Mathematics

for Caribbean Schools

Students' Book 4

Orders: please contact Hachette UK Distribution,
Hely Hutchinson Centre, Milton Road, Didcot, Oxfordshire, OX11 7HH. Telephone: +44 (0)1235 827827.
Email: education@hachette.co.uk
Lines are open from 9 a.m. to 5 p.m., Monday to Friday.
You can also order through our website:
www.hoddereducation.com

First published in 1988 by Longman Group Limited
Second edition published 2000
This edition published 2007
Published from 2015 by Hodder Education,
An Hachette UK Company
Carmelite House
50 Victoria Embankment
London EC4Y 0DZ
www.hoddereducation.com
First impression 2016

ISBN-10: 1-4058-4779-4
ISBN-13: 978-1-4058-4779-7

Set in 9.5/12 pt Stone Serif

Printed and bound by CPI Group (UK) Ltd, Croydon, CR0 4YY

Acknowledgements
The Publishers wish to acknowledge the work of J B Channon, A McLeish Smith, H C Head and M F Macrae which laid the foundations for this series.

The Publishers are grateful to the Caribbean Examinations Council for permission to reproduce examination questions as follows.

June 1986	question 7 (14b question 6) question 8 (19k question 10) question 9 (a) (i) (19s question 11) question 11(a) (22d question 7)
June 1987	question 13 (adapted) (23d question 7)
June 1988	question 14 (a) (23b question 11) question 14 (b) (23c question 8) question 6 (24 question 4)
January 1989	question 11 (22h question 9)
June 1989	question 2 (b) (19a question 40) question 2 (a) (19i question 18)
January 1990	question 7 (14b question 7) question 10 (b) (19s question 12) question 6 (22f question 18)
June 1990	question 6 (a) (23a question 6)
January 1991	question 2 (a), (b) (19s question 13) question 7(b) (22f question 17) question 3(b) (23c question 17)
June 1992	question 9 (19s question 5) question 11(b) (22e question 10) question 12 (22h question 13)
January 1996	question 13 (23c question 23) question 3 (a) (23a question 8) question 14 (23d question 10)
January 1997	question 10 (19s question 7) question 7 (b) (19s question 9) question 9 (adapted) (22g question 7) question 11 (b) (22h question 14)
January 1998	question 14 (14e question 8) question 2 (19s question 8) question 5b (22d question 12)

Preface

This is the last in a series of four volumes, intended for use primarily by students who are preparing to sit the certificate examinations held by the Caribbean Examinations Council (CXC) and by the individual countries in the Caribbean. Each volume represents material which may be covered by the average student in one year approximately (although some students may need more time) so that there is ample time for the series to be completed over a four to five year period.

It should be pointed out that the Mathematics syllabuses for the CXC Secondary Education certificate are designed to allow "... for a Basic Proficiency or 'core' syllabus which provides the minimal mathematical skills and competence necessary for any enlightened citizen in our contemporary society. At the same time additional topics have been included in the General Proficiency syllabus to meet the needs of those ... likely to work in scientific or technological fields ... requiring Mathematics beyond the secondary school level." [Excerpt from Mathematics Syllabuses (2000) Caribbean Examinations Council].

As in the first edition, the pedagogical approach links previously known concepts to the new ideas and concepts being introduced. This approach has been further promoted in this revised series by highlighting at the beginning of the chapter, the main/basic prior knowledge required to understand the new material under the heading 'Pre-requisites'. The detailed explanation of concepts, of principles and methods of working out problems, and the careful structuring of the difficulty level starting with simple examples initially – all continue to be stressed. At the end of each chapter a 'Summary' of the main points developed in the chapter, with special emphasis on new material, is included. The 'Summary' is followed by 'Practice exercises', which involve the student in not only doing but thinking before doing. In these exercises, there has been a deliberate attempt to bring together the various strands of mathematics. Key reference words are printed in bold type throughout the text.

Throughout the series, many problems have been included, both as worked examples in the teaching text to illustrate particular approaches to solving problems, and also as exercises for practice and reinforcement of the concepts. This has been done in a deliberate attempt to provide a stimulus to teachers in developing their strategies for teaching different topics; but, especially, to provide guidance to students as they work or revise on their own.

To encourage an interactive approach and to give students practice in working in teams, opportunities for group work have been afforded. Areas where such activities are relevant and would help to clarify the mathematical principles have been identified.

This revised edition seeks to incorporate changes in keeping with amendments to the syllabuses in the various Caribbean countries and that of the Caribbean Examinations Council. Attention has been given to the suggestions of Caribbean teachers whose positive reaction and responses to the series have been most encouraging. Feedback from reviews conducted throughout the Caribbean region has also informed this revised edition.

To the teacher

This edition of the series of four texts, revised with respect to content and its sequencing and to pedagogy (to a lesser extent), will be found useful in providing help and guidance in how the topics are taught and the order in which they are taken. The texts do not attempt to prescribe specific approaches to the teaching of any topic. Teachers are free to adopt or modify the suggested approaches. It is the teacher who must decide on the methodology to be used in creating the most suitable learning conditions in the classroom, in providing challenging activities which motivate the students to think and yet give them a chance to succeed in finding solutions.

It is highly recommended that teachers expose students to all the optional objectives in the CXC General Proficiency Syllabus so that students are given the opportunity to make individual decisions regarding those objectives on which they may wish to concentrate.

Adequate *time and thought* must be given to the process of trying different approaches to solving problems. The capacity to reason logically, to make valid deductions from stated data, to select the relevant information from a word problem, and to consider the reasonableness of the 'answer' are worthwhile goals to which teachers must devote attention.

In some instances, students may improve their performance by working in a group. It is suggested that after alternative strategies for tackling a problem have been discussed by the group, the actual solution(s) may be carried out in the initial group, in smaller groups, or by students working individually. This method maintains individualisation while keeping the class together as a unit. The development of positive attitudes of friendly competition, of respect for another person's suggestions, for cooperation and teamwork through such activities brings long-term benefits to the students.

Whenever the opportunity arises, teachers are urged to use, and thus reinforce, concepts taught earlier, for example, the continued application of estimation and approximation in the calculation of numerical values. In addition, it is widely accepted that learning is aided by doing. Thus concrete/practical examples and real-life applications must be provided, whenever possible, as well as the use of pictures, flow-charts and other diagrammatic representations to deepen the understanding of abstract/theoretical ideas.

In an attempt to assist the teacher and the student in quickly identifying and reviewing necessary background knowledge, we have included information to be referenced under the heading 'Pre-requisites' at the beginning of the chapters. It must be remembered that the main new ideas of each chapter are highlighted in the 'Summary' at the end of the chapter.

In addition to the 'Revision Exercises and Tests' (after Chapters 1–6 and after Chapters 7–12), the 'Revision Course' (Chapters 19–25) has been retained. The revision course relies mainly on worked examples and exercises for further practice. A revision test of multiple-choice items is also included at the end of each revision chapter.

There are in this edition at the end of the book, two 'Practice Examinations' structured in keeping with the format used by the Caribbean Examinations Council, that is, Paper 1 (an objective test of 60 items), and Paper 2 (an essay-type/open-ended question paper with a compulsory Section I and a choice of questions in Section II).

Finally, it is unfortunate that Mathematics is perceived by a large majority of students as a 'necessary evil', a subject that they have to '... get at CXC ...' in order to become employable. Teachers have a significant responsibility in helping to change this attitude, and in having students appreciate that, by acquiring the skills and techniques to solve problems in mathematics, they also acquire the tools and the ability to solve problems in the real world.

To the student

Before attempting the problems in the exercises, study and discuss the worked examples until you understand the concept. The oral/group work exercises are intended to encourage discussion. This helps to clarify lingering misunderstandings.

In particular, in solving word problems you first have to get thoroughly familiar with the problem. The next step involves translating the problem into mathematical symbols and language, for example, into an equation or an inequality, or into a graph. The next steps are applying the required mathematical operations, and finally checking the original word problem. Remember also to check that the variables are in the same units. Another useful hint is to look for patterns in similar problems and in the methods of solution. Concrete materials such as cans, coins, balls, stones and boxes are very useful aids for clearing up doubts – not for wasting time!

An electronic calculator is an excellent machine when used wisely, but you must bear in mind that it needs to be used by a clear-thinking human who fully comprehends the mathematical concepts. For example, using 0.93969262 as the value for sin 70° in a problem when other given values are stated correct to 2 significant figures indicates, among other things, a lack of appreciation of the concept of accuracy.

Nothing can replace the neat appearance of an answer that is well set out – the date, the page and exercise from which the problem has been taken, a statement of the facts given, the necessary calculations performed and the conclusions drawn which result in a final answer. This whole process helps you to think clearly.

Students are advised to ensure that before attempting a new chapter, they fully understand the required concepts as listed under 'Pre-requisites' at the start of the chapter. You are urged to make full use of the 'Revision Exercises and Tests', and the 'Revision Course', and after revision has been completed, to attempt the 'Practice Examinations' under simulated examination conditions. This will help you to identify those parts of the syllabus to which you may need to give extra time and attention.

Finally, the more a concept is applied. the clearer it becomes. Thus, PRACTICE and more practice in working out examples is an essential ingredient to success.

Althea A. Foster
E. M. Tomlinson

Master index Books 1–4

(listing reference chapters in terms of the topics defined in the Caribbean Examination Council Syllabuses)

Number theory

Book 1 Chapter 1 **Number systems (1)**
The place-value system and number bases
Chapter 3 **Number systems (2)**
Sets of whole numbers
Book 2 Chapter 1 **Number systems (3)**
Sets of numbers

Computation

Book 1 Chapter 5 **Fractions**
Chapter 9 **Decimals, fractions and percentages**
Chapter 13 **Approximation and estimation**
Chapter 20 **Ratio and rate**
Book 2 Chapter 7 **Standard form**
Large and small numbers

Sets

Book 1 Chapter 2 **Sets (1)**
Elements of sets, subsets, Venn diagrams
Book 2 Chapter 8 **Sets (2)**
Union, intersection, complement
Book 3 Chapter 6 **Sets (3)**
Equal and equivalent sets, Venn diagrams with three subsets

Measurement

Book 1 Chapter 4 **Basic units of measurement**
Chapter 14 **Plane shapes (1)**
Perimeter
Chapter 16 **Plane shapes (2)**
Area
Book 2 Chapter 6 **Measurement (1)**
Volume of solids
Chapter 11 **Scale drawing (1)**
Scale, use of scales in drawing
Chapter 17 **Measurement (2)**
Compound shapes, surface area of solids
Book 3 Chapter 11 **Measurement (3)**
Arcs and sectors of circles, similar shapes and solids

Consumer arithmetic

Book 1 Chapter 22 **Consumer arithmetic (1)**
Profit and loss, taxes
Book 2 Chapter 21 **Consumer arithmetic (2)**
Simple and compound interest, currency exchange rates
Book 3 Chapter 5 **Consumer arithmetic (3)**
Utilities, bills, investments and depreciation

Algebra

Book 1 Chapter 6 **Algebra: basic processes**
Chapter 8 **Directed numbers (1)**
Addition and multiplication
Chapter 11 **Simplifying algebraic expressions (1)**
Like terms and unlike terms, substitution
Chapter 15 **Algebraic equations (1)**
Simple linear equations, flow charts
Book 2 Chapter 3 **Directed numbers (2)**
Subtraction and division
Chapter 5 **Simplifying algebraic expressions (2)**
Brackets, indices, factors, fractions
Chapter 10 **Algebraic equations (2)**
Word problems
Chapter 16 **Inequalities (1)**
Linear inequalities
Chapter 20 **Factorisation (1)**
More factors and fractions
Book 3 Chapter 2 **Simplifying algebraic expressions (3)**
Binary operations
Chapter 3 **Algebraic equations (3)**
Formulae
Chapter 8 **Simultaneous linear equations**
Chapter 15 **Inequalities (2)**
Simple simultaneous linear inequalities, word problems
Book 4 Chapter 1 **Factorisation (2)**
Grouping, quadratic expressions
Chapter 4 **Quadratic equations**
Chapter 7 **Further algebra**
Algebraic fractions, fractional indices, algebraic identities, change of subject of formulae
Chapter 16 **Inequalities (3)**
Linear inequalities in two variables, linear programming

Relations, functions and graphs

Book 1 Chapter 19 **Relations**
Book 2 Chapter 4 **Mappings, ordered pairs**
Chapter 9 **The Cartesian plane**
Chapter 13 **Straight-line graphs**
Book 3 Chapter 7 **Linear functions**
Chapter 9 **Non-linear functions**
Book 4 Chapter 10 **Graphs (1)**
Maximum and minimum values, travel graphs

Chapter 11 **Functions**
Inverse of a function and composite functions
Chapter 15 **Graphs (2)**
Variation, exponential, inverse and cubic functions

Statistics and probability

Book 1 Chapter 18 **Statistics (1)**
Data collection and presentation
Book 2 Chapter 18 **Statistics (2)**
Mean, median, mode
Book 3 Chapter 12 **Statistics (3)**
Statistical graphs, frequency distribution
Chapter 14 **Probability (1)**
Theoretical and experimental probabilities
Book 4 Chapter 13 **Statistics and probability**
Cumulative frequency

Geometry

Book 1 Chapter 7 **Properties of solids**
Chapter 10 **Measurement of angles**
Chapter 12 **Properties of plane shapes**
Chapter 17 **Angles**
Angles between lines, angles of polygons
Chapter 21 **Symmetry**
Book 2 Chapter 2 **Geometrical constructions (1)**
Triangles, parallel and perpendicular lines
Chapter 12 **Angles (2)**
Other polygons
Chapter 14 **Solving triangles (1)**
The right-angled triangle – sides
Chapter 15 **Geometrical transformations (1)**
Congruencies
Chapter 19 **Solving triangles (2)**
The right-angled triangle – the tangent ratio, angle of elevation, angle of depression
Chapter 22 **Geometrical constructions (2)**
Constructing special angles, bisecting lines and angles
Chapter 23 **Solving triangles (3)**
The right-angled triangle – sine and cosine ratios
Book 3 Chapter 1 **Geometrical proofs (1)**
Triangles and other polygons
Chapter 4 **Geometrical proofs (2)**
Circles
Chapter 10 **Geometrical transformations (2)**
Enlargements
Chapter 13 **Scale drawing (2)**
Bearings and distances
Book 4 Chapter 5 **Geometrical proofs (3)**
Cyclic quadrilaterals and tangents
Chapter 6 **Coordinate geometry**
Parallel and perpendicular lines, midpoint and length of lines
Chapter 17 **Latitude and longitude**

Trigonometry

Book 4 Chapter 3 **Trigonometry (1)**
Trigonometric ratios of obtuse and reflex angles, positive and negative angles, and special angles
Chapter 8 **Trigonometry (2)**
The sine and cosine rules, area problems
Chapter 9 **Three-dimensional problems**
Composite solids, three-dimensional problems
Chapter 12 **Trigonometry (3)**
Trigonometric graphs

Vectors and matrices

Book 4 Chapter 2 **Matrices**
Chapter 14 **Geometrical transformations and matrices**
Combined transformations, matrices and transformations
Chapter 18 **Vectors**

Book 3: Revision course

(including a multiple-choice revision test and questions from past CXC papers in each chapter)

Chapter 16 Computation and number theory
Chapter 17 Consumer arithmetic
Chapter 18 Sets
Chapter 19 Relations, functions and graphs
Chapter 20 Algebra
Chapter 21 Measurement
Chapter 22 Geometry
Chapter 23 Statistics and probability
Chapter 24 Trigonometry

Book 4: Revision course

(including a multiple-choice revision test and questions from past CXC papers in each chapter)

Chapter 19 Sets
Chapter 20 Algebra
Chapter 21 Consumer arithmetic
Chapter 22 Relations, functions and graphs
Chapter 23 Geometry
Chapter 24 Trigonometry
Chapter 25 Matrices and vectors
Chapter 26 Statistics and probability

Contents

Chapter 1
Factorisation (2) 1
- Common factors (revision) 1
- Factorisation by grouping 1
- Factorisation of quadratic expressions 3
- Perfect squares 6
- Difference of two squares 7
- Factorisation of quadratic expressions by grouping 8

Chapter 2
Matrices 13
- Matrices 13
- Addition and subtraction of matrices 14
- Scalar multiplication 15
- Multiplication 16
- Algebra of 2×2 matrices 18
- Matrices as operators 19

Chapter 3
Trigonometry (1) 23
- Tangent, sine, cosine (revision) 23
- Trigonometric ratios of obtuse angles 23
- Trigonometric ratios of reflex angles 25
- Positive angles 26
- Negative angles 27
- Tan, sin and cos of 0°, 90°, 180°, 270°, 360° 28
- Tan, sin and cos of 45° 29
- Tan, sin and cos of 60° and 30° 30

Chapter 4
Quadratic equations 35
- Equation (revision) 35
- Quadratic equation 35
- Equations with non-rational roots 37
- Completing the square 38
- The formula for solving quadratic equations 39
- Graphical solution of quadratic equations 40
- Word problems leading to quadratic equations 44
- Simultaneous linear and quadratic equations 45
- Graphical solution of simultaneous quadratic and linear equations 46

Chapter 5
Geometrical proofs (3) 51
- Cyclic quadrilaterals 51
- Tangent to a circle 54
- Tangents from an external point 56
- Alternate segment 57

Chapter 6
Coordinate geometry 63
- Linear functions (revision) 63
- Sketching graphs of straight lines 63
- Gradients of parallel and perpendicular lines 65
- Mid-point of a line 66
- Distance between two points 67

Revision exercises and tests
Chapters 1–6 73

Chapter 7
Further algebra 79
- Simplification of fractions 79
- Multiplication and division of fractions 80
- Addition and subtraction of fractions 81
- Undefined fractions 83
- Equations with fractions 84
- Indices 86
- Algebraic identities 88
- Direct variation 89
- Inverse variation 93
- Joint variation 94
- Change of subject of formulae 95

Chapter 8
Trigonometry (2) 103
- The sine rule 103
- The cosine rule 105
- Solving triangles using the sine and cosine rules 106
- Using the cosine rule to calculate angles 107
- Using trigonometry in area problems 109

Chapter 9
Three-dimensional problems **113**
Surface area and volume of solids 113
Composite solids 116
Three-dimensional problems 121

Chapter 10
Graphs (1) **125**
Gradient of a curve 125
Maximum and minimum values 127
Travel graphs 128

Chapter 11
Functions **140**
Functional notation 140
Inverse of a function 140
Composite functions 143

Chapter 12
Trigonometry (3) **147**
Trigonometric functions 147
The sine function 147
The sine curve 148
The cosine function 149
The cosine curve 150
The tangent function 152
The tangent curve 153

Revision exercises and tests
Chapters 7–12 **156**

Chapter 13
Statistics and probability **164**
Cumulative frequency 164
Probability 167

Chapter 14
Geometrical transformations and matrices **173**
Combined transformations 173
Transformations and matrices 177

Chapter 15
Graphs (2) **186**
Exponential functions 186
Inverse functions 188
Cubic functions 190
Sketch graphs 191

Chapter 16
Inequalities (3) **196**
Inequalities in two variables 196
Linear programming 201

Chapter 17
Latitude and longitude **207**
Great circles and small circles 207
The earth, latitude and longitude 208
Distances along great circles 210
Distances along parallels of latitude 212
Miscellaneous examples 214

Chapter 18
Vectors **217**
Naming vectors (revision) 217
Addition of vectors 218
Magnitude of a vector 220
Subtraction of vectors 220
Multiplication by a scalar 221
Resultant vectors 223
Position vectors 224
Vector geometry 226
Velocity vectors 230
Vector diagrams 231

Revision course

Chapter 19
Sets **233**
Symbols, definitions and operations 233
Relations among sets 234
Intersecting sets 234

Chapter 20
Algebra **227**
Factorisation 237
Indices 240
Variation 240
Formulae 242
Equations and identities 244
Linear inequalities in two variables 245
Quadratic equations 247
Simultaneous linear and quadratic equations 250
Word problems 251
Revision test (Chapter 20) 254

Chapter 21
Consumer arithmetic **257**
Profit and loss 257
Discount 257
Mortgages and hire purchase 257
Bills and invoices 258
Investments and depreciation 258
Wages, salary and income tax 259
Exchange rates 259

Chapter 22
Relations, functions and graphs **260**
Relations and functions 260
Graphs 260
Travel graphs 263
Revision test (Chapter 22) 267

Chapter 23
Geometry **270**
Cyclic quadrilaterals 270
Tangents to circles 270
Revision test (Chapter 23) 272

Chapter 24
Trigonometry **274**
Sine, cosine, tangent 274
Solving right-angled triangles 277
Three-dimensional problems 280
Solving non-right-angled triangles 282
Trigonometric functions: the sine, cosine and tangent functions 286
Longitude and latitude 287
Revision test (Chapter 24) 291

Chapter 25
Matrices and vectors **293**
Matrices 293
Vectors 298
Geometrical transformations 301
Revision test (Chapter 25) 305

Chapter 26
Statistics and probability **307**
Cumulative frequency curve 307
Revision test (Chapter 26) 310

Practice examination I **312**
Paper 1 312
Paper 2 317

Practice examination II **321**
Paper 1 321
Paper 2 325

Mensuration tables and formulae, three-figure tables **330**

Answers **346**

Index **433**

Chapter 1

Factorisation (2)

Grouping, quadratic expressions

Pre-requisites

- common factors; expansion of algebraic expressions

Common factors (revision)

To factorise an expression is to write it as a product of its factors, that is, to do the opposite of removing brackets. Therefore, factorisation is the inverse operation of expansion.

To factorise an expression completely, we must find the highest common factor (HCF) of all the terms in the expression. In an algebraic expression, the common factor may be a number, or a letter, or may include both numbers and letters, that is, the common factor may be numeric or literal. For example,

$$5h + 10k = 5(h + 2k)$$

$$3ac + a = a(3c + 1)$$

and $$2gn - 6g^2n = 2gn(1 - 3g)$$

The common factor may be two or more literal terms joined by plus or minus signs.
If $g = [c + 5d]$ in the last expression above,then

$$2[c + 5d]n - 6[c + 5d]^2n$$
$$= 2[c + 5d]n\ (1 - 3[c + 5d])$$

The common factor is $2[c + 5d]n$.

Factorisation can be used to simplify calculations. For example, when finding the amount in simple interest calculations,

$$A = P + Prt,$$

if $P = 5000$, $r = \frac{16}{3}\%$
$t = \frac{9}{12}$,

using $A = P(1 + rt)$,
$A = 5000\ (1 + \frac{16}{300} \times \frac{9}{12})$
and the amount $= \$5200$

Exercise 1a (revision)

Factorise the following expresions.

1. $3b + 3c$
2. $3r^2 + 3rh$
3. $2m + 4m^2 - 6$
4. $7x^3 - xy - 3x$
5. $3x^2y + 5xy^2 + 2xy + y$
6. $4(a + c) + 3(a + c)$
7. $x(b + 3) + y(b + 3)$
8. $t(u + v) - t^2$
9. $m(r - h) + (r - h)^2$
10. $(x - y)(u + v) + (x - y)(p + r)$

Factorisation by grouping

By expanding the brackets, we find that

$$(a + b)(c + d) = a(c + d) + b(c + d)$$
$$= ac + ad + bc + bd$$

Therefore, to factorise $ac + ad + bc + bd$, we have to revise the steps. We group terms which have a common factor; for example, $(ac + ad)$ and $(bc + bd)$. Then we take out the common factor in each group so that the terms become $a(c + d)$ and $b(c + d)$. In each of the two products there is a factor $(c + d)$. Therefore.

$$ac + ad + bc + bd$$
$$= (ac + ad) + (bc + bd)$$
$$= a(c + d) + b(c + d)$$
$$= (c + d)(a + b)$$
$$= (a + b)(c + d) \quad \text{[commutative law]}$$

Note that, if we group $(ac + bc)$ and $(ad + bd)$, that is $c(a + b)$ and $d(a + b)$, the common factor in the two terms is $(a + b)$. Check that the final result is the same.

Example 1

Factorise $cx + cy + 2dx + 2dy$.

The terms cx and cy have c in common.
The terms $2dx$ and $2dy$ have $2d$ in common.
Grouping in pairs in this way

$$cx + cy + 2dx + 2dy = (cx + cy) + (2dx + 2dy)$$
$$= c(x + y) + 2d(x + y)$$

Hence $cx + cy + 2dx + 2dy = (x + y)(c + 2d)$

Notice that the terms may be grouped in different pairs.
The terms cx and $2dx$ have x in common
The terms cy and $2dy$ have y in common

$$\begin{aligned} cx + cy + 2dx + 2dy &= (cx + 2dx) + (cy + 2dy) \\ &= x(c + 2d) + y(c + 2d) \\ &= (c + 2d)(x + y) \end{aligned}$$

Example 2

Factorise $3a - 6b + ax - 2bx$.

$$\begin{aligned} 3a - 6b + ax - 2bx &= 3(a - 2b) + x(a - 2b) \\ &= (a - 2b)(3 + x) \end{aligned}$$

Notice that to factorise in this way, the same bracket must occur twice in the first line of working. If the given expression is to be factorised, there must be a repeated bracket. For this reason, it is often easiest to write this bracket down again immediately, as soon as it has been found. This is shown iin Example 3.

Example 3

Factorise $2x^2 - 3x + 2x - 3$.

$$2x^2 - 3x + 2x - 3 = x(2x - 3) + ?(2x - 3)$$

The terms $+2x$ and -3 in the given expression are obtained by multiplying $(2x - 3)$ by $+ 1$. Thus,

$$\begin{aligned} 2x^2 - 3x + 2x - 3 &= x(2x - 3) + 1(2x - 3) \\ &= (2x - 3)(x + 1) \end{aligned}$$

Exercise 1b

Factorise the following by grouping in pairs.

1. $ax + ay + 3bx + 3by$
2. $7a + 14b + ax + 2bx$
3. $x^2 + 5x + 2x + 10$
4. $pq + qr + ps + rs$
5. $a^2 - 9a + 3a - 27$
6. $8m - 2 + 4mn - n$
7. $5x^2 - 10x - 3x - 6$
8. $ab - bc + ad - cd$
9. $2ab - 5a + 2b - 5$
10. $3m - 1 + 6m^2 - 2m$

In many cases it is only possible to get a repeated second bracket if a negative common factor is taken. This is shown in Example 4.

Example 4

Factorise $2am - 2m^2 - 3ab + 3bm$.

$$\begin{aligned} &2am - 2m^2 - 3ab + 3bm \\ &= 2m(a - m) + ?(a - m) \end{aligned}$$

The terms $-3ab$ and $+3bm$ in the given expression are obtained by multiplying $(a - m)$ by $-3b$. Hence,

$$\begin{aligned} &2am - 2m^2 - 3ab + 3bm \\ &=2m(a - m) -3b(a - m) \\ &= (a - m)(2m - 3b) \end{aligned}$$

Exercise 1c

Factorise the following.

1. $ab + bc - am - cm$
2. $9x + 6 - 3x^2 - 2x$
3. $2ax - 2ay - 3bx + 3by$
4. $x^2 - 7x - 2x + 14$
5. $5a - 5b - ac + bc$
6. $3pq + 12pr - qy - 4ry$
7. $a^2 - 3a - 3a + 9$
8. $2ps + 5pt - 2rs - 5rt$
9. $x^2 - 6x - x + 6$
10. $3k + 1 - 3hk - h$

Sometimes the first attempt at grouping the terms does not give a common factor. In these cases, regroup the terms and try again. This is shown in Example 5.

Example 5

Factorise $cd - de + d^2 - ce$.

$$cd - de + d^2 - ce = d(c - e) + ...(\quad)$$

d^2 and ce have no common factors. Regroup the given terms.

either:

$$\begin{aligned} cd - de + d^2 - ce &= cd + d^2 - ce - de \\ &= d(c + d) - e(c + d) \\ &= (c + d)(d - e) \end{aligned}$$

or:

$$\begin{aligned} cd - de + d^2 - ce &= cd - ce + d^2 - de \\ &= c(d - e) + d(d - e) \\ &= (d - e)(c + d) \end{aligned}$$

Four terms can be grouped in pairs in three ways. If the expression factorises, two of these ways will give the required result and one will not.

Exercise 1d

Regroup and then factorise the following.

1. $6a + bm + 6b + am$
2. $pr + qs + qr + ps$
3. $15 - xy + 5y - 3x$
4. $ac - bd - bc + ad$
5. $ax - xy + x^2 - ay$
6. $ad - cm - am + cd$
7. $x^2 + 15 - 3x - 5x$
8. $8a + 15by + 12y + 10ab$
9. $3a - cb - 3b + ac$
10. $t + 6sz + 3s + 2tz$

If *all* the terms contain a common factor, it should be taken out first. This is shown in Example 6. *This should always be the rule when factorising any type of expression.*

Example 6

Factorise $2sru + 6tru - 4srv - 12trv$.

$2r$ is a factor of every term in the given expression.

$$\begin{aligned} &2sru + 6tru - 4srv - 12trv \\ &= 2r\{su + 3tu - 2sv - 6tv\} \\ &= 2r\{u(s + 3t) - 2v(s + 3t)\} \\ &= 2r(s + 3t)(u - 2v) \end{aligned}$$

Exercise 1e

Factorise the following where possible, If there are no factors, say so.

1. $mx + nx + my + ny$
2. $ax - ay + bx - by$
3. $hu + hv - ku - kv$
4. $au - bu - av + bv$
5. $am + 2bm + 2bn + an$
6. $cx - dx + 2cy - 2dy$
7. $2ce + 4df - de - 2cf$
8. $ab + 4xy - 2bx - 2ay$
9. $am - an + m - n$
10. $u + v - dv + du$
11. $a^3 + a^2 + a + 1$
12. $2mh - 3nh - 3nk + 2mk$
13. $3sx - 5ty + 5tx - 3sy$
14. $abx^2 + bxy + axy + y^2$
15. $hk - 2km + 3hn - 6mn$
16. $mn - 6xy - 3nx + 3my$
17. $2gk - 3gl + 2hk - 3hl$
18. $2fh + 4gh - fk - 2gk$
19. $3eg - 4eh - 6fg + 2fh$
20. $hl + 2kl - 3hm - 6km$
21. $3ce + 4df - 2de - 6cf$
22. $xy - 2ny - 6n^2 + 3nx$
23. $ab + 2b^2 - 2ac - 4bc$
24. $cd - ce + d^2 + de$
25. $8uv - 2v^2 + 12uw - 3vw$
26. $mn - 6pn + 3pm - 2n^2$
27. $3xy - 2ay - 3ax + 2y^2$
28. $3ab - 3bu + 3av - 3uv$
29. $ab + 6mn - 2bm - 3bn$
30. $8ce + 12de - 2cf - 3df$
31. $nuv - muv + mnu^2 - v^2$
32. $5mx - 5nx - 5my + 5ny$
33. $3ab + 3cd - bc - 9ad$
34. $6ab - 15bc - 10cd + 4ad$
35. $2amu + 2anu - 2amv - 2anv$
36. $abm^2 + 2bm - 3am - 6$
37. $4ax + 2bx + 8ay + 4by$
38. $21mn - xy - 3nx + 7my$
39. $3ax - 2a - 6bx + 2b$
40. $2am - 3m^2 + 4an - 6mn$
41. $10uv + 5u - 2v - 1$
42. $a^2m + am^2 - mn - an$
43. $2x^2y - xy^2 + 2ax - ay$
44. $1 + 3x - 5a - 15ax$
45. $2d^2x + 4dx^2y - 3dy - 6xy^2$

Factorisation of quadratic expressions

A **quadratic** expression is one in which 2 is the highest power of the unknown(s) in the expression. For example, $x^2 - 4x - 12$, $16 - a^2$, $3x^2 + 17xy + 10y^2$ are quadratic expressions.

$(x + 2)(x - 6) = x^2 - 4x - 12$

$(x + 2)$ and $(x - 6)$ are the **factors** of $x^2 - 4x - 12$. Just as in arithmetic, $5 \times 7 = 35$ where 5 and 7 are the factors of 35.

A quadratic expression may *not* have factors. In arithmetic, 13 is prime since it has no factors other than itself and 1. Similarly $x^2 + 2x - 6$ has no factors (other than itself and 1).

To **factorise** a quadratic expression is to express it as a product of its factors.
Thus $x^2 - 4x - 12$ factorises to become

$(x + 2)(x - 6)$.

Example 7 shows the steps to be followed when factorising quadratic expressions when the coefficient of the x^2 term is 1.

Example 7

Factorise the quadratic expression $x^2 + 7x + 10$.

The problem is to fill the brackets in the statement $x^2 + 7x + 10 = (\quad)(\quad)$

1st step: Look at the *first* term in the given expression, x^2. From work done in expanding brackets (as in Book 2, Chapter 5), when the first term in the expansion is x^2, x appears first in each bracket:

$x^2 + 7x + 10 = (x \quad)(x \quad)$.

2nd step: Look at the *last* term in the given expression, +10. The *product* of the last terms in the two brackets must be +10. Number pairs which have a product of +10 are:

(a) +10 and +1 (b) +5 and +2
(c) −10 and −1 (d) −5 and −2

These give four possible answers:

(a) $(x + 10)(x + 1)$
(b) $(x + 5)(x + 2)$
(c) $(x - 10)(x - 1)$
(d) $(x - 5)(x - 2)$

3rd step: Look at the coefficient of the *middle* term in the given expression, +7. The *sum* of the last terms in the two brackets must be +7. Adding the number pairs in turn:

(a) $(+10) + (+1) = +11$
(b) $(+5) + (+2) = +7$
(c) $(-10) + (-1) = -11$
(d) $(-5) + (-2) = -7$

Of these, only (b) gives +7. It follows that:
$x^2 + 7x + 10 = (x + 5)(x + 2)$

Note: 1 The answer can be checked by expanding the brackets.
2 The order of the brackets is not important.
$(x + 5)(x + 2) = (x + 2)(x + 5)$

In Example 7 both the last term and the coefficient of x were positive. It was not really necessary to consider the possibility of having negative factors of +10. Example 8 shows how the method may be shortened.

Example 8

Factorise $d^2 + 11d + 18$.

1st step: $d^2 + 11d + 18 = (d \quad)(d \quad)$

2nd step: Find two numbers such that their product is +18 and their sum is +11. Since the 18 is positive *and* the 11 is positive, consider positive factors only.

	factors of +18	*sum of factors*
(a)	+1 and +18	+19
(b)	+2 and +9	+11
(c)	+3 and +6	+9

Of these, only (b) gives the required result. Thus,

$d^2 + 11d + 18 = (d + 2)(d + 9)$

The method of Examples 7 and 8 is called a **trial and error** method. It is necessary to try various number pairs in turn, until the correct pair is found. To begin with, this will take time. With practice, you will be able to factorise quadratic expressions quite quickly.

Exercise 1f

Factorise the following quadratic expressions.

1. $x^2 + 6x + 5$
2. $x^2 + 12x + 11$
3. $a^2 + 14a + 13$
4. $b^2 + 8b + 7$
5. $y^2 + 9y + 8$
6. $z^2 + 6z + 8$
7. $c^2 + 8c + 15$
8. $d^2 + 13d + 22$
9. $n^2 + 8n + 12$
10. $r^2 + 9r + 20$
11. $s^2 + 10s + 16$
12. $t^2 + 8t + 16$

Example 9

Factorise $x^2 - 9x + 8$.

1st step: $x^2 - 9x + 8 = (x \quad)(x \quad)$

2nd step: Find two numbers such that their product is +8 and their sum is −9. List the possible pairs and find their sums:

	factors of +8	*sum of factors*
(a)	+8 and +1	+9
(b)	+4 and +2	+6
(c)	−8 and −1	−9
(d)	−4 and −2	−6

Of these, only (c) gives the required result. Thus,

$$x^2 - 9x + 8 = (x - 8)(x - 1)$$

In Example 9, the last term is positive and the coefficient of x is negative. This gives a negative sign in both brackets. Hence, it is necessary to consider only negative factors of +8.

Example 10

Factorise $t^2 - 10t + 24$.

1st step: $t^2 - 10t + 24 = (t \quad)(t \quad)$

2nd step: Find two numbers such that their product is +24 and their sum is −10. Since 24 is positive and the 10 is negative, consider negative factors only.

	factors of +24	*sum of factors*
(a)	−1 and −24	−25
(b)	−2 and −12	−14
(c)	−3 and −8	−11
(d)	−4 and −6	−10

Of these, only (d) gives the required result. Thus, $t^2 - 10t + 24 = (t - 4)(t - 6)$

Exercise 1g

Factorise the following.

1. $x^2 - 4x + 3$
2. $y^2 - 3y + 2$
3. $z^2 - 18z + 17$
4. $a^2 - 8a + 7$
5. $b^2 - 5b + 6$
6. $c^2 - 7c + 6$
7. $d^2 - 9d + 14$
8. $n^2 - 7n + 10$
9. $p^2 - 11p + 24$
10. $q^2 - 10q + 21$
11. $f^2 - 16f + 28$
12. $x^2 - 10x + 25$

So far, the given quadratic expressions have all contained a positive last term. Examples 11 and 12 show what happens when the last term is negative.

Example 11

Factorise the expression $x^2 + 2x - 15$.

1st step: $x^2 + 2x - 15 = (x \quad)(x \quad)$

2nd step: Find two numbers such that their product is −15 and their sum is +2. List the possible pairs and find their sums.

	factors of −15	*sum of factors*
(a)	−15 and +1	−14
(b)	+15 and −1	+14
(c)	−5 and +3	−2
(d)	+5 and −3	+2

Of these, only (d) gives the correct result.

$$x^2 + 2x - 15 = (x + 5)(x - 3)$$

Example 12

Factorise $x^2 - 4x - 12$.

Find two numbers such that their product is −12 and their sum is −4.

	factors of −12	*sum of factors*
(a)	−12 and +1	−11
(b)	+12 and −1	+11
(c)	−6 and +2	−4
(d)	+6 and −2	+4
(e)	−4 and +3	−1
(f)	+4 and −3	+1

Of these, only (c) gives the required result.

$$x^2 - 4x - 12 = (x - 6)(x + 2)$$

Notice that if the last term in the given expression is negative, the signs inside the brackets are different; one positive and one negative.

Exercise 1h

Factorise the following

1. $x^2 + 4x - 5$
2. $a^2 - 4a - 5$
3. $x^2 + 6x - 7$
4. $b^2 - 6b - 7$
5. $n^2 + n - 2$
6. $r^2 - 2r - 3$
7. $x^2 - 10x - 11$
8. $y^2 + 12y - 13$
9. $x^2 - 2x - 15$
10. $x^2 - 14x - 15$

11. $s^2 + 5s - 6$
12. $t^2 - 5t - 6$
13. $u^2 - u - 6$
14. $v^2 + v - 6$
15. $z^2 + z - 20$
16. $c^2 - 8c - 20$
17. $x^2 - 49$
18. $x^2 - 4$

Perfect squares

$$(a + b)^2 = (a + b)(a + b) = a^2 + ab + ab + b^2$$
$$\mathbf{(a + b)^2 = a^2 + 2ab + b^2}$$
$$(a - b)^2 = (a - b)(a - b) = a^2 - ab - ab + b^2$$
$$\mathbf{(a - b)^2 = a^2 - 2ab + b^2}$$

These results are very important and should be remembered.

Fig 1.1 gives a geometrical representation of $(a + b)^2 = a^2 + 2ab + b^2$

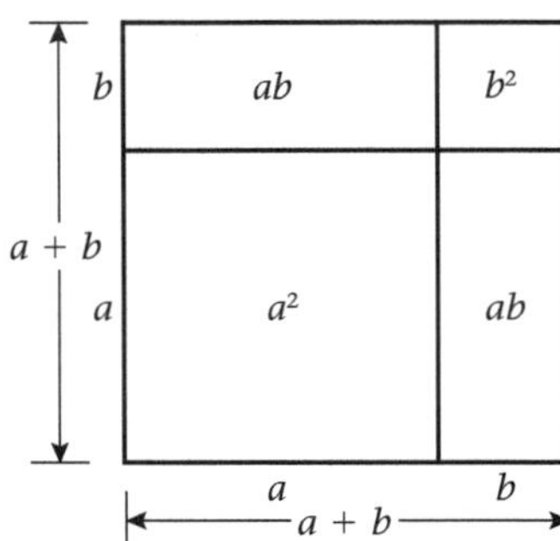

Fig. 1.1

Example 13

Expand the following

(a) $(3m + 7n)^2$ (b) $(4u - 5v)^2$

(a) $(3m + 7n)^2$
$= (3m)^2 + 2 \times 3m \times 7n + (7n)^2$
$= 9m^2 + 42mn + 49n^2$

(b) $(4u - 5v)^2$
$= (4u)^2 - 2 \times 4u \times 5v + (5v)^2$
$= 16u^2 - 40uv + 25v^2$

Notice that the squared terms are *always positive*.

Exercise 1i (Oral or written)

Expand the following.

1. $(a + 4)^2$
2. $(b - 3)^2$
3. $(5 + c)^2$
4. $(2 - d)^2$
5. $(1 + m)^2$
6. $(2n + 1)^2$
7. $(3x + y)^2$
8. $(u - 2v)^2$
9. $(5h - k)^2$
10. $(p + 4q)^2$
11. $(2a + 3d)^2$
12. $(3b - 5c)^2$
13. $(7e - 2f)^2$
14. $(10x - 1)^2$
15. $(1 + 12y)^2$
16. $(3a + 7b)^2$
17. $(c - 8d)^2$
18. $(9u + v)^2$

The expansion of a perfect square can sometimes be used to shorten the working when squaring numbers.

Example 14

Find the value of (a) 104^2 (b) 97^2

(a) $104^2 = (100 + 4)^2$
$= 100^2 + 2 \times 100 \times 4 + 4^2$
$= 10\,000 + 800 + 16$
$= 10\,816$

(b) $97^2 = (100 - 3)^2$
$= 100^2 - 2 \times 100 \times 3 + 3^2$
$= 10\,000 - 600 + 9$
$= 9409$

Exercise 1j

Find the squares of the following numbers.

1. 101
2. 99
3. 103
4. 98
5. 1001
6. 999
7. 1005
8. 996
9. 995
10. 72
11. 83
12. 79

Example 15

Factorise the following.

(a) $h^2 + 12h + 36$ (b) $25h^2 - 30hk + 9k^2$

(a) Notice that h^2 is the square of h, 36 is the square of 6 and $12h$ is twice the product of h and 6.

$h^2 + 12h + 36 = (h + 6)(h + 6)$
$= (h + 6)^2$

(b) $25h^2$ is the square of $5h$
$9k^2$ is the square of $3k$
$30hk$ is twice the product of $5h$ and $3k$
$25h^2 - 30hk + 9k^2 = (5h - 3k)^2$

Exercise 1k (Oral or written)

Give the following as the square of an expression in brackets.

1. $a^2 + 10a + 25$
2. $b^2 + 8b + 16$
3. $c^2 + 6c + 9$
4. $d^2 + 20d + 100$
5. $m^2 - 6m + 9$
6. $n^2 - 12n + 36$
7. $x^2 - 4x + 4$
8. $y^2 - 2y + 1$
9. $z^2 + 16z + 64$
10. $k^2 - 14k + 49$
11. $4 - 4b + b^2$
12. $81 + 18d + d^2$
13. $x^2 + 6xy + 9y^2$
14. $4u^2 - 12u + 9$
15. $1 - 2u + u^2$
16. $25n^2 - 30nv + 9v^2$
17. $9a^2 - 24ab + 16b^2$
18. $121 - 22y + y^2$

Difference of two squares

$(a + b)(a - b) = a^2 + ab - ab - b^2 = a^2 - b^2$

Hence

$$\boldsymbol{a^2 - b^2 = (a + b)(a - b)}$$

Fig 1.2 shows how a cardboard model can be made to demonstrate that $a^2 - b^2 = (a + b)(a - b)$.

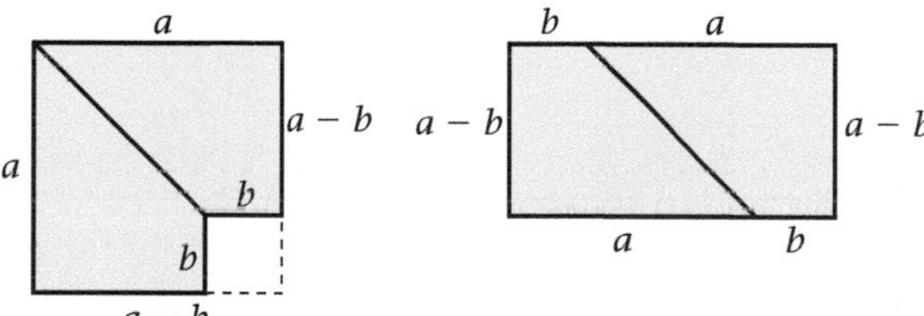

Fig 1.2

Example 16

Factorise the following
(a) $y^2 - 4$ (b) $36 - 9a^2$ (c) $25m^2 - 16n^2$

(a) $y^2 - 4 = (y)^2 - (2)^2$
$= (y + 2)(y - 2)$

(b) $36 - 9a^2 = (6)^2 - (3a)^2$
$= (6 + 3a)(6 - 3a)$

(c) $25m^2 - 16n^2 = (5m)^2 - (4n)^2$
$= (5m + 4n)(5m - 4n)$

Example 17

Factorise $5a^2 - 45$.

The two terms have the factor 5 in common. Take this our first:

$5a^2 - 45 = 5(a^2 - 9)$
$= 5(a^2 - 3^2)$
$= 5(a + 3)(a - 3)$

Exercise 1l (Oral or written)

Factorise the following.

1. $x^2 - 1$
2. $1 - y^2$
3. $4m^2 - n^2$
4. $u^2 - 16v^2$
5. $1 - a^2b^2$
6. $9 - 4c^2$
7. $4d^2 - 9e^2$
8. $3 - 3f^2$
9. $4g^2 - 4$
10. $4h^2 - 25$
11. $25k^2 - 16$
12. $49m^2 - n^2$
13. $p^2q^2 - 9$
14. $25 - u^2v^2$
15. $81 - w^2$
16. $100x^2 - 1$
17. $16y^2 - 4z^2$
18. $16h^2 - k^2$
19. $4c^2 - 49d^2$
20. $e^2 - 4f^2$
21. $36a^2 - 49b^2$
22. $5c^2 - 45d^2$
23. $x^2y^2 - z^2$
24. $100 - w^2$

The difference of two squares can sometimes be used to shorten calculations.

Example 18

Find the value of $173^2 - 127^2$.

$173^2 - 127^2 = (173 + 127)(173 - 127)$
$= 300 \times 46$
$= 13\,800$

Example 19

Fig. 1.3 shows a circular metal washer. If the diameters of the washer and its hole are 3 cm and 1 cm respectively, find the area of the washer. Use the value 3.14 for π.

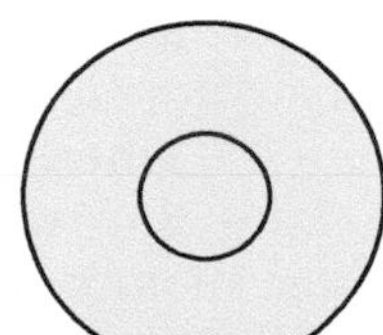

Fig. 1.3

Let the outer and inner radii of the washer be R cm and r cm respectively.

$$\begin{aligned}\text{Area of washer} &= \pi R^2 - \pi r^2 \\ &= \pi(R^2 - r^2) \\ &= \pi(R + r)(R - r)\end{aligned}$$

and $R = 1\frac{1}{2}$ and $r = \frac{1}{2}$.

$$\begin{aligned}\text{Area of washer} &= \pi(1\tfrac{1}{2} + \tfrac{1}{2})(1\tfrac{1}{2} - \tfrac{1}{2})\text{ cm}^2 \\ &= \pi \times 2 \times 1\text{ cm} \\ &= 2\pi\text{ cm}^2 \\ &= 2 \times 3.14\text{ cm}^2 \\ &= 6.28\text{ cm}^2\end{aligned}$$

Exercise 1m

In questions 1–10, use the difference of two squares to find the values of the given numerical expressions.

1. $96^2 - 4^2$
2. $118^2 - 18^2$
3. $73^2 - 71^2$
4. $98^2 - 4$
5. $103^2 - 9$
6. $52^2 - 48^2$
7. $63^2 - 37^2$
8. $57^2 - 55^2$
9. $1004^2 - 16$
10. $997^2 - 9$
11. A metal washer has an outer diameter of 14 mm and an inner diameter of 6 mm. Use the value 3.14 for π to find the area of the metal washer in mm^2.
12. A cylindrical metal pipe, 1 m long, has inner and outer radii of 3.7 cm and 2.3 cm respectively. Use the value $\frac{22}{7}$ for π to find the volume of metal in the pipe in cm^3, correct to 3 significant figures.

Factorisation of quadratic expressions by grouping

In Examples 7–12, there was only one variable in the quadratic expression. The coefficient of the quadratic term was always 1 so that the coefficient of the variable in both factors was the same, that is, 1. Therefore, the factors of the numeric term may be put in either bracket. However, there are many quadratic expressions which are not so simple. For example $2a^2 + 7a - 15$, where the coefficient of a^2 is 2, and $2x^2 + 7xy - 15y^2$ which contains more than one letter. Factors of such expressions may be found by trial and error as before. Read the following solutions to Examples 7, 9, 11 and 12 carefully. These use an alternative method, leading to factorisation by grouping.

Example 7 (alternative method)

Factorise the quadratic expression $x^2 + 7x + 10$.

1st step: Find the product of the first and last terms:

$$x^2 \times (+10) = +10x^2$$

2nd step: Find two terms such that their product is $+10x^2$ and their sum is $+7x$ (the middle term).

	factors of $+10x^2$	*sum of factors*
(a)	$+10x$ and $+x$	$+11x$
(b)	$-10x$ and $-x$	$-11x$
(c)	$+5x$ and $+2x$	$+7x$
(d)	$-5x$ and $-2x$	$-7x$

Of these, only (c) gives the required answer.

3rd step: Replace $+7x$ with $+5x$ and $+2x$ in the given expression and factorise by grouping.

$$\begin{aligned}x^2 + 7x + 10 &= x^2 + 5x + 2x + 10 \\ &= x(x + 5) + 2(x + 5) \\ &= (x + 5)(x + 2)\end{aligned}$$

Example 9 (alternative method)

Factorise the quadratic expression $x^2 - 9x + 8$.

1st step: Find the product of the first and last terms:

$$x^2 \times (+8) = +8x^2$$

2nd step: Find two terms such that their product is $+8x^2$ and their sum is $-9x$ (the middle term)

	factors of $+8x^2$	*sum of factors*
(a)	$+8x$ and $+x$	$+9x$
(b)	$-8x$ and $-x$	$-9x$
(c)	$+4x$ and $+2x$	$+6x$
(d)	$-4x$ and $-2x$	$-6x$

Of these, only (b) gives the required answer.

3rd step: Replace $-9x$ with $-8x$ and $-x$ in the given expression and factorise by grouping.

$$\begin{aligned}x^2 - 9x + 8 &= x^2 - 8x - x + 8 \\ &= x(x - 8) - 1(x - 8) \\ &= (x - 8)(x - 1)\end{aligned}$$

Example 11 (alternative method)

Factorise the quadratic expression $x^2 + 2x - 15$.

1st step: Find the product of the first and last terms: $x^2 \times (-15) = -15x^2$

2nd step: Find two terms such that their product is $-15x^2$ and their sum is $+2x$ (the middle term).

	factors of $-15x^2$	*sum of factors*
(a)	$+15x$ and $-x$	$+14x$
(b)	$-15x$ and $+x$	$-14x$
(c)	$+5x$ and $-3x$	$+2x$
(d)	$-5x$ and $+3x$	$-2x$

Of these, only (c) gives the required answer.

3rd step: Replace $+2x$ with $+5x$ and $-3x$ in the given expression and factorise by grouping.

$$\begin{aligned} x^2 + 2x - 15 &= x^2 + 5x - 3x - 15 \\ &= x(x + 5) - 3(x + 5) \\ &= (x + 5)(x - 3) \end{aligned}$$

Example 12 (alternative method)

Factorise the quadratic expression $x^2 - 4x - 12$.

1st step: Find the product of the first and last terms: $x^2 \times (-12) = -12x^2$

2nd step: Find two terms such that their product is $-12x^2$ and their sum is $-4x$ (the middle term)

	factors of $-12x^2$	*sum of factors*
(a)	$+12x$ and $-x$	$+11x$
(b)	$-12x$ and $+x$	$-11x$
(c)	$+6x$ and $-2x$	$+4x$
(d)	$-6x$ and $+2x$	$-4x$
(e)	$+4x$ and $-3x$	$+x$
(f)	$-4x$ and $+3x$	$-x$

Of these, only (d) gives the required answer.

3rd step: Replace $-4x$ with $-6x$ and $+2x$ in the given expression and factorise by grouping.

$$\begin{aligned} x^2 - 4x - 12 &= x^2 - 6x + 2x - 12 \\ &= x(x - 6) + 2(x - 6) \\ &= (x - 6)(x + 2) \end{aligned}$$

This method is recommended when factorising more difficult quadratic expressions. Consider the following examples.

Example 20

Factorise $2a^2 + 7a - 15$.

1st step: Find the product of the first and last terms: $2a^2 \times (-15) = -30a^2$

2nd step: Find two terms such that their product is $-30a^2$ and their sum is $+7a$ (the middle term).

	factors of $-30a^2$	*sum of factors*
(a)	$-30a$ and $+a$	$-29a$
(b)	$+30a$ and $-a$	$+29a$
(c)	$-15a$ and $+2a$	$-13a$
(d)	$+15a$ and $-2a$	$+13a$
(e)	$-10a$ and $+3a$	$-7a$
(f)	$+10a$ and $-3a$	$+7a$
(g)	$-6a$ and $+5a$	$-a$
(h)	$+6a$ and $-5a$	$+a$

Of these, only (f) gives the required result.

3rd step: Replace $+7a$ in the given expression with $+10a - 3a$.
Factorise by grouping.

$$\begin{aligned} 2a^2 + 7a - 15 &= 2a^2 + 10a - 3a - 15 \\ &= 2a(a + 5) - 3(a + 5) \\ &= (a + 5)(2a - 3) \end{aligned}$$

In Example 20, all 8 possibilities were written down. In practice there is no need to do this. The method can be shortened as follows.

1 Notice in the second step that the sum of the two terms is positive. It is necessary to consider only those factors in which the positive term is greater numerically than the negative term.

2 Stop when the required result is reached.

Example 21

Factorise $7 - 22x + 3x^2$.

1st step: $7 \times (+3x^2) = 21x^2$

2nd step: Find two terms such that their sum is $-22x$ and their product is $+21x^2$. Since the middle term is negative, consider negative factors only.

factors of $+21x^2$	*sum of factors*
$-21x$ and $-x$	$-22x$ (stop)

3rd step: Replace $-22x$ with $-21x - x$ in the given expression.
Factorise by grouping.

$$\begin{aligned} 7 - 22x + 3x^2 &= 7 - 21x - x + 3x^2 \\ &= 7(1 - 3x) - x(1 - 3x) \\ &= (1 - 3x)(7 - x) \end{aligned}$$

Example 22

Factorise $8x^2 - 14x - 9$.

1st step: $8x^2 \times (-9) = -72x^2$

2nd step:

factors of $-72x^2$	*sum of factors*
$-72x$ and $+x$	$-71x$
$-36x$ and $+2x$	$-34x$
$-18x$ and $+4x$	$-14x$ (stop)

3rd step:

$$\begin{aligned} 8x^2 - 14x - 9 &= 8x^2 - 18x + 4x - 9 \\ &= 2x(4x - 9) + 1(4x - 9) \\ &= (4x - 9)(2x + 1) \end{aligned}$$

Notice in Example 22 that the middle term is negative. It was necessary to consider only those factors in which the negative term was greater numerically than the positive term.

Example 23

Factorise $6a^2 + 15a + 9$.

3 is a common factor. First take out the common factor.

$6a^2 + 15a + 9 = 3(2a^2 + 5a + 3)$
$2a^2 \times (+3) = 6a^2$

factors of $+6a^2$	*sum of factors*
$+6a$ and $+a$	$+7a$
$+3a$ and $+2a$	$+5a$ (stop)

$$\begin{aligned} 6a^2 + 15a + 9 &= 3(2a^2 + 5a + 3) \\ &= 3(2a^2 + 3a + 2a + 3) \\ &= 3[a(2a + 3) + 1(2a + 3)] \\ &= 3(2a + 3)(a + 1) \end{aligned}$$

Example 24

Factorise
(a) $2a^2 + 7ab - 15b^2$ (b) $2a^2b^2 + 7ab - 15$.

Notice that these two examples have the same coefficients as the expression in Example 20.

(a) $2a^2 + 7ab - 15b^2 = (a + 5b)(2a - 3b)$

(b) $2a^2b^2 + 7ab - 15 = (ab + 5)(2ab - 3)$

Use multiplication to check the solutions to Example 24.

Exercise 1n

Factorise the following using the alternative method.

1. $a^2 + 8a + 15$
2. $b^2 - 7b + 10$
3. $c^2 + 4c - 21$
4. $d^2 - 5d - 14$
5. $e^2 + 2e - 8$
6. $w^2 + 5w + 6$
7. $x^2 + 5x - 6$
8. $y^2 - 5y + 6$
9. $z^2 - 5z - 6$
10. $2d^2 + 3d + 1$
11. $2e^2 - 3e + 1$
12. $2f^2 - f - 1$
13. $a^2b^2 + 7ab + 10$
14. $a^2 + 7ab + 10b^2$
15. $a^2b + 7ab + 10$
16. $x^2 - 2xy - 15y^2$
17. $m^2 + 10m - 24$
18. $n^2 - 10n - 24$
19. $u^2 - 10u + 24$
20. $v^2 - 11v + 24$
21. $m^2 + 4m - 21$
22. $m^2 + 4mn - 21n^2$
23. $m^2n^2 + 4mn - 21$
24. $3a^2 - 4a + 1$
25. $3b^2 + b - 2$
26. $2x^2 + 5x - 3$
27. $2y^2 - 5y - 3$
28. $2z^2 - 5z + 3$
29. $1 + 3m + 2m^2$
30. $15 - 2n - n^2$
31. $1 - 2u - 8u^2$
32. $u^2 + 2uv - 8v^2$
33. $a^2 + 5ab - 36b^2$
34. $a^2 + 9ab - 36b^2$
35. $a^2 + 16ab - 36b^2$
36. $2b^2 - 10b + 12$
37. $c^2 - 4c - 77$
38. $77 - 4d - 4d^2$
39. $3e^2 + 3e - 18$
40. $3f^2 + 2f - 1$
41. $a^2 + 4ab + 3b^2$
42. $1 + 4x + 3x^2$
43. $2g^2 - 5g + 2$
44. $2h^2 + 5h + 3$
45. $3h^2 + 7hk + 2k^2$
46. $12x^2 - 13x - 14$
47. $a^2 + 25a - 150$
48. $b^2 + 25b + 150$
49. $3c^2 - 11c + 6$
50. $3d^2 + 7d - 6$

Exercise 1o

Factorise the following.

1. $5e^2 - 9e - 2$
2. $7f^2 + 10f + 3$
3. $35 - 12a + a^2$
4. $35 - 2b - b^2$
5. $35 + 36c + c^2$
6. $35 + 30d - 5d^2$
7. $3a^2 + 5ab + 2b^2$
8. $3m^2 + 5mn - 2n^2$
9. $3u^2 + 7uv + 2v^2$
10. $6n^2 - 7n - 3$
11. $7v^2 + 22v + 3$
12. $4y^2 - 12y + 5$
13. $2h^2 - 15h - 27$
14. $2k^2 - 15k + 27$
15. $x^2y^2 - xy - 30$
16. $2u^2v^2 + uv - 6$

17 $5 - 7a - 6a^2$
18 $10p^2 - 41p - 45$
19 $10q^2 - 43q + 45$
20 $8a^2 - 17a + 9$
21 $8b^2 - 18b + 9$
22 $8c^2 - 21c - 9$
23 $8d^2 - 22d + 9$
24 $8e^2 - 49e + 75$
25 $8f^2 - 50f + 75$
26 $12a^2b^2 + 11ab - 5$
27 $12m^2 - 4mn - 5n^2$
28 $12t^2 - 11t + 2$
29 $12x^2y^2 - 11xy - 1$
30 $24p^2 + pq - 23q^2$

Summary

To **factorise** an expression is to write the expression as a product of its factors. To do this, find the HCF of the terms in the expression. It may be necessary first to **collect like** terms before finding the HCF.

Calculations may be made much simpler by first factorising an expression.

In a quadratic expression, the highest power of the variable(s) is 2, example, x^2, y^2. To **factorise** a quadratic expression is to express it as a product of its linear factors. Note, for example,

$$a^2 + 2ab + b^2 = (a + b)^2$$
$$a^2 - 2ab + b^2 = (a - b)^2$$
$$a^2 - b^2 = (a + b)(a - b)$$

Practice exercise P1.1

Factorise the following.

1 $-3g - 12rg$
2 $2k + 2k^3 + 2$
3 $3(c + 2) + 2(c + 2)^2$
4 $6x + 3yx - x^2$
5 $rd^2 - 2rd + 4dr^2$
6 $\dfrac{6y - 4}{6y + 2}$
7 $4bc + 6bcd + 2b$
8 $-5x^2 - 10x + 10$
9 $3t^3 - 3t^2 - 3st$
10 $4(l - b)h - 8(l - b)k$
11 $x(y + 1)^2 + 2b(y + 1)$
12 $8(3 - f)g - 2(3 - f)^2$
13 $3f(g + h) - 2(g + h)$
14 $(R - h)^2 + 2(R - h)$
15 $(2n^2 - 3n) + (6mn - 9m)$
16 $(x^2 + xy) + (xy + y^2)$
17 $\dfrac{5a^2 - a^2}{2a}$
18 $\dfrac{3y^2}{6y + 12}$
19 $\dfrac{9xy - 6y}{3yx + 2y}$
20 $\dfrac{6bc - 4b^3}{4b - 2bc}$

Practice exercise P1.2

Expand the following.

1 $(x + 3)(2x + 7)$
2 $3x(x - 1)$
3 $(a + 3)(b + 1)$
4 $(c - 2)(d + 1)$
5 $(a + 4)(b + 3)$
6 $(2x + 3)(y + 2)$
7 $(c - 2)(d + 5)$
7 $(x - c)(y - d)$
9 $(2f - 3)(g - 1)$
10 $(2k + 3)(2k - 1)$
11 $(3k + n)(m + 4n)$
12 $(2p + s)(p - 3s)$
13 $(3m + 2)^2$
14 $(x + 3y)(2x + y)$
15 $(d - 2r)(d + r)$
16 $(d + 2)^2$
17 $(d + 2)(d + 5)$
18 $(2 - t)(3 - t)$
19 $(3n + 1)(n - 1)$
20 $(a + 2b)(3a + b)$
21 $(3 + g)^2$
22 $(m - 2n)^2$
23 $(t - 3)^2$
24 $(2d + 1)^2$
25 $(2x - 3y)^2$
26 $(3 - 2y)^2$
27 $(-x - 3)(2x - 1)$
28 $2(3x - 2)(2x + 3)$
29 $(3x + 1)^2$
30 $(2x + 5)(2x - 5)$
31 $(n + 1)(n - 1)$
32 $(3v - 1)(3v + 1)$
33 $(x - 2y)(x + 2y)$
34 $(2x + 1)^2 + (x - 3)^2$
35 $(2x - y)^2 - (x + y)^2$
36 $(x - y)^2 - (x + y)^2$

Practice exercise P1.3

In the following expansions, write down the coefficient of (a) k^2, (b) k, (c) the constant term.

1 $(k + 4)(k + 2)$
2 $(4k + 3)(k + 1)$
3 $(k - 2)(k + 2)$
4 $(k - 3)(k - 1)$
5 $(k + 3)(k - 3)$
6 $(k + 5)(k - 5)$
7 $(k + 4)(k + 4)$
8 $(2k - 3)(2k + 3)$
9 $(4k + 1)(4k - 1)$
10 $(2k - 1)(2k + 1)$
11 $(3k + 1)(k - 3)$
12 $(2k - 1)(k + 4)$
13 $(k + 2)(k + 2)$
14 $(3k - 1)(k + 1)$
15 $(k - 3)(k - 3)$
16 $(k + 7)(k + 7)$
17 $(k + 1)(k + 1)$
18 $(k - 1)(k - 1)$
19 $(k - 4)(k - 4)$
20 $(k + 3)(k + 3)$

Practice exercise P1.4

In each of the following, add the term that will make the expression a perfect square.

1. $x^2 + 2x + \ldots$
2. $x^2 - 8x + \ldots$
3. $x^2 + \ldots + 16$
4. $\ldots + 14x + 49$
5. $x^2 + 6x + \ldots$
6. $x^2 - 6x + \ldots$

Practice exercise P1.5

Use quadratic expressions to find the value of the following without using a calculator or doing long multiplication.

1. 19^2
2. 31^2
3. $23^2 - 19^2$
4. 98^2
5. $75^2 - 70^2$
6. $83^2 - 17^2$

Practice exercise P1.6

Factorise the following.

1. $6am - 3bm + 9bn - 18an$
2. $x^2 + 5x - 3x - 15$
3. $h^2 - h - hr + r$
4. $ac + bc + 2ad + 2bd$
5. $3ux - 6uy - wx + 2wy$
6. $2rt - 3st + 2t^2 - 3rs$
6. $15 - 3v + 5w - vw$
8. $6ab^2c - 2a^2b^2 - 3abc + a^2b$
9. $6acx + 4acy - 12adx - 8ady$
10. $5y^2 - 10y - 3y + 6$

Practice exercise P1.7

Factorise the following.

1. $a^2 + 3a + 2$
2. $b^2 - 5b + 6$
3. $c^2 + 5c + 6$
4. $f^2 + 4f - 12$
5. $p^2 + 7p - 8$
6. $r^2 - 12r + 20$
7. $x^2 + 16x + 64$
8. $m^2 - 12m + 36$
9. $k^2 + 22k + 121$
10. $1 - 2y + y^2$
11. $x^2 + 14xy + 49y^2$
12. $81 + 18n + n^2$
13. $2t^2 - 7t + 6$
14. $d^2 + 6d + 5$
15. $2b^2 - 10b + 12$
16. $4h^2 - 12h - 16$
17. $g^2 - 7g + 12$
18. $j^2 + 8j + 12$
19. $3v^2 + 6v - 9$
20. $2w^2 - 10w + 12$
21. $4y^2 - 8y + 4$
22. $2y^2 + 7y + 3$
23. $x^2 - 15xy + 54y^2$
24. $2y^2 + 5y - 3$
25. $2y^2 - 5y + 3$
26. $4x^2 - 8xy + 3y^2$
27. $4x^2 - xy - 3y^2$
28. $4x^2 - 13xy + 3y^2$
29. $p^2 + 8p + 16$
30. $r^2 - 10r + 25$
31. $2t^2 + 12t + 18$
32. $d^2 + 6df + 9f^2$
33. $2b^2 - 8b + 8$
34. $3y^2 + 10y + 3$
35. $y^2 + 9y + 8$
36. $3y^2 + 6y + 3$
37. $3y^2 - 6y + 3$
38. $3x^2 - 8x - 3$
39. $6x^2 - 7x - 3$
40. $6x^2 + 7x - 3$
41. $6m^2 - 6$
42. $6m^2 - 12m + 6$
43. $6m^2 - 13m + 6$
44. $3y^2 + 10y + 3$
45. $(x - 3)^2 - 4y^2$
46. $8c^2 - 18d^2$
47. $2y^2 - 11y + 15$
48. $27x^2 - 12y^2$
49. $9y^2 - 6y + 1$
50. $x^2 - 3x - 10$

Practice exercise P1.8

Use factorisation to evaluate the following.

1. $73.41^2 - 26.59^2$
2. $60.72^2 - 0.28^2$
3. $\pi R^2 h - \pi r^2 h$, where $\pi = 3\frac{1}{7}$, $R = 17$ cm, $r = 11$ cm and $h = 9$ cm
4. $\dfrac{32}{5.4^2 - 4.6^2}$
5. $\dfrac{3.6 \times 68 + 3.6 \times 32}{6.8^2 - 3.2^2}$

Practice exercise P1.9

Use factorisation to simplify the following.

1. $\dfrac{x^3 - 16x}{x^2 - 4x}$
2. $\dfrac{x^2 - 4}{x^2 - x - 6}$
3. $\dfrac{2x^2 + 3x - 2}{x^2 - 4}$
4. $\dfrac{b + 6b^2}{1 - 36b^2}$

Chapter 2
Matrices

Pre-requisites

- Pre-requisites: directed numbers; arithmetic operations; cummutative and distributive laws

Matrices

It is quite common to store information in lists and tables. For example, Table 2.1 shows the amounts of bread, sugar and milk used by the Mills and Pitt families in one week.

Table 2.1

	Mills family	Pitt family
bread (loaves	16	15
sugar (kg)	4	$5\frac{1}{2}$
milk (bottles)	22	20

The numbers in Table 2.1 can also be written as a **matrix**:

$$\begin{pmatrix} 16 & 15 \\ 4 & 5\frac{1}{2} \\ 22 & 20 \end{pmatrix}$$

A matrix (plural matrices) is simply a set of numbers of elements arranged in a rectangular pattern. In the matrix above there are 3 **rows** and 2 **columns**; we say that the **order** of the matrix is **3 by 2**. The order of the matrix gives its size in terms of rows and columns. The number of rows is always written first.

The following are other examples of matrices:

$$\begin{pmatrix} 1 & 12 & 1 \\ 3 & -1 & 2 \end{pmatrix} \quad \begin{pmatrix} 4 \\ 7 \end{pmatrix} \quad (5, 7, 6) \quad \begin{pmatrix} 2 & 4 \\ 5 & 0 \end{pmatrix}$$

(a) (b) (c) (d)

(a) is a 2×3 matrix
(b) is a 2×1 **column matrix**
(c) is a 1×3 **row matrix**
(d) is a 2×2 **square matrix** of order 2

The elements of a row matrix are often separated by commas, but this is not essential.

Exercise 2a

1. Table 2.2 shows the amounts of bread, sugar and milk used by the Mills and Pitt families in the next week.

   Table 2.2

	Mills family	Pitt family
bread (loaves	15	10
sugar (kg)	$3\frac{1}{2}$	$4\frac{1}{2}$
milk (bottles)	18	20

   (a) Write the numbers in Table 2.2 as a matrix.
   (b) Write Mrs Mills' shopping list as a 3×1 column matrix.
   (c) Write Mrs Pitt's shopping list as a 1×3 row matrix.
   (d) Write the numbers of bottles of milk as a row matrix.

2. State the orders of the following matrices.

   (a) $\begin{pmatrix} 1 & 0 & -1 & 2 & 4 \\ 2 & 2 & 1 & -3 & 1 \end{pmatrix}$ (b) $\begin{pmatrix} x \\ y \end{pmatrix}$

   (c) $\begin{pmatrix} x & 1 & 0 \\ y & 2 & 1 \\ z & 3 & 0 \end{pmatrix}$ (d) $\begin{pmatrix} 8 & 2 \\ 9 & -3 \\ 3 & 1 \end{pmatrix}$

   (e) (4 8 10 5) (f) $\begin{pmatrix} 1 & 0 \\ 0 & 1 \end{pmatrix}$

3. Write down examples of
   (a) any 3×5 matrix
   (b) any 2×4 matrix
   (c) any 4×2 matrix
   (d) any 4×5 matrix

4. Write down any square matrix of order
   (a) 1 (b) 2 (c) 3 (d) 4

5. How many elements are in a matrix of order
   (a) 2×3 (b) 3×2 (c) 4 by 3
   (d) m by n (e) x by x (f) 1×4

6. Given the matrix $\begin{pmatrix} a & b & c & d \\ e & f & g & h \end{pmatrix}$ name the element in the
   (a) first row, first column
   (b) second row, third column
   (c) second row, second column
   (d) first row, last column.

7 $\begin{pmatrix} 9 & s & t \\ x & 0 & 5 \\ 1 & y & -2 \\ -1 & 8 & k \end{pmatrix}$

Given the above matrix, name the row and column in which the following elements appear.

(a) 5 (b) −1 (c) s (d) k
(e) 1 (f) 0 (g) t (h) y

8 Two used-car dealers have the following cars for sale: *Mike's Motors* has 11 Peugeots, 3 Fords and 5 VWs; *Pete's Cars* has 8 Peugeots and 2 VWs only. Show this information in a 2 × 3 matrix.

Addition and subtraction of matrices

The numbers in Tables 2.1 and 2.2 give the following matrices:

$$\begin{pmatrix} 16 & 15 \\ 4 & 5\frac{1}{2} \\ 22 & 20 \end{pmatrix} \text{ and } \begin{pmatrix} 15 & 10 \\ 3\frac{1}{2} & 4\frac{1}{2} \\ 18 & 20 \end{pmatrix}$$

To find the total amounts of food used by the families in the two weeks, add each number in one matrix to the corresponding number in the other matrix:

$$\begin{pmatrix} 16 & 15 \\ 4 & 5\frac{1}{2} \\ 22 & 20 \end{pmatrix} + \begin{pmatrix} 15 & 10 \\ 3\frac{1}{2} & 4\frac{1}{2} \\ 18 & 20 \end{pmatrix}$$

$$= \begin{pmatrix} 16 + 15 & 15 + 10 \\ 4\frac{1}{2} + 3\frac{1}{2} & 5\frac{1}{2} + 4\frac{1}{2} \\ 22 + 18 & 20 + 20 \end{pmatrix}$$

$$= \begin{pmatrix} 31 & 25 \\ 7\frac{1}{2} & 10 \\ 40 & 40 \end{pmatrix}$$

Matrices can be added only if they are of the same order. The resulting matrix is also of that order.

Example 1

If $A = \begin{pmatrix} 2 & 3 & 0 \\ 4 & 2 & 1 \\ 3 & 1 & -2 \end{pmatrix}$

and $B = \begin{pmatrix} 3 & -1 & 2 \\ 1 & 3 & 1 \\ 4 & 2 & -1 \end{pmatrix}$

show that A + B = B + A.

$$A + B = \begin{pmatrix} 2 & 3 & 0 \\ 4 & 2 & 1 \\ 3 & 1 & -2 \end{pmatrix} = \begin{pmatrix} 3 & -1 & 2 \\ 1 & 3 & 1 \\ 4 & 2 & -1 \end{pmatrix}$$

$$= \begin{pmatrix} 5 & 2 & 2 \\ 5 & 5 & 2 \\ 7 & 3 & -3 \end{pmatrix}$$

$$B + A = \begin{pmatrix} 3 & -1 & 2 \\ 1 & 3 & 1 \\ 4 & 2 & -1 \end{pmatrix} + \begin{pmatrix} 2 & 3 & 0 \\ 4 & 2 & 1 \\ 3 & 1 & -2 \end{pmatrix}$$

$$= \begin{pmatrix} 5 & 2 & 2 \\ 5 & 5 & 2 \\ 7 & 3 & -3 \end{pmatrix}$$

Hence A + B = B + A.

The method of subtraction follows the same pattern as that for addition. For example, if you want to find how much more food the Mills family used than the Pitt family during the first week, each number in the second matrix is taken from the corresponding number in the first matrix:

$$\begin{pmatrix} 16 & 15 \\ 4 & 5\frac{1}{2} \\ 22 & 20 \end{pmatrix} - \begin{pmatrix} 15 & 10 \\ 3\frac{1}{2} & 4\frac{1}{2} \\ 18 & 20 \end{pmatrix}$$

$$= \begin{pmatrix} 16 - 15 & 15 - 10 \\ 4\frac{1}{2} - 3\frac{1}{2} & 5\frac{1}{2} - 4\frac{1}{2} \\ 22 - 18 & 20 - 20 \end{pmatrix}$$

$$= \begin{pmatrix} 1 & 5 \\ \frac{1}{2} & 1 \\ 4 & 0 \end{pmatrix}$$

Example 2

Using the matrices A and B of Example 1, show that A − B ≠ B − A.

$$A - B = \begin{pmatrix} 2 & 3 & 0 \\ 4 & 2 & 1 \\ 3 & 1 & -2 \end{pmatrix} - \begin{pmatrix} 3 & -1 & 2 \\ 1 & 3 & 1 \\ 4 & 2 & -1 \end{pmatrix}$$

$$= \begin{pmatrix} -1 & 4 & -2 \\ 3 & -1 & 0 \\ -1 & -1 & -1 \end{pmatrix}$$

$$B - A = \begin{pmatrix} 3 & -1 & 2 \\ 1 & 3 & 1 \\ 4 & 2 & -1 \end{pmatrix} - \begin{pmatrix} 2 & 3 & 0 \\ 4 & 2 & 1 \\ 3 & 1 & -2 \end{pmatrix}$$

$$= \begin{pmatrix} 1 & -4 & 2 \\ -3 & 1 & 0 \\ 1 & 1 & 1 \end{pmatrix}$$

Hence A − B ≠ B − A. (However, it can be seen that A − B = −(B − A).)

Exercise 2b

Combine the following matrices where possible.

1. $\begin{pmatrix} 1 & 2 \\ 3 & 0 \end{pmatrix} + \begin{pmatrix} 2 & -1 \\ 4 & 5 \end{pmatrix}$
2. $\begin{pmatrix} 3 & 1 \\ 2 & 0 \\ 4 & 7 \end{pmatrix} + \begin{pmatrix} 1 & 2 \\ 3 & 1 \\ 2 & -3 \end{pmatrix}$
3. $\begin{pmatrix} 3 & -1 & -2 \\ 2 & 3 & 1 \end{pmatrix} - \begin{pmatrix} 2 & -2 & 3 \\ 1 & 3 & 1 \end{pmatrix}$
4. $(2, 1, 3) - (3, 4)$
5. $\begin{pmatrix} 4 \\ 3 \\ 1 \end{pmatrix} - \begin{pmatrix} 6 \\ 1 \\ 3 \end{pmatrix}$
6. $\begin{pmatrix} 1.3 & 4.2 \\ 3.1 & 6.2 \end{pmatrix} + \begin{pmatrix} 7.1 & -3.2 \\ -2.9 & 4.3 \end{pmatrix}$
7. $\begin{pmatrix} 3 \\ 1 \end{pmatrix} + \begin{pmatrix} -1 \\ 3 \\ 2 \end{pmatrix}$
8. $\begin{pmatrix} 1 & 3 & 1 \\ 2 & 4 & 1 \end{pmatrix} + \begin{pmatrix} 3 & -2 & 1 & 5 \\ 4 & 2 & 3 & 2 \end{pmatrix}$
9. $\begin{pmatrix} 2 & 1 & 3 \\ 1 & 3 & 2 \\ 4 & 1 & 1 \end{pmatrix} + \begin{pmatrix} 7 & -1 & 0 \\ 2 & 3 & -2 \\ 1 & 4 & 6 \end{pmatrix} + \begin{pmatrix} 3 & -2 & -1 \\ 4 & -1 & 3 \\ 2 & 3 & 1 \end{pmatrix}$
10. $\begin{pmatrix} 3 & 2 \\ 1 & 0 \\ -1 & 1 \end{pmatrix} - \begin{pmatrix} 2 & 3 \\ -1 & -2 \\ -1 & 3 \end{pmatrix}$
11. $\begin{pmatrix} 2 & 1.3 \\ 3 & 2.1 \end{pmatrix} - \begin{pmatrix} 4.2 & 3 \\ 2.1 & 4 \end{pmatrix}$
12. $\begin{pmatrix} 2 \\ 7 \end{pmatrix} + \begin{pmatrix} 0 \\ -9 \end{pmatrix} - \begin{pmatrix} 5 \\ -4 \end{pmatrix}$
13. $(1, 5, 8) + (6, -4, 0) - \begin{pmatrix} 2 \\ 4 \\ 3 \end{pmatrix}$
14. $\begin{pmatrix} 2 & 8 \\ 9 & 3 \end{pmatrix} + \begin{pmatrix} 1 & 5 \\ 2 & 9 \end{pmatrix} - \begin{pmatrix} 3 & 7 \\ 6 & 5 \end{pmatrix}$
15. $\begin{pmatrix} 4 & -1 \\ 2 & 2 \end{pmatrix} - \begin{pmatrix} -2 & 9 \\ 6 & -8 \end{pmatrix} + \begin{pmatrix} -5 & 0 \\ 4 & 1 \end{pmatrix}$

Scalar multiplication

Example 3

If $A = \begin{pmatrix} 2 & x \\ y & 0 \end{pmatrix}$, find 3A.

$$\begin{aligned} 3A &= A + A + A \\ &= \begin{pmatrix} 2 & x \\ y & 0 \end{pmatrix} + \begin{pmatrix} 2 & x \\ y & 0 \end{pmatrix} + \begin{pmatrix} 2 & x \\ y & 0 \end{pmatrix} \\ &= \begin{pmatrix} 4 & 2x \\ 2y & 0 \end{pmatrix} + \begin{pmatrix} 2 & x \\ y & 0 \end{pmatrix} \\ &= \begin{pmatrix} 6 & 3x \\ 3y & 0 \end{pmatrix} \end{aligned}$$

In Example 3 the working can be shortened by multiplying each element of A by 3:

$$\begin{aligned} 3A &= 3\begin{pmatrix} 2 & x \\ y & 0 \end{pmatrix} = \begin{pmatrix} 3 \times 2 & 3 \times x \\ 3 \times y & 3 \times 0 \end{pmatrix} \\ &= \begin{pmatrix} 6 & 3x \\ 3y & 0 \end{pmatrix} \end{aligned}$$

The number multiplying the matrix is called a **scalar**. To multiply a matrix by a scalar, multiply each element of the matrix by the scalar.

Example 4

Find x if

$$5\begin{pmatrix} 4 & 3 \\ 7 & 3 \end{pmatrix} - 2\begin{pmatrix} 1 & 4 \\ 3 & 2 \end{pmatrix} = \begin{pmatrix} w & x \\ y & z \end{pmatrix}$$

$$\begin{pmatrix} 20 & 15 \\ 35 & 15 \end{pmatrix} - \begin{pmatrix} 2 & 8 \\ 6 & 4 \end{pmatrix} = \begin{pmatrix} w & x \\ y & z \end{pmatrix}$$

$$\begin{pmatrix} 18 & 7 \\ 29 & 11 \end{pmatrix} = \begin{pmatrix} w & x \\ y & z \end{pmatrix}$$

Since the final matrices are equal, corresponding elements must be equal: $x = 7$.

Notice that two matrixes are **equal** if they are of the same order and their corresponding elements are equal.

Exercise 2c

1. If $A = \begin{pmatrix} 3 \\ 2 \end{pmatrix}$, $B = \begin{pmatrix} 2 & 1 \\ 0 & 1 \end{pmatrix}$, $C = (1, 2, 8)$, $D = \begin{pmatrix} 6 & 8 \\ 2 & 14 \end{pmatrix}$, find
   (a) 3A (b) 4B (c) −2C
   (d) $\frac{1}{2}$D (e) D + 3B (f) D − 3B
2. Find n if
   $$n\begin{pmatrix} 1 & 2 \\ 1 & 0 \end{pmatrix} + 2\begin{pmatrix} 0 & 1 \\ 3 & 5 \end{pmatrix} = \begin{pmatrix} 3 & 8 \\ 9 & 10 \end{pmatrix}$$
3. Find the matrix M which satisfies
   (a) $7M = 3\begin{pmatrix} -4 & 5 \\ 0 & 1 \end{pmatrix} + \begin{pmatrix} -2 & 6 \\ 0 & 4 \end{pmatrix}$
   (b) $\begin{pmatrix} 9 & 1 \\ 2 & 6 \end{pmatrix} + M = \begin{pmatrix} 3 & 7 \\ 6 & 8 \end{pmatrix} - M$
4. Find x and y if
   $$5\begin{pmatrix} 1 & 2 \\ 0 & -1 \end{pmatrix} + \begin{pmatrix} w & x \\ y & z \end{pmatrix} = 3\begin{pmatrix} 3 & 4 \\ -1 & 7 \end{pmatrix}$$
5. Find p and q if
   $$3\begin{pmatrix} p & 5 \\ 4 & 8 \end{pmatrix} - \begin{pmatrix} -2 & 7 \\ 0 & 2 \end{pmatrix} = 2\begin{pmatrix} -5 & 4 \\ 6 & q \end{pmatrix}$$

Multiplication

Row and column matrices

Look at the foods given in Table 2.1 on page 13. If bread costs 30c per loaf, sugar 90c per kg and milk 45c per bottle, then the row matrix (30, 90, 45) represents their respective costs. The total cost of the Mills' food is the following matrix product:

$$(30,\ 90,\ 45)\begin{pmatrix}16\\4\\22\end{pmatrix}$$
$$= (30\times16 + 90\times4 + 45\times22)$$
$$= (480 + 360 + 990)$$
$$= (1830)$$

The Mills' food costs 1830 cents, or \$18.30.

The total cost of the Pitts' food is the product:

$$(30,\ 90,\ 45)\begin{pmatrix}15\\5\frac{1}{2}\\20\end{pmatrix}$$
$$= (30\times15 + 90\times5\tfrac{1}{2} + 45\times20)$$
$$= (450 + 495 + 900)$$
$$= (1845)$$

The Pitts' food costs 1845 cents, or \$18.45.

The two calculations above can be combined as a single matrix product:

$$(30,\ 90,\ 45)\begin{pmatrix}16 & 15\\4 & 5\frac{1}{2}\\22 & 20\end{pmatrix} = (1830,\ 1845)$$

Notice that each element in the resulting matrix is made up by multiplying the elements of the row matrix by the elements of the columns in turn. It follows that for multiplication to be possible there must be as many columns in the first matrix as there are rows in the second matrix.

Example 5

Find the product $(2,\ 2,\ -4,\ 5)\begin{pmatrix}-3\\7\\-9\\-8\end{pmatrix}$

$$(2,\ 2,\ -4,\ 5)\begin{pmatrix}-3\\7\\-9\\-8\end{pmatrix}$$
$$= 2\times(-3) + 2\times7 + (-4)\times(-9) + 5\times(-8)$$
$$= -6 + 14 + 36 - 40$$
$$= 4$$

Example 6

A hotel has 8 single rooms and 14 double rooms. The costs per night of single and double rooms are \$70 and \$80 respectively. Use a matrix method to show how much money the hotel makes per night when full.

The row matrix \$(70, 80) represents the costs.

The column matrix $\begin{pmatrix}8\\14\end{pmatrix}$ represents the numbers of rooms.

Total income per night when full

$$= \$(70,\ 80)\begin{pmatrix}8\\14\end{pmatrix}$$
$$= \$(70\times8 + 80\times14)$$
$$= \$(560 + 1120)$$
$$= \$1680$$

Exercise 2d

1. Calculate the following products.

   (a) $(2,\ 3)\begin{pmatrix}5\\6\end{pmatrix}$ (b) $(4,\ -3)\begin{pmatrix}-3\\9\end{pmatrix}$

   (c) $(1,\ 2,\ 1)\begin{pmatrix}0\\0\\1\end{pmatrix}$ (d) $(2,\ 3,\ 1)\begin{pmatrix}5\\6\\7\end{pmatrix}$

   (e) $(1,\ 2,\ -1)\begin{pmatrix}\frac{1}{2}\\2\\2\frac{1}{2}\end{pmatrix}$ (f) $(6,\ 8,\ -3,\ \frac{1}{2})\begin{pmatrix}2\\-3\\-5\\-6\end{pmatrix}$

2. Do the following multiplications, *where possible.*

   (a) $(3,\ 6,\ 4)\begin{pmatrix}12\\10\\15\end{pmatrix}$ (b) $(3,\ 6,\ 4)\begin{pmatrix}11\\9\end{pmatrix}$

   (c) $(1,\ 1,\ 0)\begin{pmatrix}1\\0\end{pmatrix}$ (d) $(1,\ 0)\begin{pmatrix}0\\1\end{pmatrix}$

3. One night the hotel in Example 6 had guests in only 5 single rooms and 3 double rooms. Use a matrix method to find the income of the hotel for that night.

4. Given that bread costs 80c per loaf, sugar 100c per kg and milk 50c per bottle, find the costs for the Mills and Pitt families of the foodstuffs given in Table 2.1 on page 13.

5. A cinema has 400 seats upstairs and 600 seats downstairs, each seat costing \$2.00 and \$1.20 respectively. Use a matrix method to compare the income of the cinema on a night when it is full with that when only 305 upstairs and 420 downstairs were sold.

General matrix multiplication

Look at the following matrix multiplication.

$$(30, 40, 20)\begin{pmatrix}16 & 15\\ 4 & 5\frac{1}{2}\\ 22 & 20\end{pmatrix} = (1080, 1070)$$

If the row matrix is changed to (32, 50, 25), check that

$$(30, 50, 25)\begin{pmatrix}16 & 15\\ 4 & 5\frac{1}{2}\\ 22 & 20\end{pmatrix} = (1262, 1255)$$

These two results can be combined as a single matrix product:

$$\begin{pmatrix}30 & 40 & 20\\ 32 & 50 & 25\end{pmatrix}\begin{pmatrix}16 & 15\\ 4 & 5\frac{1}{2}\\ 22 & 20\end{pmatrix} = \begin{pmatrix}1080 & 1070\\ 1262 & 1255\end{pmatrix}$$

Notice, in this case that a 2×3 matrix multiplies a 3×2 matrix to give a 2×2 matrix product:

$$(2 \times \boxed{3) \times (3} \times 2) \rightarrow (2 \times 2)$$

Example 7

Multiply $\begin{pmatrix}3 & 2\\ 4 & 3\\ 6 & 0\\ 3 & 1\end{pmatrix}$ by $\begin{pmatrix}6 & 0\\ 1 & -1\end{pmatrix}$

$$\begin{pmatrix}3 & 2\\ 4 & 3\\ 6 & 0\\ 3 & 1\end{pmatrix}\begin{pmatrix}6 & 0\\ 1 & -1\end{pmatrix} = \begin{pmatrix}20 & -2\\ 27 & -3\\ 36 & 0\\ 19 & -1\end{pmatrix}$$

method:

The first element of the first row of the product is given by:

$$(3\ \ 2)\begin{pmatrix}6\\ 1\end{pmatrix} = 3 \times 6 + 2 \times 1 = 18 + 2 = 20$$

The second element of the first row is given by:

$$(3\ \ 2)\begin{pmatrix}0\\ -1\end{pmatrix} = 3 \times 0 + 2 \times (-1) = 0 + (-2) = -2$$

The first element of the second row is given by:

$$(4\ \ 3)\begin{pmatrix}6\\ 1\end{pmatrix} = 4 \times 6 + 3 \times 1 = 24 + 3 = 27$$

... and so on.

Notice in Example 7 that a 4×2 matrix multiplies a 2×2 matrix to give a 4×2 product:

$$(4 \times \boxed{2) \times (2} \times 2) \rightarrow (4 \times 2)$$

In order for it to be possible to multiply two matrices the first matrix must have the same number of columns as the second matrix has of rows. The product will have the same number of rows as the first matrix and the same number of columns as the second matrix.

Hence a $p \times q$ matrix will multiply a $q \times r$ matrix to give a $p \times r$ product:

$$(p \times \boxed{q) \times (q} \times r) \rightarrow (p \times r)$$

Example 8

If $A = \begin{pmatrix}3 & 2\\ 1 & 4\end{pmatrix}$ and $B = \begin{pmatrix}2 & -1\\ 0 & 3\end{pmatrix}$ show that $AB \neq BA$.

$$\begin{aligned}AB &= \begin{pmatrix}3 & 2\\ 1 & 4\end{pmatrix}\begin{pmatrix}2 & -1\\ 0 & 3\end{pmatrix}\\ &= \begin{pmatrix}3 \times 2 + 2 \times 0 & 3 \times (-1) + 2 \times 3\\ 1 \times 2 + 4 \times 0 & 1 \times (-1) + 4 \times 3\end{pmatrix}\\ &= \begin{pmatrix}6 & 3\\ 2 & 11\end{pmatrix}\end{aligned}$$

$$\begin{aligned}BA &= \begin{pmatrix}2 & -1\\ 0 & 3\end{pmatrix}\begin{pmatrix}3 & 2\\ 1 & 4\end{pmatrix}\\ &= \begin{pmatrix}2 \times 3 + (-1) \times 1 & 2 \times 2 + (-1) \times 4\\ 0 \times 3 + \ \ 3 \times 1 & 0 \times 2 + \ \ 3 \times 4\end{pmatrix}\\ &= \begin{pmatrix}5 & 0\\ 3 & 12\end{pmatrix}\end{aligned}$$

Hence $AB \neq BA$.

Exercise 2e

Find the following matrix products.

1. $\begin{pmatrix}3 & 2\\ 1 & 4\end{pmatrix}\begin{pmatrix}2\\ 3\end{pmatrix}$
2. $\begin{pmatrix}4 & 1\\ 3 & 2\end{pmatrix}\begin{pmatrix}3 & 1\\ 2 & 1\end{pmatrix}$
3. $\begin{pmatrix}3 & 1\\ 2 & 1\end{pmatrix}\begin{pmatrix}4 & 1\\ 3 & 2\end{pmatrix}$
4. $\begin{pmatrix}11 & 2\\ 5 & 1\end{pmatrix}\begin{pmatrix}1 & -2\\ -5 & 11\end{pmatrix}$
5. $\begin{pmatrix}2 & -1\\ -3 & 6\end{pmatrix}\begin{pmatrix}6 & 1\\ 3 & 2\end{pmatrix}$
6. $\begin{pmatrix}2 & 3 & 1\\ 4 & 2 & 1\\ 3 & 1 & 1\end{pmatrix}\begin{pmatrix}2\\ 1\\ 3\end{pmatrix}$
7. $\begin{pmatrix}2 & 2\\ 0 & -4\end{pmatrix}\begin{pmatrix}2 & 2\\ 0 & -4\end{pmatrix}$
8. $\begin{pmatrix}1 & 2 & -1\\ 3 & 1 & -2\end{pmatrix}\begin{pmatrix}1 & 2\\ -2 & 1\\ -3 & 0\end{pmatrix}$
9. $\begin{pmatrix}1 & 2\\ -2 & 1\end{pmatrix}\begin{pmatrix}1 & 2 & -1\\ 3 & 1 & -2\end{pmatrix}$
10. $\begin{pmatrix}1 & 0\\ 3 & 1\end{pmatrix}\begin{pmatrix}1 & 4 & -5\\ 1 & 7 & -2\end{pmatrix}$

Algebra of 2 × 2 matrices

Null (zero) matrix

If any 2 × 2 matrix $\begin{pmatrix} a & b \\ c & d \end{pmatrix}$ is pre-multiplied or post-multiplied by the matrix $\begin{pmatrix} 0 & 0 \\ 0 & 0 \end{pmatrix}$, the matrix $\begin{pmatrix} a & b \\ c & d \end{pmatrix}$ is reduced to the matrix $\begin{pmatrix} 0 & 0 \\ 0 & 0 \end{pmatrix}$.

Check that:

$$\begin{pmatrix} 0 & 0 \\ 0 & 0 \end{pmatrix}\begin{pmatrix} a & b \\ c & d \end{pmatrix} = \begin{pmatrix} 0 & 0 \\ 0 & 0 \end{pmatrix}$$

and

$$\begin{pmatrix} a & b \\ c & d \end{pmatrix}\begin{pmatrix} 0 & 0 \\ 0 & 0 \end{pmatrix} = \begin{pmatrix} 0 & 0 \\ 0 & 0 \end{pmatrix}$$

$\begin{pmatrix} 0 & 0 \\ 0 & 0 \end{pmatrix}$ is called the **null matrix** or **zero matrix**.

Identity matrix

If any 2 × 2 matrix $\begin{pmatrix} a & b \\ c & d \end{pmatrix}$ is pre-multiplied or post-multiplied by the matrix $\begin{pmatrix} 1 & 0 \\ 0 & 1 \end{pmatrix}$, the matrix $\begin{pmatrix} a & b \\ c & d \end{pmatrix}$ remains unchanged.

Check that:

$$\begin{pmatrix} 1 & 0 \\ 0 & 1 \end{pmatrix}\begin{pmatrix} a & b \\ c & d \end{pmatrix} = \begin{pmatrix} a & b \\ c & d \end{pmatrix}$$

and

$$\begin{pmatrix} a & b \\ c & d \end{pmatrix}\begin{pmatrix} 1 & 0 \\ 0 & 1 \end{pmatrix} = \begin{pmatrix} a & b \\ c & d \end{pmatrix}$$

$\begin{pmatrix} 1 & 0 \\ 0 & 1 \end{pmatrix}$ is called the **identity matrix**. It is given the symbol I, that is, $I = \begin{pmatrix} 1 & 0 \\ 0 & 1 \end{pmatrix}$.

Inverse of a 2 × 2 matrix

If $\begin{pmatrix} p & q \\ r & s \end{pmatrix}$ is a matrix such that

$$\begin{pmatrix} p & q \\ r & s \end{pmatrix}\begin{pmatrix} a & b \\ c & d \end{pmatrix} = I = \begin{pmatrix} a & b \\ c & d \end{pmatrix}\begin{pmatrix} p & q \\ r & s \end{pmatrix}$$

then $\begin{pmatrix} p & q \\ r & s \end{pmatrix}$ is the **inverse** of $\begin{pmatrix} a & b \\ c & d \end{pmatrix}$.

Given that

$$\begin{pmatrix} p & q \\ r & s \end{pmatrix}\begin{pmatrix} a & b \\ c & d \end{pmatrix} = \begin{pmatrix} 1 & 0 \\ 0 & 1 \end{pmatrix}$$

then
$$\begin{aligned} pa + qc &= 1 && (1)\\ pb + qd &= 0 && (2)\\ ra + sc &= 0 && (3)\\ rb + sd &= 1 && (4) \end{aligned}$$

$$\begin{aligned} (1)\times d:&\quad pad + qcd = d && (5)\\ (2)\times c:&\quad pbc + qcd = 0 && (6)\\ (5)-(6):&\quad pad - pbc = d\\ &\quad p(ad - bc) = d \end{aligned}$$

$$p = \frac{d}{ad - bc}$$

Similarly $q = \dfrac{-b}{ad - bc}$

and $r = \dfrac{-c}{ad - bc}$

and $s = \dfrac{a}{ad - bc}$

$$\begin{pmatrix} p & q \\ r & s \end{pmatrix} = \begin{pmatrix} \dfrac{d}{ad - bc} & \dfrac{-b}{ad - bc} \\ \dfrac{-c}{ad - bc} & \dfrac{a}{ad - bc} \end{pmatrix} = \frac{1}{ad - bc}\begin{pmatrix} d & -b \\ -c & a \end{pmatrix}$$

Hence, the inverse of any 2 × 2 matrix $\begin{pmatrix} a & b \\ c & d \end{pmatrix}$ is $\dfrac{1}{ad - bc}\begin{pmatrix} d & -b \\ -c & a \end{pmatrix}$

$ad - bc$ is called the **determinant** of the matrix $\begin{pmatrix} a & b \\ c & d \end{pmatrix}$.

To find the inverse of a 2 × 2 matrix:

1 interchange the top left-hand and bottom right-hand elements;

2 multiply the other two elements by −1;

3 divide the resulting matrix by the determinant of the original matrix.

Example 9

Find the inverse of

(a) $\begin{pmatrix} 3 & -2 \\ 4 & 1 \end{pmatrix}$ (b) $\begin{pmatrix} 2 & 5 \\ 3 & 8 \end{pmatrix}$ (c) $\begin{pmatrix} 6 & 3 \\ 2 & 1 \end{pmatrix}$.

(a) The determinant of $\begin{pmatrix} 3 & -2 \\ 4 & 1 \end{pmatrix}$ is

$3 \times 1 - 4 \times (-2) = 3 + 8 = 11$

Its inverse is $\dfrac{1}{11}\begin{pmatrix} 1 & 2 \\ -4 & 3 \end{pmatrix}$

Check: $\dfrac{1}{11}\begin{pmatrix} 1 & 2 \\ -4 & 3 \end{pmatrix}\begin{pmatrix} 3 & -2 \\ 4 & 1 \end{pmatrix}$

$$= \frac{1}{11}\begin{pmatrix} 11 & 0 \\ 0 & 11 \end{pmatrix} = \begin{pmatrix} 1 & 0 \\ 0 & 1 \end{pmatrix}$$

(b) The determinant of $\begin{pmatrix} 2 & 5 \\ 3 & 8 \end{pmatrix}$ is

$2 \times 8 - 3 \times 5 = 16 - 15 = 1$

Its inverse is $\frac{1}{1}\begin{pmatrix} 8 & -5 \\ -3 & 2 \end{pmatrix} = \begin{pmatrix} 8 & -5 \\ -3 & 2 \end{pmatrix}$

Check: $\begin{pmatrix} 8 & -5 \\ -3 & 2 \end{pmatrix}\begin{pmatrix} 2 & 5 \\ 3 & 8 \end{pmatrix}$

$= \begin{pmatrix} 1 & 0 \\ 0 & 1 \end{pmatrix}$

(c) The determinant of $\begin{pmatrix} 6 & 3 \\ 2 & 1 \end{pmatrix}$ is

$6 \times 1 - 2 \times 3 = 0$

The inverse of the given matrix would contain the fraction $\frac{1}{0}$. Since division by 0 is impossible,

it follows that $\begin{pmatrix} 6 & 3 \\ 2 & 1 \end{pmatrix}$ has no inverse.

Notice in Example 9 that a matrix whose determinant is zero has no inverse. Such matrices are called **singular** matrices.

Exercise 2f

Find the inverses of the following matrices, where possible. Use multiplication to check each result.

1. $\begin{pmatrix} 6 & 3 \\ 1 & 2 \end{pmatrix}$
2. $\begin{pmatrix} 5 & 3 \\ 2 & 3 \end{pmatrix}$
3. $\begin{pmatrix} 4 & 3 \\ 2 & 1 \end{pmatrix}$
4. $\begin{pmatrix} 4\frac{1}{2} & 3 \\ 5 & 3 \end{pmatrix}$
5. $\begin{pmatrix} 2 & 3 \\ 1 & 2 \end{pmatrix}$
6. $\begin{pmatrix} 3 & 2 \\ 2 & 1 \end{pmatrix}$
7. $\begin{pmatrix} 6 & 14 \\ 3 & 7 \end{pmatrix}$
8. $\begin{pmatrix} -5 & 3 \\ -1 & 4 \end{pmatrix}$
9. $\begin{pmatrix} 2 & -9 \\ 8 & -6 \end{pmatrix}$
10. $\begin{pmatrix} -5 & -4 \\ -2 & 0 \end{pmatrix}$
11. $\frac{1}{2}\begin{pmatrix} 4 & 1 \\ 16 & 3 \end{pmatrix}$
12. $\frac{1}{6}\begin{pmatrix} 0 & -3 \\ 2 & 0 \end{pmatrix}$

Matrices as operators

Simultaneous linear equations

If $\begin{pmatrix} 9 & 4 \\ 2 & 1 \end{pmatrix}\begin{pmatrix} x \\ y \end{pmatrix} = \begin{pmatrix} 17 \\ 4 \end{pmatrix}$

then $\begin{pmatrix} 9x + 4y \\ 2x + y \end{pmatrix} = \begin{pmatrix} 17 \\ 4 \end{pmatrix}$

or $9x + 4y = 17$ (1)

and $2x + y = 4$ (2)

Hence the simultaneous equations (1) and (2) can be written as a single matrix equation:

$$\begin{pmatrix} 9 & 4 \\ 2 & 1 \end{pmatrix}\begin{pmatrix} x \\ y \end{pmatrix} = \begin{pmatrix} 17 \\ 4 \end{pmatrix} \quad (3)$$

In (3) the matrix $\begin{pmatrix} 9 & 4 \\ 2 & 1 \end{pmatrix}$ multiplies $\begin{pmatrix} x \\ y \end{pmatrix}$.

Multiplication is an arithmetical operation; we say that the matrix acts as an **operator**.

The inverse of $\begin{pmatrix} 9 & 4 \\ 2 & 1 \end{pmatrix}$ is $\begin{pmatrix} 1 & -4 \\ -2 & 9 \end{pmatrix}$.

Note that the determinant of $\begin{pmatrix} 9 & 4 \\ 2 & 1 \end{pmatrix}$ is $9 \times 1 - 4 \times 2 = 1$.

Pre-multiply both sides of (3) by this inverse matrix:

$$\begin{pmatrix} \mathbf{1} & \mathbf{-4} \\ \mathbf{-2} & \mathbf{9} \end{pmatrix}\begin{pmatrix} 9 & 4 \\ 2 & 1 \end{pmatrix}\begin{pmatrix} x \\ y \end{pmatrix} = \begin{pmatrix} \mathbf{1} & \mathbf{-4} \\ \mathbf{-2} & \mathbf{9} \end{pmatrix}\begin{pmatrix} 17 \\ 4 \end{pmatrix}$$

$$\begin{pmatrix} 1 & 0 \\ 0 & 1 \end{pmatrix}\begin{pmatrix} x \\ y \end{pmatrix} = \begin{pmatrix} 1 \\ 2 \end{pmatrix}$$

$$\begin{pmatrix} x \\ y \end{pmatrix} = \begin{pmatrix} 1 \\ 2 \end{pmatrix}$$

that is $x = 1$

and $y = 2$

Example 10

Solve the equations $3x - 4y = 1$ and $7x + y = 23$.

$3x - 4y = 1$

$7x + y = 23$

$$\begin{pmatrix} 3 & -4 \\ 7 & 1 \end{pmatrix}\begin{pmatrix} x \\ y \end{pmatrix} = \begin{pmatrix} 1 \\ 23 \end{pmatrix} \quad (1)$$

The determinant of $\begin{pmatrix} 3 & -4 \\ 7 & 1 \end{pmatrix}$ is:

$3 \times 1 - 7 \times (-4) \quad 31$

Its inverse is $\frac{1}{31}\begin{pmatrix} 1 & 4 \\ -7 & 3 \end{pmatrix}$.

Pre-multiply both sides of (1) by this inverse:

$$\frac{\mathbf{1}}{\mathbf{31}}\begin{pmatrix} \mathbf{1} & \mathbf{4} \\ \mathbf{-7} & \mathbf{3} \end{pmatrix}\begin{pmatrix} 3 & -4 \\ 7 & 1 \end{pmatrix}\begin{pmatrix} x \\ y \end{pmatrix} = \frac{\mathbf{1}}{\mathbf{31}}\begin{pmatrix} \mathbf{1} & \mathbf{4} \\ \mathbf{-7} & \mathbf{3} \end{pmatrix}\begin{pmatrix} 1 \\ 23 \end{pmatrix}$$

$$\frac{1}{31}\begin{pmatrix} 31 & 0 \\ 0 & 31 \end{pmatrix}\begin{pmatrix} x \\ y \end{pmatrix} = \frac{1}{31}\begin{pmatrix} 93 \\ 62 \end{pmatrix}$$

$$\begin{pmatrix} 1 & 0 \\ 0 & 1 \end{pmatrix}\begin{pmatrix} x \\ y \end{pmatrix} = \begin{pmatrix} 3 \\ 2 \end{pmatrix}$$

$$\begin{pmatrix} x \\ y \end{pmatrix} = \begin{pmatrix} 3 \\ 2 \end{pmatrix}$$

that is $x = 3$

and $y = 2$

Check: $3 \times 3 - 4 \times 2 = 1$ and $7 \times 3 + 2 = 23$.

Exercise 2g

Use the matrix method to solve the following pairs of simultaneous equations.

1. $6x + 11y = 29$, $x + 2y = 5$
2. $7x + 4y = 29$, $4x + 2y = 16$
3. $2x + y = 3$, $x + 2y = 9$
4. $21x - 2y = 15$, $13x - y = 10$
5. $4x - 2y = 9$, $x + y = 3$
6. $x - 3y = 2$, $2x + 4y = -1$
7. $2a - 3b = 3$, $a + b = 4$
8. $\frac{1}{2}u - \frac{1}{3}v = 4$, $\frac{1}{3}u + \frac{1}{2}v = 7$
9. $15s - 6t = 4$, $6s + 18t = 5$
10. $6c - d = -2$, $10c - 3d = -10$

Further examples of the use of matrices as operators are given in Chapter 14.

Exercise 2h

1. Evaluate as a single matrix $\begin{pmatrix} -3 & -8 \\ 6 & 2 \end{pmatrix} + 2\begin{pmatrix} -2 & 1 \\ -3 & 5 \end{pmatrix}$.

2. If $M = \begin{pmatrix} 2 & -6 \\ -1 & 4 \end{pmatrix}$,
   (a) find the value of the determinant of M,
   (b) hence write down the inverse of M.

3. Find the value of the determinant of the matrix $\begin{pmatrix} -2 & -4 \\ 5 & 3 \end{pmatrix}$. Hence write down the inverse of the matrix.

4. The value of the determinant of the matrix $\begin{pmatrix} 5 & -2 \\ -4 & x \end{pmatrix}$ is 7.
   (a) Find the value of x.
   (b) Hence write down the inverse of the matrix.

5. Find the value of k for which the matrix $\begin{pmatrix} 4 & k-2 \\ 8 & 6 \end{pmatrix}$ is a singular matrix.

6. If $A = \begin{pmatrix} 3 \\ -1 \\ 2 \end{pmatrix}$ and $B = (-2 \quad 5 \quad 0)$, evaluate
   (a) AB, (b) BA.

7. Express each of the following as a single matrix.
   (a) $\begin{pmatrix} 1 & 5 \\ 4 & 0 \\ 2 & 3 \end{pmatrix}\begin{pmatrix} -1 \\ 3 \end{pmatrix}$
   (b) $\begin{pmatrix} 1 & 4 \\ 0 & 1 \end{pmatrix}\begin{pmatrix} 0 & 0 & 3 & 3 \\ 0 & 3 & 3 & 0 \end{pmatrix}$

8. Given that $\begin{pmatrix} 3 & 2 & 4 \\ 6 & 0 & 1 \end{pmatrix}\begin{pmatrix} 4 \\ m \\ -3 \end{pmatrix} = \begin{pmatrix} 10 \\ 7n \end{pmatrix}$ find the values of m and n.

9. P is a 2×2 matrix such that $\begin{pmatrix} 3 & 0 \\ 0 & 3 \end{pmatrix} P - P = \begin{pmatrix} -2 & 0 \\ 2 & 4 \end{pmatrix}$.
   Find the matrix P.

10. $M = \begin{pmatrix} 3 & 0 \\ 0 & 2 \end{pmatrix}$ and $N = \begin{pmatrix} -1 & 1 \\ 0 & 3 \end{pmatrix}$.
    (a) Find $M - 2N$.
    (b) Find the values of p and q if $M\begin{pmatrix} 4 \\ p \end{pmatrix} = N\begin{pmatrix} q \\ 6 \end{pmatrix}$.

11. Given that
    $\begin{pmatrix} 2c & 1 \\ -3 & 5 \end{pmatrix} + \begin{pmatrix} 3 & -1 \\ 0 & -2 \end{pmatrix} = \begin{pmatrix} 2 & 0 \\ 0 & -1 \end{pmatrix}\begin{pmatrix} 1 & a \\ 3 & b \end{pmatrix}$
    find the value of a, b and c.

12. Find a and b if $\begin{pmatrix} 3 & 7 \\ b & a \end{pmatrix}\begin{pmatrix} a & -7 \\ -2 & 3 \end{pmatrix} = I$
    where I is the identity matrix.

13. $P = \begin{pmatrix} 2 & 0 \\ 1 & 3 \end{pmatrix}$ and $R = \begin{pmatrix} 2 & -\frac{1}{2} \\ r & 1 \end{pmatrix}$.
    (a) Evaluate P^2
    (b) Find the value of r which makes R a singular matrix.
    (c) Find the value of r which makes $PR = \frac{1}{2}\begin{pmatrix} 8 & -2 \\ -8 & -5 \end{pmatrix}$

14. (a) Write down the inverse of the matrix $\begin{pmatrix} 3 & -4 \\ 5 & 7 \end{pmatrix}$
    (b) Hence find x and y if $\begin{pmatrix} 3 & -4 \\ 5 & 7 \end{pmatrix}\begin{pmatrix} x \\ y \end{pmatrix} = \begin{pmatrix} 10 \\ 3 \end{pmatrix}$

15. Express the simultaneous equations
    $3y = -5x + 3$
    $2y = -3x + 1$
    as a single matrix equation in the form
    $\begin{pmatrix} a & b \\ c & d \end{pmatrix}\begin{pmatrix} x \\ y \end{pmatrix} = \begin{pmatrix} p \\ q \end{pmatrix}$
    where a, b, c, d, p, q, are integers.
    Hence find the values of x and y.

Summary

A set of numbers (or **elements**) arranged in a rectangular pattern of rows and columns is termed a **matrix**. The **order** of the matrix is given as $\boldsymbol{m \times n}$, m and n being the number of rows and columns respectively.

Examples of matrices are **column matrix, row matrix** and **square matrix.**

Matrices can be added (or subtracted) **only if they are of the same order** – corresponding elements are added (or subtracted) to result in a matrix of the same order.

If A and B are two matrices of the same order, then

$$A + B = B + A$$

but $A - B \neq B - A$ in general.

To multiply a matrix by a **scalar**, multiply each element of the matrix by the scalar.

To multiply two matrices, **the number of columns in the first matrix P must equal the number of rows in the second matrix, Q.** The resulting matrix has as many rows as the first matrix and as many columns as the second matrix.

In general $PQ \neq QP$.

If two matrices P and Q both have the same number of rows and columns (square matrices) then multiplication is always possible.

$\begin{pmatrix} 0 & 0 \\ 0 & 0 \end{pmatrix}$ is called the **null** or **zero** matrix

$\begin{pmatrix} 1 & 0 \\ 0 & 1 \end{pmatrix}$ is called the **identity** matrix, I.

The **inverse** of a matrix $\begin{pmatrix} a & b \\ c & d \end{pmatrix}$ is

$\dfrac{1}{ad - bc}\begin{pmatrix} d & -b \\ -c & a \end{pmatrix}$, where $(ad - bc)$ is called the **determinant**. If the determinant equals zero, the matrix has no inverse and is called a **singular** matrix.

The product of a matrix and its inverse is the identity matrix.

Two simultaneous linear equations may be written as a single matrix equation, that is

$$ax + by = m \qquad (1)$$
$$cx + dy = n \qquad (2)$$

may be written as

$$\begin{pmatrix} a & b \\ c & d \end{pmatrix}\begin{pmatrix} x \\ y \end{pmatrix} = \begin{pmatrix} m \\ n \end{pmatrix} \qquad (3)$$

The matrix $\begin{pmatrix} a & b \\ c & d \end{pmatrix}$ acts as an operator in equation (3). Equations (1) and (2) may be solved by multiplying both sides of equation (3) by the inverse of this matrix.

Practice exercise P2.1

1. State the orders of the following matrices.

   (a) $\begin{pmatrix} 5 \\ 4 \end{pmatrix}$ (b) $\begin{pmatrix} 8 & -1 & 6 \\ 5 & 2 & -3 \\ 4 & 9 & 3 \end{pmatrix}$

   (c) $\begin{pmatrix} 8.2 & 5.8 & 3.7 \end{pmatrix}$ (d) $\begin{pmatrix} 5 & -2 \\ 2 & 8 \\ -3 & 1 \end{pmatrix}$

2. If $P = \begin{pmatrix} 0.5 & 3 \\ 0.9 & 0.3 \end{pmatrix}$ and $Q = \begin{pmatrix} 0.2 & -1 \\ 0.7 & 3 \end{pmatrix}$, evaluate

   (a) $P + Q$, (b) $P - Q$, (c) $3P$, (d) $-2Q$.

3. Why it is not possible to combine these matrices?

   $\begin{pmatrix} 5 & 9 \\ -2 & -1 \end{pmatrix} + \begin{pmatrix} 2 & 6 \end{pmatrix}$

4. Evaluate as a single matrix.

   $\begin{pmatrix} 2 & 1 \\ -3 & -4 \end{pmatrix} + 2\begin{pmatrix} -1 & 1 \\ 0 & 3 \end{pmatrix}$

5. Find n if

   $n\begin{pmatrix} 3 & 3 \\ 4 & 9 \end{pmatrix} + 2\begin{pmatrix} 7 & -1 \\ 0 & 4 \end{pmatrix} = \begin{pmatrix} 20 & 4 \\ 8 & 26 \end{pmatrix}$.

6. Find x and y if

   $3\begin{pmatrix} 1 & 0 \\ 4 & 8 \end{pmatrix} - \begin{pmatrix} w & x \\ y & z \end{pmatrix} = \begin{pmatrix} 0 & -2 \\ -2 & 5 \end{pmatrix}$.

Practice exercise P2.2

1. Calculate the following products.

   (a) $\begin{pmatrix} 2 & 6 \end{pmatrix}\begin{pmatrix} 3 \\ 5 \end{pmatrix}$ (b) $\begin{pmatrix} 5 & 11 & 3 \end{pmatrix}\begin{pmatrix} 6 \\ 0 \\ -1 \end{pmatrix}$

   (c) $\begin{pmatrix} 7 & 3 \end{pmatrix}\begin{pmatrix} 0.2 \\ 3.2 \end{pmatrix}$ (d) $\begin{pmatrix} 8 & 5 & 9 \end{pmatrix}\begin{pmatrix} x \\ y \\ z \end{pmatrix}$

2. An apple weighs 0.25 kg, a banana weighs 0.2 kg and an orange weighs 0.3 kg. Use a matrix method to find the weight of fruit salad made by a restaurant when it uses 13 apples, 21 bananas and 27 oranges.

3. If $A = \begin{pmatrix} 2 & 3 \\ -3 & 5 \end{pmatrix}$ and $B = \begin{pmatrix} -1 & 2 \\ 4 & 2 \end{pmatrix}$, evaluate
   (a) AB, (b) BA.

4. Find the following matrix products.
   (a) $\begin{pmatrix} 3 & 7 \\ 0 & 7 \end{pmatrix} \begin{pmatrix} 8 \\ 4 \end{pmatrix}$
   (b) $\begin{pmatrix} -6 & 8 \\ 2 & -5 \end{pmatrix} \begin{pmatrix} 9 & -3 \\ -1 & -7 \end{pmatrix}$
   (c) $\begin{pmatrix} 9 & 8 \\ 3 & 2 \end{pmatrix} \begin{pmatrix} -2 & 8 & -1 \\ -5 & 7 & 2 \end{pmatrix}$

Practice exercise P2.3

1. Find the inverses of the following matrices.
   (a) $\begin{pmatrix} -3 & 4 \\ -2 & -3 \end{pmatrix}$ (b) $\begin{pmatrix} 6 & 2 \\ 5 & 2 \end{pmatrix}$
   (c) $\begin{pmatrix} -2 & 8 \\ 1 & -5 \end{pmatrix}$ (d) $\begin{pmatrix} 9 & 4 \\ 7 & 2 \end{pmatrix}$
   (e) $\begin{pmatrix} 2 & 7 \\ 1 & 6 \end{pmatrix}$ (f) $\begin{pmatrix} 8 & -3 \\ 2 & -2 \end{pmatrix}$

2. The value of the determinant of the matrix $\begin{pmatrix} 3 & -4 \\ -3 & k \end{pmatrix}$ is 6.
   (a) Find the value of k.
   (b) Hence find the inverse of the matrix.

3. Given that $\begin{pmatrix} 4 & -4 \\ 3 & p \end{pmatrix}$ is a singular matrix, find the value of p.

Practice exercise P2.4

Use the matrix method to solve the following pairs of simultaneous equations.

1. $x + 4y = 39$
   $3x + 5y = 54$
2. $3x + 5y = 46$
   $6x - 3y = -12$
3. $5x - 4y = -27$
   $8x - y = 0$
4. $9x + 5y = 42$
   $6x - 8y = -6$
5. $5x + 5y = 5$
   $4x - 2y = -38$
6. $x - 6y = -20$
   $4x - 2y = -36$
7. $3x - 7y = -41$
   $9x + y = 53$
8. $8x - 5y = -33$
   $3x - 3y = -18$

Practice exercise P2.5

1. (a) Express the simultaneous equations
   $2x + 5y = 14$
   $3x - 8y = -41$
   as a single matrix equation.
   (b) Hence find the values of x and y.

2. (a) Find the inverse of the matrix $\begin{pmatrix} 3 & -5 \\ 2 & -3 \end{pmatrix}$.
   (b) Hence find x and y if $\begin{pmatrix} 3 & -5 \\ 2 & 3 \end{pmatrix} \begin{pmatrix} x \\ y \end{pmatrix} = \begin{pmatrix} 18 \\ 13 \end{pmatrix}$.

3. A student bought 5 pencils and 7 pens for a total cost of \$375.00.
   Another student bought 5 pencils and 2 pens for a total of \$155.00.
   (a) Using x to represent the cost of a pencil and y to represent the cost of a pen, write two equations to represent the information above.
   (b) Using a matrix equation, find the cost of a pen and the cost of a pencil.

4. A scientist weighs 9 identical blocks of metal A and 5 identical blocks of metal B. The total weight is 799 grams.
   He weighs 3 of the blocks of metal A and 14 of the blocks of metal B. The total weight is 661 grams.
   (a) Using a to represent the weight of one block of metal A and b to represent the weight of one block of metal B, write two equations to represent this information.
   (b) Using a matrix equation, find the weight of one block of metal A and the weight of one block of metal B.

Chapter 3

Trigonometry (1)

Trigonometric ratios of obtuse and reflex angles, positive and negative angles, and special angles

Pre-requisites

- trigonometric ratios of acute angles; Cartesian axes; surds

Tangent, sine, cosine (revision)

In Fig. 3.1 $\triangle$ABC is any triangle, right-angled at A.

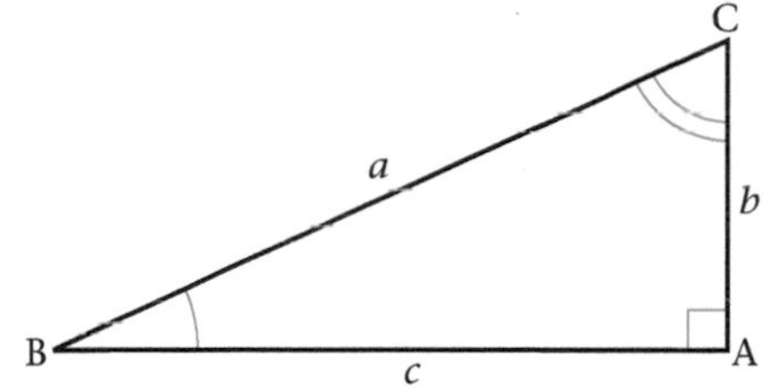

Fig. 3.1

$$\tan B = \frac{b}{c}, \tan C = \frac{c}{b} \quad \left(tan: \frac{opp}{adj}\right)$$

$$\sin B = \frac{b}{a}, \sin C = \frac{c}{a} \quad \left(sin: \frac{opp}{hyp}\right)$$

$$\cos B = \frac{c}{a}, \cos C = \frac{b}{a} \quad \left(cos: \frac{adj}{hyp}\right)$$

Note that, in right-angled $\triangle$ABC,

$\sin B = \cos C$, and $\cos B = \sin C$

In $\triangle$ABC, $\widehat{B}$ and $\widehat{C}$ are complementary angles (i.e. $\widehat{B} + \widehat{C} = 90°$). If $\widehat{B} = \theta$, then $\widehat{C} = 90° - \theta$ (Fig. 3.2).

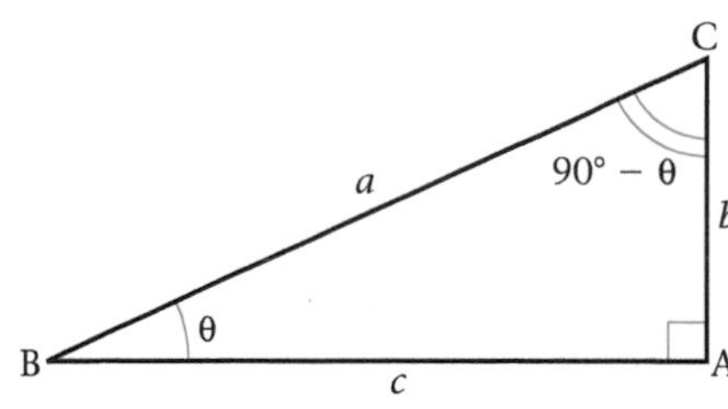

Fig. 3.2

$$\therefore \sin \theta = \cos (90° - \theta) = \frac{b}{a}$$

and $\cos \theta = \sin (90° - \theta) = \frac{c}{a}$

$$\tan (90° - \theta) = \frac{c}{b} = \frac{1}{\tan \theta}$$

Trigonometric ratios of obtuse angles

The trigonometric ratios of an acute angle in a right-angled triangle have already been defined.

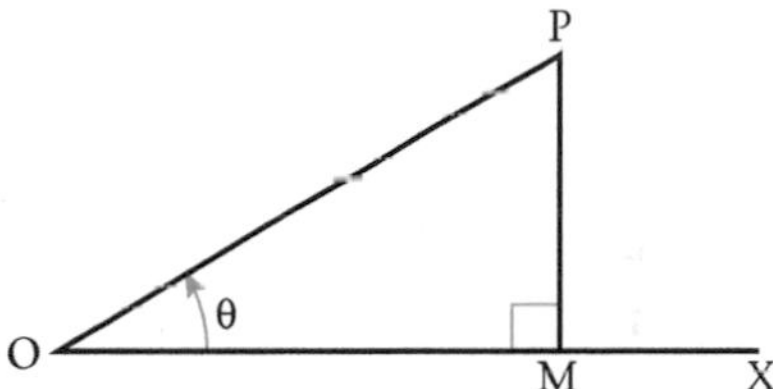

Fig. 3.3

If OX in Fig. 3.4 is kept fixed and OP allowed to rotate *anticlockwise*, there will come a stage when θ becomes obtuse (Fig. 3.5). Name the obtuse angle θ_2.

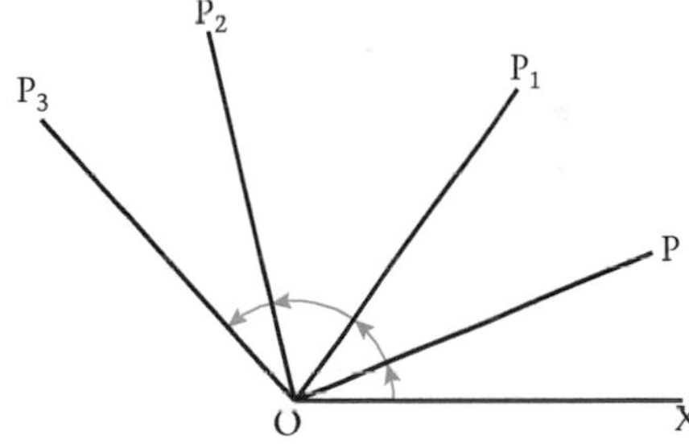

Fig. 3.4

When θ is obtuse, it is no longer in a right-angled triangle. It is therefore impossible to define $\sin \theta_2$, $\cos \theta_2$ and $\tan \theta_2$ in terms of the ratios of the hypotenuse, adjacent and opposite sides of a right-angled triangle (Fig. 3.5).

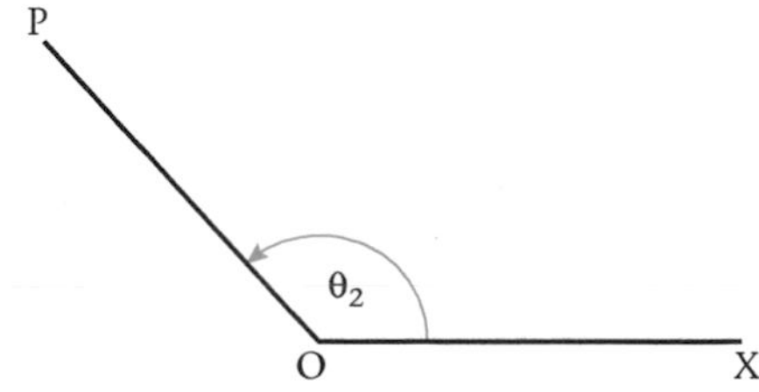

Fig. 3.5

It is necessary, therefore, to define the trigonometric ratios in such a way as to be suitable for obtuse angles as well as acute angles. Fig. 3.6 shows acute angle θ_1 and obtuse angle θ_2 within Cartesian axes Ox, Oy.

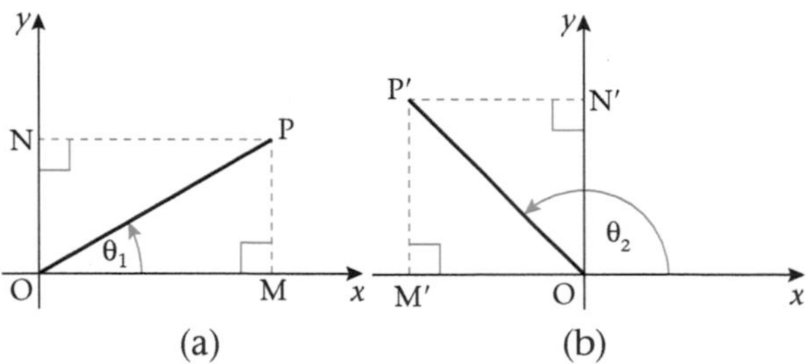

Fig. 3.6

Notice that the lettering of Fig. 3.6(a) is very much the same as that of Fig. 3.3.

In Fig. 3.6(a)

$$\sin\theta_1 = \frac{MP}{OP} = \frac{ON}{OP}$$

$$\cos\theta_1 = \frac{OM}{OP}$$

$$\tan\theta_1 = \frac{MP}{OM} = \frac{ON}{OM}$$

In Fig. 3.6 OM is called the **projection of OP on Ox**. On is the projection of OP on Oy. This makes it possible to define the trigonometric ratios in a new way:

$$\sin\theta = \frac{\text{projection of OP on Oy}}{\text{OP}}$$

$$\cos\theta = \frac{\text{projection of OP on Ox}}{\text{OP}}$$

$$\tan\theta = \frac{\text{projection of OP on Oy}}{\text{projection of OP on Ox}}$$

With these definitions in Fig. 13.6(b),

$$\sin\theta_2 = \frac{ON'}{OP'};\ \cos\theta_2 = \frac{OM'}{OP'};\ \tan\theta_2 = \frac{ON'}{OM'}$$

The ratios are now the same for both parts of Fig. 3.6. In Fig. 3.7, O is the centre of the circle and the origin of axes Ox, Oy.

$\widehat{POM} = \widehat{QOL} = \theta$ (acute)

In Fig. 3.7,

Let $\widehat{MOQ} = \theta_2 = 180° - \theta$ (obtuse)
OP = OQ (radii, both taken to be positive lengths)

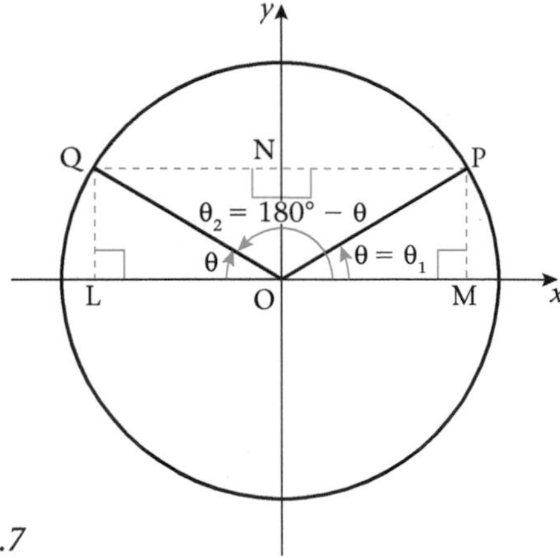

Fig. 3.7

ON is a positive length since it is on the positive part of Oy, OM is a positive length since it is on the positive part of Ox. OL is a negative length since it is on the negative part of Ox.

From the symmetry of the figure, OL = −OM. Hence

$$\sin\theta_2 = \sin(180° - \theta) = \frac{ON}{OQ} = \frac{ON}{OP} = \sin\theta$$

$$\cos\theta_2 = \cos(180° - \theta) = \frac{OL}{OQ} = \frac{-OM}{OP} = -\cos\theta$$

$$\tan\theta_2 = \tan(180° - \theta) = \frac{ON}{OL} = \frac{ON}{-OM} = -\tan\theta$$

The following numerical examples show how to use the above statements.

$$\sin 160° = \sin(180 - 20)° = \sin 20° = 0.342$$
$$\cos 160° = \cos(180 - 20)° = -\cos 20° = -0.940$$
$$\tan 160° = \tan(180 - 20)° = -\tan 20° = -0.364$$

Exercise 3a

Use tables or a calculator to find the values of the following.

1. sin 110°
2. cos 110°
3. tan 110°
4. sin 153°
5. sin 98°
6. cos 142°
7. tan 93°
6. cos 128°
9. sin 156°30′
10. tan 173.5°
11. cos 161.4°
12. tan 131°42′
13. cos 103°6′
14. sin 118°42′
15. sin 178.4°
16. tan 92°40′

Example 1

Find the values of θ lying between 0° and 180° in each of the following.

(a) $\cos\theta = 0.287$ (b) $\sin\theta = 0.936$
(c) $\cos\theta = -0.822$ (d) $\tan\theta = -2.164$

(a) $\cos\theta = 0.287$
From tables or a calculator, $\theta = 73.3°$
Since $\cos\theta$ is positive, θ is acute.

(b) $\sin\theta = 0.936$
From tables or a calculator, $\theta = 69.4°$
But $\sin 69.4° = \sin(180 - 69.4)°$
$= \sin 110.6°$
$\theta = 69.4°$ or $\theta = 110.6°$

(c) $\cos\theta = -0.822$
Since $\cos\theta$ is negative, θ is obtuse.
Note that the obtuse angle may be obtained directly from some calculators when the negative value of the trigonometric ratio is input.
First find the acute angle whose cosine is 0.822.
From tables or a calculator, $0.822 = \cos 34.7°$
$\Rightarrow -0.822 = \cos(180 - 34.7°)$
$= \cos 145.3°$
$\Rightarrow \theta = 145.3°$

(d) $\tan\theta = -2.164$
Since $\tan\theta$ is negative, θ is obtuse.
Note that the obtuse angle may be obtained directly from some calculators when the negative value of the trigonometric ratio is input.
From tables or a calculator,
$2.164 = \tan 65.2°$
$\Rightarrow \theta = 180° - 65.2° = 114.8°$

Exercise 3b

Find the values of θ lying between 0° and 180° in each of the following. Give the answers in degrees to 1 or 2 d.p. where appropriate.

1. $\cos\theta = 0.809$
2. $\cos\theta = -0.809$
3. $\tan\theta = 3.732$
4. $\tan\theta = -3.732$
5. $\sin\theta = 0.920$
6. $\tan\theta = -1.963$
7. $\cos\theta = -0.940$
7. $\sin\theta = 0.423$
9. $\cos\theta = -0.139$
10. $\tan\theta = -0.625$
11. $\tan\theta = -2.106$
12. $\tan\theta = -0.315$
13. $\cos\theta = -0.426$
14. $\sin\theta = 0.833$
15. $\tan\theta = -0.724$
16. $\sin\theta = 0.959$

Trigonometric ratios of reflex angles

As OP continues to rotate anticlockwise, the angle θ becomes θ_3, a reflex angle between 180° and 270°, and θ_4, a reflex angle between 170° and 360°.

In Fig. 3.8,

$\theta_3 = M\hat{O}V = 180° + \theta$
$\theta_4 = M\hat{O}W = 360° - \theta$
and OV = OW = OP (radius)

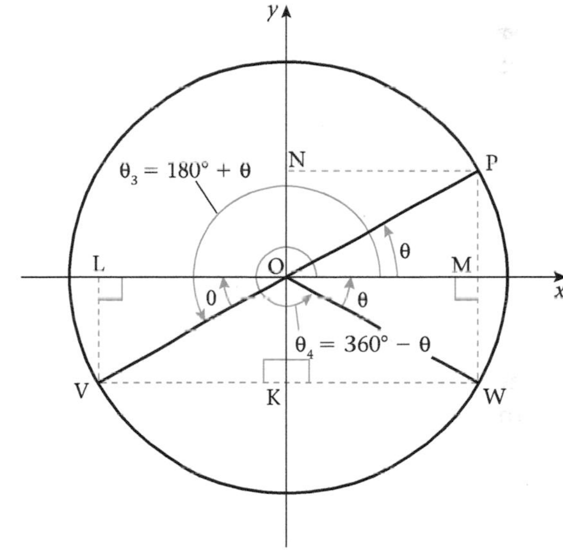

Fig. 3.8

Note that OV and OW are taken as positive lengths. OL is a negative length since it is on the negative part of Ox. OK is a negative length since it is on the negative part of Oy.
From symmetry, OL = −OM and OK = −ON

Hence, by a method similar to that used for obtuse angles,

$$\sin\theta_3 = \sin(180° + \theta) = \frac{OK}{OV} = \frac{-ON}{OP} = -\sin\theta$$

$$\cos\theta_3 = \cos(180° + \theta) = \frac{OL}{OV} = \frac{-OM}{OP} = -\cos\theta$$

$$\tan\theta_3 = \tan(180° + \theta) = \frac{OK}{OL} = \frac{-ON}{-OM} = \tan\theta$$

and

$\sin \theta_4 = \sin (360° - \theta) = \frac{OK}{OW} = \frac{-ON}{OP} = -\sin \theta$

$\cos \theta_4 = \cos (360° - \theta) = \frac{OM}{OW} = \frac{OM}{OP} = \cos \theta$

$\tan \theta_4 = \tan (360° - \theta) = \frac{OK}{OM} = \frac{-ON}{OM} = -\tan \theta$

The following numerical examples show you how to use the above statements.

$\sin 230° = \sin (180° + 50°) = -\sin 50° = -0.765$
$\cos 230° = \cos (180° + 50°) = -\cos 50° = -0.643$
$\tan 230° = \tan (180° + 50°) = \tan 50° = 1.192$
and
$\sin 320° = \sin (360° - 40°) = -\sin 40° = -0.643$
$\cos 320° = \cos (360° - 40°) = \cos 40° = 0.766$
$\tan 320° = \tan (360° - 40°) = -\tan 40° = -0.839$

Exercise 3c

Use tables or a calculator to find the values, correct to 3 significant figures, of the following:

1. $\sin 250°$
2. $\cos 250°$
3. $\tan 250°$
4. $\cos 233°$
5. $\sin 315°$
6. $\tan 330°$
7. $\sin 290.6°$
8. $\cos 280°36'$
9. $\cos 210.4°$
10. $\tan 194.2°$
11. $\tan 240°12'$
12. $\sin 342°42'$

Example 2

Find the values of θ between 180° and 360° in each of the following:

(a) $\cos \theta = 0.342$
(b) $\sin \theta = -0.842$
(c) $\tan \theta = 3.53$
(d) $\cos \theta = -0.681$
(e) $\tan \theta = -0.725$

(a) $\cos \theta = 0.342$
From tables or a calculator, $\theta = 70.0°$
Since $\cos \theta$ is positive, θ lies between 270° and 360°
$\theta = 360° - 70° = 290°$

(b) $\sin \theta = -0.842$
From tables or a calculator,
when $\sin \theta = 0.842$, $\theta = 57.3°$
Since $\sin \theta$ is negative, θ lies between 180° and 270° or between 270° and 360°
$\theta = 180° + 57.3° = 237.3°$
or $\theta = 360° - 57.3° = 302.7°$

(c) $\tan \theta = 3.53$
From tables (or a calculator), $\theta = 74.2°$
Since $\tan \theta$ is positive, θ lies between 180° and 270°
$\theta = 180° + 74.2° = 254.2°$

(d) $\cos \theta = -0.681$
From tables or a calculator,
when $\cos \theta = 0.681$, $\theta = 47.1°$
Since $\cos \theta$ is negative, θ lies between 180° and 270°
$\theta = 180° + 47.1° = 227.1°$

(e) $\tan \theta = -0.725$
From tables or a calculator,
when $\tan \theta = 0.725$, $\theta = 35.9°$
Since $\tan \theta$ is negative, θ lies between 270° and 360°
$\theta = 360° - 35.9° = 324.1°$

Exercise 3d

Find the values of between 180° and 360° in each of the following, giving your answer in degrees to 1 decimal place.

1. $\sin \theta = -0.255$
2. $\cos \theta = 0.342$
3. $\tan \theta = 2.504$
4. $\cos \theta = -0.746$
5. $\sin \theta = -0.931$
6. $\tan \theta = -0.583$
7. $\tan \theta = -2.906$
8. $\cos \theta = -0.286$
9. $\cos \theta = 0.621$
10. $\tan \theta = 1.942$

Positive angles

The x-axis and the y-axis divide the Cartesian plane into four regions, called **quadrants**. As shown in Fig. 3.9, the quadrants are numbered from 1 to 4, Q_1 to Q_4 in Fig. 3.9.

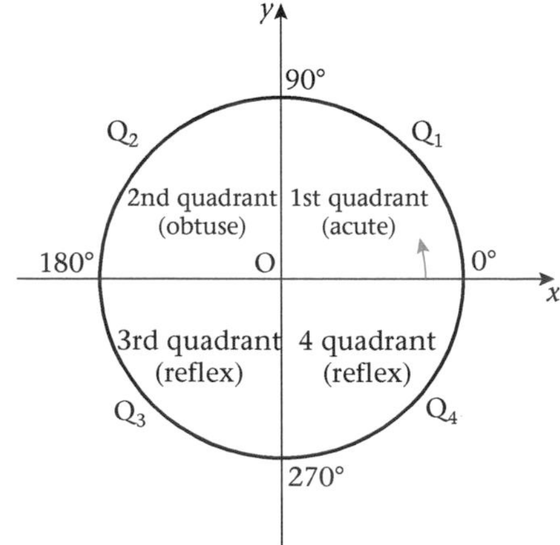

Fig. 3.9 Positive angles

When a line OP starts at the positive x-axis in the first quadrant and rotates anticlockwise through 360°, the angles made by the rotating line OP and the positive x-axis are **positive angles**. (See Fig. 3.9.)

The angles in the first quadrant are acute angles, (angles between 0° and 90°). As described in the earlier sections of this chapter, in the first quadrant all the trigonometric ratios are positive.

In the second quadrant, the angles made by OP with the positive x-axis are obtuse (angles between 90° and 180°) and only the sine (sin) ratio is positive.

In the third quadrant, only the tangent (tan) ratio of the reflex angles (those angles between 180° and 270°) is positive.

In the fourth quadrant, only the cosine (cos) ratio of the reflex angles (those angles between 270° and 360° is positive.

Negative angles

When a line OP starts at the positive x-axis in the first quadrant and rotates clockwise through 360°, the angles made by the rotating line OP and the positive x-axis are **negative angles** (Fig. 3.10).

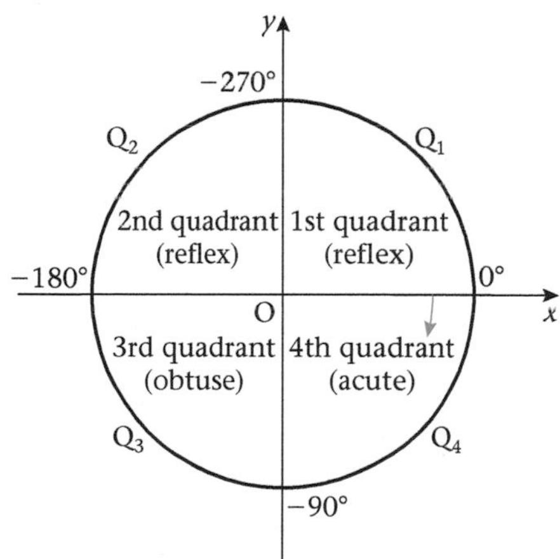

Fig. 3.10 Negative angles

When we use the same method that we used for positive angles to find the trigonometric ratios of negative angles between 0° and −360°, we see that in each of the four quadrants the trigonometric ratios of negative angles have the same values as the ratios of positive angles in the same quadrant.

Hence, for an angle of (−40°) which lies in the fourth quadrant, the trigonometric ratios are

$$\sin(-40°) = -\sin 40°$$
$$\cos(-40°) = \cos 40°$$
$$\tan(-40°) = -\tan 40°$$

and for an angle of (−140°), which lies in the third quadrant,

$$\sin(-140°) = -\sin 40°$$
$$\cos(-140°) = -\cos 40°$$
$$\tan(-140°) = \tan 40°$$

and for an angle of (−220°), which lies in the second quadrant,

$$\sin(-220°) = \sin 40°$$
$$\cos(-220°) = -\cos 40°$$
$$\tan(-220°) = -\tan 40°$$

and for an angle of (−320°), which lies in the first quadrant,

$$\sin(-320°) = \sin 40°$$
$$\cos(-320°) = \cos 40°$$
$$\tan(-320°) = \tan 40°$$

Exercise 3e

Use tables or a calculator to find the values, correct to 3 significant figures, of the following:

1. sin −50°
2. cos −250°
3. tan −150°
4. cos −33°
5. sin −135°
6. tan −330°
7. sin −290.6°
8. cos −200°36′
9. cos −310.4°
10. tan −14.2°

Notice that for angles between 0° and −90°, and between 0° and 90°, for example,

$$\sin(-40°) = -\sin 40°$$
$$\cos(-40°) = \cos 40°$$
$$\tan(-40°) = -\tan 40°$$

and for angles between −90° and −180°, and between 90° and 180°, for example,

$$\sin(-140°) = -\sin 40° = -\sin(140°)$$
$$\cos(-140°) = -\cos 40° = \cos(140°)$$
$$\tan(-140°) = \tan 40° = -\tan(140°)$$

and for angles between −180° and −270°, and between 180° and 270°, for example,

$$\sin(-220°) = \sin 40° = -\sin(220°)$$
$$\cos(-220°) = -\cos 40° = \cos(220°)$$
$$\tan(-220°) = -\tan 40° = -\tan(220°)$$

and for angles between $-270°$ and $-360°$, and between $270°$ and $360°$, for example,

$\sin(-320°) = \sin 40° = -\sin(320°)$
$\cos(-320°) = \cos 40° = \cos(320°)$
$\tan(-320°) = \tan 40° = -\tan(320°)$

Hence,
$\mathbf{\sin(-\theta) = -\sin\theta;\ \cos(-\theta) = \cos\theta;}$
$\mathbf{\tan(-\theta) = -\tan\theta}$

Fig. 3.11 summarises the values of the sine, cosine and tangent ratios in the four quadrants for both positive and negative angles.

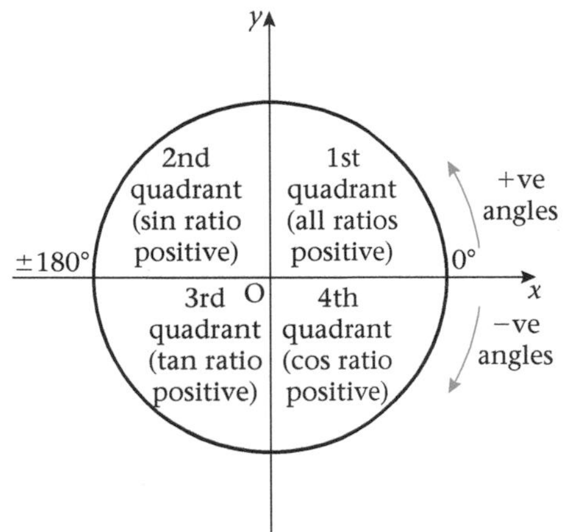

Fig. 3.11 Values of trigonometric ratios

Note that for all angles,

$$\tan\theta = \frac{\sin\theta}{\cos\theta}$$

Exercise 3f

Find all the values of between $-360°$ and $360°$ in each of the following, giving your answer in degrees to 1 decimal place.

1. $\sin\theta = -0.255$
2. $\cos\theta = 0.342$
3. $\tan\theta = 2.504$
4. $\cos\theta = -0.746$
5. $\sin\theta = -0.931$
6. $\tan\theta = -0.583$
7. $\tan\theta = -2.906$
8. $\cos\theta = -0.286$
9. $\cos\theta = 0.621$
10. $\tan\theta = 1.942$

Tan, sin and cos of 0°, 90°, 180°, 270°, 360°

When the rotating line OP lies along Ox, the positive x-axis in the first quadrant, as in Fig. 13.12, the angle θ is equal to zero degrees ($\theta = 0°$). Then the projection of OP on Ox is OP itself, and the projection of OP on Oy is zero units in length.

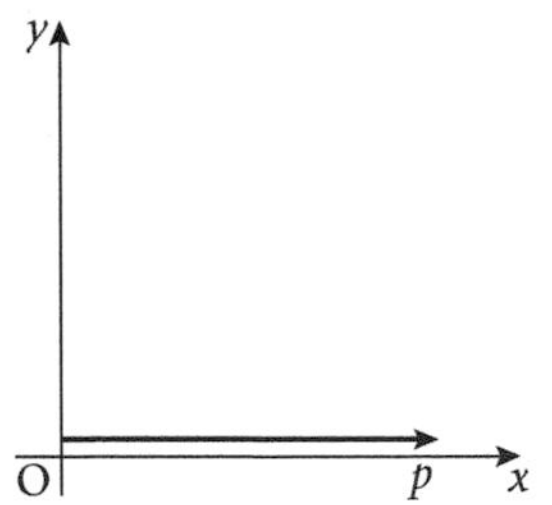

Fig. 3.12

Thus

$$\sin 0° = \frac{0}{\text{OP}} = 0$$

$$\cos 0° = \frac{\text{OP}}{\text{OP}} = 1$$

$$\tan 0° = \frac{0}{\text{OP}} = 0$$

Similarly, when the line OP lies along Oy, the positive y-axis, as in Fig. 3.13, $\theta = 90°$. Then the projection of OP on Oy is OP itself, and the projection of OP on Ox is zero units in length.

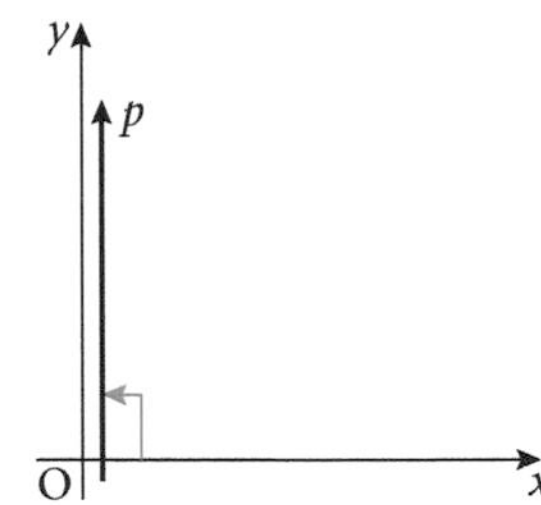

Fig. 3.13

Then

$$\sin 90° = \frac{\text{OP}}{\text{OP}} = 1$$

$$\cos 90° = \frac{0}{\text{OP}} = 0$$

$$\tan 90° = \frac{\text{OP}}{0} = \text{is undefined}$$

since division by zero is undefined.

When the rotating line OP lies along he negative x-axis in the third quadrant as in Fig. 3.14, the angle θ is equal to 180 degrees ($\theta = 180°$). Then the projection of OP on the x-axis is equal to $-$OP, and the projection of OP on Oy is zero units in length.

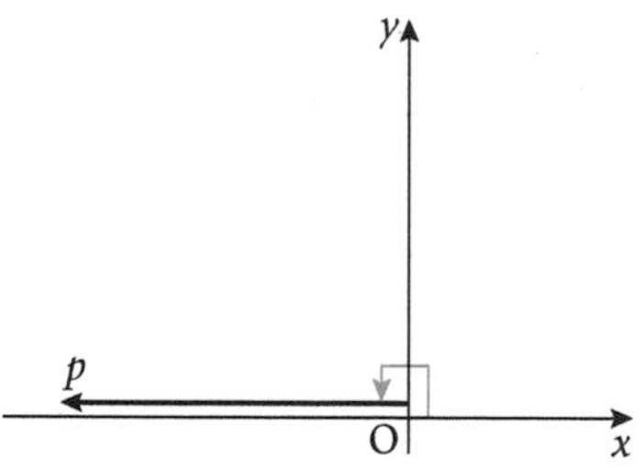

Fig. 3.14

Thus

$$\sin 180° = \frac{0}{OP} = 0$$

$$\cos 180° = \frac{OP}{OP} = -1$$

$$\tan 180° = \frac{0}{-OP} = 0$$

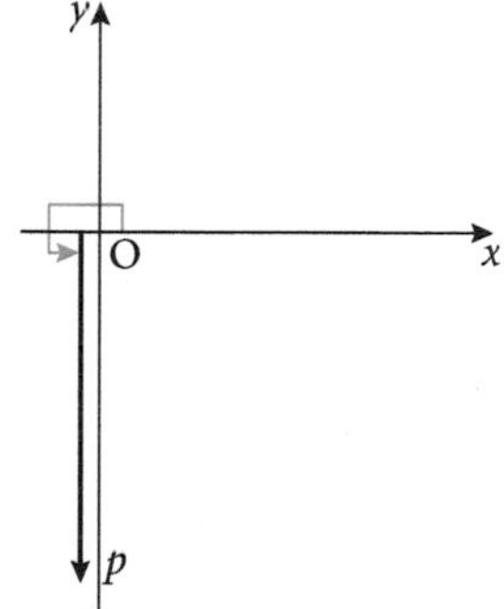

Fig. 3.15

Similarly, when the line OP lies along the negative y-axis in the third quadrant as in Fig. 3.15, $\theta = 270°$. Then the projection of OP on the negative y-axis is equal to $-$OP, and the projection of OP on Oy is zero units in length.

$$\sin 270° = \frac{-OP}{OP} = -1$$

$$\cos 270° = \frac{0}{OP} = 0$$

$$\tan 270° = \frac{-OP}{0} = \text{is undefined}$$

since division by zero is undefined.

When the rotating line OP has turned through 360°, the line OP again lies along Ox, the positive x-axis in the first quadrant as in Fig. 3.16 and $\theta = 360°$. Then the projection of OP on Ox is OP itself, and the projection of OP on Oy is zero units in length.

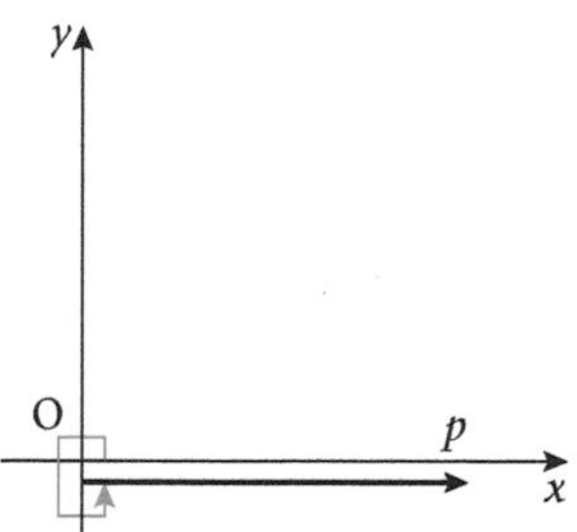

Fig. 3.16

Thus

$$\sin 360° = \frac{0}{OP} = 0$$

$$\cos 360° = \frac{OP}{OP} = 1$$

$$\tan 360° = \frac{0}{OP} = 0$$

Tan, sin and cos of 45°

In Fig. 3.17, △ABC is right-angled at B and AB = BC = 1 unit.

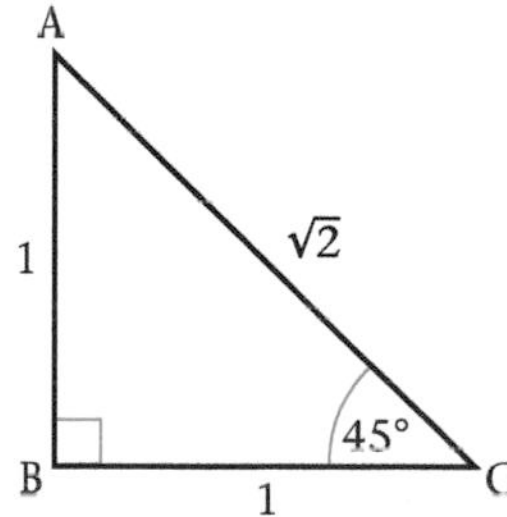

Fig. 3.17

$AC^2 = 1^2 + 1^2$ (*Pythagoras' theorem*)

$\quad = 2$

AC $= \sqrt{2}$ units

Since AB = BC, $\widehat{A} = \widehat{C}$ (*isosceles* △)

But $\widehat{A} + \widehat{C} = 90°$ (*sum of angles of* △)

Hence $\widehat{A} = \widehat{C} = 45°$

$$\mathbf{\tan 45° = \frac{1}{1} = 1}$$

$$\mathbf{\sin 45° = \frac{1}{\sqrt{2}}}$$

$$\mathbf{\cos 45° = \frac{1}{\sqrt{2}}}$$

Any triangle with angles of 45°, 45°, 90° has sides whose lengths are in the ratio $1 : 1 : \sqrt{2}$.

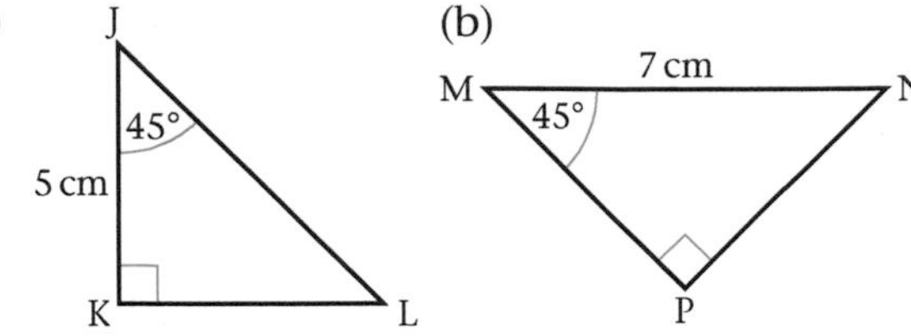

Fig. 3.18

For example, in Fig. 3.18(a), JK = 5 cm. Therefore KL = 5 cm and JL = $5\sqrt{2}$ cm.

In Fig. 3.18(b), MN = 7 cm.

$$\cos 45° = \frac{MP}{7} = \frac{1}{\sqrt{2}}$$

$$MP = \frac{7}{\sqrt{2}}\text{ cm}$$

Similarly, $NP = \frac{7}{\sqrt{2}}$ cm

Hence MP and NP are found by dividing MN by $\sqrt{2}$.

Note: To write a length correct to a given degree of accuracy, to 2 significant figures say, we must first rationalise the denominator, for example:

$$MP = \frac{7}{\sqrt{2}}\text{ cm}$$

$$= \frac{7\sqrt{2}}{\sqrt{2}\cdot\sqrt{2}}\text{ cm}$$

$$= \frac{7\sqrt{2}}{2}\text{ cm}$$

$$= \frac{7 \times 1.41}{2}\text{ cm}$$

$$= 4.94\text{ cm}$$

$$= 4.9\text{ cm to 2 s.f.}$$

Tan, sin and cos of 60° and 30°

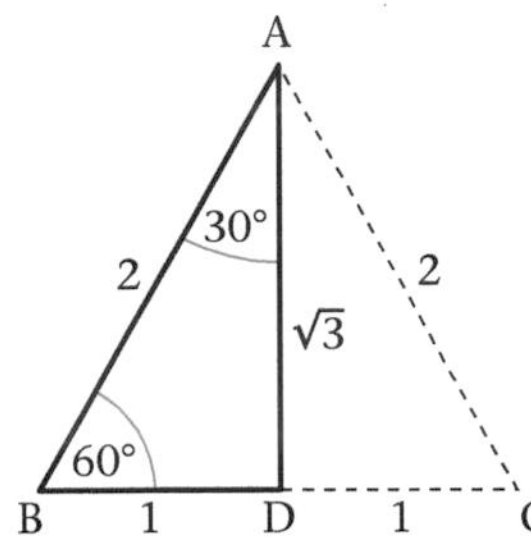

Fig. 3.19

In Fig 3.19, ABC is an equilateral triangle with sides of 2 units in length. AD is an altitude.

BD = DC = 1 unit (AD *bisects* BC)

In △ABD,

$AB^2 = AD^2 + BD^2$ (*Pythagoras' theorem*)

$2^2 = AD^2 + 1^2$

$AD^2 = 2^2 - 1^2 = 4 - 1 = 3$

$AD = \sqrt{3}$ units

Since $\widehat{B} = 60°$ (*equilateral* △)

$$\tan 60° = \frac{\sqrt{3}}{1} = \sqrt{3}$$

$$\sin 60° = \frac{\sqrt{3}}{2}$$

$$\cos 60° = \frac{1}{2}$$

Since $B\widehat{A}D = 30°$ (*sum of angles of* △ABD)

$$\tan 30° = \frac{1}{\sqrt{3}}$$

$$\sin 30° = \frac{1}{2}$$

$$\cos 30° = \frac{\sqrt{3}}{2}$$

Note: since 60° and 30° are complementary angles,
sin 60° = cos 30° and **cos 60° = sin 30°**

Any triangles with angles of 30°, 60°, 90° has sides whose lengths are in the ratio $1 : \sqrt{3} : 2$.

For example, in Fig. 3.20,

(a) If QS = 3 cm
then QR = 6 cm and RS = $3\sqrt{3}$ cm

(b) If TV = 8 cm,
then UV = 4 cm and RS = $4\sqrt{3}$ cm

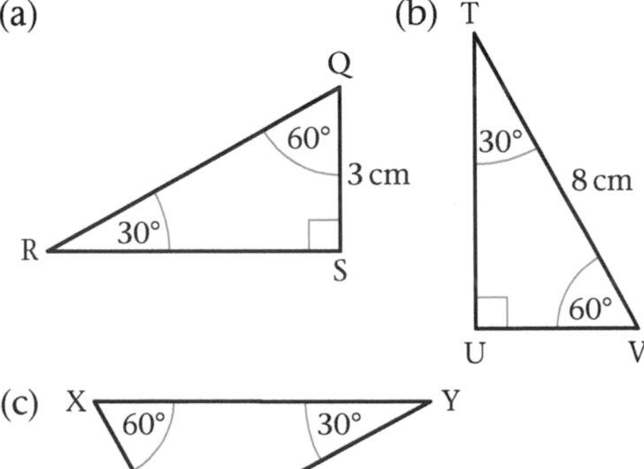

Fig. 3.20

(c) $\tan 30° = \frac{XZ}{YZ}$

$$\frac{1}{\sqrt{3}} = \frac{XZ}{5}$$

$$XZ = \frac{5}{\sqrt{3}}\text{ cm}$$

$$\sin 30° = \frac{XZ}{XY} = \frac{1}{2}$$

$$XY = 2 \times XZ$$

$$= 2 \times \frac{5}{\sqrt{3}}\text{ cm} = \frac{10}{\sqrt{3}}\text{ cm}$$

Notice that XZ : ZY : XY

$$= \frac{5}{\sqrt{3}} : 5 : \frac{10}{\sqrt{3}}$$

$$= 5 : 5\sqrt{3} : 10$$

$$= 1 : \sqrt{3} : 2$$

Note: $XZ = \frac{5}{\sqrt{3}}\text{ cm} = \frac{5\sqrt{3}}{3}\text{ cm}$

and $XY = \frac{10\sqrt{3}}{3}\text{ cm}$

Example 3

In Fig. 3.21 if BC = 4 cm, find AD.

Either: Using the ratios of the sides of the $\triangle$s:
In $\triangle$ABC with angles 30°, 60°, 90°,
if BC = 4 cm, then AC = $4\sqrt{3}$ cm

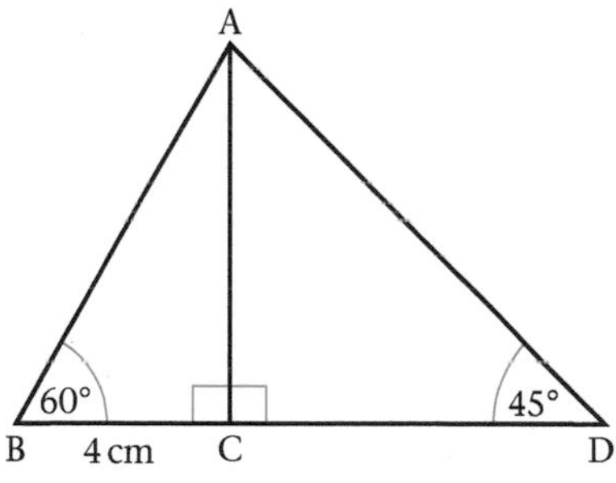

Fig. 3.21

In $\triangle$ACD with angles of 45°, 45°, 90°,
If AC = $4\sqrt{3}$ cm, then AD = $4\sqrt{3}$ cm $\times \sqrt{2}$
$= 4\sqrt{6}$ cm

Or: Using trigonometrical ratios of 60° and 45°:

In $\triangle$ABC, $\tan 60° = \frac{AC}{BC}$

$$\sqrt{3} = \frac{AC}{4}$$

$$AC = 4\sqrt{3}\text{ cm}$$

In $\triangle$ACD, $\sin 45° = \frac{AC}{AD}$

$$\frac{1}{\sqrt{2}} = \frac{4\sqrt{3}}{AD}$$

$$AD = 4\sqrt{3} \times \sqrt{2}\text{ cm} = 4\sqrt{6}\text{ cm}$$

Example 4

In Fig. 3.22, if PX = 24 m, find PQ.

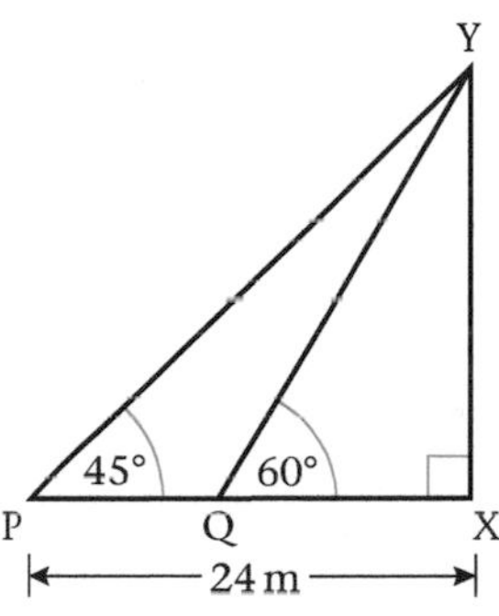

Fig. 3.22

Either: using the ratios of the sides of the $\triangle$s:
In $\triangle$PXY with angles of 45°, 45°, 90°,
if PX = 24 m, then XY = 24 m
In $\triangle$QXY with angles of 30°, 60°, 90°
if XY = 24 m, then $QX = \frac{24}{\sqrt{3}}\text{ m} = \frac{24\sqrt{3}}{3}$
$= 8\sqrt{3}$ m

Hence PQ = $(24 - 8\sqrt{3})$ m
$= 8(3 - \sqrt{3})$ m

Or: Using the trigonometrical ratios of 45° and 60°:

In $\triangle$PXY, $\tan 45° = \frac{XY}{PX}$

$$1 = \frac{XY}{24}$$

$$XY = 24\text{ m}$$

In $\triangle$QXY, $\tan 60° = \frac{XY}{QX}$

$$\sqrt{3} = \frac{24}{QX}$$

$$QX = \frac{24}{\sqrt{3}}$$

$= 8\sqrt{3}$ m (as before)

Hence PQ = $8(3 - \sqrt{3})$ m (as before)

Example 5

In Fig. 3.23, if AB = 6 cm, calculate x and y. (Leave the answers in surd form with rational denominators.)

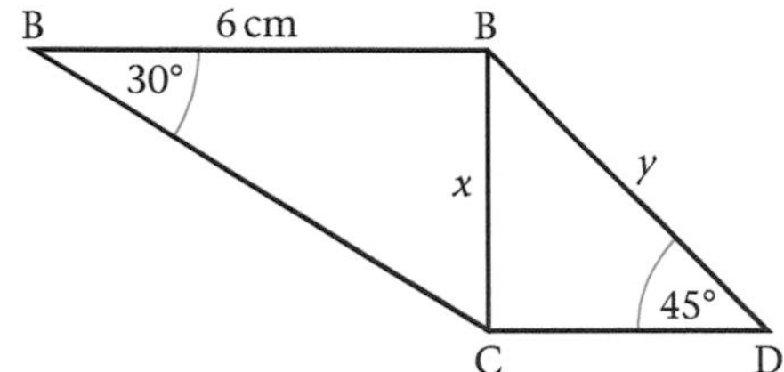

Fig. 3.23

In △ABC,

$$\tan 30° = \frac{x}{6}$$

$$\frac{x}{6} = \frac{1}{\sqrt{3}}$$

$$x = \frac{6}{\sqrt{3}} \text{ cm}$$

$$= \frac{6\sqrt{3}}{3}$$

$$= 2\sqrt{3} \text{ cm}$$

In △BCD,

$$\sin 45° = \frac{x}{y}$$

$$\frac{1}{\sqrt{2}} = \frac{2\sqrt{3}}{y}$$

$$y = 2\sqrt{3} \times \sqrt{2} \text{ cm} = 2\sqrt{6} \text{ cm}$$

Exercise 3g

In each part of Fig. 3.24, calculate the lengths marked x and y. All dimensions are in cm. Give answers in surd form with rational denominators.

1
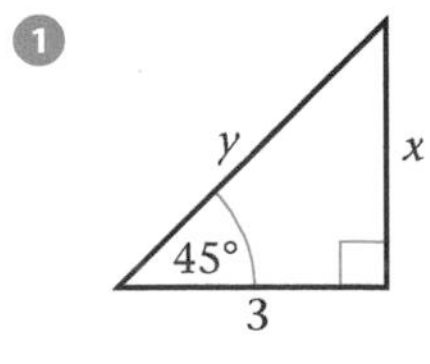

2
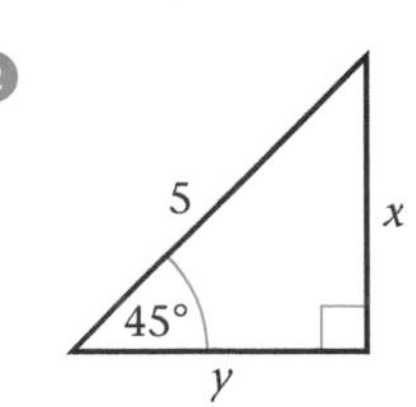

3
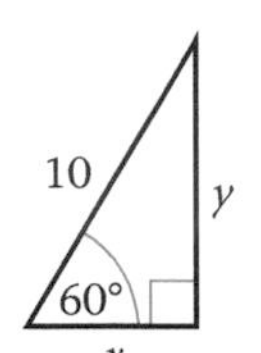

4
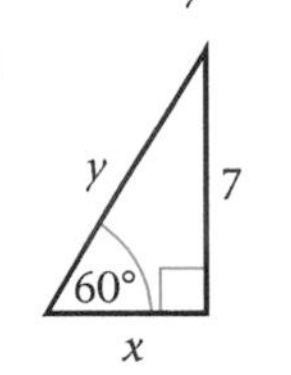

5
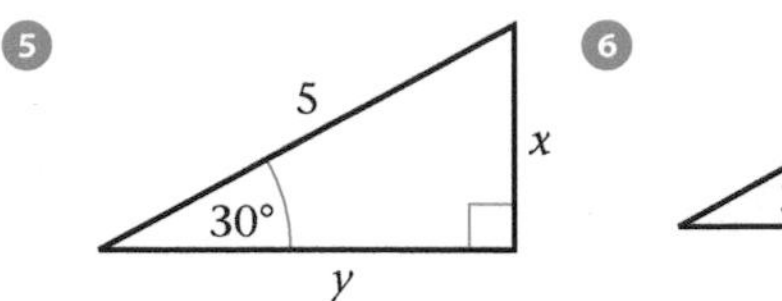

6
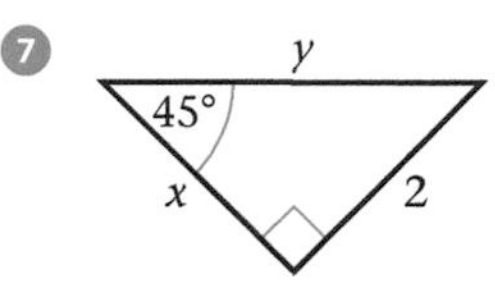

7
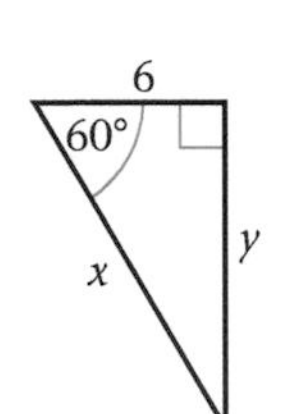

8
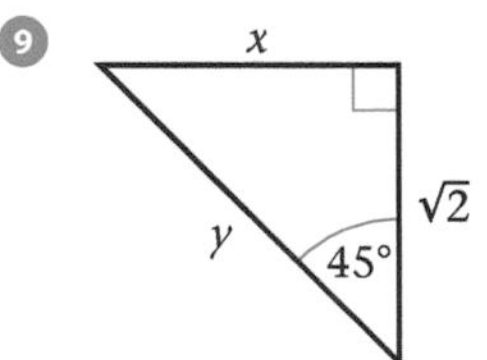

9
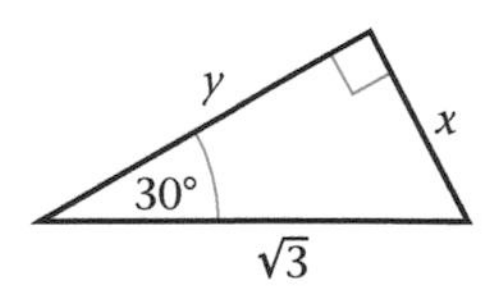

10
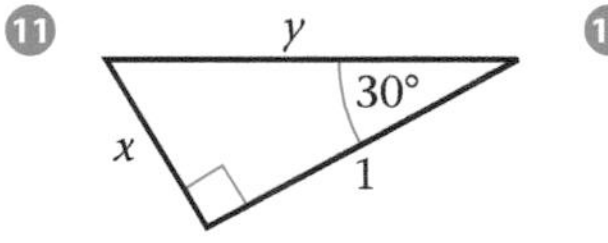

11
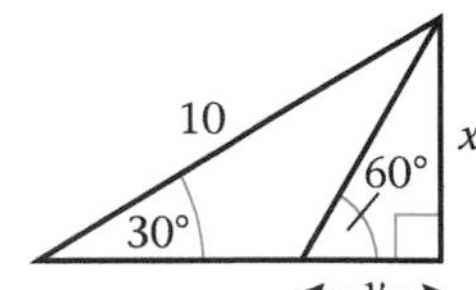

12
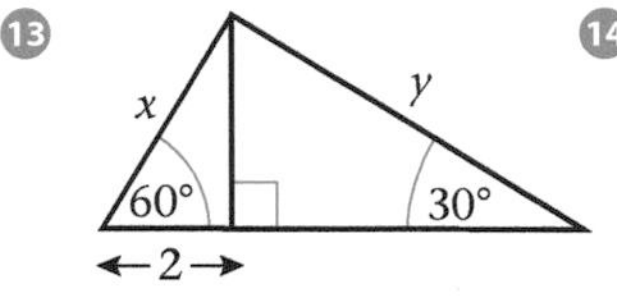

13
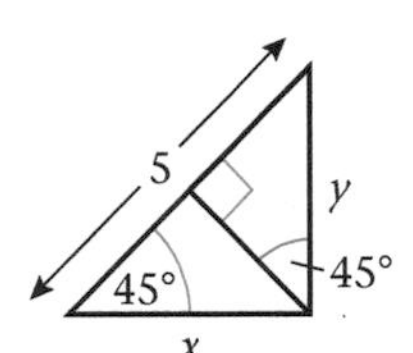

14
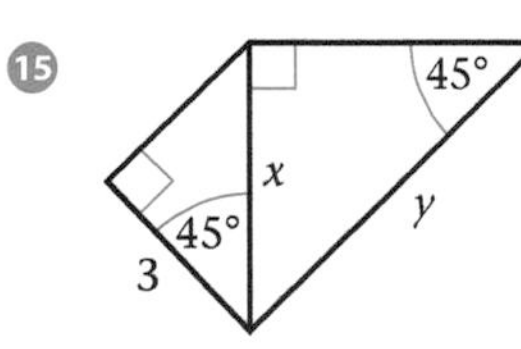

15
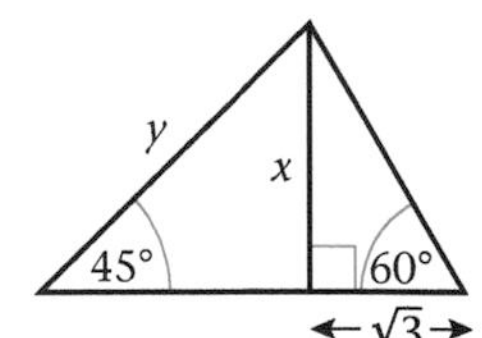

16
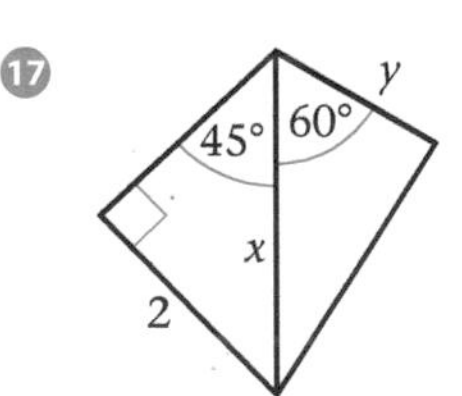

17
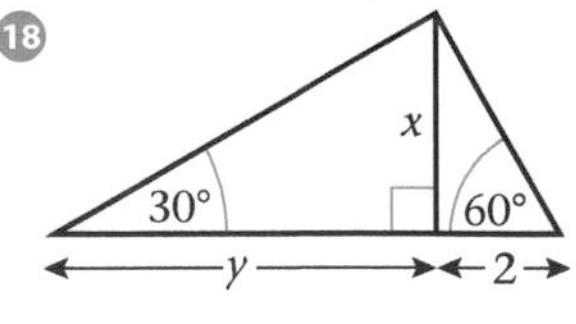

18
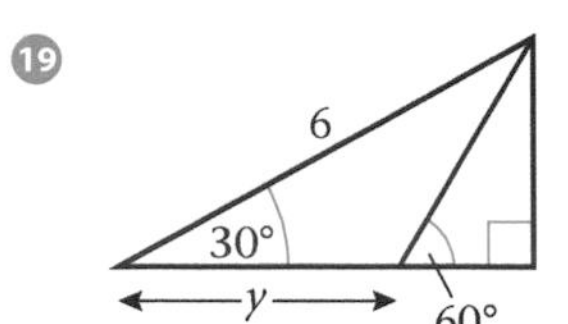

19
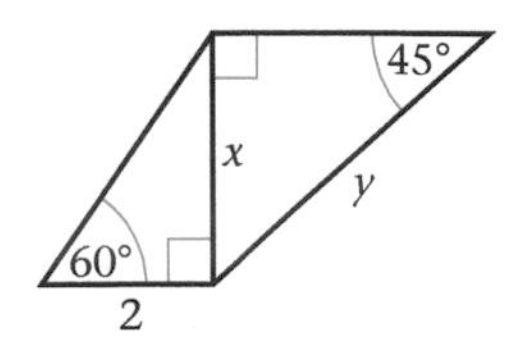

20

Fig. 3.24

Example 6

From the top of a tower 80 m high, two boats are seen in a direction due south. The angles of depression of the boats from the top of the tower are 45° and 30°. Find the distance between the boats.

First, draw a sketch. In Fig. 3.25, HT represents the tower; A and B are the positions of the boats.

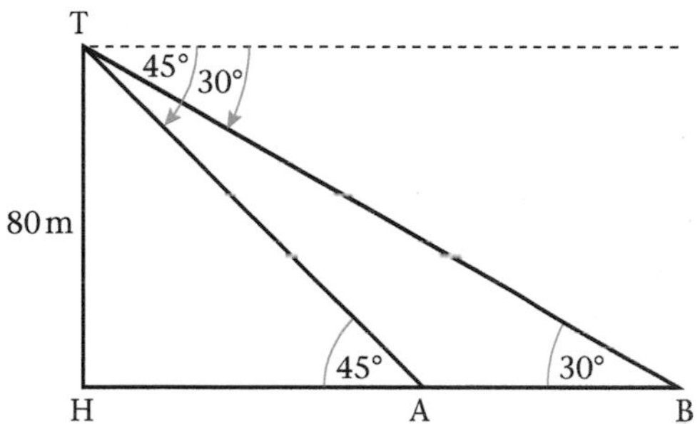

Fig. 3.25

AB is the distance between the boats.

$$AB = HB - HA$$

In △THB,

$$\tan 30° = \frac{80}{HB}$$

$$\frac{1}{\sqrt{3}} = \frac{80}{HB}$$

$$HB = 80\sqrt{3}\text{ m}$$

In △THA,

$$\tan 45° = \frac{80}{HA}$$

$$1 = \frac{80}{HA}$$

$$HA = 80\text{ m}$$

$$\text{Hence } AB = 80\sqrt{3} - 80\text{ m}$$

$$= 80(\sqrt{3} - 1)\text{ m}$$

Notice the importance of drawing a sketch.

Exercise 3h

Draw a sketch in each question. Leave the answers in surd form with rational denominators.

1. From the top of a tower, the angle of depression of a car is 30°. If the tower is 20 m high, how far is the car from the foot of the tower?
2. The angle of elevation of X from Y is 30°. If XY = 40 m, how high is X above Y?
3. In △ABC, AB = 6 cm, $\widehat{A}$ = 90° and $\widehat{B}$ = 30°. AD is an altitude. Find CD.
4. In the isosceles triangle ACB, AB = AC = 4 cm, $B\widehat{A}C$ = 30° and CN is an altitude. Find BN.
5. XYZ is an isosceles triangle with XY = XZ = 6 cm and $Y\widehat{X}Z$ = 120°. Calculate the length of YZ.
6. A cone has a circular base, a perpendicular height of 21 cm, and a semi-vertical angle of 30°. Calculate the slant height of the cone. Find the area of its base. (Take π to be $\frac{22}{7}$.)
7. When the angle of elevation of the sun is 30°, the shadow of a vertical tower is 20 m longer than when the elevation of the sun is 60°. Find the height of the tower.
8. Two huts and a radio mast are on level ground such that one hut is due east of the mast and the other is due west of it. From the top of the mast the angles of depression of the huts are 60° and 45° respectively. If the mast is 150 m high, find the distance between the huts.
9. A and B are two points on level ground, both due south of a flag-pole. The angle of elevation of the top of the flag-pole is 60° from A and 45° from B. If A is 20 m from the foot of the flag-pole, find AB.
10. The top of a building 24 m high is observed from the top and from the bottom of a tree (which is vertical). The angles of elevation are found to be 45° and 60° respectively. By a suitable calculation find the height of the tree.

Summary

If θ is any acute angle, then we have for the trigonometric ratios of

complementary angles

$\sin(90° - \theta) = \cos\theta$; $\cos(90° - \theta) = \sin\theta$; and $\tan(90° - \theta) = \frac{1}{\tan\theta}$

obtuse angles
$\sin(180° - \theta) = \sin\theta$
$\cos(180° - \theta) = -\cos\theta$
$\tan(180° - \theta) = -\tan\theta$

reflex angles
$\sin(180° + \theta) = -\sin\theta$
$\cos(180° + \theta) = -\cos\theta$
$\tan(180° + \theta) = \tan\theta$

and

$\sin(360° - \theta) = -\sin\theta$
$\cos(360° - \theta) = \cos\theta$
$\tan(360° - \theta) = -\tan\theta$

Also for any negative angle θ
$\sin(-\theta) = -\sin\theta$; $\cos(-\theta) = \cos\theta$;
$\tan(-\theta) = -\tan\theta$

θ	0°	30°	45°	60°	90°	180°	270°	360°
sin	0	$\frac{1}{2}$	$\frac{1}{\sqrt{2}}$	$\frac{\sqrt{3}}{2}$	1	0	−1	0
cos	1	$\frac{\sqrt{3}}{2}$	$\frac{1}{\sqrt{2}}$	$\frac{1}{2}$	0	−1	0	1
tan	0	$\frac{1}{\sqrt{3}}$	1	$\sqrt{3}$	undefined	0	undefined	0

Practice exercise P3.1

Use tables or a calculator to find the values, correct to 3 significant figures, of these.

1. $\sin -80°$
2. $\cos -140°$
3. $\tan -260°$
4. $\sin -253°$
5. $\cos -58°$
6. $\tan -164°$

Practice exercise P3.2

Find all the values of θ between 0° and 360° in each of the following, giving your answer in degrees to the nearest whole number.

1. $\sin\theta = 0.966$
2. $\cos\theta = 0.225$
3. $\tan\theta = 2.356$
4. $\sin\theta = -0.993$
5. $\cos\theta = -0.545$
6. $\tan\theta = -0.649$

Practice exercise P3.3

Draw a sketch for each question. Leave the answers in surd form with rational denominators.

1. Given that $\sin\theta = \frac{\sqrt{3}}{2}$, then the possible values of θ are
   A 60° and 30°, B 60° and 120°,
   C 30° and 150°, D 60° and 300°.
2. In triangle PQR, PQ = a cm, PR = 12 cm, $Q\hat{P}R = 30°$ and $P\hat{Q}R = 90°$. Show that $a = 6\sqrt{3}$ cm.
3. From the top of a building 28 m high, the angle of depression of a car is 30°. Calculate the distance of the car from the foot of the building.
4. In triangle ABC, AC = 15 cm, $C\hat{A}B = 30°$ and $A\hat{B}C = 90°$.
   (a) Calculate the length of BC.
   (b) Show that the length of AB is $\frac{15\sqrt{3}}{2}$ cm.
5. P is a point on the ground 10.5 m from the foot of a flag-pole. The angle of elevation of the top of the flag-pole from P is 30°. Calculate the height of the flag-pole.
6. P and Q are two points on level ground and due south of a building. The angle of elevation of the top of the building is 60° from P and 45° from Q. If Q is 54 m from the foot of the building, calculate
   (a) the height of the building,
   (b) the distance of P from the foot of the building.
7. Triangle PQR is an equilateral triangle with sides of length 13 cm. Calculate the height of the triangle.
8. The angle of depression of a boat from a cliff 24 m high is 60°. Find the distance, measured in a straight line, of the boat from the top of the cliff.
9. A child 120 cm tall is flying a kite. The angle of elevation of the kite from the child is 45°. The kite is 6 m vertically above the ground. How long is the kite string?
10. ABCD is a rectangle with AB = 15 cm, $C\hat{A}B = 60°$ and $A\hat{B}C = 90°$.
    Calculate
    (a) BC, (b) AC, (c) the acute angle between the diagonals.

Chapter 4

Quadratic equations

Pre-requisites

- expansion of algebraic expressions; linear equations; factorisation of quadratic expressions

Equation (revision)

An equation is a statement about an unknown quantity called a variable. The variable is usually defined by a letter. There are two sides in an equation – the left-hand side (LHS) and the right-hand side (RHS).

To solve an equation is to find the value of the unknown (or the variable) which makes the two sides of the equation equal, that is, the value which makes the statement true.
If the power of the variable is 1, the equation is a linear equation, for example,

$$3m + 12 = 14$$

and the solution (or root) of the equation is $m = 4$

Quadratic equation

A **quadratic equation** is one in which 2 is the highest power of the variable(s) (or unknown(s)) in the equation. We say that it is an equation of the **second degree**. For example,

$$m^2 - 5m - 14 = 0$$

is an equation of the second degree, or a quadratic equation.

In order to solve a quadratic equation, we use the fact that, if the product of two numeric quantities is zero, then one of the quantities (or possibly both of them) must be zero. For example,

$$3 \times 0 = 0; -5 \times 0 = 0; \text{ and } 0 \times 0 = 0$$

Given that $p \times q = 0$
then *either* $p = 0$
or $q = 0$

If $3pq = 0$
then $pq = 0$ (since $3 \neq 0$)
and *either* $p = 0$ or $q = 0$

Suppose $p = (a + 2)$ and $q = (a + 3)$, and $pq = 0$
then $a + 2 = 0$ or $a + 3 = 0$
that is, $a = -2$ or $a = -3$

Example 1

Solve the equation $(b - 2)(b - 5) = 0$.

If $(b - 2)(b - 5) = 0$.
then either $b - 2 = 0$ or $b - 5 = 0$
$b = 2$ or $b = 5$

Example 2

Solve the equation $4r(r - 7) = 0$.

If $4r(r - 7) = 0$
then either $r = 0$ or $r - 7 = 0$ $(4 \neq 0)$
$r = 0$ or $r = 7$

Example 3

Solve the equation $(t - 3)^2 = 0$.

If $(t - 3)^2 = 0$
then $(t - 3)(t - 3) = 0$
$t - 3 = 0$ twice
$t = 3$ twice

The solutions of the equations in Examples 1, 2 and 3 are called the **roots** of the equations. Notice that in Example 3 there are repeated roots. The meaning of these will be explained on page 40.

Exercise 4a

Solve the following equations.

1. $(a + 2)(a - 4) = 0$
2. $b(b + 6) = 0$
3. $3d(d - 7) = 0$
4. $6m(m + 3) = 0$
5. $(6 - n)(4 + n) = 0$
6. $0 = (5 + u)(3 - u)$
7. $(v - 2)(v + 2) = 0$
8. $(x + 5)(x - 5) = 0$
9. $(9 - f)^2 = 0$
10. $(1 + z)^2 = 0$

Example 4

Solve the equation $(3a + 2)(2a - 7) = 0$.

If $(3a + 2)(2a - 7) = 0$, then

$$\begin{array}{llcl} \text{either} & 3a + 2 = 0 & \text{or} & 2a - 7 = 0 \\ & 3a = -2 & \text{or} & 2a = 7 \\ & a = -\frac{2}{3} & \text{or} & a = \frac{7}{2} \\ & \text{i.e. } a = -\frac{2}{3} & \text{or} & 3\frac{1}{2} \end{array}$$

Check: By substitution:

If $a = -\frac{2}{3}$

$$\begin{aligned}(3a + 2)(2a - 7) &= (-2 + 2)(-1\tfrac{1}{3} - 7) \\ &= 0 \times (-8\tfrac{1}{3}) = 0\end{aligned}$$

If $a = 3\frac{1}{2}$,

$$\begin{aligned}(3a + 2)(2a - 7) &= (10\tfrac{1}{2} + 2)(7 - 7) \\ &= 12\tfrac{1}{2} \times 0 = 0\end{aligned}$$

Exercise 4b

Solve the following equations. Check the results by substitution.

1. $(3a + 10)(3a - 12) = 0$
2. $(4b - 3)^2 = 0$
3. $(2c + 1)^2 = 0$
4. $(1 - 2d)(2 + 3d) = 0$
5. $(3e + 5)^2 = 0$
6. $5m(3m - 4) = 0$
7. $(9 + 3n)(5 - 7n) = 0$
8. $(11 - 4x)^2 = 0$
9. $3y(2y + 9) = 0$
10. $5a(15 + 4a) = 0$

In the quadratic equation $m^2 - 5m - 14 = 0$, the left-hand side (LHS) of the equation can be factorised and written as a product of two linear factors, that is,

$$m^2 - 5m - 14 = (m + 2)(m - 7)$$

Then we can write $(m + 2)(m - 7) = 0$
so that $(m + 2) = 0$ or $(m - 7) = 0$
that is, either $m = -2$, or $m = 7$

Notice that before the solutions of a quadratic equation can be found, the terms in the equation must be arranged, where necessary, so that the right-hand side (RHS) of the equation is zero. The quadratic expression on the left-hand side (LHS) is then factorised.

Example 5

Solve the equation $y^2 + 4y + 21 = 0$.

$$\begin{aligned} y^2 + 4y - 21 &= 0 \\ (y + 7)(y - 3) &= 0 \end{aligned}$$

$$\begin{array}{llcl} \text{Either} & y + 7 = 0 & \text{or} & y - 3 = 0 \\ & y = -7 & \text{or} & y = 3 \end{array}$$

Example 6

Solve the equation $n^2 = 9$.

First rearrange the equation:

$$\begin{aligned} n^2 &= 9 \\ n^2 - 9 &= 0 \end{aligned}$$

Factorise the left-hand side (LHS) (difference of two squares):

$$(n - 3)(n + 3) = 0$$

$$\begin{array}{llcl} \text{Either} & n - 3 = 0 & \text{or} & n + 3 = 0 \\ & n = 3 & \text{or} & n = -3 \\ & n = \pm 3 & & \end{array}$$

Example 7

Solve the equation $x^2 = 4x + 5$.

Rearrange the equation to give a quadratic expression on the LHS and 0 (zero) on the RHS:

$$\begin{aligned} x^2 &= 4x + 5 \\ x^2 - 4x - 5 &= 0 \\ (x - 5)(x + 1) &= 0 \end{aligned}$$

$$\begin{array}{llcl} \text{Either} & x - 5 = 0 & \text{or} & x + 1 = 0 \\ & x = 5 & \text{or} & x = -1 \end{array}$$

Exercise 4c

Solve the following quadratic equations.

1. $a^2 - 3a + 2 = 0$
2. $b^2 + 5b + 6 = 0$
3. $c^2 - c - 2 = 0$
4. $d^2 + 2d - 3 = 0$
5. $e^2 - 7e + 10 = 0$
6. $m^2 - 4m = 0$
7. $n^2 + 5n = 0$
8. $p^2 + 7p + 12 = 0$
9. $q^2 + 2q - 8 = 0$
10. $x^2 - 2x + 1 = 0$
11. $y^2 - 5y + 4 = 0$
12. $a^2 - 9a = 0$
13. $b^2 - 9 = 0$
14. $c^2 = 25$
15. $u^2 - 8u - 9 = 0$
16. $v^2 + 2v - 35 = 0$
17. $x^2 - 6x + 9 = 0$
18. $y^2 + 8y + 16 = 0$
19. $z^2 - 4z = 0$
20. $z^2 - 4 = 0$

Example 8

Solve the equation $4y^2 + 5y - 21 = 0$.

$$4y^2 + 5y - 21 = 0$$
$$(y + 3)(4y - 7) = 0$$

either $y + 3 = 0$ or $4y - 7 = 0$

$y = -3$ or $4y = 7$

$y = -3$ or $y = \frac{7}{4}$

i.e. $y = -3$ or $1\frac{3}{4}$

Check: By substitution:

If $y = -3$,

$4y^2 + 5y - 21 = 36 - 15 - 21 = 0$

If $y = 1\frac{3}{4}$,

$4y^2 + 5y - 21 = 4 \times \frac{7}{4} \times \frac{7}{4} + 5 \times \frac{7}{4} - 21$

$= \frac{49}{4} + \frac{35}{4} - 21 = 0$

Example 9

Solve the equation $a^2 - 3a = 0$.

$$a^2 - 3a = 0$$
$$a(a - 3) = 0$$

either $a = 0$ or $a - 3 = 0$

i.e. $a = 0$ or 3

Example 10

Solve the equation $m^2 = 16$.

Rearrange the equation.

If $m^2 = 16$

then $m^2 - 16 = 0$

$(m - 4)(m + 4) = 0$

either $m - 4 = 0$ or $m + 4 = 0$

i.e. $m = +4$ or $m = -4$

i.e. $m = \pm 4$

Note: when $m^2 = 16$

$m = \pm\sqrt{16}$

i.e. $m = \pm 4$

Example 11

Solve the equation $2x^2 = 3x + 5$.

Rearrange the equation to give a quadratic expression on the LHS and 0 (zero) on the RHS.

$$2x^2 = 3x + 5$$
$$2x^2 - 3x - 5 = 0$$
$$(2x - 5)(x + 1) = 0$$

either $2x - 5 = 0$ or $x + 1 = 0$

$2x = 5$ or $x = -1$

$x = \frac{5}{2}$ or $x = -1$

i.e. $x = 2\frac{1}{2}$ or -1

Notice, in Examples 10 and 11, that where necessary the given equation should be arranged to give a quadratic expression on the left-hand side and zero on the other side.

Exercise 4d

Solve the following quadratic equations.

1. $h^2 - 15h + 54 = 0$
2. $k^2 - 15k - 54 = 0$
3. $2m^2 - 5m = 0$
4. $2m^2 - 5m + 3 = 0$
5. $2m^2 - 5m - 3 = 0$
6. $3n^2 + n = 0$
7. $a^2 + a = 90$
8. $b^2 - b = 72$
9. $3x^2 + 4x + 1 = 0$
10. $9h^2 = 6h - 1$
11. $3d^2 - 5d - 2 = 0$
12. $4e^2 - 20e + 25 = 0$
13. $9f^2 + 12f + 4 = 0$
14. $4u^2 = 11u - 3$
15. $b^2 + 7b = 44$
16. $7m^2 = 3m$
17. $2p^2 - 11p + 5 = 0$
18. $6y^2 = y + 1$
19. $6h^2 + 13h - 5 = 0$
20. $16t^2 = 49$
21. $4r^2 - 49 = 0$
22. $8s^2 + 14s = 15$
23. $6x^2 = 7x + 20$
24. $63z = 49 + 18z^2$

Equations with non-rational roots

If $m^2 = 16$, then $m = \pm 4$. (See Example 10.)

Hence, if $(x - 3)^2 = 16$

then $x - 3 = \pm 4$

and $x = 3 \pm 4$

that is, $x = 3 + 4$ or $x = 3 - 4$

$x = 7$ or $x = -1$

Example 12

Solve the equation $(x + 3)^2 = 7$.

If $(x + 3)^2 = 7$

then $x + 3 = \pm\sqrt{7}$

$x = -3 \pm \sqrt{7}$

Since $\sqrt{7}$ is a non-rational number, the roots of the equation $(x + 3)^2 = 7$ are non-rational. The roots may be found approximately by writing $\pm\sqrt{7}$ as ± 2.65.

Exercise 4e

Solve the following equations. If an equation has non-rational roots, leave the answer in the form given in Example 12.

1. $(x - 2)^2 = 9$
2. $(x - 7)^2 = 4$
3. $(x + 3)^2 = 4$
4. $(x + 2)^2 = 25$

5 $(x - 1)^2 = 2$
6 $(x + 4)^2 = 3$
7 $(x - 2)^2 = \frac{1}{4}$
8 $(x - 6)^2 = 36$
9 $(x - 4)^2 = 10$
10 $(x + 5)^2 = \frac{1}{9}$
11 $(x - 8)^2 = 3$
12 $(x - 1)^2 = \frac{9}{25}$
13 $(x + 1)^2 = 2\frac{1}{4}$
14 $(x + 7)^2 = 6$
15 $(x + \frac{1}{3})^2 = \frac{4}{9}$
16 $(x + 9)^2 = 3$
17 $(x - 6)^2 = 5$
18 $(x - 2\frac{1}{2})^2 = 6\frac{1}{4}$
19 $(x + 10)^2 = 8$
20 $(x - 6)^2 = 2\frac{7}{9}$

Completing the square

In order to solve a quadratic equation, it is sometimes useful to write the quadratic expression on the left-hand side (LHS) as a perfect square. However, this becomes necessary when the quadratic expression on the LHS does not factorise and the roots are non-rational. The following method gives the formula for finding the roots of any quadratic equation.

Remember that in a quadratic expression which is a perfect square, there must be three terms. If the variable in the quadratic expression is x, these terms are: a term in x^2, a term in x, and a constant (numeric) term. Let us first consider a quadratic expression which is a perfect square, when the coefficient of the x^2-term is 1.

Suppose $x^2 + 6x + k$ is a perfect square and that it is equal to $(x + a)^2$.

Let $x^2 + 6x + k = (x + a)^2$

Then $x^2 + 6x + k = x^2 + 2ax + a^2$

By comparing the coefficients of x,

$2a = 6$
$a = 3$

By comparing the constant terms,

$k = a^2$
$k = 3^2 = 9$

Check: $x^2 + 6x + 9 = (x + 3)^2$

There is a definite relation between the coefficients of the x^2-term and of the x-term, and the constant term. In the expression $x^2 + 6x + k$, the coefficient of x is 6, half of 6 is 3 and the square of 3 is 9. **Hence, the constant term is the square of half of the coefficient of x**, that is, $k = 9$.

***Note* that the coefficient of x^2 is 1.**

Example 13

What must be added to $d^2 - 5d$ to make it into a perfect square? Factorise the result.

The coefficient of d is -5. Half of -5 is $-\frac{5}{2}$.
$(-\frac{5}{2})^2 = +\frac{25}{4}$
$\frac{25}{4}$ must be added.
$d^2 - 5d + \frac{25}{4}$ is a perfect square.
$d^2 - 5d + \frac{25}{4} = (d - \frac{5}{2})^2$

Notice that the constant term in the bracketed expression to be squared is half the coefficient of the d-term, that is $(d - \frac{5}{2})$.

Example 14

Add a term to $n^2 + 1\frac{1}{2}n$ to make the expression a perfect square. Express the result as the square of a bracketed expression.

The coefficient of n is $+1\frac{1}{2}$.
Half of $+1\frac{1}{2} = \frac{1}{2} \times \frac{3}{2} = \frac{3}{4}$
$(\frac{3}{4})^2 = \frac{9}{16}$
$\frac{9}{16}$ must be added.
$n^2 + 1\frac{1}{2}n + \frac{9}{16} = (n + \frac{3}{4})^2$

Notice that the constant term is the bracketed expression to be squared is half the coefficient of the n-term, that is $(n + \frac{3}{4})$

Examples 13 and 14 may be checked by squaring out the bracket in the final result.

Exercise 4f

In the following, add a term making the given expression into a perfect square. Write the result as the square of a bracketed expression.

1 $a^2 + 8a$
2 $b^2 + 10b$
3 $c^2 - 4c$
4 $d^2 - 6d$
5 $x^2 + 5x$
6 $y^2 - 3y$
7 $z^2 - 7z$
8 $m^2 + 2m$
9 $n^2 - n$
10 $u^2 - \frac{1}{2}u$
11 $h^2 + \frac{2}{3}h$
12 $k^2 - 1\frac{1}{3}k$
13 $g^2 - 4\frac{2}{3}g$
14 $a^2 + \frac{3}{5}a$
15 $b^2 - \frac{4}{5}b$
16 $m^2 - 8m$
17 $m^2 - 8mn$
18 $a^2 - 6ad$
19 $x^2 + 10xy$
20 $m^2 + 3mn$

Example 15

Solve the equation $x^2 - 8x + 3 = 0$.

The LHS does not factorise, so the equation is rearranged leaving the terms in x on the LHS. A constant to make the LHS a perfect square is added to both the LHS and the RHS.

$$x^2 - 8x + 3 = 0$$

Subtract 3 from both sides.

$$x^2 - 8x = -3$$

Add $\left(\frac{-8}{2}\right)^2$, that is, 16 to both sides.

$$x^2 - 8x + 16 = -3 + 16$$
$$(x - 4)^2 = 13$$

Take the square root of both sides of the equation.

$$x - 4 = \pm\sqrt{13}$$
$$x = 4 \pm \sqrt{13}$$

Note:

1 The same constant (positive or negative) must always be added to both sides of an equation.

2 Both the positive and negative values of the square root on the RHS are used.

Example 16

Solve the equation $a^2 + 3a - 2 = 0$.

$$a^2 + 3a - 2 = 0$$

The LHS does not factorise; rearrange the terms in the equation.

$$a^2 + 3a = 2$$

Add to both sides the square of $\frac{3}{2}$.

$$a^2 + 3a + \left(\tfrac{3}{2}\right)^2 = 2 + \tfrac{9}{4}$$
$$= \frac{8 + 9}{4}$$
$$\left(a + \tfrac{3}{2}\right)^2 = \tfrac{17}{4}$$
$$a + \tfrac{3}{2} = \pm\sqrt{\tfrac{17}{4}}$$
$$= \pm\frac{\sqrt{17}}{2}$$
$$a = -\tfrac{3}{2} \pm \tfrac{\sqrt{17}}{2}$$
$$= \frac{-3 \pm \sqrt{17}}{2}$$

If the LHS of the equation factorises, use the method of factorisation rather than completing the square.

For example,

$$x^2 - 7x + 10 = 0$$
$$(x - 5)(x - 2) = 0$$
$$x = 5 \text{ or } 2$$

rather than

$$x^2 - 7x + 10 = 0$$
$$x^2 - 7x = -10$$
$$x^2 - 7x + \left(\tfrac{7}{2}\right)^2 = -10 + \tfrac{49}{4}$$
$$= \frac{-40 + 49}{4}$$
$$\left(x - \tfrac{7}{2}\right)^2 = \tfrac{9}{4}$$
$$x - \tfrac{7}{2} = \pm\tfrac{3}{2}$$
$$x = \tfrac{7}{2} \pm \tfrac{3}{2}$$
$$= \tfrac{10}{2} \text{ or } \tfrac{4}{2}$$
$$= 5 \text{ or } 2$$

Exercise 4g

Solve the following equations. Factorise where possible. Otherwise, solve by completing the square, leaving the answers in the form of Examples 15 and 16.

1. $a^2 + 4a - 21 = 0$
2. $b^2 - b - 12 = 0$
3. $c^2 - 4c - 2 = 0$
4. $d^2 + 2d - 2 = 0$
5. $n^2 + 4n + 4 = 0$
6. $p^2 - 10p + 15 = 0$
7. $q^2 + 10q + 22 = 0$
8. $t^2 - 6t + 9 = 0$
9. $m^2 + 6m + 7 = 0$
10. $y^2 - 3y + 1 = 0$
11. $z^2 - 5z + 6 = 0$
12. $h^2 + 5h + 4 = 0$
13. $k^2 - 5k + 2 = 0$
14. $g^2 + 5g + 2 = 0$
15. $x^2 - 8x - 1 = 0$
16. $a^2 - a - 1 = 0$
17. $b^2 + b - 3 = 0$
18. $y^2 + 7y - 30 = 0$
19. $x^2 - 10x + 25 = 0$
20. $n^2 - 12n + 1 = 0$

The formula for solving quadratic equations

When the coefficient of the x^2-term in a quadratic equation is not equal to $+1$, we can apply a formula to find the roots of the equation. The following steps show how we get the formula.

The general form of a quadratic equation in x is $ax^2 + bx + c = 0$, where a, b and c are constants. The roots of this equation are found by completing the square.

$$ax^2 + bx + c = 0$$

1st step: Make the coefficient of the x^2 term equal to 1.

$$x^2 + \frac{b}{a}x + \frac{c}{a} = 0$$

2nd step: Put the constant term on the RHS of the equation.

$$x^2 + \frac{b}{a}x = \frac{c}{a}$$

3rd step: Add a constant equal to half the coefficient of the x-term squared to both sides of the equation.

$$x^2 + \frac{b}{a}x + \left(\frac{b}{2a}\right)^2 = -\frac{c}{a} + \left(\frac{b}{2a}\right)^2$$
$$= -\frac{c}{a} + \frac{b^2}{4a^2}$$

4th step: Write the LHS as a perfect square and simplify the RHS.

$$\left(x + \frac{b}{2a}\right)^2 = \frac{b^2 - 4ac}{4a^2}$$

5th step: Take the square root of both sides of the equation.

$$x + \frac{b}{2a} = \pm\sqrt{\frac{b^2 - 4ac}{4a^2}}$$
$$= \pm\frac{\sqrt{b^2 - 4ac}}{2a}$$

6th step: Write the equation with x only on the LHS and simplify the RHS.

$$x = -\frac{b}{2a} \pm \frac{\sqrt{b^2 - 4ac}}{2a}$$

$$\boldsymbol{x = \frac{-b \pm \sqrt{b^2 - 4ac}}{2a}}$$

Example 17

Find, correct to 2 decimal places, the roots of the equation $3x^2 - 5x - 7 = 0$.

Comparing $3x^2 - 5x - 7 = 0$
with $ax^2 + bx + c = 0$:
$a = 3, b = -5, c = -7$.

$$x = \frac{-(-5) + \sqrt{(-5)^2 - 4 \times 3 \times (-7)}}{2 \times 3}$$
$$= \frac{5 \pm \sqrt{25 + 84}}{6}$$
$$= \frac{5 \pm \sqrt{109}}{6}$$
$$= \frac{5 \pm 10.44}{6}$$
$$= \frac{15.44}{6} \text{ or } \frac{-5.44}{6}$$
$$\simeq 2.57 \text{ or } -0.91$$

Exercise 4h

Use the formula to solve the following equations. Give the roots correct to 2 decimal places where necessary. Use factorisation to check the results of the first ten.

1. $x^2 + 5x + 6 = 0$
2. $x^2 - 5x + 4 = 0$
3. $x^2 - 4x - 5 = 0$
4. $2x^2 + 5x + 3 = 0$
5. $3x^2 - 4x + 1 = 0$
6. $3x^2 - 5x - 2 = 0$
7. $5x^2 - 3x - 2 = 0$
8. $4x^2 + 7x - 2 = 0$
9. $6x^2 + 13x + 6 = 0$
10. $3x^2 - 13x - 10 = 0$
11. $x^2 + 3x + 1 = 0$
12. $x^2 - 2x - 4 = 0$
13. $2x^2 + 7x - 3 = 0$
14. $3x^2 - 5x - 3 = 0$
15. $5x^2 - 6x - 3 = 0$
16. $5x^2 + 8x - 2 = 0$
17. $3x^2 + 7x + 3 = 0$
18. $3x^2 - 12x + 10 = 0$
19. $3x^2 - 8x + 2 = 0$
20. $5x^2 + 3x - 3 = 0$

Graphical solution of quadratic equations

Example 18

Solve the equation $x^2 + 2x - 3 = 0$ graphically.

The LHS of the given equation, $x^2 - 2x - 3$, is a function of x.
Let $y = x^2 - 2x - 3$

Table 4.1 is a table of values for the function $y = x^2 - 2x - 3$.

Table 4.1

x	−2	−1	0	1	2	3	4	5
x^2	4	1	0	1	4	9	16	25
$-2x$	4	2	0	−2	−4	−6	−8	−10
-3	−3	−3	−3	−3	−3	−3	−3	−3
$x^2 - 2x - 3$	5	0	−3	−4	−3	0	5	12

Fig. 4.1 is the graph of the function $x^2 - 2x - 3$.

We know that $y = x^2 - 2x - 3$, so the solutions of $x^2 - 2x - 3 = 0$ are the values of x where $y = 0$, the points at which $y = 0$ lie on the x-axis. In Fig. 4.1, the curve cuts the x-axis at points A and B.

At A, $x = 3$

At B, $x = -1$

Hence the roots of the equation $x^2 - 2x - 3 = 0$ are $x = 3$ or -1.

Check: By factorisation:

$$x^2 - 2x - 3 = 0$$
$$(x - 3)(x + 1) = 0$$
$$x = 3 \text{ or } -1$$

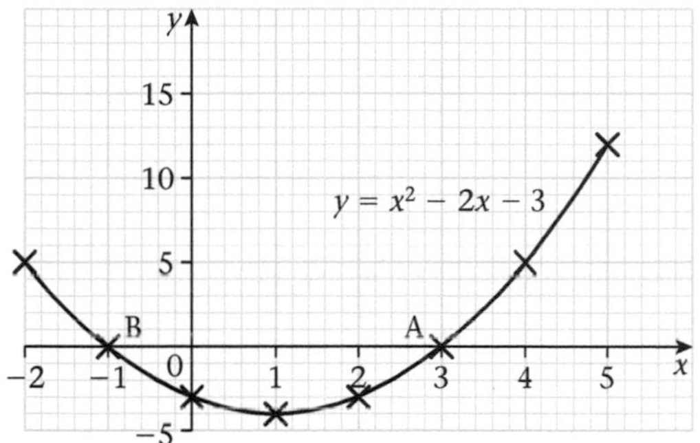

Fig. 4.1

For the equation given in Example 18, the method of factorisation is much easier than the graphical method. However, when the expression cannot be factorised, the graphical method can be used to give approximate roots as in Example 19.

Example 19

Find the roots of the equation $3x^2 + x - 7 = 0$.

$3x^2 + x - 7$ does not factorise.
Let $y = 3x^2 + x - 7$ and construct a table of values (Table 4.2).

Table 4.2

x	-3	-2	-1	0	1	2	3
$3x^2$	27	12	3	0	3	12	27
$+x$	-3	-2	-1	0	1	2	3
-7	-7	-7	-7	-7	-7	-7	-7
y	17	3	-5	-7	-3	7	23

Fig. 4.2 is the graph of $y = 3x^2 + x - 7$.

From the graph, $y = 0$ when
$x \simeq 1.4$
or $x \simeq -1.7$

Hence the approximate roots of the equation are 1.4 and -1.7.

The accuracy of the results depends on the scale used to draw the graph. With the scale in Fig. 4.2 results are correct to one decimal place only. In practice a larger scale should be used, such as 2 cm to one unit on the x-axis and 2 cm to 2 units on the y-axis.

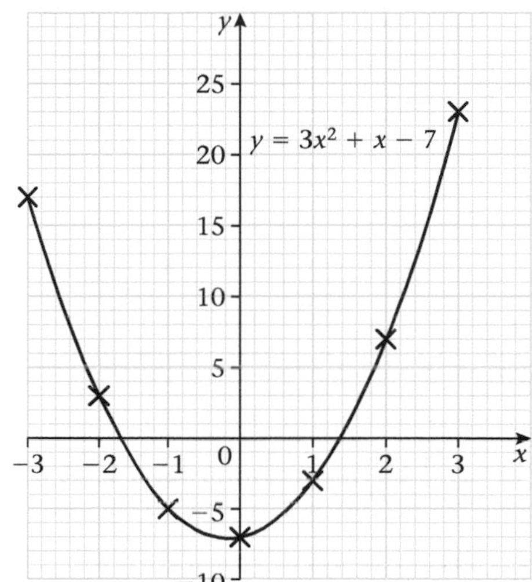

Fig. 4.2

Example 20

Draw a graph to find the roots of the equation $4x^2 - 20x + 25 = 0$.

Let $y = 4x^2 - 20x + 25$ and construct a table of values (Table 4.3).

Table 4.3

x	0	1	2	3	4	5
$4x^2$	0	4	16	36	64	100
$-20x$	0	-20	-40	-60	-80	-100
$+25$	25	25	25	25	25	25
y	25	9	1	1	9	25

Fig. 4.3 is the graph of $y = 4x^2 - 20x + 25$.

The curve does not *cut* the x-axis. It appears to *touch* the x-axis where $x = 2.5$.

This result can be checked by factorisation:

$$4x^2 - 20x + 25 = 0$$
$$(2x - 5)(2x - 5) = 0$$
$$\text{i.e. } (2x - 5)^2 = 0$$
$$x = 2\tfrac{1}{2} \text{ (twice)}$$

When the curve touches the x-axis, the roots are said to be **coincident**.

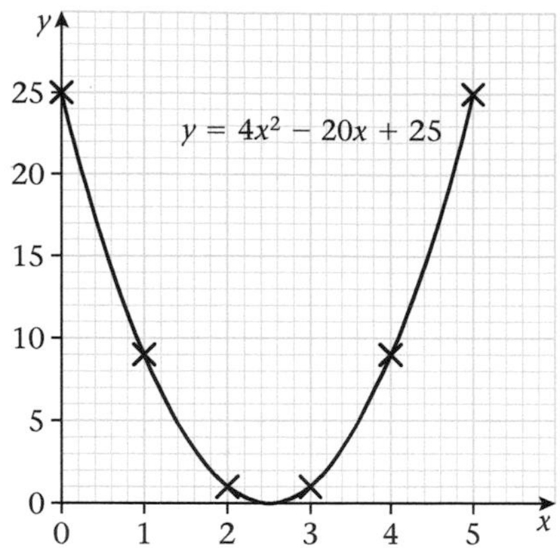

Fig. 4.3

Remember that the curve of a quadratic function is usually in one of three positions with respect to the x-axis (Fig. 4.4). [See also Book 3, Chapter 9].

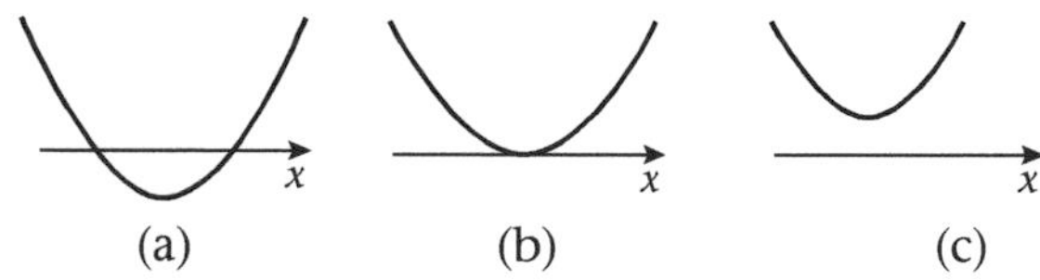

Fig. 4.4

In Fig. 4.4(a) the curve cuts the x-axis at two clear points. These two points give the roots of the related equation.

In Fig. 4.4(b), the two points are coincident, i.e. the points are so close together that the curve touches the x-axis at one point. This corresponds to an equation which has one repeated root.

In Fig. 4.4(c) the curve does not cut the x-axis. The roots of the related equation are said to be **imaginary**. For example, in Fig. 4.5 the roots of the equation $x^2 - 5x + 8 = 0$ are imaginary since the graph of $y = x^2 - 5x + 8$ does not cut the x-axis.

Exercise 4i

1. Fig. 4.6 is the graph of $y = x^2 + 2x - 3$.

   Use Fig. 4.6 to write down the roots of the equation $x^2 + 2x - 3 = 0$. Check the result by factorisation.

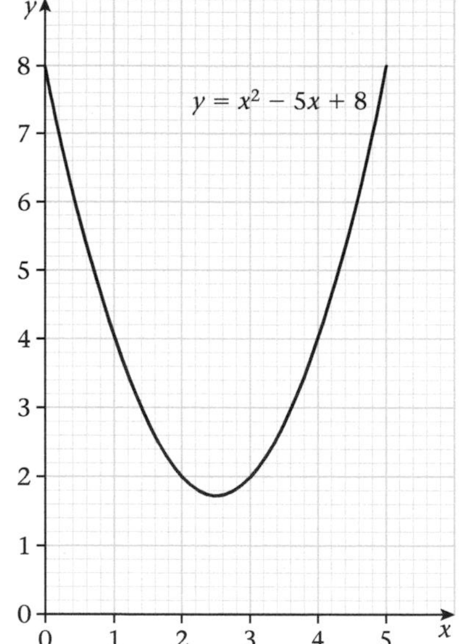

Fig. 4.5

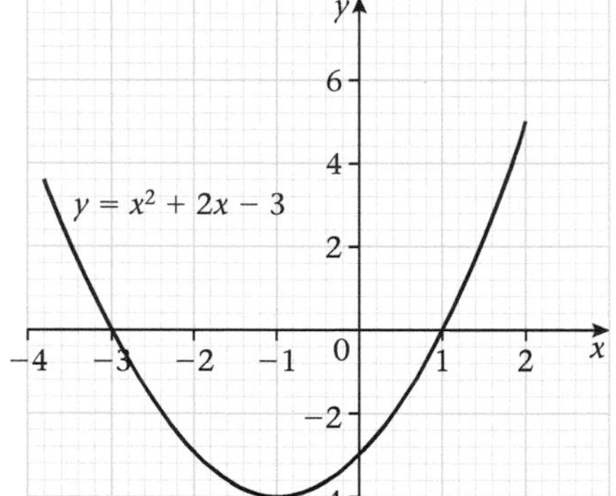

Fig. 4.6

2. Use Fig. 4.7 to write down the roots of the equation $2 - 3x - 2x^2 = 0$. Use factorisation to check the result.

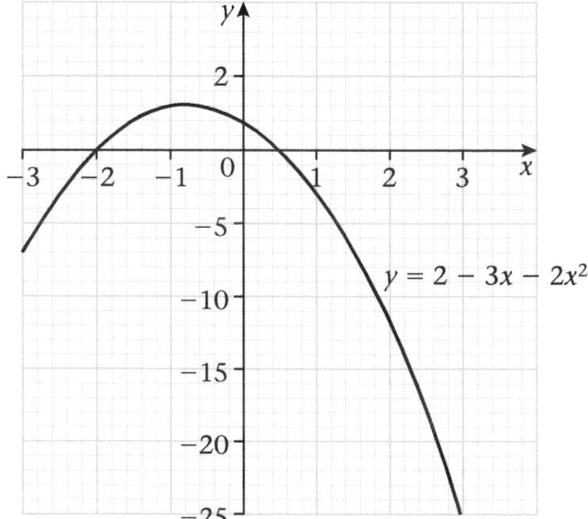

Fig. 4.7

3 Use Fig. 4.8 to write down the roots of the equation $3x - x^2 = 0$. Check the result by factorisation.

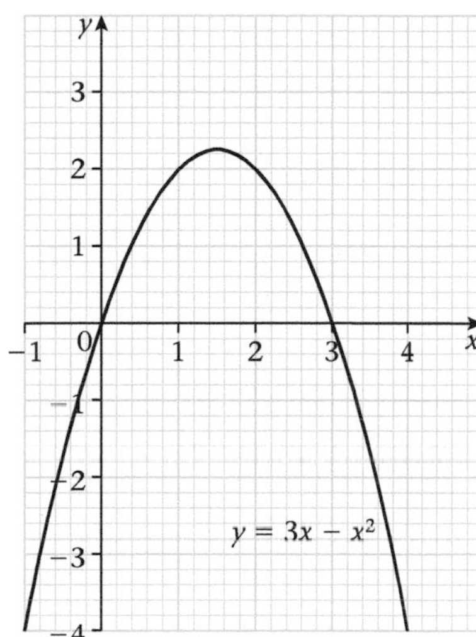

Fig. 4.8

4 Table 4.4 is a table of values for $y = x^2 + x - 8$ from $x = -4$ to $x = +3$.

Table 4.4

x	−4	−3	−2	−1	0	1	2	3
y	4	−2	−6	−8	−8	−6	−2	4

Use Table 4.4 to solve the equation $x^2 + x - 8 = 0$ graphically.

5 Table 4.5 is a table of values for $y = 3x^2 + 10x + 6$ from $x = -4$ to $x = 1$.

Table 4.5

x	−4	−3	−2	−1	0	1
y	14	3	−2	−1	6	19

Use Table 4.5 to solve the equation $3x^2 + 10x + 6 = 0$ graphically.

6 Each curve in Fig. 4.9 corresponds to a quadratic equation. Find the roots of these equations where possible.

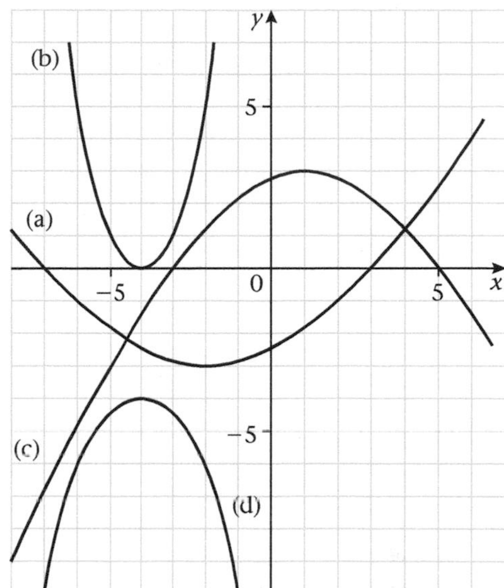

Fig. 4.9

7 (a) Given that $y = x^2 + 3x - 2$, copy and complete Table 4.6.

Table 4.6

x	−5	−4	−3	−2	−1	0	1	2
x^2	25	16	9	4				
$+3x$	−15	−12	−9	−6				
-2	−2	−2	−2	−2	−2			
y	8	2	−2	−4				

(b) Hence draw a graph to find the roots of the equation $x^2 + 3x - 2 = 0$.

8 (a) Given that $y = 4x^2 - 12x + 9$, copy and complete Table 4.7.

Table 4.7

x	−1	0	1	2	3	4
$4x^2$	4			16		64
$+12x$	12			−24		−48
-9	9			9		9
y	25			1		25

(b) Hence draw a graph and find the roots of the equation $4x^2 - 12x + 9 = 0$.

9 (a) Draw the graph of the function $x^2 + 2x - 2$ from $x = -4$ to $x = +2$.

(b) Hence find the approximate roots of the equation $x^2 + 2x - 2 = 0$.

10 (a) Draw the graph of the function $11 + 8x - 2x^2$ from $x = -2$ to $x = +6$.
(b) Hence find the approximate roots of the equation $2x^2 - 8x - 11 = 0$.

Word problems leading to quadratic equations

Example 21

Find two numbers whose difference is 5 and whose product is 266.

Let the smaller number be x.
Then the larger number is $x + 5$.
Their product is $x(x + 5)$.

Hence $x(x + 5) = 266$

$x^2 + 5x - 266 = 0$

$(x - 14)(x + 19) = 0$

$x = 14$ or -19

The other number is $14 + 5$ or $-19 + 5$
i.e. 19 or -14.

The two numbers are 14 and 19, or -19 and -14.
Check: $14 \times 19 = 266$ and $-19 \times -14 = 266$

Example 22

A man's age is 4 times his son's age. 5 years ago the product of their ages was 175. Find their present ages.

Let the son's age be x years.
Then the father's age is $4x$ years.
5 years ago, the son's age was $(x - 5)$ years and the father's age was $(4x - 5)$years.
The product of their ages was $(x - 5)(4x - 5)$.

Hence $(x - 5)(4x -5) = 175$

$4x^2 - 25x + 25 = 175$

$4x - 25x - 150 = 0$

$(4x + 15)(x - 10) = 0$

$x = 10$ or $-\frac{15}{4}$

However, $-\frac{15}{4}$ is not sensible for an age (ages cannot be negative). Therefore the son is 10 years old and the father is 40 years old.
Check: $(40 - 5)(10 - 5) = 35 \times 5 = 175$

Notice in Example 22 that one of the roots does not give a sensible answer. Disregard any root which is not sensible in terms of the given question.

Exercise 4j

In each problem, form a quadratic equation and then solve it.

Note: in this exercise all the quadratic equations can be solved by factorisation.

1 Find two numbers which differ by 4 and whose product is 45.

2 The width of a classroom is 4 metres less than the length. Its area is 45 m^2. Find the dimensions of the classroom.

3 Two numbers have a difference of 3. The sum of their squares is 89. Find the numbers.

4 Two square rooms have a total floor area of 89 m^2. One room is 3 m longer each way than the other. Find the dimensions of the two rooms.

5 A girl is 6 years younger than her oldest brother. The product of their ages is 135. Find their ages.

6 The ages of two brothers are 11 and 8 years. In how many years' time will the product of their ages be 208?

7 Find the number which, when added to its square, makes 90.

8 Twice the square of a certain whole number added to 3 times the number makes 90. Find the number.

9 The area of a rectangle is 60 cm^2. The length is 11 cm more than the width. Find the width.

10 A rectangular plot measures 12 m by 5 m. A path of constant width runs along one side and one end. If the total area of the plot and the path is 120 m^2, find the width of the path. (*Hint*: Make a sketch and let the width of the path be x m.)

11 A rectangular piece of cardboard measures 17 cm by 14 cm. Strips of equal width are cut off one side and one end. The area of the remaining piece is 108 cm^2. Find the width of the strips removed.

12 A man is 37 years old and his son's age is 8. How many years ago was the product of their ages 96?

13. A certain number is subtracted from 18 and from 13. The product of the two numbers obtained is 66. Find the first number.
14. Find two consecutive numbers whose product is 156.
15. Find two consecutive even numbers whose product is 224.
16. The square of a certain number is 22 less than 13 times the original number. Find the number.

Simultaneous linear and quadratic equations

Given a pair of simultaneous equations in two unknowns, of which one is linear and the other is quadratic, the most usual method of solving the equations is by **substitution**. Use one of the equations, generally the linear equation, to write one of the unknowns in terms of the other. Then, by substitution, a quadratic equation in one unknown is obtained. This equation may be solved by one of the usual methods.

Example 23

Solve the equations $3x + y = 10$, $2x^2 + y^2 = 19$.

$$3x + y = 10 \qquad (1)$$
$$2x^2 + y^2 = 19 \qquad (2)$$

From (1), $y = 10 - 3x$

Substitute $10 - 3x$ for y in (2)

$$2x^2 + (10 - 3x)^2 = 19$$
$$2x^2 + 100 - 60x + 9x^2 = 19$$
$$11x^2 - 60x + 81 = 0$$
$$(x - 3)(11x - 27) = 0$$
$$x = 3 \text{ or } \tfrac{27}{11}$$

When $x = 3$, $y = 10 - 3 \times 3$
$= 10 - 9 = 1$

When $x = \frac{27}{11}$, $y = 10 - 3 \times \frac{27}{11}$
$= 10 - \frac{81}{11} = \frac{29}{11}$

The solutions are $x = 3$ and $y = 1$
or $x = 2\frac{5}{11}$ and $y = 2\frac{7}{11}$

The solutions can be given more neatly as ordered pairs: $(3, 1)$, $(2\frac{5}{11}, 2\frac{7}{11})$.

In Example 23, y, one of the unknowns, is expressed in terms of the other unknown from the linear equation. A substitution is made in the quadratic equation to give an equation in one of the unknowns only. Notice also that when the values of x were found, the corresponding values of y were found by substitution in the *linear* equation. The result can be checked by substituting each pair of values in the given quadratic equation.

Example 24

Solve the equations $3x + 4y = 11$, $xy = 2$.

$$3x + 4y = 11 \qquad (1)$$
$$xy = 2 \qquad (2)$$

From (2), $x = \dfrac{2}{y}$

Substitute $\dfrac{2}{y}$ for x in (1)

$$3\left(\frac{2}{y}\right) + 4y = 11$$
$$6 + 4y^2 = 11y$$
$$4y^2 - 11y + 6 = 0$$
$$(y - 2)(4y - 3) = 0$$
$$y = 2 \text{ or } \tfrac{3}{4}$$

When $y = 2$, $x = \frac{2}{2} = 1$

When $y = \frac{3}{4}$, $x = \dfrac{2}{\frac{3}{4}} = \frac{8}{3} = 2\frac{2}{3}$

The solutions are $(1, 2)$, $(2\frac{2}{3}, \frac{3}{4})$.

In Example 24, one unknown is expressed in terms of the other from the quadratic equation. A substitution is made in the linear equation to give an equation in one of the unknowns only. This method will always lead to a solution. The next example shows that it is sometimes possible to use another method.

Example 25

Solve the equations $3x - y = 3$, $9x^2 - y^2 = 45$.

$$9x^2 - y^2 = 45$$
$$(3x - y)(3x + y) = 45$$

Hence, $3(3x + y) = 45$ (since $3x - y = 3$)

$\therefore$ $3x + y = 15$ (1)

$3x - y = 3$ (2)

Add: $6x = 18$

$x = 3$

Subtracting (2) from (1)

$2y = 12$

$y = 6$

$x = 3$ and $y = 6$

The method of Example 25 can be used when one of the given equations can be expressed in factors, one of which is part of the linear equation.

Exercise 4k

1. Use the method of Example 23 to solve the following.

   (a) $3x^2 - 4y = -1$, $2x - y = 1$ (b) $x^2 + y^2 = 34$, $x - y = 2$

   (c) $2x^2 + y^2 = 19$, $x + 3y = 0$ (d) $9y^2 + 8x = 12$, $2x + 3y = 4$

   (e) $xy + 3x = 3$, $3x + y = 7$ (f) $3x^2 - xy = 0$, $2y - 5x = 1$

2. Use the method of Example 24 to solve the following.

   (a) $3x + y = 25$, $xy = 8$ (b) $3x - 4y = 2$, $xy = 2$

   (c) $2x - 5y = 7$, $xy = 6$ (d) $3x + 4y = 7$, $2xy + 3 = 0$

3. In each of the following, factorise the expression on the LHS of the first equation. Use the method of Example 25 to solve the equations.

   (a) $x^2 - y^2 = 27$, $x + y = 3$ (b) $x^2 - 4y^2 = 9$, $x + 2y = 1$

   (c) $4x^2 - y^2 = 15$, $2x - y = 5$ (d) $4x^2 - 9y^2 = 19$, $2x + 3y = 1$

4. A woman is q years old while her son is p years old. The sum of their ages is equal to twice the difference of their ages. The product of their ages is 675.

   Write down the equations connecting their ages and solve the equations in order to find the ages of the woman and her son.

5. The product of two numbers is 12. The sum of the larger number and twice the smaller number is 11. Find the two numbers.

Graphical solution of simultaneous quadratic and linear equations

Example 26

Use a graphical method to solve the equations $y = 2x^2 + x - 2$, $y = 4 - 2x$ simultaneously.

Table 4.8

$y = 2x^2 + x - 2$

x	−3	−2	−1	0	1	2
y	13	4	−1	−2	1	8

$y = 4 - 2x$

x	−3	0	2
y	10	4	0

Method: Draw the graphs of both equations on the same axes. The graphs are drawn from the data in Table 4.8.

Fig. 4.10 shows both graphs draw on the same axes.

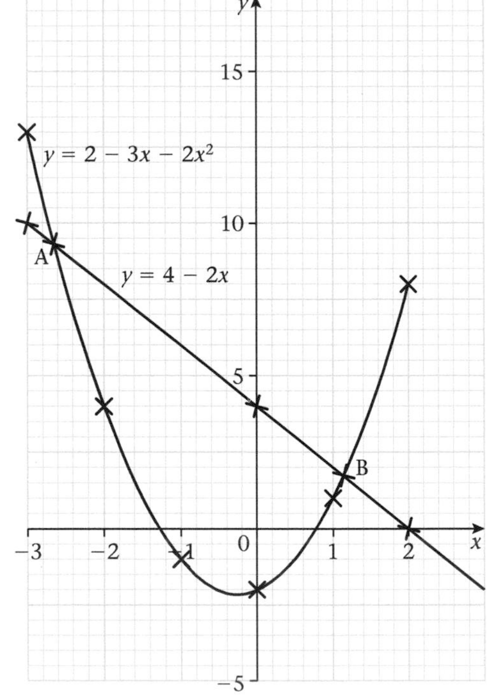

Fig. 4.10

From the graph,

At A, $y = 9.2$, when $x = -2.6$

At B, $y = 1.7$, when $x = 1.1$

The approximate solutions to the given simultaneous equations are $(-2.6, 9.2)$ and $(1.1, 1.7)$.

Example 27

The curve in Fig. 4.11 is the graph of

$y = 4x^2 - 9x - 1$

Use the graph to solve the following equations.

(a) $4x^2 - 9x - 1 = 7$

(b) $4x^2 - 9x + 3 = 0$

(c) $4x^2 - 9x - 1 = 1\frac{1}{4}x + 1\frac{1}{4}$

(a) $4x^2 - 9x - 1 = 7$ at the points where the line $y = 7$ cuts the curve. See construction (a) on Fig. 4.11.
At these points, $x = 2.9$ or $x = -0.7$.

(b) $4x^2 - 9x + 3 = 0$
Subtract 4 from both sides
$4x^2 - 9x - 1 = -4$
In construction (b) on Fig. 4.11, the line $y = -4$ cuts the curve in two points.
At these points, $x = 0.4$ or $x = 1.8$.

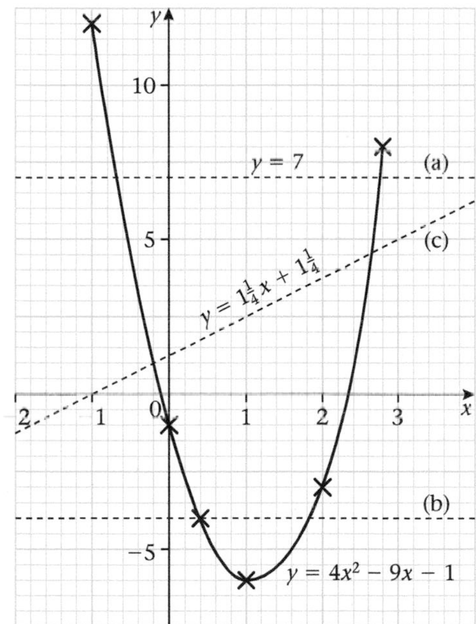

Fig. 4.11

(c) Construction (c) shows the graph of the line $y = 1\frac{1}{4}x + 1\frac{1}{4}$, If $y = 4x^2 - 9x - 1$ and $y = 1\frac{1}{4}x + 1\frac{1}{4}$, then, at the points where the line cuts the curve, $4x^2 - 9x - 1 = 1\frac{1}{4}x + 1\frac{1}{4}$.
At these points, $x = -0.2$ or $x = 2.8$

(*Note*: The results in this example are accurate to 1 d.p. only.)

Exercise 4l

1 The graphs of $y = 2x^2 - 3x - 7$ and $y = 2x - 1$ are given in Fig. 4.12.
Use Fig. 4.12 to answer the following.

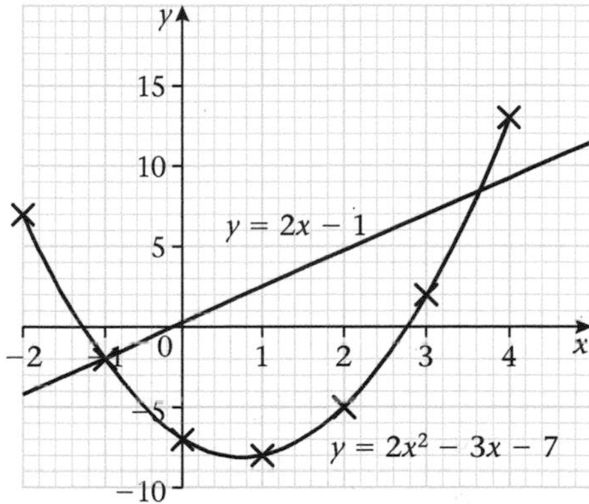

Fig. 4.12

(a) What are the roots of the equation $2x^2 - 3x - 7 = 0$?

(b) What are the solutions of the simultaneous equations $y = 2x^2 - 3x - 7$ and $y = 2x - 1$?

(c) What are the solutions of the equation $2x^2 - 3x - 7 = 2x - 1$?

(d) For what values of x is $2x^2 - 3x - 7 = 5$?

(e) What are the roots of the equation $2x^2 - 3x - 4 = 0$?

2 Use Fig. 4.5 on page 42 to solve the equations
(a) $x^2 - 5x + 8 = 6$, (b) $x^2 - 5x + 3 = 0$.

3 Use Fig. 4.8 on page 43 to solve the simultaneous equations $y = x$, $y = 3x - x^2$.

4 Use Fig. 4.8 on page 43 to solve the equations
(a) $3x - x^2 = 1$, (b) $x^2 - 3x - 2 = 0$.

5 Draw the graph of $y = 3x(4 - x)$ for values of x ranging from -2 to $+6$. On the same graph draw the line $y = 5(x - 2)$. Use a scale of 1 cm to 1 unit on the x-axis and 1 cm to 5 units on the y-axis. From your graph,

(a) find the roots of the equation $3x(4 - x) = 0$

(b) deduce the roots of the equation $3x(4 - x) = 5(x - 2)$.

Summary

An equation in which the highest power of the variable(s) is 2, that is, an equation of the **second degree**, is a **quadratic equation**. The values of the variable(s) which make the equation true are called the **roots** of the equation.

When a quadratic expression can be factorised, then the factors of the corresponding quadratic equation can be written as the product of two linear expressions, each factor being equal to zero. The value of the variable in each of the two linear equations gives the two roots of the quadratic equation.

When a **quadratic expression** cannot be factorised, there are other methods of solving the corresponding quadratic equation. These include:

(a) **completing the square** from first principles;

(b) applying the **formula** derived from completing the square for the general quadratic equation, $ax^2 + bx + c = 0$, that is, using

$$x = \frac{-b \pm \sqrt{b^2 - 4ac}}{2a};$$

(c) the graphical method: writing the corresponding quadratic function, y, and finding on the graph of $y = f(x)$ the value(s) of x when $y = 0$ where the curve crosses the x-axis.

The two common methods of solving simultaneous equations in two variables when one of the simultaneous equations is a linear equation and the other is a quadratic equation are

(a) by **substitution** for one of the variables in terms of the other: the value of the second variable is found by substituting the value found for the first variable in one of the original equations.

(b) by a **graphical method** in which the points of intersection of the two graphs plotted (in this case a straight line and a curve) give the values of the variables.

Practice exercise P4.1

Solve the following equations.

1. $(x + 1)(x + 3) = 0$
2. $x(x - 1) = 0$
3. $(x + 2)^2 = 0$
4. $(x - 5)^2 = 0$
5. $p^2 + 7p = 8$
6. $r^2 = 12r - 20$
7. $(4 + x)(x - 5) = 0$
8. $x^2 - 3x = 10$

Practice exercise P4.2

In the following, add a term making the given expression into a perfect square. Write the result as the square of a bracketed expression. Where necessary, make the coefficient of the quadratic term equal to +1.

1. $d^2 + 6d$
2. $b^2 - 5b$
3. $a^2 + 3a$
4. $-x^2 + 16x$
5. $3v^2 + 6v$
6. $10w - 2w^2$
7. $2t^2 - 7t$
8. $-4y^2 - 8y$

Practice exercise P4.3

Write each of the following as a perfect square plus or minus a constant.

1. $p^2 + 7p - 8$
2. $m^2 - 12m + 25$
3. $c^2 + 5c + 6$
4. $x^2 + 11x - 14$
5. $2b^2 - 10b + 12$
6. $12r - 20 - r^2$
7. $12 - f^2 - 4f$
8. $4c^2 - 9 - 6c$
9. $1 - 2y - 2y^2$
10. $16 - 4h^2 + 12h$

Practice exercise P4.4

Use the method of completing the square to solve the following equations, giving your answer in surd form.

1. $r^2 = 2r + 5$
2. $t^2 + 6t - 3 = 0$
3. $d^2 + 4d = 1$
4. $b^2 - b + 8 = 9$
5. $k^2 = 6k - 6$
6. $y^2 + 4y = 6$
7. $2y^2 + 5y - 20 = 0$
8. $x^2 + 16x + 14 = 0$
9. $m^2 = 12m - 25$
10. $4c^2 - 9 = 6c$

Practice exercise P4.5

Use the formula to solve the equations given in Exercise P 4.4, writing your answers correct to 2 decimal places where necessary.

Practice exercise P4.6

1. (a) A boy gave the solution set of the quadratic equation $x^2 + 7x + 12 = 0$ as $\{3, 4\}$. How can you tell *immediately that this is not the correct set?*

   (b) A girl was given the equation $x^2 + 2\frac{1}{6}x - \frac{5}{6} = 0$ to solve. She guessed from a sketch graph that $\frac{1}{3}$ was one of the roots. She then calculated that the other root must be $\frac{5}{2}$. Was her calculation correct? Give a reason for your answer.

2. In the general quadratic equation $ax^2 + bx + c = 0$, the roots are given by $x = \dfrac{-b \pm \sqrt{b^2 - 4ac}}{2a}$.
   The expression $b^2 - 4ac$ is called the *discriminant* because it discriminates between three different cases.
   If $b^2 > 4ac$ there are two distinct roots
   If $b^2 < 4ac$ there are no roots that belong to the set of *real* numbers
   If $b^2 = 4ac$ there is one repeated root and the quadratic function is a perfect square.
   Use the discriminant to decide which of the following equations have two roots, one repeated root or no real roots.

   (a) $x^2 + 3x + 5 = 0$

   (b) $x^2 + 2x + 1 = 0$

   (c) $x^2 + x - 2 = 0$

3. The distance, s metres, travelled by a train in t seconds is given by the equation $s = ut + \frac{1}{2}at^2$, where u metres per second is the initial speed of the train and a metres per second per second is its acceleration. The train enters a station that is 140 m long at a speed of 7 m/s and travels through with an acceleration of 3.5 m/s². Find the time it will take for the train to pass through the station.

Practice exercise P4.7

(a) Draw, on graph paper, the graph of each of the following quadratic functions, $y = f(x)$.

(b) Use the graph to find the roots of the equation $f(x) = 0$. If necessary, write your answers correct to 1 decimal place.

(c) Check your answers by using the formula.

1. $y = x^2 + 3x - 4$
2. $y = x^2 - 2x - 4$
3. $y = x^2 + 5x + 4$
4. $y = x^2 - 4x$
5. $y = 2x^2 + 3x - 6$

Practice exercise P4.8

1. (a) Given that $y = x^2 - 3x - 2$, copy and complete Table 4.9.

   Table 4.9

x	−2	−1	0	1	2	3	4
x^2							
$-3x$							
-2							
y							

   (b) Draw a graph of the function.

   (c) Find the roots of the following equations giving your answers correct to 1 decimal place where necessary.
   (i) $x^2 - 3x - 2 = 0$
   (ii) $x^2 - 3x - 2 = -3$
   (iii) $2 = x^2 - 3x - 2$

2. The incomplete table below (Table 4.10) gives corresponding values for the function $y = px^2 + 2x + 4$, for $-2 \leqslant x \leqslant 4$ where p is a constant.

   Table 4.10

x	−2	−1	0	1	2	3	4
y	−4	1	4				

   (a) Use the table to find the value of the constant p.

   (b) Copy and complete the table of values.

   (c) Using a scale of 2 cm to 1 unit on both axes, draw the graph of the function for the given domain.

   (d) Use your graph to obtain the roots of these equations.
   (i) $px^2 + 2x + 4 = 0$
   (ii) $px^2 + 2x + 4 = 2$

3 A stone is projected vertically upwards. Its position after t seconds is given by the ordered pair (t, s), where s metres is the distance moved. The set of pairs representing the motion is $\{(t, s): s = 18t - 3t^2\}$.
 (a) Using a scale of 2 cm to 0.5 unit on the t-axis and 2 cm to 5 units on the s-axis, draw the graph of the relation for the domain $0 \leqslant t \leqslant 6, t \in \mathbb{Q}$.
 (b) From your graph, estimate
  (i) the times between which the stone is in the air,
  (ii) the times between which the stone is higher than 25 m,
  (iii) the times at which the stone is 20 m high.

Practice exercise P4.9

In each problem, form a quadratic equation and then solve it.

1 The sum of a number and its square is 90. If the number is greater than zero, find the number.

2 Divide 28 into two parts so that the product of the parts is 192.

3 The area of a rectangle is $180\,\text{cm}^2$. Its length is reduced by 5 cm and its breadth by 2 cm so that it becomes a square. Find the length of a side of the square.

4 Andy subtracted the same number from 15 and from 21. The product of his answers is 135. Find the possible numbers.

5 A fruit vendor sold n oranges at $(2n - 15)$ cents each and received \$42.50. Write an equation to show this information and find the number of oranges she sold.

6 Twice a whole number subtracted from 3 times the square of the number leaves 133. Find the number.

7 When 12 times a positive number is added to twice its square, the result is 270. Find the number.

8 The length of a plank is five times its width. When the width is reduced by 1 cm and the length is also reduced by 1 cm, the area is $352\,\text{cm}^2$. Find the original length and width of the plank.

Practice exercise P4.10

Solve these simultaneous equations algebraically.

1 $xy = 15$, $4x - y = 7$

2 $x^2 - y^2 = -3$, $x + y = 3$

3 $x^2 + 2y^2 = 3$, $x - 3y = 2$

4 $x^2 + y^2 = 25$, $2x + y = 5$

5 $xy = 30$, $3x + y = 21$

6 $x^2 + 2xy = 8$, $x + 2y = 2$

7 $2x + y = 1$, $4x^2 + y^2 = 61$

8 $2x^2 - y^2 = -2$, $3x + y = 1$

9 $x^2 - y^2 = -7$, $x + y = 2$

10 $x^2 - 3x = y$, $y = 2 - 2x$

Use a graphical method to solve each pair of simultaneous equations.

11 $x^2 - 3x = y$ and $y = 2 - 2x$

12 $y = x^2 - 2x - 5$ and $y = 2x - 3$

Practice exercise P4.11

Write the equation in the form $ax^2 + bx + c = 0$, that is satisfied by the x-coordinates of the points of intersection of the following graphs.

1 $y = x^2 - 3x - 6$ and $y = 3x - 2$

2 $y = 3x^2 + 8x - 1$ and $y = 2(3x - 1)$

3 $y = 5 - 3x - 4x^2$ and $3x + y - 7 = 0$

Practice exercise P4.12

1 A girl is 24 years younger than her mother. The product of their ages is 180 years. Write the algebraic expressions to represent this information and calculate their ages.

2 Ann paid \$480 for some tins of paint. If she had paid \$8 less for each tin of paint, she could have bought 2 more tins of paint for the same amount of money. Find
 (a) the number of tins of paint she bought,
 (b) the cost of one tin.

3 Given the equation of the quadratic curve $y = x^2 + 3x - 2$ and the equation of the straight line $y = 5x - 3$, show that the straight line is a tangent to the curve
 (a) by solving the equations simultaneously,
 (b) graphically.

Geometrical proofs (3)

Cyclic quadrilaterals and tangents

Pre-requisites
- angle properties of a circle

Cyclic quadrilaterals

The vertices of a **cyclic quadrilateral** lie on the circumference of a circle. In Fig. 5.1, ABCD and PQRS are cyclic quadrilaterals.

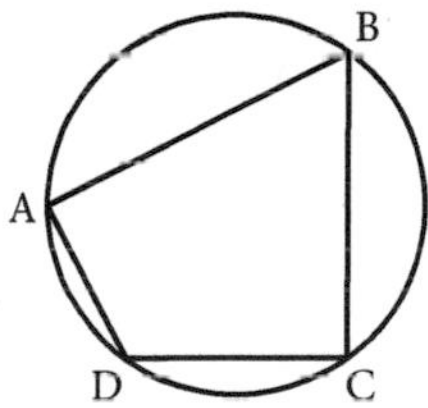

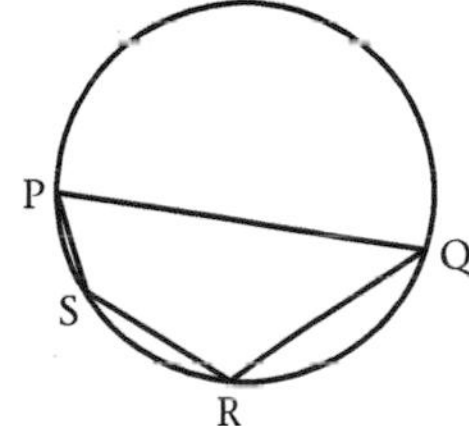

Fig. 5.1

Opposite angles of a cyclic quadrilateral lie in opposite segments of a circle (Fig. 5.2).

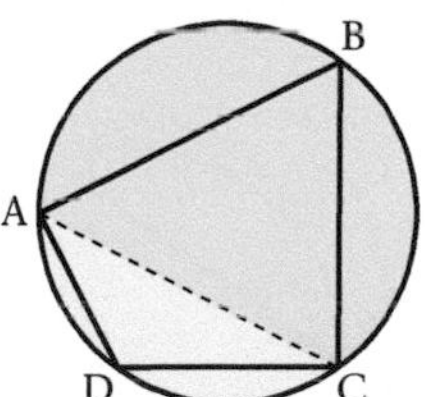

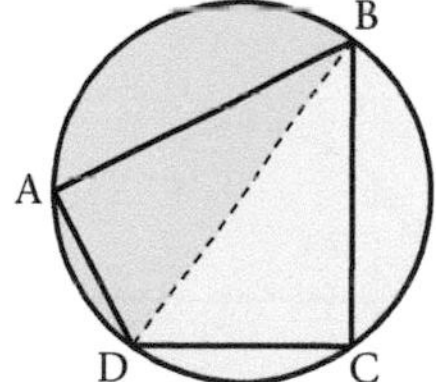

Fig. 5.2

Example 1

Show that the opposite angles of a cyclic quadrilateral are supplementary.

Quadrilateral ABCD, Fig. 5.3 is a cyclic quadrilateral with O as centre of the circle. We want to show that $B\hat{C}D + B\hat{A}D = 180°$.

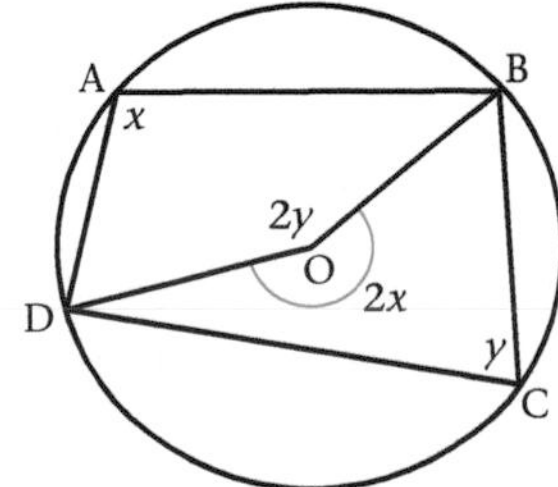

Fig. 5.3

$B\hat{O}D = 2y$	(*angle at centre* = 2 × *angle at circumference*)
reflex $B\hat{O}D = 2x$	(*same reason*)
$\therefore 2x + 2y = 360°$	(*angles at a point*)
$\therefore x + y = 180°$	
$\therefore B\hat{A}D + B\hat{C}D - 180°$	

The opposite angles of a cyclic quadrilateral are supplementary.

or

Angles in opposite segments are supplementary.

Example 2

ABCD is a cyclic quadrilateral (Fig. 5.4) with DC produced to X. Show that $x_2 = x_1$.

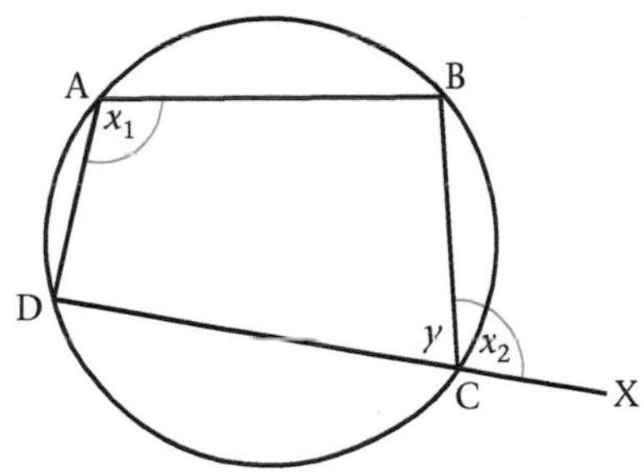

Fig. 5.4

In Fig. 5.4.

$x_1 + y = 180°$	(*opp. angles of cyclic quad.*)
$x_2 + y = 180°$	(*straight angle*)
$\therefore \quad x_1 = x_2$	(= 180° − y)
$\therefore \quad B\hat{C}X = B\hat{A}D$	

The exterior angle of a cyclic quadrilateral is equal to the interior opposite angle.

Example 3

In Fig. 5.5, CE is a diameter of circle ABCDE. If $A\hat{B}C = 127°$, find $A\hat{C}E$.

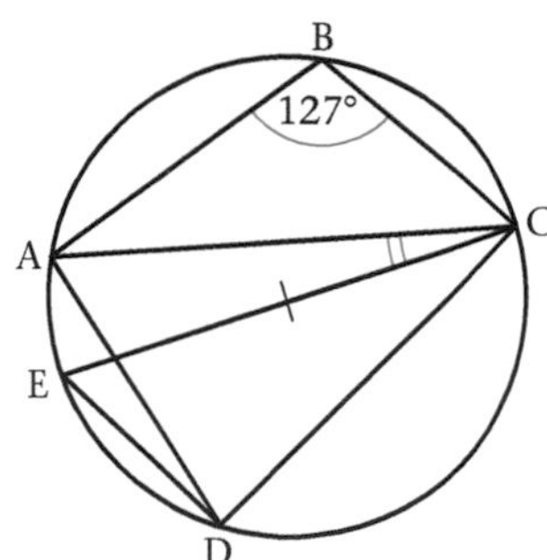

Fig. 5.5

$A\hat{D}C = 180° - 127°$ (*opp. angles of cyclic quad.*)

$= 53°$

$E\hat{D}C = 90°$ (*angle in semicircle*)

$\therefore E\hat{D}A = 90° - 53°$

$= 37°$

$\therefore A\hat{C}E = 37°$ (*in same segment as* $E\hat{D}A$)

Example 4

In Fig. 5.6, P, Q, R, S are points on a circle centre O. QP is produced to X. If $X\hat{P}S = 77°$ and $P\hat{S}O = 68°$, find $P\hat{Q}O$.

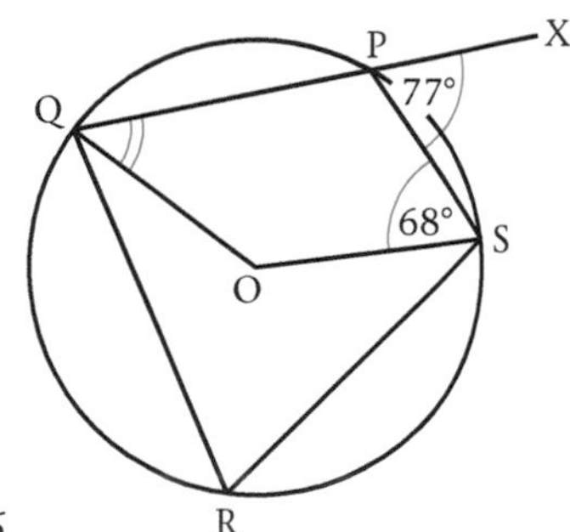

Fig. 5.6

$Q\hat{R}S = 77°$ (*= ext. angle of cyclic quad.*)

$\therefore Q\hat{O}S = 2 \times 77°$ (*angle at centre = 2 × angle at circumference*)

$= 154°$

$Q\hat{P}S = 180° - 77°$ (*straight angle*)

$= 103°$

In quad. PQOS,

$P\hat{Q}O = 360° - 154° - 103° - 68°$ (*angle sum of quad.*)

$\therefore P\hat{Q}O = 35°$

Exercise 5a

1 Use Fig. 5.7 to answer the following.

(a) (i) Which arc subtends $A\hat{B}E$?
(ii) Write down an angle equal to $A\hat{B}E$.

(b) (i) Which arc subtends $B\hat{E}C$?
(ii) Write down two angles equal to $B\hat{E}C$.

(c) (i) Which arc subtends $E\hat{A}C$?
(ii) Write down an angle equal to $E\hat{A}C$.

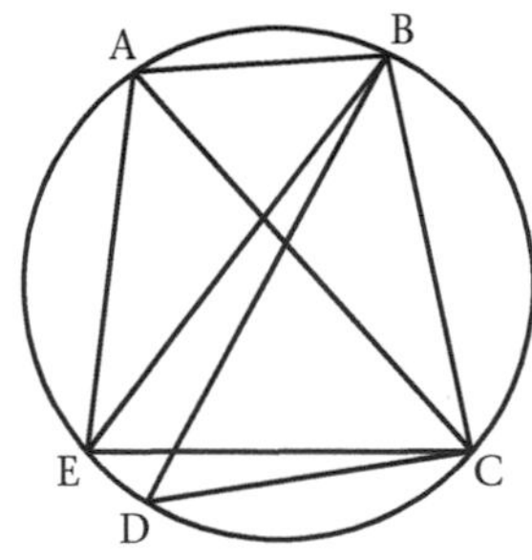

Fig. 5.7

2 Find the marked angle in each part of Fig. 5.8. Where a point O is given it is the centre of the circle. (*Hint*: make a careful copy of each figure and write in the angles as they are found. Some constructions may be necessary.)

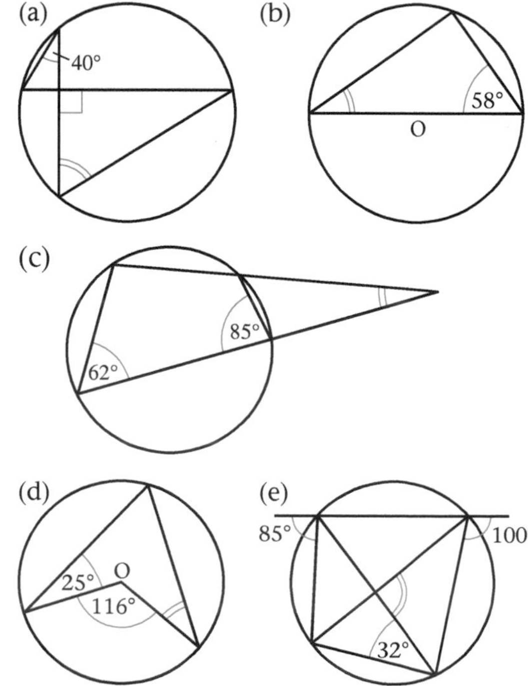

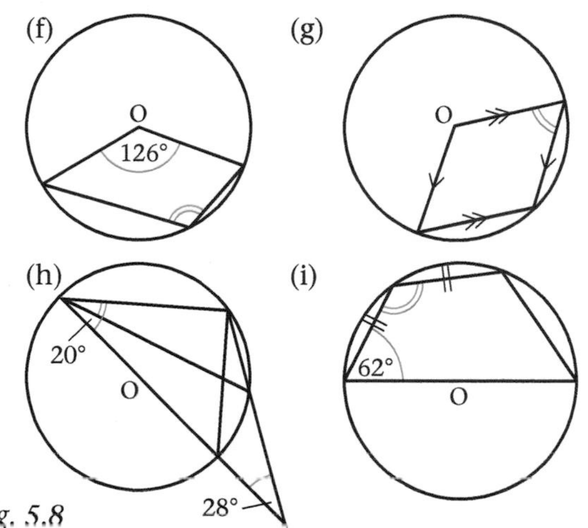

Fig. 5.8

3 In Fig. 5.9, AB is a diameter of semicircle ABCD. If $A\hat{B}D = 16°$, calculate $B\hat{C}D$. (*Hint*: Join CA.)

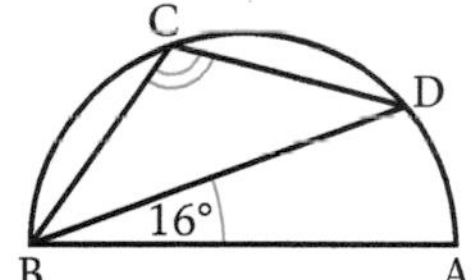

Fig. 5.9

4 In Fig. 5.10, A, B, C, D are points on a circle such that $A\hat{B}C = 102°$. CD is produced to E so that $A\hat{E}D = 47°$. Calculate $E\hat{A}D$.

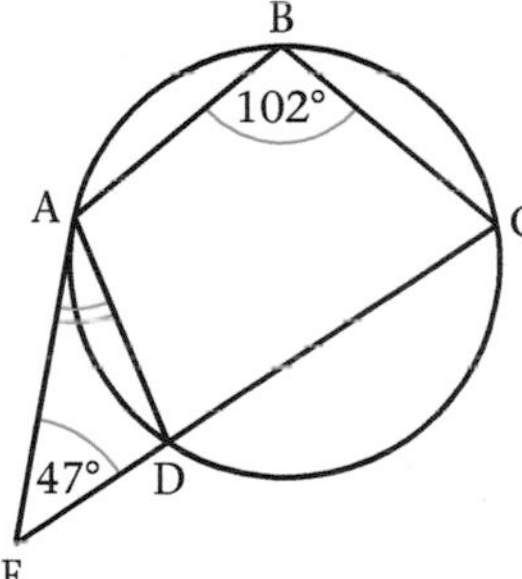

Fig. 5.10

5 In Fig. 5.11, ABCD is a cyclic quadrilateral such that AB∥DC. If $A\hat{C}D = 35°$ and $D\hat{B}C = 72°$, calculate the sizes of the angles of the trapezium.

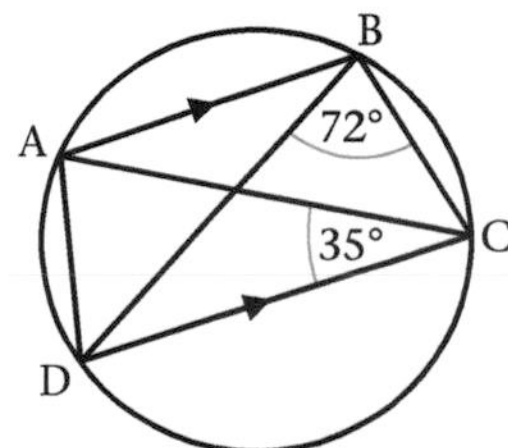

Fig. 5.11

6 In Fig. 5.12, $P\hat{Q}R = 80°$ and $S\hat{R}T = 20°$. What size is $P\hat{X}R$?

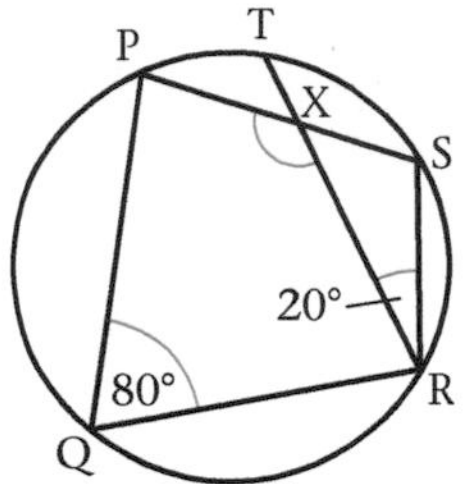

Fig. 5.12

7 In Fig. 5.13, O is the centre of circle MLY. If $O\hat{L}Y = 50°$ and $O\hat{M}Y = 15°$, calculate $M\hat{O}L$.

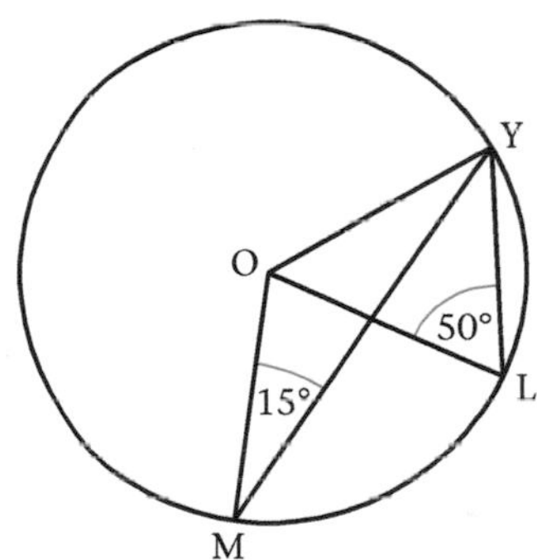

Fig. 5.13

8 In Fig. 5.14, PQRS is a cyclic quadrilateral such that $Q\hat{P}R = 18°$ and $R\hat{Q}S = 42°$. If $P\hat{S}R = 78°$, calculate the angles of △PQS. What kind of triangle is △PQS?

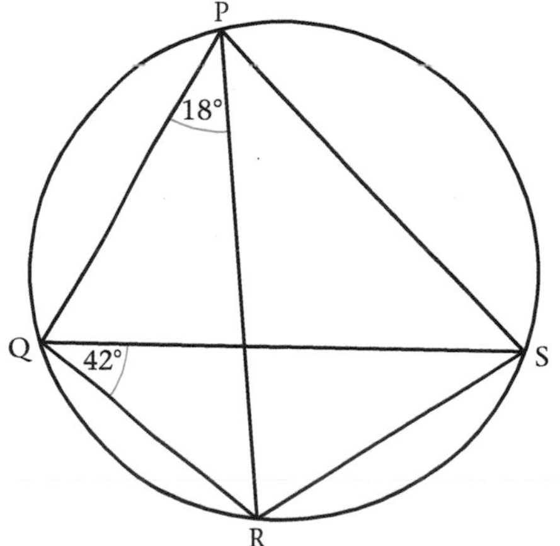

Fig. 5.14

9 In Fig. 5.15, calculate the value of x, giving a reason for each step in your answer.

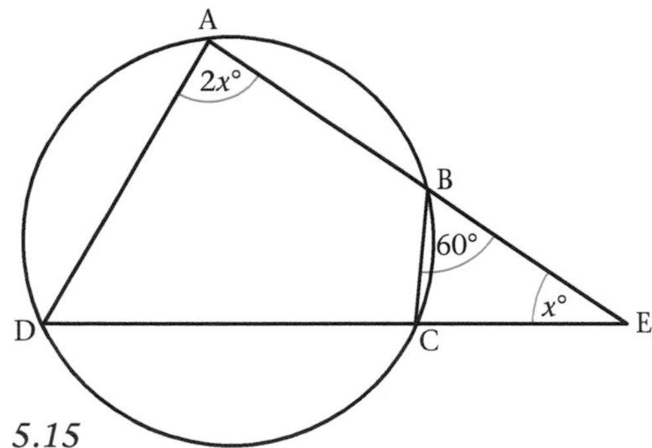

Fig. 5.15

10 In Fig. 5.16, P, Q, R, S are points on a circle such that $P\hat{Q}S = P\hat{R}Q$.

(a) Prove that PS = PQ.

(b) Hence, if SQ is a diameter of the circle and QR is produced to Z, determine $S\hat{Q}P$ and $S\hat{R}Z$, giving reasons.

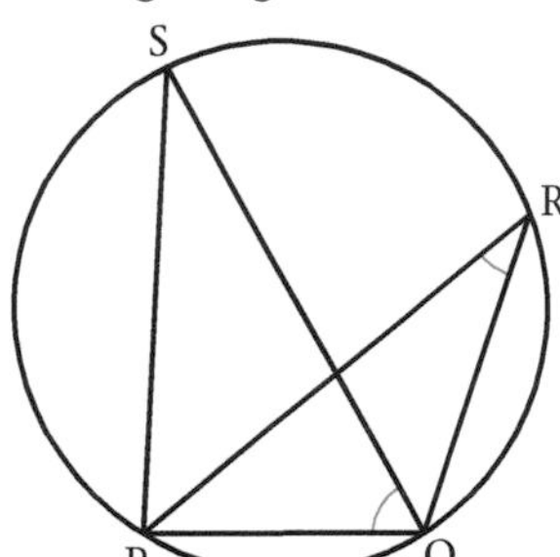

Fig. 5.16

Tangent to a circle

In Fig. 5.17, a circle, centre O, is cut by a straight line MN at the two points X and Y. △OXY is isosceles and $O\hat{X}Y = O\hat{Y}X$.

Hence $O\hat{X}M = O\hat{Y}N$.

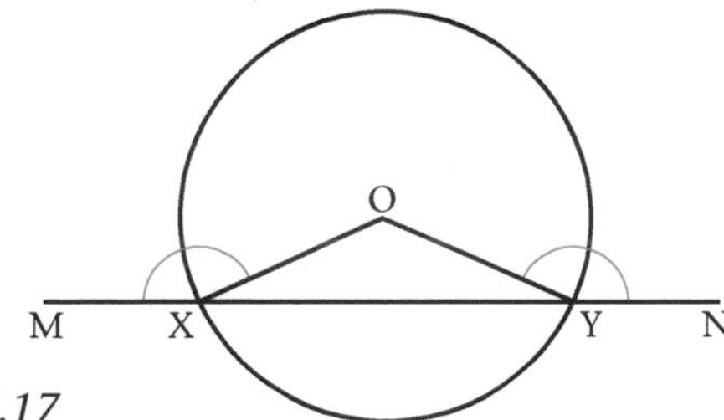

Fig. 5.17

Fig. 5.18 shows what happens to the positions of X and Y if MN moves downwards. X and Y occupy new positions such as X_1, Y_1 and X_2, Y_2, etc., becoming closer to each other.

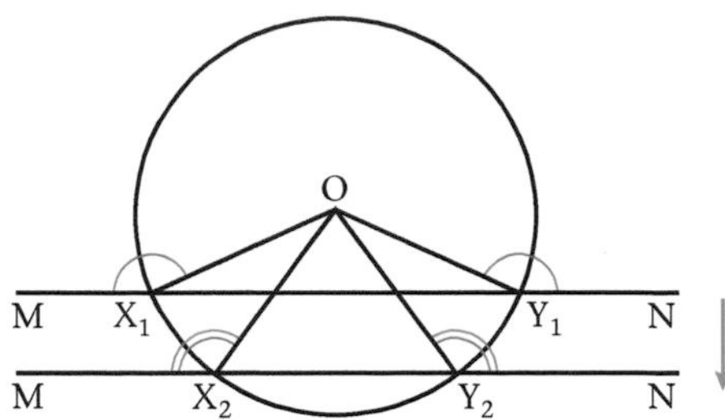

Fig. 5.18

Eventually the points X and Y will coincide at a single point T as in Fig. 5.19. Similarly, the radii OX and OY will coincide to become one radius, OT.

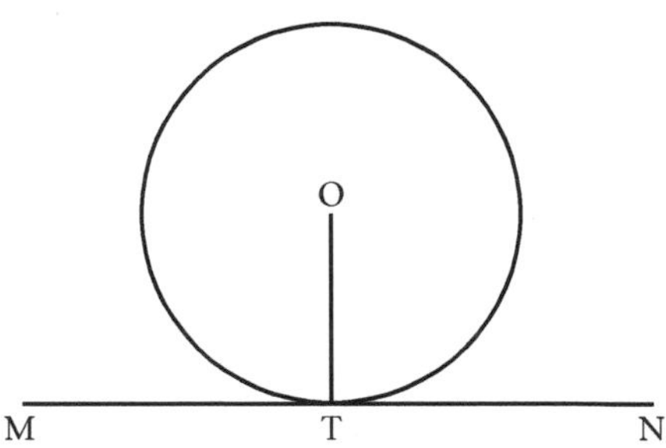

Fig. 5.19

Since $O\hat{X}M = O\hat{Y}N$ (Fig. 5.17), it follows that in Fig. 5.19

$$O\hat{T}M = O\hat{T}N$$

and since MTN is a straight line,

$$O\hat{T}M = O\hat{T}N = 90°$$

Hence OT ⊥ MN.

In Fig. 5.19, the line MN is said to be a **tangent** to the circle. The tangent *does not cut* the circle; it *touches* the circle.

Remember the following:

A tangent to a circle is perpendicular to the radius drawn to its point of contact.

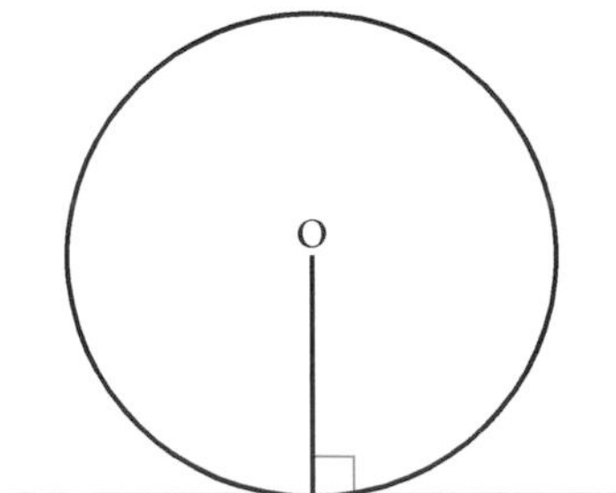

Fig. 5.20

The perpendicular to a tangent at its point of contact passes through the centre of the circle.

Example 5

TA is a tangent at A to a circle, centre O. AB is a chord. If $B\widehat{A}T = x°$, show that $B\widehat{O}A = 2x°$.

Fig. 5.21 is a sketch of the circle.

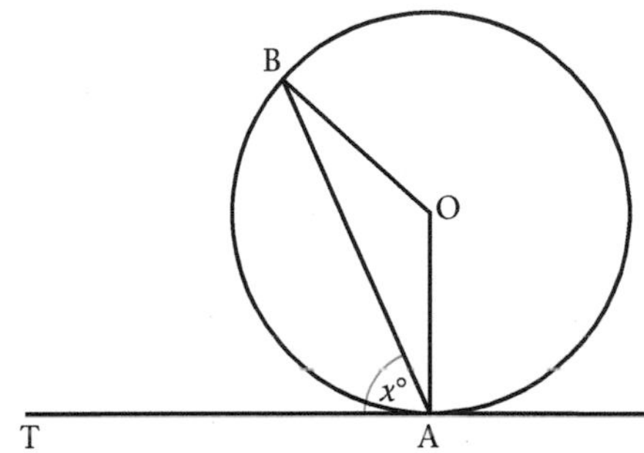

Fig. 5.21

If $B\widehat{A}T = x°$

then $B\widehat{A}O = (90 - x)°$ *(radius ⊥ tangent)*

and $A\widehat{B}O = (90 - x)°$ *(isos. △AOB)*

∴ $B\widehat{O}A = 180° - 2(90° - x)°$ *(sum of angles of △ABO)*

$= 180° - 180° + 2x°$

$= 2x°$

Exercise 5b

1. Calculate the size of angle α in each part of Fig. 5.22. In each figure, O is the centre of the circle.

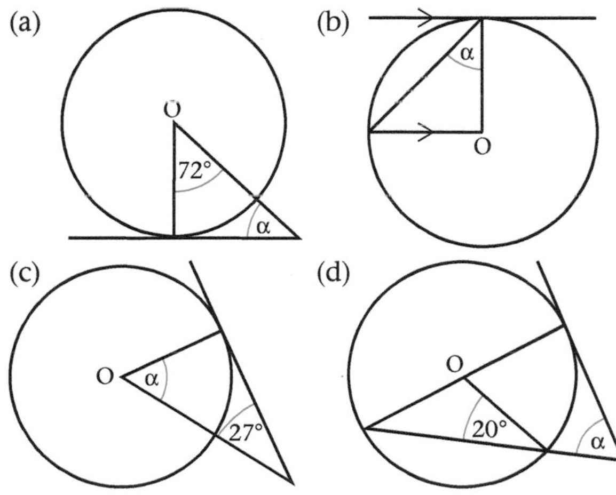

Fig. 5.22

2. Calculate the size of θ in each part of Fig. 5.23. In each figure, O is the centre of the circle.

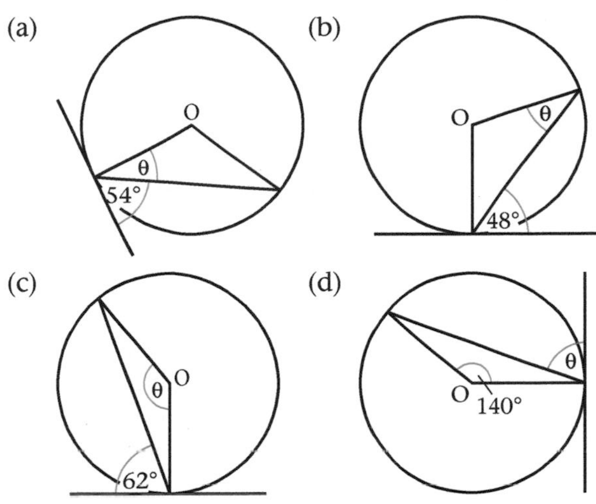

Fig. 5.23

3. Calculate OA in each part of Fig. 5.24. In each figure, O is the centre of the circle and the dimensions are in cm.

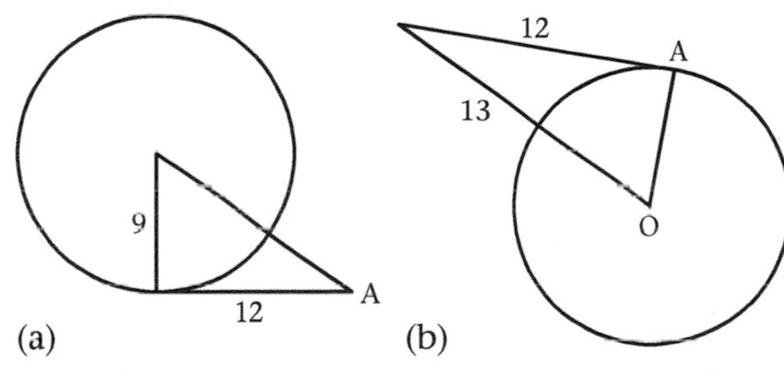

Fig. 5.24

4. In Fig. 5.25, O is the centre of the circle and TA is a tangent at A.

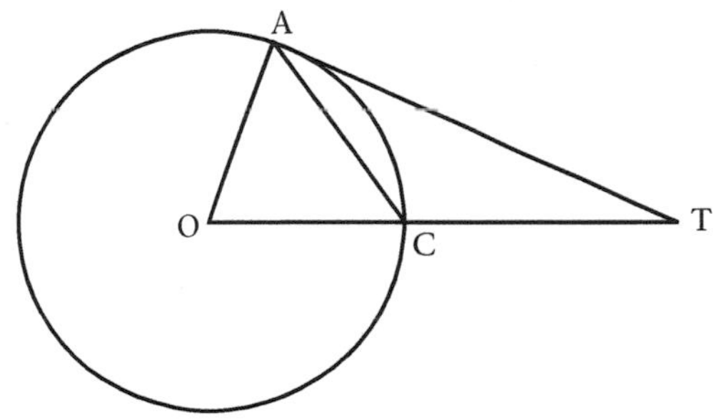

Fig. 5.25

Use Fig. 5.25 to answer the following.

(a) If $A\widehat{O}C = 86°$, calculate $C\widehat{A}T$.
(b) If $A\widehat{T}O = 36°$, calculate $A\widehat{C}O$.
(c) If $O\widehat{A}C = 69°$, calculate $A\widehat{T}C$.
(d) If $A\widehat{C}T = 122°$, calculate $C\widehat{A}T$.

5. The tangent from a point T touches a circle at R. If the radius of the circle is 2.8 cm, and T is 5.3 cm from the centre, calculate TR.

6. A point P is 6.5 cm from the centre of a circle, and the length of the tangent from P is 5.6 cm. Calculate the radius of the circle.

7. AB is a chord and O is the centre of a circle. If $A\hat{O}B = 78°$, calculate the obtuse angle between AB and the tangent at B.

8. Two circles have the same centre and their radii are 15 cm and 17 cm. A tangent to the inner circle at P cuts the outer circle at Q. Calculate PQ.

9. In Fig. 5.26 AC is a tangent to the circle, centre O, and $B\hat{C}A = 90°$.

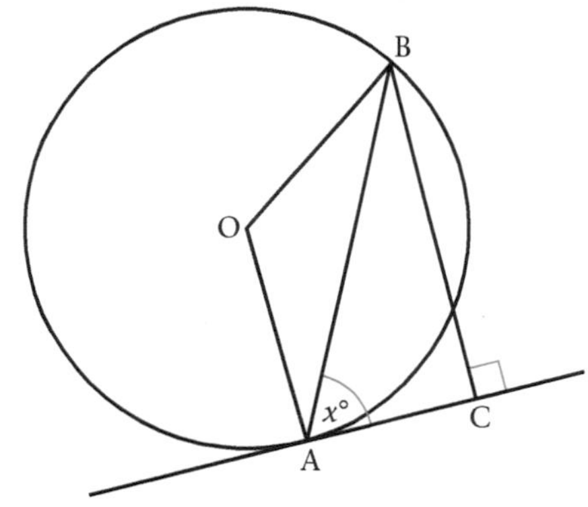

Fig. 5.26

   (a) If $B\hat{A}C = x°$, find (i) $O\hat{B}A$, (ii) $A\hat{B}C$ in terms of x.
   (b) Hence show that AB bisects $O\hat{B}C$.

10. AD is a diameter of a circle, AB is a chord and AT is a tangent.
   (a) State the size of $A\hat{B}D$.
   (b) If $B\hat{A}T$ is an acute angle of $x°$, find the size of $D\hat{A}B$ in terms of x.
   (c) Hence prove that $B\hat{A}T = A\hat{D}B$.

Tangents from an external point

Example 6

In Fig. 5.27, T is a point outside the circle, centre O and TA, TB are tangents to the circle. Show that TA = TB.

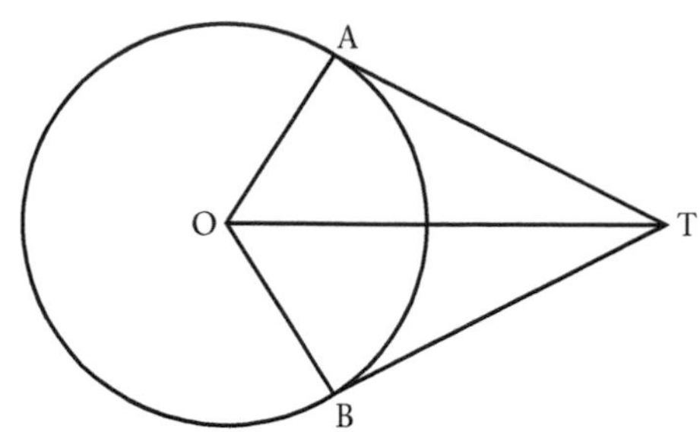

Fig. 5.27

In $\triangle$s OAT and OBT,

$\hat{A} = \hat{B} = 90°$ (*radius ⊥ tangent*)
OA = OB (*radii*)
OT = OT (*common side*)
$\therefore \triangle OAT \equiv \triangle OBT$
$\therefore$ TA = TB

The tangents to a circle from an external point are equal.

Notice also that in Fig. 5.27 $A\hat{O}T = B\hat{O}T$ and $A\hat{T}O = B\hat{T}O$.
Hence the line joining the external point to the centre of the circle bisects the angle between the tangents and the angle between the radii drawn to the points of contact of the tangents, i.e. OT is on the line of symmetry of Fig. 5.27.

Example 7

In Fig. 5.28 O is the centre of the circle and TA and TB are tangents. If $A\hat{T}O = 39°$, calculate $T\hat{B}X$.

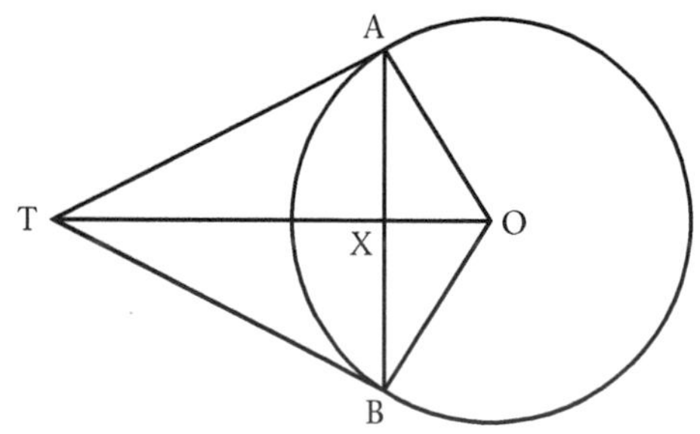

Fig. 5.28

In $\triangle$TAX,

$A\hat{X}T = 90°$ (*symmetry of Fig. 5.28*)
$\therefore T\hat{A}X = 180° - (90° + 39°)$ (*sum of angles of △*)
$= 180° - 129°$
$= 51°$
$\therefore T\hat{B}X = 51°$ (*symmetry*)

In Example 7 there are many ways of showing that $T\hat{B}X = 51°$, e.g. by noticing that $A\hat{T}X$ is the semi-vertical angle of isosceles triangle ATB.

Example 8

X, Y, Z are three points on a circle centre O as shown in Fig. 5.29. The tangents at X and Y meet at T. If $X\hat{T}Y = 58°$, calculate $X\hat{Z}Y$.

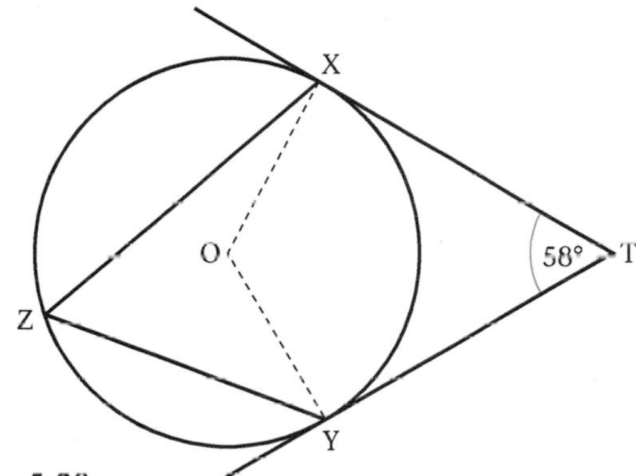

Fig. 5.29

Join OX and OY.
In quadrilateral TXOY,

$O\hat{X}T = O\hat{Y}T = 90°$ (*radius ⊥ tangent*)
$\therefore O\hat{X}T + O\hat{Y}T = 180°$
$\therefore X\hat{T}Y + X\hat{O}Y = 180°$ (*angles of quad.*)
$\therefore X\hat{O}Y = 180° - 58°$
$= 122°$
$\therefore X\hat{Z}Y = \frac{1}{2}$ of $122°$ (*angle at centre = 2 × angle at circumference*)
$= 61°$

Exercise 5c

1. Use Fig. 5.28 to answer the following.
   (a) If $A\hat{T}O = 36°$, calculate $A\hat{B}O$.
   (b) If $A\hat{B}T = 57°$, calculate $A\hat{O}T$.
   (c) If $B\hat{T}O = 44°$, calculate $T\hat{A}X$.
   (d) If AB = 18 cm and TB = 15cm, calculate TX.

2. In Fig. 5.28 show that TAOB is a cyclic quadrilateral.

3. In Fig. 5.28, if $A\hat{O}T = 47°$, calculate $A\hat{B}O$.

4. A, B, C are three points on a circle, centre O, such that $B\hat{A}C = 37°$. The tangents at B and C meet at T. Calculate $B\hat{T}C$.
   (*Hint*: Make a sketch and join BO and CO.)

5. P, Q, R are three points on a circle, centre O. The tangents at P and Q meet at T. If $P\hat{T}O = 62°$, calculate $P\hat{R}Q$.

6. In Fig. 5.30 AB is a diameter of circle ABC, centre O. TA and TC are tangents.

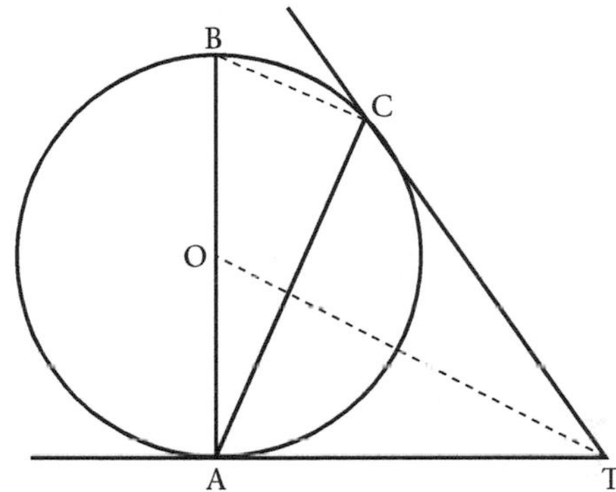

Fig. 5.30

   If $A\hat{T}C = 2x$, show that $B\hat{A}C = x$.

7. In Fig. 5.30 show that BC is parallel to OT.

8. A quadrilateral PQRS is such that a circle can be drawn inside it to touch all four sides. Show that PQ + RS = PS + QR.

9. A, B, C are three points on a circle. The tangents at A and B meet at T, and BC||TA. Show that AB bisects $T\hat{B}C$.

10. O is the centre of a circle, and two tangents from a point T touch the circle at A and B. BT is produced to C. If $A\hat{O}T = 67°$, calculate $A\hat{T}C$.

Alternate segment

In both parts of Fig. 5.31 SAT is a tangent to the circle at A. The chord AB divides the circle into two segments APB and AQB.

In Fig. 5.31(a) the segment APB is the **alternate segment** to $T\hat{A}B$, i.e. it is on the other side of AB from $T\hat{A}B$. Similarly, in Fig. 5.31(b), segment AQB is the alternate segment to $S\hat{A}B$.

Sometimes the word **opposite** is used instead of alternate.

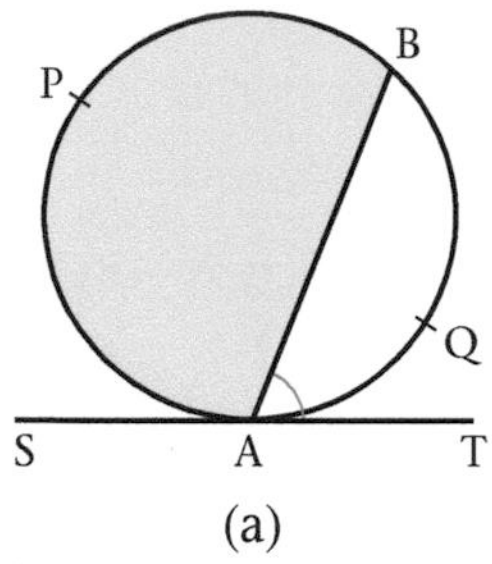

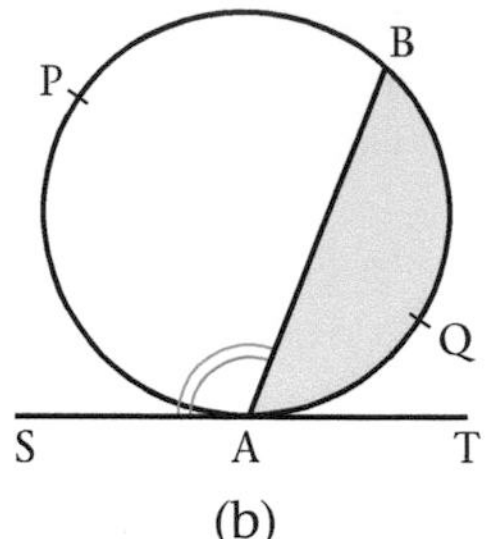

Fig. 5.31

Example 9

In Fig. 5.32, AD is the diameter of the circle with SAT a tangent at A and chord AB dividing the circle into two segments, APB and AQB.

Show that $T\hat{A}B = A\hat{P}B$
and $S\hat{A}B = A\hat{Q}B$

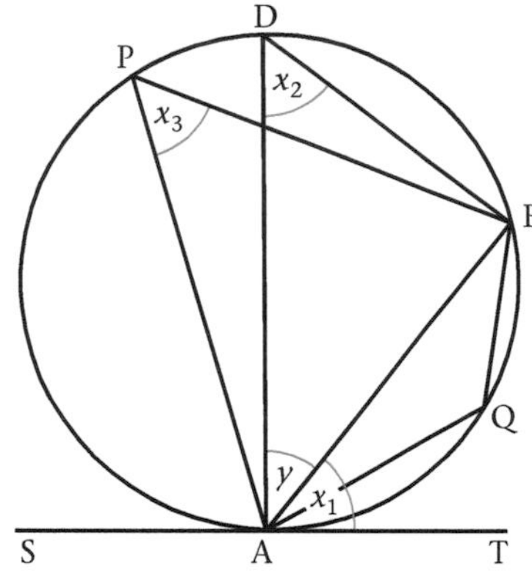

Fig. 5.32

With the lettering of Fig. 5.32

$x_1 + y = 90°$ (*tangent ⊥ radius*)
also $A\hat{B}D = 90°$ (*angle in semicircle*)
$\therefore\ x_2 + y = 90°$ (*sum of angles of △*)
$\therefore\ x_1 = x_2$
and $x_2 = x_3$ (*angles in same segment*)
$\therefore\ T\hat{A}B = A\hat{P}B$
Also $S\hat{A}B = 180° - x_1$ (*angles on str. line*)
$= 180° - x_3$ ($x_1 = x_3$ proved)
$= A\hat{Q}B$ (*opp angles of cyclic quad.*)

If a straight line touches a circle, and from the point of contact a chord is drawn, the angles which the chord makes with the tangent are equal to the angles in the alternate segments.

Example 10

In Fig. 5.33, PQX is a tangent to the circle QRS. Calculate $S\hat{Q}X$.

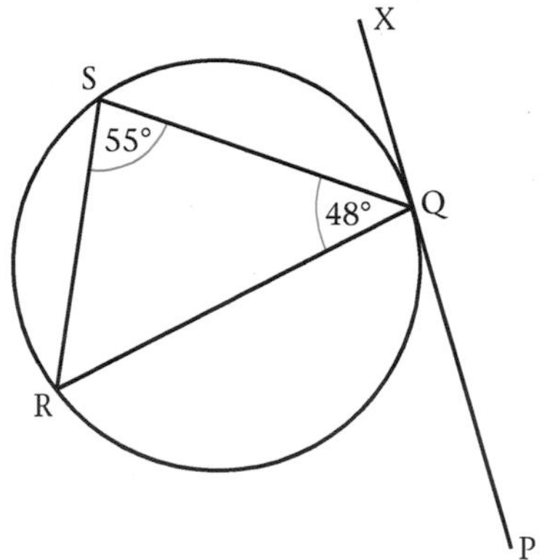

Fig. 5.33

In △QRS,

$Q\hat{R}S = 180° - (55° + 48°)$ (*sum of angles of △*)
$= 180° - 103°$
$= 77°$
$\therefore\ S\hat{Q}X = 77°$ (*alternate segment*)

Example 11

In Fig. 5.34 PT is a tangent to circle ABCT, BA = BT and $A\hat{T}P = 82°$. Calculate $B\hat{C}T$.

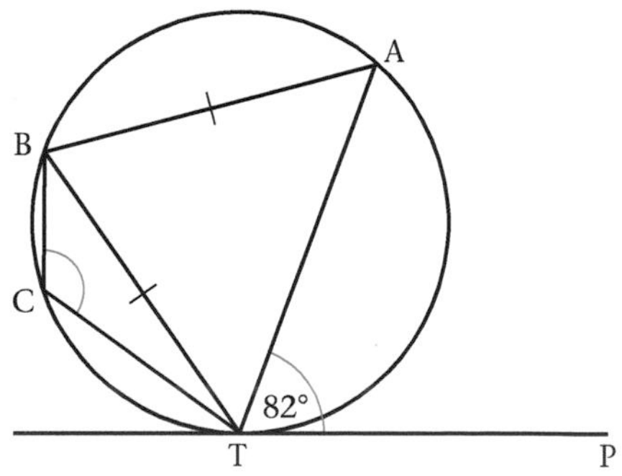

Fig. 5.34

$A\hat{B}T = 82°$ (*alternate segment*)
In △ABT,
$B\hat{A}T = \frac{1}{2}(180° - 82°)$ (*sum of angles of isos. △*)
$= \frac{1}{2} \times 98°$
$= 49°$
$\therefore\ B\hat{C}T = 180° - 49°$ (*opp. angles of cyclic quad.*)
$= 131°$

Exercise 5d

1. In Fig. 5.35 XYZ is a tangent to circle ABCY.
   (a) Name two angles equal to $A\hat{Y}Z$.
   (b) Name two angles equal to $C\hat{Y}X$.
   (c) Name an angle equal to $B\hat{Y}Z$.
   (d) Name an angle equal to $B\hat{Y}X$.

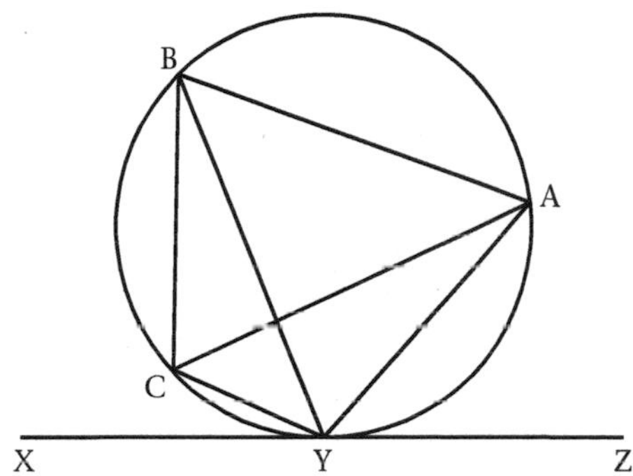

Fig. 5.35

   (e) If $A\hat{Y}Z = 58°$ what is the size of $A\hat{C}Y$?
   (f) If $B\hat{C}Y = 112°$ what is the size of $B\hat{Y}Z$?
   (g) If $B\hat{C}Y = 125°$ what is the size of $B\hat{Y}X$?
   (h) If $B\hat{Y}Z = 100°$ what is the size of $B\hat{A}Y$?

2. In Fig. 5.36 TAX and TBY are tangents to the circle and C is a point on the major arc AB.

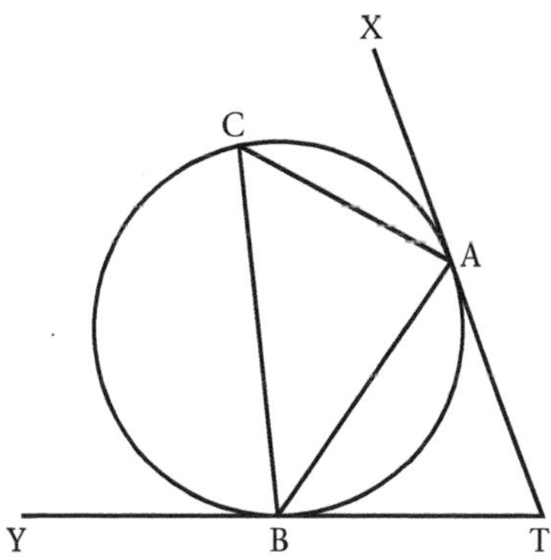

Fig. 5.36

   (a) If $A\hat{T}B = 68°$, calculate $A\hat{C}B$.
   (b) If $A\hat{B}C = 33°$, $B\hat{A}C = 83°$, calculate $A\hat{T}B$.
   (c) If $A\hat{C}B = 59°$, $C\hat{B}Y = 78°$, calculate $C\hat{A}X$.
   (d) If $C\hat{A}X = 65°$, $C\hat{B}Y = 76°$, calculate $A\hat{T}B$.
   (e) If $A\hat{B}C = 48°$, $A\hat{T}B = 72°$, calculate $B\hat{A}C$.

3. In Fig. 5.37, XYZ is a tangent to the circle at Y. Name an angle equal to $Y\hat{Q}P$.

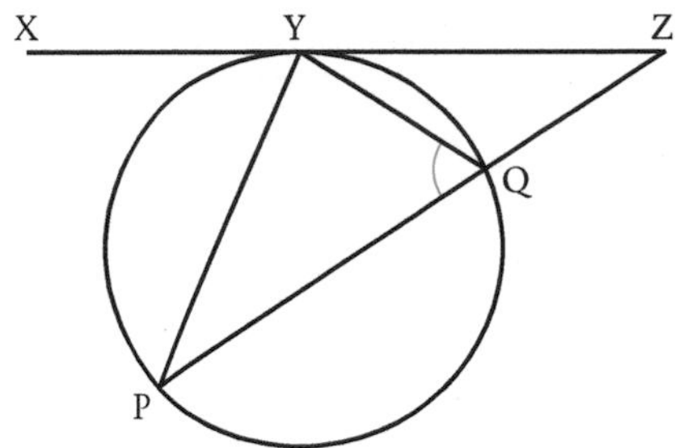

Fig. 5.37

4. In Fig. 5.38, TY is a tangent to the circle TVS. If $S\hat{V}T = 48°$ and VS = ST, what is the size of $V\hat{T}Y$?

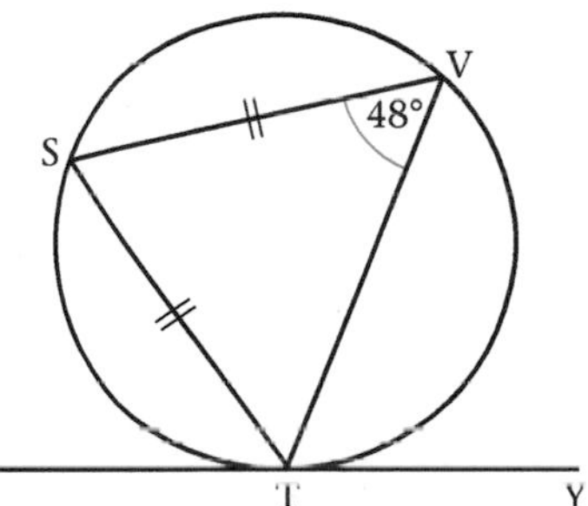

Fig. 5.38

5. In Fig. 5.39, the tangents from T touch a circle at A and B and BC is a chord parallel to TA. If $B\hat{A}T = 54°$, calculate $B\hat{A}C$.

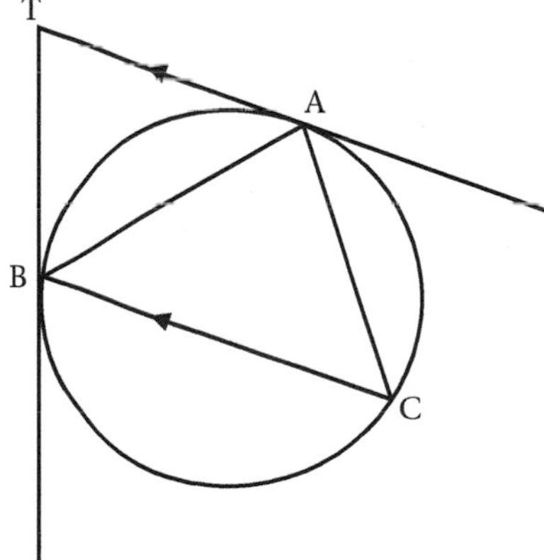

Fig. 5.39

6. In Fig. 5.39, if $A\hat{T}B = 82°$, calculate the angles of $\triangle ABC$.

7. In Fig. 5.40, TS is a tangent to circle PQRS. If PR = PS and $P\hat{Q}R = 117°$, calculate $R\hat{S}T$.

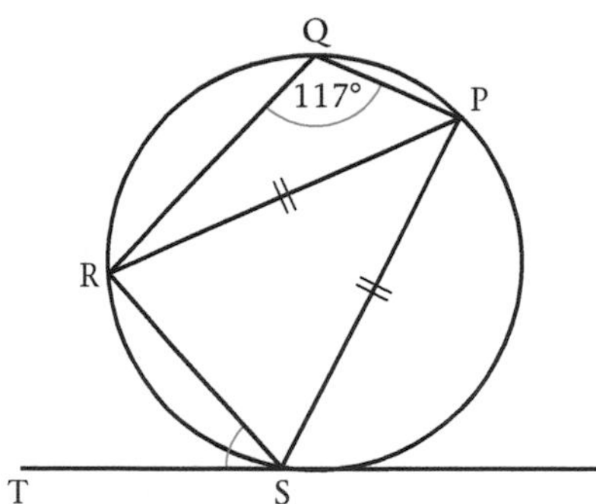

Fig. 5.40

8. In Fig. 5.41, if $A\hat{C}B = 37°$ and $A\hat{T}B = 42°$, calculate $A\hat{B}T$ and $A\hat{E}B$.

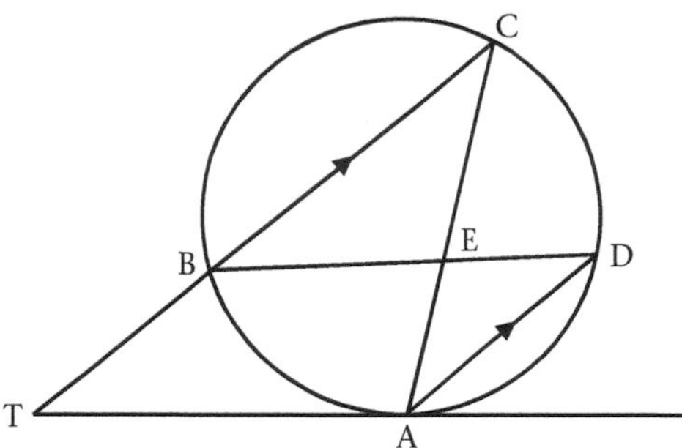

Fig. 5.41

9. AB is a chord of a circle and the tangents at A and B meet at T. C is a point on the minor arc AB. If $A\hat{T}B = 54°$ and $C\hat{B}T = 23°$, calculate $C\hat{A}T$.

10. A, B, C are three points on a circle. The tangent at C meets AB produced at T. If $A\hat{C}T = 103°$, $A\hat{T}C = 43°$, calculate the angles of $\triangle ABC$.

11. The angles of a triangle are 40°, 60°, 80°, and a circle touches its sides at P, Q, R. Calculate the angles of $\triangle PQR$.

12. AT is a tangent to the circle ABCD, $B\hat{A}C = 64°$ and $C\hat{A}T = 72°$. Calculate $B\hat{C}A$ and $C\hat{D}A$.

Summary

Opposite angles of a **cyclic quadrilateral** are supplementary. The exterior angle of a cyclic quadrilateral is equal to the interior opposite angle.

A **tangent** is a line which touches a circle.

A right angle is formed at the point where the tangent meets the radius of the circle.

Tangents to a circle from an external point are equal.

If from the point of contact a chord of the circle is drawn, then the angle between the tangent and the chord is equal to the angle in the **alternate segment**.

Practice exercise P5.1

In the circle ABCD, centre O (Fig. 5.42), let $B\hat{A}D = a$.

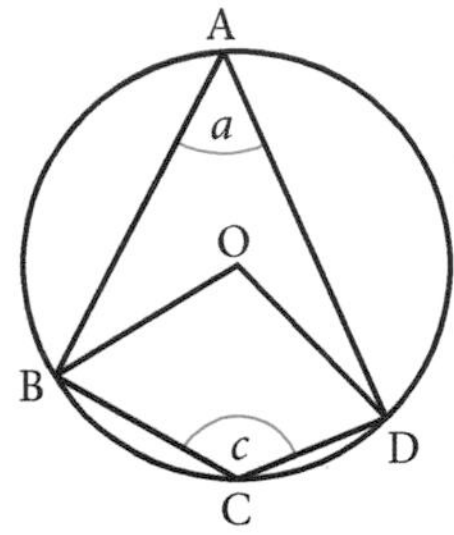

Fig. 5.42

1. Find the reflex angle $B\hat{O}D$ in terms of a.
2. Show that $B\hat{A}D + B\hat{C}D = 180°$, drawing any necessary construction lines and giving the reason for each statement.
3. Copy and complete the proof to show that $B\hat{C}D + B\hat{O}D = 180° + B\hat{A}D$.

   Proof

   $B\hat{O}D = 2B\hat{A}D$

   because ....

   $B\hat{C}D = 180° -$ ....

   because ....

   Adding the two equations gives

   ....

Practice exercise P5.2

In each part of Fig. 5.43, O is the centre of the circle. Find the sizes of the marked angles. Give the reason for each step in your calculations.

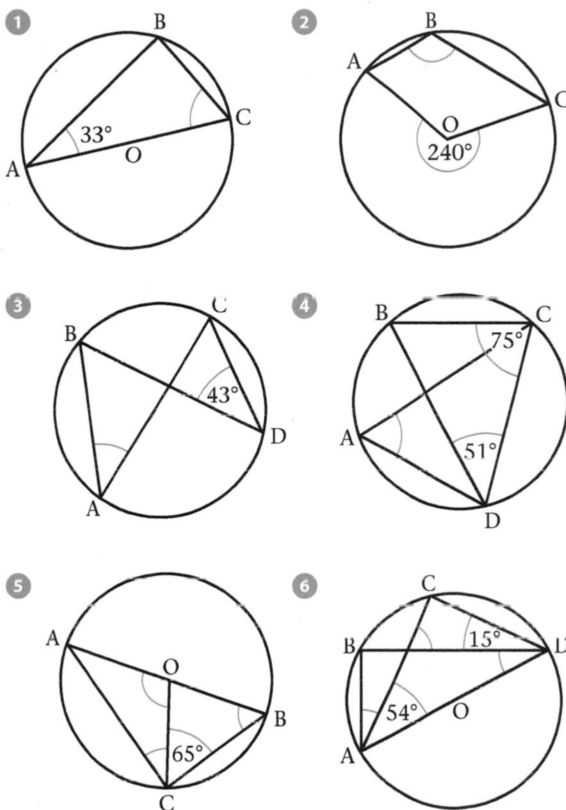

Fig. 5.43

Practice exercise P5.3

In each part of Fig. 5.44, O is the centre of the circle. Find the sizes of the marked angles. Give the reason for each step in your calculations.

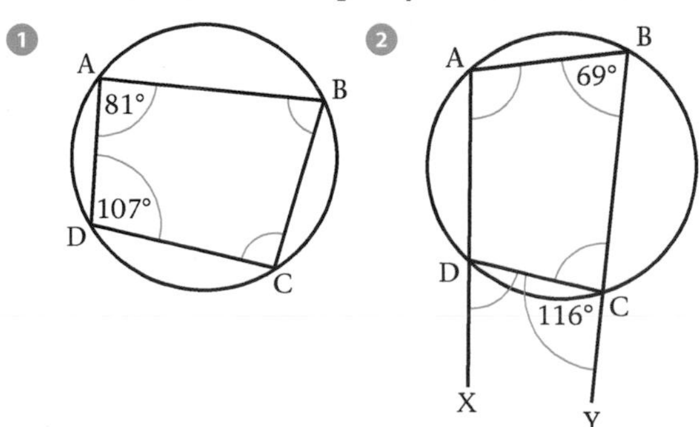

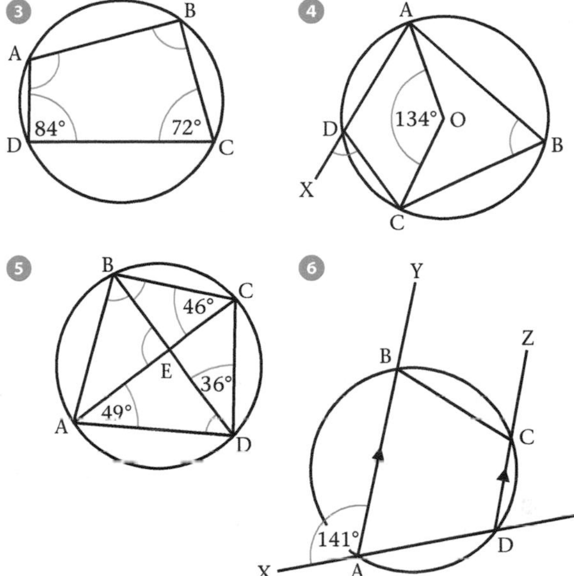

Fig. 5.44

Practice exercise P5.4

In each part of Fig. 5.45, calculate the sizes of the marked angles. Give the reason for each statement. In each case O is the centre of the circle.

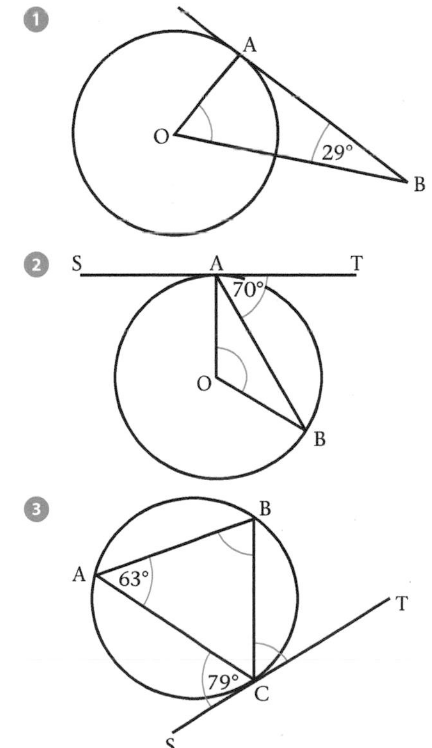

4

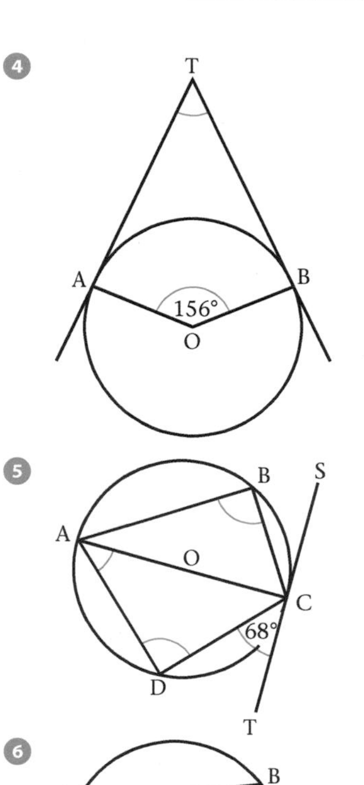

5

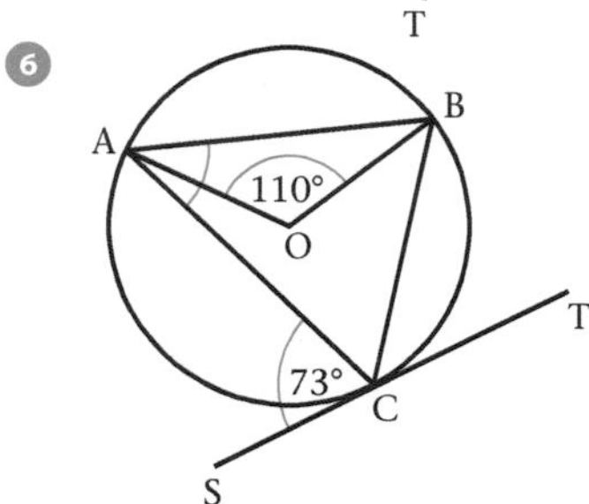

6

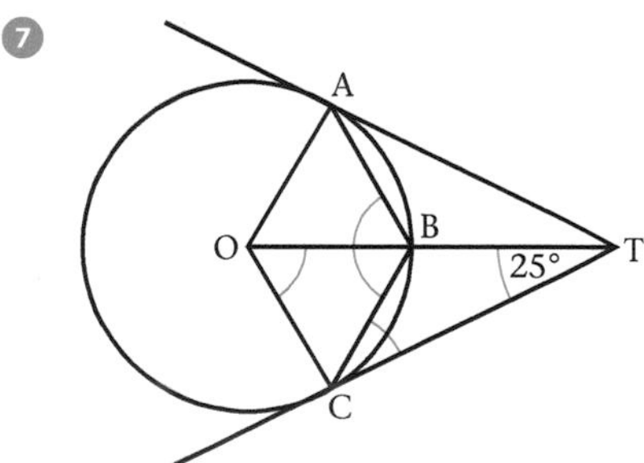

7

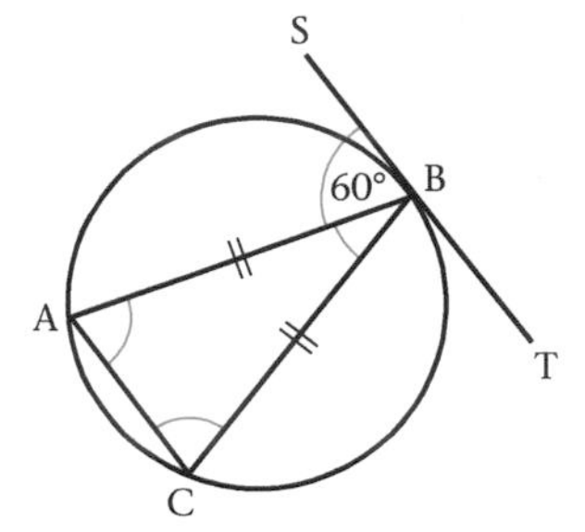

8

Fig. 5.45

Practice exercise P5.5

1 In Fig. 5.46, AD is a tangent to the circle.

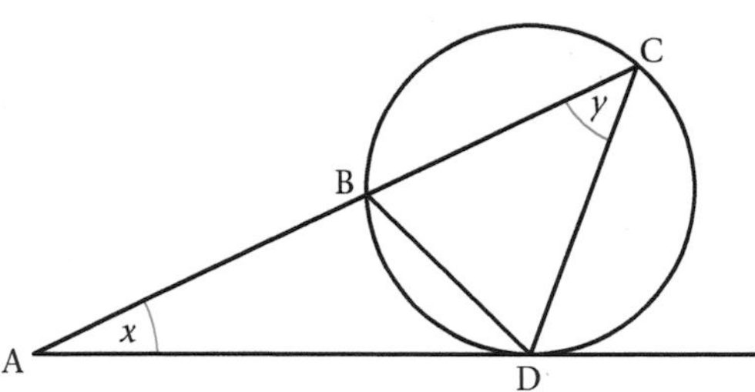

Fig. 5.46

(a) Find $A\hat{B}D$ in terms of x and y.
(b) Show that triangles ABD and ACD are similar.

2 Using the information in Fig. 5.47, show that $C\hat{B}D = \frac{1}{2}(x + y)$.

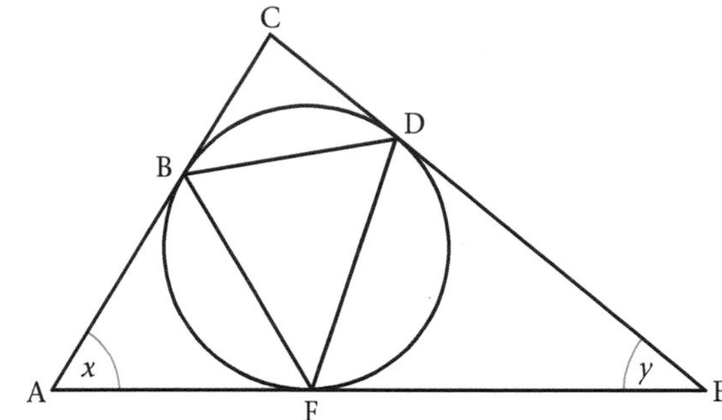

Fig. 5.47

3 In Fig. 5.48, OD and OE are tangents and AB is parallel to OE. Prove that $A\hat{O}E = A\hat{C}B$.

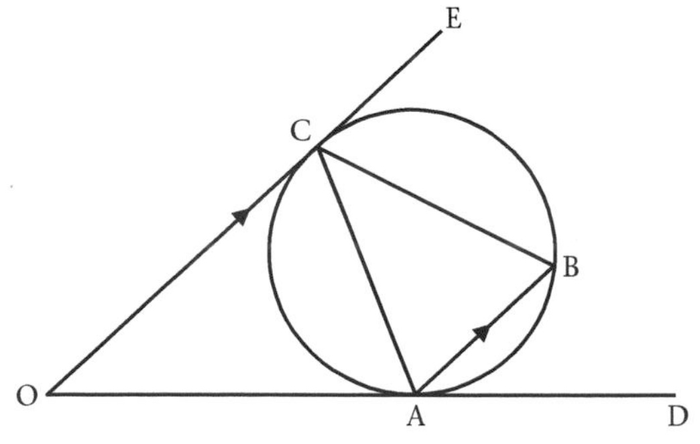

Fig. 5.48

Chapter 6

Coordinate geometry

Parallel and perpendicular lines, mid-point and length of lines

Pre-requisites

- Cartesian coordinates and graphs; gradient of a straight line; equation of a straight line

Linear functions (revision)

In Book 3 you studied the graphs of linear functions, the equation and the gradient of the straight line.

Example 1

Write down (i) the gradients, (ii) the equations of the straight lines (a) and (b) in Fig 6.1.

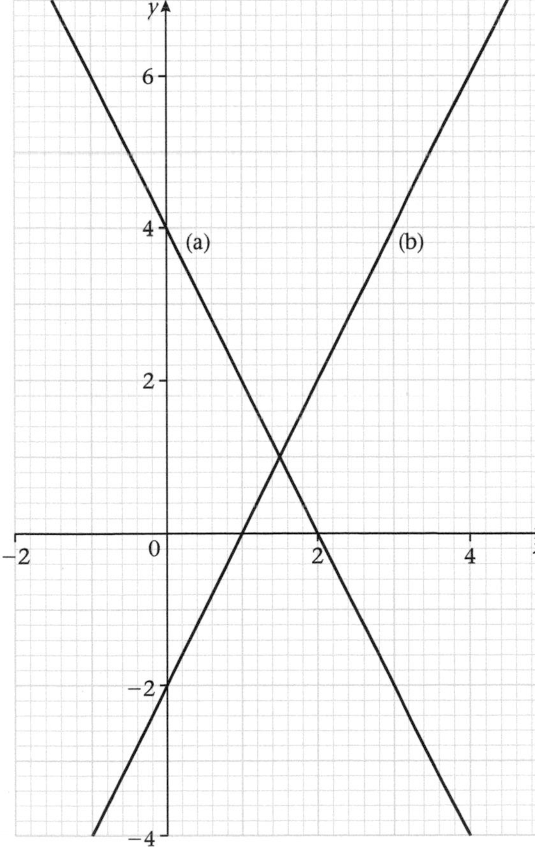

Fig. 6.1

(a) (i) gradient $= -\frac{4}{2} = -2$

(ii) y intercept $= 4$

$\therefore$ Equation of line is given by $y = -2x + 4$

$y + 2x = 4$

(b) (i) Gradient $= \frac{4}{2} = 2$

(ii) y intercept $= -2$

$\therefore$ Equation of line is given by

$y + 2x \pm 2 = 0$

Sketching graphs of straight lines

Sometimes you will be asked to sketch rather than draw the graph of a straight line. This can be done if the gradient and one point on the line are known.

Example 2

Make a rough sketch of the line whose equation is $2x + 4y = 9$.

First: Rearrange the equation to make y the subject.

$$2x + 4y = 9$$
$$4y = -2x + 9$$
$$y = -\tfrac{1}{2}x + 2\tfrac{1}{4}$$

Since the gradient of the line is $-\frac{1}{2}$ then

$$\frac{\text{increase in } y}{\text{increase in } x} = -\frac{1}{2}$$

Hence when x increases by 2 units, y changes by -1 unit.

Second: Find a point on the line. The simplest point is usually that where $x = 0$. When $x = 0$, $y = 2\frac{1}{4}$. $(0, 2\frac{1}{4})$ is a point on the line.

Fig. 6.2 is a rough sketch of the line.

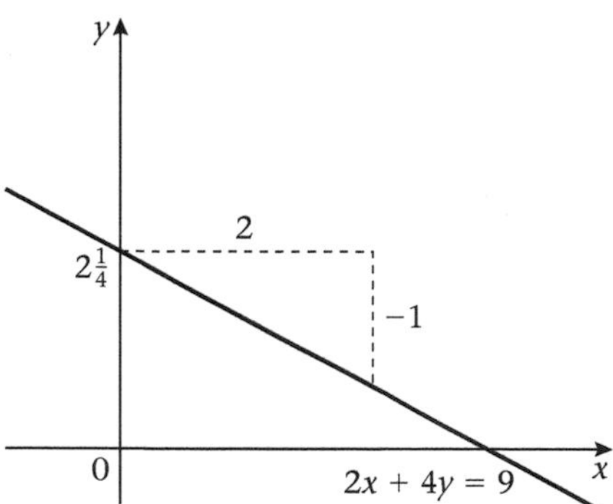

Fig. 6.2

Notice that the axes x and y should always be shown on a rough sketch.

An alternative method of sketching a straight line is to find the two points where the line crosses the axes.

Example 3

Sketch the graph of the line whose equation is $4x - 3y = 12$.

When $x = 0$, $-3y = 12$
$y = -4$

The line crosses the y-axis at $(0, -4)$
When $y = 0$, $4x = 12$
$x = 3$

The line crosses the x-axis at $(3, 0)$.
Fig. 6.3 is a rough sketch of the line.

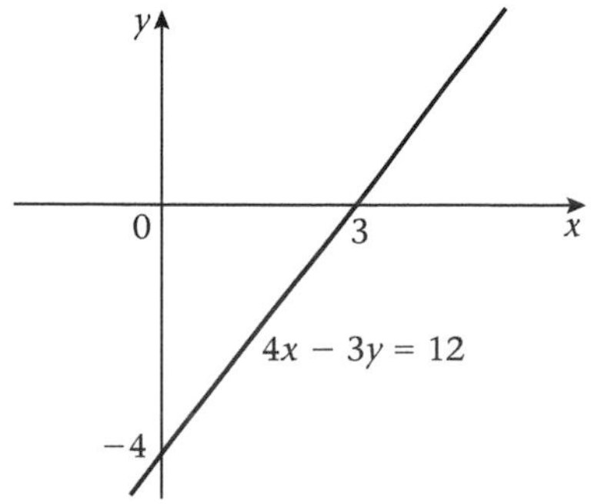

Fig. 6.3

Any line which is parallel to the x-axis has a **zero gradient**. The equations of such lines are always in the form $y = c$, where c may be any number. Fig. 6.4 shows the graphs of $y = 5$ and $y = -3$.

Notice that the equation of the x-axis is $y = 0$.

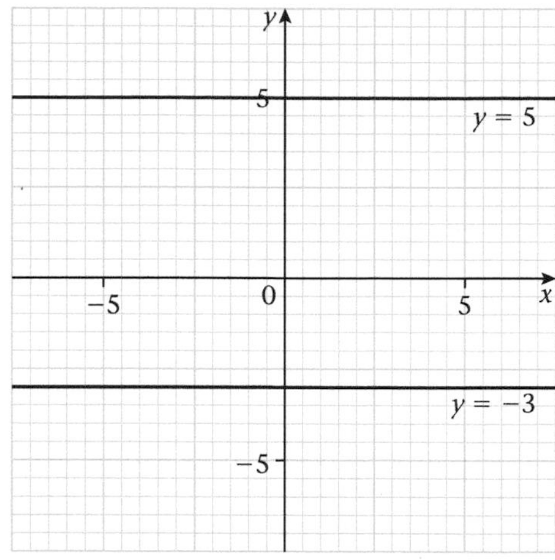

Fig. 6.4

The gradient of a line which is parallel to the y-axis is undefined (i.e. cannot be found). The equations of such lines are always in the form $x = a$, where a may be any number. Fig. 6.5 shows the graphs of the lines $x = 2$ and $x = -4$.

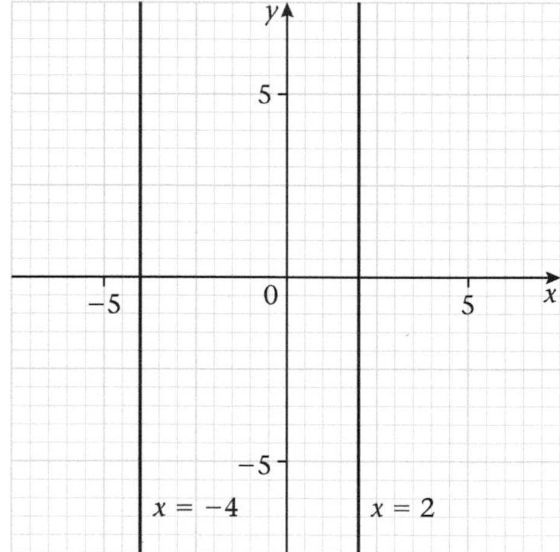

Fig. 6.5

Notice that the equation of the y-axis is $x = 0$.

Exercise 6a

1. Sketch the lines which pass through the following points with the given gradients.
   (a) point $(2, 1)$, gradient 3
   (b) point $(5, 0)$, gradient -2
   (c) point $(1, -3)$, gradient -3
   (d) point $(-4, -2)$, gradient $\frac{2}{3}$
   (e) point $(5, -2)$, gradient $-\frac{4}{3}$

2. Write down the gradients of the lines represented by the following equations. Hence sketch the graphs of the lines.
   (a) $y = 2x + 3$
   (b) $y = \frac{1}{3}x$
   (c) $y = \frac{5}{4}x - 2$
   (d) $3x + 7y = 5$
   (e) $4x - 7y = 7$

3. Find the coordinates of the points where the lines represented by the following equations cross the axes. Hence sketch the graphs of the lines.
   (a) $y = 2x - 2$
   (b) $y = \frac{1}{3}x + 1$
   (c) $3x - 5y = 30$
   (d) $4x + 3y = 2$
   (e) $8x + 5y = 4$

4. Write down the gradients of the lines represented by the sketches in Fig. 6.6.

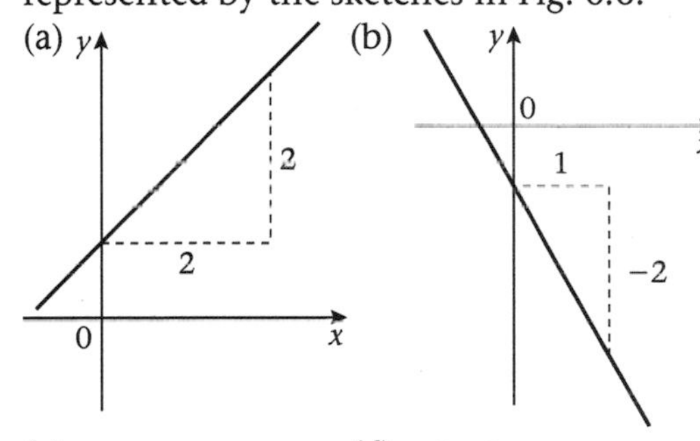

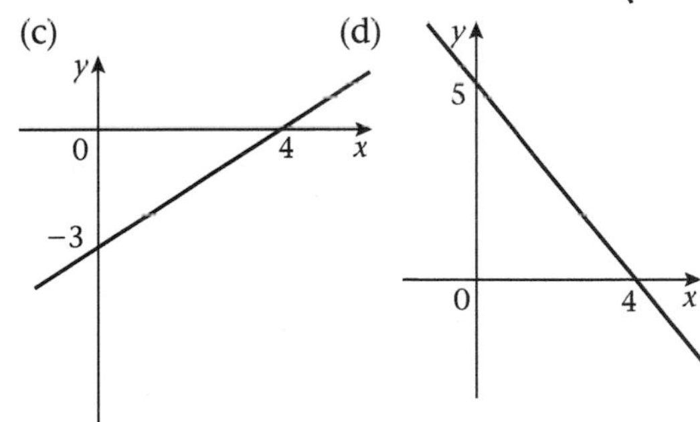

Fig. 6.6

Gradients of parallel and perpendicular lines

Example 4

Using the same coordinate axes, sketch the graphs of the straight lines

(a) $y = 3x + 2$
(b) $y = 3x - 2$
(c) $y = 3x + 6$

What do you notice?

The graphs are shown on Fig. 6.7. The straight lines are all parallel.

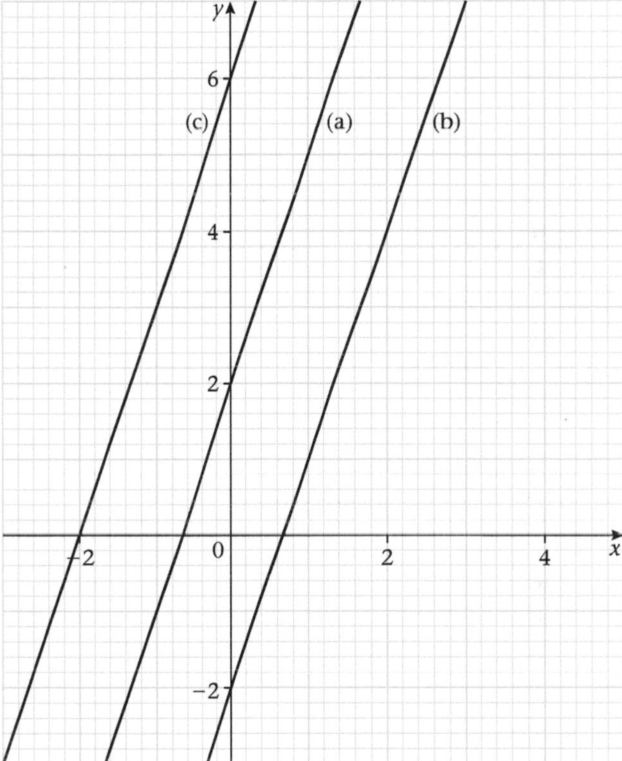

Fig. 6.7

Example 5

Using the same coordinate axes, sketch the graphs of the straight lines

(a) $y = 2x - 1$
(b) $2y - 4x = 3$

Fig. 6.8 shows the graphs of these two straight lines. You will notice that the lines are parallel.

Parallel lines have the same gradient.

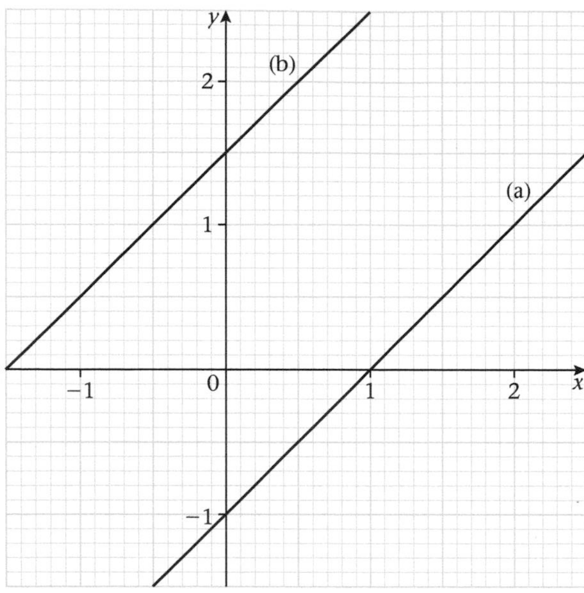

Fig. 6.8

Example 6

Using the same coordinate axes, sketch the graphs of the straight lines

$y = 2x - 1$

$2y + x = 2$

What can you say about the angle between the lines?

Fig. 6.9 shows the graphs of the two lines.

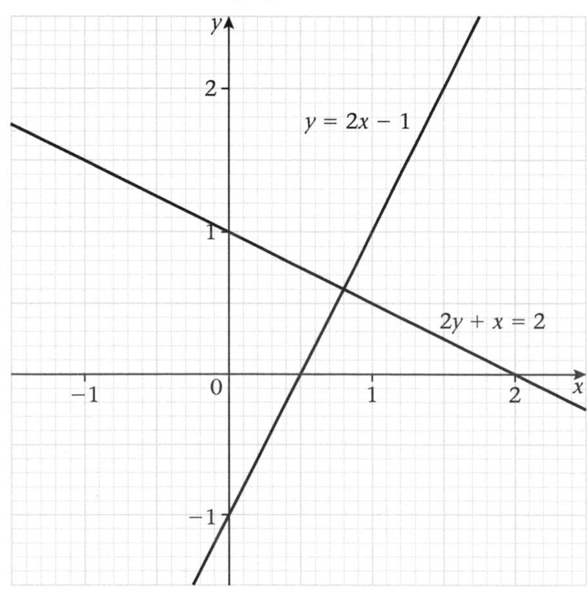

Fig. 6.9

The angle between the two lines is a right angle.

The line $y = 2x - 1$ is perpendicular to the line $2y + x = 2$.
The gradient of the line $y = 2x - 1$ is 2.

The gradient of the line $2y + x = 2$ is $-\frac{1}{2}$ (check by rearranging the equation).
The product of the gradients is

$2 \times -\frac{1}{2} = -1$

When lines are perpendicular, the product of the gradients of the lines equals −1, i.e. $mm_1 = -1$ where m, m_1 are the gradients of the lines.

Exercise 6b

State which of the following pairs of lines are (i) perpendicular, (ii) parallel.

1. $y = x + 5$, $\quad y = -x + 5$
2. $y = -3x + 2$, $\quad y + 3x = 7$
3. $2y - 6 = 5x$, $\quad 3 - 5y = 2x$
4. $2y + 7 = 3x$, $\quad 2y - 3x = 0$
5. $2y = x + 3$, $\quad y - \frac{x}{2} = 4$
6. $x = 3y + 5$, $\quad y + 3x = 2$

Mid-point of a line

Example 7

Find the coordinates of the mid-point of the line joining A(2, 1) and B(8, 5).

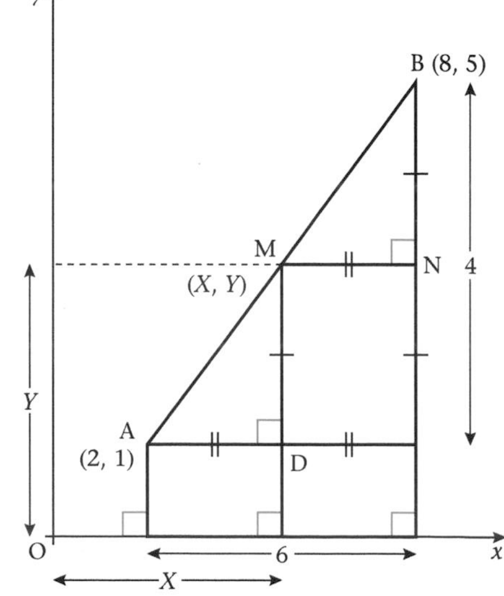

Fig. 6.10

In Fig. 6.10, let the mid-point of AB be M(X, Y).

Δs AMD, MBN are congruent (corresponding sides and angles are equal).
Then AD = MN and MD = BN
Hence, from Fig. 6.10, the mid-point is at M(5, 3).

In Fig. 6.11, M(X, Y) is the mid-point of the line joining the points P(a, b) and Q(c, d).

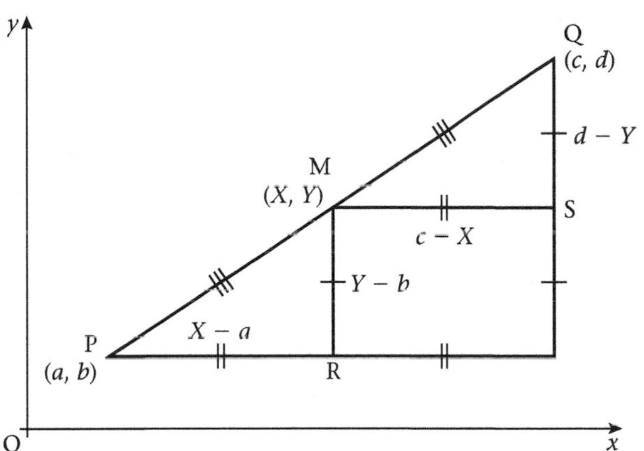

Fig. 6.11

Then PR = $X - a$, MS = $c - X$
and MR = $Y - b$, QS = $d - Y$

Δs MPR, QMS are congruent so that PR = MS and MR = QS, hence,

$$X - a = c - X$$
$$2X = a + c$$
$$X = \frac{a + c}{2}$$

Similarly

$$Y - b = d - Y$$

and $$Y = \frac{b + d}{2}$$

Hence the mid-point of the straight line joining (a, b) to (c, d) is

$$\left(\frac{a + c}{2}, \frac{b + d}{2}\right)$$

Using this formula in Example 7:

$$X = \frac{2 + 8}{2} = 5$$

and $$Y = \frac{1 + 5}{2} = 3$$

Example 8

Find the coordinates of M, the mid-point of the line joining P(−1, 4) and Q(5, −2),

From Fig. 6.12, M is the point (2, 1)

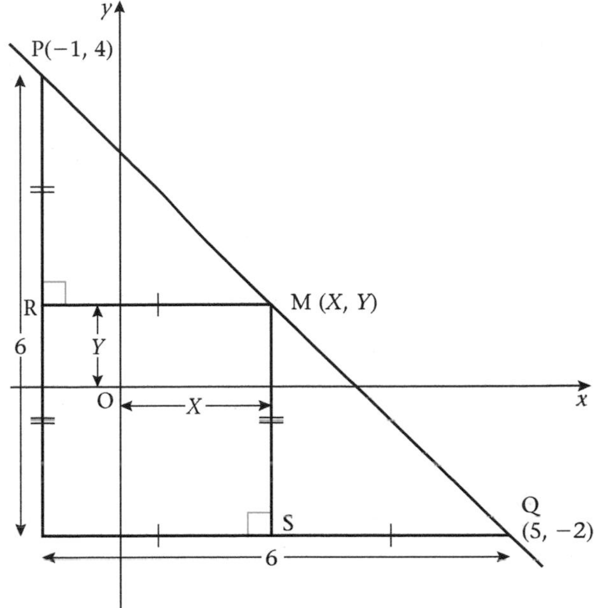

Fig. 6.12

or using $X = \frac{a + c}{2}$ and $Y = \frac{b + d}{2}$

$$X_M = \frac{-1 + 5}{2} \quad \text{and} \quad Y_M = \frac{4 + (-2)}{2}$$
$$= 2 \qquad\qquad = 1$$

M is the point (2, 1).

Exercise 6c

Find the coordinates of the mid-point of the line joining the following pairs of points.

1. (3, 4) and (1, 2)
2. (2, −5) and (3, 6)
3. (−3, 4) and (−1, 2)
4. (2, 5) and (−3, 6)
5. (3, 4) and (−1, −2)
6. (−2, −5) and (3, −6)

Distance between two points

Example 9

Find the distance between A(7, 5) and B(3, 2).

In Fig. 6.13, using Pythagoras' theorem

$$(AB)^2 = (AF)^2 + (BF)^2$$
$$= (4)^2 + (3)^2$$
$$= 25$$
$$AB = 5 \text{ units}$$

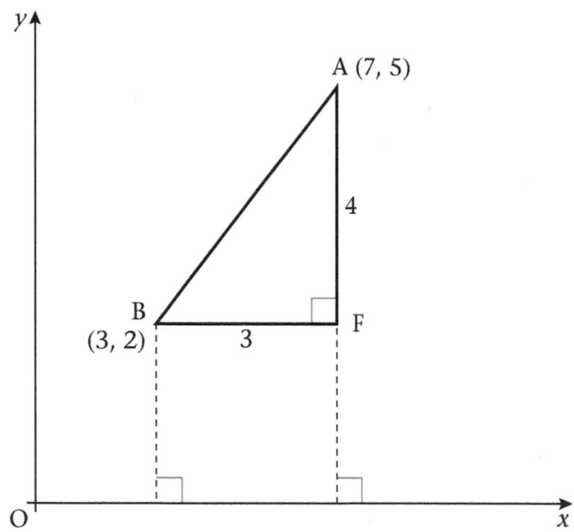

Fig. 6.13

In Fig. 6.14, the coordinates of the points P and Q are (a, b) and (c, d) respectively, Let the length of PQ be l. Then $PR = c - a$, and $QR = d - b$. Using Pythagoras' theorem

$$(PQ)^2 = (PR)^2 + (QR)^2$$
$$l^2 = (c - a)^2 + (d - b)^2$$
$$l = \sqrt{(c - a)^2 + (d - b)^2}$$

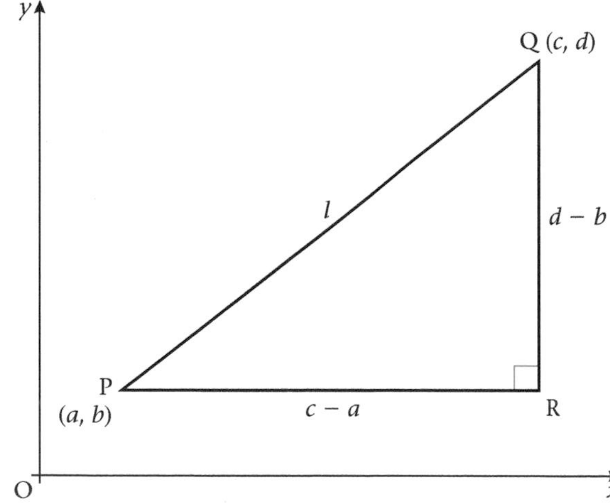

Fig. 6.14

Using this formula in Example 9,

$$AB = \sqrt{(7 - 3)^2 + (5 - 2)^2} = \sqrt{4^2 + 3^2}$$
$$= \sqrt{16 + 9} = \sqrt{25} = 5 \text{ units}$$

Example 10

Find the distance between each pair of points (a) (7, 2) and (1, 6), (b) (3, −3) and (−2, 5).

Using $l = \sqrt{(c - a)^2 + (d - b)^2}$

(a) $l = \sqrt{(7 - 1)^2 + (2 - 6)^2}$
$= \sqrt{6^2 + (-4)^2} = \sqrt{36 + 16}$
$= \sqrt{52} = 2\sqrt{13}$
$= 2(3.61) = 7.22$ units

(b) $l = \sqrt{(3 - (-2))^2 + (-3 - 5)^2}$
$= \sqrt{5^2 + (-8)^2} = \sqrt{89}$
$= 9.43$ units

Exercise 6d

Find the distance between the points in each of the following pairs.

1. (3, 4) and (1, 2)
2. (2, −5) and (3, 6)
3. (−3, 4) and (−1, 2)
4. (2, 5) and (−3, 6)
5. (3, 4) and (−1, −2)
6. (−2, −5) and (3, −6)

Example 11

P and Q are the points (4, 7) and (−1, −5) on a straight line. Find (a) the coordinates (X, Y) of the mid-point, M, of PQ, (b) the distance PQ.

Either

From Fig. 6.15, M is the point $(\frac{3}{2}, 1)$

and $(PQ)^2 = 5^2 + 12^2 = 169$

$PQ = 13$ units

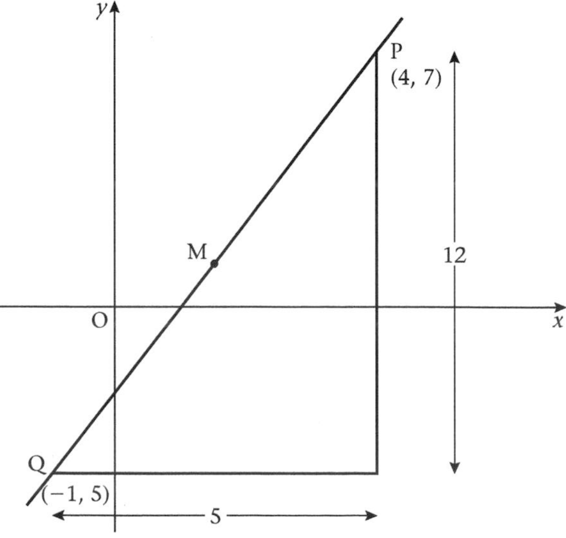

Fig. 6.15

or

using $X = \dfrac{a+c}{2}$ and $Y = \dfrac{b+d}{2}$

$$X_M = \frac{4 + (-1)}{2} = \frac{3}{2}$$

$$Y_M = \frac{7 + (-5)}{2} = 1$$

using $l = \sqrt{(c-a)^2 + (d-b)^2}$

$$PQ = \sqrt{(4 - (-1))^2 + (7 - (-5))^2}$$
$$= \sqrt{(4+1)^2 + (7+5)^2}$$
$$= \sqrt{5^2 + 12^2} = \sqrt{169} = 13 \text{ units}$$

Exercise 6e

1. Calculate, for the line joining each of the following pairs of points,
   (i) the gradient of the line,
   (ii) the equation of the line,
   (iii) the coordinates of the mid-point,
   (iv) the length of the line.
   (a) (3, 2) and (7, 4)
   (b) (3, −8) and (−1, 4)
   (c) (−4, −1) and (4, 1)
   (d) (−6, 8) and (2, −7)
   (e) (−3, −2) and (−1, 6)
   (f) (−3, −2) and (−7, −4)

2. Find the length of the hypotenuse of a right-angled triangle PQR when the coordinates of the vertices are P(−2, 7), Q(3, 2) and R(0, 1).

3. The coordinates of the vertices of ABC are A(3, 0), B(9, 8) and C(9, −2). (a) Show that the triangle is isosceles. (b) Find the area of △ABC.

4. The vertices of △ABC are A(7, 7), B(−4, 3) and C(2, −5), Calculate the length of (a) the longest side of △ABC, (b) the line AM, where M is the mid-point of the side opposite A.

5. A straight line cuts the x-axis at P and the y-axis at R. The gradient of the line RP is $-\frac{3}{2}$ and the line passes through the point (2, 3). Find (a) the equation of the line RP, (b) the intercept on the y-axis, (c) the area of △ROP, where O is the origin.

Summary

Given that (x_1, y_1) and (x_2, y_2) are the coordinates of any two points P and Q on a straight line, m, the gradient of the line, is given by

$$m = \frac{y_2 - y_1}{x_2 - x_1}$$

The coordinates of the **mid-point** of PQ are

$$\left(\frac{(x_1 + x_2)}{2}, \frac{(y_1 + y_2)}{2}\right)$$

The length of PQ is given by

$$l = \sqrt{(x_1 - x_2)^2 + (y_1 - y_2)^2}$$

Given two straight lines with gradients m_1 and m_2, then if $m_1 = m_2$ the lines are parallel.
if $m_1 m_2 = -1$ the lines are perpendicular to each other.

Practice exercise P6.1

Find the gradient of each of the lines joining the following pairs of points.

1. (3, 5) and (5, 9)
2. (4, 2) and (7, 1)
3. (−3, 1) and (3, −2)
4. (2, 3) and (−1, −1)
5. (2, 5) and (6, 5)
5. (2, −1) and (2, 3)
7. (2, −1) and (2, 1)
6. (3, 4) and (4, −2)
9. (0, 6) and (−2, −2)
10. (1, −2) and (0, 2)

Practice exercise P6.2

Write down the gradients of the lines represented by the sketches in Fig. 6.16.

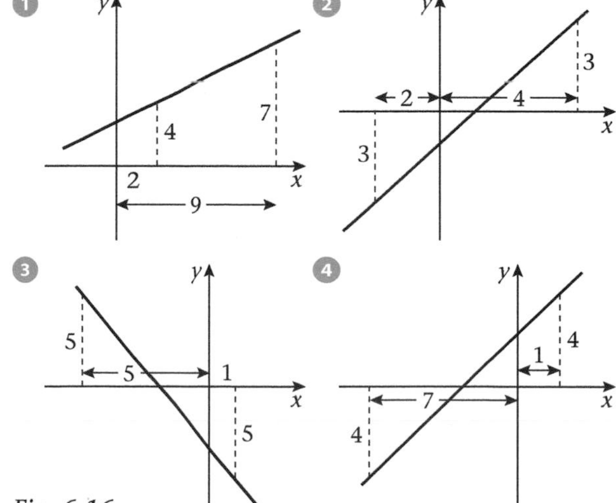

Fig. 6.16

Practice exercise P6.3

1. Show that the line passing through (1, 7) and (2, 10) has a gradient equal to 3.
2. Find OP, the intercept on the y-axis of the line of gradient 3 that passes through the point (1, 7).
3. Show that the line of gradient 3 through (2, 10) intersects the y-axis at the same point P.

Practice exercise P6.4

1. Draw on the same axes the straight lines that pass through each of the following pairs of points.
   (a) (5, 3) and (1, −5)
   (b) (2, 6) and (1, 4)
   (c) (−2, −3) and (2, 5)
2. For each of the lines in question 1, find the point of intersection with the y-axis.
3. For each of the lines in question 1, calculate the gradient of the line.
4. What do you notice about the lines in question 1?

Practice exercise P6.5

1. Draw on the same axes each pair of straight lines passing through the following pairs of points.
   (a) (5, 3) and (1, −5)
   and (0, 2) and (6, −1)
   (b) (−1, −3) and (2, 6)
   and (4, 5) and (1, 6)
2. Calculate the gradient of the line joining each pair of points in question 1.
3. For each pair of lines in question 1, state the relationship between the gradients of the two lines.
4. What do you notice about each pair of lines in question 1?

Practice exercise P6.6

1. Using Fig. 6.17, find for each line
   (i) the gradient,
   (ii) the intercept on the y-axis.
2. Are there any parallel lines in Fig. 6.17? Give a reason for your answer.
3. Are there any perpendicular lines in Fig. 6.17? Give a reason for your answer.

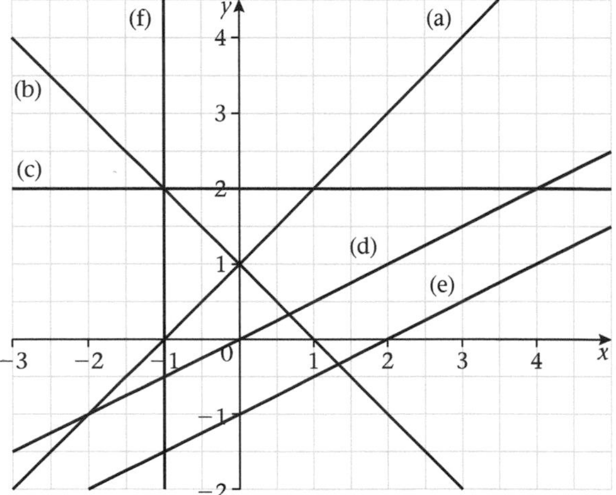

Fig. 6.17

Practice exercise P6.7

1. Draw x- and y-axes using a scale of 1 cm : 1 unit on each axis.
   On the same axes draw the line that passes through each of the following pairs of points.
   (a) (−1, −1) and (3, 3)
   (b) (3, 4) and (3, −1)
   (c) (2, −1) and (4, 1)
   (d) (1, 0) and (−2, 3)
   (e) (−2, −4) and (4, −4)
   (f) (0, 5) and (5, 0)
2. For each of the lines in question 1,
   (i) calculate the gradient,
   (ii) state the intercept on the y-axis.
3. What can you say about the lines in question 1?

Practice exercise P6.8

1. (a) Draw the graph of the line $4x + 2y = 5$.
   (b) Calculate the gradient by using two points on the line.
2. (a) Write down the gradient of the line, corresponding to each of the following relations.
   (i) $y = 7x$ (ii) $y - 5x = 0$
   (iii) $2y + x = 0$ (iv) $\frac{y}{x} = m$
   (b) Which of the lines (i), (ii) or (iii) has the greatest gradient and which has the least?
   (c) Sketch each of the lines.

Practice exercise P6.9

Write the equation of the line passing through each of the following pairs of points.

1. (2, 5) and (−2, 6)
2. (3, −2) and (−1, 1)
3. (4, −1) and (0, 2)
4. (−1, 3) and (−2, −3)
5. (0, 0) and (3, 7)
6. (−4, −3) and (3, −6)
7. (7, 2) and (−9, 7)
8. (3, 5) and (6, 8)
9. (0, 0) and (3, −7)
10. (−1, 4) and (5, −2)

Find the equation of the line which passes through the point

11. (4, 8) and has a gradient of 3.
12. (0, 0) and has a gradient of $2\frac{1}{2}$.
13. (−2, 6) and has a gradient of −1.
14. (0, −5) and has a gradient of −2.
15. (3, −1) and has a gradient of 3.

Practice exercise P6.10

1. The equation of the line ℓ is $y = 6 - \frac{1}{2}x$.
   (a) (i) State the gradient of the line ℓ.
   (ii) State the intercept on the y-axis.

   A line L passes through the point $(-\frac{1}{2}, 6)$ and is perpendicular to the line ℓ.
   (b) Find the equation of the line L.
2. The line ℓ passes through the points (0, 0) and (7, −2).
   (a) Find the equation of the line ℓ.

   The line L is parallel to the line ℓ and crosses the x-axis at (8, 0).
   (b) (i) Find the equation of the line L.
   (ii) Find the coordinates of the point where L crosses the y-axis.

Practice exercise P6.11

For each of the following straight lines, write down
(a) the gradient of the line,
(b) the intercept on the y-axis.

1. $y = 2x$
2. $y = 2x + 3$
3. $2y = 4x - 3$
4. $2y + x = 0$

Practice exercise P6.12

Find the coordinates of the mid-point of the line joining each of the following pairs of points.

1. (−1, 5) and (3, 3)
2. (3, 4) and (1, 2)
3. (2, 5) and (−3, 6)
4. (0, 1) and (−2, 3)
5. (−3, 4) and (−1, 2)
6. (0, 5) and (5, 0)
7. (2, 5) and (−2, 6)
8. (3, −2) and (−1, 1)
9. (4, −1) and (0, 2)
10. (−1, 3) and (−2, −3)

Practice exercise P6.13

1. In the triangle TOF shown in Fig 6.18, O is the origin and OF = X.

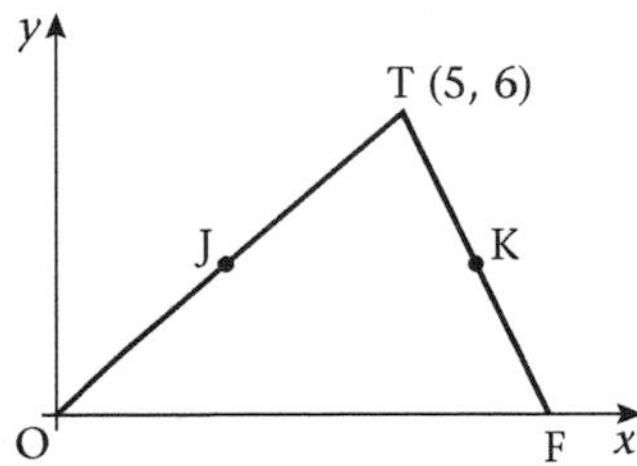

Fig. 6.18

   (a) find, in terms of X where necessary, the coordinates of the mid-point
   (i) J of TO, (ii) K of TF.

(b) Comment on the y-coordinates of J and K.
(c) Find the coordinates of the mid-points of TO and TF when the coordinates of T are (x_1, y_1).
(d) What can you deduce about the line joining J and K?

2 A parallelogram OABC is drawn as in Fig. 6.19, with coordinates O(0, 0), A(4, 0) and B(6, 3).

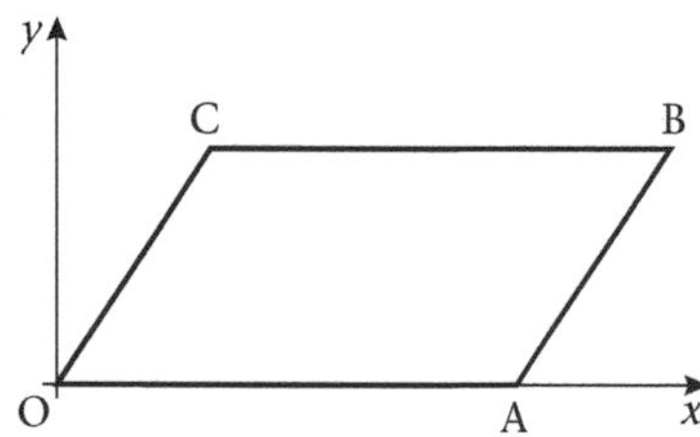

Fig. 6.19

(a) Find the coordinates of C.
(b) Work out the coordinates of the mid-point of
(i) OB, (ii) CA.
(c) Comment on the special property of a parallelogram illustrated.
(d) If in another parallelogram OABC, the coordinates of A are (5, 0) and of C are (1, 3), find the coordinates of B.
(e) In a third parallelogram OABC, the coordinates of A are (6, 0) and the coordinates of the point of intersection of the diagonals are (2, 2). Find the coordinates of B and of C.

Practice exercise P6.14

Find the length of the line joining each of the following pairs of points.

1 (−1, −1) and (3, 3)
2 (3, 4) and (3, −1)
3 (2, −1) and (4, 1)
4 (1, 0) and (−2, 3)
5 (−2, −4) and (4, −4)

Practice exercise P6.15

For the line joining each of the following pairs of points, calculate
(a) the gradient of the line,
(b) the equation of the line,
(c) the coordinates of the midpoint,
(d) the length of the line.

1 (−1, 4) and (3, −8)
2 (−3, 4) and (4, −12)
3 (8, 7) and (2, 5)
4 (2, 5) and (4, −1)
5 (5, 3) and (1, −5)
6 (a) Prove that the distance of the point (x_1, y_1) from the origin is $\sqrt{x_1^2 + y_1^2}$.
(b) (i) Explain why points which satisfy the equation $\sqrt{x_1^2 + y_1^2} = 10$ lie on a circle.
(ii) State the coordinates of the centre of the circle.
(iii) Find the radius of the circle.

Chapters 1–6

Revision exercise 1 (Chapters 1, 4)

1 The values of p and q are connected by the equation $y = x^2 + px + q$. When $x = -1$, $y = 0$ and when $x = -2$, $y = 0$. Find the values of p and q.

2 In each of the following, regroup the given terms then factorise.
(a) $3ay - 2bx + 2xy - 3ab$
(b) $ds + rt - dt - rs$
(c) $pq - pr + 8q - 8r$
(d) $5x + ky + 5y + kx$

3 Factorise the following.
(a) $x^2 - 3x - 4$
(b) $6y^2 + 11y + 3$
(c) $3a^2 - 11a + 8$
(d) $2d^2 + 7d - 15$

4 Solve the following equations.
(a) $(d - 5)(3d - 2) = 0$
(b) $(2m - 1)(m + 4) = 0$

5 Factorise the following.
(a) $a^2 + 3a$
(b) $9p^2 - 12pq + 4q^2$
(c) $16x^2 - 1$
(d) $2y^2 + 8y + 8$

6 Solve the following equations.
(a) $x^2 - 5x = 0$
(b) $2a^2 + 11a + 5 = 0$
(c) $6t^2 - 5t - 4 = 0$
(d) $x^2 + \frac{3}{2}x - 1 = 0$

7 Solve the equation $2x^2 + 6x + 1 = 0$,
(a) by completing the square,
(b) by formula.

8 Table R1 gives corresponding values of x and y for the relation $y = 2x^2 - 5x - 6$.

Table R1

x	−2	−1	0	1	2	3	4
y	12	1	−6	−9	−8	−3	6

(a) Use scales of 2 cm to 1 unit on the x-axis and 1 cm to 1 unit on the y-axis and draw the graph of $y = 2x^2 - 5x - 6$.
(b) Use your graph to find (i) the least value of y, (ii) the solutions of $2x^2 - 5x - 6 = 0$.

9 Twice a certain whole number subtracted from 3 times the square of the number leaves 133. Find the number.

10 The base of a triangle is 3 cm longer than its corresponding height. If the area is 44 cm^2, find the length of its base.

Revision test 1 (Chapters 1, 4)

1 $(2x + 3)$ is a factor of $6x^2 + x - 12$. The other factor is
A $(2x - 3)$ B $(3x + 4)$
C $(3x - 4)$ D $(4x - 9)$

2 Find the roots of the equation $x^2 + 12x - 28 = 0$. The greater of the two roots is
A 14 B 2 C −2 D −14

3 Find the value of k such that $x^2 - \frac{3}{4}x + k$ is a perfect square.
A $\frac{3}{2}$ B $\frac{3}{8}$ C $\frac{9}{64}$ D $\frac{9}{16}$

4 If $m^2 + km + \frac{64}{9}$ is a perfect square, then $k =$
A $\frac{2}{3}$ B $\frac{4}{3}$ C $\frac{8}{3}$ D $\frac{16}{3}$

5 The straight line $x + y = 2$ and the curve $x^2 + y^2 = 10$ intersect at the point $(3, -1)$. They also intersect at the point
A $(-3, 1)$ B $(-3, -1)$
C $(-1, 3)$ D $(1, -3)$

6 Factorise the following, simplifying where possible.
(a) $3a^2 + a(2a + b)$
(b) $(5x - 2y)(a - b) - (2x - y)(a - b)$
(c) $a^2 + 7a + 6$
(d) $2b^3 - 22b^2 + 56b$
(e) $15c^2 + 31c + 10$
(f) $8d^2 + 37d - 15$

7 Solve the following quadratic equations.
(a) $x^2 - 4x = 0$
(b) $y^2 - 10y + 16 = 0$
(c) $3x^2 + 14x + 8 = 0$
(d) $2x^2 - 11x = 21$

8 Find x.
(a) $(x - 3)^2 = 1\frac{7}{9}$ (b) $2x^2 - 18 = 0$
(c) $(x + 5)^2 = 8$ (d) $x^2 + 2x = 1$
(Use square-root tables or a calculator where necessary.)

9 The area of a rectangular plot of land is 216 m². Its length is one and a half times its breadth. Find the length and breadth of the plot of land.

10 Using a graphical method, or otherwise, solve the simultaneous equations $y = x^2$, $y = 1 + x$.

Revision exercise 2 (Chapters 2, 3)

1 $A = \begin{pmatrix} 2 \\ -1 \end{pmatrix}$ and $B = \begin{pmatrix} -3 \\ 4 \end{pmatrix}$. Find 2A + B.

2 $P = \begin{pmatrix} 2 & -1 \\ 0 & 1 \end{pmatrix}$ and $Q = \begin{pmatrix} -1 & 0 \\ 3 & 1 \end{pmatrix}$.
Show that 3(P + Q) = 3P + 3Q.

3 Find the following.
(a) $\begin{pmatrix} 3 & 4 \\ -1 & 2 \end{pmatrix}\begin{pmatrix} -2 \\ 3 \end{pmatrix}$
(b) $\begin{pmatrix} 2 & 3 \\ 1 & -2 \end{pmatrix}\begin{pmatrix} -1 & 0 \\ 2 & 1 \end{pmatrix}$

4 In each part of Fig. R1 calculate the lengths marked x and y, giving the answers in surd form with rational denominators. Assume that all dimensions are cm.

(a)
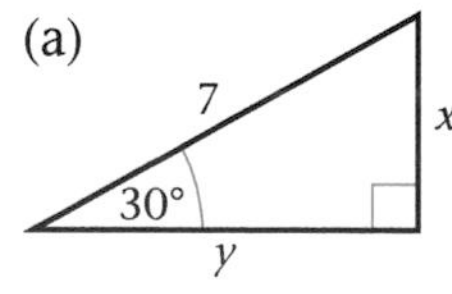

(b)
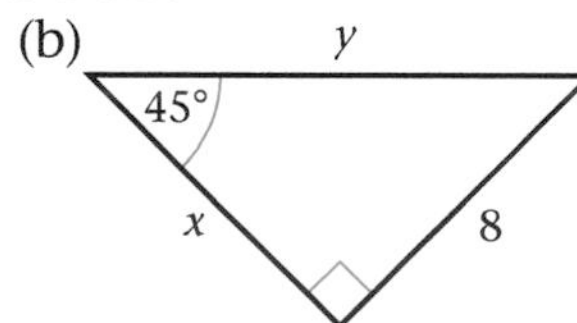

(c)
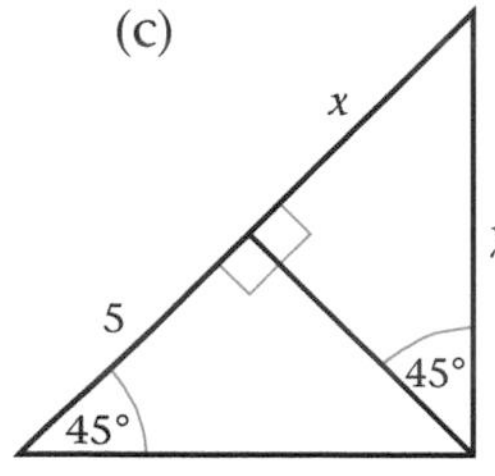

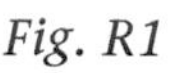

(d)

Fig. R1

5 Three teams L, M and N play in a competition. Each team plays 6 matches. L wins 3 matches and loses 1. M loses 2 and draws 2. Write down, in a matrix, the results for the three teams.

6 $P = \begin{pmatrix} 3 & a \\ b & -1 \end{pmatrix}$ and $Q = \begin{pmatrix} -2 & 2 \\ 1 & 3 \end{pmatrix}$.
If $P + Q = \begin{pmatrix} 1 & 0 \\ 0 & 2 \end{pmatrix}$, find the values of a and b.

7 Use 3-figure tables or a calculator to find, to 2 decimal places, the value of
(a) sin 138.2° (b) cos 151.7°
(c) cos 263° 24′ (d) sin 282°36′
(e) tan 325.4° (f) tan 220° 48′

8 Find the values of x for $-360° < x° < 360°$, if
(a) $\sin x = -0.530$
(b) $\cos x = 0.225$
(c) $\tan x = \pm 2.747$

9 A chord of a circle subtends an angle of 60° at the centre of a circle radius 7 cm. Find the perimeter of the minor segment of the circle.

10 In an equilateral triangle ABC, AD is perpendicular to BC, Prove that $AD^2 = \frac{3}{4}BC^2$.

Revision test 2 (Chapters 2, 3)

1 The order of the matrix $\begin{pmatrix} 1 & 2 \\ -1 & 0 \end{pmatrix}$ is
A 2 B 1 × 2
C 2 × 1 D 2 × 2

2 A diagonal of a rectangle is 4 cm long and makes an angle of 60° with one side. The length, in cm, of a longer side of the rectangle is
A $2\sqrt{2}$ B $2\sqrt{3}$ C 4 D $4\sqrt{2}$

3 If $P = \begin{pmatrix} 1 & 2 \\ -2 & 0 \end{pmatrix}$, 3P =
A $\begin{pmatrix} 3 & 2 \\ -6 & 0 \end{pmatrix}$ B $\begin{pmatrix} 3 & 6 \\ -2 & 0 \end{pmatrix}$
C $\begin{pmatrix} 3 & 2 \\ -6 & 0 \end{pmatrix}$ D $\begin{pmatrix} 3 & 6 \\ -6 & 0 \end{pmatrix}$

4 $X = \begin{pmatrix} 3 & -1 \\ 2 & 1 \end{pmatrix}$ and $Y = \begin{pmatrix} 0 & 4 \\ -3 & -1 \end{pmatrix}$.
Then X − Y =

A $\begin{pmatrix} 3 & 3 \\ -1 & 0 \end{pmatrix}$ B $\begin{pmatrix} 3 & 13 \\ -3 & 7 \end{pmatrix}$

C $\begin{pmatrix} 3 & -5 \\ 5 & 2 \end{pmatrix}$ D $\begin{pmatrix} 0 & -4 \\ -6 & -1 \end{pmatrix}$

5 If $\cos\theta = -0.5$ and $0° \leqslant \theta \leqslant 180°$, then $\theta =$
A 30° B 60° C 120° D 150°

6 Given that $\tan(180° - \theta) = \frac{5}{12}$
(a) find the value of $\cos\theta$,
(b) show that $\sqrt{1 + \tan^2\theta} = \frac{13}{5}$.

7 $M = \begin{pmatrix} 1 & 3 \\ 2 & 0 \end{pmatrix}$ and $N = \begin{pmatrix} 0 & -1 \\ 1 & 2 \end{pmatrix}$
Find (a) M + 2N (b) MN

8 In Fig. R2, $A\hat{B}D = A\hat{E}C = 90°$ and AB = 10 cm.
(a) Calculate BC, AC, CD, CE and $C\widehat{A}D$.
(b) Hence write down a numerical expression for sin 15°.
(Leave answers in terms of $\sqrt{2}$ and $\sqrt{3}$ where necessary).

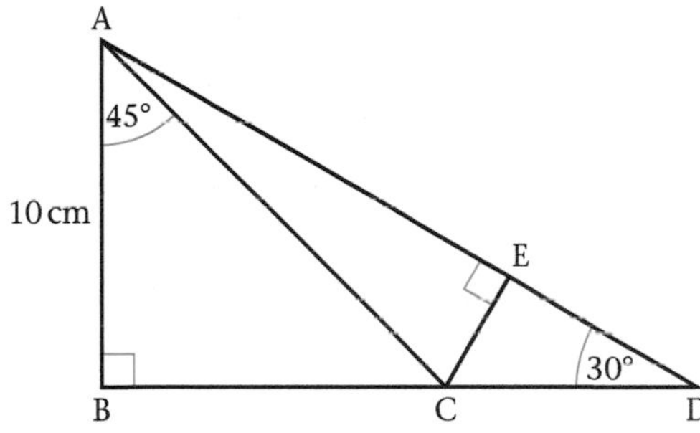

Fig. R2

9 (a) Given that
$$2x - y = 7$$
and $x + 3y = 0$,
write the above equations as a single matrix equation in the form
$$P\begin{pmatrix} x \\ y \end{pmatrix} = N.$$
(b) Find the value of the determinant of P.
(c) Write down the inverse of P.
(d) Use the inverse of P to find the values of x and y that satisfy the given equations.

10 A pendulum consists of a mass hanging at the end of a string 18 cm long. Find the vertical height through which the mass rises and falls as the pendulum swings through 30° on each side of the vertical. (Use the value 1.7 for $\sqrt{3}$.)

Revision exercise 3 (Chapters 5, 6)

1 For each of the following pairs of lines, state whether they are parallel or perpendicular
(a) $y = 2x + 3, y = 2x - 9$
(b) $y = 4x - 1, y = 3 - \frac{x}{4}$
(c) $y = 3 - 2x, y + 2x - 8 = 0$
(d) $3x - 4y + 9 = 0, 8x + 6y - 9 = 0$

2 For each of the following pairs of points, calculate
(i) the distance between the points,
(ii) the mid-point of the line joining the points,
(iii) the gradient of the line joining the points.
(a) (2, 9) and (8, 3)
(b) (−3, 4) and (−4, 3)
(c) (−6, 2) and (8, −5)

3 In Fig. R3, chord SR is parallel to diameter PQ. If angle $P\widehat{Q}R = 56°$, find a, b, c, d, e, giving reasons for your answers.

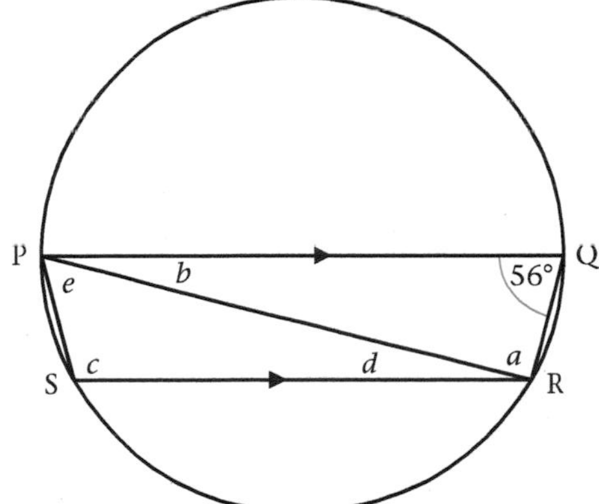

Fig. R3

4 In Fig. R4, TA and TD are tangents to the circle ABCD at A and D.
(a) Name three angles equal to $T\widehat{A}D$.
(b) Name an angle equal to $C\widehat{A}T$.
(c) If $T = z°$, express $A\widehat{C}D$ in terms of z.

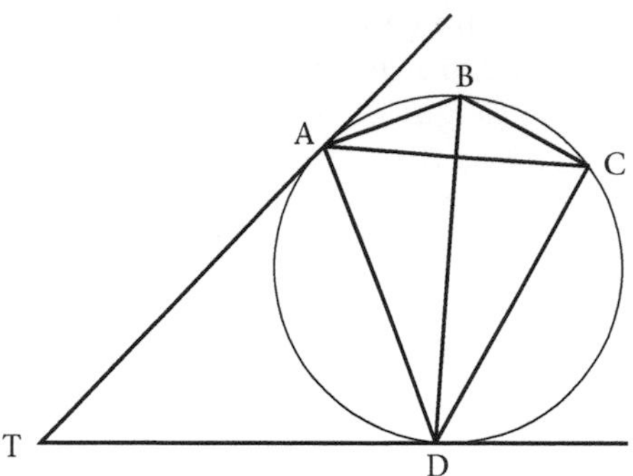

Fig. R4

5 In Fig. R4, if BD is a diameter of the circle and $B\widehat{D}C = 29°$, calculate $C\widehat{A}D$.

6 In Fig. R5, QR is a tangent to the circle and CA is parallel to QR. Show that triangle ACQ is an isosceles triangle.

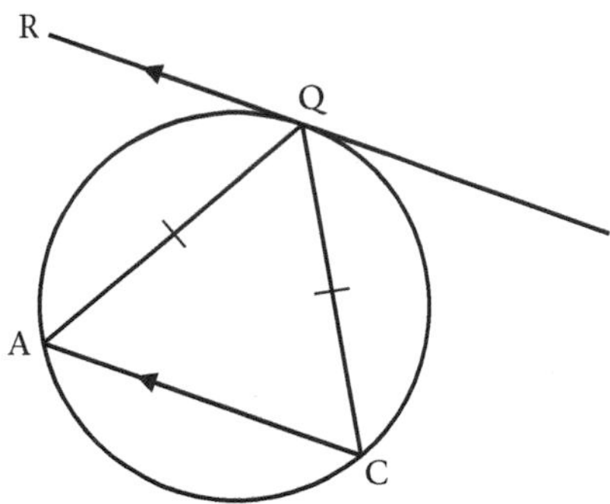

Fig. R5

7 The coordinates of the vertices of the triangle ABC are A(2, 2), B(8, 6) and C(12, 0).
 (a) Show that triangle ABC is a right-angled triangle.
 (b) Calculate the length of the hypotenuse of the triangle.

8 The line $y = 2x - 1$ intersects at P(1, 1) with a line passing through the point Q(9, −3).
 (a) Show that the line PQ is perpendicular to the line $y = 2x - 1$.
 (b) Calculate the equation of the line PQ.

9 In Fig. R6, AT is a tangent to circle QRST, TR is a diameter, QS||AT and $A\widehat{T}Q = 73°$.
 (a) Calculate the angles of quadrilateral QRST.
 (b) What kind of quadrilateral is QRST?

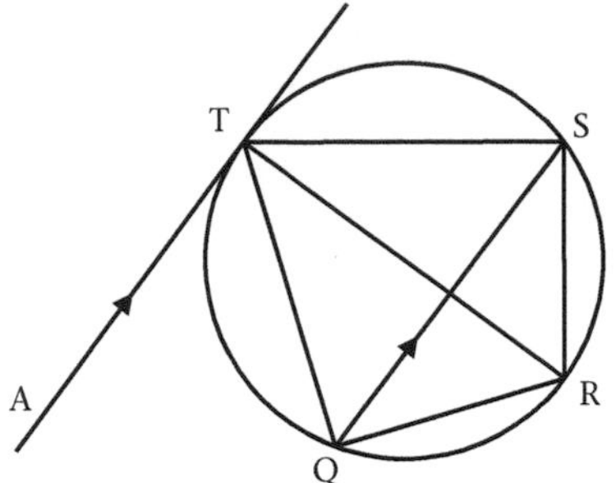

Fig. R6

Revision test 3 (Chapters 5, 6)

1 The gradient of the line through the points (−10, −6) and (11, 8) is
 A −2 B $\frac{2}{3}$ C $\frac{3}{2}$ D 2

2 The coordinates of P and Q are (−3, 2) and (5, 7). M is the mid-point of PQ. The coordinates of M are
 A $(-1\frac{1}{2}, \frac{5}{2})$ B $(-1, \frac{9}{2})$
 C $(1, \frac{5}{2})$ 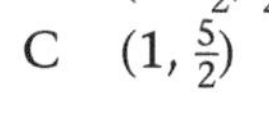 D $(1, \frac{9}{2})$

3

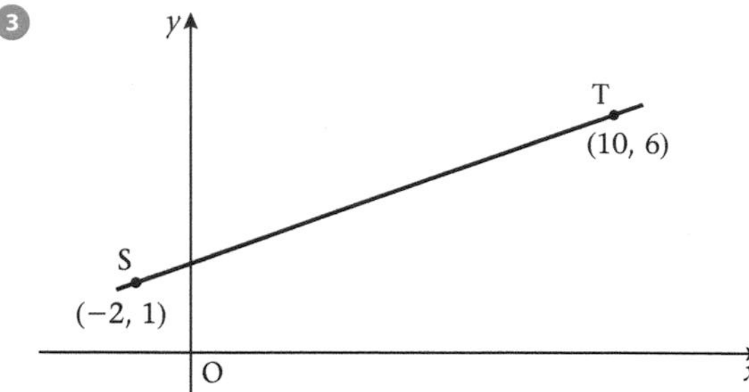

Fig. R7

From Fig. R7, the distance ST is
 A 11 B 13 C 15 D 16

In Fig. R8, XY is a tangent to circle PQRST at T.

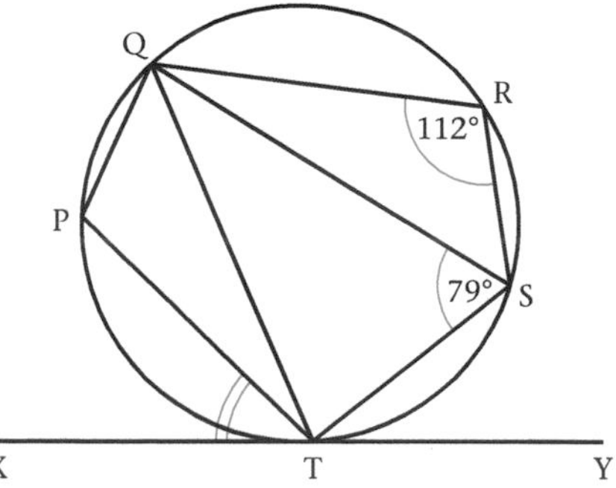

Fig. R8

Use Fig. R8 to answer questions 4 and 5.

4 Which of these angles at Q equals $P\hat{T}X$?
A $P\hat{Q}T$ B $T\hat{Q}S$ C $S\hat{Q}R$ D $T\hat{Q}R$

5 If $Q\hat{R}S = 112°$, and $Q\hat{S}T = 79°$, then $S\hat{T}Y$ is
A 11° B 22° C 33° D 44°

6 Line l passes through the points (0, 0) and (7, −2). Line m is parallel to l and crosses the y-axis at (0, 8). Find
(a) the equation of line l,
(b) the equation of m,
(c) the coordinates of the point where m crosses the x-axis.

7 Draw the graph of $y = 5 - \frac{1}{2}x^2$ for values of x from −4 to 4. Estimate the gradient at the points where $x = 2, -1, -3$.

8 In Fig. R9, O is the centre of the circle and $O\hat{D}B = 25°$. Calculate, giving reasons, the size of $O\hat{B}F$, $O\hat{B}D$ and $B\hat{F}D$.

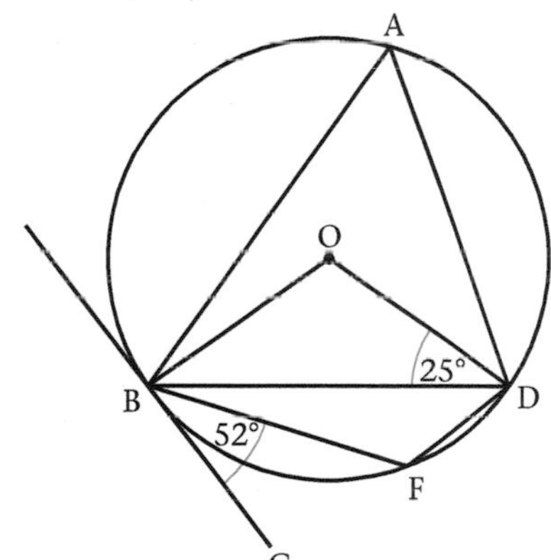

Fig. R9

9 In Fig. R10, TP and TQ are tangents to the circle centre O. Calculate, giving reasons, the size of angle $P\hat{T}Q$.

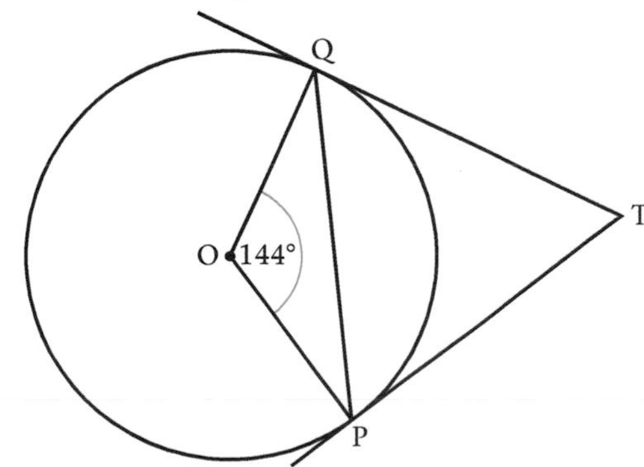

Fig. R10

10 In Fig. R11, PQ is a tangent to circle ABCT at T. TB is a diameter and AB||TC.
(a) If $C\hat{A}B = a$, express $C\hat{T}P$ and $A\hat{T}Q$ in terms of a.
(b) Hence or otherwise show that AC and BT intersect at the centre of the circle.

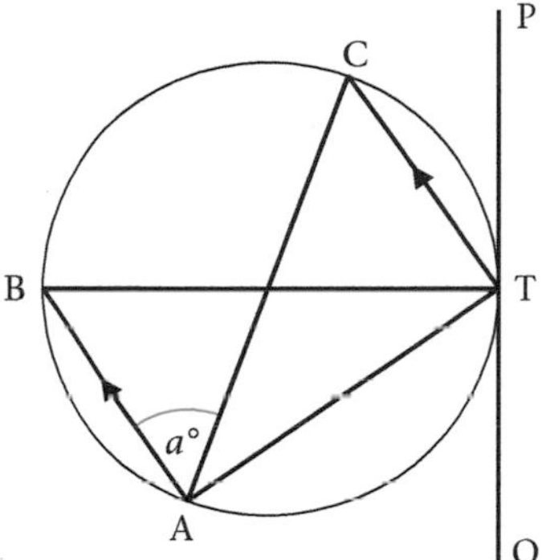

Fig. R11

General revision test A (Chapters 1–6)

1 The number of elements in a 6 by 2 matrix is
A 3 B 4 C 8 D 12

2 Which of the following are factors of $6a^2 - 2ab - 3ab + b^2$?
I $(3a - b)$, II $(2a + b)$, III $(2a - b)$
A I only B I and II only
C I and III only D II and III only

3 Given that $\cos(90° + \theta) = \frac{1}{2}$ then
A $\cos\theta = -\frac{1}{2}$ B $\sin\theta = -\frac{1}{2}$
C $\tan\theta = -\frac{1}{2}$ D $\tan\theta = -1$

4 If $x^2 - 10x - 24 = 0$, then $x = 12$ or
A −2 B −1 C 1 D 2

5 The equation of the straight line passing through (6, −2) and with gradient 5 is
A $2y + 6x = 5$ B $y - 5x = 6$
C $2y = 6x + 5$ D $y = 5x - 32$

6 If $\cos\hat{M} = -0.264$, then $\hat{M} =$
A −15.3° B −74.7°
C −105.3° D −164.7°

7 If $P = \begin{pmatrix} 4 \\ -2 \end{pmatrix} - \begin{pmatrix} -3 \\ -1 \end{pmatrix}$, then P as a single matrix is

A $\begin{pmatrix} 1 \\ -3 \end{pmatrix}$ B $\begin{pmatrix} 7 \\ -1 \end{pmatrix}$ C $\begin{pmatrix} 7 \\ -3 \end{pmatrix}$ D $\begin{pmatrix} -1 \\ 3 \end{pmatrix}$

8 In Fig. R12, MT is a tangent to circle TXYZ.

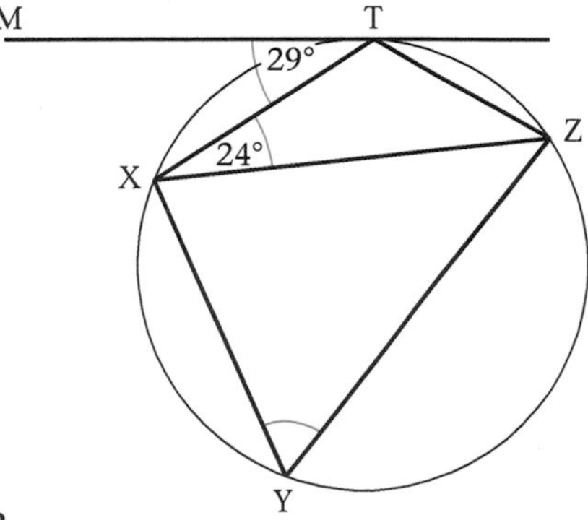

Fig. R12

If $M\hat{T}X = 29°$ and $T\hat{X}Z = 24°$, then $X\hat{Y}Z =$

A 37° B 53° C 61° D 66°

9 In Fig. R13, RT is a tangent to the circle, PQRS, centre O, and chord QR is produced to Z. Which of the following statements is **not necessarily true**?

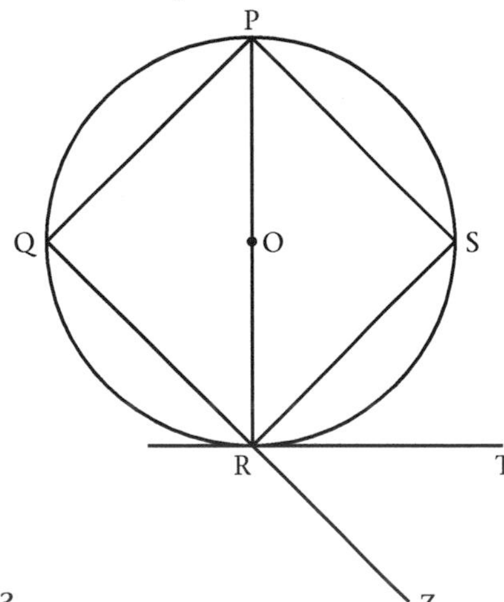

Fig. R13

A $S\hat{R}Z = Q\hat{P}S$ B $S\hat{R}T = R\hat{P}S$
C $S\hat{R}T = P\hat{Q}R$ D $P\hat{R}T = P\hat{S}R$

10 Factorise
(a) $x^3 + x^2 + x + 1$
(b) $2 - b^3 - 2b^2 + b$
(c) $x^2 + px - qx - pq$

11 If $3M = \begin{pmatrix} 0 & -2 \\ 3 & 1 \end{pmatrix} - \begin{pmatrix} -3 & 1 \\ 3 & -2 \end{pmatrix}$ find (a) M, (b) the inverse of M.

12 Find the simplest factors of the following
(a) $3c^3 - 15c$ (b) $4x^2 - 20x + 25$
(c) $2y^4 - 32$

13 Using the values 1.4 for $\sqrt{2}$ and 1.7 for $\sqrt{3}$, find the lengths marked x and y in each part of Fig. R14.
(a) (b)

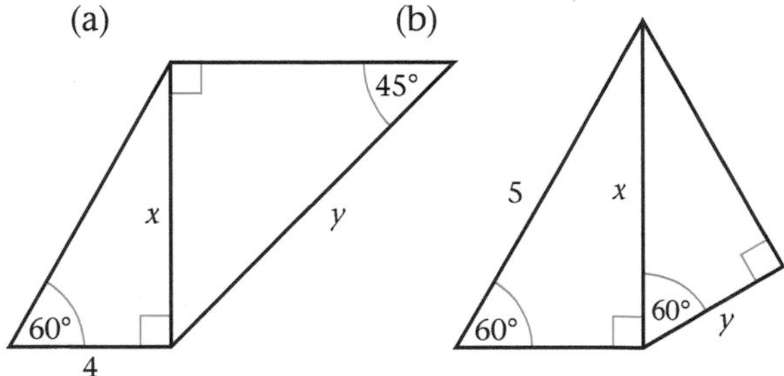

Fig. R14

14 Solve the following equations, giving answers correct to 1 d.p. where necessary.
(a) $2x^2 - x - 21 = 0$
(b) $x^2 + 14x + 8 = 0$
(c) $2p^2 + 9p - 30 = 0$

15 The lines $y = ax + 2$ and $y = mx - 2$ are perpendicular to each other at the point $(-2, 1)$. Find the value of a and hence state the equation of each line.

16 In Fig. R15, XPY is a tangent to the circle, PR is a diameter which cuts chord SQ at A. Angles $P\hat{A}S$ and $P\hat{Q}A$ are 111° and 48° respectively.
Calculate (a) $A\hat{P}Q$ (b) $Y\hat{P}Q$ (c) $S\hat{P}Q$

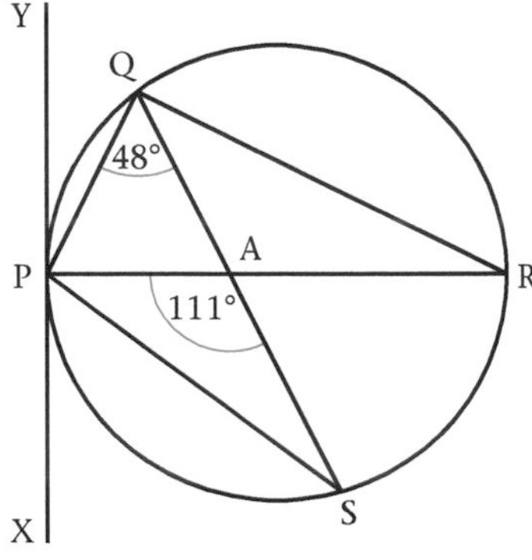

Fig. R15

Chapter 7

Further algebra

Algebraic fractions; fractional indices; algebraic identities; variation; change of subject of formulae

Pre-requisites

- ratios; linear formulae; common factors; algebraic fractions; integral indices

Simplification of fractions

Lowest terms

When simplifying algebraic fractions, always fully factorise the numerators and denominators. It may then be possible to divide the numerator and denominator by any factors which they have in common.

Example 1

Reduce $\frac{6m^2u^2 - 4mu^3}{9m^3u - 4mu^3}$ to its lowest terms.

$$\frac{6m^2u^2 - 4mu^3}{9m^3u - 4mu^3}$$

$$= \frac{2mu^2(3m - 2u)}{mu(9m^2 - 4u^2)} \quad \text{(taking out common factors)}$$

$$= \frac{2mu^2(3m - 2u)}{mu(3m + 2u)(3m - 2u)} \quad \text{(difference of two squares)}$$

$$= \frac{2u}{3m + 2u} \quad \text{(dividing numerator and denomiantor by } mu(3m - 2u))$$

Example 2

Simplify $\frac{a^2 + ax - 6x^2}{2x^2 + ax - a^2}$.

$$\frac{a^2 + ax - 6x^2}{2x^2 + ax - a^2} = \frac{(a - 2x)(a + 3x)}{(2x - a)(x + a)}$$

$$= -\frac{a + 3x}{x + a}$$

In Example 2, notice that $a - 2x = -(2x - a)$

so that $\frac{a - 2x}{2x - a} = \frac{-(2x - a)}{(2x - a)} = -1$.

In general

$$\frac{c}{-d} = -\frac{c}{d} \text{ and } \frac{-c}{d} = -\frac{c}{d}$$

so that

$$\frac{c}{-d} = \frac{-c}{d} = -\frac{c}{d}.$$

Remember also that $\frac{-c}{-d} = \frac{c}{d}$ since two negative quantities, divided one by the other, give a positive result.

In the same way,

$$\frac{a - m}{2m - a} = \frac{a - m}{-(a - 2m)} = -\frac{a - m}{a - 2m}$$

$$\frac{a - m}{2m - a} = \frac{-(m - a)}{2m - a} = -\frac{m - a}{2m - a}$$

$$\frac{a - m}{2m - a} = \frac{-(m - a)}{-(a - 2m)} = \frac{m - a}{a - 2m}$$

Hence if the sign of the numerator *or* the denominator is changed, the sign of the fraction is changed. However, if the sign of *both* the numerator *and* the denominator are changed, the sign of the fraction is unchanged. Because of this, there will sometimes be alternative answers to those given in the examples and exercises in this chapter. (Notice that to change the sign of a term is equivalent to multiplying it by -1.) Remember that the line between the numerator and the denominator behaves like a bracket.

Example 3

Simplify $\frac{a^2 - 5a + 6}{2 - 3a + a^2}$.

$$\frac{a^2 - 5a + 6}{2 - 3a + a^2} = \frac{(a - 2)(a - 3)}{(2 - a)(1 - a)}$$

$$= -\frac{a - 3}{1 - a} \text{ or } \frac{a - 3}{a - 1}$$

$$\text{or } \frac{3 - a}{1 - a} \text{ or } -\frac{3 - a}{a - 1}$$

Exercise 7a

Simplify the following fractions. If there is no simpler form, say so.

1. $\frac{mnu}{nuv}$
2. $\frac{8x^2z}{10xyz}$
3. $\frac{u+m}{u+n}$
4. $\frac{ab+ac}{ad+ae}$
5. $\frac{a^2+ab}{a^2+ac}$
6. $\frac{a^2+b^2}{a+b}$
7. $\frac{h^2-hk}{hk}$
8. $\frac{5d^2nv^3}{15d^3n^2v^4}$
9. $\frac{c^2-cd}{d^2-cd}$
10. $\frac{a^2-b^2}{b^2-ab}$
11. $\frac{x^2+xy}{x^2-y^2}$
12. $\frac{28c^2d^2e^2}{35ce^4}$
13. $\frac{x^2-4x}{x^2-4}$
14. $\frac{c^2-2cd+d^2}{c^2-cd}$
15. $\frac{m^2+2mn+n^2}{m^2-n^2}$
16. $\frac{c^2-2c-15}{c^2-3c-10}$
17. $\frac{d^2-9}{d^2-7d+12}$
18. $\frac{xy-y^2}{(x-y)^2}$
19. $\frac{xy-y^2}{(x+y)^2}$
20. $\frac{h^2+k^2}{(h+k)^2}$
21. $\frac{h^2-k^2}{(h-k)^2}$
22. $\frac{x^2+xy-6y^2}{x^2-3xy+2y^2}$
23. $\frac{15+2x-x^2}{x^2-25}$
24. $\frac{9a^2-m^2}{m^2-2am-3a^2}$
25. $\frac{8-2a-a^2}{2a^2-3a-2}$
26. $\frac{a^2+am-an-mn}{a^2+am+an+mn}$
27. $\frac{(u+w)^2-v^2}{(v+w)^2-u^2}$
28. $\frac{a^2+(b+c)^2}{c^2-(a-b)^2}$
29. $\frac{a(b+c)+(b+c)^2}{b^2-c^2+ab-ac}$
30. $\frac{b^2+ac-ab-bc}{c^2-ac+ab-bc}$

Multiplication and division of fractions

Factorise fully first, then divide the numerator and denominator by any common factors.

Example 4

Simplify $\frac{a^2+2a-3}{a^2-16} \times \frac{a+4}{a^2+8a+15}$.

Given expression

$$= \frac{\cancel{(a+3)}(a-1)}{(a-4)\cancel{(a+4)}} \times \frac{\cancel{a+4}}{(a+5)\cancel{(a+3)}}$$

$$= \frac{a-1}{(a-4)(a+5)}$$

The answer should be left in the form given. Do not multiply out the brackets.

Example 5

Simplify $\frac{m^2-a^2}{m^2+bm+am+ab} \div \frac{m^2-2am+a^2}{cm+bc}$.

To divide by a fraction, multiply by its reciprocal.
Given expression

$$= \frac{m^2-a^2}{m^2+bm+am+ab} \times \frac{cm+bc}{m^2-2am+a^2}$$

$$= \frac{\cancel{(m-a)}\cancel{(m+a)}}{\cancel{(m+b)}\cancel{(m+a)}} \times \frac{c\cancel{(m+b)}}{\cancel{(m-a)}(m-a)}$$

$$= \frac{c}{m-a}$$ (dividing above and below by $(m-a)$, $(m+a)$, $(m+b)$)

Example 6

Simplify

$$\frac{a^2+ab}{a^2-2ab+b^2} \div \frac{a+3b}{a+2b} \times \frac{ab-a^2}{a^2+3ab+2b^2}$$

Given expression

$$= \frac{a^2+ab}{a^2-2ab+b^2} \times \frac{a+2b}{a+3b} \times \frac{ab-a^2}{a^2+3ab+2b^2}$$

$$= \frac{a\cancel{(a+b)}}{\cancel{(a-b)}(a-b)} \times \frac{\cancel{a+2b}}{a+3b} \times \frac{a\cancel{(b-a)}^{-1}}{\cancel{(a+b)}\cancel{(a+2b)}}$$

$$= -\frac{a^2}{(a-b)(a+3b)}$$ (dividing above and below by $(a+b)$, $(a+2b)$, $(a-b)$)

In Example 6, notice that $(a-b)$ divides into $(b-a)$ to give -1. This is because $-1 \times (a-b) = (b-a)$.

Exercise 7b

Simplify the following.

1. $\frac{18ab}{15bc} \times \frac{20cd}{24de}$
2. $\frac{12dn^3}{15cd^3} \div \frac{9c^3n}{10c^2d^2}$
3. $\frac{m+n}{m} \times \frac{mn}{3m+3n}$
4. $\frac{uv}{3u-6v} \times \frac{4u-8v}{u^2v}$
5. $\frac{2a-2b+2c}{8bc} \times \frac{10abc}{5a-5b+5c}$
6. $\frac{18m^2u}{16n^3v^2} \div \frac{24m}{15nu^3} \times \frac{8n^2v^3}{30m^3u}$
7. $\frac{a^2-4}{a^2-3a+2} \div \frac{a}{a-1}$
8. $\frac{m^2-9}{m^2-m-6} \times \frac{m^2+2m}{m^2}$

9 $\dfrac{a^2 - b^2}{ab + a^2} \times \dfrac{2a^3}{ab - a^2}$

10 $\dfrac{3d^2 - 12}{9d^2} \times \dfrac{6d^3}{4d + 8}$

11 $\dfrac{a^2 - b^2}{a^2 + ab} \div \dfrac{2a - 2b}{ab}$

12 $\dfrac{n^2 - 9}{n^2 - n} \times \dfrac{n^2 - 3n + 2}{n^2 + n - 6}$

13 $\dfrac{m^2 - n^2}{m^2 - 2mn + n^2} \div \dfrac{m^2 + mn}{n^2 - mn}$

14 $\dfrac{a^2 - ab - 6b^2}{a^2 + ab - 6b^2} \times \dfrac{a^2 - ab - 2b^2}{a^2 - 2ab - 3b^2}$

15 $\dfrac{5abc^2 - 10abcd}{3b^2c^2 - 6b^2cd} \times \dfrac{12bc^2d}{10acd}$

Addition and subtraction of fractions

Example 7

Simplify $2 + \dfrac{6a^2 + 2b^2}{3ab} - \dfrac{4a - b}{2b}$.

The denominators are $3ab$ and $2b$. The LCM of $3ab$ and $2b$ is $6ab$. Express each fraction in the expression with a common denominator of $6ab$. Expand all the brackets in the numerator. Collect like terms. Simplify the expression by factorising fully and dividing the numerator and denominator by any common factors.

$$2 + \frac{6a^2 + 2b^2}{3ab} - \frac{4a - b}{2b}$$

$$= \frac{2 \times 6ab}{6ab} + \frac{2(6a^2 + 2b^2)}{6ab} - \frac{3a(4a - b)}{6ab}$$

$$= \frac{12ab + 2(6a^2 + 2b^2) - 3a(4a - b)}{6ab}$$

$$= \frac{12ab + 12a^2 + 4b^2 - 12a^2 + 3ab}{6ab}$$

$$= \frac{15ab + 4b^2}{6ab}$$

$$= \frac{b(15a + 4b)}{6ab} = \frac{15a + 4b}{6a}$$

Example 8

Simplify $\dfrac{3}{m^2 + mn - 2n^2} - \dfrac{2}{m^2 - 4mn + 3n^2}$.

Factorise the denominators so that their LCM can be used as the common denominator.

Given expression

$$= \frac{3}{(m - n)(m + 2n)} - \frac{2}{(m - n)(m - 3n)}$$

The LCM of the denominators is $(m - n)(m + 2n)(m - 3n)$.

Given expression

$$= \frac{3(m - 3n) - 2(m + 2n)}{(m - n)(m + 2n)(m - 3n)}$$

$$= \frac{3m - 9n - 2m - 4n}{(m - n)(m + 2n)(m - 3n)}$$

$$= \frac{m - 13n}{(m - n)(m + 2n)(m - 3n)}$$

Example 9

Simplify $\dfrac{x + 4}{x^2 - 3x} - \dfrac{x - 1}{9 - x^2}$.

$$\frac{x + 4}{x^2 - 3x} - \frac{x - 1}{9 - x^2}$$

$$= \frac{x + 4}{x(x - 3)} - \frac{x - 1}{(3 - x)(3 + x)}$$

$$= \frac{x + 4}{x(x - 3)} + \frac{x - 1}{(x - 3)(3 + x)}^{\dagger}$$

$$= \frac{(x + 4)(x + 3) + x(x - 1)}{x(x - 3)(x + 3)}$$

$$= \frac{x^2 + 7x + 12 + x^2 - x}{x(x - 3)(x + 3)}$$

$$= \frac{2x^2 + 6x + 12}{x(x - 3)(x + 3)}$$

$$= \frac{2(x^2 + 3x + 6)}{x(x - 3)(x + 3)}$$

†Notice that the sign in front of the fraction is changed since $(3 - x) = -(x - 3)$. This gives an LCM of $x(x - 3)(x + 3)$.

Example 10

Simplify $\dfrac{3a - 5m}{a^2 - 5am + 6m^2} + \dfrac{1}{a - 2m} - \dfrac{2}{a - 3m}$.

Given expression

$$= \frac{3a - 5m}{(a - 2m)(a - 3m)} + \frac{1}{a - 2m} - \frac{2}{a - 3m}$$

$$= \frac{3a - 5m + (a - 3m) - 2(a - 2m)}{(a - 2m)(a - 3m)}$$

$$= \frac{3a - 5m + a - 3m - 2a + 4m}{(a - 2m)(a - 3m)}$$

$$= \frac{2a - 4m}{(a - 2m)(a - 3m)}$$

$$= \frac{2(a - 2m)}{(a - 2m)(a - 3m)}$$

$$= \frac{2}{a - 3m}$$

Exercise 7c

Simplify the following.

1. $\dfrac{3}{2ab} + \dfrac{4}{3bc}$
2. $5 - \dfrac{a - b}{c}$
3. $\dfrac{3a - b}{5ab} - \dfrac{2b + 3c}{6bc} + \dfrac{3c - 2a}{15ac}$
4. $\dfrac{3}{2(x + y)} - \dfrac{1}{3(x + y)}$
5. $\dfrac{6}{a - 2b} + \dfrac{4}{2b - a}$
6. $3 + \dfrac{2b}{a - b}$
7. $2 - \dfrac{x}{x + 2y}$
8. $\dfrac{7u}{2u + 3v} - 3$
9. $\dfrac{3a}{2a + b} - \dfrac{b}{4a + 2b}$
10. $\dfrac{3mn}{2m^2 + 2n^2} + \dfrac{5mn}{3m^2 + 3n^2}$
11. $\dfrac{1}{4x - 2y} - \dfrac{1}{y - 2x}$
12. $\dfrac{u^2 - v^2}{uv} + \dfrac{v}{u} - \dfrac{3uv - u^2}{v^2}$
13. $\dfrac{2}{a + 1} + \dfrac{3}{a + 2}$
14. $\dfrac{3x}{x - 1} - \dfrac{4}{x + 2}$
15. $\dfrac{3}{m - n} + \dfrac{m + 3n}{(m - n)^2}$
16. $\dfrac{a + b}{(a - 2b)^2} - \dfrac{1}{2b - a}$
17. $\dfrac{3c}{c^2 - d^2} - \dfrac{3d}{d^2 - c^2}$
18. $\dfrac{4m - 9n}{16m^2 - 9n^2} + \dfrac{1}{4m - 3n}$

Example 11

Given that $x : y = 9 : 4$, evaluate $\dfrac{8x - 3y}{x - \frac{3}{4}y}$.

Method 1 If $x : y = 9 : 4$, then $\frac{x}{y} = \frac{9}{4}$.

Divide numerator and denominator of $\dfrac{8x - 3y}{x - \frac{3}{4}y}$ by y.

$$\frac{8x - 3y}{x - \frac{3}{4}y} = \frac{8\left(\frac{x}{y}\right) - 3}{\frac{x}{y} - \frac{3}{4}}$$

Substitute $\frac{9}{4}$ for $\frac{x}{y}$ in the expression.

Value of expression

$$= \frac{8 \times \frac{9}{4} - 3}{\frac{9}{4} - \frac{3}{4}} = \frac{18 - 3}{1\frac{1}{2}} = \frac{15}{1\frac{1}{2}}$$

$$= 15 \div \tfrac{3}{2} = 15 \times \tfrac{2}{3} = 10$$

Method 2 If $x : y = 9 : 4$, then $\frac{x}{y} = \frac{9}{4}$

that is, $4x = 9y$ and $x = \frac{9}{4}y$

Substitute $\frac{9y}{4}$ for x in the expression.

$$\frac{8x - 3y}{x - \frac{3}{4}y} = \frac{8\left(\frac{9y}{4}\right) - 3y}{\frac{9y}{4} - \frac{3y}{4}}$$

$$= \frac{18y - 3y}{\frac{6y}{4}} = 15y \times \frac{4}{6y} = 10$$

Example 12

If $x = \frac{2a + 3}{3a - 2}$, express $\frac{x - 1}{2x + 1}$ in terms of a.

Substitute $\frac{2a + 3}{3a - 2}$ for x in the given expression.

$$\frac{x - 1}{2x + 1} = \frac{\frac{2a + 3}{3a - 2} - 1}{2 \times \frac{2a + 3}{3a - 2} + 1}$$

Multiply the numerator and denominator by $(3a - 2)$. In this way, the value of the given expression does not change.

$$\frac{x - 1}{2x + 1} = \frac{(2a + 3) - (3a - 2)}{2(2a + 3) + (3a - 2)}$$

$$= \frac{2a + 3 - 3a + 2}{4a + 6 + 3a - 2} = \frac{-a + 5}{7a + 4}$$

Exercise 7d

1. If $\frac{x}{y} = \frac{3}{4}$, evaluate $\frac{2x - y}{2x + y}$.
2. Given $p : q = 9 : 5$, evaluate $\frac{15p - 2q}{5p + 16q}$.
3. If $a : b = 5 : 3$, evaluate $\frac{6a + b}{a - \frac{1}{3}b}$.
4. If $a = \frac{d + 1}{d - 1}$, express $\frac{a + 1}{a - 1}$ in terms of d.
5. If $x = \frac{a + 3}{2a - 1}$, express $\frac{2x + 1}{3x - 1}$ in terms of a.
6. If $x = \frac{3m - 5}{3m + 5}$, express $\frac{x - 1}{x + 1}$ in terms of m.
7. If $x = \frac{3w - 1}{w + 2}$, express $\frac{2x - 3}{3x + 1}$ in terms of w.
8. If $X = \frac{2a + 3}{3a - 2}$, express $\frac{X - 1}{2X + 1}$ in terms of a.

Undefined fractions

Let $V = \frac{1}{x - 1}$

When $x = 1$, $V = \frac{1}{1 - 1} = \frac{1}{0}$

Division by zero is impossible. The fraction $\frac{1}{x - 1}$ is said to be **undefined** when $x = 1$.

If the denominator of a fraction has the value zero, the fraction will be undefined. If an expression contains an undefined fraction, the whole expression is undefined.

Example 13

Find the values of x for which the following fractions are not defined.

(a) $\frac{3}{x + 2}$ (b) $\frac{2x + 13}{3x - 12}$ (c) $\frac{x^2 - 2x + 3}{(x + 3)(x - 8)}$

(a) $\frac{3}{x + 2}$ is undefined when $x + 2 = 0$.

If $x + 2 = 0$ then $x = -2$

The fraction is not defined when $x = -2$.

(b) $\frac{2x + 13}{3x - 12}$ is undefined when $3x - 12 = 0$.

If $3x - 12 = 0$ then $3x = 12$

$x = 4$

The fraction is undefined when $x = 4$.

(c) $\frac{x^2 - 2x + 3}{(x + 3)(x - 8)}$ is undefined

when $(x + 3)(x - 8) = 0$

If $(x + 3)(x - 8) = 0$

then either $x + 3 = 0$ or $x - 8 = 0$

i.e. either $x = -3$ or $x = 8$

The fraction is undefined when $x = -3$ or 8.

If part of an expression is undefined, then the whole expression is not defined.

Example 14

Find the values of x for which the expression $\frac{a}{x} - \frac{b}{x^2 + 6x - 7}$ is not defined.

$$\frac{a}{x} - \frac{b}{x^2 + 6x - 7} = \frac{a}{x} - \frac{b}{(x - 1)(x + 7)}$$

The expression is not defined if any of its fractions has a denominator of 0.

$\frac{a}{x}$ is undefined when $x = 0$.

$\frac{b}{(x - 1)(x + 7)}$ is undefined when $(x - 1)(x + 7) = 0$.

If $(x - 1)(x + 7) = 0$

then either $(x - 1) = 0$ or $(x + 7) = 0$

i.e. either $x = 1$ or $x = -7$

The expression is not defined when $x = 0$, or 8. 1 or -7.

Exercise 7e

Find the values of x for which the following expressions are not defined.

1. $\frac{7}{x-3}$
2. $\frac{2x}{4-x}$
3. $\frac{3x+2}{x+7}$
4. $\frac{3+x}{x}$
5. $\frac{6x}{2x-5}$
6. $\frac{y}{20-3x}$
7. $\frac{2a}{x(x+2)}$
8. $\frac{3x+1}{(x+4)(x+3)}$
9. $\frac{7x^2}{(x+1)(x-1)}$
10. $\frac{4}{(x-6)(x-6)}$

Example 15

(a) For what value(s) of x is the expression $\frac{x^2+15x+50}{x-5}$ not defined?

(b) Find the value(s) of x for which the expression is zero.

(a) The expression is not defined when its denominator is zero, i.e. when $x - 5 = 0$
so $x = 5$

(b) Let $\frac{x^2+15x+50}{x-5} = 0$

Multiply both sides by $x - 5$.

$x^2 + 15x + 50 = 0$

$(x + 5)(x + 10) = 0$

either $x + 5 = 0$ or $x + 10 = 0$

i.e. either $x = -5$ or $x = -10$

The expression is zero when $x = -5$ or $x = -10$.

Exercise 7f

Find the values of x for which the following expressions are not defined.

1. $\frac{x^2+12x+36}{x^2-3x-10}$
2. $\frac{x^2-3x-10}{x^2+12x+36}$
3. $\frac{18}{x}+\frac{x^2+1}{x^2-9}$
4. $\frac{a}{x-2}+\frac{b}{x^2-2x}-\frac{c}{x+2}$

Equations with fractions

Example 16

Solve the equation $\frac{1}{3a-1} = \frac{2}{a+1} - \frac{3}{8}$.

The LCM of the denominators is $8(3a - 1)(a + 1)$. To clear fractions, multiply the terms on both sides of the equation by $8(3a - 1)(a + 1)$.

If $\frac{1}{3a-1} = \frac{2}{a+1} - \frac{3}{8}$

$$\left(\frac{1}{3a-1}\right)[8(3a-1)(a+1)]$$

$$= \left(\frac{2}{a+1} - \frac{3}{8}\right)[8(3a-1)(a+1)]$$

then $\frac{1}{\cancel{3a-1}} \times 8(\cancel{3a-1})(a+1)$

$$= \frac{2}{\cancel{a+1}} \times 8(3a-1)(\cancel{a+1})$$

$$- \frac{3}{\cancel{8}} \times \cancel{8}(3a-1)(a+1)$$

i.e. $8(a + 1) = 16(3a - 1) - 3(3a - 1)(a + 1)$

$8a + 8 = 48a - 16 - 3(3a^2 + 2a - 1)$

$8a + 8 = 48a - 16 - 9a^2 - 6a + 3$

$8a + 8 - 48a + 16 + 9a^2 + 6a - 3 = 0$

$9a^2 - 34a + 21 = 0$

$(a - 3)(9a - 7) = 0$

$\therefore a = 3$ or $9a = 7$

$\therefore a = 3$ or $\frac{7}{9}$

Check: If $a = 3$,

$$\text{LHS} = \frac{1}{3a-1} = \frac{1}{9-1} = \frac{1}{8}$$

$$\text{RHS} = \frac{2}{a+1} - \frac{3}{8} = \frac{2}{4} - \frac{3}{8}$$

$$= \frac{1}{2} - \frac{3}{8} = \frac{1}{8}$$

If $a = \frac{7}{9}$,

$$\text{LHS} = \frac{1}{3a-1} = \frac{1}{\frac{7}{3}-1} = \frac{1}{\frac{4}{3}} = \frac{3}{4}$$

$$\text{RHS} = \frac{2}{a+1} - \frac{3}{8} = \frac{1}{1\frac{7}{9}} - \frac{3}{8}$$

$$= \frac{18}{16} - \frac{3}{8} = \frac{9}{8} - \frac{3}{8} = \frac{3}{4}$$

Example 17

Solve the equation $\frac{3}{x^2-5x+6} = \frac{2}{x^2-x-6}$.

Factorise the denominators of the fractions.

$$\frac{3}{(x-2)(x-3)} = \frac{2}{(x-3)(x+2)}$$

Multiply both sides by $(x - 2)(x - 3)(x + 2)$.
Then $3(x + 2) = 2(x - 2)$

$$3x + 6 = 2x - 4$$
$$3x - 2x = -4-6$$
$$x = -10$$

Check: If $x = -10$,

$$\text{LHS} = \frac{3}{x^2 - 5x + 6} = \frac{3}{100 + 50 + 6} = \frac{3}{156} = \frac{1}{52}$$

$$\text{RHS} = \frac{3}{x^2 - x - 6} = \frac{2}{100 + 10 - 6} = \frac{2}{104} = \frac{1}{52}$$

Exercise 7g

Solve the following equations.

1. $\dfrac{4}{w + 3} - \dfrac{3}{w + 2} - 0$
2. $\dfrac{3}{2b - 5} - \dfrac{4}{b - 3} = 0$
3. $\dfrac{3}{a} = a - 2$
4. $\dfrac{7}{3} + \dfrac{2}{e} = e$
5. $m = \dfrac{8}{3m + 2}$
6. $3x - 2 = \dfrac{4}{x - 1}$
7. $\dfrac{x - 2}{x + 4} = x$
8. $\dfrac{a - 4}{7} = \dfrac{2}{3a - 1}$
9. $\dfrac{3n}{2n - 1} = n$
10. $\dfrac{2}{d - 2} = \dfrac{3d}{4d + 12}$
11. $\dfrac{4n - 3}{6n + 1} = \dfrac{2n - 1}{3n + 4}$
12. $\dfrac{2m + 3}{2m + 5} - \dfrac{m - 1}{m - 2} = 0$
13. $\dfrac{3}{c + 2} - \dfrac{2}{2c - 3} = \dfrac{1}{7}$
14. $\dfrac{3}{x - 4} = \dfrac{2}{x - 1} - 4$
15. $\dfrac{1}{2a - 5} + \dfrac{7}{9} = \dfrac{2}{a + 5}$
16. $\dfrac{2}{d + 3} = \dfrac{3}{2d - 1} - \dfrac{4}{15}$
17. $\dfrac{11}{m + 3} = \dfrac{5}{2m} - \dfrac{1}{m - 4}$
18. $\dfrac{1}{2n - 3} + \dfrac{1}{2n + 1} - \dfrac{1}{n - 1} = 0$
19. $\dfrac{2}{u + 2} = \dfrac{2}{u + 1} - \dfrac{2}{u + 4}$
20. $\dfrac{4a - 1}{a + 4} - 2 = \dfrac{2a - 1}{a + 2}$

Compare Examples 16 and 17 with Examples 7, 8, 9 and 10. In Examples 16 and 17, both sides of the **equations** are multiplied by the LCM of the denominators. Hence every denominator becomes 1 and the fractions are cleared. This is valid since both sides of the equation still balance.

However, in Examples 7, 8, 9 and 10, the common denominators must stay in the given expressions, so that the **expressions** have the same values. This is an important difference between **solving equations** with fractions and **simplifying expressions** with fractions.

Exercise 7h

1. Reduce $\dfrac{3}{x} - \dfrac{x}{2} + 5$ to a single fraction.
2. Simplify $\dfrac{3x + 2}{3} - \dfrac{x - 1}{4} - \dfrac{5}{12}$.
3. Simplify $\dfrac{3}{2x - 4} + \dfrac{2}{6 - 3x}$.
4. Simplify $\left(\dfrac{2}{x} - \dfrac{5}{y}\right) \div \dfrac{4}{xy}$.
5. If $a = \dfrac{2m + 1}{2m - 1}$, express $\dfrac{2a + 1}{2a - 1}$ in terms of m.
6. If $x : y = 9 : 1$, evaluate $\dfrac{11x + y}{x + y}$.
7. If $A = \dfrac{3x + 2}{x + 3}$,
   (a) for what value of x is A undefined?
   (b) for what range of values of x is $A < 2$?
8. (a) Simplify $\dfrac{5}{a + 4} - \dfrac{2}{a - 2}$.
   (b) Solve $\dfrac{5}{a + 4} - \dfrac{2}{a - 2} = 0$.
9. (a) If k is a constant not equal to zero, find the value(s) of x for which the expression $\dfrac{k}{x} + \dfrac{b}{x - 3} + \dfrac{c}{x(x - 3)}$ is not defined.
   (b) If x is not equal to any of the values obtained in (a), find the value of k such that $\dfrac{k}{x} + \dfrac{2}{x - 3} - \dfrac{6}{x(x - 3)} = 0$.
10. Simplify $\dfrac{y^2 + 2yz + z^2}{y^2 - z^2}$.
11. Simplify $\dfrac{3}{a^2 - 3a + 2} \div \dfrac{2}{2a^2 - 5a + 2}$ and state the value of a for which the simplified expression is not defined.

12 Simplify

$$\frac{2b}{a^2 - b^2} + \frac{a}{b^2 - ab}$$

13 Simplify $\dfrac{3y}{x^2 - xy - 2y^2} - \dfrac{2y}{x^2 - 2xy} + \dfrac{2x + y}{x^2 + xy}$.

14 Solve the equation

$$\frac{2}{2x - 3} + \frac{3}{2x + 3} = \frac{2}{4x^2 - 9}.$$

Indices

In Books 2 and 3, it was shown that the following laws of indices are true for integral values of a and b and for all values of x.

1 $x^a \times x^b = x^{a+b}$

2 $x^a \div x^b = x^{a-b}$

3 $x^0 = 1$

4 $x^{-a} = \dfrac{1}{x^a}$

5 $(x^a)^b = x^{ab}$

Fractional indices

$\sqrt{x}$ is short for the **square root** of x.

$\sqrt{1}\times \sqrt{x} = x$

Let $\sqrt{x} = x^p$

then, $x^p \times x^p = \sqrt{x} \times \sqrt{x} = x^1$

By the laws of indices

$x^p \times x^p = x^{p+p} = x^{2p}$

or

$x^p \times x^p = (x^p)^2 = x^{p \times 2} = x^{2p}$

$x^{2p} = x^1$

Therefore $2p = 1$

and $p = \frac{1}{2}$

Thus $\sqrt{x} = x^{\frac{1}{2}}$

Similarly, $\sqrt[3]{x}$ is short for the **cube root** of x. For example, $\sqrt[3]{8} = 2$ since $2 \times 2 \times 2 = 8$ and $\sqrt[3]{-27} = -3$ since $-3 \times -3 \times -3 = -27$.

$\sqrt[3]{x} \times \sqrt[3]{x} \times \sqrt[3]{x} = x$

Let $\sqrt[3]{x} = x^r$

then, $x^r \times x^r \times x^r = x^1$

Also

$x^r \times x^r \times x^r = x^{r+r+r} = x^{3r}$

or

$x^r \times x^r \times x^r = (x^r)^3 = x^{r \times 3} = x^{3r}$

$x^{3r} = x^1$

Therefore $3r = 1$

and $r = \frac{1}{3}$

Thus $\sqrt[3]{x} = x^{\frac{1}{3}}$

$x^{\frac{1}{2}} = \sqrt{x}$ and $x^{\frac{1}{3}} = \sqrt[3]{x}$

In general, $x^{\frac{1}{a}} = \sqrt[a]{x}$.

Also $x^{\frac{2}{3}} = x^{2 \times \frac{1}{3}} = (x^2)^{\frac{1}{3}} = \sqrt[3]{x^2}$

or $x^{\frac{2}{3}} = x^{\frac{1}{3} \times 2} = (x^{\frac{1}{3}})^2 = (\sqrt[3]{x})^2$

In general, $x^{\frac{a}{b}} = \sqrt[b]{x^a}$ or $(\sqrt[b]{x})^a$.

Example 18

Simplify (a) $9^{\frac{1}{2}}$, (b) $8^{-\frac{2}{3}}$, (c) $4 \times 4^{-\frac{1}{2}}$, (d) $(\frac{16}{81})^{-\frac{3}{4}}$.

(a) $9^{\frac{1}{2}} = \sqrt{9}\ \pm 3$

or

$9^{\frac{1}{2}} = ((\pm 3)^2)^{\frac{1}{2}} = (\pm 3)^{2 \times \frac{1}{2}} = \pm 3$

(b) $8^{-\frac{2}{3}} = \dfrac{1}{8^{\frac{2}{3}}} = \dfrac{1}{(\sqrt[3]{8})^2} = \dfrac{1}{2^2} = \dfrac{1}{4}$

or

$8^{-\frac{2}{3}} = (2^3)^{-\frac{2}{3}} = (2)^{3 \times \frac{-2}{3}} = 2^{-2}$

$= \dfrac{1}{2^2} = \dfrac{1}{4}$

(c) $4 \times 4^{-\frac{1}{2}} = 4^{1 + (-\frac{1}{2})} = 4^{\frac{1}{2}} = \sqrt{4} = \pm 2$

or

$4 \times 4^{-\frac{1}{2}} = 4^{1 + (-\frac{1}{2})} = 4^{\frac{1}{2}}$

$= ((\pm 2)^2)^{\frac{1}{2}} = \pm 2$

(d) $\left(\frac{16}{81}\right)^{-\frac{3}{4}} = \dfrac{1}{\left(\frac{16}{81}\right)^{\frac{3}{4}}} = \left(\frac{81}{16}\right)^{\frac{3}{4}}$

$= \left(\sqrt[4]{\dfrac{81}{16}}\right)^3 = \left(\dfrac{\pm 3}{\pm 2}\right)^3 = \pm\dfrac{27}{8}$

or

$\left(\frac{16}{81}\right)^{-\frac{3}{4}} = \left(\frac{81}{16}\right)^{\frac{3}{4}} = \left(\left(\pm\frac{3}{2}\right)^4\right)^{\frac{3}{4}}$

$= \left(\pm\frac{3}{2}\right)^{4 \times \frac{3}{4}} = \left(\pm\frac{3}{2}\right)^3 = \pm\frac{27}{8}$

Notice the following in Example 18:

(i) The square root of a positive number may be positive or negative.
$\sqrt{9} = \pm 3$. ± 3 means $+3$ or -3.

(ii) The cube root of a positive number is positive; $\sqrt[3]{8} = 2$
the cube root of a negative number is negative; $\sqrt[3]{-27} = -3$

(iii) $\sqrt[4]{n}$ means the 4th root of n, i.e. the number which multiplied by itself 4 times gives n. For example $\sqrt[4]{16} = \pm 2$ since $2 \times 2 \times 2 \times 2 = 16$ and $-2 \times -2 \times -2 \times -2 = 16$.

(iv) A number raised to a negative power is equivalent to the reciprocal of the number raised to a positive power of the same numerical value. For example, $(\frac{3}{4})^{-2} = (\frac{4}{3})^2 = \frac{16}{9}$.

Example 19

Rewrite the following expressions with positive indices only.

(a) pq^{-2} (b) $\left(\frac{2a}{b}\right)^{-1}$

(a) $pq^{-2} = \frac{p}{q^2}$

Note that the index -2 refers to q only.

(b) $\left(\frac{2a}{b}\right)^{-1} = \frac{b}{2a}$

Exercise 7i

Rewrite the following expressions using positive indices only.

1. a^{-2}
2. b^{-1}
3. $c^{-\frac{2}{3}}$
4. xy^{-1}
5. $(xy)^{-1}$
6. $a^{-2}b^3$
7. ab^{-3}
8. $(ab)^{-3}$
9. $2x^{-\frac{1}{2}}$
10. $3y^{-\frac{2}{3}}$
11. $\left(\frac{a}{3b}\right)^{-2}$
12. $\left(\frac{1}{n}\right)^{-\frac{1}{3}}$

Note that $\sqrt[x]{n}$ means the xth root of n;
if x is even, n must be positive since any number, whether positive or negative, raised to an even power is always positive;
if x is odd and n is positive, the result must be positive;
if x is odd and n is negative, the result must be negative.
When x and n are mixed numbers, these numbers must be changed to improper fractions first.

Example 20

Simplify (a) $\left(2\frac{1}{4}\right)^{1\frac{1}{2}}$, (b) $\left(\frac{8}{50}\right)^{-\frac{1}{2}}$, (c) $\sqrt{\frac{72a^3b^{-2}}{2a^5b^{-6}}}$.

(a) Change mixed numbers to fractions.

$$\left(2\tfrac{1}{4}\right)^{1\frac{1}{2}} = \left(\tfrac{9}{4}\right)^{\frac{3}{2}} = \left(\sqrt{\tfrac{9}{4}}\right)^3 = \left(\pm\tfrac{3}{2}\right)^3 = \pm\tfrac{27}{8} = \pm3\tfrac{3}{8}$$

(b) Reduce the given fraction to its lowest terms.

$$\left(\tfrac{8}{50}\right)^{-\frac{1}{2}} = \left(\tfrac{4}{25}\right)^{-\frac{1}{2}} = \left(\tfrac{25}{4}\right)^{\frac{1}{2}} = \sqrt{\tfrac{25}{4}} = \pm\tfrac{5}{2} = \pm2\tfrac{1}{2}$$

(c)
$$\begin{aligned}\sqrt{\frac{72a^3b^{-2}}{2a^5b^{-6}}} &= \sqrt{36 \times a^{3-5} \times b^{-2-(-6)}} \\ &= \sqrt{36 \times a^{-2} \times b^4} \\ &= (36)^{\frac{1}{2}} \times (a^{-2})^{\frac{1}{2}} \times (b^4)^{\frac{1}{2}} \\ &= \pm6 \times a^{-1} \times b^2 \\ &= \pm\frac{6b^2}{a}\end{aligned}$$

Exercise 7j

Simplify the following.

1. $2a \times (3a)^2$
2. $4^{\frac{1}{2}}$
3. $125^{\frac{1}{3}}$
4. $\sqrt[3]{2^6}$
5. $8^{\frac{2}{3}}$
6. 2^{-2}
7. 3^{-3}
8. $9^{-\frac{1}{2}}$
9. $(25a^2)^{\frac{1}{2}}$
10. $2a^{-1}$
11. $4^{\frac{3}{2}}$
12. $2^{-2} \times 2^3$
13. 10^{-2}
14. $\sqrt{1\frac{9}{16}}$
15. $3a^{-2}$
16. $(3a)^{-2}$
17. $\sqrt{3^4}$
18. $(a^2)^{-\frac{1}{2}}$
19. $\left(\frac{1}{9}\right)^{-1}$
20. $3^{\frac{1}{2}} \times 3^{\frac{3}{2}}$
21. $\left(\frac{1}{27}\right)^{-\frac{2}{3}}$
22. $3^{\frac{1}{2}} \times 3^{-\frac{3}{2}}$
23. $0.04^{\frac{1}{2}}$
24. $2a^{-1} \times (3a)^2$
25. $16^{-\frac{3}{2}}$
26. $2^{\frac{1}{2}} \times 2^{\frac{3}{2}}$
27. $125^{-\frac{2}{3}}$
28. $3^x \times 3^{-x}$
29. $0.027^{\frac{2}{3}}$
30. $2a \times 3a^{-2}$
31. $2a \times (3a)^{-2}$
32. $\left(\frac{8}{27}\right)^{-\frac{2}{3}}$
33. $\sqrt[3]{4^{1.5}}$
34. $\sqrt[3]{8a^{-6}}$
35. $0.125^{-\frac{1}{3}}$
36. $4a^3b \times 3ab^{-2}$
37. $\sqrt{(125^2)^{-\frac{1}{3}}}$
38. $(2x)^{\frac{1}{2}} \times (2x^3)^{\frac{3}{2}}$

Example 21

Solve the following equations.

(a) $x^{\frac{1}{3}} = 4$ (b) $2a^{-\frac{1}{2}} = -14$ (c) $8^x = 32$

(a) *Either*: $x^{\frac{1}{3}} = 4$ $\sqrt[3]{x} = 4$

Cube both sides

$$\left(\sqrt[3]{x}\right)^3 = 4^3 = 4 \times 4 \times 4$$
$$x = 64$$

or:
$$\begin{aligned}x^{\frac{1}{3}} &= 4 \\ \left(x^{\frac{1}{3}}\right)^3 &= 4^3 \\ x^{\frac{1}{3}\times 3} &= 64 \\ x &= 64\end{aligned}$$

(b) $2a^{-\frac{1}{2}} = -14$

Divide both sides by 2

$$\begin{aligned}a^{-\frac{1}{2}} &= -7 \\ \frac{1}{a^{\frac{1}{2}}} &= -7 \\ a^{\frac{1}{2}} &= -\tfrac{1}{7} \\ \sqrt{a} &= -\tfrac{1}{7}\end{aligned}$$

Square both sides

$$a = \tfrac{1}{49}$$

(c) $8^x = 32$

Express 8 and 32 as powers of 2

$$\begin{aligned}(2^3)^x &= 2^5 \\ 2^{3x} &= 2^5\end{aligned}$$

Equate the powers of 2

$$\begin{aligned}3x &= 5 \\ x &= \tfrac{5}{3} = 1\tfrac{2}{3}\end{aligned}$$

Exercise 7k

Solve the following equations.

1. $x^{\frac{1}{2}} = 2$
2. $x^{\frac{1}{3}} = 3$
3. $a^{-1} = 2$
4. $a^{-2} = 9$
5. $2x^3 = 54$
6. $x^{-\frac{1}{2}} = 5$
7. $n^{-\frac{2}{3}} = 9$
8. $2r^{-3} = -16$
9. $5x = 40x^{-\frac{1}{2}}$
10. $5^x = 25$

Algebraic identities

Consider

$$x + 5 = 7 \qquad (1)$$

The algebraic statement (1) is called an equation. $x = 2$ is called the solution of equation (1). For that particular value, and *only* that value, of the variable x, the equation is true.

Consider

$$x + 5 = 3x + 7 - 2(x + 1) \qquad (2)$$

Whatever value is substituted for x, the left-hand side (LHS) of the algebraic statement equals the right-hand side (RHS) in (2). An algebraic statement which is true for all values of the variable, such as (2), is called an **algebraic identity**.

In algebraic identities the terms on the LHS may be rearranged and expressed in the format of the RHS, and vice versa. Examples of such algebraic statements include:

(a) expansion of algebraic terms in brackets, such as

$$(x + 3)(x - 2) = x^2 + x - 6$$

and $(x + a)^3 = x^3 + 3ax^2 + 3a^2x + a^3$

(b) factorisation, for example

$$c^2 - 9 = (c - 3)(c + 3)$$

(c) simplification of fractions, for example

$$\frac{3}{y - 1} + \frac{1}{y + 2} = \frac{4y + 5}{(y - 1)(y + 2)}$$

The statements of algebraic laws, such as the laws of indices, are also examples of algebraic identities:

$$a^x \times a^y = a^{x+y}$$
$$a^x \div a^y = a^{x-y}$$
$$(a^x)^y = a^{xy}$$

are true for all values of x and y.

To **prove an identity** is to rearrange the terms so that the terms on both sides are in the same format. This may be done by changing only one side of the identity or by changing both sides to a new but common format.

Notice that when proving algebraic identities, the commutative, associative and distributive properties are often used.

Example 22

Prove the following identities.

(a) $(a + c)(a - 2c) = a^2 - ac - 2c^2$

(b) $(xy^3)^2x^2 = (x^2y^2)^2y^2$

(a) LHS $= (a + c)(a - 2c)$
$= a^2 - 2ac + ac - 2c^2$
$= a^2 - ac - 2c^2$
= RHS

(b) LHS $= (xy^3)^2\, x^2$
$= (x^2\, y^{3 \times 2})x^2$
$= x^{2+2}\, y^{3\times2}$
$= x^4\, y^6$
$= (x^4\, y^4)\, y^2$
$= (x^2\, y^2)^2\, y^2$
= RHS

Example 23

For each of the following (i) state whether it is an equation E or an identity I; (ii) either solve the equation E or prove the identity I.

(a) $3a + 1 = 7$

(b) $ah + \frac{1}{2}(b - a)h = \frac{1}{2}(a + b)h$

(c) $(2xy)^3(x^3y)^2 = 8x^9y^5$

(d) $\dfrac{y + 4}{2} + \dfrac{3}{y} = \dfrac{-1}{2}$

(a) (i) E
(ii) $3a + 1 = 7$
$3a = 7 - 1$
$a = 6-3$
$a = 2$

(b) (i) I
(ii) LHS $= ah + \frac{1}{2}(b - a)h$
$= ah + \frac{1}{2}bh - \frac{1}{2}ah$
$= \frac{1}{2}(a + b)h$
= RHS

(c) (i) I

(ii) $\text{LHS} = (2xy)^3(x^3y)^2$
$= 2^3\, x^3y^3\, x^{3\times 2}\, y^2$
$= 8\, x^{3+6}\, y^{3+2}$
$= 8\, x^9\, y^5$
$= \text{RHS}$

(d) (i) E

(ii) $\frac{y+4}{2} + \frac{3}{y} = \frac{-1}{2}$

$$\frac{y(y+4) + 3(2)}{2y} = \frac{-1(y)}{2y}$$
$$y(y+4) + 6 = -y$$
$$y^2 + 4y + y + 6 = 0$$
$$y^2 + 5y + 6 = 0$$
$$(y+3)(y+2) = 0$$
$$y + 3 = 0 \quad \text{or} \quad y + 2 = 0$$
$$y = -3 \quad \text{or} \quad y = -2$$

Example 24

If $p(x + 3) = x - q$ for all values of x, find the values of p and q.

$p(x + 3) = x - q$
$px + 3p = x - q$

In an identity, like terms must be identical on both sides. Hence, $px = x$ and $3p = -q$ that is, $p = 1$ and $-q = 3(1)$.
Hence $p = 1$ and $q = -3$.

Exercise 7I

1. For each of the following (i) indicate whether the statement is an equation (E) or an identity (I), (ii) either solve the equation or prove the identity.
   (a) $3(x + 2) - 4 = 2(1 - x) + 5x$
   (b) $\frac{x - \frac{1}{2}}{x} + \frac{1}{2x} = 1$
   (c) $3a^3b^2\,(2a)^2b^2 = 12a^5\,b^4$
   (d) $2^{x+1} = 32$
   (e) $x - 1 - \frac{4}{2 + x} = x + 3$

2. Given that the following are true for all values of x, find the values of p and q in each case.
   (a) $p(x + 1) + q\,(x + 2) = 3qx + 2$
   (b) $\frac{p}{x - 2} + \frac{q}{x - 1} = \frac{q - 2px + 8}{x^2 - 3x + 2}$
   (c) $(x + p)\,(x + q) = x^2 + 2x - 3$

Direct variation

Fig. 7.1 shows a new pencil cut into a number of pieces.

Fig. 7.1

The mass of each piece is proportional to its length. The ratio of mass to length is the same for all the pieces.

If a person walks at a steady speed, the distance travelled is proportional to the time taken.

These are both examples of **direct proportion**, or **direct variation**. In the first example, the mass, M, varies directly with the length, L. In the second, the distance, D, varies directly with the time, T.

The symbol $\propto$ means 'varies with' or 'is proportional to'. The statements in the previous paragraph are written:

$M \propto L$
$D \propto T$

$D \propto T$ means that the ratio $\frac{D}{T}$ is **constant** (i.e. stays the same), so that $\frac{D}{T} = k$, where k is a constant.

Example 25

If $D \propto T$ and $D = 80$ when $T = 5$, find (a) the relationship between D and T, (b) the value of T when $D = 56$.

(a) If $D \propto T$
then $D = kT$, where k is a constant.
$D = 80$ when $T = 5$,
hence $80 = k \times 5$
$k = \frac{80}{5}$
$= 16$
The relationship between D and T is $D = 16T$.

(b) $D = 16T$
When $D = 56$
$56 = 16T$
$T = \frac{56}{16} = \frac{7}{2} = 3\tfrac{1}{2}$

Fig. 7.2 is a **sketch graph** of the relation $D \propto T$.

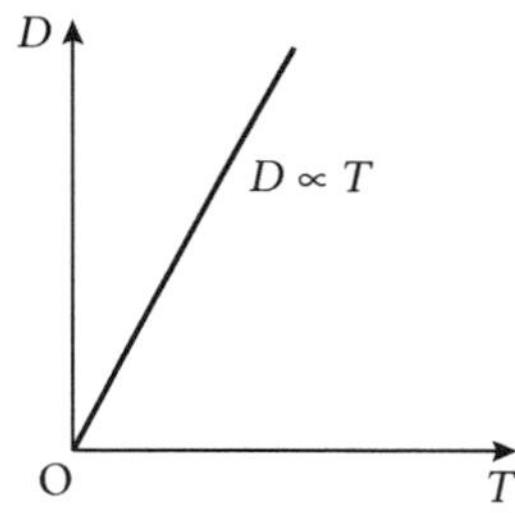

Fig. 7.2

Since $D = 16T$, the graph is a straight line of gradient 16 passing through the origin.
The graph of the relation between any two linear quantities which vary directly is always a straight line through the origin.

Example 26

Table 7.1 shows the extension E cm in an elastic string when it is pulled by a force of T newtons.

Table 7.1

T	5	8	11
E	7.5		16.5

(a) Show that E is directly proportional to T.
(b) Find the value of E when $T = 8$.

(a) By calculation:
If $E \propto T$
then $E = kT$, where k is a constant,
or $\frac{E}{T} = k$
i.e. if $E \propto T$, then $\frac{E}{T}$ should have a constant value for the results given in Table 7.1.
When $T = 5$, $E = 7.5$

$$\frac{E}{T} = \frac{7.5}{5} = 1.5$$

When $T = 11$, $E = 16.5$

$$\frac{E}{T} = \frac{16.5}{11} = 1.5$$

Since $\frac{E}{T} = 1.5$ in both cases, E is directly proportional to T.

Graphically:
Plot the given values (Fig. 7.3).

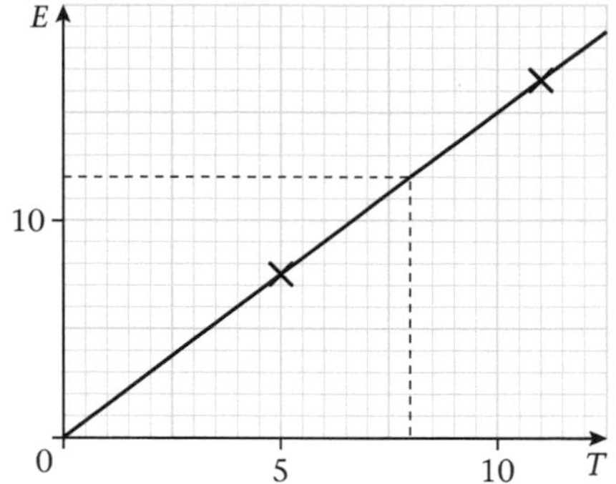

Fig. 7.3

Since the graph of E against T is a straight line passing through the origin, the gradient of the graph is a constant so that $\frac{E}{T}$ is a constant. Hence, $E \propto T$.

(b) By calculation:
$E = 1.5T$
When $T = 8$
$E = 1.5 \times 8 = 12$
or, from the graph:
when $T = 8$, $E = 12$.

Exercise 7m

1. If 1 m of wire has a mass of x g, what will be the mass of 25 m of the same wire?
2. If 1 jar of coffee costs $\$y$, what will be the cost of 4 jars of coffee?
3. If a man cycles 15 km in 1 hour, how far will he cycle in t hours if he keeps up the same rate?
4. Two variables A and B have corresponding values as shown in Table 7.2.

Table 7.2

A	6	10	12
B	34		68

(a) Either graphically, or by calculation, show that $B \propto A$.
(b) Find the value of B when $A = 10$.

5 The number of dollars ($) exchanged for a number of pounds sterling (£) is given in Table 7.3.

Table 7.3

£	2	4	6	8	10
$	3.20	6.40	9.60	12.80	16.00

(a) Show that $\$ \propto £$.
(b) Find the value of £7 in dollars.
(c) Find the value of $8 in £.

6 If $C \propto n$ and $C = 28$ when $n = 4$, find the formula connecting C and n.

7 If $d \propto s$ and $d = 120$ when $s = 30$, find the formula connecting d and s.

8 If $a \propto b$ and $a = 2.4$ when $b = 3$, find the relationship between a and b.

9 $x \propto y$ and $x = 30$ when $y = 12$. Find (a) the formula connecting x and y, (b) x when $y = 10$, (c) y when $x = 14$.

10 $P \propto Q$ and $P = 4.5$ when $Q = 12$. Find (a) the relationship between P and Q, (b) P when $Q = 16$, (c) Q when $P = 2.4$.

11 y varies directly with x.
(a) If $y = 9$ when $x = 45$, find the equation that connects x and y.
(b) Find y when $x = 40$.
(c) Find x when $y = 10$.

12 If $p \propto q$ and $p = 0.7$ when $q = 0.028$, find the relationship between p and q.

13 $x \propto y$ and $x = 17\frac{1}{2}$ when $y = 10\frac{1}{2}$.
(a) Find the equation that connects x and y.
(b) Find x when $y = 12$.

14 The mass of a plastic disc is proportional to its area. A disc of area 180 cm² has a mass of 200 g. If a similar disc has a mass of 250 g, what is its area?

15 The height (H cm) of liquid in a tube and the volume (V cm³) of the liquid are as given in Table 7.4.

Table 7.4

V	3	6	12
H	2.6	5.2	10.4

(a) Show that $H \propto V$.
(b) Find the law of variation in the form $H = kV$.
(c) Find V when $H = 6.5$.
(d) Find H when $V = 4.5$.

Direct variation between non-linear quantities

Expressions stating the relation between quantities which vary directly are not always in linear form. For example, the mass, m, of a cardboard square is directly proportional to its area, A.

$$m \propto A$$

However, $A = x^2$

where x is the length of a side of a square, so it follows that $m \propto x^2$.

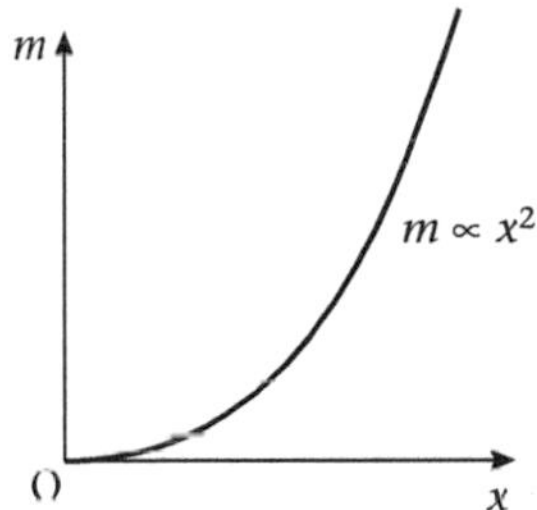

Fig. 7.4

$m \propto x^2$ is an example of direct proportion in which one of the variables is in quadratic form. Fig. 7.4 is a sketch of $m \propto x^2$.

Notice that the curve is similar to the graph of $y = x^2$.

Similarly, the volumes of spheres are directly proportional to the cubes of their radii:

$$V = \frac{4}{3}\pi r^3$$

so that $V \propto r^3$, since $\frac{4}{3}\pi$ is a constant.

Fig. 7.5 is a sketch of the graph of $V \propto r^3$.

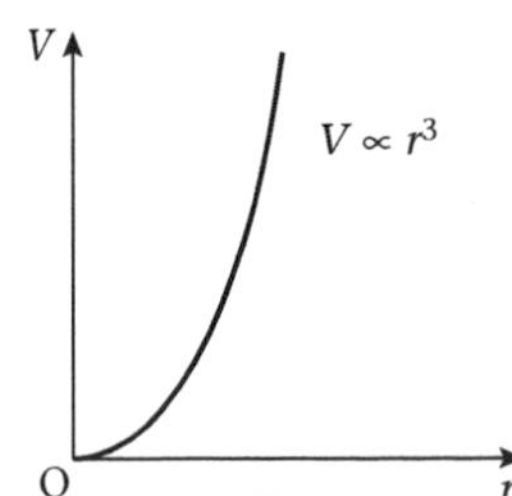

Fig. 7.5

Notice that the curve in Fig. 7.5 rises more steeply than that of Fig. 7.4.

Example 27

$y \propto \sqrt{x}$ and $y = 4\frac{1}{2}$ when $x = 9$. (a) Find the relationship between x and y. (b) Find y when $x = 25$. (c) Find x when $y = 6$.

(a) If $y \propto \sqrt{x}$
then $y = k\sqrt{x}$, where k is a constant.
When $x = 9$, $y = 4\frac{1}{2}$

$$4\tfrac{1}{2} = k\sqrt{9}$$
$$= 3k$$
$$k = \frac{4\frac{1}{2}}{3} = \frac{3}{2}$$
$$y = \tfrac{3}{2}\sqrt{x}$$

(b) When $x = 25$

$$y = \tfrac{3}{2}\sqrt{25} = 7\tfrac{1}{2}$$

(c) When $y = 6$

$$6 = \tfrac{3}{2}\sqrt{x}$$
$$\sqrt{x} = \tfrac{12}{3} = 4$$
$$x = 16$$

The next example shows how a variation problem may be solved when the value of the constant k is not required.

Example 28

x is directly proportional to the square of y. What is the percentage change in x if y increases by 20%?

$x \propto y^2$

Let $x = ky^2$ (1)

Then $k = \dfrac{x}{y^2}$

Let x become X if y increases to $\dfrac{120}{100}y$.

Then $X = k\left(\dfrac{120}{100}y\right)^2$ (2)

Either, $X = \left(\dfrac{x}{y^2}\right)\left(\dfrac{144}{100}y^2\right)$ *since* $k = \dfrac{x}{y^2}$

$$= \frac{144}{100}x$$

or,
dividing (2) by (1),

$$\frac{X}{x} = \frac{k\left(\frac{120}{100}y\right)^2}{ky^2} = \left(\frac{120}{100}\right)^2 = \frac{144}{100}$$
$$X = \frac{144}{100}x$$

Hence x increases by 44%.

Exercise 7n

1. A particle moves in such a way that its displacement, s metres, at time t seconds is given by the relation $s = at^2$, where a is a constant. Calculate a if $s = 32$ when $t = 4$.
2. $A \propto B^3$ and $A = 32$ when $B = 4$.
   (a) Find the formula connecting A and B.
   (b) Find A when $B = 6$.
   (c) Find B when $A = 13.5$.
3. P varies directly as the square root of Q and $P = 10$ when $Q = 16$.
   (a) Find the equation in P and Q.
   (b) Find P when $Q = 9$.
   (c) Find Q when $P = 1\frac{7}{8}$.
4. y varies directly with x^2. Table 7.5 shows some corresponding values of x and y.

   Table 7.5

x	-2	-1	$\frac{1}{2}$
y	16	4	

   Find the relation between x and y and complete the table.
5. The power, P watts, used in an electric circuit is proportional to the square of the current, C amps. When the current is 4 amps the circuit uses 500 watts. Find the current when the circuit uses 2420 watts.
6. The distance of the horizon from an observer varies directly with the square root of the height of the observer above ground level. At a height of 8 metres the horizon is 10 km away. Find the distance of the horizon from an observer at a height of 98 metres.
7. x varies directly with the square of y. Find the percentage change in x if y is (a) increased by 10%, (b) decreased by 10%.
8. x is directly proportional to the square root of y. Calculate the percentage change in x if y is increased by 44%.
9. If $V \propto R^3$, calculate the percentage increase in V if R increases by 20%.
10. If $W \propto D^2$, calculate the percentage decrease in W if D decreases by 15%.

Inverse variation

Fig. 7.6 shows a circle cut into (a) 5 equal sectors, (b) 12 equal sectors.

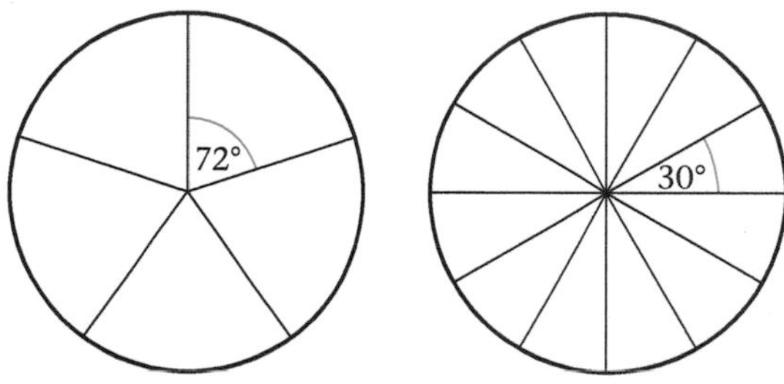

Fig. 7.6

If the circle is cut into a greater number of sectors, it is seen that the greater the number of sectors, the smaller the angle of each sector.

If a pot of tea is shared between some people, the *greater* the number of people, the *less* tea each will receive.

These are examples of **inverse proportion**, or **inverse variation**. In the first example, the size of the angle, θ, **varies inversely** with the number of sectors, n. In the second, the volume of tea received, V, is inversely proportional to the number of people, n. These statements are written:

$$\theta \propto \frac{1}{n} \quad \text{and} \quad V \propto \frac{1}{n}$$

These equations may be written as

$$\theta = \frac{k}{n} \quad \text{and} \quad V = \frac{K}{n} \quad \text{so that}$$

$$k = \theta n \quad \text{and} \quad K = Vn$$

where k and K are constants.

Example 29

If V varies inversely with n and $V = 220$ when $n = 6$, find V when $n = 8$.

If $V \propto \frac{1}{n}$ then $V = \frac{k}{n}$, where k is a constant.

$V = 220$ when $n = 6$

$$220 = \frac{k}{6} \quad \Leftrightarrow \quad k = 6 \times 220$$

and $V = \dfrac{6 \times 220}{n}$

When $n = 8$,

$$V = \frac{6 \times 220}{8}$$

$$= 165$$

Fig. 7.7 is a sketch graph of the relation $V \propto \frac{1}{n}$.

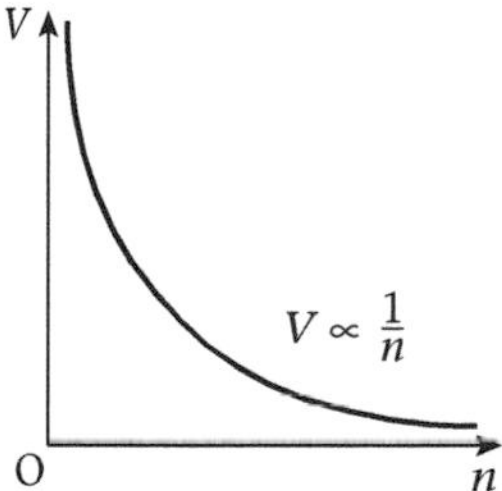

Fig. 7.7

Notice that the curve approaches both axes but does not reach them. This is because division by 0 is undefined.

Example 30

The number of spherical glass beads which can be made from a given volume of glass varies inversely with the cube of the diameter of the beads. When the diameter is 2 mm, the number of beads is 2700. How many beads of diameter 3 mm can be made from the glass?

Let N = number of beads, d = diameter of the beads. From the first sentence,

$$N \propto \frac{1}{d^3} \text{ or } N = \frac{k}{d^3}, \text{ where } k \text{ is a constant.}$$

From the second sentence,

$$2700 = \frac{k}{2^3}$$

$$k = 8 \times 2700$$

So $\quad N = \dfrac{2700 \times 8}{d^3}$

When $d = 3$,

$$N = \frac{2700 \times 8}{3^3}$$

$$= \frac{2700 \times 8}{27}$$

$$= 800$$

800 beads can be made.

Exercise 7o

1. If d varies inversely as t, use the symbol $\propto$ to show a connection between d and t.

2. A piece of string is cut into n pieces of equal length l.
   (a) Does n vary directly or inversely with l?
   (b) Use the symbol $\propto$ to show a connection between n and l.

3. A rectangle has a constant area, A. Its length is l and its breadth is b.
   (a) Write a formula for l in terms of A and b.
   (b) Write a formula for b in terms of A and l.
   (c) Does l vary inversely or directly with b?

4. If $R \propto \frac{1}{T}$ and $T = 8$ when $R = 4$, find the relationship between R and T.

5. If y varies inversely as x, and $y = 2$ when $x = 3$, find y when $x = 6$.

6. P is inversely proportional to Q and $P = 5$ when $Q = 4$. Calculate the value of Q when $P = 25$.

7. The electrical resistance R of a wire varies inversely as the square of the radius r, Use a constant k to show the relation between R and r.

8. P varies inversely as the square root of v and $P = 4.5$ when $v = 25$. Find v when $P = 15$.

9. The variables X and Y are connected by the relation 'Y varies inversely as X'. Table 7.6 shows the values of Y for some selected values of X.

   Table 7.6

X	10	20	30	40
Y	12	6	?	3

   Calculate the missing value of Y.

10. The length of wire that can be made from a mass of copper is inversely proportional to the square of the diameter of the wire. When the diameter is 3 mm the length of the wire is 1.8 km. Find the length of the wire when its diameter is 1.2 mm.

Joint variation

The mass, M, of a coin of radius r and thickness h depends on the volume, V, of metal in the coin.
i.e. $M \propto V$
or $M \propto \pi r^2 h$ (since $V = \pi r^2 h$)
or $M \propto r^2 h$ (since π is a constant)
$M \propto r^2 h$ means that the mass of the coin **varies jointly** with the square of the radius *and* the thickness. This is an example of **joint variation**.

Example 31

The mass of a wire varies jointly with its length and the square of its diameter. 500 m of wire of diameter 3 mm has a mass of 31.5 kg. What is the mass of 1 km of wire of diameter 2 mm?

Let M = mass in kg, d = diameter in mm and L = length in m.

Then, from the first sentence,
$M \propto Ld^2$
or $M = kLd^2$, where k is a constant.

From the second sentence,
$$31.5 = k \times 500 \times 3^2 \qquad (1)$$

From the third sentence,
$$M = k \times 1\,000 \times 2^2 \qquad (2)$$

Dividing (2) by (1),
$$\frac{M}{31.5} = \frac{1000 \times 2^2}{500 \times 3^2} = \frac{2 \times 4}{9} = \frac{8}{9}$$
$$M = \tfrac{8}{9} \times 31.5 = 28$$
Hence the mass is 28 kg.

Notice that it was not necessary to find k. Notice also that it is possible to mix dimensions, such as mm and m, so long as this is done *consistently*.

Example 32

If $P \propto \frac{1}{V}$ and $V \propto R^2$, show that $P \propto \frac{1}{R^2}$.

If $V \propto R^2$,
then $V = kR^2$ where k is a constant.
If $P \propto \frac{1}{V}$, substituting kR^2 for V
then $P \propto \frac{1}{kR^2}$
that is $P \propto \frac{1}{R^2}$, since k is a constant.
The method in Example 32 is to eliminate the variable which is not required.

Exercise 7p

1. $x \propto yz$. when $y = 2$ and $z = 3$, $x = 30$.
   (a) Find the relation between x, y and z.
   (b) Find x when $y = 4$ and $z = 6$.
2. $x \propto \frac{y}{z}$. $x = 27$ when $y = 9$ and $z = 2$.
   (a) Find the relation between x, y and z.
   (b) Find x when $y = 14$ and $z = 12$.
3. $p \propto \frac{q}{r^2}$ and $p = 3\frac{1}{3}$ when $q = 5$ and $r = 3$.
   (a) Find the equation connecting p, q and r.
   (b) Find p when $q = 9$ and $r = 1.2$.
4. The height (h) of a cone varies directly as its volume (V) and inversely as the square of its radius (r). Use the constant k to show the relationship between h, V and r.
5. $A \propto BC$ and $A = 6$ when $B = 4$ and $C = 9$.
   (a) Find A when $B = 3$ and $C = 10$.
   (b) Find C if $A = 20$ and $B = 15$.
   (c) By what percentage does A change if B is increased by 10% and C is decreased by 10%?
6. x varies directly with the square of y and with z. When $y = 2$ and $z = 3$, $x = 4\frac{1}{2}$.
   (a) Find x when $y = 5$ and $z = 4$.
   (b) Find y when $x = 21$ and $z = 3\frac{1}{2}$.
   (c) What happens to x if y is doubled and z halved?
7. $x \propto y$ and $y \propto z^3$. How does x vary with z?
8. $x \propto y^2$ and $y \propto z^2$. How does x vary with z?
9. $A \propto BC$ and $B \propto \frac{1}{C^2}$. How does A vary with C?
10. The mass, m, of a roller varies jointly with its length, l, and the square of its diameter, d. A roller of diameter 20 mm is 5 cm long and has a mass of 0.29 kg. Calculate the mass, in kg, of a roller 30 mm in diameter and 2 cm long.

Change of subject of formulae

The letter I is the **subject** of the formula

$$I = \frac{PRT}{100}$$

The subject stands on its own. Its value can be found directly by substituting the values of the other letters in the formula.

It is often necessary to **change the subject** of a formula. This means to rearrange the order of the letters in the formula so that one of the other letters becomes the subject. For example, to find the value of R if the values of I, P and T are given, make R the subject of the formula before calculating R.

To do this, think of the formula as an equation. Solve the equation for the letter which is to become the subject. The following examples show how different formulae can be rearranged in order to change the subject.

Example 33

Make x the subject of the formula $a = b(1 - x)$.

$$a = b(1 - x)$$

Clear brackets.

$$a = b - bx$$

Rearrange to give terms in x on one side of the equation.

$$bx = b - a$$

Divide both sides by b.

$$x = \frac{b - a}{b}$$

Example 34

Make P, R and T in turn the subject of the formula $I = \frac{PRT}{100}$.

$$I = \frac{PRT}{100}.$$

Multiply both sides by 100

$$100I = \text{PRT} \qquad (1)$$

Dividing both sides of equation (1) by RT gives

$$P = \frac{100I}{RT} \qquad (2)$$

Dividing both sides of equation (1) by PT gives

$$R = \frac{100I}{PT} \qquad (3)$$

Dividing both sides of equation (1) by PR gives

$$T = \frac{100I}{PR} \qquad (4)$$

Equations (2), (3) and (4) show P, R and T respectively as subjects of the given formula.

Example 35

Make x the subject of the following.

(a) $y = x - 9$ (b) $N = 7x$ (c) $\frac{x}{a} = 8$

(d) $\frac{h}{x} = k$ (e) $y = 2x + 1$

(a) $y = x - 9$
Add 9 to both sides,
$y + 9 = x$
$x = y + 9$

(b) $N = 7x$
Divide both sides by 7
$\frac{N}{7} = x$ or $x = \frac{N}{x}$

(c) $\frac{x}{a} = 8$
Multiply both sides by a
$x = 8a$

(d) $\frac{h}{x} = k$
Multiply both sides by x
$h = kx$
Divide both sides by k
$\frac{h}{k} = x$ or $x = \frac{h}{k}$

(e) $y = 2x + 1$
Subtract 1 from both sides
$y - 1 = 2x$
Divide both sides by 2
$\frac{y - 1}{2} = x$
$x = \frac{y - 1}{2}$

Exercise 7q

In each question a formula is given. A letter is printed in heavy type after it. Make that letter the subject of the formula. If more than one letter is given, make each the subject in turn.

1. $y = x + 8$ **x**
2. $y = x - 3$ **x**
3. $b = a + c$ **a, c**
4. $y = 3x$ **x**
5. $y = \frac{x}{4}$ **x**
6. $b = ac$ **a, c**
7. $n = 5ax$ **a, x**
8. $\frac{x}{y} = 9$ **x**
9. $\frac{y}{x} = 2$ **x**
10. $y = 6x + 11$ **x**
11. $b = 5a - c$ **a**
12. $x + y = 13$ **x, y**
13. $2p - q = 0$ **q, p**
14. $2x - y = d$ **x, y**
15. $p = 4d$ **d**
16. $T = \theta + 273$ θ
17. $W = VI$ **V, I**
18. $A = \pi rl$ **r, l**
19. $V = lbh$ **l, b, h**
20. $A = 2\pi rh$ **r, h**

Example 36

Given $3x - 2y = 8$, (a) express x in terms of y and find x when $y = 11$, (b) obtain a formula for y and find y when $x = 2$.

(a) 'Express x in terms of y' means 'make x the subject of the formula'.
$3x - 2y = 8$
Add $2y$ to both sides
$3x = 8 + 2y$
Divide both sides by 3
$x = \frac{8 + 2y}{3}$
When $y = 11$,
$x = \frac{8 + 2 \times 11}{3} = \frac{8 + 22}{3}$
$= \frac{30}{3}$
$x = 10$

(b) 'Obtain a formula for y' means 'make y the subject of the formula'.
$3x - 2y = 8$
Add $2y$ to both sides
$3x = 8 + 2y$
Subtract 8 from both sides
$3x - 8 = 2y$
Divide both sides by 2
$\frac{3x - 8}{2} = y$
When $x = 2$
$y = \frac{3 \times 2 - 8}{2} = \frac{6 - 8}{2} = \frac{-2}{2}$
$y = -1$

Exercise 7r

1. If $y = 2x - 9$, (a) express x in terms of y, (b) find x when $y = 5$.

2. If $3x + y = d$, (a) express x in terms of y and d, (b) find x when $d = 1$ and $y = 13$.

3. A cylinder of radius r cm and height h cm has a curved surface area A cm^2, where $A = 2\pi rh$.
   (a) Obtain a formula for h.
   (b) Find the value of h when $A = 93$, $r = 2.5$ and $\pi = 3.1$.

4. A triangle of base b cm and height h cm has an area A cm^2, where $A = \frac{1}{2}bh$.
   (a) Express b in terms of A and h.
   (b) Hence find the value of b when $A = 135$ and $h = 18$.

5. The wage, w dollars, of a woman who works r hours of overtime is given by the formula $w = 3r + 59$.
   (a) Make r the subject of this formula.
   (b) Hence find the number of hours of overtime worked by a woman whose total wage is \$83.

6. (a) Make P the subject of the simple interest formula $I = \dfrac{PRT}{100}$.
   (b) Hence find the principal which makes an interest of \$297.50 in 7 years at a rate of 5% per annum.

7. (a) Make R the subject of the simple interest formula $I = \dfrac{PRT}{100}$.
   (b) Hence find the rate of interest if \$162 makes an interest of \$21.60 in 4 years.

8. If $p = \dfrac{c}{d}$,
   (a) express d in terms of p and c,
   (b) find d when $p = 3$ and $c = 5.7$.

9. Make x the subject of the relation $y \propto \dfrac{1}{\sqrt{x}}$.

10. A quantity $(y - k)$ varies inversely as the square of x. Make y the subject of an equation in x, k and h, where k and h are constants.

Example 37

Make x the subject of the formula $a = \dfrac{b + x}{b - x}$.

$$a = \frac{b + x}{b - x}$$

Clear fractions. Multiply both sides by $(b - x)$.

$$a(b - x) = b + x$$

Clear brackets.

$$ab - ax = b + x$$

Collect terms in x on one side of the equation.

$$ab - b = x + ax$$

Take x outside a bracket.

$$ab - b = x(1 + a)$$

Divide both sides by $(1 + a)$.

$$\frac{ab - b}{1 + a} = x \quad \Rightarrow \quad x = \frac{b(a - 1)}{1 + a}$$

Example 38

Make x the subject of the formula $b = \frac{1}{2}\sqrt{a^2 - x^2}$.

$$b = \tfrac{1}{2}\sqrt{a^2 - x^2}.$$

Clear fractions.

$$2b = \sqrt{a^2 - x^2}$$

Square both sides.

$$(2b)^2 = a^2 - x^2$$
$$4b^2 = a^2 - x^2$$

Rearrange to have the term in x on one side of the equation.

$$x^2 = a^2 - 4b^2$$

Take the square root of both sides.

$$x = \sqrt{a^2 - x^2}$$

The general method of Examples 37 and 38 is to treat the formula as an equation and the new subject as the unknown of the equation. There are many different formulae and it is not possible to give general rules for changing their subject. However, the following points should be remembered.

1 Begin by clearing fractions, brackets and root signs.

2 Rearrange the formula so that all the terms which contain the new subject are on one side of the equals sign and the rest on the other. Do not try to place the subject on the left-hand side if it comes more naturally on the right (see Example 37).
3 If more than one term contains the subject, take it outside a bracket.
4 Divide both sides by the bracket, then simplify as far as possible.

Exercise 7s

Make x the subject of the following.

1 $x + a = b$
2 $a - x = b$
3 $ax = b$
4 $ax + bx = c$
5 $ax + b = x$
6 $\frac{a}{x} = b$
7 $\frac{a}{x} + b = c$
8 $\frac{x}{a} + b = c$
9 $\frac{x}{a} + \frac{x}{b} = 1$
10 $\frac{a}{x} + \frac{b}{x} = 1$
11 $a(x + b) = c$
12 $ax = b(c + x)$
13 $a(b - x) = cx$
14 $\frac{x}{2a} + \frac{x}{3a} = b$
15 $x(a - b) = b(c - x)$
16 $\frac{a}{b - x} = c$
17 $a = \frac{2b + 3x}{3b - 2x}$
18 $\sqrt{x} = a$
19 $\sqrt{2x} = a$
20 $2\sqrt{x} = a$
21 $\sqrt{\frac{x}{2}} = a$
22 $\frac{\sqrt{x}}{2} = a$
23 $a\sqrt{x} = b$
24 $\sqrt{ax} = b$
25 $\sqrt[3]{\frac{x}{a}} = b$
26 $x^2 = a^4$
27 $x^2 = a$
28 $\sqrt{x + a} = b$
29 $\sqrt{x} + a = b$
30 $\sqrt{x^2 + a^2} = b$
31 $\sqrt{x^2 + a^2} = 3a$
32 $\frac{a}{x} - 1 = \frac{b}{2x}$
33 $a\sqrt{x - 1} = b$
34 $a\sqrt{x} - 1 = b$
35 $(ax - b)(bx + a) = (bx^2 + a)a$
36 $\frac{a}{a - x} = \frac{b}{b + x}$
37 $\frac{b}{b - x} = \frac{b}{a + x}$
38 $a(a^2 - x) = b(b^2 - x)$
39 $\frac{x}{x + a} - \frac{a}{x - b} = 1$
40 $\frac{x^2}{a^2} + \frac{y^2}{b^2} = 1$

Example 39

If $u = 1 - \frac{3v}{vt - w}$, express t in terms of the other letters.

$$u = 1 - \frac{3v}{vt - w},$$

Clear fractions.

$$u(vt - w) = 1(vt - w) - 3v$$

Clear brackets.

$$uvt - uw = vt - w - 3v$$

Collect terms in t.

$$uvt - vt = uw - w - 3v$$

Take t outside a bracket.

$$t(uv - v) = uw - w - 3v$$

$$t = \frac{uw - w - 3v}{uv - v}$$

Example 40

The period of a compound pendulum is given by $T = 2\pi\sqrt{\left(\frac{h^2 + k^2}{gh}\right)}$. Express k in terms of T, h and g, taking π^2 as 10.

$$T = 2\pi\sqrt{\left(\frac{h^2 + k^2}{gh}\right)}.$$

$$T^2 = 4\pi^2\left(\frac{h^2 + k^2}{gh}\right)$$

$$\frac{T^2}{4\pi^2} = \frac{h^2 + k^2}{gh}$$

$$\frac{T^2 gh}{4\pi^2} = h^2 + k^2$$

$$k^2 = \frac{T^2 gh}{4\pi^2} - h^2$$

$$k = \sqrt{\frac{T^2 gh}{4\pi^2} - h^2}$$

Using the value 10 for π^2,

$$k = \sqrt{\left(\frac{T^2 gh}{40} - h^2\right)}$$

Exercise 7t

In each question a formula is given. A letter is printed in heavy type after it. Make the letter the subject of the formula. If more than one letter is given, make each letter the subject in turn.

1. $c = 2\pi r$ **r**
2. $P = aW + b$ **W**
3. $P = \dfrac{N+2}{D}$ **N, D**
4. $A = P + \dfrac{PRT}{100}$ **T, P**
5. $v^2 = u^2 + 2as$ **s, u**
6. $s = \dfrac{n}{2}(a + l)$ **n, l**
7. $S = 2\pi r(r + h)$ **h**
8. $k = \dfrac{brt}{v - b}$ **b**
9. $S = 4\pi r^2$ **r**
10. $V = \pi h^2\left(r - \dfrac{h}{3}\right)$ **r**
11. $L = \dfrac{Wh}{a(W + P)}$ **W**
12. $\dfrac{L}{E} = \dfrac{2a}{R - r}$ **R**
13. $T = \dfrac{mu^2}{K} - 5mg$ **K**
14. $D = \sqrt{\dfrac{3h}{2}}$ **h**
15. $t = \dfrac{3p}{r} + s$ **r**
16. $\dfrac{a}{p} - \dfrac{b}{q} = c$ **q**
17. $H = \dfrac{m(v^2 - u^2)}{2gx}$ **v**
18. $R = \sqrt{\dfrac{ax - P}{Q + bx}}$ **x**
19. $T = 2\pi\sqrt{\dfrac{I}{MH}}$ **M**
20. $A = \frac{1}{2}m(v^2 - u^2)$ **u**

Exercise 7u

1. The volume of a square-based pyramid is given by $V = \dfrac{a^2h}{3}$, where h is the height of the pyramid and a is the length of one of the base edges.
   (a) Make a the subject of the formula.
   (b) Find a when $V = 162$ and $h = 24$.

2. The volume V of a cone of height h and base radius r is $\frac{1}{3}\pi r^2 h$.
   (a) Obtain a formula for r in terms of V, π and h.
   (b) Calculate the base radius of a cone of height 14 cm and volume $91\frac{2}{3}\,\text{cm}^3$ using the value $\frac{22}{7}$ for π.

3. The length of the hypotenuse, h, in a right-angled triangle is given by the formula $h = \sqrt{(a^2 + b^2)}$ where a and b are the lengths of the other two sides of the triangle.
   (a) Make a the subject of this formula.
   (b) Hence find a if $h = 34$ and $b = 16$.

4. (a) Make d the subject of the formula
   $$S_n = \frac{n}{2}[2a + (n - 1)d]$$
   (b) Hence find d if $S_{32} = 56$ and $a = 25$.
   (*Note*: S_{32} is the value of S_n when $n = 32$.)

5. The energy E possessed by an object of mass m kg travelling at a height h m with a velocity v m/s is given by $E = \dfrac{mv^2}{2} + mgh$ joules.
   (a) Express v in terms of the other letters.
   (b) If the energy of a 20 kg mass at a height of 15 m is 4900 joules and $g = 9.8$, how fast is the mass moving?

6. The formula $V = \frac{1}{3}\pi r^2(2r + h)$ gives the volume V of a solid consisting of a cone of height h and base radius r attached to a hemisphere of the same radius.
   (a) Change the subject of the formula to h.
   (b) Hence calculate the height of the cone, if the solid has a volume of $55\,\text{cm}^3$ and the common radius is $2\frac{1}{2}$ cm. (Use the value $\frac{22}{7}$ for π.)

7. If a wire L metres long is stretched tightly between two points at the same level d metres apart, the sag in the middle of the wire is s metres, where $s = \sqrt{\dfrac{3d(L - d)}{8}}$.
   (a) Change the subject of the formula to L.
   (b) How long is the wire if $d = 16$ and $s = 0.6$?

8 The formula $T = \frac{4b^2}{21}(d - \frac{3}{5}b)$ gives the approximate mass, T tonnes, of a ship d metres long and b metres wide.
(a) Make d the subject of the formula.
(b) Calculate the length of a 4500 tonne ship if it is 20 m wide, giving your answer to the nearest 5 m.

Summary

It is important to distinguish between **simplifying an algebraic expression** and **solving an algebraic equation.**

In simplifying an expression, the value of the given expression must not change. However, the form of the expression changes; for example, by collecting like terms; and, when there are fractions, by using a common denominator and/or by dividing both numerator and denominator by common factors.

In solving an equation we have to find the particular value(s) of the variable(s) which make the LHS and RHS of the equation equal. Note that in solving an equation, it is often necessary to first simplify algebraic expressions.

In an **algebraic identity**, the **commutative, associative** and **distributive** properties are generally used in proving the statement to be true for **all** values of the variable(s) in the identity.

If x **varies directly** with y^n, where $n = 1, 2, 3,$ ... then $x \propto y^n$

so that $x = ky^n$ where k is a constant.

If x **varies inversely** with y^n, where $n = 1, 2, 3,$ ... then $x \propto \frac{1}{y^n}$

so that $x = \frac{k}{y^n}$ where k is a constant.

In the formula for the area of a circle, the area A is given by the formula, $A = \pi r^2$. Then A is called the **subject** of the formula.

To **change the subject of the formula** is to rearrange the terms so that another letter is placed on the LHS; for example

$$r = \sqrt{\frac{A}{\pi}}$$

Practice exercise P7.1

Simplify the following.

1 $\frac{x+4}{2} + \frac{x+4}{3}$

2 $\frac{x+4}{2} - \frac{x+4}{3}$

3 $\frac{x+4}{2} \times \frac{x+4}{3}$

4 $\frac{x+4}{2} \div \frac{x+4}{3}$

5 $\frac{3}{x+3} + \frac{4}{x+4}$

6 $\frac{3}{x+3} - \frac{4}{x+4}$

7 $\frac{3}{x+3} \times \frac{4}{x+4}$

8 $\frac{3}{x+3} \div \frac{4}{x+4}$

Practice exercise P7.2

Simplify the following.

1 $\frac{y-4}{2} - \frac{y}{6} + \frac{1-y}{3}$

2 $\frac{a+3}{3} \times \frac{2a-1}{2}$

3 $\frac{w+4}{w+1} + \frac{2w-1}{w+2}$

4 $\frac{1}{2} - \frac{3-m}{m+6}$

5 $\frac{2d-3}{d+3} - \frac{4-d}{d-2}$

6 $\frac{3v+5}{3} + 2v$

7 $y - \frac{2y-3}{3}$

8 $\frac{p-3}{2p+1} \div \frac{3-p}{p-2}$

9 $\frac{3}{abc} - \frac{2}{a^2b^2} + \frac{1}{b^2c^2}$

10 $\frac{x+2}{2} \times \frac{1-x}{3}$

11 $\frac{x-4}{x+2} - \frac{3}{5}$

12 $\frac{m^2-3m}{2m} - \frac{3n^2+n}{6n}$

13 $\frac{4}{y-2} - \frac{3}{y+3}$

14 $\frac{h-4}{h+3} - \frac{5}{6} + \frac{1-h}{h+1}$

15 $\frac{3y-1}{2y} - \frac{y}{6} + \frac{1-3y}{3}$

Practice exercise P7.3

1 If $v = -1$, $x = -2$, $y = 3$, find the value of the following.

(a) vx
(b) vy
(c) $3x^2$
(d) $(3x)^2$
(e) $3v + 2x$
(f) $2y + 4v - x$
(g) $vx^2 + y^2$
(h) $(vx - y)^2$
(i) $\frac{x+y}{2}$
(j) $\frac{2v-y}{5}$
(k) $\frac{xy(v-3)}{4vy}$
(l) $\frac{(vx-y)^2}{v-xy}$

2. If $P = 2h + 2w$, evaluate
   (a) P when $h = 12$ and $w = 9$,
   (b) h when $P = 34$ and $w = 7$.
3. Evaluate $4y^2 + 3y - 2$ when $y =$
   (a) 2, (b) -2, (c) 0, (d) -3, (e) -1.
4. Find the value of $x^2y - 4y^2 + 3xy^2 - x^2$ when $x = 3$ and $y = -2$.
5. Evaluate the following given that $r = -3$, $s = 2$, and $t = -1$.
   (a) $r^2 - 5tr - 4t^2$ (b) $3srt - rs^2$
   (c) $\dfrac{tsr}{5tr - 3sr}$
6. Using $F = \dfrac{M}{8} + 2$ and $T = \dfrac{M}{4F}$, write T in terms of F only.
7. The formula for the circumference of a circle is $C = 2\pi r$. Find the circumference of the base of a cylindrical tank if its radius is 0.7 metres.
8. \$2700 is the interest, I, paid on a principal, P, of \$20 000 after 3 years. Calculate R, the rate paid, using the formula $I = \dfrac{PRT}{100}$, where T is time.
9. (a) A woman has \$$x$ in an account. She adds \$$d$ every week for n weeks. She then has \$$y$ in the account. Write the formula connecting y and x.
   (b) Use the formula to find n when $x = 560$, $d = 35$ and $y = 980$.

Practice exercise P7.4

In each question a formula is given. A letter is printed in heavy type after it. Make that letter the subject of the formula. If more than one letter is given, make each the subject in turn.

1. $K = C + 273$ **C**
2. $P = 2l + 2b$ **b**
3. $t = \frac{1}{2}s + r$ **s, r**
4. $I = \dfrac{PRT}{100}$ **T, R, P**
5. $v = u + at$ **u, t**
6. $s = ut + \frac{1}{2}a^2t$ **u, t**
7. $p = \dfrac{2n - 3}{n + 5}$ **n**

Practice exercise P7.5

Simplify the following.

1. $4x^3 + 4x^3$
2. $4x^3 \times 4x^3$
3. $4x^3 \div 4x^3$
4. $(4x^3)^3$

Practice exercise P7.6

Simplify the following.

1. $a^4 \times 4a^3$
2. $p^6r^5q^5 \div pq^2$
3. $3d^2 \times 5bd^3$
4. $3xy^4 \times 2yz^3$
5. $9h^2k^5 \div 3h^5jk^3$
6. $12ab^5c^5 \div 3b^5c^3$
7. $(2a^3f)^3$
8. $3(hkn^2)^3 \times (3hk)^2$

Practice exercise P7.7

Express the following in the form x^n.

1. $\dfrac{1}{x^5}$
2. $\sqrt[5]{x}$
3. $\dfrac{1}{\sqrt{x}}$
4. $(\sqrt{x})^5$

Practice exercise P7.8

Simplify the following.

1. $9^{-\frac{1}{2}}$
2. $(\frac{1}{8})^{\frac{1}{3}}$
3. $(16b^6)^{\frac{1}{2}}$
4. $\sqrt{1\frac{7}{9}}$
5. $2x^{-1} \times 3x^3$
6. $2x^{-1} \times (3x)^3$
7. $36^{-\frac{3}{4}}$
8. $(\frac{1}{27})^{\frac{1}{3}}$
9. $(16k^4)^{\frac{3}{4}}$
10. $(\frac{3}{8})^{-1}$
11. $a^{-3} \times c^0$
12. $(-3f^2)^3$
13. $x^3y^2 \times xy^{-1}$
14. $(r^3 \div 3r)^{\frac{2}{3}}$
15. $(a^3x^2y)^3 \div 2xy^2$
16. $2h^{-3} \times 3h^{-2}$
17. $(2y)^{-1} \times (2y)^3$
18. $(4v)^{-2} \times (2v)^4$
19. $(3h)^{\frac{1}{2}} \times h^{-\frac{1}{2}}$
20. $m^{\frac{1}{3}} \div m^{-\frac{3}{2}}$

Practice exercise P7.9

(a) State whether each of the following is an equation or an identity.
(b) Either solve the equation or prove the identity.

1. $3x + 4 = 5 - 2x$
2. $3y + 1 - z = 3 + 3(z + y) - 2(1 + 2z)$
3. $2d(d + 2) = d^2 + d(d + 1)$
4. $(c + 2)(c - 3) = c^2 - c - 6$

5. $5h - 2 = 3h + 4$
6. $2(x + 3y) + x = 3x + 2y$
7. $2 + \frac{m}{3} = 2m - \frac{4}{3}$
8. $\frac{2t - 1}{2t} - \frac{t + \frac{1}{2}}{1 + t} = -\frac{1}{2t(1 - t)}$
9. $8(y^4)^{\frac{3}{4}} + 1 = 8y^3 - \frac{1}{4}y$
10. $2(f + 4)(2 - f) - 3f + 2$
    $= (2 - 3f)(1 - 2f) - 8(f^2 - 2)$
11. $a^2 + 5a + 1 = (a + 2)^2 + 2a$
12. $12b^4c^3 \div 3bc^3 = 32$

Practice exercise P7.10

Show that each of the following statements is an identity.

1. $3x + 4 \equiv \frac{1}{2}(8 + 6x)$
2. $3a - b + 1 - b \equiv b + 5 + 3(a - b) - 4$
3. $d(2d + 1) \equiv d^2 + d(d + 1)$
4. $(x + y)(2y + 1) \equiv 3y^2 + x + y(2x - y + 1)$
5. $3(x + 1) - 2(x - 1) \equiv 5 + x$
6. $2(a - 2b) + b \equiv 3(a - b) - a$
7. $m^2 + n(m + 2) \equiv m(m + n) + 2n$
8. $(j + k)(k + j) \equiv 2(k^2 + j^2 + kj) - (j^2 + k^2)$
9. $\frac{x^2 + 4x + 3}{x^2 + 3x} \equiv \frac{x + 1}{x}$
10. $m(n - p) + n(p - m) + p(m - n) = 0$

Practice exercise P7.11

1. If $c(2x - 3) + 2d \equiv 4x + 2$, where c and d are constants, find c and d.
2. If $(x + p)^2 - (y + q)^2 \equiv (x + y + 1)(x - y - 3)$, where p and q are constants, find p and q.

Practice exercise P7.12

1. An arc of length l in a circle of radius r, circumference C and area A, subtends an angle θ at the centre of the circle. Using the appropriate equations, state the variation connecting
   (a) C and r, (b) A and r,
   (c) A and C, (d) l, r and θ.
2. The volume V of a sphere varies as the cube of its radius r.
   Write the variation connecting V and r.
3. The time of a swing of a pendulum varies as the square root of its length. If the time of a swing of a 2-metre pendulum is 2.1 seconds, find the time of a 3.5-metre pendulum.
4. If $y \propto \frac{1}{x}$ and $x = \frac{1}{a}$ when $y = \frac{1}{b}$, find the value of y when $x = c$.

Chapter 8

Trigonometry (2)

The sine and cosine rules; area problems

Pre-requisites

- properties of polygons; geometrical constructions; trigonometric ratios

The sine rule

In any triangle $\triangle ABC$, the angles are usually denoted by the capital letters, A, B, C and the sides opposite these angles by a, b, c respectively.

Consider triangle ABC shown in Fig. 8.1(a) (acute angled) and Fig. 8.1(b) (obtuse-angled).

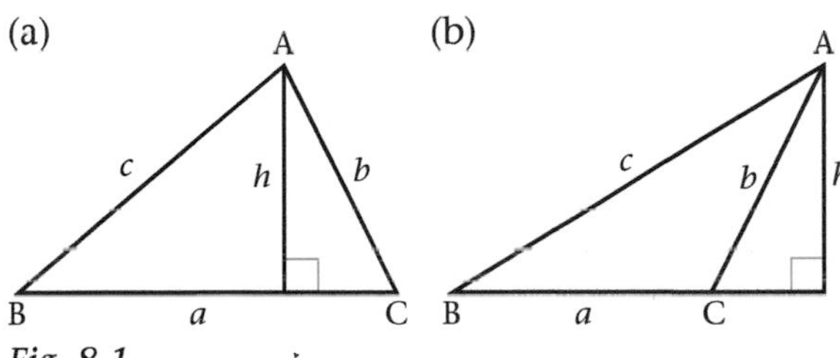

Fig. 8.1

To show that

$$\frac{a}{\sin A} = \frac{b}{\sin A} = \frac{c}{\sin A}$$

Draw the perpendicular from A to BC in Fig. 8.1(a), and to BC produced in Fig. 8.1(b).

In Fig. 8.1(a) and (b),

$$\sin B - \frac{h}{c} \qquad (1)$$

In Fig. 8.1(a),

$$\sin C = \frac{h}{b} \qquad (2)$$

In Fig. 8.1(b),

$$\sin (180° - C) = \frac{h}{b}$$

$$\Rightarrow \sin C = \frac{h}{b} \qquad [\sin (180° - \theta) = \sin \theta] \quad (2)$$

From (1), $h = c \sin B$
From (2), $h = b \sin C$

$$\Rightarrow c \sin B = b \sin C$$

$$\Rightarrow \frac{b}{\sin B} = \frac{c}{\sin C}$$

Similarly, by drawing a perpendicular from C to AB, you can show that

$$\frac{a}{\sin A} = \frac{b}{\sin B}$$

$$\Rightarrow \mathbf{\frac{a}{\sin A} = \frac{b}{\sin B} = \frac{c}{\sin C}}$$

This formula is used for solving triangles which are not right-angled and in which either **two angles and any side** are given or **two sides and the angle opposite one of them** are given.

Example 1

In $\triangle ABC$, B = 39°, C = 82°, a = 6.73 cm. Find c.

First draw a sketch of the information (Fig. 8.2).

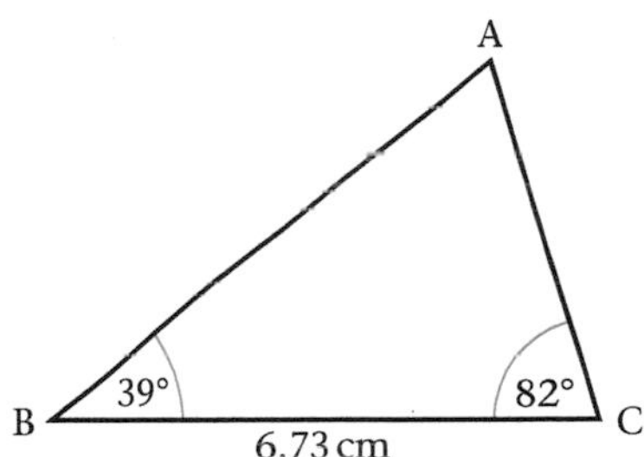

Fig. 8.2

In Fig. 8.2,

$$A = 180° - (39° + 82°)$$
$$= 59°$$

$$\frac{c}{\sin C} = \frac{a}{\sin A}$$

$$\frac{c}{\sin 82°} = \frac{6.73}{\sin 59°}$$

$$\Rightarrow \quad c = \frac{6.73 \times \sin 82°}{\sin 59°} \text{ cm}$$

$$= 7.78 \text{ cm}$$

working:

No	Log
6.73	0.828
sin 82°	$\bar{1}$.996
	0.834
sin 59°	$\bar{1}$.933
7.78°	0.891

In Example 1, when three-figure tables (logarithms and logarithms of sines) are used, the working is set at the side as shown. When a calculator is used care must be taken to input the correct digits, and the values in the correct sequence.

When using the sine rule to calculate angles, the formula

$$\frac{a}{\sin A} = \frac{b}{\sin B} = \frac{c}{\sin C}$$

may be rearranged as

$$\frac{\sin A}{a} = \frac{\sin B}{b} = \frac{\sin C}{c}$$

Example 2

Find the remaining angles of △ABC in which a = 12.5 cm, c = 17.7 cm and C = 116°.
First make a sketch of the information.

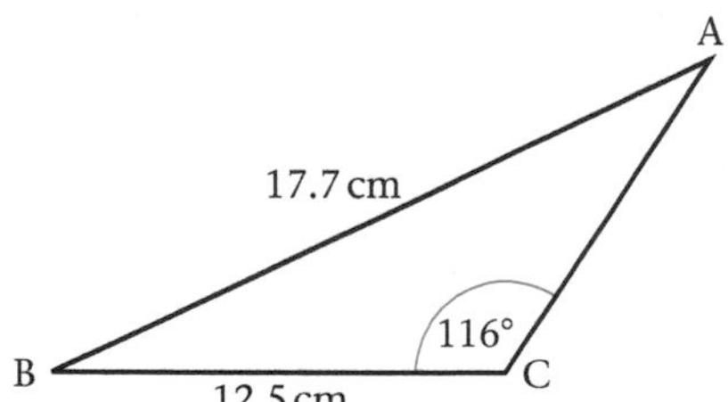

Fig. 8.3

In Fig. 8.3, c and C are known. Since a is also given, A can be found using the arrangement of the sine rule which gives the unknown first:

$$\frac{\sin A}{a} = \frac{\sin C}{c}$$

$$\frac{\sin A}{12.5} = \frac{\sin 116°}{17.7}$$

$$\Rightarrow \quad \sin A = \frac{12.5 \times \sin 116°}{17.7}$$

$$= \frac{12.5 \times \sin 64°}{17.7}$$

$$\Rightarrow \quad A = 39.4° \text{ or } (180 - 39.4°)$$

$$= 39.4° \text{ or } 140.6°$$

working:

No	Log
12.5	1.097
sin 64°	$\bar{1}$.954
	1.051
17.7	1.248
sin 39.4°	$\bar{1}$.803

But C is obtuse, therefore A cannot be obtuse.

$\Rightarrow$ A = 39.4° (or 39°24′)
B = 180° − (39.4 + 116)°
= 180° − 155. 4°
= 24.6° (or 24°36′)

Example 3

In △ABC, a = 7.1 cm, b = 9.5 cm and B = 63°18′. Solve the triangle completely.
To solve a triangle *completely* means to calculate all the unknown sides and angles.
Make a sketch of the given information.

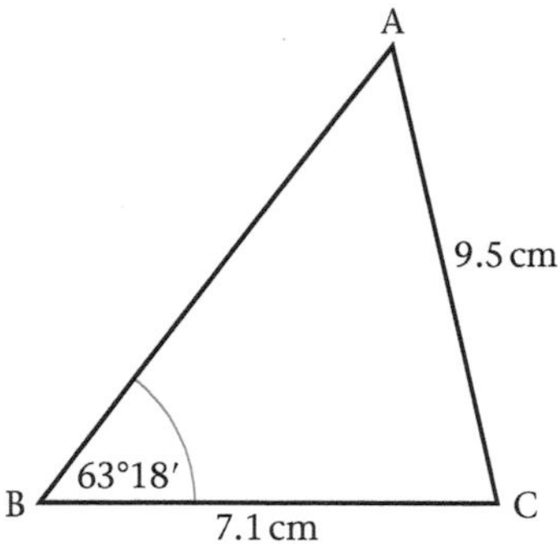

Fig. 8.4

$$63°18' = 63\tfrac{18}{60}° = 63\tfrac{3}{10}° = 63.3°$$

$$\frac{\sin A}{a} = \frac{\sin B}{b}$$

$$\frac{\sin A}{7.1} = \frac{\sin 63.3°}{9.5}$$

$$\Rightarrow \quad \sin A = \frac{7.1 \times \sin 63.3°}{9.5}$$

$$A = 41.8° \text{ or } (180° - 41.8°)$$

$$= 41.8° \text{ or } 138.2°$$

working:

No	Log
7.1	0.851
sin 63.3°	$\bar{1}$.951
	0.802
9.5	0.978
sin 41.8°	$\bar{1}$.824

But $a < b$
$\Rightarrow$ A < B

Note that in a triangle, smaller angles are opposite shorter sides.

∴ A = 41.8°
$\Rightarrow$ C = 180° − (63.3 + 41.8)°
= 180° − 105.1° = 74.9°

$$\frac{c}{\sin C} = \frac{b}{\sin B}$$

$$\frac{c}{\sin 74.9°} = \frac{9.5}{\sin 63.3°}$$

$$\Rightarrow c = \frac{9.5 \times \sin 74.9°}{\sin 63.3°} \text{ cm} = 10.3 \text{ cm}$$

working:

No	Log
9.5	0.978
sin 74.9°	$\bar{1}$.985
	0.963
sin 63.3°	$\bar{1}$.951
10.3	1.012

A = 41.8°, C = 74.9°, c = 10.3 cm
or
A = 41°48′, C= 74°54′, c = 10.3 cm

Exercise 8a

1

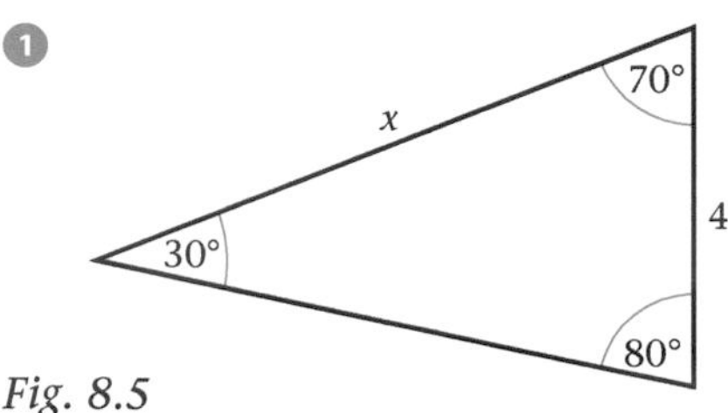

Fig. 8.5

Using the information given in Fig. 8.5 write down, but do not solve, an equation that can be used to find x.

2 In $\triangle$ABC, A = 29°, B = 36°, b = 15.8 cm. Find a.

3 In $\triangle$ABC, A = 54°12′, B = 71°30′, a = 12.4 cm. Find b.

4 In $\triangle$ABC, B= 104.3°, C = 31.3°, a = 29.0 cm. Calculate c.

5 In $\triangle$PQR, P = 83°, p = 285 m, r = 216 m. Calculate R.

6 In $\triangle$ABC, C = 53°, b = 3.56 m, c = 4.28 m. Calculate B.

7 In $\triangle$ABC, A = 115°, a = 65 m, b = 32 m. Solve the triangle completely.

8 In $\triangle$ABC, B = 25°36′, C = 124°24′, c = 39.2 m. Solve the triangle completely.

The cosine rule

Consider the $\triangle$ABC (acute-angled and obtuse-angled) given in Fig. 8.6.

(a)

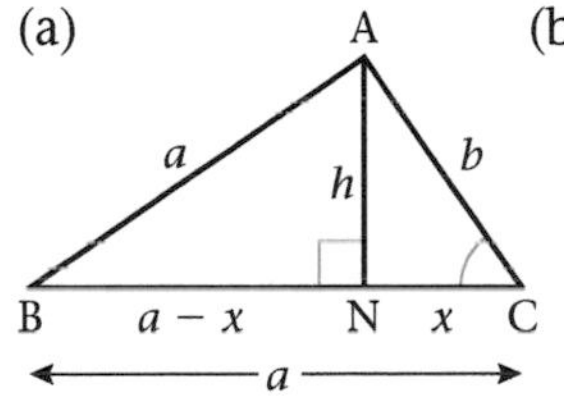

(b)

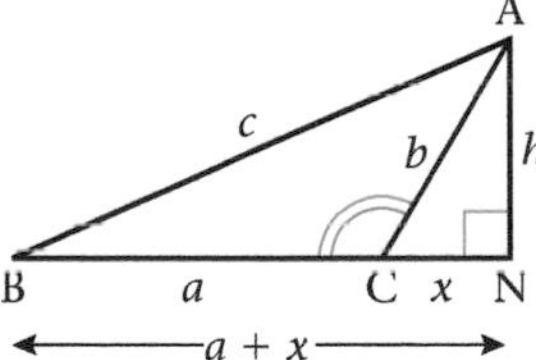

Fig. 8.6

To show that

$$c^2 = a^2 + b^2 - 2ab \cos C$$

Draw the perpendicular from A to BC (produced if necessary).

In Fig. 8.6 (a), with $\widehat{C}$ acute,

$$c^2 = (a - x)^2 + h^2 \qquad \textit{(Pythagoras)}$$
$$= a^2 - 2ax + x^2 + h^2$$
$$= a^2 - 2ax + b^2 \qquad (\textit{In } \triangle ACN,\ x^2 + h^2 = b^2)$$
$$= a^2 + b^2 - 2ab \cos C \qquad (\textit{In } \triangle ACN,\ \frac{x}{b} = \cos C,\ x = b \cos C)$$

In Fig. 8.6(b), with $\widehat{C}$ obtuse,

$$c^2 = (a + x)^2 + h^2 \qquad \textit{(Pythagoras)}$$
$$= a^2 + 2ax + x^2 + h^2$$
$$= a^2 + 2ax + b^2 \qquad (\textit{In } \triangle ACN,\ x^2 + h^2 = b^2)$$
$$\text{but } \frac{x}{b} = \cos A\widehat{C}N \qquad (\textit{in } \triangle ACN)$$
$$= \cos (180° - C)$$
$$= -\cos C$$
$$x = -b \cos C$$
$$\therefore\ c^2 = a^2 + b^2 + 2a(-b \cos C) \quad \textit{(substituting for x)}$$
$$= a^2 + b^2 - 2ab \cos C$$

In either case, $\mathbf{c^2 = a^2 + b^2 - 2ab \cos C}$

Similarly, $\mathbf{b^2 = a^2 + c^2 - 2ac \cos B}$

and $\mathbf{a^2 = b^2 + c^2 - 2bc \cos A}$

This formula is for solving triangles which are not right-angled in which **two sides and the included angle** are given.

Example 4

Find AB in Fig. 8.7.

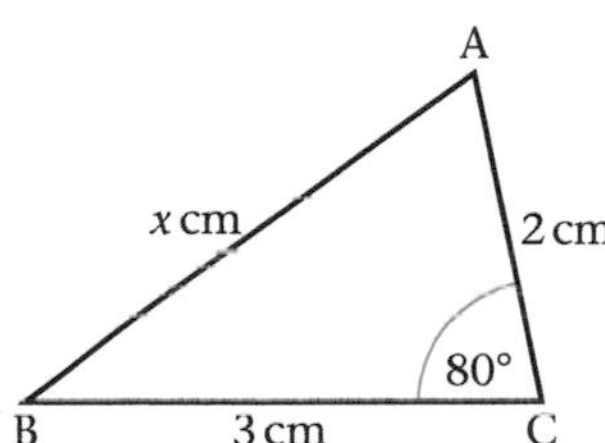

Fig. 8.7

By the cosine rule,

$$x^2 = 2^2 + 3^2 + 2 \times 2 \times 3 \times \cos 80°$$
$$= 4 + 9 - 12 \times 0.174$$
$$= 13 - 2.088$$
$$= 10.912 = 10.9 \text{ to 3 s.f.}$$
$$x = \sqrt{10.9}$$
$$x = 3.30 = 3.3 \text{ to 2 s.f.}$$
$$AB \simeq 3.3 \text{ cm}$$

Example 5

Find y in Fig. 8.8.

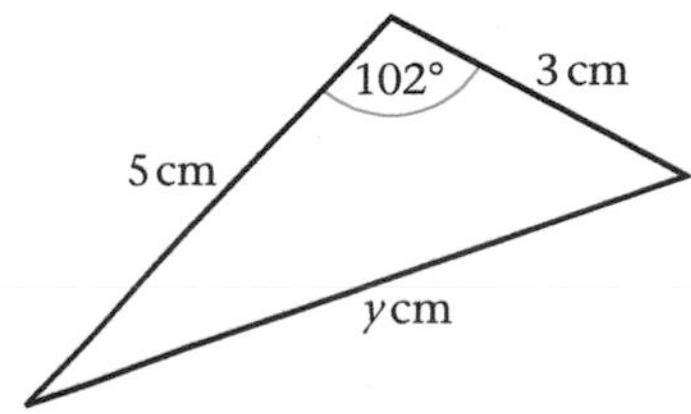

Fig. 8.8

$$\begin{aligned} y^2 &= 5^2 + 3^2 - 2 \times 5 \times 3 \cos 102^\circ \\ &= 25 + 9 - 30\,(-\cos 78^\circ) \\ &= 34 + 30 \cos 78^\circ \\ &= 34 + 30 \times 0.208 \\ &= 34 + 6.24 = 40.24 \\ &= 40.2 \text{ to 3 s.f.} \\ y &= \sqrt{40.2} \\ &= 6.34 \end{aligned}$$

Notice that in Example 4, $x^2 < 2^2 + 3^2$, and in Example 5, $y^2 > 5^2 + 3^2$.

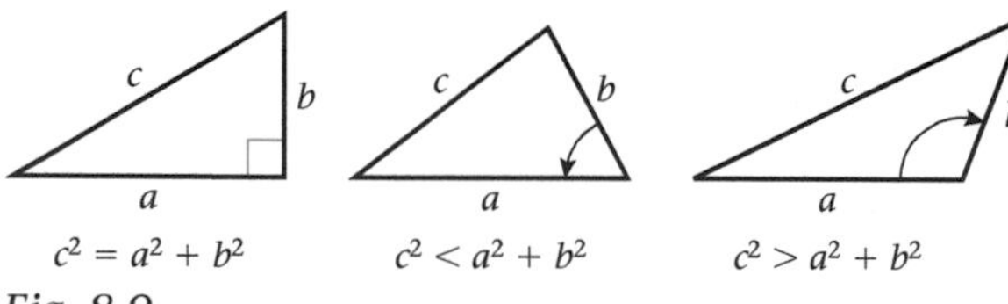

Fig. 8.9

Fig. 8.9 shows that
$c^2 = a^2 + b^2$ when c is opposite a *right* angle
$c^2 < a^2 + b^2$ when c is opposite an *acute* angle
$c^2 > a^2 + b^2$ when c is opposite an *obtuse* angle

Example 6

In $\triangle$ABC, $c = 8.44$ m, $a = 7.92$ m *and* B = 151.3°. Calculate AC.

First, make a sketch of the data (Fig. 8.10).

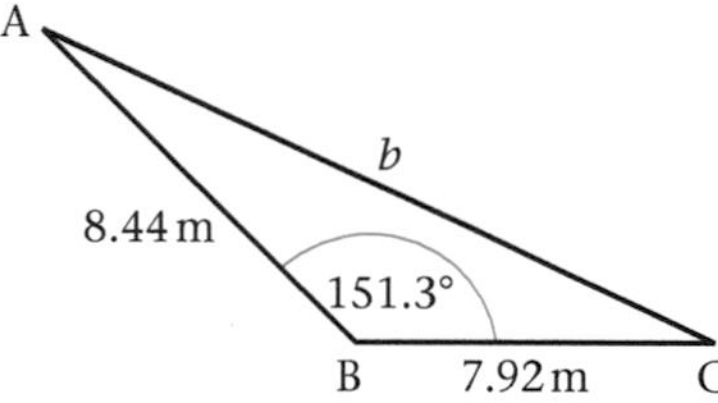

Fig. 8.10

By the cosine rule,

$$\begin{aligned} b^2 &= 8.44^2 + 7.92^2 - 2 \times 8.44 \times 7.92 \times \cos 151.3^\circ \\ &= 71.23 + 62.73 + 2 \times 8.44 \times 7.92 \times \cos 28.7^\circ \\ &= 133.96 + 16.9 \times 7.92 \times \cos 28.7^\circ \\ &= 133.96 + 117 \\ &= 250.96 \\ &= 251 \text{ to 3 s.f.} \\ b &= \sqrt{251} \\ &= 15.8 \end{aligned}$$

AC = 15.8 m

working:

No	Log
16.9	1.228
7.92	0.899
cos 28.7°	1.943
117	2.070

Notice the use of tables of squares, logarithms and square roots in Example 6, but note that a calculator may be used.

Exercise 8b

Calculate the length of the side opposite the given angle in each of the $\triangle$s ABC given.

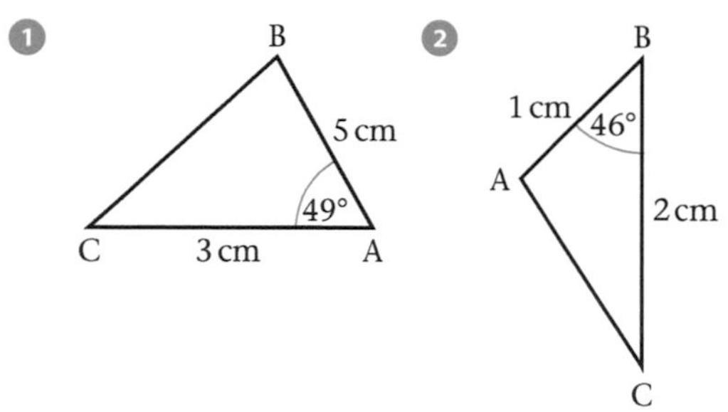

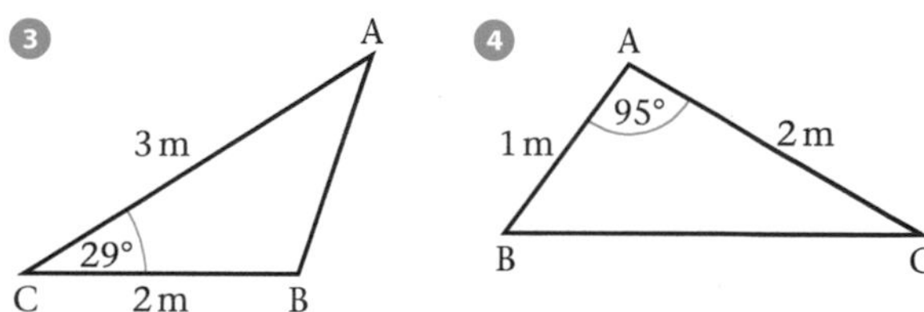

Fig. 8.11

5. A = 120°, $b = 7$ cm, $c = 12$ cm
6. B = 54°, $c = 4$ cm, $a = 5$ cm
7. C = 13°, $a = 10$ m, $b = 15$ m
8. B = 135.5°, $c = 8$ cm, $a = 5$ cm

Solving triangles using the sine and cosine rules

Example 7

In $\triangle$ABC, $a = 6.7$ cm, $c = 2.3$ cm and B = 46.6°. Find b, A and C.

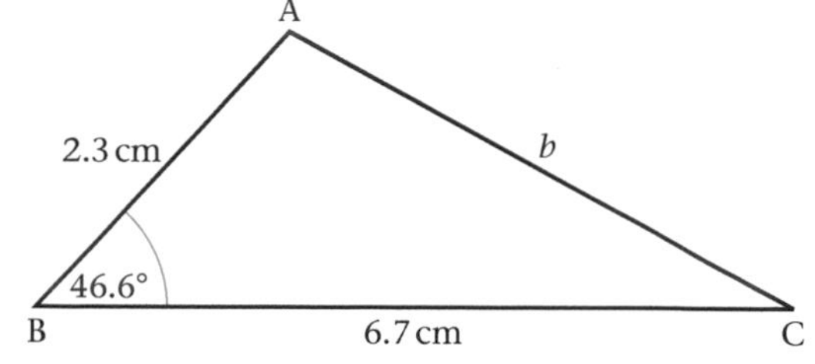

Fig. 8.12

In Fig. 8.12, using the cosine rule,

$b^2 = 2.3^2 + 6.7^2 - 2 \times 2.3 \times 6.7 \times \cos 46.6°$
$= 5.29 + 44.89 - 4.6 \times 6.7 \times \cos 46.6°$
$= 50.18 - 21.2$
$= 29.0$ to 3 s.f.
$b = \sqrt{29.0} = 5.39$ cm

working:

No	Log
4.6	0.663
6.7	0.826
cos 46.6°	$\bar{1}.837$
21.2	1.323

Using the sine rule,

$$\frac{\sin C}{2.3} = \frac{\sin 46.6°}{5.39}$$

$$\sin C = \frac{2.3 \sin 46.6°}{5.39}$$

$C = 18.05°$
$A = 180° - (46.6 + 18.05)°$
$= 180° - 64.65° = 115.35°$
$b = 5.4$ cm, $C = 18.1°$, $A = 115.4°$, correct to 1 d.p.

Notice the following points in Example 7.

1 The cosine rule is used to find b.
2 The sine rule is used to find one of the remaining angles.
3 The *smaller* of the two unknown angles is found first, since this must be an acute angle. If the sine rule had been used to find A, log sin A would have been $\bar{1}.956$, which gives A = 64.6° *or* 115.4°. To avoid this ambiguity, always find the smaller angle first, since this angle must be acute.
4 Rounding off numbers only takes place at the *final* answer stage. Do *not* use rounded numbers at intermediate stages of the calculation.

Exercise 8c

Calculate the unknown side and angles in each of the △s ABC given.

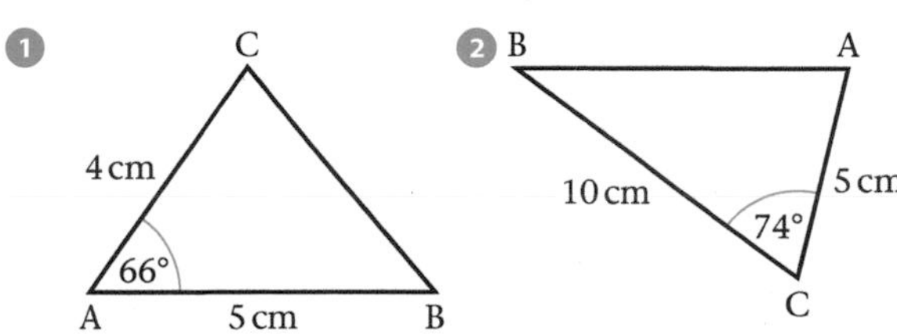

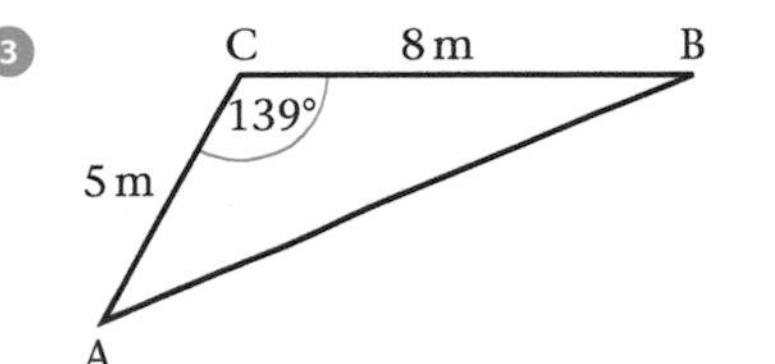

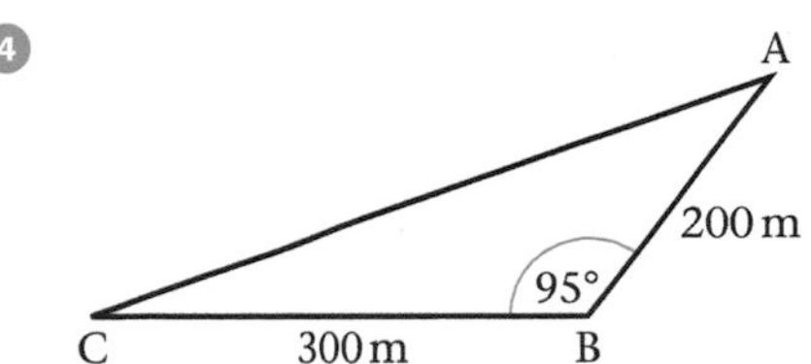

Fig. 8.13

5 $A = 58.1°$, $b = 10$ m, $c = 8.5$ m

6 $B = 126°$, $c = 5.6$ cm, $a = 5$ cm

7 $C = 25.7°$, $b = 3.5$ cm, $a = 6$ cm

8 $A = 140.2°$, $b = 45$ m, $c = 24$ m

Using the cosine rule to calculate angles

The formula $a^2 = b^2 + c^2 - 2bc \cos A$ can be rearranged with cos A as the subject of the formula:

$$\cos A = \frac{b^2 + c^2 - a^2}{2bc}$$

Similarly, $$\cos B = \frac{c^2 + a^2 - b^2}{2ca}$$

and $$\cos C = \frac{a^2 + b^2 - c^2}{2ab}$$

This formula can be used to calculate the angles of a triangle in which *all three sides* are given.

Example 8

Calculate the angles of a triangle with sides of length 4 m, 5 m and 7 m.

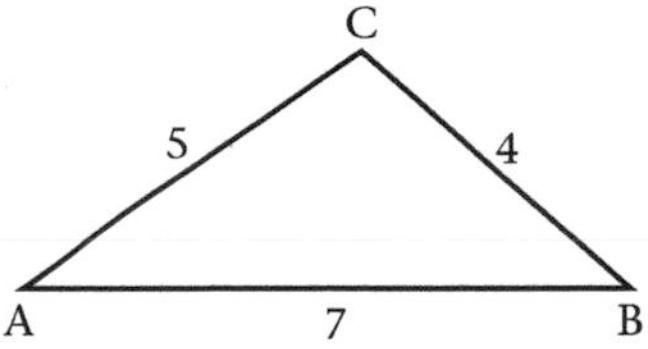

Fig. 8.14

With the lettering of Fig. 8.14,

$$\cos A = \frac{5^2 + 7^2 - 4^2}{2 \times 5 \times 7} = \frac{58}{70} = 0.829$$

$\Rightarrow \quad A = 34°$

$$\cos B = \frac{4^2 + 7^2 - 5^2}{2 \times 4 \times 7} = \frac{40}{56} = \frac{5}{7} = 0.714$$

$\Rightarrow \quad B = 44.4°$

$$\cos C = \frac{4^2 + 5^2 - 7^2}{2 \times 4 \times 5} = \frac{-8}{40} = -\frac{1}{5} = 0.200$$

$\Rightarrow \quad C = 180° - 78.4° = 101.6°$

Check:
$A + B + C = 34 + 44.4° + 101.6° = 180°$

In questions such as Example 8, it is advisable to use the cosine formula to find every angle, then to check the results by addition.

Example 9

Calculate the angles of triangles which have sides (a)40 m, 500 m, 700 m, (b)2.8 cm, 4.2 cm, 5.6 cm.

In both cases, the calculation is simplified by considering similar triangles (i.e. equiangular) which have less complex numbers as sides.

(a) 400 : 500 : 700 = 4 : 5 : 7
Solve the triangle with sides of 4, 5 and 7 units, as in Example 8.

(b) 2.8 : 4.2 : 5.6 = 28 : 42 : 56
= 2 : 3 : 4
Solve the triangle with sides of 2, 3 and 4 units. This example is left as an exercise.

Example 10

The sides of a parallelogram are 7 cm and 10 cm and one of its diagonals is 15 cm. Use the cosine formula to find the length of the other diagonal.

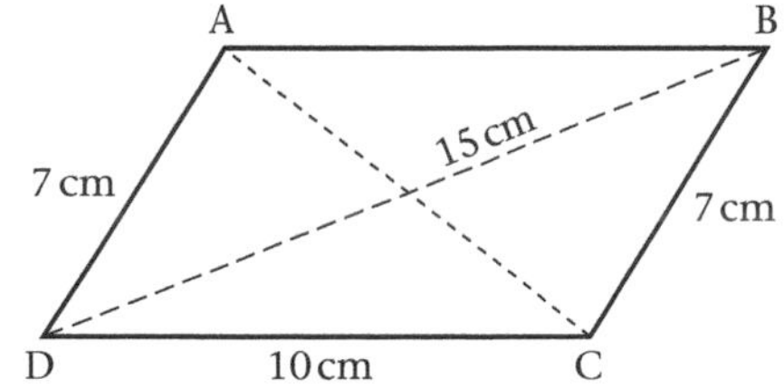

Fig. 8.15

In △BDC,

$$\cos C = \frac{10^2 + 7^2 - 15^2}{2 \times 10 \times 7} = \frac{149 - 225}{140} = \frac{76}{140}$$

In parallelogram ABCD,

$\widehat{ADC} + \widehat{DCB} = 180°$ *(adjacent angles of* $\|^{gm}$*)*

$\Rightarrow \quad \widehat{ADC} = 180° - \widehat{DCB}$

$\Rightarrow \quad \cos \widehat{ADC} = \cos(180° - \widehat{DCB})$

$\Rightarrow \quad \cos \widehat{ADC} = -\cos \widehat{DCB} = \frac{76}{140}$

In △ADC,

$AC^2 = 7^2 + 10^2 - 2 \times 7 \times 10 \times \cos \widehat{ADC}$
$= 49 + 100 - 140 \times \frac{76}{140} = 149 - 76 = 73$
$AC = \sqrt{73}\text{ cm} = 8.54\text{ cm}$

The other diagonal is approximately 8.54 cm long.

Notice in Example 10 that it was not necessary to find the value of C in degrees.

Exercise 8d

1. In △ABC in Fig. 8.16, m is the mid-point of BC. (a) Calculate cos B in nABC. (b) Hence calculate AM. (*Note*: Do not find B in degrees.)

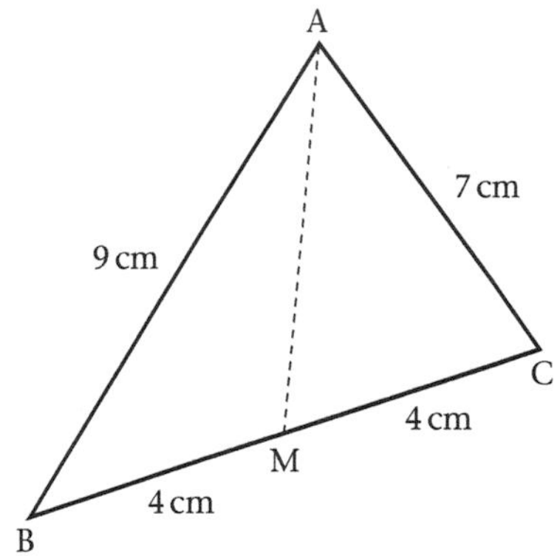

Fig. 8.16

2. With the data of Fig. 8.17, calculate (a) $\widehat{ABC}$, (b) AC, all lengths being given in cm.

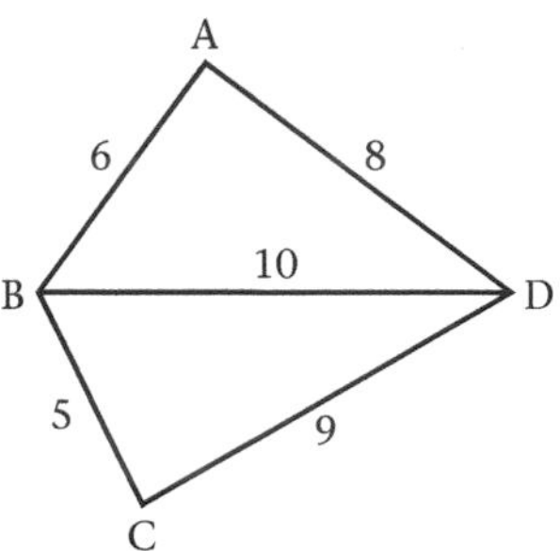

Fig. 8.17

3. In Fig 8.18, find x and θ. All lengths are given in cm.

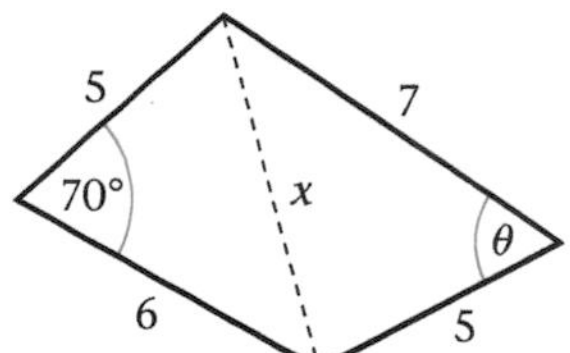

Fig. 8.18

4. In Fig. 8.19, PQRS is a cyclic quadrilateral, PQ = 7 cm, QR = 8 cm and PR = 7.5 cm.

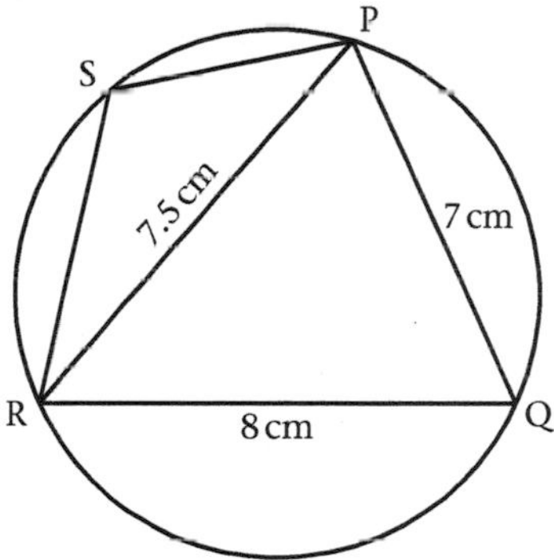

Fig. 8.19

(a) Calculate $P\widehat{S}R$.

(b) Hence if SR = SP, calculate $S\widehat{P}R$.

5. In a $\triangle$ABC, AB = 8 cm, BC = 4 cm, CA = 5 cm and BC is produced to P so that CP = 4 cm. Use the cosine rule to find cos ACB. Hence find AP.

6. Given Fig. 8.20, find cos P in $\triangle$PQR. Hence find QX.

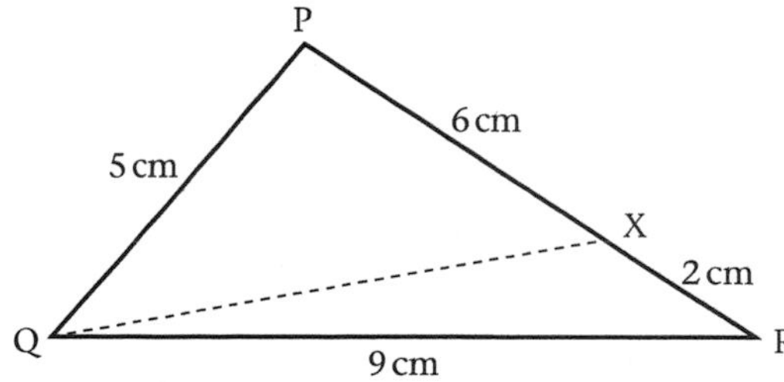

Fig. 8.20

7. The sides of a parallelogram are 3 cm and 5 cm and include an angle of 144°. Find the lengths of the diagonals of the parallelogram.

8. Calculate y in Fig. 8.21. (First find cos θ, but do not work out θ.)

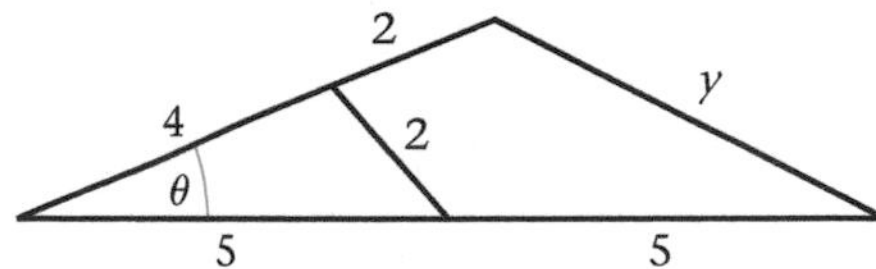

Fig. 8.21

9. In $\triangle$PQR, $p : q : r = \sqrt{3} : 1 : 1$. Calculate the ratio $\widehat{P} : \widehat{Q} : \widehat{R}$ in its simplest form.

10. In Fig. 8.22 ABCD is a trapezium with sides of lengths as shown.

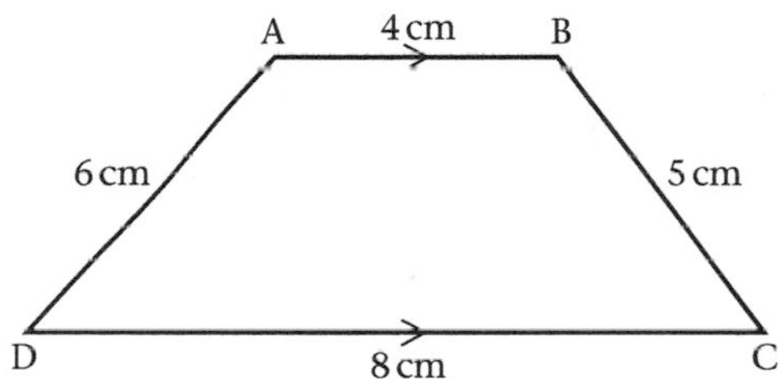

Fig. 8.22

Copy Fig. 8.22 and draw a line BX parallel to AD to cut CD in X. Hence calculate (a) C, (b) BD.

Using trigonometry in area problems

Example 11

Find the area of $\triangle$ABC to the nearest cm² if BA = 6 cm, BC = 7 cm and $\widehat{B}$ = 34°.

Let the height of the triangle be x cm.

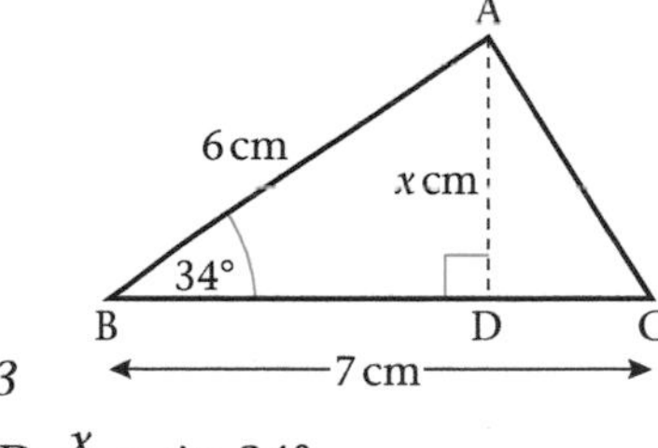

Fig. 8.23

In $\triangle$ABD, $\frac{x}{6} = \sin 34°$

$$x = 6 \sin 34° = 6 \times 0.559 = 3.354$$

$$\begin{aligned}\text{Area of } \triangle ABC &= \tfrac{1}{2} \times BC \times AD \\ &= \tfrac{1}{2} \times 7 \times 3.354\,\text{cm}^2 \\ &= \tfrac{1}{2} \times 23.478\,\text{cm}^2 \\ &= 11.739\,\text{cm}^2 \\ &= 12\,\text{cm}^2 \text{ to the nearest cm}^2\end{aligned}$$

Example 12

Calculate the area of parallelogram PQRS if QR = 5 cm, RS = 6 cm, $Q\widehat{R}S = 118°$.

In Fig. 8.24 QD is the height of the parallelogram.

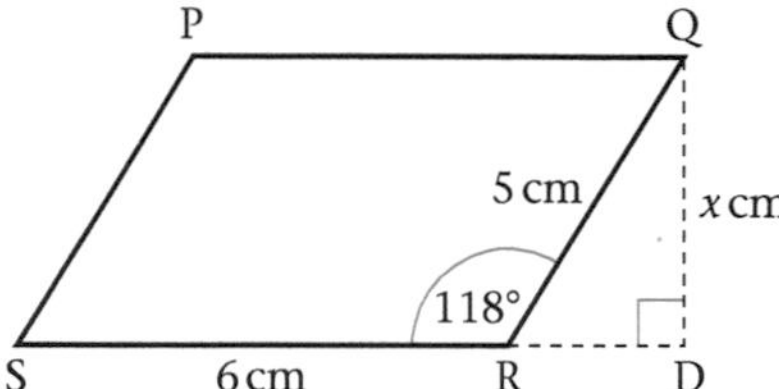

Fig. 8.24

Let QD be x cm.
In $\triangle$QRD, $Q\widehat{R}D = 180° - 118° = 62°$

$$\frac{x}{5} = \sin 62°$$
$$x = 5 \sin 62°$$
$$= 5 \times 0.883 = 4.415$$

$$\text{Area of PQRS} = \text{SR} \times \text{QD}$$
$$= 6 \times 4.415\ \text{cm}^2$$
$$= 26.490\ \text{cm}^2$$
$$= 26\ \text{cm}^2 \text{ to the nearest cm}^2$$

In Examples 11 and 12, since the data of the questions are given in whole numbers of cm and degrees, it is appropriate to give the results to the nearest whole number of cm^2.

Example 13

Calculate the area of the trapezium in Fig. 8.25.

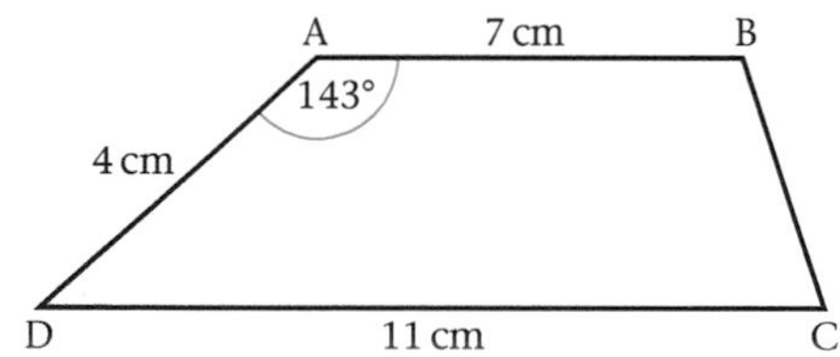

Fig. 8.25

Construct the height AP of the trapezium as in Fig. 8.26.

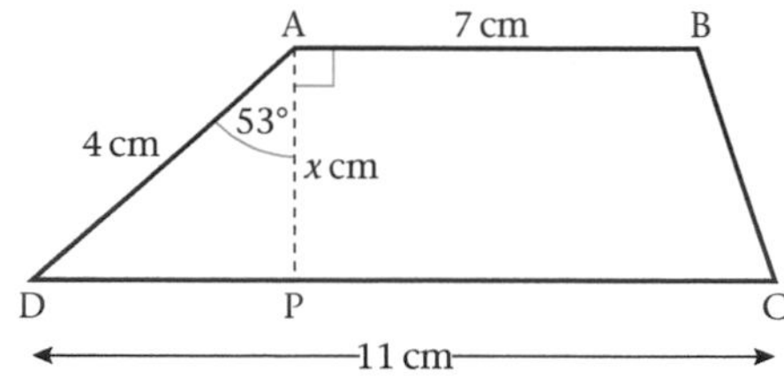

Fig. 8.26

In $\triangle$ADP, $D\widehat{A}P = 143° - 90° = 53°$

$$\frac{x}{4} = \cos 53°$$
$$x = 4 \cos 53°$$
$$= 4 \times 0.602 = 2.408$$

$$\text{Area of ABCD} = \tfrac{1}{2}(\text{AB} + \text{DC}) \times \text{AP}$$
$$= \tfrac{1}{2}(7 + 11) \times 2.408\ \text{cm}^2$$
$$= \tfrac{1}{2} \times 18 \times 2.408\ \text{cm}^2$$
$$= 9 \times 2.408\ \text{cm}^2$$
$$= 21.672\ \text{cm}^2$$
$$= 22\ \text{cm}^2 \text{ to 2 s.f.}$$

Notice in Examples 11, 12 and 13 that rounding off is only done at the *last* stage of the working. Do not round off at an earlier stage.

Exercise 8e

Give all answers to a suitable degree of accuracy.

1. Find the areas of the triangles in Fig. 8.27. All dimensions are in cm.

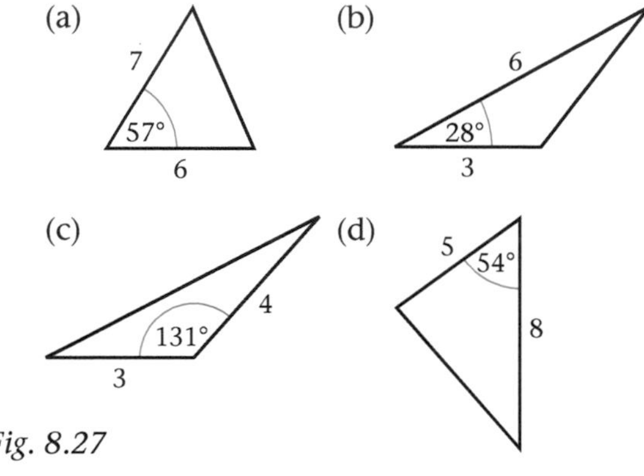

Fig. 8.27

2. Find the areas of the parallelograms in Fig. 8.28. All dimensions are in cm.

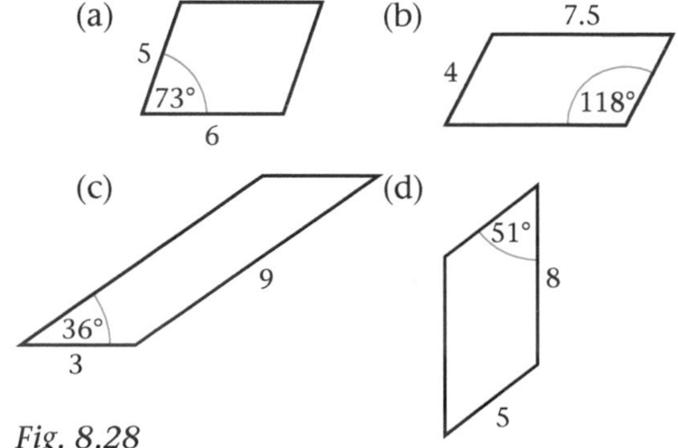

Fig. 8.28

3. Find the areas of the trapeziums in Fig. 8.29. All dimensions are in metres.

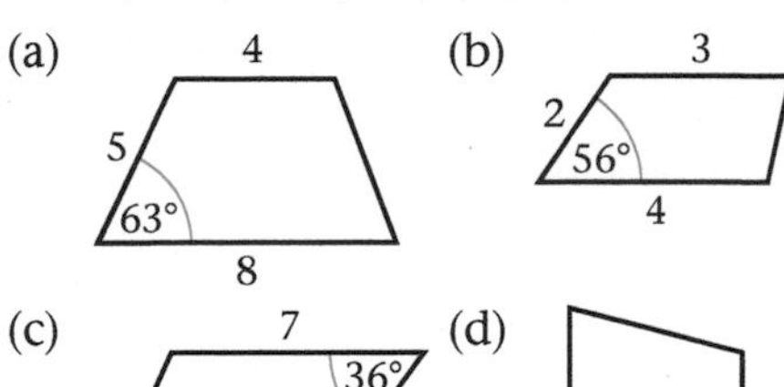

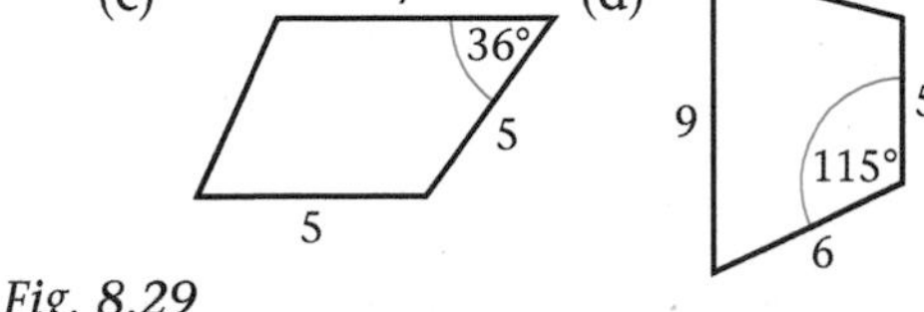

Fig. 8.29

4. In Fig. 8.30, PQRS is a parallelogram, HSR is a straight line and $\text{H}\widehat{\text{P}}\text{Q} = 90°$. If HQ = 10 cm and PQ = 6 cm, what is the area of the parallelogram?

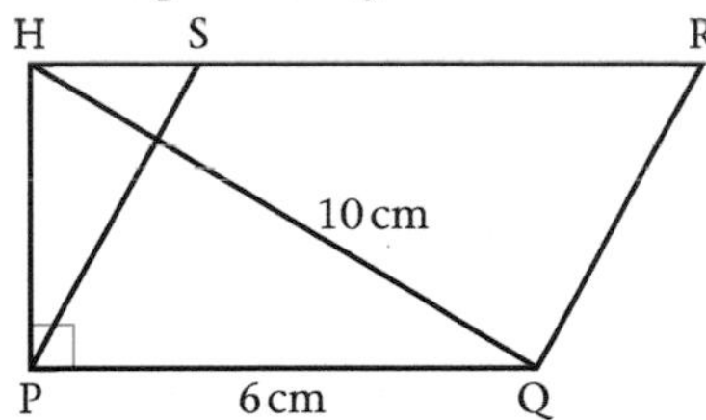

Fig. 8.30

5. Calculate the area of each shape in Fig. 8.31. correct to 2 s.f. All dimensions are in metres.

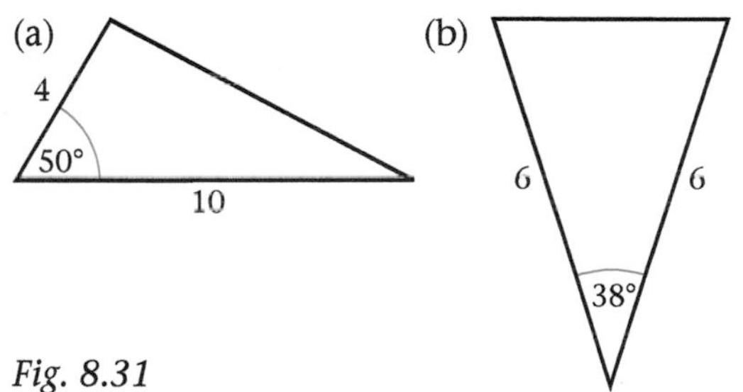

Fig. 8.31

6. PQRS is a trapezium in which PQ||SR, PQ = 16 cm, QR = 12 cm, RS = 8 cm and $\text{P}\widehat{\text{Q}}\text{R} = 30°$ (Fig. 8.32). Calculate the area of PQRS.

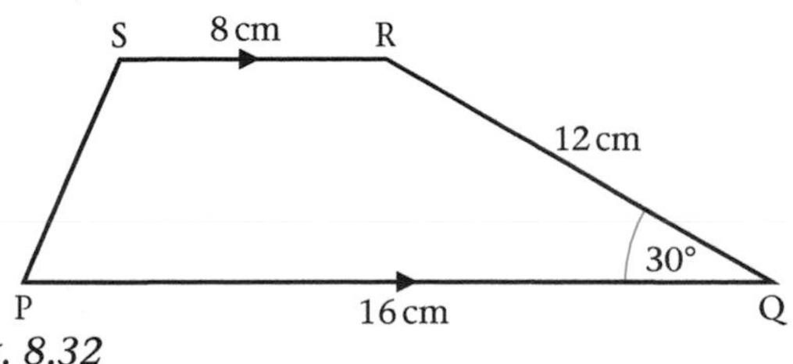

Fig. 8.32

7. The area of a parallelogram ABCD is 43 cm². AB = 7 cm, BC = 9 cm and $\text{A}\widehat{\text{B}}\text{C} < 90°$. Calculate $\text{A}\widehat{\text{B}}\text{C}$ to the nearest degree.

Summary

Solution of triangles

The sine and cosine formulae can be used to find unknown sides and angles of non-right-angled triangles.

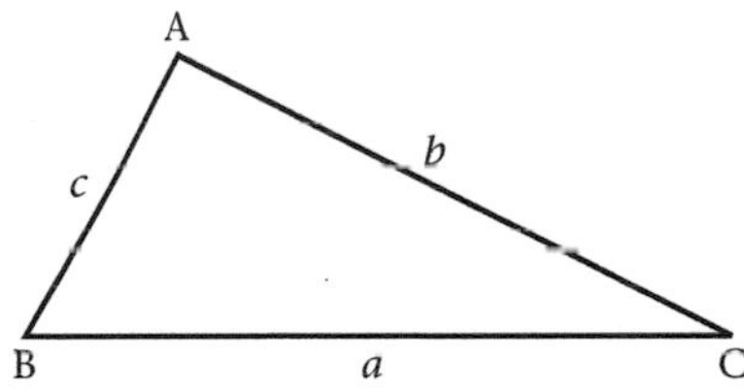

Fig. 8.33

With the lettering of Fig. 8.33.

sine formula

$$\frac{a}{\sin A} = \frac{b}{\sin B} = \frac{c}{\sin C} \quad \textit{or}$$

$$\frac{\sin A}{a} = \frac{\sin B}{b} = \frac{\sin C}{c}$$

cosine formula

$$a^2 = b^2 + c^2 - 2bc\cos A \quad \textit{or} \quad \cos A = \frac{b^2 + c^2 - a^2}{2bc}$$

$$b^2 = a^2 + c^2 - 2ac\cos B \quad \textit{or} \quad \cos B = \frac{a^2 + c^2 - b^2}{2ac}$$

$$c^2 = a^2 + b^2 - 2ab\cos C \quad \textit{or} \quad \cos C = \frac{a^2 + b^2 - c^2}{2ab}$$

Trigonometry is often used in finding the area of shapes.

triangle **parallelogram**

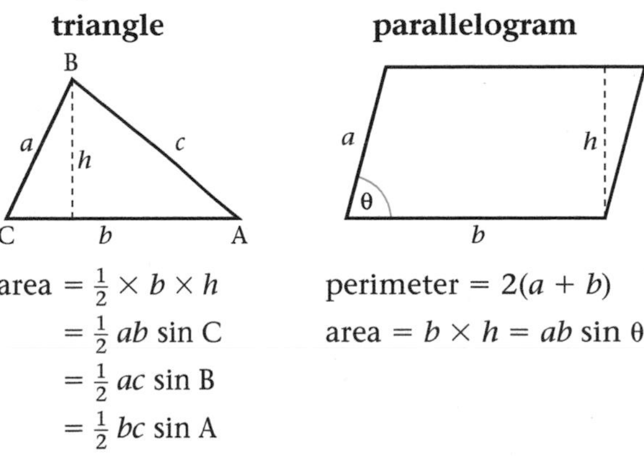

area $= \frac{1}{2} \times b \times h$

$= \frac{1}{2}ab\sin C$

$= \frac{1}{2}ac\sin B$

$= \frac{1}{2}bc\sin A$

perimeter $= 2(a + b)$

area $= b \times h = ab\sin\theta$

Practice exercise P8.1

Give the size of angles to the nearest whole number and the lengths of sides to 1 decimal place.

1. In △ABC, A = 55°, B = 69°, b = 14.9 cm. Find a.
2. In △XYZ, X = 86°, x = 17.7 m, y = 7.3 m. Calculate Y.
3. In △PQR, P = 14°, Q = 77°, r = 9.8 cm. Solve the triangle completely.
4. In △ABC, A = 109°, b = 24.3 m, c = 38.6 m. Find a.
5. In △XYZ, X = 64°, y = 6.4 mm, z = 4.9 mm. Find x.
6. In △PQR, p = 4.1 cm, q = 5.3 cm, r = 8.1 cm. Find P.
7. In △ABC, a = 18.7 cm, b = 15.5 cm, c = 12.2 cm. Solve the triangle completely.
8. In △ABC, A = 129°, b = 9 cm, c = 4.9 cm. Solve the triangle completely.

Practice exercise P8.2

Give your answers to a sensible degree of accuracy.

1. The diagonals of a parallelogram are of lengths 10 cm and 15 cm. The acute angle between the diagonals is 69°. Calculate the lengths of the sides of the parallelogram.
2. A boat sails 12 km from a harbour on a bearing of 210° and then 16.5 km on a bearing of 330°. Calculate the distance of the boat from the harbour.
3. The bearings of B and C from a point T are 042° and 162°. B is 170 km due north of C. Find the distances of B and C from point T.
4. Towns A, B and C are three towns. Town A is 30 km from town B on a bearing of 222° and town C is 50 km due south of town B. Calculate
   (a) the bearing of town B from town A,
   (b) the distance from town A to town C,
   (c) the bearing of town C from town A.

Practice exercise P8.3

Give the size of angles to the nearest whole number and the lengths of sides and areas of shapes to 1 decimal place.

1. In △ABC, A = 53°, b = 2.6 cm, c = 13.2 cm. Find the area of △ABC.
2. Parallelogram ABCD has sides AB = 15.2 cm, BC = 13.8 cm and B = 102°. Calculate the area of parallelogram ABCD.
3. In triangle PQR, PQ = 16.5 cm, QR = 19.4 cm and the area of triangle PQR is 145.6 cm^2. Calculate
   (a) the size of angle Q,
   (b) the length of PR.
4. A triangle KLM has LM = 17.1 cm, K = 41°, M = 47°. Find the area of triangle KLM.
5. PQRS is a quadrilateral with PQ = 4.2 cm, QR = 3.8 cm, RS = 5.6 cm, Q = 65° and S = 53°. Calculate
   (a) the area of triangle PQR,
   (b) the length of PR,
   (c) angle P,
   (d) the area of the quadrilateral PQRS.
6. In triangle PQR, PQ = 6.5 cm, QR = 5.1 cm, RP = 2.2 cm. Calculate
   (a) the size of the angles of the triangle,
   (b) the area of the triangle PQR.

Chapter 9

Three-dimensional problems

Composite solids; three-dimensional problems

Pre-requisites

- properties of solids; areas and volumes; trigonometric ratios; bearings

Surface area and volume of solids

Formulae for the areas and volumes of common solids already found in earlier books of this course are given below.

Prisms

In general,

volume = areas of constant cross-section × perpendicular height
= area of base × height (Fig. 9.1)

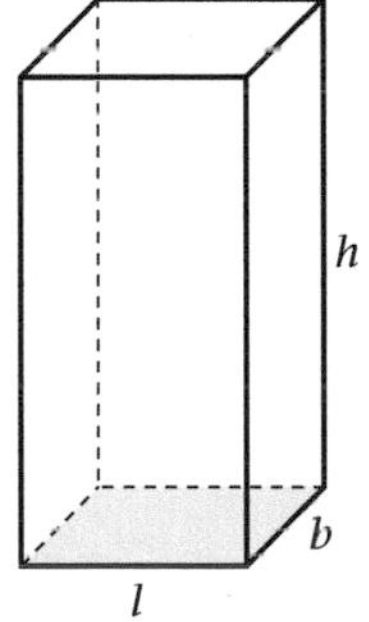

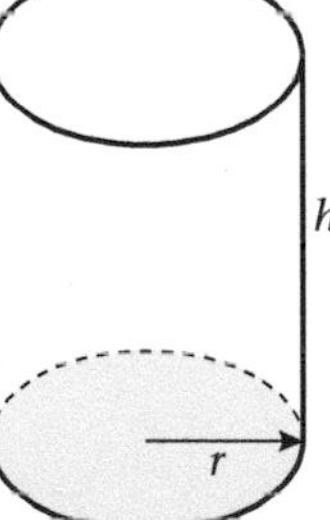

Fig. 9.1

cuboid

volume = lbh

surface area = $2(lb + lb + bh)$

cylinder

volume = $\pi r^2 h$

curved surface area = $2\pi rh$

total surface area = $2\pi rh + 2\pi r^2$
= $2\pi r(h + r)$

Pyramid and cone

In general,

volume = $\frac{1}{3}$ × base area × height (Fig. 9.2)

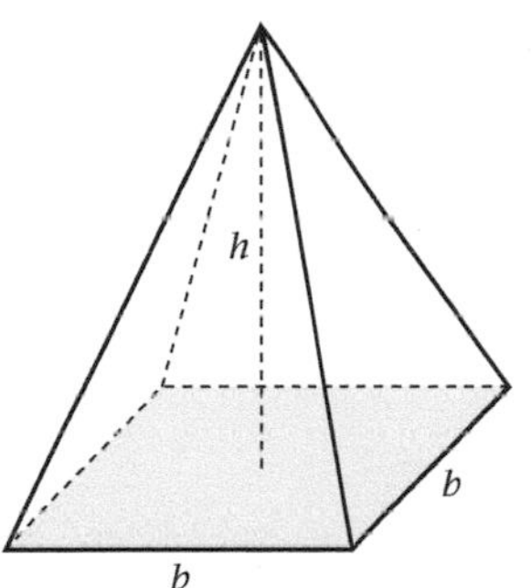

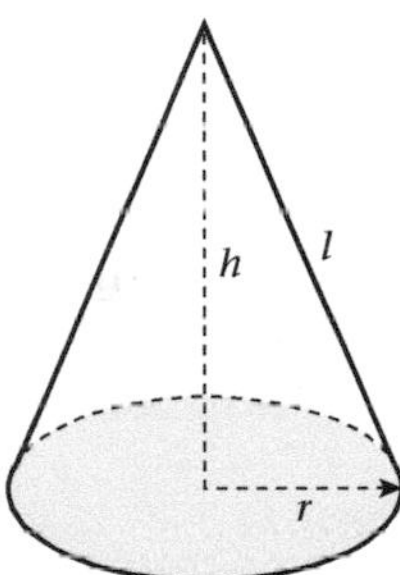

Fig. 9.2

square-based pyramid

volume = $\frac{1}{3}b^2h$

cone

volume = $\frac{1}{3}\pi r^2 h$

curved surface area = πrl

total surface area = $\pi rl + \pi r^2$
= $\pi r(l + r)$

Example 1

A car petrol tank is 0.8 m long, 25 cm wide and 20 cm deep. How many litres of petrol can it hold?

Working in cm,

$$\text{volume of tank} = 80 \times 25 \times 20\,\text{cm}^3$$
$$1\text{ litre} = 1000\,\text{cm}^3$$
$$\text{capacity of tank} = \frac{80 \times 25 \times 20}{1000}\text{ litres}$$
$$= 40\text{ litres}$$

The tank can hold 40 litres of petrol.

Example 2

A circular metal sheet 48 cm in diameter and 2 mm thick is melted and recast into a cylindrical bar 6 cm in diameter. How long is the bar?

Radius of sheet $= \frac{48}{2}$ cm $= 24$ cm
Radius of bar $= \frac{6}{2}$ cm $= 3$ cm
Let the bar be x cm long.

$$\text{Then its volume} = \pi \times 3^2 \times x \text{ cm}^3$$
$$\text{Volume of circular sheet} = \pi \times 24^2 \times \tfrac{1}{5} \text{ cm}^3$$
$$\text{Hence } \pi \times 3^2 \times x = \pi \times 24^2 \times \tfrac{1}{5}$$
$$x = \frac{\pi \times 24^2 \times \frac{1}{5}}{\pi \times 3^2}$$
$$= \frac{576}{9 \times 5}$$
$$= \frac{64}{5} = 12.8$$

The bar is 12.8 cm long.

Notice in Example 2 that no numerical value of π was needed. Never substitute a value for π unless it is necessary.

Example 3

Fig. 9.3 shows a wooden block in the form of a prism. PQRS is a trapezium with PQ||SR, PQ = 7 cm, PS = 5 cm and SR = 4 cm. If the block is 12 cm long, calculate its volume.

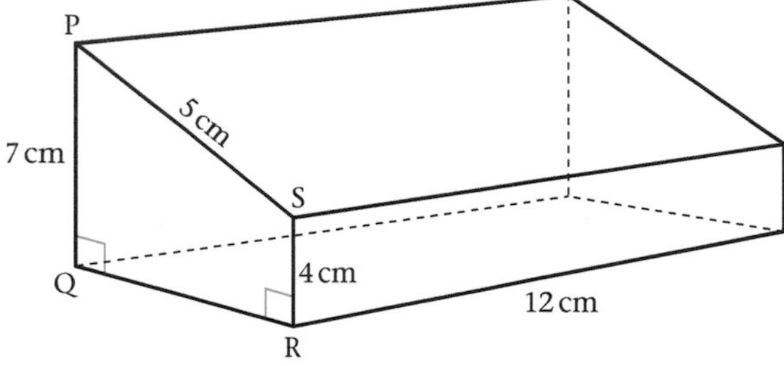

Fig. 9.3

$$\text{Volume of block} = (\text{area of PQRS} \times 12) \text{ cm}^3$$
$$\text{area of block} = (\tfrac{1}{2}(7 + 4) \times \text{QR}) \text{ cm}^2$$

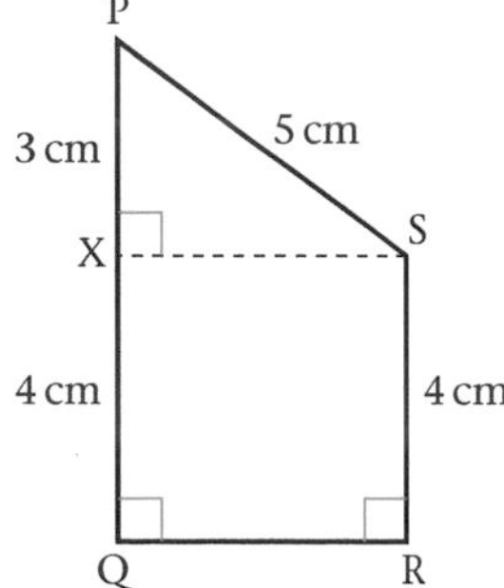

Fig. 9.4

With the construction of Fig. 9.4,
SX = QR
But SX = 4 cm (*the sides of* $\triangle$PXS *form a* 3 : 4 : 5 *pythanorean triple*)

$$\text{Volume of block} = (\tfrac{1}{2}(7 + 4) \times 4 \times 12) \text{ cm}^3$$
$$= (11 \times 2 \times 12) \text{ cm}^3$$
$$= 264 \text{ cm}^3$$

Exercise 9a

Use the value of $3\frac{1}{7}$ for π where necessary.

1. Calculate the volumes of the solids in Fig. 9.5. All lengths are in cm.

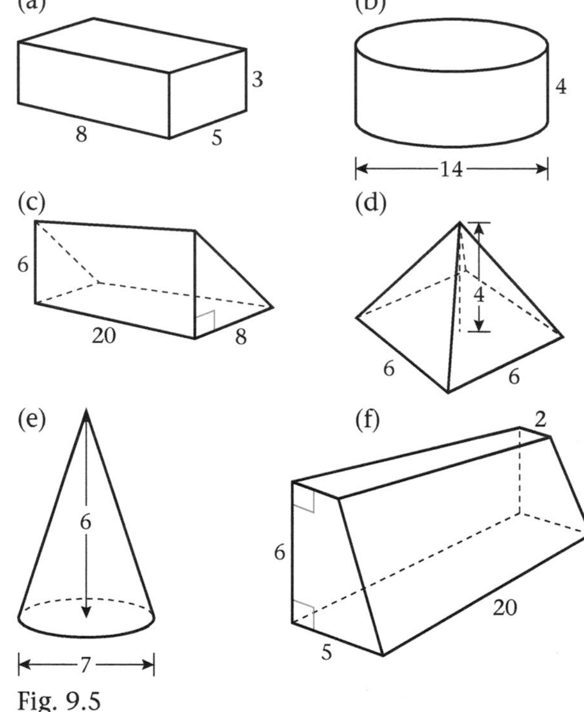

Fig. 9.5

2. Calculate the total surface areas of the solids in parts (a), (b), (c) of Fig. 9.5.

3. A water tanks is 1.2 m square and 1.35 m deep. It is half full of water. How many times can a 9-litre bucket be filled from the tank?

4. $2\frac{1}{2}$ litres of oil are poured into a container whose cross-section is a square of side $12\frac{1}{2}$ cm. How deep is the oil in the container?

5 The diagrams in Fig. 9.6 show the cross-sections of steel beams. All dimensions are in cm. Calculate the volumes, in cm^3, of 5-metre lengths of the beams.

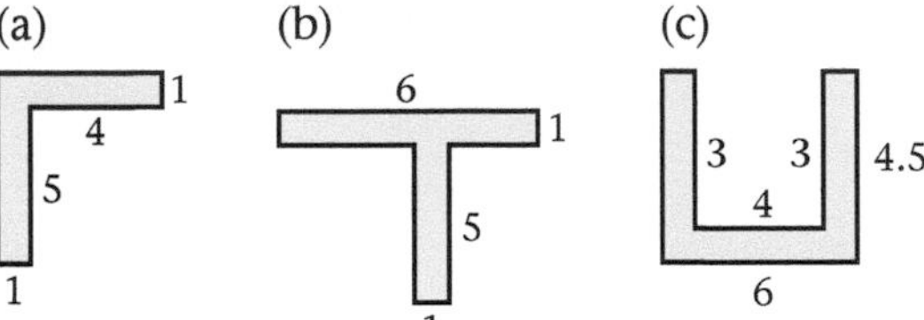

Fig. 9.6

6 Fig. 9.7 shows the cross-section of a steel rail, dimensions being given in cm. Calculate the mass, in tonnes, of a 20-metre length of the rail if the mass of 1 cm^3 of the steel is 7.5 g.

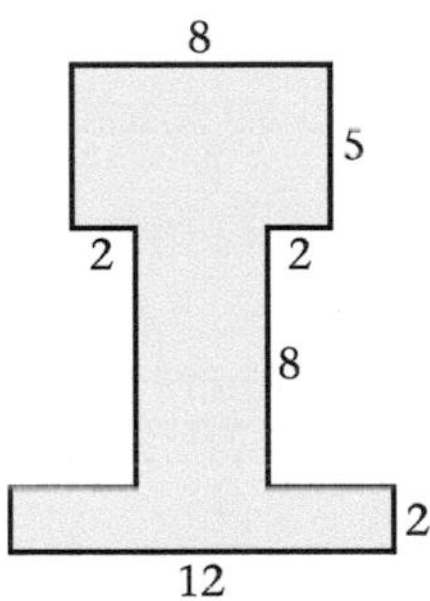

Fig. 9.7

7 Fig. 9.8 shows the cross-section of a ruler.

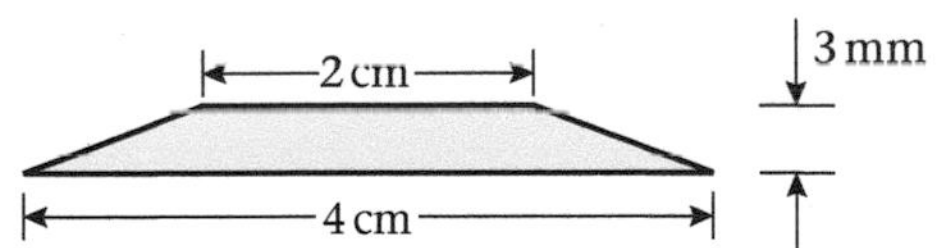

Fig. 9.8

(a) Calculate the volume of the ruler in cm^3 if it is 30 cm long.
(b) If the ruler is made of plastic and has a mass of 45 g, calculate the density of the plastic in g/cm^3.
(c) Find the mass, to the nearest g, of the ruler if it is made of wood of density 0.7 g/cm^3.

8 Calculate, in terms of π, the total surface area of a solid cylinder of radius 3 cm and height 4 cm.

9 A paper label just covers the curved surface of a cylindrical tin of diameter 12 cm and height $10\frac{1}{2}$ cm. Calculate the area of the paper label.

10 A cylindrical tin full of engine oil has a diameter of 12 cm and a height of 14 cm. The oil is poured into a rectangular tin 16 cm long and 11 cm wide. Calculate the depth of the oil in the tin.

11 A cylindrical shoe polish tin is 10 cm in diameter and 3.5 cm deep.
(a) Calculate the capacity of the tin in cm^3.
(b) When full, the tin contains 300 g of polish. Calculate the density of the polish in g/cm^3 correct to 2 d.p.

12 A wire of circular cross-section has a diameter of 2 mm and a length of 350 m. If the mass of the wire is 6.82 kg, calculate its density in g/cm^3.

13 A measuring cylinder of radius 3 cm contains water to a height of 49 cm. If this water is poured into a similar cylinder of radius 7 cm, calculate the height of the water column.

14 A solid metal cylinder, 8 cm in diameter and 8 cm long, is to be made into discs 4 cm in diameter and 5 mm thick. Assuming no wastage, calculate the number of discs that can be made.

15 A paper cone has a base diameter of 8 cm and a height of 3 cm.
(a) Calculate the volume of the cone in terms of π.
(b) Make a sketch of the cone and hence use Pythagoras' theorem to calculate its slant height.
(c) Calculate the curved surface area of the cone in terms of π.
(d) If the cone is cut and opened out into a sector of a circle, calculate the angle of the sector.

Sphere

Fig. 9.9 represents a solid sphere of radius r.

Volume $= \frac{4}{3}\pi r^3$

Curved surface area $= 4\pi r^2$

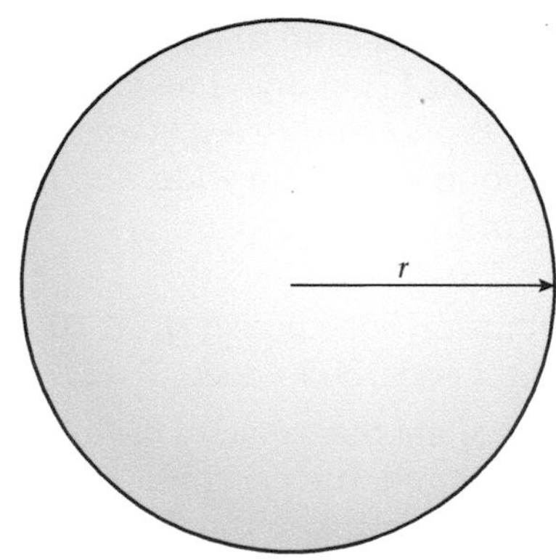

Fig. 9.9

The proof of these formulae is beyond the scope of this course.

Example 4

A solid sphere has a radius of 5 cm and is made of metal of density 7.2 g/cm³. Calculate the mass of the sphere in kg.

Volume of sphere $= (\frac{4}{3} \times 5^3)\,\text{cm}^3$

$= \frac{4 \times \pi \times 125}{3}\,\text{cm}^3$

$= \frac{500\pi}{3}\,\text{cm}^3$

Mass of sphere $= \frac{500\pi}{3} \times 7.2\,\text{g}$

$= \frac{500\pi \times 7.2}{3 \times 1000}\,\text{kg} = \frac{7.2\pi}{3 \times 2}\,\text{kg}$

$= 1.2\pi\,\text{kg} = 1.2 \times 3.14\,\text{kg} = 3.768\,\text{kg}$

$= 3.77$ kg to 3 s.f.

Example 5

Calculate the total surface area of a solid hemisphere of radius 6.8 cm. Use the value 0.497 for log π.

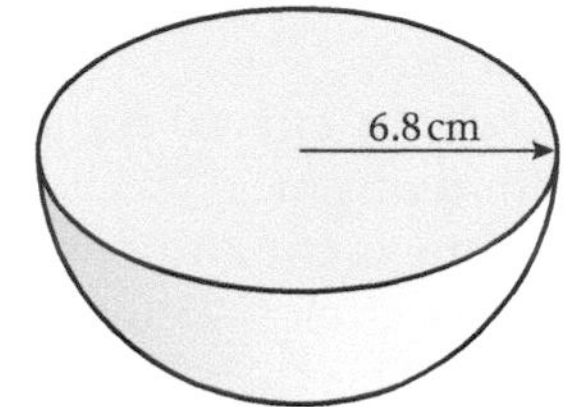

Fig. 9.10

In Fig. 9.10, total surface area

$=$ curved surface area $+$ plane surface area

$= 2\pi r^2 + \pi r^2$

$= 3\pi r^2$

When $r + 6.8$,

total surface area

$= 3 \times \pi \times 6.8^2\,\text{cm}^2$

$= 437\,\text{cm}^2$

working:

No	Log
6.8^2	0.833×2 $= 1.666$
π	0.497
3	0.477
437	2.640

Exercise 9b

Use the value 3.14 for π or 0.497 for log π, whichever is more convenient.

1. Calculate the volume and surface area to 3 s.f. of each of the following.
   (a) A sphere, radius 10 cm
   (b) A sphere, diameter 16 cm
   (c) A hemisphere, radius 2 cm
   (d) A hemisphere, diameter 9 cm
2. The diameter of an iron ball used in 'putting the shot' is 12 cm. If the density of iron is 7.8 g/cm³, calculate the mass of the ball in kg to 3 s.f.
3. A cylinder and sphere both have the same diameter and the same volume. If the height of the cylinder is 36 cm, find their common radius.
4. A metal sphere 6 cm in diameter is melted and cast into balls of diameter $\frac{1}{2}$ cm. Calculate the number of smaller balls made.
5. A sphere has a volume of 1000 cm³.
   (a) Use tables to calculate its radius correct to 3 s.f.
   (b) Hence calculate the surface area of the sphere.

Composite solids

Many **composite solids** can be made by joining basic solids together.

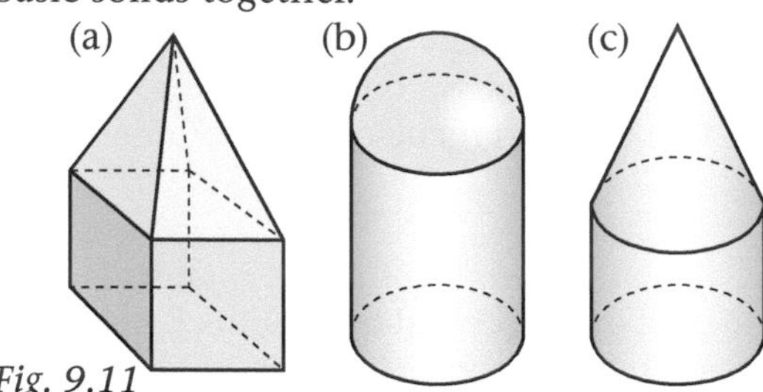

Fig. 9.11

In Fig. 9.11, the composite solids are made as follows:
(a) a cube and a square-based pyramid,
(b) a cylinder and hemisphere,
(c) a cylinder and cone.

Example 6

Fig. 9.12 represents a gas tank in the shape of a cylinder with a hemispherical top. The internal height and diameter are 1 m and 30 cm respectively. Calculate the capacity of the tank to the nearest litre.

With the lettering of Fig. 9.12,
volume of tank
= volume of cylinder + volume of hemisphere
$= \pi r^2 h + \frac{2}{3}\pi r^3$

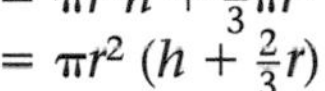

$= \pi r^2 (h + \frac{2}{3}r)$

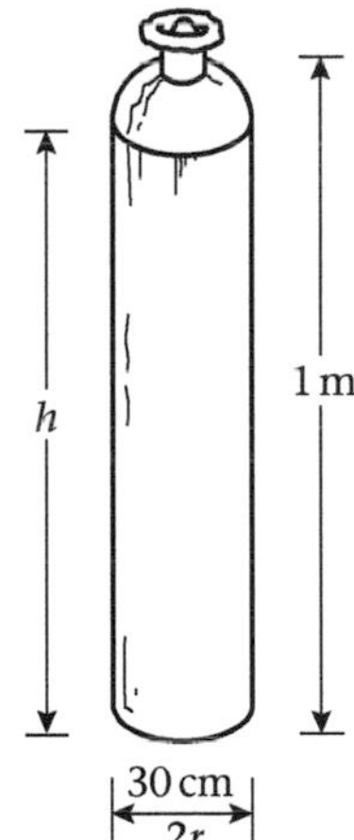

Fig. 9.12

In Fig. 9.12
$r = 15$
$h = 100 - 15 = 85$
volume of tank
$= \pi 15^2 (85 + \frac{2}{3} \times 15)\ \text{cm}^3$
$= 225\pi (85 + 10)\ \text{cm}^3$
$= 225\pi \times 95\ \text{cm}^3$
capacity in litres
$= \frac{225 \times \pi \times 95}{100} = 67.1$
= 67 to the nearest litre

working:

No	Log
225	2.352
π	0.497
95	1.978
	4.827
1000	3.000
67.1	1.827

Some solids are formed by removing a part of an original solid. If a cone or pyramid, standing on a horizontal table, is cut through parallel to the table, the top part is a smaller cone or pyramid. The other part is called a **frustum** (Fig. 9.13).

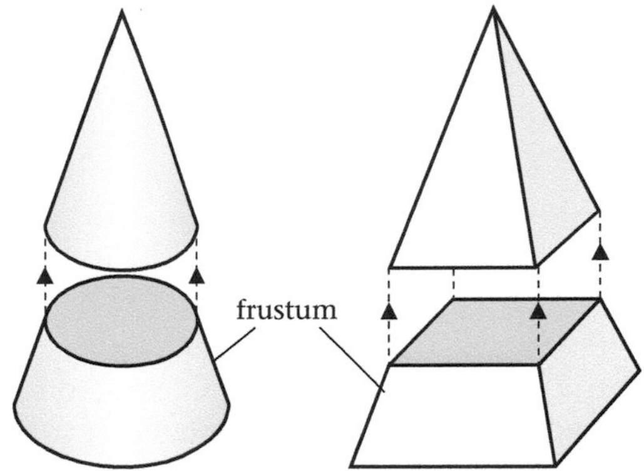

Fig. 9.13

To find the volume or surface of a frustum, it is necessary to consider the frustum as a complete cone (or pyramid) with the smaller cone removed. Read Examples 7 and 8 carefully.

Example 7

Find the capacity in litres of a bucket 24 cm in diameter at the top, 16 cm in diameter at the bottom and 20 cm deep.

Complete the cone of which the bucket is a frustum, i.e. add a cone of height x cm and base diameter 16 cm as in Fig. 9.14

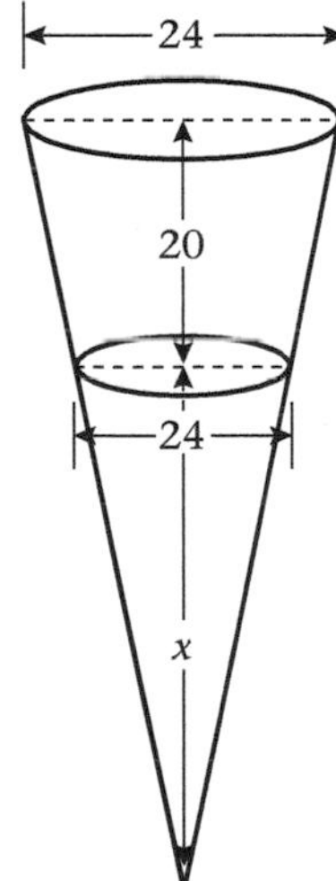

Fig. 9.14

By similar triangles,

$\frac{x}{8} = \frac{x + 20}{12}$
$12x = 8x + 160$
$4x = 160$
$x = 40$

Volume of bucket

$= (\frac{1}{3}\pi\ 12^2 \times 60 - \frac{1}{3}\pi 8^2 \times 40)\ \text{cm}^3$

$= \frac{1}{3}\pi\ (8640 - 2560)\ \text{cm}^3$

$= \frac{1}{3}\pi \times 6080\ \text{cm}^3$

$= 6366\ \text{cm}^3$

Capacity of bucket = 6.37 litres to 3 s.f.

Example 8

Find, in cm^3, the area of material required for a lampshade in the form of a frustum of a cone of which the top and bottom diameters are 20 cm and 30 cm respectively, and the vertical height is 12 cm.

Complete the cone of which the lampshade is a frustum as in Fig. 9.15.

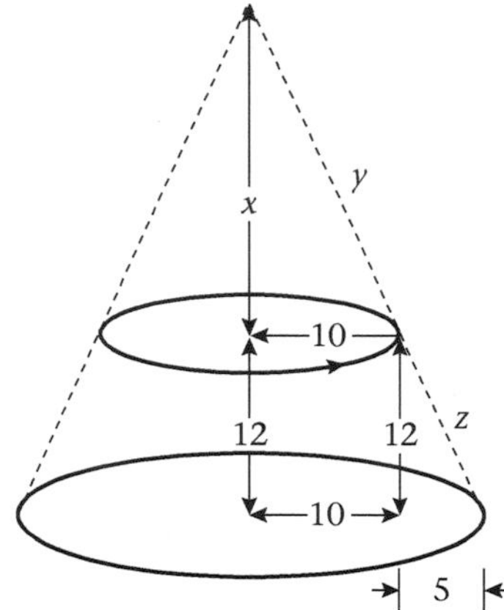

Fig. 9.15

With the lettering of Fig. 9.15, by similar triangles,

$\frac{x}{10} = \frac{12}{5}$ or $\frac{x}{10} = \frac{x + 12}{15}$

$x = 24$

By Pythagoras' theorem, $y = 26$ and $z = 13$.

Surface area of frustum

$= (\pi \times 15 \times 39 - \pi \times 10 \times 26)\ \text{cm}^2$

$= 13\pi(45 - 20)\ \text{cm}^2$

$= (13\pi \times 25)\ \text{cm}^2$

$= 1021\ \text{cm}^2$

Area of material required = 1020 cm^2 to 3 s.f.

Hollow objects such as boxes and pipes, have space inside.

The volume of material in a hollow object is found by subtracting the volume of the space inside from the volume of the object as if it were solid.

Example 9

Fig. 9.16 represents an open rectangular box made of wood 1 cm thick. If the external dimensions of the box are 42 cm long, 32 cm wide and 15 cm deep, calculate the volume of wood in the box.

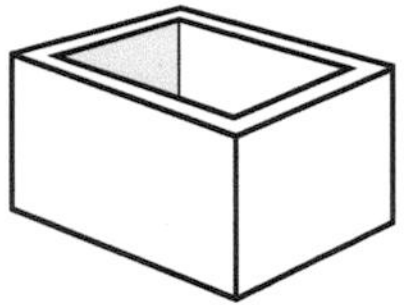

Fig. 9.16

Note that the internal length and width are 2 cm less, but the internal depth is only 1 cm less than the external dimensions.

⇒ the internal measurements of the box are 40 cm long, 30 cm wide and 14 cm deep.

External volume $= (42 \times 32 \times 15)\ \text{cm}^3$
$= 20\,160\ \text{cm}^3$

Internal volume $= (40 \times 30 \times 14)\ \text{cm}^3$
$= 16\,800\ \text{cm}^3$

Volume of wood $= 20\,160\ \text{cm}^3 - 16\,800\ \text{cm}^3$
$= 3360\ \text{cm}^3$

Example 10

Find the mass of a cylindrical iron pipe 2.1 m long and 12 cm in external diameter, if the metal is 1 cm thick and of density 7.8 g/cm^3. Take π to be $\frac{22}{7}$.

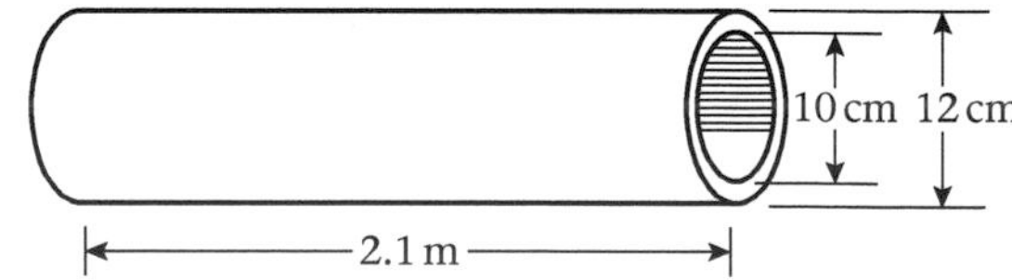

Fig. 9.17

Volume of outside cylinder $= (\pi \times 6^2 \times 210)\ \text{cm}^3$

Volume of inside cylinder $= (\pi \times 5^2 \times 210)\ \text{cm}^3$

Volume of iron

$= (\pi \times 6^2 \times 210 - \pi \times 5^2 \times 210)\ \text{cm}^3$

$= 210\pi(6^2 - 5^2)\ \text{cm}^3$

$= 210\pi(6 - 5)(6 - 5)\ \text{cm}^3$

$= (210\pi \times 11)\ \text{cm}^3$

Mass of iron

$= 210\pi \times 11 \times 7.8\,\text{g}$

$= \dfrac{210 \times 22 \times 11 \times 7.8}{7 \times 1000}\,\text{kg}$

$= 56.6\,\text{kg}$ to 3 s.f.

(The working of the final line should be checked using tables.)

Example 11

Calculate, to one place of decimals, the volume in cm^3 of the metal in a hollow sphere 10 cm in external diameter, the metal being 1 mm thick. Take log π to be 0.497.

Outer radius = 5 cm
Inner radius = 4.9 cm

Volume of metal

$= (\frac{4}{3}\pi \times 5^3 - \frac{4}{3}\pi \times 4.9^3)\,\text{cm}^3$

$= \frac{4}{3}\pi(125 - 4.9^3)\,\text{cm}^3$

$= \frac{4}{3}\pi(125 - 117)\,\text{cm}^3$

$= \dfrac{4 \times \pi \times 8}{3}\,\text{cm}^3$

$= \dfrac{32\pi}{3}\,\text{cm}^3$

$= 33.5\,\text{cm}^3$ to 1 d.p. (calculator accuracy)

working:

No	Log
4.9^3	0.690 × 3
117	2.070

(If tables are used for final calculation there will be no need to round off.)

In Examples 6, 7, 8, 10 and 11, notice how factorisation simplifies the calculation.

Example 12

A right circular cylinder of height 12 cm and radius 4 cm is filled with water. A heavy circular cone of height 9 cm and base-radius 6 cm is lowered, with vertex downwards and axis vertical, into the cylinder until the cone rests on the rim of the cylinder. Find (a) the volume of water which spills over from the cylinder, and (b) the height of the water in the cylinder after the cone has been removed.

Fig. 9.18 shows the position of the cone and cylinder.

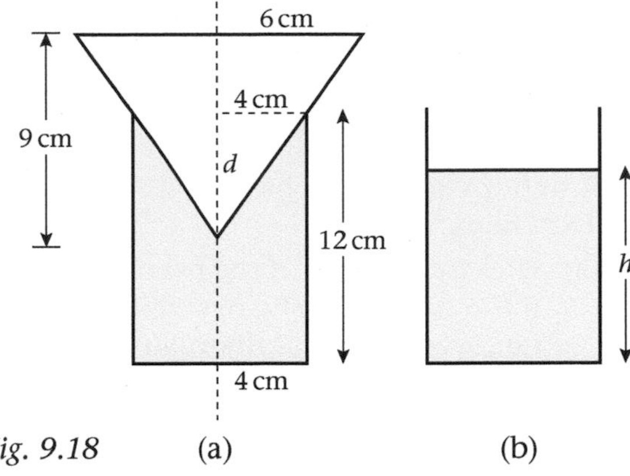

Fig. 9.18 (a) (b)

(a) Let the cone be immersed to a depth d cm. By similar triangles,

$\dfrac{d}{4} = \dfrac{9}{6}$

$d = \dfrac{9 \times 4}{6} = 6$

Volume of water which spills over

$= \text{volume displaced by end of cone}$

$= (\frac{1}{3}\pi \times 4^2 \times d)\,\text{cm}^3$

$= (\frac{1}{3} \times \frac{22}{7} \times 16 \times 6)\,\text{cm}^3$

$= \frac{704}{7}\,\text{cm}^3 = 100\frac{4}{7}\,\text{cm}^3$

(b) Let the height of the water after the cone has been removed be h cm.
Volume of water in Fig. 9.18(a)
= volume of water in Fig. 9.18(b).

$$\pi \times 4^2 \times 12 - \tfrac{1}{3} \times \pi \times 4^2 \times 6 = \pi \times 4^2 \times h$$
$$\pi 4^2(12 - \tfrac{1}{3} \times 6) = \pi 4^2 h$$
$$12 - 2 = h$$
$$h = 10$$

The height of the water will be 10 cm.

Exercise 9c

Use the value $\frac{22}{7}$ for π or 0.497 for log π, whichever is more convenient.

1. An open rectangular box has internal dimensions 2 m long, 20 cm wide and 22.5 cm deep. If the box is made of wood 2.5 cm thick, calculate the volume of the wood in cm^3.
2. An open concrete tank is internally 1 m wide, 2 m long and 1.5 m deep, the concrete being 10 cm thick. Calculate (a) the capacity of the tank in litres, and (b) the volume of concrete used in m^3.

3. A frustum of a cone has top and bottom diameters of 14 cm and 10 cm respectively and a depth of 6 cm. Find the volume of the frustum in terms of π.

4. A right pyramid on a base 10 m square is 15 m high.
   (a) Find the volume of the pyramid.
   (b) If the top 6 m of the pyramid are removed, calculate the volume of the remaining frustum.

5. Fig. 9.19 shows the plan of a foundation which is of uniform width 1 m.

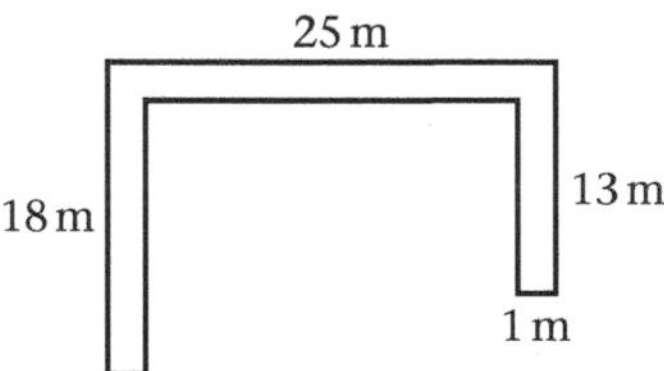

Fig. 9.19

If 1 m^3 of earth has a mass of $1\frac{1}{2}$ tonnes, calculate the mass of earth removed when digging the foundation to a depth of $1\frac{1}{2}$ m.

6. A cast iron pipe has a cross-section as shown in Fig. 9.20, the iron being 1 cm thick. The mass of 1 cm^3 of cast iron is 7.2 g. Calculate the mass of a 2-metre length of the pipe.

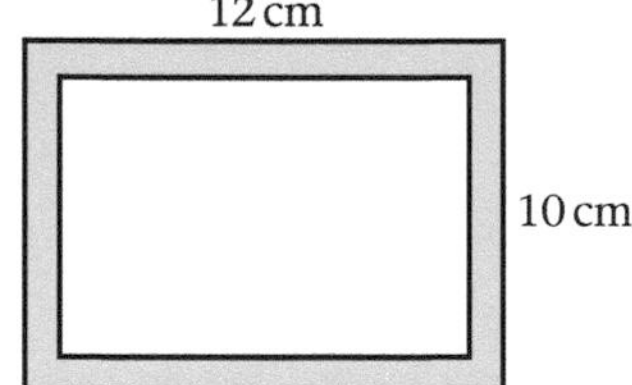

Fig. 9.20

7. A fish tank is in the shape of an open glass cuboid 30 cm deep with a base 16 cm by 17 cm, these measurements being external. If the glass is 0.5 cm thick and its density is 3 g/cm^3, find (a) the capacity of the tank in litres, and (b) the mass of the tank in kg.

8. Fig. 9.21 shows a storage tank made from a cylinder with a hemispherical end. Use the dimensions in Fig. 9.21 to calculate the volume of the tank in m^3.

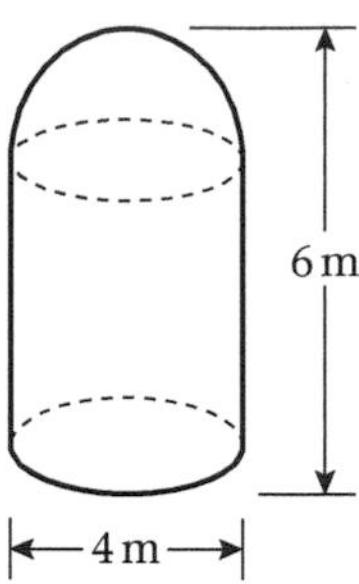

Fig. 9.21

9. Fig. 9.22 shows a cylindrical casting of height 8 cm and external and internal diameters 9 cm and 5 cm respectively. Calculate the volume of metal in the casting.

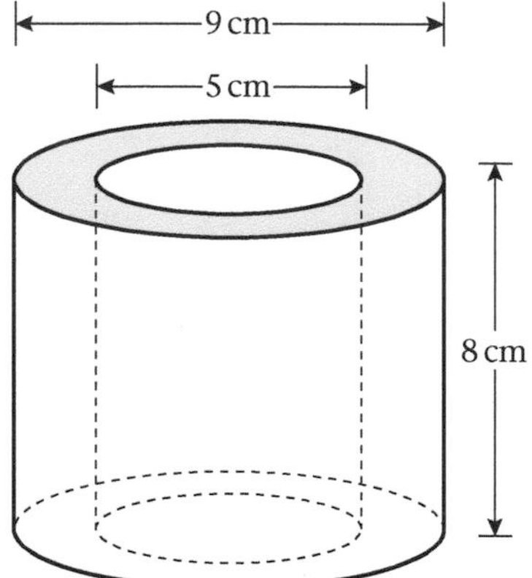

Fig. 9.22

10. A rectangular box, 18 cm by 12 cm by 6 cm, contains six tennis balls, each of diameter 6 cm. Calculate the percentage of the volume of the box occupied by the tennis balls.

11. Calculate the approximate mass in kg of a 2 m length of cylindrical clay pipe of external and internal diameters 15 cm and 12 cm. The density of clay is 1.3 g/cm^3.

12. A lampshade like that of Fig. 9.15 has a height of 12 cm and upper and lower diameters of 10 cm and 20 cm. Calculate
    (a) the area of material required to cover the curved surface of the lampshade.
    (b) the volume of the lampshade.
    (Give both answers in terms of π.)

13 The volume of a right circular cone is 5 litres. Calculate the volumes of the two parts into which the cone is divided by a plane parallel to the base, one-third of the way down from the vertex to the base. Give your answers to the nearest cm^3.

14 A storage container is in the form of a frustum of a right pyramid 4 m square at the top and 2.5 m square at the bottom. If the container is 3 m deep, calculate its capacity in m^3.

15 The cone in Fig. 9.23 is exactly half full of water by volume. Calculate the depth of water in the cone.

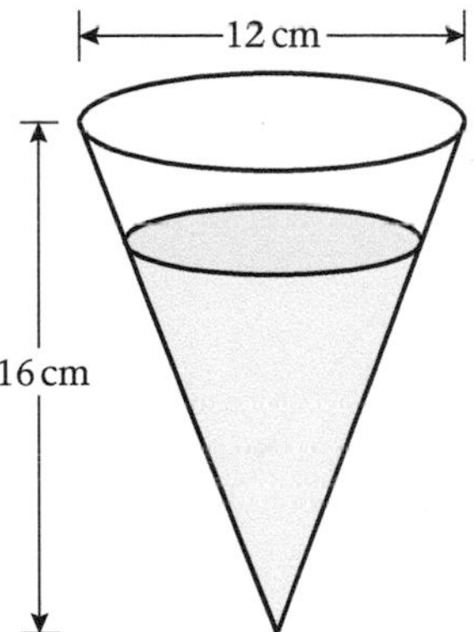

Fig. 9.23

16 A conical funnel 12 cm deep and 15 cm in diameter is full of liquid. It is emptied into a cylindrical tin 10 cm in diameter. Calculate the height of the liquid in the tin.

17 A cylindrical tin of internal diameter 8 cm contains water to a depth of 6 cm. How far does the water level rise when a heavy ball of diameter 6 cm is placed in the tin?

18 A solid cube of side 8 cm is dropped into a cylindrical tank of radius 7 cm. Calculate the rise in the water level if the original depth of water was 9 cm.

19 An iron ball, 6 cm in diameter, is placed in a cylindrical tin 12 cm in diameter. Water is poured into the tin until its depth is 8 cm. If the ball is now removed, find how far the water level drops.

20 A solid aluminium casting for a pulley consists of three discs, each $1\frac{1}{2}$ cm thick, of diameters 4 cm, 6 cm and 8 cm. A central hole 2 cm in diameter is drilled out as in Fig. 9.24. If the density of aluminium is 2.8 g/cm^3, calculate the mass of the casting.

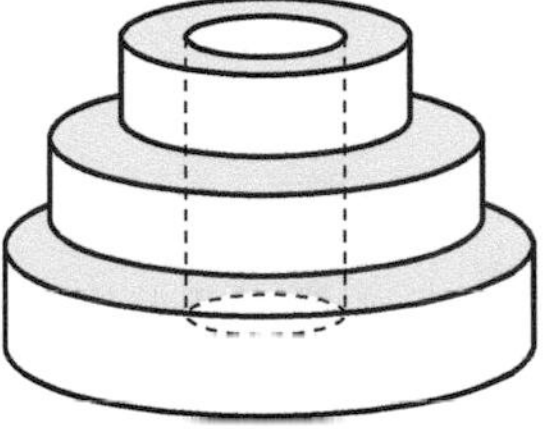

Fig. 9.24

Three-dimensional problems

Example 13

A 216° sector of a circle of radius 5 cm is bent to form a cone. Find the radius of the base of the cone and its vertical angle.

In Fig. 9.25, the radius of the base of the cone is r cm and the vertical angle is 2α.

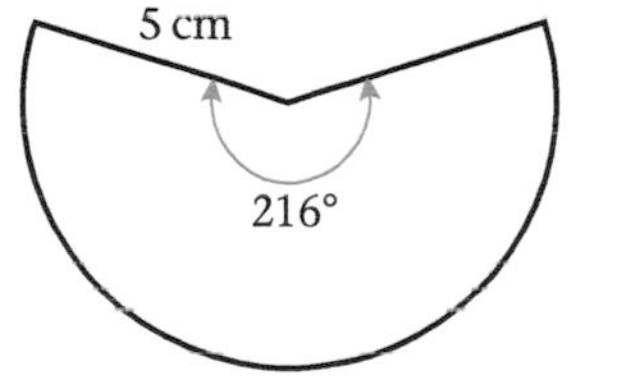

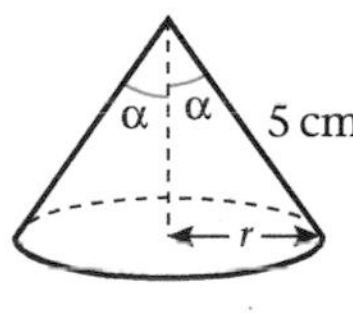

Fig. 9.25

Circumference of base of cone
= length of arc of sector

$$2\pi r = \frac{216}{360} \times 2\pi \times 5$$

$$r = \frac{216}{360} \times 5 = 3$$

$$\sin\alpha = \tfrac{3}{5} \times 0.600$$

$$\alpha = 36.9°$$

$$2\alpha = 73.8°$$

Radius of base = 3 cm
Vertical angle = 73.8°

Example 14

A semicircle of radius 80 mm is folded into a cone. Calculate (a) the angle of slope of the curved surface, that is, the slant edge of the cone, (b) the height of the cone to the nearest mm.

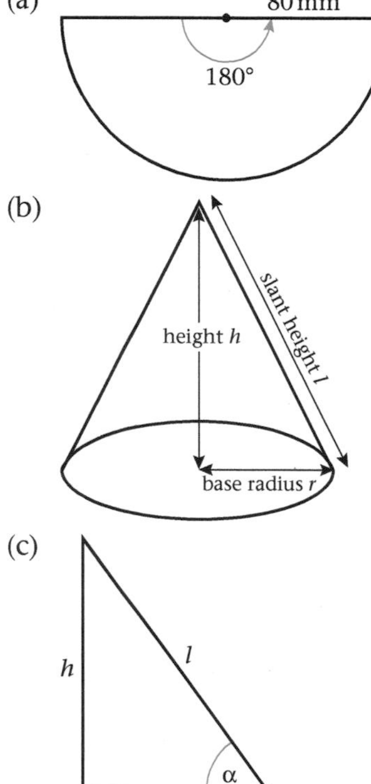

Fig. 9.26

Since the arc of the semicircle of Fig. 9.26(a) becomes the circumference of the base of the cone of Fig. 9.26(b),

then, $\frac{1}{2}(2\pi \times 80) \times 2\pi \times$ base radius of cone

base radius of cone, $r = 80\,\text{mm} \times \frac{1}{2} \times 40\,\text{mm}$

slant height of cone, $l = 80\,\text{mm}$ (radius of semicircle)

(a) In Fig. 9.26(c)

$$\frac{r}{l} = \cos\alpha \quad (\alpha \text{ is angle of slope})$$

$$\frac{40}{80} = \cos\alpha = \frac{1}{2}$$

$$\alpha = 60°$$

(b) $h^2 = l^2 - r^2$

$$= 80^2 - 40^2$$

$$h = \sqrt{4800} = 40\sqrt{3} = 69.2\,\text{mm}$$

$$= 69\,\text{mm to nearest mm}$$

Example 15

A tower is on a bearing 036° from a point Y and 322° from a point X. If X is 200 m due east of Y and the angle of elevation of the top of the tower from Y is 12°, calculate (a) the distance of the tower from Y, (b) the height of the tower to 2 s.f.

(a)

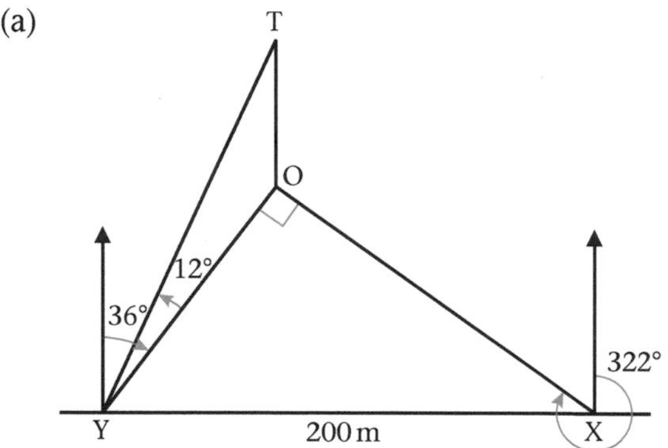

(b)

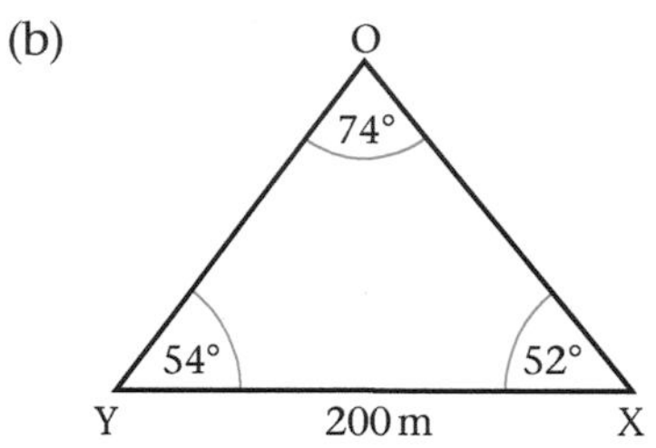

(c)

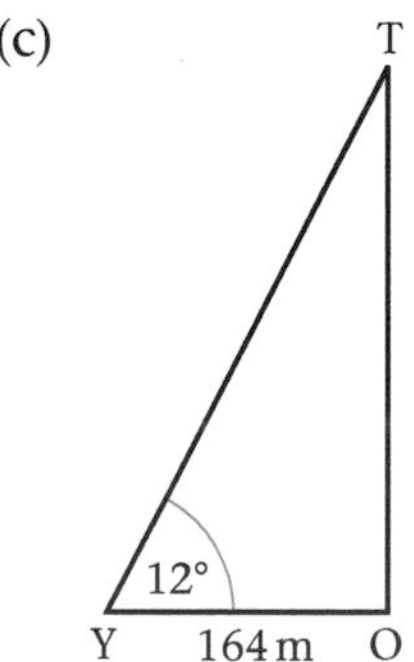

Fig. 9.27

In Fig. 9.27(a) △OYX is in a horizontal plane

$O\hat{Y}X = 90° - 36° - 54°$

$O\hat{X}Y = 322° - 270° - 52°$

$Y\hat{O}X = 180° - (54 + 52)°$

$= 74°$

(a) For the distance of the tower from Y, in $\triangle OYX$ (Fig. 9.27(b))

$$\frac{OY}{\sin 52^\circ} = \frac{200}{\sin 74^\circ}$$

$$OY = \frac{200 \sin 52^\circ}{74^\circ}$$

$$= 164\,\text{m}$$

working:

No	Log
200	2.301
sin 52°	$\bar{1}.897$
	2.198
sin 74°	$\bar{1}.983$
164	2.215

(b) For the height of the tower

$$\frac{TO}{164} = \tan 12^\circ$$

$$TO = 164 \tan 12^\circ$$

$$= 164 \times 0.213$$

$$= 34.9\,\text{m}$$

$$= 35\,\text{m to 2 s.f.}$$

Exercise 9d

1. From points A and B on level ground, the angles of elevation of the top of a building are 25° and 37° respectively. If A is 5.7 m from the foot of the building, calculate, to the nearest metre, the distances of the top of the building from A and B.

2. A 210° sector of a circle of radius 5 cm is folded to form a cone. Calculate the angle of slope of the curved surface of the cone.

3. A point Y is 34 m due east of a point X. The bearings of a point O, the foot of a flag-pole, from X and Y are 018° and 320° respectively. If the height of the flag-pole is 12 m, calculate (a) the distance of the flag-pole from Y, (b) the angle of elevation of the top of the flag-pole from Y.

4. The angle of elevation of the top of a tower, T from a point P on the ground is 65°. The points P is 15 m from the foot of the tower and 28 m from a point Q due east of P. If the point Q is 18 m from the foot of the tower, calculate (a) the height of the tower, giving your answer to 1 d.p., (b) the angle of elevation of the top of the tower from Q.

5. Fig. 9.28 shows the net of a pyramid on a square base BDFH of side 3 cm with slant edges of length 5 cm.

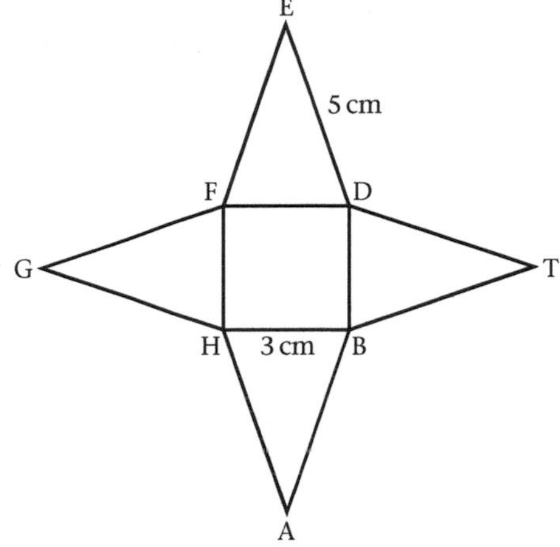

Fig. 9.28

Calculate (a) the height of the pyramid, (b) the angle of slope of the edge AB.

6. Two students are standing at the opposite ends of the diagonal AB of a rectangular playing field, 30 m long by 24 m wide. They both see an aeroplane which is flying at a constant height of 750 m above the playing field. Calculate the angle of elevation of the aeroplane as seen from B when it is immediately above A. (Give your answer to the nearest degree.)

Summary

Volume of a pyramid = area of constant cross-section × perpendicular height

Volume of pyramid = $\frac{1}{3}$ × area of base × height

Volume of cone = $\frac{1}{3}\pi r^2 h$

Practice exercise P9.1

Unless otherwise instructed, give your answers correct to 3 significant figures where necessary.

1. Calculate the volume and surface area of a pyramid of height of 14 cm with a rectangular base measuring 9 cm by 12 cm.

2. Calculate the volume and surface area of a prism of length 16 cm and with a base in the shape of a triangle with a base of 3 cm and a height of 7 cm.
3. Calculate the volume and surface area of a solid cylinder of height 7 cm and radius 2.4 cm, giving your answer in terms of π.
4. Calculate the volume and surface area of a sphere of radius 13 m.
5. Calculate the volume and surface area of a pyramid with vertical height 7.4 cm and a rectangular base measuring 3.4 cm by 5.7 cm.
6. A cylindrical tin of radius 10 cm contains water to a depth of 8 cm.
   (a) Calculate the volume, in litres, of water in the tin.
   (b) If the tin will be just filled when an additional 2.5 litres of water is poured in the tin, calculate the height of the tin.
7. An inverted cone has a base radius of 9 cm. 1.018 litres of water completely fills the cone. Calculate the height of the cone giving your answer to the nearest centimetre.
8. A sphere of diameter 12 cm is placed inside a cube of side 12 cm. What volume of air within the cube is not occupied?
9. A solid metal bar in the shape of a cuboid 7 cm by 12 cm and 2 m long is melted down and made into a solid cone with a base radius of 10 cm. How tall is the cone?
10. A solid metal cylinder with radius 9 cm and height 10 cm is melted down and made into solid spheres of radius 5 mm. How many spheres can be made?
11. A hemispherical bowl of height 7 cm has a capacity of 1 litre. Calculate the radius of the bowl.
12. Calculate the volume and curved surface area of a cone with a base diameter of 9.2 cm and a height of 2.7 cm.

Practice exercise P9.2

Where necessary, give your answers to 3 significant figu s.

1. Find the volume of a frustum of a cone of height 14 cm with a base radius of 12 cm and a top radius of 5 cm.
2. Calculate the volume of metal used to make a cylindrical pipe 22 m long, 3 cm thick and with an internal radius of 17 cm.
3. An open rectangular box has external dimensions 33 cm long, 13 cm wide and 6 cm deep. It is made of plastic 7 mm thick. Calculate the volume of plastic used to make the box.

Practice exercise P9.3

Unless otherwise instructed, give your answers correct to 3 significant figures where necessary.

1. An electric pole is planted so that a length of 5.4 m is above the surface of the ground. The pole is held firm by two lengths of wire each attached at one end at 1.5 m from the top of the pole and the other end buried in the ground 5 m from the foot of the pole. The angle made by the two wires at the pole is 168°. Calculate the distance between the two ends of the wire buried in the ground.
2. A 173° sector of a circle of radius 6 cm is folded to form a cone. Calculate
   (a) the angle of the slope of the curved surface of the cone,
   (b) the height of the cone to the nearest millimetre.
3. A lighthouse 30 m tall is on a bearing of 55° from ship A. The angle of elevation of the top of the lighthouse from ship A is 13°.
   (a) How far from the foot of the lighthouse is ship A?
   Ship B is on a bearing of 80° from ship A and the angle of elevation of the top of the lighthouse from ship B is 25°.
   (b) Find the distance between the two ships.

Chapter 10

Graphs (1)

Maximum and minimum values, travel graphs

Pre-requisites

- linear functions; quadratic functions; areas of quadrilaterals; approximation and estimation

Gradient of a curve

Earlier, in Book 3 and revised in Chapter 6, you saw that if a graph connects x and x its gradient is the rate of change of y compared with x. The gradient of the straight line in Fig. 10.1 is the same at every point.

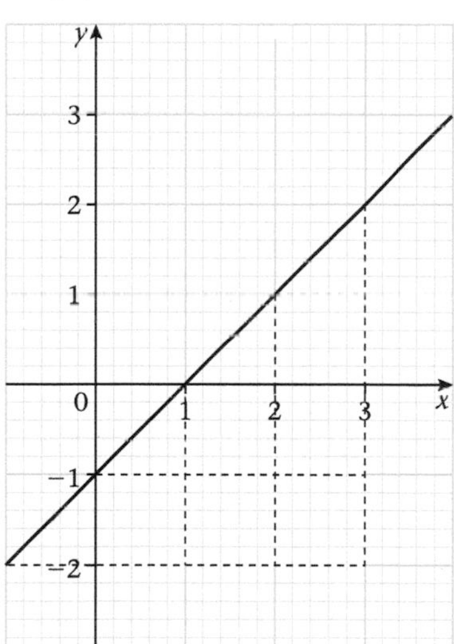

Fig. 10.1

The gradient at any particular point on a curve is defined as being the gradient of the tangent to the curve at that point.

In Fig. 10.2 the gradient of the curve at point P is the gradient of the tangent TP, i.e. tan θ. The tangent is drawn by placing a ruler against the curve at P and drawing a line, taking care that the 'angles' between the line and the curve at both sides of the point appear equal.

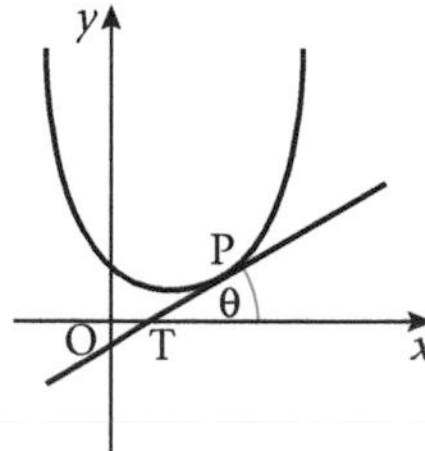

Fig. 10.2

Notice that the gradient of a straight line is the same at any point on the line, but that the gradient of a curve changes from point to point.

Example 1

Fig. 10.3 is the graph of the curve $y = x^2 + 2x - 3$. Use the given tangent to find the gradient of the curve at P.

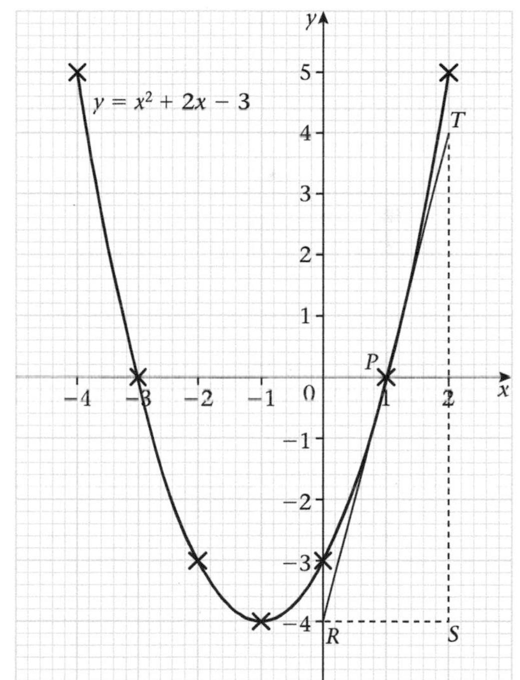

Fig. 10.3

Gradient of curve at P

$= \text{gradient of TP}$

$= \frac{\text{ST}}{\text{RS}} = \frac{8}{2} = 4$

Example 2

Fig. 10.4 is the graph of the curve $y = 2 + x - x^2$ for values of x from -2 to 3. Use the given tangents to find the gradient of the curve at (a) P, (b) Q.

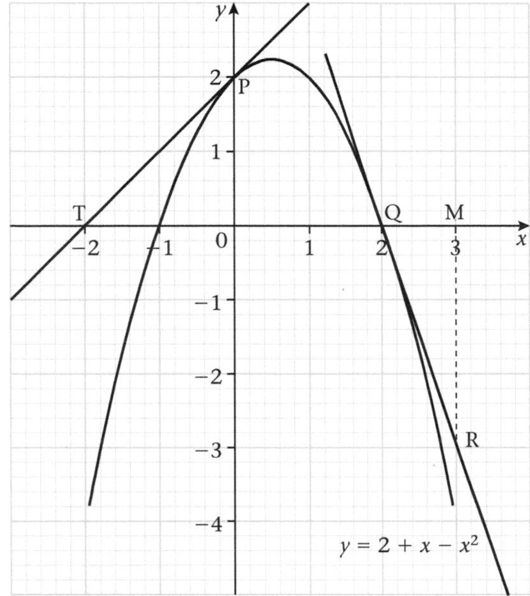

Fig. 10.4

(a) Gradient of the curve at P
  = gradient of tangent TP
  $= \frac{OP}{TO} = \frac{2}{2} = 1$

(b) Gradient of the curve at Q
  = gradient of tangent QR
  $= \frac{-MR}{QM} = \frac{-3}{1} = -3$

Notice that $\triangle$s TOP and QMR were used to find the gradients of the tangents. Any suitable right-angled triangles could have been used. In this case the intercepts TO and QM were of convenient lengths.

Example 3

Draw the graph of $y = \frac{1}{4}x^2$ for values of x from -2 to 3. Find the gradient of the curve at the point where x has the value (a) 3, (b) -2.

Table 10.1 is the table of values.

$y = \frac{1}{4}x^2$

Table 10.1

x	-2	-1	0	1	2	3
x^2	4	1	0	1	4	9
y	1	$\frac{1}{4}$	0	$\frac{1}{4}$	1	$2\frac{1}{4}$

Fig. 10.5 shows the required graph.

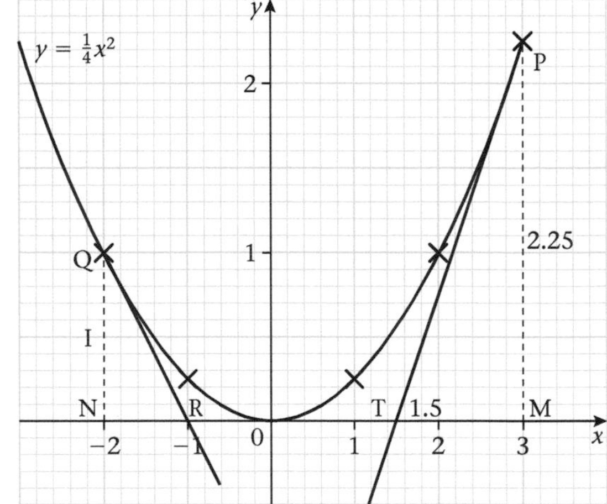

Fig. 10.5

Using the tangents drawn to the curve where $x = 3$ and $x = -2$:

(a) Gradient of curve where $x = 3$
  = gradient of tangent PT
  $= \frac{MP}{TM} = \frac{2.25}{1.5} = \frac{2\frac{1}{4}}{1\frac{1}{2}} = \frac{9}{6} = 1\frac{1}{2}$

  When $x = 3$, the gradient of the curve is $1\frac{1}{2}$ (i.e. at P, y is increasing $1\frac{1}{2}$ times as fast as x).

(b) Gradient of curve where $x = -2$
  = gradient of tangent QR
  $= \frac{-QN}{NR} = \frac{-1}{1} = -1$

When $x = -2$, the gradient of the curve is -1 (i.e. at Q, y is *decreasing* at the same rate as x is increasing).

Notice the following points:

1 In Example 2 the lengths MP, TM, QN, NR are measured according to the *scales* of the axes.

2 In Examples 1 and 2 the method of drawing tangents using a ruler can give inaccurate results. Gradients found by this method must only be taken as approximate.

Maximum and minimum values

Fig. 10.6 shows the tangents drawn at the **turning points** of two quadratic functions. In each case the tangent is parallel to the x-axis.

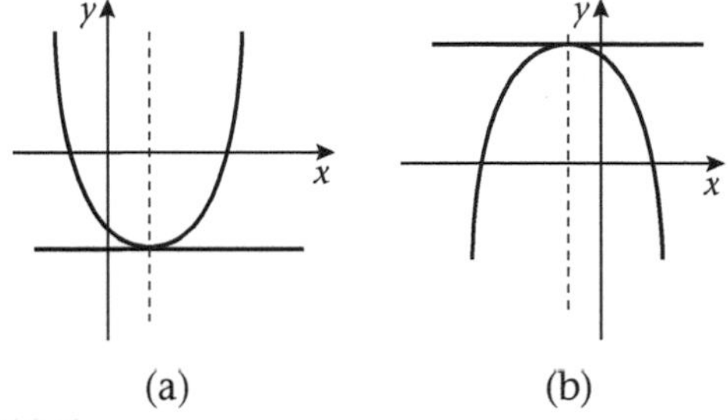

(a) (b)

Fig. 10.6

Hence the gradient at a turning point is zero. In Fig. 10.6(a) the turning point corresponds to the **minimum** value of the function. In Fig. 10(b) the turning point corresponds to the **maximum** value of the function. In each figure the line of symmetry of the curve is shown by a broken line. Note that the line of symmetry passes through the maximum or minimum value of the function.

Exercise 10a

1. Write down the equation of the line of symmetry of the curve in (a) Fig. 10.4, (b) Fig. 10.5.

2. Fig. 10.7 is the graph of the function $y = x^2 - 6x + 4$.

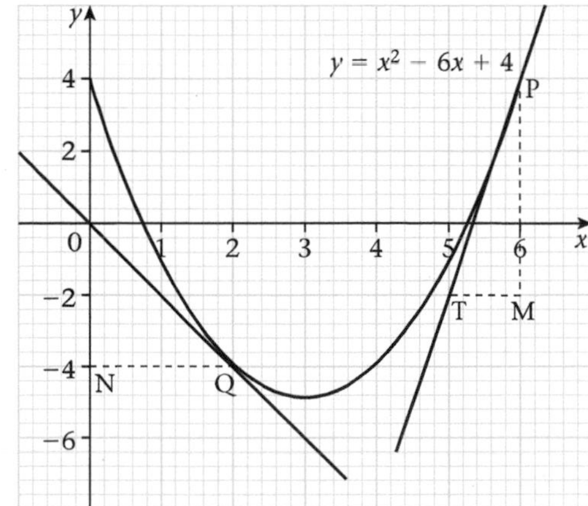

Fig. 10.7

(a) Use the given tangents to find the gradient of the curve (i) at P, (ii) at Q.
(b) Find the minimum value of the function.
(c) Write down the equation of the line of symmetry of the curve.

3. Fig. 10.8 is the graph of $y = 3 - 2x - x^2$.

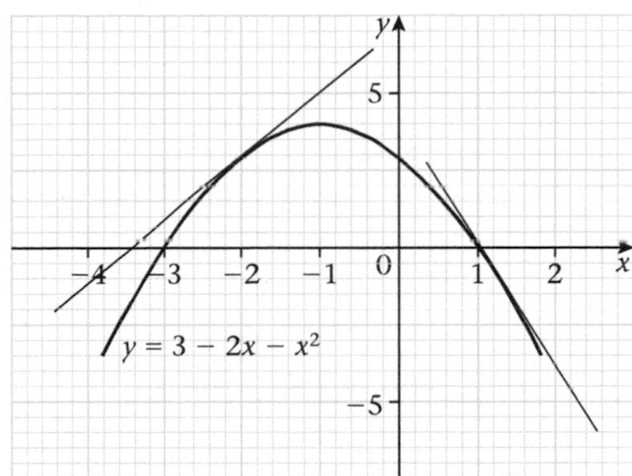

Fig. 10.8

(a) Use the given tangents to find the gradient of the curve (i) when $x = -2$, (ii) when $x = 1$.
(b) What is the maximum value of $3 - 2x - x^2$?
(c) Write down the equation of the line of symmetry of the curve.

4. (a) Copy and complete Table 10.2 for the relation $y = 3x - x^2$.

Table 10.2

x	−1	0	1	2	3	4
$3x$	−3		3			
$-x^2$	−1		−1			
y	−4		2			

(b) Draw the graph of $y = 3x - x^2$ from $x = -2$ to $x = 4$, using a scale of 2 cm to 1 unit on both axes.
(c) Find the gradient of the curve at (i) $x = 0$, (ii) $x = 2$.
(d) Write down the equation of the line of symmetry of the curve.
(e) Find the maximum value of $3x - x^2$.

5 Draw the graph of $y = \frac{1}{5}x^2$ for values of x from -4 to 4. Use a scale of 1 cm to 1 unit on the x-axis and 2 cm to 1 unit on the y-axis. Find the gradient at the point where (a) $x = 3$, (b) $x = 1.5$, (c) $x = -2$.

6 Copy and complete Table 10.3 giving values for the function $y = 2x^2 - 4x + 3$ from $x = -2$ to $x = 4$.

Table 10.3

x	-2	-1	0	1	2	3	4
$2x^2$	8		0	2			
$-4x$	8		0	-4			
$+3$	3		3	3			
y	19		3	1			

Draw the graph of $y = 2x^2 - 4x + 3$, using a scale of 2 cm to 1 unit on the x-axis and 1 cm to 2 units on the y-axis.
From your graph, find

(a) the equation of the line of symmetry of the curve,

(b) the gradient of the curve at $x = 3$,

(c) the minimum value of y.

7 Draw the graph of $y = x^2 - 4x$ from $x = -1$ to $x = 5$. Use a scale of 2 cm to 1 unit on both axes. Find the gradient at the point where (a) $x = 4$, (b) $x = 2$, (c) $x = 0$.

8 Draw the graph of $y = 5x - 2x^2$ from $x = -1$ to $x = 4$. Use a scale of 2 cm to 1 unit on the x-axis and 1 cm to 1 unit on the y-axis.

(a) Find the gradient of the curve at the point where $x =$ (i) 0, (ii) 1, (iii) 3.

(b) Write down the equation of the line of symmetry of the curve.

9 Draw the graph of $y = x^2 - 3x + 2$ for values of x from -1 to 4. Find the gradient at the point where x has the value (a) $2\frac{1}{2}$, (b) $1\frac{1}{2}$, (c) 0, (d) $-\frac{1}{2}$.

10 Draw the graph of $y = 1 + x - x^2$ from $x = -2$ to $x = 3$. Find the gradient at the point where x has the value (a) $2\frac{1}{2}$, (b) $1\frac{1}{2}$, (c) $\frac{1}{2}$, (d) -1.

Travel graphs

Distance–time graphs

Fig. 10.9 is a graph representing the journey of a cyclist who travels from town A to town B.

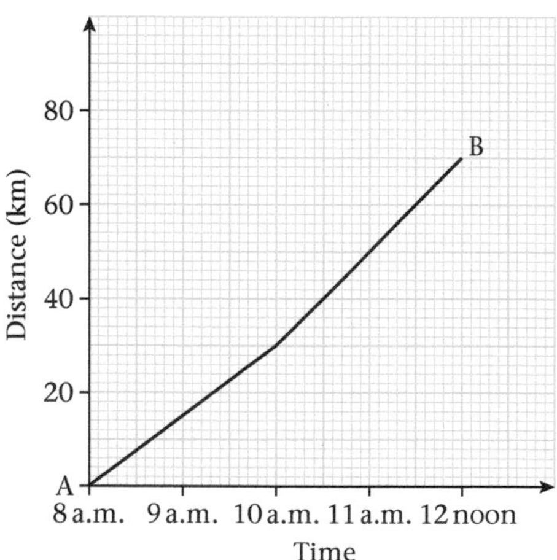

Fig. 10.9

The graph shows that the cyclist travelled a distance of 30 km in the first 2 hours and 40 km in the second two hours.

The cyclist travelled a total distance of 70 km in 4 hours.

The graph in Fig. 10.10 represents the journey of a motorist between two towns P and Q.

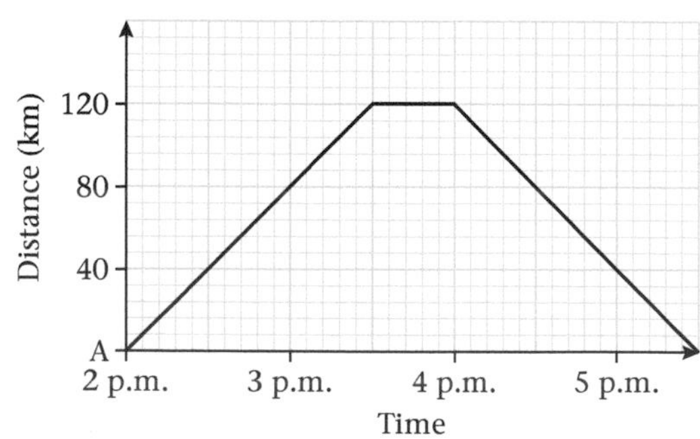

Fig. 10.10

The graph shows that the motorist drove steadily for $1\frac{1}{2}$ hours for the distance of 120 km from P to Q. She rested for $\frac{1}{2}$ hour and then drove back to P in $1\frac{1}{2}$ hours.

Example 4

Fig. 10.11 is a graph represnting the journeys of a pedestrian X and a motorist Y. X walks steadily towards a village. Y drives to the village, stays for a while and then returns.

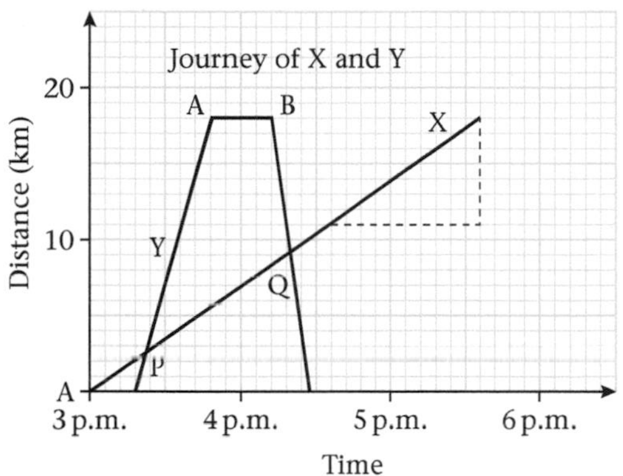

Fig. 10.11

(a) What was X's average walking speed?
(b) How many times did y stay in the village?
(c) At what speed did Y drive back from the village?
(d) How far did X walk between the two times that Y passed him?

(a) From the graph it can be seen that X travelled a total distance of 18 km in 2.6 h so that
either:

$$\text{X's average speed} = \frac{\text{total distance travelled}}{\text{total time taken}} = \frac{18\text{ km}}{2.6\text{ h}} \simeq 6.9\text{ km/h}$$

or:
Draw a horizontal line 1 hour at a convenient place. Read off the corresponding vertical line. (See the dotted lined in Fig. 10.11.) The vertical line represents the distance travelled in 1 hour, and is 7 km.
Hence X's speed $\simeq$ 7 km/h.

(b) On the time axis, 1 small square represents $\frac{1}{10}$ hour (6 min). In Fig. 10.11 the time that Y stays in the village is shown by the horizontal line, 4 small squares long.
$\therefore$ Time that Y stays in village
$$\simeq 4 \times 6\text{ min} \simeq 24\text{ min}$$

(c)
$$\text{Y's return speed} = \frac{\text{distance travelled}}{\text{time taken}}$$
$$\simeq \frac{18}{0.25}\text{ km/h}$$
$$\simeq 18 \div \tfrac{1}{4}\text{ km/h}$$
$$\simeq 18 \times 4\text{ km/h}$$
$$\simeq 72\text{ km/h}$$

(d) In Fig. 10.11 the points P and Q represent the times and positions of X and Y when Y passed X.
At Q, distance travelled by X $\simeq$ 9 km
At P, distance travelled by X $\simeq 2\frac{1}{2}$ km
Distance travelled by X between P and Q
$$\simeq (9 - 2\tfrac{1}{2})\text{ km}$$
$$\simeq 6\tfrac{1}{2}\text{ km}$$

Notice the following points:

1 In travel graphs, time is always given on the horizontal axis.
2 Answers obtained from graphs are not usually exact. However, accuracy can be improved by drawing graphs to a larger scale.
3 Speed is the rate of change of distance with time.

In the sketch graph of Fig. 10.12, PQR is any right-angled triangle drawn on the travel graph. Between P and Q the time changes from P to R and the distance changes from R to Q.

$$\text{Speed along PQ} = \frac{\text{change of distance}}{\text{change of time}} = \frac{RQ}{PR}$$

$$\text{Notice also, } \tan\theta = \frac{RQ}{PR}$$

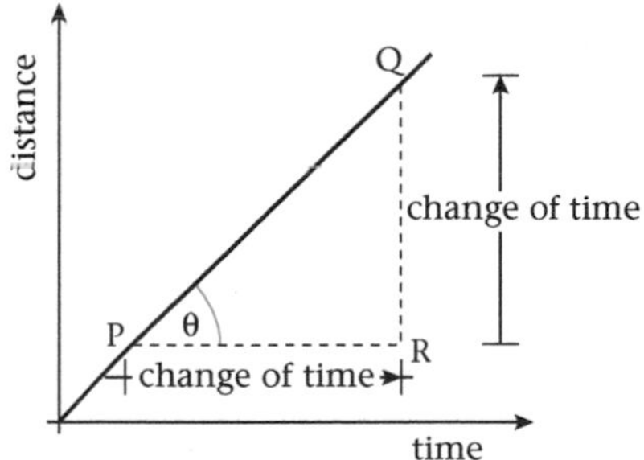

Fig. 10.12

The value of tan θ is called the **gradient** of the line PQ. In a time–distance graph, the gradient of a line always gives a measure of the speed on that part of the graph. Since θ can be taken at any point on the line, any convenient right-angled triangle can be used to find the gradient (as in Example 2, part (a)).

Exercise 10b

Most of the questions in this exercise are suitable for class discussion.

1. Use Fig. 10.11 to answer the following.
   (a) How far had X walked when Y started towards the village?
   (b) At what speed did Y drive towards the village?
   (c) What was the time between Y leaving the village and X arriving at the village?
   (d) How far had X walked when Y completed his journey?

2. Fig. 10.13 is a time–distance graph showing the distance covered in 2 hours at a speed of 48 km/h.
   (a) What is the time and distance covered at point A?
   (b) To the nearest km, what distance is covered in (i) 1 h 42 min, (ii) $\frac{3}{4}$ h?
   (c) Find the time it takes to travel (i) 60 km, (ii) 34 km.

3. Fig. 10.14 is the travel graph of a cyclist who stopped once on a journey of 15 km.
   (a) How long did the cyclist stop for?
   (b) How far had the cyclist travelled after 48 min?
   (c) What was the average speed for the whole journey?

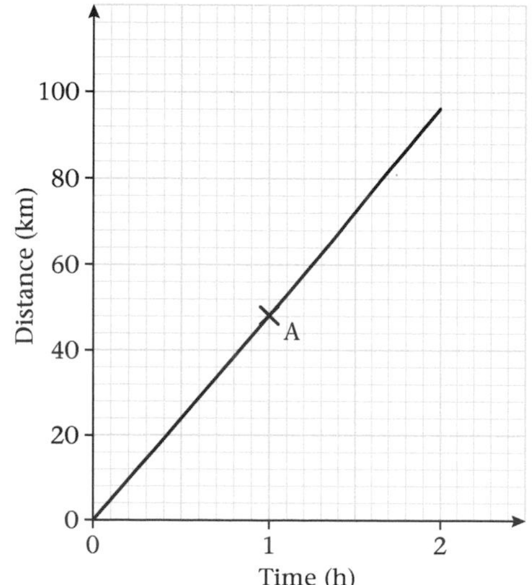

Fig. 10.13

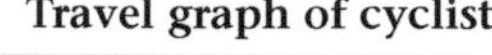

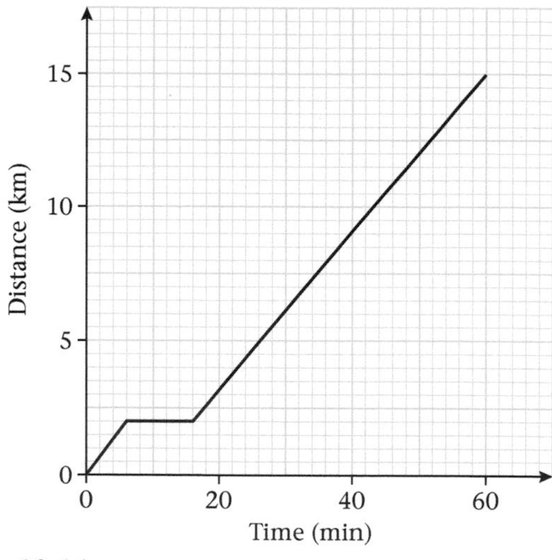

Fig. 10.14

   (d) Neglecting the stop, what was the average *cycling* speed?
   (e) How long would the journey have taken without the stop?

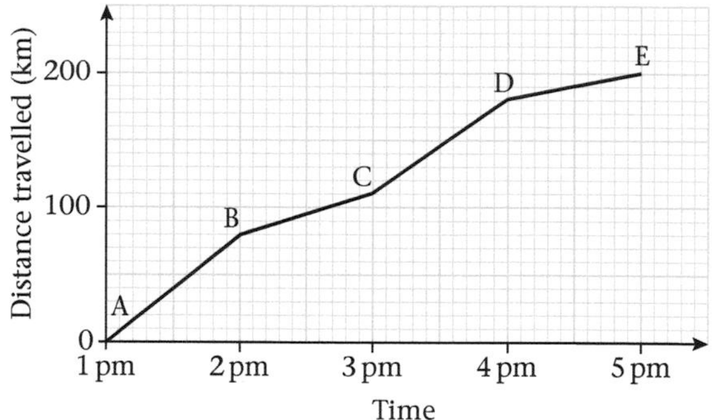

Fig. 10.15

4. Fig. 10.15 shows the journey of a motorist.
   (a) Did the motorist stop at any time?
   (b) At what time had the motorist completed half the distance?
   (c) How far had the motorist travelled by 3.30 p.m.?
   (d) What was the average speed for the *whole* journey?
   (e) What was the speed between stages A and B of the journey?
   (f) What was the speed between stages B and C of the journey?

(g) What was the speed between stages C and D of the journey?
(h) What was the speed between stages D and E of the journey?

5 Two soldiers, A and B, march backwards and forwards outside the gate of a military barracks. They cross in front of the gate. Fig. 10.16 is a graph of their movements.
(a) What is the greatest distance between a soldier and the gate?
(b) How far does each soldier march in 1 min?
(c) Calculate their marching speed in km/h.
(d) After $\frac{1}{2}$ hour, how many times will they have passed each other?

6 Fig. 10.17 shows the outcome of a 100 m race between A and B.
(a) Who won the race?
(b) What distance did he win by?
(c) How far apart were the runners after 6 seconds?
(d) What was A's speed (in m/s)?

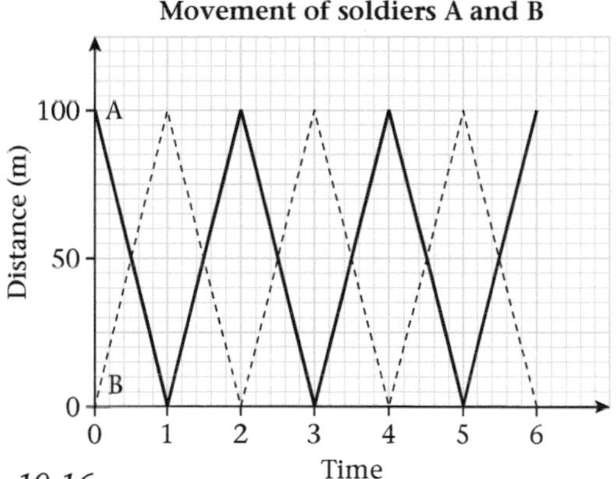

Fig. 10.16

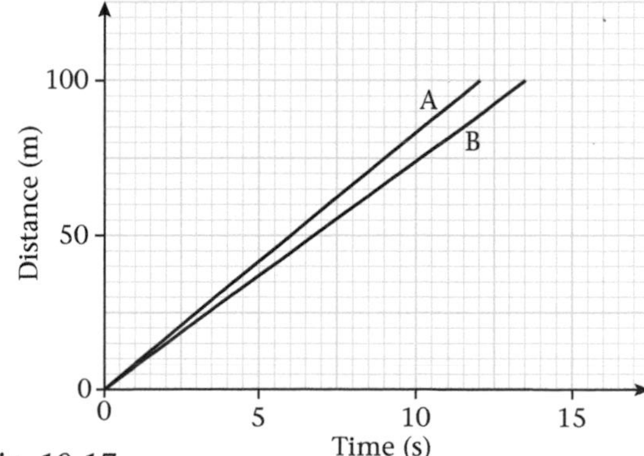

Fig. 10.17

Drawing distance–time graphs

Example 5

A cyclist leaves home at 09:00 and rides at a steady 12 km/h to a place 20 km away. She spends 45 minutes there, then returns at 16 km/h. At what time does she get home again?

The following have been given.

1 The time the cyclist started.
2 The cyclist's speed for the first hour.
3 The distance of the outward journey.
4 The rest time.
5 The cyclist's speed for the first hour of the return journey.

All these can be represented on a graph in order to find the time the cyclist returned home.

Fig. 10.18 is a graph of the cyclist's journey.

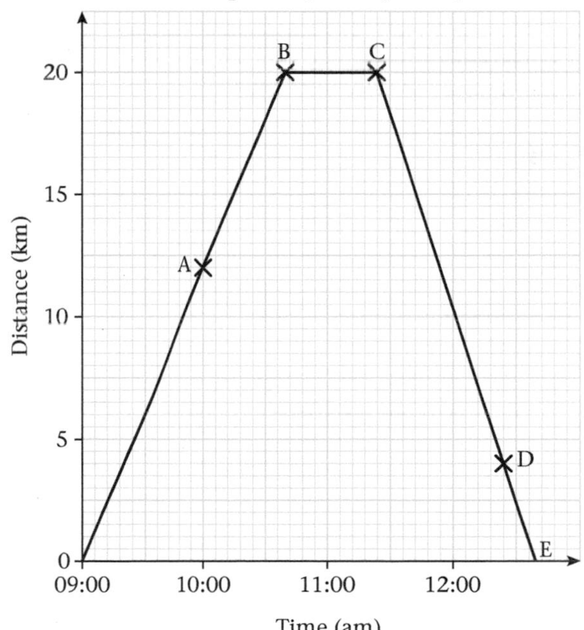

Fig. 10.18

Method:

1 Choose suitable scales, place time on the horizontal axis and mark 09:00 at the origin.

2 In one hour the cyclist travels 12 km. Plot the point A at (10:00, 12 km). Join the origin to A and produce it to B (20 km from home).
3 Mark a point C at the same horizontal level but 45 minutes beyond B. C represents the starting point for the journey home. (Note: between B and C time increases, but distance stays the same.)
4 Plot a point D, 1 hour and 16 km from C. The distance is measured downwards a the cyclist returns home. Join CD and produce to cut the time axis at E.
5 E gives the time of arrival. This is approximately 12:39 (each small square represents 6 min on this scale).

Example 6

A motorist starts from A at 11:00 and plans to arrive at B, 100 km away, at 13:00. After $\frac{1}{2}$ hour he has a puncture which takes 18 min to mend. how fast must he then travel so that he still arrives at B at 13:00?

Fig. 10.19 is the travel graph of the motorist's journey.

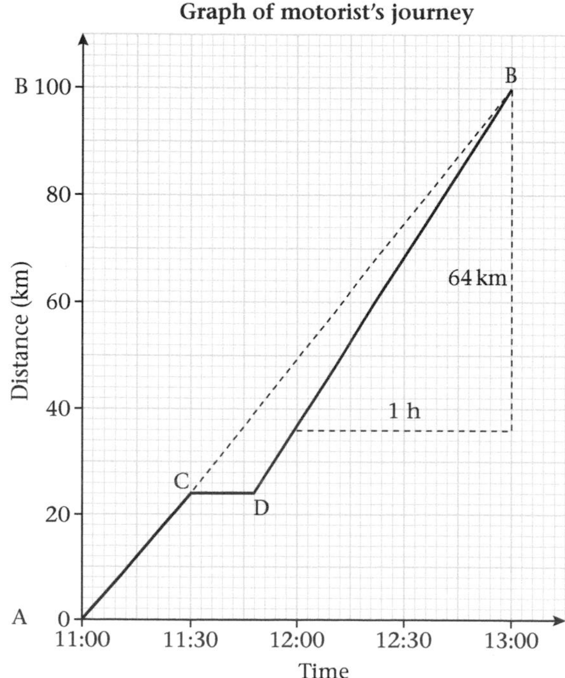

Fig. 10.19

Method:

1 Choose suitable scales. Place 11:00 at the origin A. Mark a point B at (13:00, 100 km). AB represents the motorist's journey if he had not had a puncture.
2 Mark a point C on AB at 11:30. Draw the line CD horizontally 1.2 cm long (representing 18 min on the scale in Fig. 10.19).
3 Join DB. Then ACDB represents the motorist's actual journey.
4 To find the speed between D and B, draw a horizontal line one hour long at any convenient place. Read off the corresponding vertical distance. (See the dotted lines to the right of DB in Fig. 10.19.) 64 km corresponds to 1 hour.
5 The motorist must travel at 64 km/h (approximately) to reach point B on time.

Exercise 10c

1 (a) Using scales of 2 cm to 10 s on the horizontal axis and 2 cm to 50 m on the vertical axis, draw on the same axes, the graphs of the following world sprint records (1984):

distance (m)	*time (s)*
100	9.9
200	19.7
400	43.8

(b) Which record represents the fastest speed?

2 A car averages 68 km/h.

(a) Using scales of 2 cm to 10 min and 2 cm to 10 km, draw a time–distance graph from 0 to 30 min.

(b) Read off the distance covered in (i) 11 min, (ii) 25 min.

(c) Read off the time taken to travel (i) 10 km, (ii) 29 km.

3 A man sets out at 10:00 to walk 25 km. He walks steadily at 6 km/h, but sits down for 12 min after each hour's walking. Draw a travel graph and hence find the time when he completes his journey.

4 At 10:00 a girl starts walking to a town 8 km away. She walks at 6 km/h. She rests for $\frac{1}{4}$ hour at the town and then returns at 16 km/h on a bicycle. Draw a travel graph and hence find the time when she gets home again.

5 Three cars, A, B, C, start one after the other in that order, at 5 min intervals, travelling at 90, 120, 150 km/h respectively. How long after the start does B pass A, C pass A, C pass B?

6 Two women start at 08:00 and travel towards one another from places 32 km apart. One cycles at 20 km/h and the other walks at 5 km/h. Draw the graphs of their journeys on the same axes and hence find
(a) the time when they pass each other,
(b) the times when they are 5 km apart.

7 At 09:00 a woman starts walking from K to C 32 km away at a steady 6 km/h. She sits down to rest for $\frac{1}{2}$ hour at 11:00. A bus which averages 30 km/h starts from K in the same direction at 11:15. Draw travel graphs of the woman and the bus on the same axes. Hence find (a) the time, and (b) the distance from K when the bus passes the woman.

8 Two men travel to a village 12 m away. The first walks steadily at 6 km/h without stopping. The second starts 30 min later and runs at 10 km/h, but takes a 30 min rest after 1 hour's running. Using scales of 2 cm to 30 min on the time axis and 1 cm to 1 km on the distance axis, draw a travel graph of their journeys. Hence find which man reaches the village first and by how many minutes.

9 X can run 100 m in 11.7 s and Y can run the same distance in 12.3 s. Use a graphical method to find how many metres start X should give Y in a 100 m race if they are to finish together.

10 Cecil and Kevin live 30 km apart. They arrange to meet at a point half-way between their houses at 12:00. Cecil starts at 10:30 and cycles at 10 km/h. After 5 km he has a puncture which delays him for 10 min. Find graphically Cecil's speed for the last 10 km if he arrived at the meeting point on time.

Speed–time graphs

Fig. 10.20 is a graph showing how the speed of a car varies with time over a short journey.

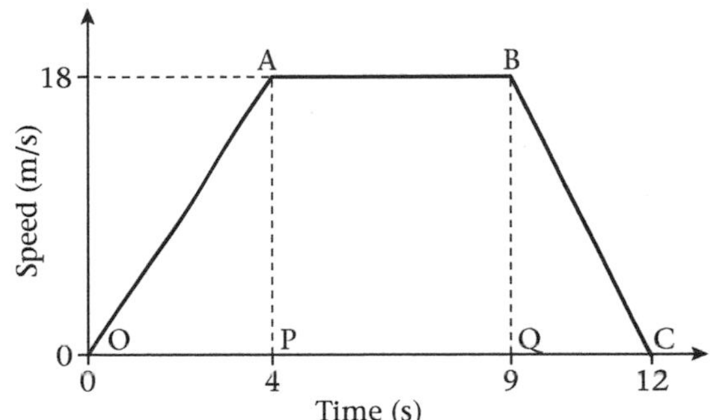

Fig. 10.20

The journey is in three stages, OA, AB, BC.

Stage OA

During the first 4 seconds the car speed up, or **accelerates**, uniformly from rest, 0 m/s, to 18 m/s.

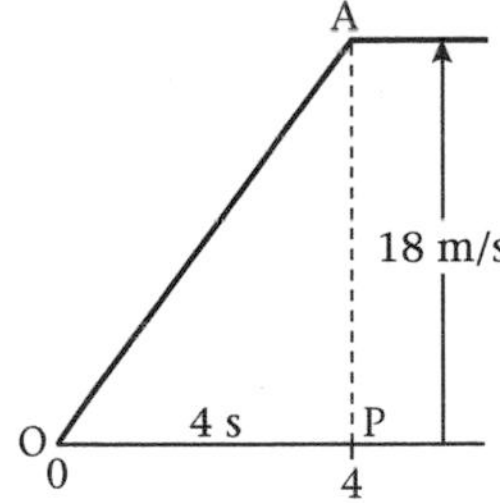

Fig. 10.21

The gradient of the graph during this stage gives the rate of change of speed, or **acceleration**, of the car.

Acceleration between O and A

$= \text{gradient of OA}$

$= \frac{\text{PA}}{\text{OP}}$

$= \frac{18\text{ m/s}}{4\text{ s}}$

$= 4\frac{1}{2}\text{ m/s per second}$

The car is accelerating at $4\frac{1}{2}$ m/s per second. m/s^2 is short for m/s per second.

Distance travelled during first stage

$= \text{average speed} \times \text{time}$

$= \frac{(0 + 18)}{2}\,\text{m/s} \times 4\,\text{s}$

$= 36\,\text{m}$

Alternatively, notice in Fig. 10.21

area of $\triangle OAP = \frac{1}{2} \times 4\,\text{s} \times 18\,\text{m/s}$
$= 36\,\text{m}$

Hence the area under the graph represents the distance travelled.

Stage AB

During the second stage the car travels with a constant speed of 18 m/s for 5 seconds.

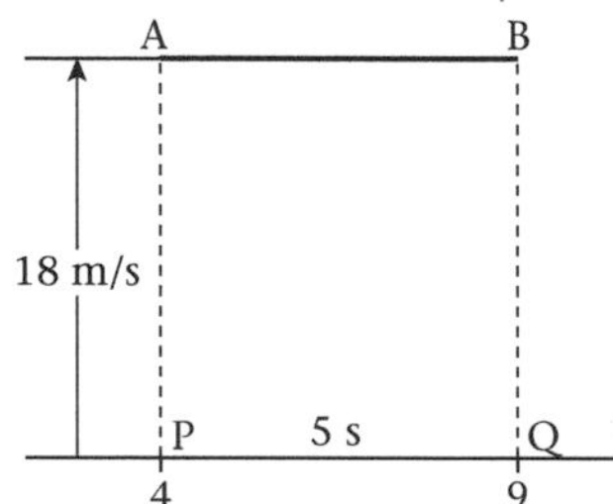

Fig. 10.22

The speed of the car does not change
⇒ acceleration between A and B = 0
Distance travelled during second stage

either $= \text{average speed} \times \text{time}$
$= 18\,\text{m/s} \times 5\,\text{s}$
$= 90\,\text{m}$

or $= \text{area under AB}$
$= 18\,\text{m/s} \times 5\,\text{s}$
$= 90\,\text{m}$

Stage BC

During the final stage the car slows down, or **decelerates**, uniformly from 18 m/s to rest, 0 m/s.

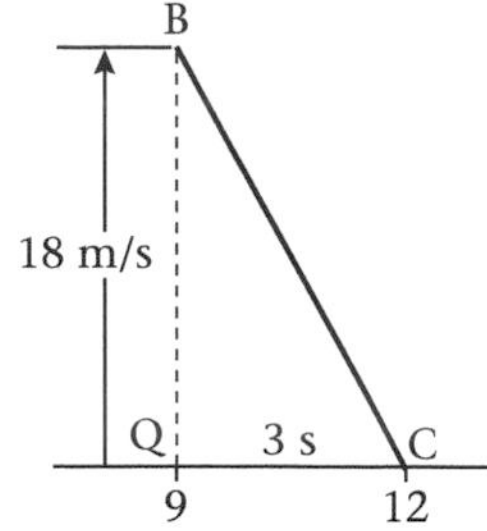

Fig. 10.23

The gradient of the graph during this final stage gives the rate of change of speed. Since the gradient is negative there is a negative acceleration, or **deceleration.**

Acceleration between B and C

$= \text{gradient of BC}$

$= \frac{BQ}{QC}$

$= \frac{-18\,\text{m/s}}{3\,\text{s}}$

$= -6\,\text{m/s per second}$

The car is decelerating at $6\,\text{m/s}^2$.
Distance travelled during final stage

either $= \text{average speed} \times \text{time}$

$= \frac{(18 + 0)}{2}\,\text{m/s} \times 3\,\text{s}$

$= 27\,\text{m}$

or $= \text{area under BC}$
$= \frac{1}{2} \times 3\,\text{s} \times 18\,\text{m/s}$
$= 27\,\text{m}$

Notice the following:

1. **Acceleration** is the rate of change of speed with time. The gradient of a speed–time graph gives the acceleration of the object under consideration. **Deceleration** is the decrease of speed with time.
2. The area under a speed–time graph represents the distance travelled by the object under consideration.

Example 7

During a journey, a car accelerates uniformly for 40 seconds. Its speed, v km/h, is given at 10-second intervals in Table 10.4.

Table 10.4

t(s)	0	10	20	30	40
v(m/s)	9	12	15	18	21

Find (a) its acceleration in m/s per second, (b) the distance travelled during the whole 40 seconds.

Fig. 10.24 is a graph of the data in Table 10.4.

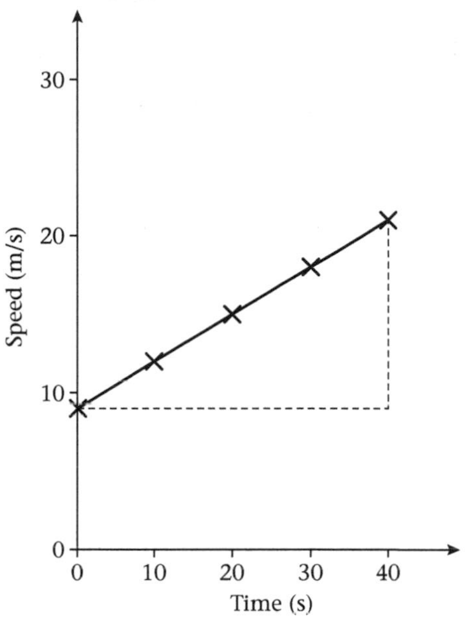

Fig. 10.24

(a) Acceleration = gradient of the graph

$$= \frac{(21 - 9)\,\text{m/s}}{40\,\text{s}}$$

$$= \frac{12\,\text{m/s}}{40\,\text{s}}$$

$$= 0.3\,\text{m/s per second}$$

(b) The area under the graph represents the distance travelled.
Area under graph
$= \frac{1}{2}(9 + 21) \times 40\,\text{m}$
$= \frac{1}{2} \times 30 \times 40\,\text{m}$
$= 600\,\text{m}$

Example 8

Fig. 10.25 is the speed–time graph of a car journey.

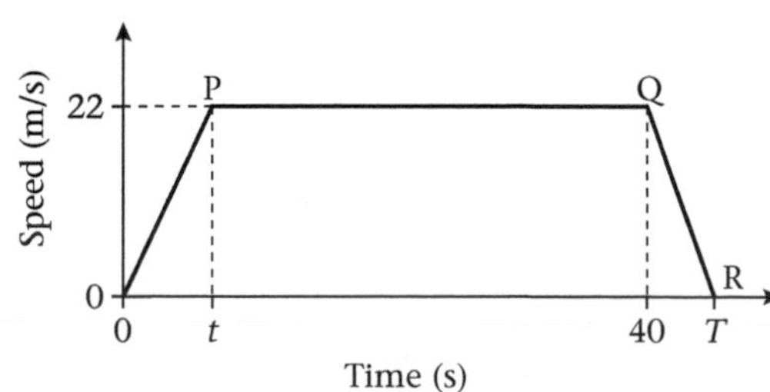

Fig. 10.25

The car starts from rest and accelerates at $2\frac{3}{4}\,\text{m/s}^2$ for t seconds until its speed is 22 m/s. It then travels at this speed until, 40 seconds after starting, its brakes bring it uniformly to rest. The total journey is 847 m long and takes T seconds.

Calculate (a) the value of t,
(b) the distance travelled during the first t seconds,
(c) the value of T,
(d) the final deceleration.

(a) Initial acceleration = gradient of OP

$$\text{Hence } 2\tfrac{3}{4} = \frac{22}{t}$$

$$\Rightarrow t = \frac{22}{2\frac{3}{4}} = 8$$

(b) Distance travelled during first t seconds

$= \text{area under OP}$

$= \frac{1}{2} \times 8 \times 22\,\text{m}$

$= 88\,\text{m}$

(c) Total distance travelled

$= \text{area of trapezium OPQR}$

$= \frac{1}{2}(\text{PQ} + T) \times 22 \text{ metres}$

$= (32 + T)\,11 \text{ metres}$

Hence $847 = (32 + T)11$

$$\Rightarrow 32 + T = \frac{847}{11} = 77$$

$$\Rightarrow T = 77 - 32 = 45$$

(d) Acceleration during last stage

$= \text{gradient of QR}$

$$= \frac{-22}{T - 40}$$

$$= \frac{-22}{5}\,\text{m/s}^2$$

$$= -4.4\,\text{m/s}^2$$

The final deceleration is $4.4\,\text{m/s}^2$.

Example 9

In Fig. 10.26 OABC is the speed–time graph of a journey.

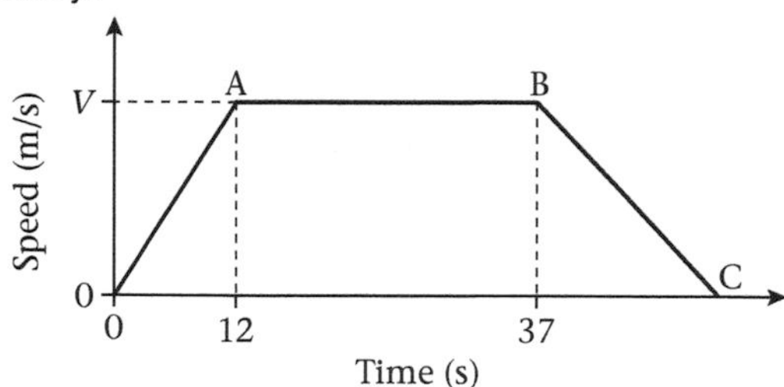

Fig. 10.26

(a) If 868 m is covered in the first 37 seconds, calculate the top speed V. (b) If the final deceleration is $3\frac{1}{2}$ m/s^2, calculate the total time for the journey. (c) Find the average speed for the whole journey.

(a) Distance travelled in the first 37 seconds

$= \text{area under OAB}$

$= \frac{1}{2}(\text{AB} + 37) \times V$ metres

where AB $= 37 - 12 = 25$

Hence $868 = \frac{1}{2}(25 + 37)V$

$868 = \frac{1}{2} \times 62V$

$\Rightarrow V = \frac{868}{31} = 28$

The top speed is 28 m/s.

(b) If the deceleration from B to C takes t seconds, then

$\frac{V}{t} = 3\frac{1}{2}$

$\frac{28}{t} = 3\frac{1}{2}$

$\Rightarrow t = \frac{28}{3\frac{1}{2}} = 8$

Total time taken $= 37\text{ s} + 8\text{ s} = 45\text{ s}$

(c) Average speed for the whole journey

$= \frac{\text{total distance travelled}}{\text{total time taken}}$

$= \frac{\frac{1}{2}(45 + 25)28}{45}$ m/s

$= \frac{35 \times 28}{45}$ m/s

$= \frac{7 \times 28}{9}$ m/s $= 21\frac{7}{9}$ m/s

Exercise 10d

1. In your own words describe the journeys shown in the speed–time graphs in Fig. 10.27.
2. Calculate the acceleration in each part of Fig. 10.27.

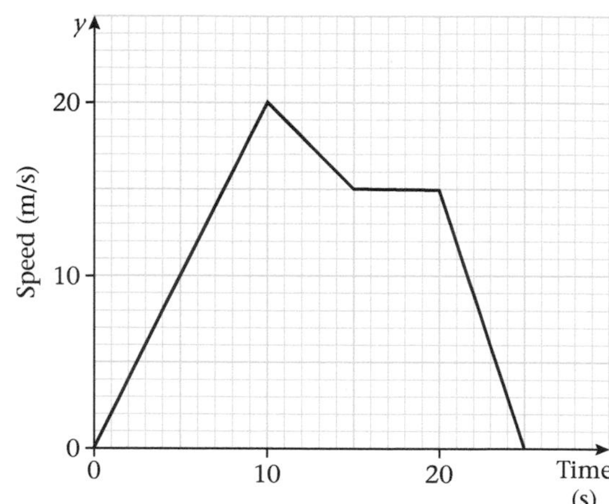

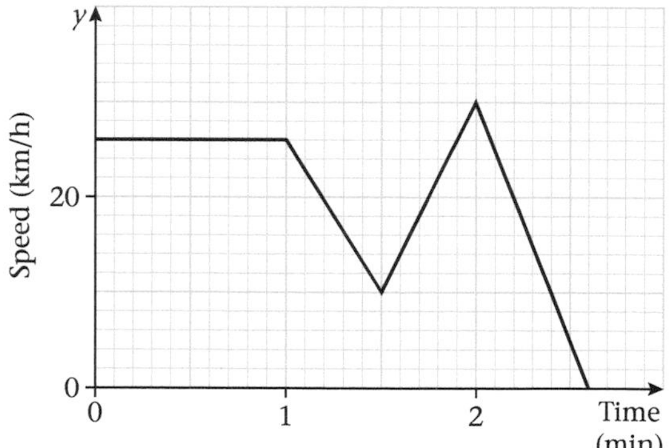

Fig. 10.27

3. Find the total distance travelled for each journey in Fig. 10.27 (Answer in metres for part (a), in km for part (b).)
4. Table 10.5 gives the speeds, v m/s, of an object at 10-second intervals.

Table 10.5

t(s)	0	10	20	30	40
v(m/s)	9	16	23	30	37

Find (a) the acceleration and (b) the distance travelled throughout the 40 seconds.

5 An object starts from rest and accelerates uniformly to a speed of 18 m/s. It travels at this speed for 10 seconds and then comes to rest after a further 3 seconds. If the object travelled for a total time of 15 seconds, find

(a) the time during which the object accelerates,
(b) the acceleration during this time,
(c) the total distance travelled by the object.

6 Fig. 10.28 is the speed–time graph of an object which accelerates from rest to a speed of v m/s then decelerates to rest, taking 54 seconds altogether.

(a) If the journey is 810, find the value of v.
(b) Hence find the deceleration, given that the initial acceleration is $1\frac{2}{3}/s^2$.

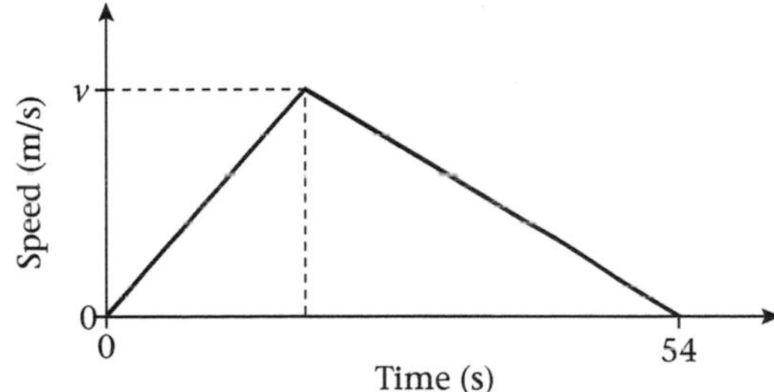

Fig. 10.28

7 Fig. 10.29 is the graph of a car which starts from rest in 1st gear and then changes up to 2nd gear.

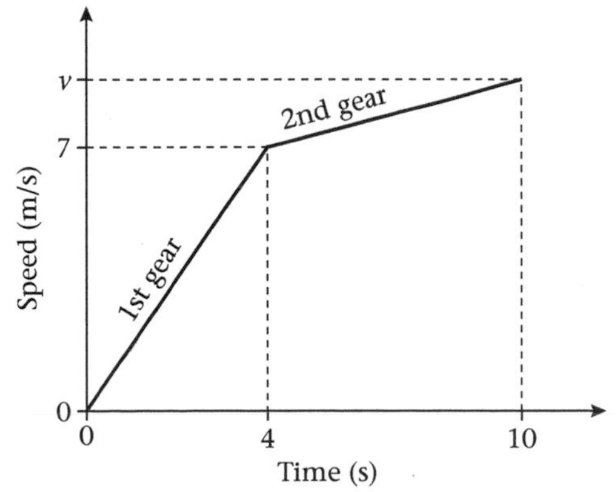

Fig. 10.29

(a) Find the acceleration and distance covered in 1st gear.
(b) If the car travels 54 m in 2nd gear, find the value of v and the acceleration in that gear.

8 A car is travelling initially at 30 m/s. Between midday and 12:05 it decelerates at 0.8 m/s per min. It then accelerates at 0.65 m/s per min for 12 min, after which it maintains a constant speed. Graphically or otherwise find the speed of the car at the following times.

(a) 12:01 (b) 12:04
(c) 12:05 (d) 12:07
(e) 12:11 (f) 12:19

9 Fig. 10.30 is the speed–time graph of an object which travels at a constant speed of 48 m/s for 5 seconds and then slows down uniformly, coming to rest after a further 3 seconds.

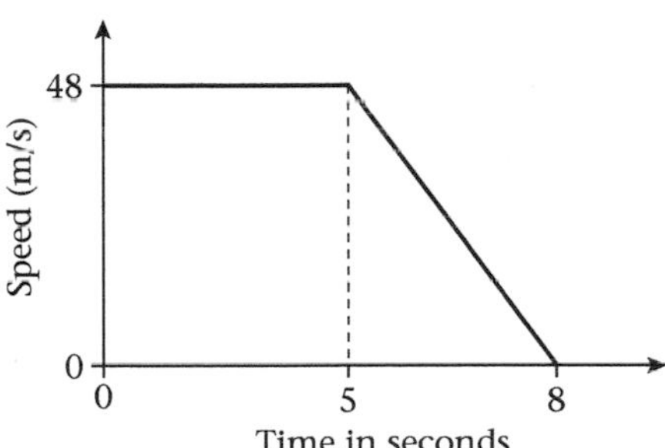

Fig. 10.30

Calculate (a) the speed of the object after 6 seconds, (b) the total distance travelled during the 8 seconds, (c) the averages speed of the object during the 8 seconds.

10 Fig. 10.31 on the following page is a travel graph showing the motion of an object which has a starting speed of 26 m/s. It decelerates at 4 m/s^2 for the first 3 seconds, travels at a constant speed for the next 5 seconds, and finally accelerates for 2 seconds until its speed is 37 m/s.

Find (a) its speed after 7 seconds, (b) its speed after 9 seconds, (c) its average speed over the whole 10 seconds.

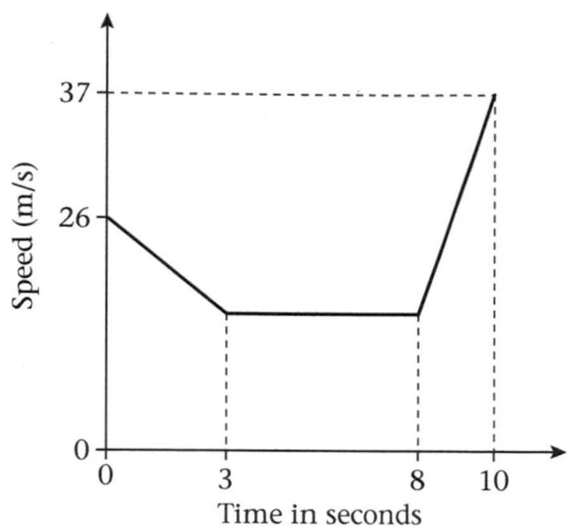

Fig. 10.31

Summary

The **gradient of a curve** at a point (x, y) is the gradient of the tangent to the curve at that point. At the **turning points** of a curve the tangent is parallel to the x-axis. Hence, at a turning point the gradient of a curve is zero.

In **speed–time** and **distance–time** graphs, time is always plotted on the horizontal axis. Constant speed is indicated by a line parallel to the time axis on the speed–time graph. Constant speed means zero acceleration.

Deceleration is negative acceleration, therefore the slope of the line will be negative.

Total distance travelled may be calculated from the area under the curve of a speed–time graph.

Practice exercise P10.1

1 (a) Copy and complete Table 10.6 for the function $y = \frac{1}{2}x^2$.

Table 10.6

x	-4	-3	-2	-1	0	1	2	3	4
x^2	16	9							
y	8	4.5							

(b) Using scales of 2 cm : 1 unit on each axis, draw the graph of the function $y = \frac{1}{2}x^2$ for $-4 \leqslant x \leqslant 4$.

(c) Draw tangents to the curve at
(i) $x = -3$, (ii) $x = -1$,
(iii) $x = 0$, (iv) $x = 2$.

(d) Use the tangents to find the gradient of the curve at
(i) $x = -3$, (ii) $x = -1$,
(iii) $x = 0$, (iv) $x = 2$.

2 (i) Copy the graphs in Fig. 10.32 and Fig. 10.33 and use the tangents shown to estimate the gradient of the curve at the marked points.

(ii) By drawing a suitable tangent, find the coordinates of the turning point of the curve and state whether it is a maximum or a minimum point.

(iii) Draw and write down the equation of the line of symmetry of the curve.

(a)

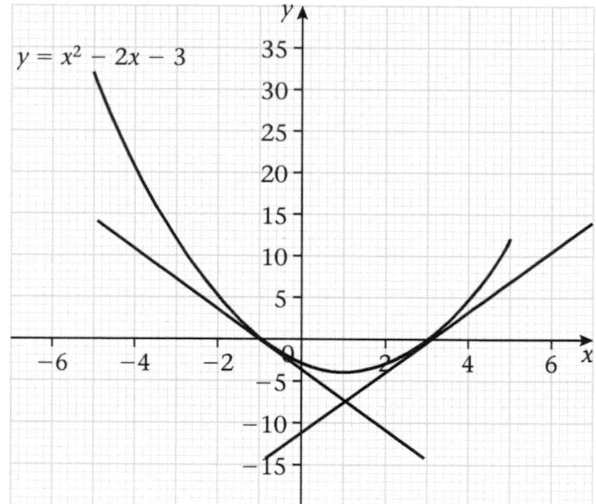

Fig. 10.32

(b)

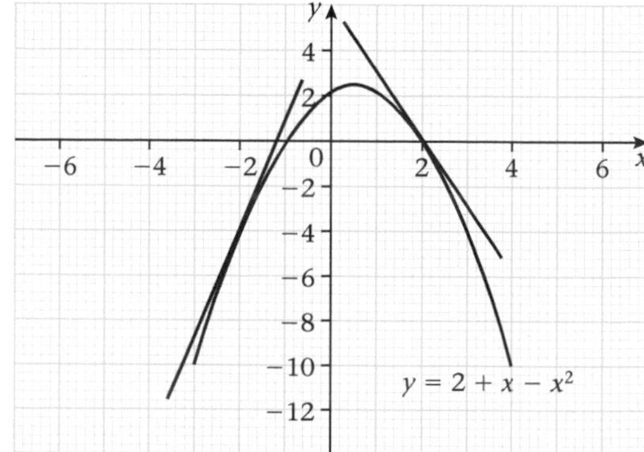

Fig. 10.33

3 (a) Copy and complete Table 10.7 for the function $y = x^2 - 4x + 1$.

Table 10.7

x	−4	−3	−2	−1	0	1	2	3	4	5
x^2		9		1	0	1		9		
$-4x$		12		4				−12		
$+1$		1		1				1		
y		22		6				−2		

(b) Using scales of 1 cm : 1 unit on the x-axis and 2 cm : 5 units on the y-axis, draw the graph of the function $y = x^2 - 4x + 1$ for $-4 \leqslant x \leqslant 5$.

(c) Use the graph to find the roots of the equation $x^2 - 4x + 1 = 0$.

(d) Draw tangents to the curve at
(i) $x = -3$, (ii) $x = -1$,
(iii) $x = 0$, (iv) $x = 2$.

(e) Use the tangents to estimate the gradient of the curve at
(i) $x = -3$, (ii) $x = -1$,
(iii) $x = 0$, (iv) $x = 2$.

(f) Find the turning point of the curve. State whether the turning point is a maximum or a minimum.

Practice exercise P10.2

1 Describe the information represented in the graph in Fig. 10.34.

(a)

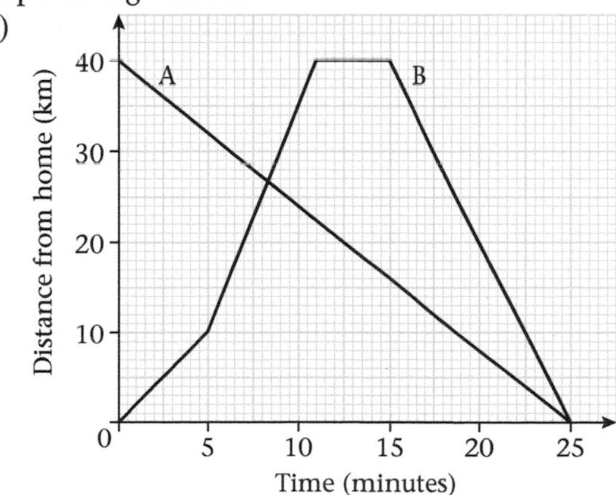

Fig. 10.34

2 The graph in Fig. 10.35 shows the relationship between the distance of an object measured from a point O and the corresponding times.

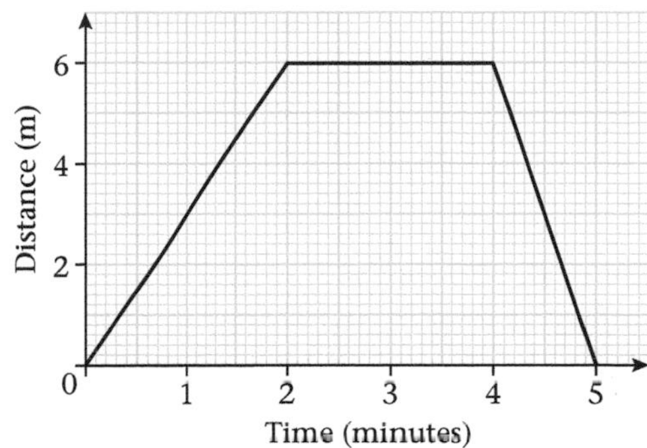

Fig. 10.35

(a) Describe the information given by each straight line (OA, AB, ...) of the graph.

(b) Calculate
(i) the total distance travelled,
(ii) the average speed for the whole journey.

Chapter 11

Functions

Inverse of a function, composite functions

Pre-requisites

- relations and mappings; flow charts; linear and non-linear functions

A **function** may be defined as:

(a) a relation between two sets in which each member of the first set is connected to one and only one member of the second set.

(b) a set of ordered pairs if the first element appears only once in any ordered pair in the set.

Functional notation

Functions are usually denoted by small (lower case) letters such as f, g and h. The set of ordered pairs

$(-1, -5), (0, -2), (1, 1), (2, 4)$

shows that for each x value the image is $3x - 2$ so that x is mapped onto $3x - 2$. The function f may be written as

$f: x \to 3x - 2$ (read 'f is the function that maps x to $3x - 2$')

$f(x) = 3x - 2$ ($f(x)$ if the image of each element x)

$y = f(x) = 3x - 2$ (y is a function of x)

Note that $f(x)$ is read simply as 'f of x' and does *not* mean 'f times x'.

Example 1

Given that a function $g(x) = 2x - 3$,

(a) find the value of $g(2)$;

(b) find the element in the domain for which the element in the range is 0;

(c) list the set of ordered pairs for integral values of x in the domain $-2 \leqslant x \leqslant 4$.

$g(x) = 2x - 3$

(a) If $x = 2$

$g(2) = 2(2) - 3 = 1$

(b) If $g(x) = 0$

$2x - 3 = 0$

$2x = 3$

$x = \frac{3}{2}$

(c) $\{(-2, -7), (-1, -5), (0, -3), (1, -1), (2, 1), (3, 3)\}$

Exercise 11a

1. Given that a function is defined by $f(x) = 3x - \frac{2}{x}$, evaluate
   (a) $f(1)$, (b) $f(5)$, (c) $f(0)$,
   (d) $f(-1)$, (e) $f(b)$, (f) $f(-b)$,
   (g) $f(-x)$, (h) $f(2x)$, (i) $f(-2x)$,

2. If $h(x) = x^3 + 3x - 1$, evaluate $h(5) - h(2)$.

3. If $g(x) = \dfrac{2x - 3}{x - 2}$ find x when $g(x)$ is
   (a) 1, (b) -3, (c) 0, (d) x.

4. Given that $h: x \to (x - 1)^2$,
   (a) find the value of
   (i) $h(2)$, (ii) $h(0)$, (iii) $h(a)$;
   (b) find x when $h(x) =$
   (i) 0, (ii) 4, (iii) $x^2 - 3$.

5. Given that $f(x) = cx^2 - 2$
   and $f(3) = 25$
   find the value of c.

6. Write each of the following in the form $y = f(x)$.
   (a) $2x + 5y = 1$
   (b) $\frac{x}{3} + \frac{y}{2} = -2$
   (c) $x^2 + x - y = 3$
   (d) $3(x - 1) + 2(y + 3) = 0$
   (e) $x^2 + xy - 4 + y = 3$

Inverse of a function

Given $f: x \to 3x - 2$ then $\{(1, 1), (2, 4), (3, 7), (4, 10)\}$ is a set of ordered pairs for the function f.

If there is a function g that is satisfied by the ordered pairs (1, 1), (4, 2), (7, 3), (10, 4), then the function g is the **inverse of the function** f. This is written

$g = f^{-1}$ (read 'f minus one')

Note that the order of the numbers in the ordered pairs is significant.

Since each element in the domain of a function must be mapped onto just one element in the range (in an arrow diagram only one arrow leaves each element in the domain and only one arrow goes to each element in the range), and since all elements in the domain must be mapped onto an element in the range, *only* one-to-one functions have inverses.

Example 2

Describe, giving reasons, which of the functions in Fig. 11.1 have inverses.

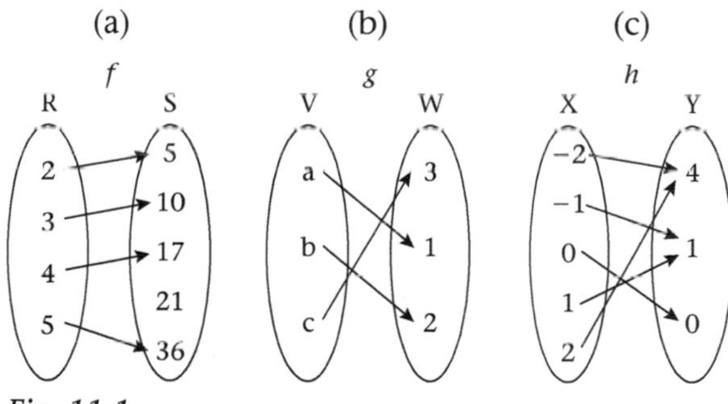

Fig. 11.1

(a) f maps all the elements in R onto elements in S; but there is an element 21 in S that cannot be mapped onto a corresponding element in R. Since in the reverse process *all* the elements in the set S cannot be mapped onto the set R, no inverse function f^{-1} exists.

(b) g maps V onto W so that each member of V is mapped onto one, and only one, member of W; hence, each element of W can be mapped onto one, and only one, corresponding element of V; that is, g and its inverse g^{-1} exist.

(c) h maps X onto Y so that only one arrow leaves each element in the set X; thus, this mapping is a function. However, for the mapping from Y onto X, the elements 4 and 1 in the set Y are each mapped onto two elements in the set X so that this mapping is not a function. Hence, there cannot exist an inverse function, h^{-1}.

The conditions for any functions f and g to be inverse functions are

1. f and g must each be a one-to-one mapping,
2. the range of f must be the domain of g and the range of g must be the domain of f,
3. all the elements in the domain of f must be mapped onto all the elements in the domain of g.

An algebraic expression denotes a function by defining the operations to map the elements in the domain onto the range. In order to represent the inverse function by an algebraic expression, the operations must be reversed and performed in the reverse order.

Example 3

Given that $f(x) = 3x - 2$, derive the expression for the inverse function $f^{-1}(x)$.

Method 1
In the function $f(x) = 3x - 2$, x is first multiplied by 3 and then 2 is subtracted.
The reverse operations are add 2 and then divide by 3, that is

$$f^{-1}(x) = \frac{x + 2}{3}$$

Method 2
Write the function in terms of another variable, usually y, that is

$$y = 3x - 2$$

Interchange the variable x and y, so that the equation becomes

$$x = 3y - 2$$

and then write the equation in terms of the 'new' y

$$x + 2 = 3y$$

$$y = \frac{x + 2}{3}$$

$$\text{i.e. } f^{-1}(x) = \frac{x + 2}{3}$$

Example 3 may be illustrated graphically by using a **flow chart.** In a flow chart each operation is put in a rectangle or box. The boxes are linked by arrows showing the sequence or order of the operations (Fig 11.2).

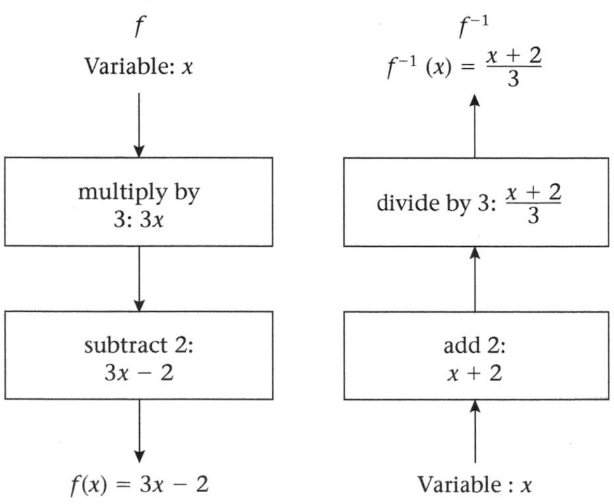

Fig. 11.2

Example 4

Given $f(x) = \dfrac{5x - 3}{x + 4}$,

find (a) $f^{-1}(x)$, (b) $f^{-1}(2)$

(a) Let $y = \dfrac{5x - 3}{x + 4}$

Write $x = \dfrac{5y - 3}{y + 4}$ (interchange x and y)

$$xy + 4x = 5y - 3$$
$$xy - 5y = -4x - 3 = -(4x + 3)$$
$$y(x - 5) = -(4x + 3)$$
$$y = \frac{-(4x + 3)}{x - 5} = \frac{4x + 3}{5 - x}$$
$$f^{-1}(x) = \frac{4x + 3}{5 - x}$$

(b) $f^{-1}(2) = \dfrac{8 + 3}{5 - 2} = \dfrac{11}{3} = 3\frac{2}{3}$

Example 5

Given that $f : x \rightarrow \dfrac{3x + 2}{x + 2}$,

(a) evaluate $f(1)$ and $f(-3)$,
(b) determine $f^{-1}(x)$,
(c) find the value of x if (i) $f(x) = -2$, (ii) $f(x) = x$, (iii) $f(x)$ is undefined.

$$f(x) = \frac{3x + 2}{x + 2}$$

(a) $f(1) = \dfrac{3(1) + 2}{(1) + 2} = \dfrac{5}{3}$

$$f(-3) = \frac{3(-3) + 2}{(-3) + 2}$$
$$= \frac{-9 + 2}{-3 + 2} = \frac{-7}{-1} = 7$$

(b) Let $y = \dfrac{3x + 2}{x + 2}$

$$x = \frac{3y + 2}{y + 2}$$

Then $xy + 2x = 3y + 2$

$$xy - 3y = 2 - 2x$$
$$y(x - 3) = 2 - 2x = 2(1 - x)$$
$$y = \frac{2(1 - x)}{x - 3}$$

that is, $f^{-1}(x) = \dfrac{2(1 - x)}{x - 3}$

(c) (i) $\dfrac{3x + 2}{x + 2} = -2$

$$3x + 2 = -2x - 4$$
$$5x = -6$$
$$x = \frac{-6}{5}$$

(ii) $\dfrac{3x + 2}{x + 2} = x$

$$3x + 2 = x^2 + 2x$$
$$x^2 - x - 2 = 0$$
$$(x + 1)(x - 2) = 0$$
$$x = -1 \quad \text{or} \quad x = 2$$

(iii) $f(x) = \dfrac{3x + 2}{x + 2}$,

$f(x)$ is undefined when $x + 2 = 0$, that is, when $x = -2$.

The inverse of a function can also be used to solve an equation.

Example 6

(a) Find the inverse of $f(x) = \dfrac{x - 5}{x + 3}$.

(b) Using the inverse in (a) solve the equation

$$\frac{x - 5}{x + 3} = 3.$$

(a) Let $y = \frac{x - 5}{x + 3}$

Write $x = \frac{y - 5}{y + 3}$

$$xy + 3x = y - 5$$
$$xy - y = -3x - 5$$
$$y(x - 1) = -(3x + 5)$$
$$y = \frac{3x + 5}{1 - x}$$
$$f^{-1}(x) = \frac{3x + 5}{1 - x}$$

(b) To solve $\frac{x - 5}{x + 3} = 3$,

note that if $f(a) = b$ then $f^{-1}(b) = a$. Hence

$$f^{-1}(3) = \frac{9 + 5}{1 - 3} = -7$$

The solution of the given equation is $x = -7$.

Exercise 11b

1. Given the following functions,
   (i) state whether the inverse function exists, giving reasons,
   (ii) if so, describe the inverse function.
   (a) f is the function 'times 2'
   (b) g is the function 'is a town in'
   (c) h is the function 'is the volume of'

2. The function f is defined by $f: x \mapsto x + 5$
   (a) Calculate the values of f when x is -2, -1, 0, 1, 2.
   (b) Does the inverse function exist in the given domain? If it does, define the expression and verify it.
   (c) Draw an arrow diagram to illustrate your answer.

3. For each of the following functions
   (i) state a suitable domain and range,
   (ii) find an expression for the inverse function, if it exists:

   (a) $f(x) = 5x$ (b) $f(x) = \frac{x}{2} + 1$

   (c) $f(x) = \frac{x - 3}{4}$ (d) $f(x) = (3x + 1)^2$

   (e) $f(x) = \frac{2x + 5}{x - 3}$

4. If f is a function defined by
   $f(x) = \frac{5x - 3}{x - 4}$,
   (a) write down an expression for $f^{-1}(x)$,
   (b) find the value of $f^{-1}(-3)$,
   (c) state the value of x for which $f(x)$ is undefined.

5. By finding the inverse of
   $f(x) = \frac{3x + 2}{x + 3}$,
   solve the equation
   $$\frac{3x + 2}{x + 3} = 2$$

Composite functions

Let f and g be two functions such that f maps the element x in the domain C onto w in the range H, and g maps the element w in the domain H onto y in the range D. Hence, a function is now defined which maps the element x in the domain C onto y in the range D. This may be represented graphically as in Fig. 11.3.

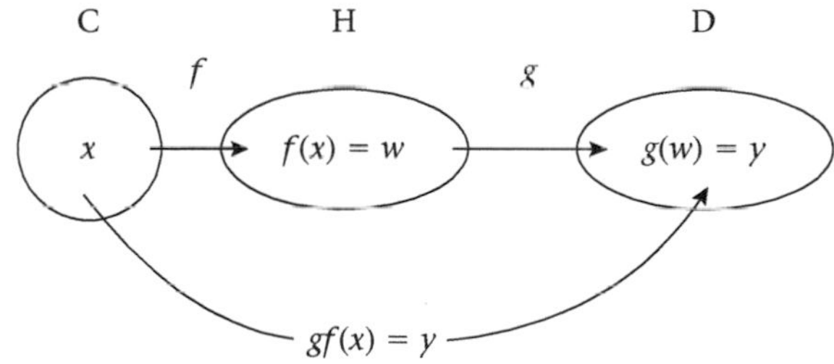

Fig. 11.3

This defined function is written as $gf(x)$ and is called a **composite function**. Note that f is the function that comes first. The elements in the domain of $gf(x)$ are those that are in both the domain and the range of f.

The composite function $gf(x)$ means apply the function f first and then apply the function g to the result. What would the composite function $fg(x)$ mean?

Example 7

A composite function is defined as $h(x) = x^2 + 3$. Write $h(x)$ as two simpler functions, f and g, indicating clearly the order of f and g in the composite function h.

In the function h, first square the variable x and then add 3.

Hence, let $f(x) = x^2 = w$

then $g(w) = w + 3 = x^2 + 3$

so that $h(x) = g(w) = g(f(x))$

that is $h(x) = gf(x)$

Example 8

Given that $f(x) = 2x + 1$

and $g(x) = 3x - 2$

express the composite functions (a) $fg(x)$ as a single function $h(x)$, (b) $gf(x)$ as a single function $k(x)$.

(a) $fg(x) = f(3x - 2) = 2(3x - 2) + 1$
$= 6x - 4 + 1$
$h(x) = 6x - 3$

(b) $gf(x) = g(2x + 1) = 3(2x + 1) - 2$
$= 6x + 3 - 2$
$k(x) = 6x + 1$

Notice that in this example $fg(x)$ is not equal to $gf(x)$, that is the composite of two functions may not be commutative.

Example 9

Use the functions f and g in Example 8 to calculate

(a) $(fg)^{-1}$, (b) $(gf)^{-1}$, (c) $f^{-1}g^{-1}$, (d) $g^{-1}f^{-1}$.

(a) $fg(x) = 6x - 3$

$(fg)^{-1}(x) = \frac{x + 3}{6}$

(b) $gf(x) = 6x + 1$

$(gf)^{-1}(x) = \frac{x - 1}{6}$

(c) $f(x) = 2x + 1$

$f^{-1}(x) = \frac{x - 1}{2}$

and $g(x) = 3x - 2$

$g^{-1}(x) = \frac{x + 2}{3}$

$f^{-1}g^{-1}(x) = f^{-1}\left(\frac{x + 2}{3}\right)$

$= \frac{\frac{x + 2}{3} - 1}{2} = \frac{x + 2 - 3}{6}$

$= \frac{x - 1}{6}$

(d) $g^{-1}f^{-1}(x) = g^{-1}\left(\frac{x - 1}{2}\right)$

$= \frac{\frac{x - 1}{2} + 2}{3} = \frac{x - 1 + 4}{6}$

$= \frac{x + 3}{6}$

Notice that Example 9 illustrates that

$(fg)^{-1} = g^{-1}f^{-1}$

and $(gf)^{-1} = f^{-1}g^{-1}$

Exercise 11c

1 The arrow diagram in Fig. 11.4 shows the domain and range for the functions f and g.

(a) Find $gf(1)$, $gf(2)$, $gf(3)$.

(b) State the domain and range of gf.

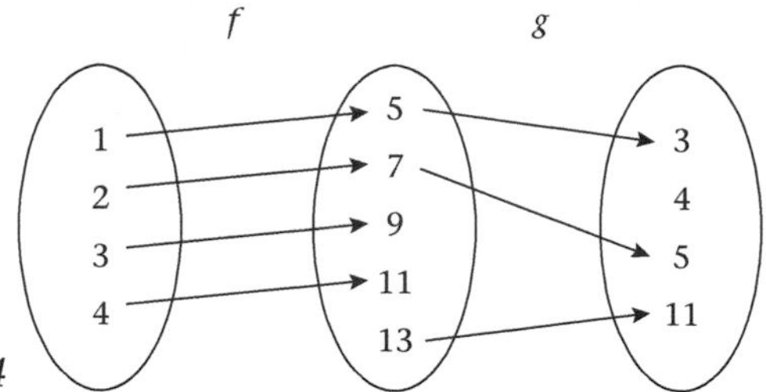

Fig. 11.4

2 f and g are functions on the set of integers defined by the expressions

$f(x) = x + 2$ and $g(x) = 3x - 1$.

(a) Complete Table 11.1.

Table 11.1

x →	$f(x)$ →	$g(f(x))$
−2	0	−1
−1		
0		
1		
2		

(b) Show that $gf(x) = 3x + 5$.

(c) Draw an arrow diagram to show the domain and range of $gf(x)$.

3 (a) Find the inverses f^{-1} and g^{-1} of the functions $f : x \rightarrow 1 - 2x$ and $g : x \rightarrow 3x$.

(b) Write down the expressions for

(i) gf, (ii) $f^{-1}g^{-1}$, (iii) $(gf)^{-1}$.

4. f, g and h are functions defined by the expressions

   $f : x \rightarrow 3x \qquad g : x \rightarrow \frac{x}{4} + 1 \qquad h : x \rightarrow x^2$

   (a) Write down expressions for

   (i) f^{-1} (ii) g^{-1} (iii) h^{-1}
   (iv) fg (v) gf (vi) fh
   (vii) $(hf)^{-1}$ (viii) $(gh)^{-1}$. (ix) $(hg)^{-1}$

   (b) Show that

   (i) $(hf)^{-1} = f^{-1}h^{-1}$
   (ii) $h^{-1}g^{-1} = (gh)^{-1}$

5. If $f(x) = x^2 - 8x + 15$, find

   (a) $f(3)$, $f(-2)$; (b) x if $f(x) = 0$;
   (c) x if $f(x) = 8$.

Summary

A function is a one-to-one or many-to-one relation between the elements of two sets A and B and is usually denoted by a lower case letter.

$$f : x \rightarrow y, \quad f(x) = y, \quad y = f(x)$$

are functional notations.

If f : A $\rightarrow$ B and g : B $\rightarrow$ A, f and g are **inverse functions** only if

(a) f and g are one-to-one functions,
(b) the range of A is the domain of B and the domain of A is the range of B,
(c) all the elements in A are the domain of f and all the elements in B are the domain of g.

Given that f and g are functions with domains A and B, then the **composite** of f and g is denoted by the functions $gf(x)$, that is, $g(f(x))$, with a domain on the set of elements that are both in the domain and range of A.

Practice exercise P11.1

1. State which of the arrow diagrams in Fig. 11.5 illustrate functions, giving reasons for your statement on each relation.

(a)

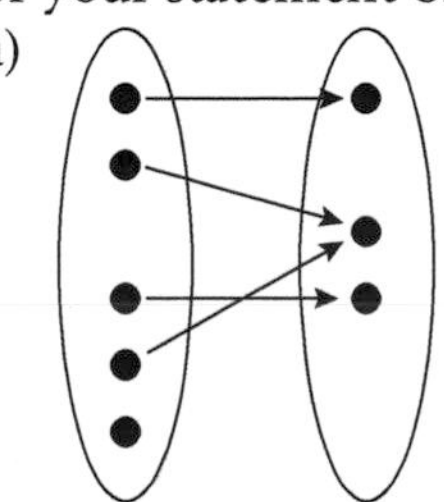

(b)

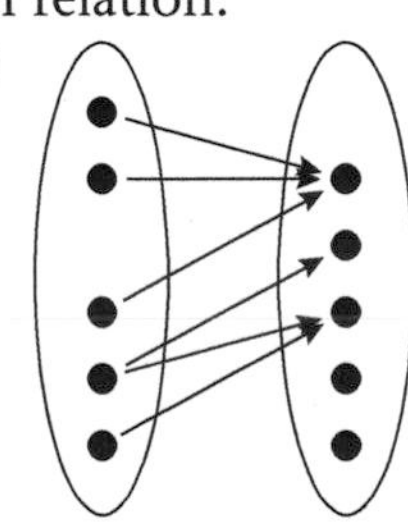

(c) (d)

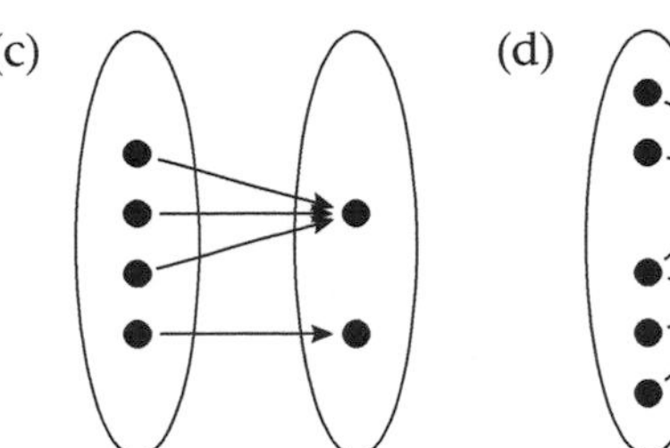

Fig. 11.5

2. (i) Draw up a table of values for each of the following functions.
   (ii) Write the functions as ordered pairs.

   (a) $y = x^2 - 2x + 4$, where $x = -3, -2, -1, 0, 1, 2, 3$

   (b) $h(x) = \frac{2x + 1}{x - 3}$ where $-2 < x \leqslant 3$

3. For each relation shown below

   (i) list the ordered pairs,
   (ii) state whether the relation is a function or not, giving your reason,
   (iii) if the relation is a function, say whether it is one-to-one or many-to-one.

	domain	***relation***	***range***
(a)	$\{-3, -1, 0, 1, 2, 3\}$	$x \rightarrow -x + 1$	$\{4, 2, -1, -3\}$
(b)	$\{15, 12, 9, 6, 3\}$	is a multiple of	$\{5, 4, 3, 2\}$
(c)	$\{-3, -2, -1, 0, 1, 2\}$	$x \rightarrow x^2 - 2$	$\{7, -1, -2, 2\}$
(d)	$\{2, 3, 4, 5, 6\}$	is a prime factor of	$\{10, 12, 14, 16, 18, 20\}$

Practice exercise P11.2

1. If $h : x \rightarrow (x - 2)^2$,

   (a) evaluate
   (i) $h(1)$,
   (ii) $h(-1)$,
   (iii) $-h(w)$,

   (b) find x when
   (i) $h(x) = 0$,
   (ii) $h(x) = x$,
   (iii) $h(x) = 2x^2 - 5$.

2. The function g is defined by

$$g : x \rightarrow \frac{x - 5}{2x + 4}$$

(a) Calculate the values of g when x is -3, -2, -1, 0, 1, 2, 3, stating any value of x for which the function is undefined.
(b) List the set of ordered pairs for integral values of x in the domain $-1 < x \leq 3$.
(c) Draw an arrow diagram to illustrate your answer.

3. N = {1, 2, 3, 4} and L = {m, n}.
A relation P maps each odd number to m and each even number to n.
(a) List P as a set of ordered pairs.
(b) Draw an arrow diagram.
(c) Is P a function from N to L? Give a reason for your answer.

Practice exercise P11.3

1. Given the functions,
(a) f is the function 'is the capital city of',
(b) g is the function 'is 5 more than',
(c) h is the function 'is a street in',

(i) state whether an inverse function exists, giving reasons for your answer,
(ii) if the inverse function exists, describe it.

2. Use a flow chart to illustrate the operations that define
(a) the function $y = g(x)$, where $2y - 3x = 4$,
(b) the inverse of the function y.

3. (a) By interchanging the variables in the function $f(x) = \frac{2x - 5}{x + 3}$, derive the inverse of the function.
(b) State any value(s) of x for which the function is undefined.
(c) Use the inverse function to find the solution of the equation $\frac{2x - 5}{x + 3} = 1$.

4. For each of the following functions,
(i) find the value for integral values of x in the domain $-2 < x < 4$,
(ii) if it exists, derive the inverse function,
(iii) state the value of x, if any, for which the function is undefined.

(a) $f(x) = -2x + 5$
(b) $f(x) = 2x^2 - x - 2$
(c) $f(x) = -x^2 - 3x - 1$
(d) $f(x) = \frac{2 - x}{2x + 1}$

Practice exercise P11.4

1. Given the two functions $f(x)$ and $g(x)$, find
(i) $gf(x)$, (ii) $fg(x)$, where
(a) $f(x) = 4x$ and $g(x) = x + 1$,
(b) $f(x) = x - 3$ and $g(x) = x^2 + 1$.

2. Given the function $f : x \rightarrow x + 5$,
(a) find $f(3)$ and $ff(3)$ and express $ff(x)$ in its simplest form,
(b) find $f^{-1}(x)$ and the value of $f^{-1}(3)$ and $f^{-1}f^{-1}(3)$,
(c) express $f^{-1}f^{-1}(x)$ in its simplest form.

3. (a) Find two simple functions f and g such that $gf(x) \rightarrow x^2 + 5$.
(b) Find $fg(x)$ and the value of $gf(2)$ and $fg(2)$.

4. (a) Find the inverses f^{-1} and g^{-1} of the functions $f : x \rightarrow 2 - 3x$ and $g : x \rightarrow 2x$.
(b) Find expressions for
(i) gf, (ii) $f^{-1}g^{-1}$, (iii) $(gf)^{-1}$.

5. Given the function $f(x) = 2x - 5$,
(a) express $f^{-1}(x)$ in terms of x,
(b) find $f(7)$, $f^{-1}(7)$, $f^{-1}f(7)$, $ff^{-1}(7)$ and $f^{-1}f^{-1}(7)$,
(c) express $f^{-1}f(x)$ and $ff^{-1}(x)$ in their simplest forms and comment on the answers.

6. Given the two functions $f(x) = 2x + 1$ and $g(x) = x^2 - 1$,
(a) find two values of x such that $fg(x) = 17$,
(b) find two values of x such that $gf(x) = 8$.

Chapter 12

Trigonometry (3)

Trigonometric graphs

Pre-requisites

- trigonometric ratios; non-linear functions

Trigonometric functions

In Chapter 3, the values of the trigonometric ratios of positive angles between 0° and 360° were calculated. The sign of the value of the ratios was found by rotating a line OP *anticlockwise*, starting from the positive x-axis. The radius, OP, of the circle was of length +1 unit.

Thus, it was was possible to find the trigonometric ratios of acute angles (θ_1, between 0° and 90°), of obtuse angles (θ_2, between 90° and 180°), and of reflex angles (θ_3, between 180° and 270° and θ_4, between 270° and 360°).

Similarly, the values of the trigonometric ratios of negative angles between 0° and −360° were determined, but, starting from the positive x-axis, the line OP was rotated *clockwise*.

Since the trigonometric functions, which give the relation between any angle, θ, and the values of the trigonometric ratios, may be defined by the above method, the trigonometric functions are also called circular **functions**.

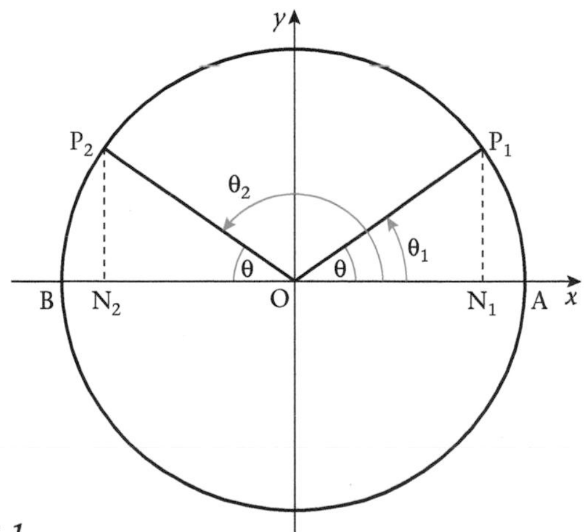

Fig. 12.1

The sine function

In Fig. 12.1, O is the centre of the circle and the origin of axes Ox and Oy. OP_1 and OP_2 are radii, each of length 1 unit.

$$A\widehat{O}D_1 = \theta_1 = \theta$$

and $\quad A\widehat{O}P_2 = \theta_2 = (180° - \theta)$

where $\quad B\widehat{O}P_2 = \theta$

As P moves around the circle in an anticlockwise direction, P_1N_1, the distance of P_1 from the horizontal axis, represents the sine of angle θ.

That is, $\quad P_1N_1 = \sin\theta$

Since $\quad A\widehat{O}P_1 = B\widehat{O}P_2$

then $\quad P_2N_2 = P_1N_1$

so that, as in Chapter 3,

$\sin\theta_2 = \sin(180° - \theta) = \sin\theta$

In Fig. 12.2, $\quad A\widehat{O}P_3 = \theta_3 = (180° + \theta)$

$\quad A\widehat{O}P_4 = \theta_4 = (360° - \theta)$

and $\quad P_3N_2 = P_4N_4 = -P_1N_1$

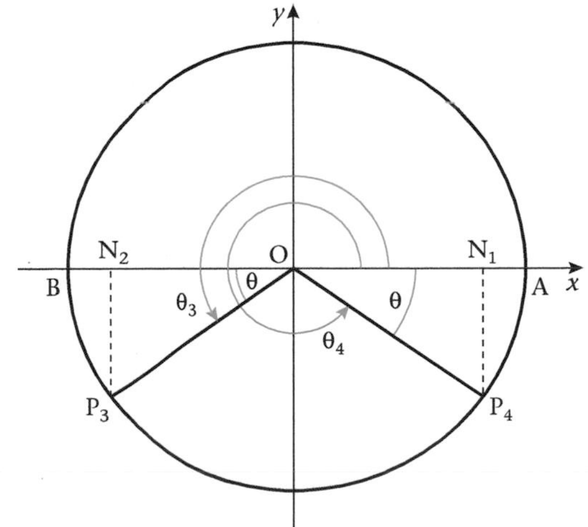

Fig. 12.2

Hence, for angles greater than 180°, as shown in Fig. 12.2,

$\sin \theta_3 = \sin(180° + \theta) = -\sin \theta$
$\sin \theta_4 = \sin(360° - \theta) = -\sin \theta$

Example 1 (Revision)

Express the following as ratios of actute angles.
(a) sin 120°, sin 240°, sin 300°
(b) sin 145°, sin 215°, sin 325°

(a) $\sin 120° = \sin(180 - 120)° = \sin 60°$
$\sin 240° = \sin(180 + 60)° = -\sin 60°$
$\sin 300° = \sin(360 - 60)° = -\sin 60°$

(b) $\sin 145° = \sin(180 - 35)° = \sin 35°$
$\sin 215° = \sin(180 + 35)° = -\sin 35°$
$\sin 325° = \sin(360 - 35)° = -\sin 35°$

As in Chapter 3, and shown in Fig. 12.1 and Fig. 12.2,

$\sin 0° = 0$
$\sin 90° = 1$ $\quad \sin 270° = -1$
$\sin 180° = 0$ $\quad \sin 360° = 0$
$\sin(180° - \theta) = \sin \theta$
$\sin(180° + \theta) = -\sin \theta$
$\sin(360° - \theta) = -\sin \theta$

The sine curve

Using the values of the sine ratio for 0°, 90°, 180°, 270°, and 360°, and the values of the sine ratio for 30° and 60° (using an equilateral triangle – see Chapter 3), and for 45° (using an isosceles right-angled triangle – see Chapter 3), a table of values for the trigonometric function

$y = \sin \theta°, \quad \text{for } 0 \leqslant \theta \leqslant 360$

may be drawn up as follows (Table 12.1).

Table 12.1

θ	0	30	45	60	90	120	135	150	180
sin θ°	0	0.5	0.71	0.87	1.0	0.87	0.71	0.5	0

θ	210	225	240	270	300	315	330	360
sin θ°	−0.5	−0.71	−0.87	−1.0	−0.87	−0.71	−0.5	0

From Chapter 3, $\sin(-\theta) = -\sin \theta$ for *all* values of θ. The table of values for

$y = \sin \theta°, \quad \text{for } -360 \leqslant \theta \leqslant 0$

is shown in Table 12.2.

Table 12.2

θ	0	−30	−45	−60	−90	−120	−135	−150	−180
sin θ°	0	−0.5	−0.71	−0.87	−1.0	−0.87	−0.71	−0.5	0

θ	−210	−225	−240	−270	−300	−315	−330	−360
sin θ°	0.5	0.71	0.87	1.0	0.87	0.71	0.5	0

Fig. 12.3 is the graph of the function

$y = \sin \theta°, \quad \text{for } -360 \leqslant \theta \leqslant 360$

This graph is called the **sine curve**.

From the graph in Fig. 12.3, note that

1 the maximum value of the sine function is 1,
2 the minimum value of the sine function is −1,
3 the values of the sine function between −360° and 0° are repeated between 0° and 360°; such a function, when the values of the function are repeated after a constant period, is called a **periodic function**. For the sine curve, the period is 360°.

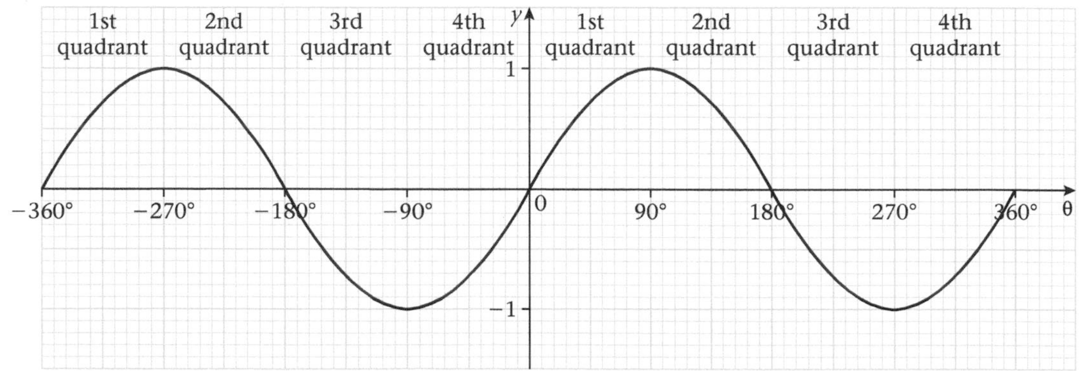

Fig. 12.3

Example 2

Given that $y = \sin x°$, for $0 \leqslant x \leqslant 180$, copy and complete the table below.

Table 12.3

x	0	30	45	60	90	120	135	150	180
$\sin x°$	0	0.5	0.71	0.87					

Using a scale of 1 cm to represent 30° on the x-axis and 2 cm to represent 1 unit on the y-axis, draw the graph of $y = \sin x°$, for $0 \leqslant x \leqslant 180$. Hence, state the solutions of the equation $\sin x° =$ for 0.65, for $0 < x < 180$.

The required table is Table 12.4 and the graph is Fig. 12.4.

Table 12.3

x	0	30	45	60	90	120	135	150	180
$\sin x°$	0	0.5	0.71	0.87	1.0	0.87	0.71	0.5	0

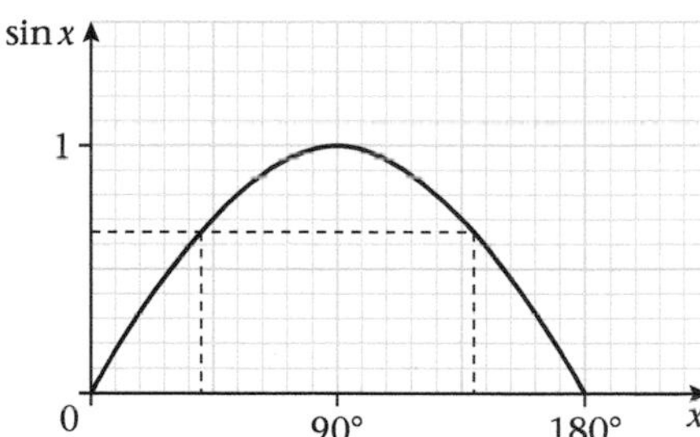

Fig. 12.4

From the graph, when $\sin x° = 0.65$,

$x° \simeq 40°, 140°$.

Exercise 12a

1. Write down the values of the following using tables or a calculator.
   (a) $\sin 63°$ (b) $\sin 139°$
   (c) $\sin 224°$ (d) $\sin 330°$
   (e) $\sin(-32°)$ (f) $\sin(-140°)$
2. Using the $\sin\theta$ table, on page 335, write down the acute angle for which $\sin\theta = 0.7$.
3. Using the sine curve in Fig. 12.3, or otherwise, write down all the angles between $-360°$ and $360°$ that satisfy the equation $\sin\theta = 0.7$.
4. Given that $\sin x = 153°$ and x is acute, using Fig. 12.3, or otherwise, find x.
5. What values between 0° and 180° satisfy the equation $\sin x = 0.67$?
6. What values between $-360°$ and 360° satisfy the equation $\sin x = -0.88$?
7. Simplify
   (a) $\dfrac{\sin(180° - \theta)}{\cos\theta}$ (b) $\dfrac{\sin(180° + \theta)}{\cos\theta}$
8. Using a scale of 1 cm to represent 30° on the x-axis and 2 cm to represent 1 unit on the y-axis, draw the graph of $y = 2\sin x - 1$ for values of x between 0° and 360°.
9. From the graph drawn in question 8, find the values of x which satisfy the equations
   (a) $2\sin x - 1 = 0.5$,
   (b) $2\sin x - 1 = -0.6$.
10. Using the graph of question 8 and drawing a suitable straight line graph, solve the equation $2\sin x = -1$.
11. Given that $y = \sin 2x$, copy and complete the table for $0 \leqslant x \leqslant 90$.

Table 12.5

x	0	30	45	60	90	15	22.5	75
$2x$	0	60	90	120	180	30		
$\sin 2x°$	0	0.87	0.71					

Using a scale of 2 cm to represent 30° on the x-axis and 2 cm to represent 1 unit on the y-axis, draw the graph of

$y = \sin 2x°$, for $0 \leqslant x \leqslant 90$

Hence, state the solutions of the equation

$\sin 2x° = 0.5$, for $0 \leqslant x \leqslant 90$

[Hint: Note 'extra' values of x are added to the tables as required.]

The cosine function

In Figs. 12.5 and 12.6, as in Fig. 12.1 and 12.2, O is the centre of the circle and the origin of axes Ox, and Oy. OP is a radius of unit length.

As P moves around the circle in an anticlockwise direction, N, the foot of the perpendicular from

P to the x-axis moves from A to B, and back again to A. ON represents the cosine of the angle θ.

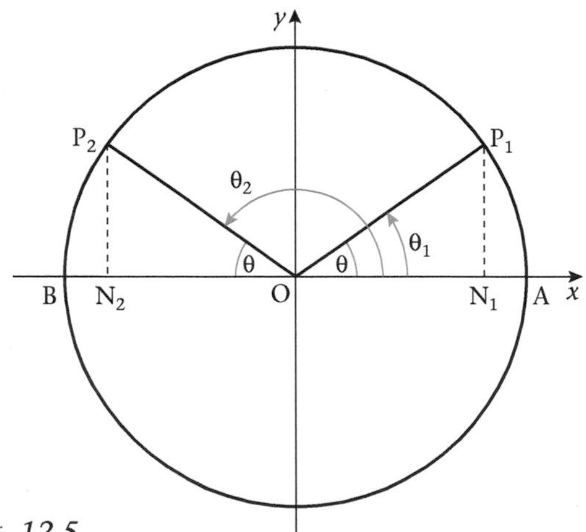

Fig. 12.5

In Fig. 12.5, $OP_1 = OP_2 = OP$

$A\widehat{O}P_1 = \theta_1 = \theta$

and $\cos\theta_1 = ON_1 = \cos\theta$

Also, $A\widehat{O}P_2 = \theta_2 = (180° - \theta)$

where $B\widehat{O}P_2 = \theta$

and $ON_2 = -ON_1$

so that $\cos\theta_2 = -\cos\theta$

In Fig. 12.6, $OP_3 = OP_4 = OP_1$

$A\widehat{O}P_3 = \theta_3 = (180° + \theta)$

and $ON_2 = -ON_1$

so that $\cos\theta_3 = -\cos\theta$

Also $A\widehat{O}P_4 = \theta_4 = (360° - \theta)$

so that $\cos\theta_4 = \cos\theta$

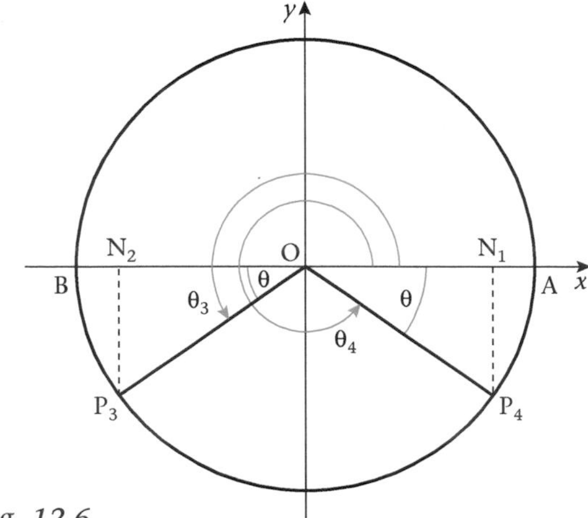

Fig. 12.6

As in Chapter 3, and shown in Fig. 12.5 and Fig. 12.6,

$$\cos(180 - \theta) = -\cos\theta$$
$$\cos(180 + \theta) = -\cos\theta$$
$$\cos(360 - \theta) = \cos\theta$$
$$\cos 0° = -1$$
$$\cos 90° = 1$$
$$\cos 180° = -1$$
$$\cos 27° = 0$$
$$\cos 360° = 1$$

Example 3 (Revision)

Express the following as ratios of acute angles.
(a) cos 150°, cos 210°, cos 330°
(b) cos 108°, cos 252°, cos 288°

(a) $\cos 150° = (\cos 180 - 30)° = -\cos 30°$
$\cos 210° = (\cos 180 + 30)° = -\cos 30°$
$\cos 330° = (\cos 360 - 30)° = \cos 30°$

(b) $\cos 108° = (\cos 180 - 72)° = -\cos 72°$
$\cos 252° = (\cos 180 + 72)° = -\cos 72°$
$\cos 288° = (\cos 360 - 72)° = \cos 72°$

The cosine curve

A table of values for the trigonometric function

$$y = \cos\theta°, \text{ for } 0 \leqslant \theta \leqslant 360$$

may be drawn up, using the values of the cosine ratio of the 'special' angles (Table 12.6).

Table 12.6

θ	0	30	45	60	90	120	135	150	180
cos θ°	1.0	0.87	0.71	0.5	0	−0.5	−0.71	−0.87	−1.0

θ	210	225	240	270	300	315	330	360
cos θ°	−0.87	−0.71	−0.5	0	0.5	0.71	0.87	1.0

Since $\cos(-\theta) = \cos\theta$, for all values of θ, the values for $-360 \leqslant \theta < 0$ are shown in Table 12.7.

Table 12.7

θ	−30	−45	−60	−90	−120	−135	−150	−180
cos θ°	0.87	0.71	0.5	0	−0.5	−0.71	−0.87	−1.0

θ	−210	−225	−240	−270	−300	−315	−330	−360
cos θ°	−0.87	−0.71	−0.5	0	0.5	0.71	0.87	1.0

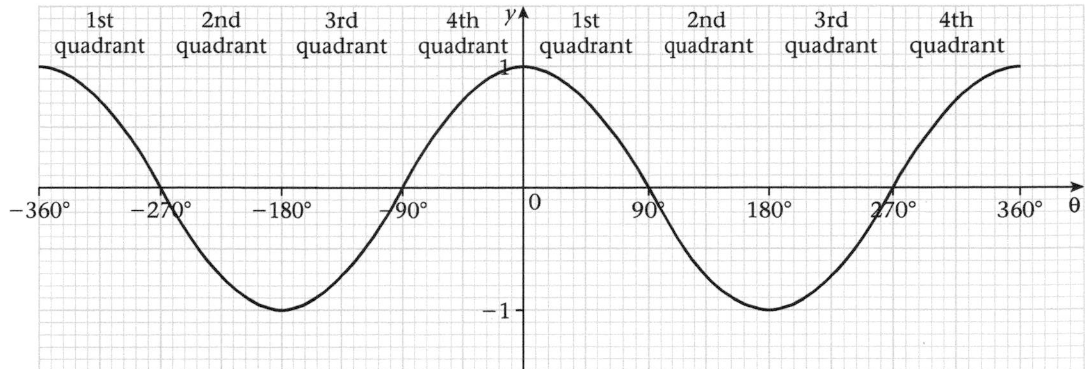

Fig. 12.7

Fig. 12.7 is the graph of the function

$y = \cos \theta°$, for $-360 \leqslant \theta \leqslant 360$

This graph is called the cosine curve.

From the graph in Fig. 12.7, note that as for the sine curve

1 the maximum value of the cosine function is 1,
2 the minimum value of the cosine function is −1,
3 the values of the cosine function between −360° and 0° are repeated between 0° and 360°; such a function, when the values of the function are repeated after a constant period, is called a **periodic function**. For the cosine curve, the period is 360°;

but

4 the same values of the sine function and the cosine function occur at values of angles that differ by 90°, that is,

$\cos \theta = \sin (90 + \theta)$

Example 4

Given that $y = \cos x$, for $0 \leqslant \theta \leqslant 180$, copy and complete the table below.

Table 12.8

x	0	30	45	60	90	120	135	150	180
$\cos x°$	1.0	0.87	0.71	0.5					

Using a scale of 1 cm to represent 45° on the x-axis and 2 cm to represent 1 unit on the y-axis, draw the graph of $y = \cos x$, for $0 \leqslant \theta \leqslant 180$.

Hence, state the solutions of the equation

$\cos x° = 0.65$, for $0 < x < 180$

Table 12.9

x	0	30	45	60	90	120	135	150	180
$\cos x°$	1.0	0.87	0.71	0.5	0	−0.5	−0.71	−0.87	−1.0

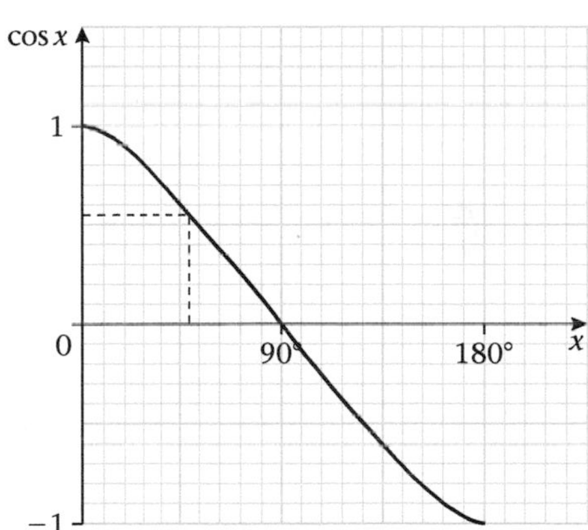

Fig. 12.8

From the graph, when $\cos x = 0.65$,

$x = 50°$, for $0 \leqslant x \leqslant 180$.

Example 5

On the same axes draw the graphs of $\sin \theta$ and $\cos \theta$ for $0° \leqslant \theta \leqslant 180°$. Hence, find the values of θ for which $\sin \theta = \cos \theta$.

From the graph in Fig. 12.9, the required values are at points where the curves intersect, i.e. $\theta = 45°, 225°$.

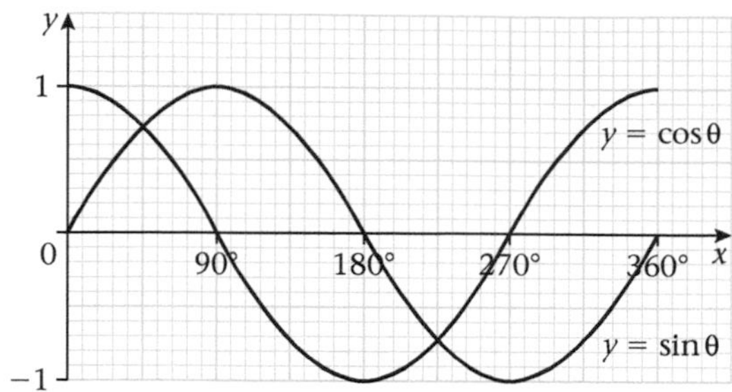

Fig. 12.9

Exercise 12b

1. Write down the values of the following using tables or a calculator.
   (a) cos 63° (b) cos 139°
   (c) cos 224° (d) cos 330°
   (e) cos (−32°) (f) cos (−135°)
2. Using the cos θ table, on page 336, or a calculator, write down the acute angle for which cos θ = 0.4.
3. Using the cosine curve in Fig. 12.7, or otherwise, write down all the angles between −360° and 360° that satisfy the equation cos θ = 0.7.
4. If cos x° = −cos 155° and x is acute, find x.
5. What values between 0° and 180° satisfy the equation cos x = 0.67?
6. What values between −360° and 360° satisfy the equation cos x = −0.88?
7. Simplify
   (a) $\dfrac{\sin(180° - \theta)}{\cos(180° - \theta)}$ (b) $\dfrac{\sin(180° + \theta)}{\cos(180° + \theta)}$
7. Given that $y = \cos x°$, for $-90 \leqslant x \leqslant 90$, copy and complete the following table.

Table 12.10

x	−90	−60	−45	−30	0	30	45	60	90
cos x°		0.5		0.87			0.71		

Using a scale of 1 cm to represent 30° on the x-axis and 2 cm to represent 1 unit on the y-axis, draw the graph of

$y = \cos x°$, for $-90 \leqslant x \leqslant 90$

Hence, state the solutions of the equation

$\cos x° = 0.5$, for $-90 \leqslant x \leqslant 90$

9. Using a scale of 2 cm to represent 30° on the x-axis and 2 cm to represent 1 unit on the y-axis, draw the graph of $y = 2\cos x + 1$ for values of x between 0° and 180°.
10. From the graph drawn in question 9, find the values of x which satisfy the equations
    (a) $2\cos x + 1 = 0.5$,
    (b) $2\cos x + 1 = -0.6$.
11. Using the graph of question 9, and drawing a suitable straight line graph, solve the equation $-\cos x = -2.0$.

The tangent function

In Figs. 12.1 and 12.2, 12.5 and 12.6,

sin θ = PN (the y-coordinate)
cos θ = PN (the x-coordinate)

$$\text{but } \tan\theta = \text{the } \frac{y\text{-coordinate}}{\text{the } x\text{-coordinate}} = \frac{PN}{ON} = \frac{\sin\theta}{\cos\theta}$$

The tangent of an angle θ is then defined as $\dfrac{\sin\theta}{\cos\theta}$.

By extension,

$$\tan(180° - \theta) = \frac{\sin(180° - \theta)}{\cos(180° - \theta)} = \frac{\sin\theta}{-\cos\theta} = -\tan\theta$$

$$\tan(180° + \theta) = \frac{\sin(180° + \theta)}{\cos(180° + \theta)} = \frac{-\sin\theta}{-\cos\theta} = \tan\theta$$

$$\tan(360° - \theta) = \frac{\sin(360° - \theta)}{\cos(360° - \theta)} = \frac{-\sin\theta}{\cos\theta} = -\tan\theta$$

Example 6

Express the following as ratios of acute angles.
(a) tan 135°, tan 225°, tan 315°
(b) tan 107°, tan 253°, tan 287°

(a) tan 135° = (tan 180 − 45)° = −tan 45°
tan 225° = (tan 180 + 45)° = tan 45°
tan 315° = (tan 360 − 45)° = −tan 45°

(b) tan 107° = (tan 180 − 73)° = −tan 73°
tan 253° = (tan 180 + 73)° = tan 73°
tan 287° = (tan 360 − 73)° = −tan 73°

The tangent curve

Using the same angles as for the sine and cosine curves, the corresponding values for the tangent function are given in Tables 12.11 and 12.12.

Table 12.11

θ	0	30	45	60	90	120	135	150	180
$\tan\theta$	0	0.58	1.0	1.73	—	−1.73	−1.0	−0.58	0

θ	210	225	240	270	300	315	330	360
$\tan\theta$	0.58	1.0	1.73	—	−1.73	−1.0	−0.58	0

Table 12.12

θ	−30	−45	−60	−90	−120	−135	−150	−180
$\tan\theta$	−0.58	−1.0	−1.73	—	1.73	1.0	0.58	0

θ	−210	−225	−240	−270	−300	−315	−330	−360
$\tan\theta$	−0.58	−1.0	−1.73	—	1.73	1.0	0.58	0

Since the cosine ratios of ±90° and ±270° are zero and division by zero is undefined, the tangent ratio of these angles is undefined. Hence, in Fig. 12.10, the graph of the trigonometric function $y = \tan\theta$ is not continuous at these values of θ.

From the graph in Fig. 12.10, note that the values of the tangent function between −90° and 90° are repeated between 90° and 270°; the tangent function is therefore another **periodic function**. For the tangent curve, the period is 180°.

Exercise 12c

1. Write down the values of the following using tables or a calculator.
   (a) tan 53° (b) tan 139°
   (c) tan 254° (d) tan 318°
   (e) tan (−64°) (f) tan (−138°)
2. Using the tan θ curve, write down all the angles between 0° and 360° for which
   (a) $\tan\theta = 1.73$ (b) $\tan\theta = -0.7$
3. Given that $\tan x = \tan 215°$ and x is acute, find x.
4. If $\tan x = -\tan 132°$ and x is acute, find x.
5. What values between −180° and 180° satisfy the equation $\tan x = 0.577$?
6. What values between −360° and 360° satisfy the equation $\tan x = -0.88$?
7. Simplify
   (a) $\dfrac{\tan(180° - \theta)}{\tan\theta}$ (b) $\dfrac{\tan(180° + \theta)}{\tan\theta}$

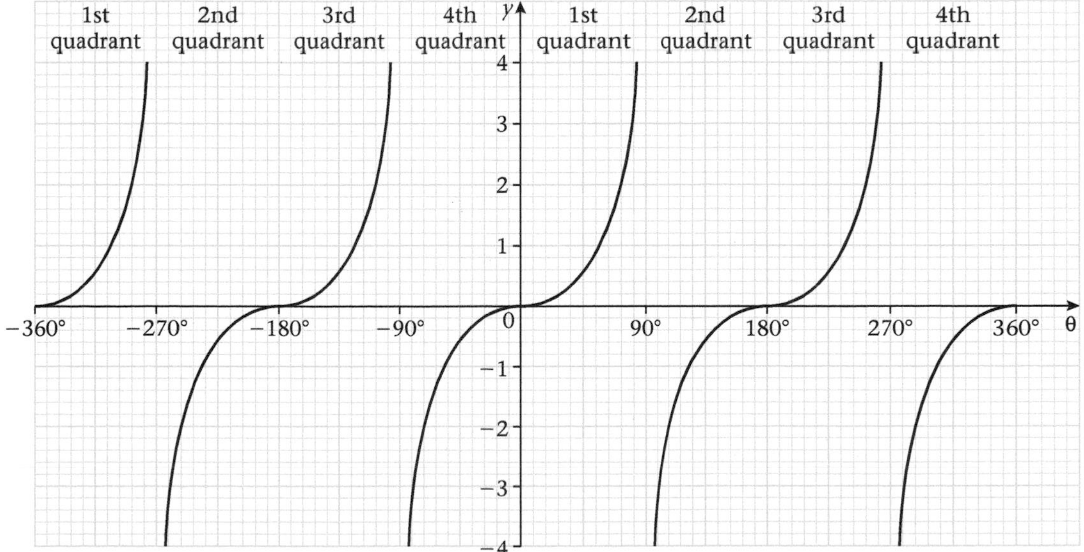

Fig. 12.10

8. Using a scale of 1 cm to represent 30° on the x-axis and 2 cm to represent units on the y-axis, draw the graph of $y = \tan x° + 1$ for values of x between 0° and 360°.

9. From the graph drawn in question 8, find the values of x which satisfy the equations (a) $\tan x + 1 = 0.5$, (b) $\tan x + 1 = -0.6$.

10. Using the graph of question 8 and drawing a suitable straight line graph, solve the equation $\tan x = -3$.

4. If $\sin\theta = -\sin 203°$ and θ is acute, find θ.

5. If $\cos\theta = \cos 309°$ and θ is acute, find θ.

6. If $\tan\theta = -\tan 107°$ and θ is acute, find θ.

7. Find the values of θ between 0° and 360° for which $\sin\theta = \cos\theta$.

8. What values between 0° and 360° satisfy the equation $\sin x = 0.799$?

9. What values between 0° and 360° satisfy the equation $\cos x = -0.602$?

10. What values between 0° and 360° satisfy the equation $\tan x = 2.475$?

Summary

Angles in terms of θ (acute)	Ratios in terms of θ		
	sin	cos	tan
θ	$\sin\theta$	$\cos\theta$	$\tan\theta$
$180 - \theta$	$\sin\theta$	$-\cos\theta$	$-\tan\theta$
$180 + \theta$	$-\sin\theta$	$-\cos\theta$	$\tan\theta$
$360 - \theta$	$-\sin\theta$	$\cos\theta$	$-\tan\theta$

The graphs of the named trigonometric functions are shown on pages 148, 151 and 153 for $-360° \leqslant \theta \leqslant 360°$.

Practice exercise P12.1

1. Using the sine curve in Fig. 12.3, or otherwise, write down all the angles between 0° and 360 that satisfy the equations
(a) $\sin\theta = 0.5$, (b) $\sin\theta = \frac{\sqrt{3}}{2}$,
(c) $\sin\theta = 0.92$, (d) $\sin\theta = 0.79$,
(e) $\sin\theta = 0.74$, (f) $\sin\theta = 0.36$.

2. Using the cosine curve in Fig. 12.7, or otherwise, write down all the angles between 0° and 360° that satisfy the equations
(a) $\cos\theta = 0.45$, (b) $\cos\theta = 0.71$,
(c) $\cos\theta = -0.5$, (d) $\cos\theta = 0.10$.

3. Using the tangent curve in Fig. 12.10, or otherwise, write down all the angles between 0 and 360° that satisfy the equation $\tan\theta = 0.45$.

Practice exercise P12.2

1. Simplify $\frac{\sin(180 - x)}{\sin(180 + x)}$.

2. Simplify $\frac{\cos(180 - x)}{\cos x}$.

3. Simplify $\frac{\tan(180 + x)}{\tan(180 - x)}$.

Practice exercise P12.3

1. Using a scale of 2 cm to represent 30° on the x-axis and 2 cm to represent 1 unit on the y-axis, draw the graph of $y = 2\sin x + 1$ for values of x between 0° and 180°.

2. From the graph drawn in question 1 find the values of x which satisfy the equations
(a) $2\sin x + 1 = 1$,
(b) $2\sin x + 1 = 1.5$.

3. Using the graph drawn in question 1 and drawing a suitable straight line, solve the equation $2\sin x = 2$.

4. Using a scale of 2 cm to represent 30° on the x-axis and 2 cm to represent 1 unit on the y-axis, draw the graph of $y = 2\cos x - 1$ for values of x between 0° and 180°.

5. From the graph drawn in question 4 find the values of x which satisfy the equations
(a) $2\cos x - 1 = -1$,
(b) $2\cos x - 1 = 0.6$.

6 Using the graph drawn in question 4 and drawing a suitable straight line, solve the equation $2 \cos x = -1$.

7 Using a scale of 2 cm to represent 30° on the x-axis and 2 cm to represent 5 units on the y-axis, draw the graph of $y = 2 \tan x - 1$ for values of x between 0° and 180°.

8 From the graph drawn in question 7 find the values of x which satisfy the equations
(a) $2 \tan x - 1 = -3$,
(b) $2 \tan x - 1 = 5$.

9 Using the graph drawn in question 7 and drawing a suitable straight line, solve the equation $2 \tan x = 0$.

Revision exercises and tests

Chapters 7–12

Revision exercise 4 (Chapters 7, 9)

1. Simplify

   (a) $\dfrac{x^2 - y^2}{(x - y)^2}$ (b) $\left(1 - \frac{3}{x}\right) \div \left(x - \frac{9}{x}\right)$

2. A frustum of a pyramid is 16 cm square at the bottom, 6 cm square at the top and 12 cm high. Find the volume of the frustum.

3. $V = kr^3$ where k is a constant. $V = 120$ when $r = 2$. Find the value of k and hence find V when $r = 6$.

4. The length of a cuboid is twice its width. The perimeter of a face which includes the length and width is 36 cm. Calculate the length of the cuboid. If the height of the cuboid is 8 cm, calculate its volume.

5. Simplify the following.

   (a) $\left(\dfrac{y^{-3} \times y^2}{y^{-7}}\right)^{\frac{1}{2}}$ (b) $(x^{1\frac{1}{3}})^3 \times \dfrac{1}{x^4}$

6. A boy is flying a kite at the end of a string 5 m long. The angle of elevation of the kite from the point, P, where the boy is standing is 52°. Calculate the height of the kite above the ground. If the angle of elevation of the kite from another point, X, is 35° calculate the distance from X to the kite.

7. A cylindrical container of radius 8 cm and height 21 cm is full of water. A spherical ball of radius $3\frac{1}{2}$ cm is dropped into the container and the water overflows into a cylindrical jug of radius 5 cm. Calculate the height of water in the jug.

8. The surface area of a closed cone of height h and base radius r is given by

   $A = \pi r^2 + \pi r\sqrt{h^2 + r^2}$

   (a) Show that $h = \dfrac{A}{\pi r}\sqrt{1 - \dfrac{2\pi r^2}{A}}$.

   (b) Hence calculate the height of a cone of surface area $157\frac{1}{7}$ cm² and base radius 3 cm. (Use the value $3\frac{1}{7}$ for π.)

Revision test 4 (Chapters 7, 9)

1. Which one of the following is *not* an identity?

   A $(x + 1)^2 - 2x = x^2 + 1$
   B $3(x + 3) = 2(x + 5) - 1 + x$
   C $\tan x = \dfrac{\sin x}{\cos x}$
   D $4x^2y = \dfrac{(2xy)^3}{xy^2}$

2. The volume of a sphere of radius 3 cm is

   A 36π cm³
   B 12π cm²
   C 18π cm³
   D 36π cm²

3. p is directly proportional to the square of q and $p = 7$ when $q = 4$. The value of q when $p = 28$ is

   A 6 B 8 C 16 D 64

4. Which one of the following is *not* true given that $x * y = xy(x + y)$?

   A $2 * 3 = 3 * 2$
   B $(2 * 3) - 2 = 2 - (2 * 3)$
   C $2(2 * 3) = (2 * 3)2$
   D $(2 * 3) * 2 = 2 * (2 * 3)$

5. The number of faces, edges and vertices in a triangular-based pyramid are respectively

   A 3, 3, 1,
   B 3, 4, 3,
   C 4, 6, 3,
   D 4, 6, 4.

6. The right triangular prism in Fig. R16 has AB = 18 cm, AD = AE = 10 cm and ED = 16 cm. Calculate the volume of the prism.

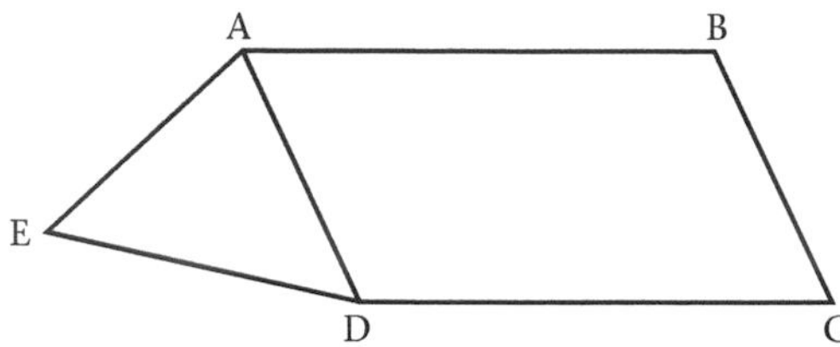

Fig. R16

7 A football just fits inside a cube of edge 28 cm (Fig. R17). Use the value $\frac{22}{7}$ for π to calculate the volume of the cube not occupied.

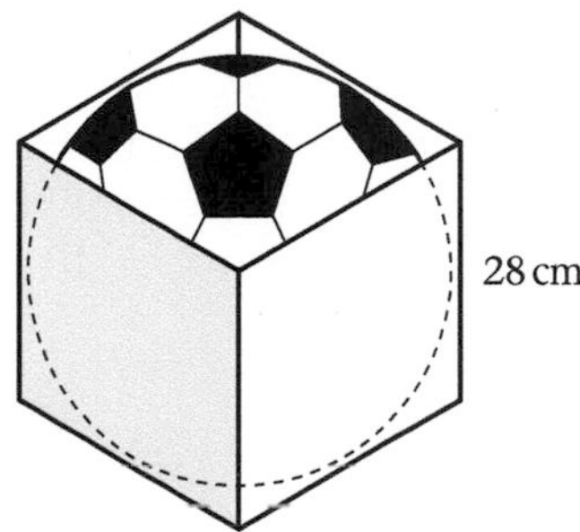

Fig. R17

8 Solve $\frac{1}{3-a} - \frac{1}{3} + \frac{4}{2a-5} = 0$.

9 (a) For what value(s) of x is the expression $\frac{2x+11}{x^2+x-20}$ not defined?
(b) For what value(s) of x is the expression zero?

10 (a) Simplify $\frac{5}{2b+2} - \frac{2b+1}{b^2-2b-3} - \frac{1}{3-b}$.
(b) Solve $\frac{5}{2b+2} - \frac{2b+1}{b^2-2b-3} = \frac{1}{3-b}$.

Revision exercise 5 (Chapters 8, 12)

1 In $\triangle XYZ$, $x = 3.5$ cm, $y = 4.0$ cm, $z = 5.5$ cm. Find, correct to the nearest degree, the angles of the triangle.

2 In $\triangle ABC$, $b = 5.2$ cm, $c = 4.5$ cm and $\hat{C} = 35°30'$. Solve $\triangle ABC$ completely, giving final answers correct to 1 d.p.

3 In $\triangle PQR$, PR = 7 cm and QR = 8 cm. T is the foot of the perpendicular from Q to PR and QT = 4 cm.
(a) Make a sketch of $\triangle PQR$.
(b) Calculate the area of $\triangle PQR$.
(c) Show that $P\hat{R}Q = 30°$.
(d) Hence, find the length of QP, to 2 significant figures.

4

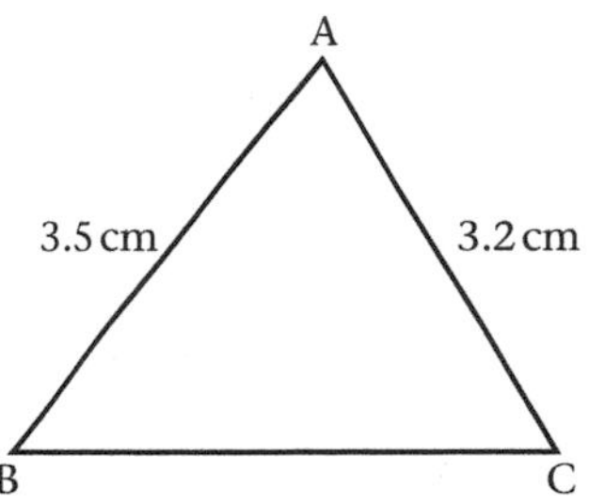

Fig. R18

In Fig. R18, AB = 3.5 cm, and AC = 3.2 cm. Given that the area of $\triangle ABC = 3.36\text{ cm}^2$, find (a) $B\hat{A}C$, (b) the length of the perpendicular from B to AC.

5 The acute angles of a rhombus are 76°. If the shorter diagonal is of length 8 cm, calculate (a) the length of the sides of the rhombus to the nearest mm, (b) the area of the rhombus.

6

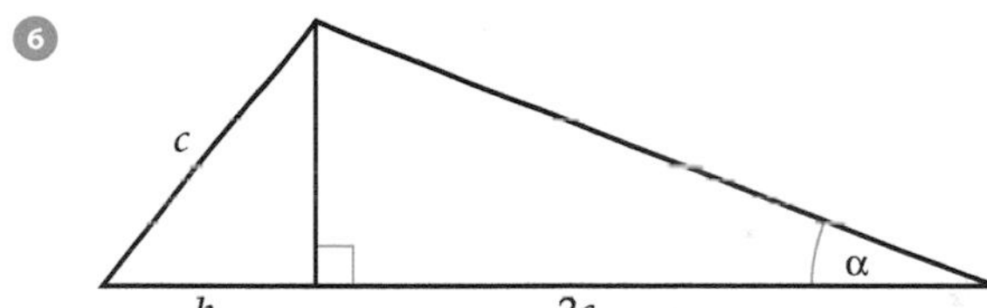

Fig. R19

Using Pythagoras' theorem, show that in Fig. R19.

$$\sin^2 \alpha = \frac{c^2 - b^2}{5c^2 - b^2}$$

7 (a) Taking 2 cm to represent 30° on the x-axis, and 2 cm to represent 0.5 unit on the y-axis, draw the graph of $y = \sin x + 1$, for $0° \leqslant x \leqslant 180°$.
(b) On the same axes and using the same scales, draw the graph of $y = 2 \cos x + 1$.

8 (a) Show that the equation $\tan x - 2 = 0$ may be solved from the graphs drawn in question 7, and state the solution(s) for the given domain.
(b) By drawing a straight line, solve the equation $\sin x = 0.65$ for the given domain.

9 In Fig. R20, calculate the value of
(a) d and (b) α

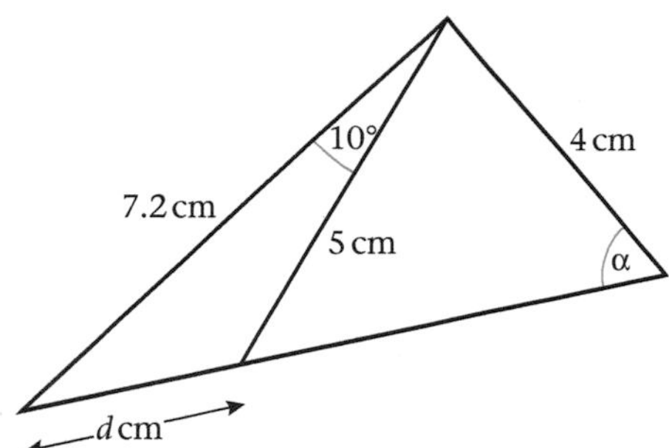

Fig. R20

10 In Fig. R21, AB is a chord of a circle of radius 10 cm, and centre at O. M is the mid-point of AB and MN ⊥ AB.

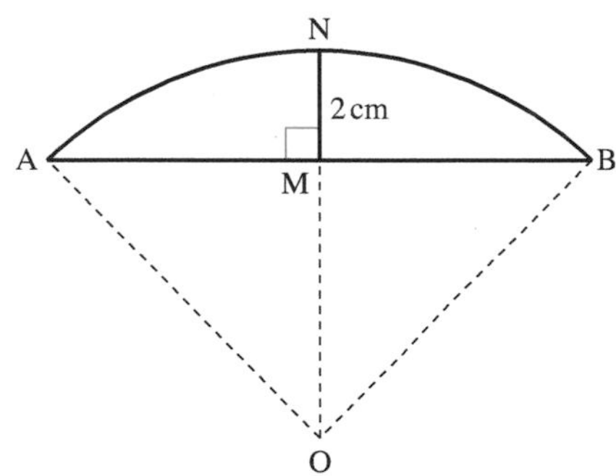

Fig. R21

If MN = 2 cm, calculate (a) the angle that arc ANB subtends at the centre of the circle O, (b) the difference in length between AB and arc ANB to the nearest mm, (c) the area of △AOB, (d) the area of the segment ANBM.

Revision test 5 (Chapters, 8, 12)

Fig. R22 refers to questions 1 and 2.

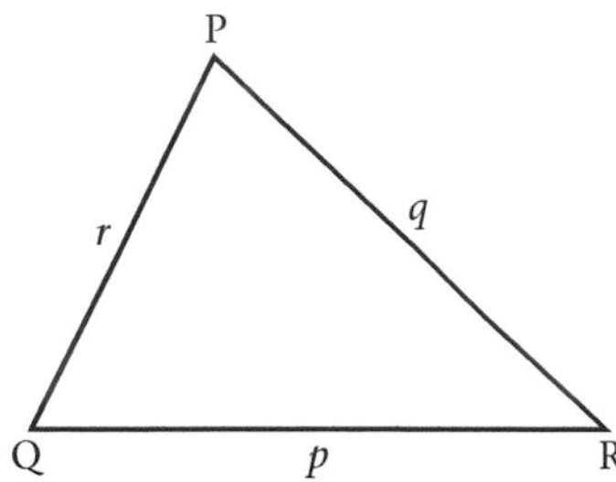

Fig. R22

1 In △PQR, $r =$

A $\dfrac{q \sin P}{\sin R}$ B $\dfrac{q \sin R}{\sin P}$

C $\dfrac{p \sin P}{\sin R}$ D $\dfrac{p \sin R}{\sin P}$

2 In Fig. R22, which of the following give(s) the perpendicular distance of P from QR?
I $r \sin \hat{Q}$, II $q \sin \hat{R}$, III $p \sin R$.

A I only
B III only
C I and II only
D I and III only

3 In △ABC, $a = 6$, $b = 3$ and $c = 4$. The value of cos A is

A $-\frac{11}{24}$ B $-\frac{11}{22}$ C $\frac{11}{12}$ D $\frac{11}{24}$

4 The graph of the function $y = \sin 2x$ for $0° \leqslant x \leqslant 90°$ is represented by

A
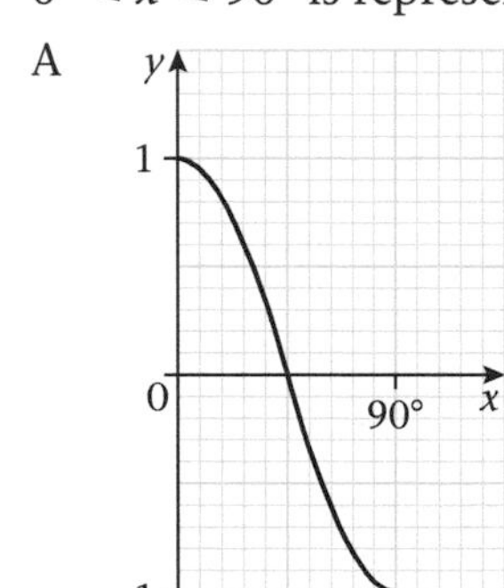

B
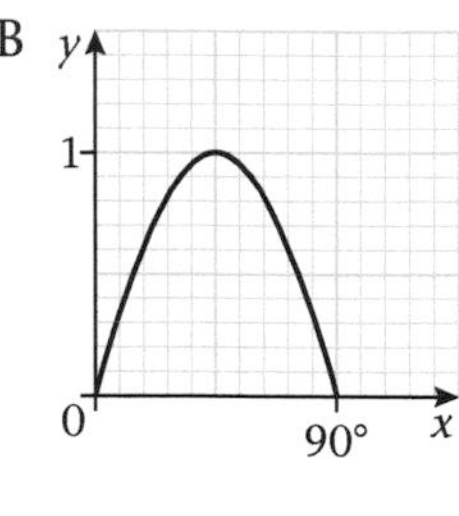

C
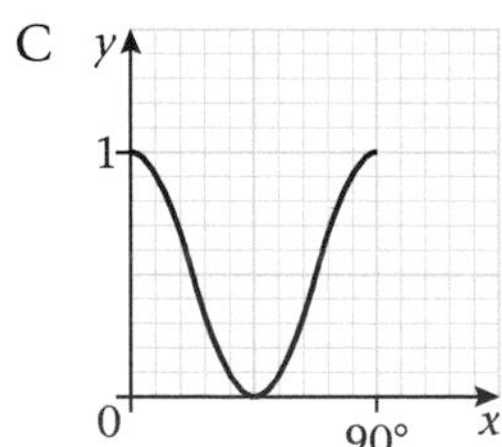

D
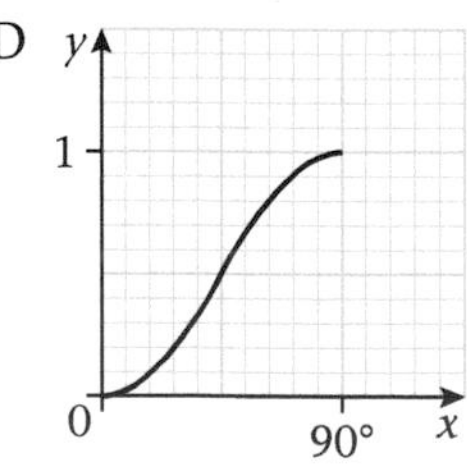

Fig. R23

5 In which of the following cases can two different triangles be drawn?

A $p = 5, q = 9, \hat{R} = 46°$
B $p = 6, r = 5, \hat{R} = 50°$
C $p = 3, q = 4, r = 6$
D $p = 6, \hat{R} = 64°, \hat{Q} = 72°$

6 Find the area of the shapes in Fig. R24.

(a)

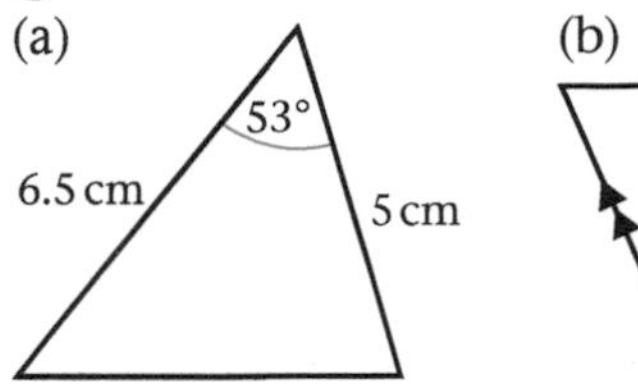

(b)

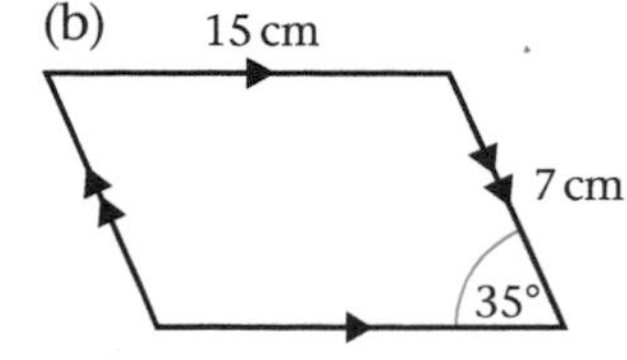

Fig. R24

7 A camper travelled from the campsite P for a distance of 2 km on a bearing of 035° to a point R. He then continued to another fixed position T on a bearing of 110° until he was 5 km away on a straight path from the campsite.

(a) Draw a sketch showing the location of P, R and T.

(b) Calculate (i) the distance RT, (ii) the bearing of T from P.

8 ABCD is a trapezium which AB//DC, and BC ⊥ DC. AB = 3.1 cm, DC = 5.3 cm and the area of ABCD is 14.7 cm².

(a) Show that the perpendicular distance between AB and DC is 3.5 cm.

(b) Calculate the size of $A\hat{D}C$.

9 Taking 1 cm to represent 30° on the x-axis, and 2 cm to represent 1 unit on the y-axis, draw the graph of the function $y = 2\cos\theta - 1$, for $0° \leqslant \theta \leqslant 360°$

10 Using the graph drawn in question 9, find the values of θ for $0° \leqslant \theta \leqslant 360°$ which satisfy the equation $2\cos\theta - 1 = \frac{1}{2}$.

Revision exercise 6 (Chapters 10, 11)

1 Fig. R25 shows the distance–time graph for a cyclist who travels from home to a town A where he stops for some time before going on to town B and then travels home.

(a) What is the distance from home to town A?

(b) What is the distance between towns A and B?

(c) How long did it take the cyclist from home to town A?

(d) How long did the cyclist stay at town A?

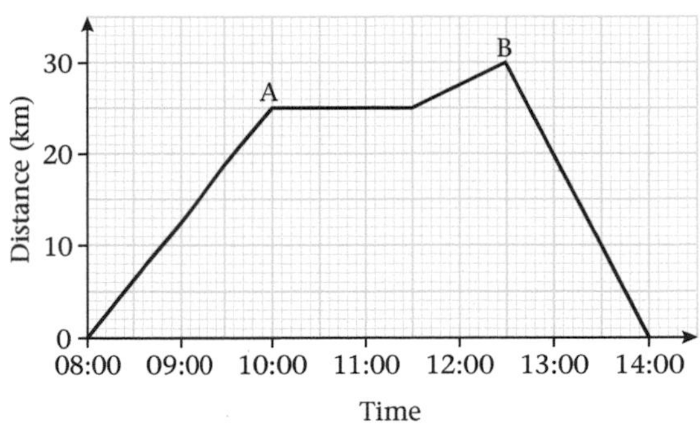

Fig. R25

(e) How long was the return journey?

(f) How long did the cyclist take for the whole journey?

(g) What was the average speed of the cyclist for the entire journey?

2 A cyclist started from and cycled for 1 hour at 15 km/h, stopped for 30 minutes and then continued to cycle for $\frac{1}{2}$ hour at 12 km/h. Using a scale of 1 cm to represent $\frac{1}{2}$ hour on the time axis and 1 cm to represent 5 km on the distance axis, draw the graph to represent the cyclist's journey. What is the total distance travelled by the cyclist?

3 A man took 40 min to run 8 km from town A to town B. He spent 25 min in town B. He then left town B in a taxi and arrived back at town A 10 min later. Represent this information on a graph using a scale of 1 cm to 5 min on the time axis and 1 cm to 1 km on the distance axis.

4 Using the graph drawn in question 3, or otherwise, find, in km/h,

(a) the average running speed of the man,

(b) the average speed of the taxi,

(c) the average speed of the man for the whole journey from A to B and back to A again (including the stop at B).

5 Fig. R26 is the graph of the function $y = x^2 + 3x$.

(a) Use a tangent to estimate the gradient of the curve at $x = 0.25$.

(b) State the minimum value of the function.

(c) Write down the equation of the line of symmetry of the curve.

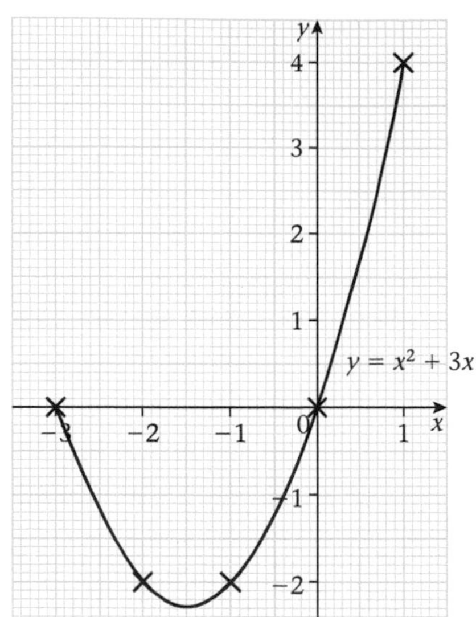

Fig. R26

6 Draw the graph of $y = x^2 - 3x + 2$ from $x = -1$ to 4, using 2 cm to represent 1 unit on each axis. Using the graph,
   (a) estimate the minimum value of the function,
   (b) write down the equation of the line of symmetry,
   (c) find the gradient of the curve at $x = 3$.

7 Fig. R27 is the speed–time graph of a train journey. Describe the four parts of the journey in your own words.

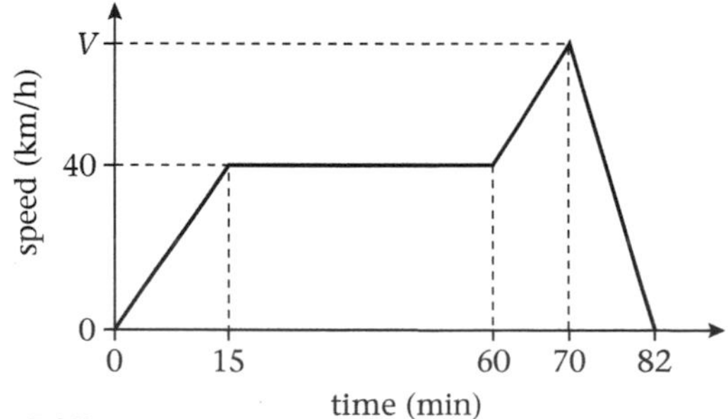

Fig. R27

8 Use Fig. R27 to answer the following.
   (a) Find the initial acceleration in km/h².
   (b) If the final deceleration is 270 km/h², find the maximum speed of the train, V km/h.
   (c) Hence find the total distance travelled by the train.

9 Given $f: x \to 2x - \frac{1}{2}$, $g: x \to \frac{2x + 1}{3}$ find (i) $fg(x)$, (ii) $gf(x)$ (iii) $(gf)^{-1}$, (iv) $(fg)^{-1}$.

10 Given $h: x \to \frac{3x - 2}{3 - x}$, find
   (i) the value of x for which h is undefined,
   (ii) $h^{-1}(x)$.

   Hence solve the equation $\frac{3x - 2}{3 - x} = 2$.

Revision test 6 (Chapters, 10, 11)

Fig. R28 is a graph showing the amount of water in a container during a period of $\frac{1}{4}$ hour. Use the graph to answer questions 1, 2 and 3.

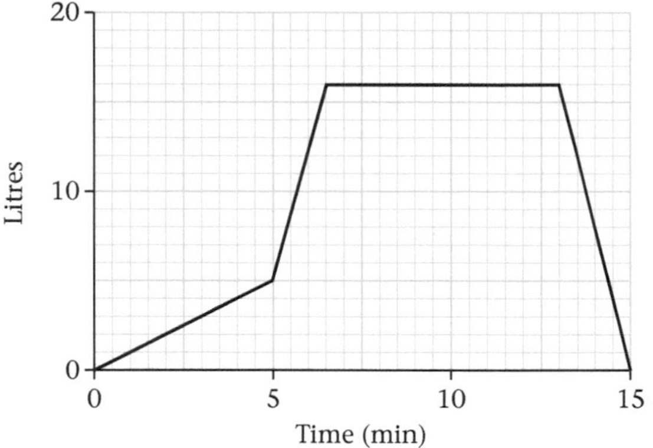

Fig. R28

1 For how long was the container being filled with water?

A	$1\frac{1}{2}$ min	B	5 min
C	$6\frac{1}{2}$ min	D	8 min

2 What was the greatest amount of water in the container?

A 10 *l* B 13 *l* C 16 *l* D 18 *l*

3 (a, b) is a point on a velocity–time curve. The gradient of the tangent to the curve at this point is a measure of
   A the acceleration after a seconds
   B the acceleration after b seconds
   C the distance after a seconds
   D the distance after b seconds

4 If $f: x \to 3(x^2 + 1) - \frac{2}{x}$, $f(-2) =$

A −13 B −11 C 14 D 16

5 A motorist travels for $\frac{1}{2}$ hour at 48 km/h, $\frac{3}{4}$ hour at 60 km/h and 15 minutes at 72 km/h. The average speed for the entire journey is

A 120 km/h
B 80 km/h
C 60 km/h
D 58 km/h

6 Line l passes through the points (0, 0) and (7, −2). Line m is parallel to l and crosses the y-axis at (0, 8). Find (a) the equation of l, (b) the equation of m, (c) the coordinates of the point where m crosses the x-axis.

7 An object moves in a straight line with velocity v m/s after t seconds such that $v = 3t^2 + 2t$.

(a) Copy and complete Table R2.

Table R2

t	0	1	2	3	4	5
v						

(b) With scales of 2 cm to 1 second on the t-axis and 2 cm to 20 m/s on the v-axis, draw a graph to show the relation between v and t.

(c) Use your graph to find (i) the acceleration after 2 s, (ii) the distance travelled after 4 s, (iii) the distance travelled during the 4th second.

8 (a) Given that

$f: x \rightarrow 3x + 8$

$g: x \rightarrow \frac{4x}{5} - 3$

Calculate $fg(10)$.

(b) Given that

$h: x \rightarrow \frac{2x - 5}{x - 3}$

(i) State the values of x for which $h(x) = 0$ and $h(x)$ is undefined.

(ii) Determine $h^{-1}(x)$. Hence solve the equation

$\frac{2x - 5}{x - 3} = 3$

General revision test B (Chapters 7–12)

1 $\frac{x^2 + 2x - 3}{x - 3} \div \frac{x - 1}{x - 3} =$

A $\frac{x + 2}{x - 1}$
B $x + 2$
C $x + 3$
D $(x - 1)(x + 3)$

2 If $P = 2(l + b)$, b expressed in terms of P and l is

A $2(l + P)$
B $P - 2l$
C $\frac{P - 2}{l}$
D $\frac{P}{2} - 1$

3 Each sloping edge of a pyramid, on a square base of side 8 cm, is 5 cm long. The surface area, in cm^2, of the triangular faces is

A 160, B 96, C 80, D 48.

4 Fig R29 is the graph of the function $y = f(x)$ for $0° \leq x \leq 180°$, where $f(x) =$

A $2 \cos \frac{x}{2}$
B $\cos x + 1$
C $2 \cos x$
D $2 \cos 2x$

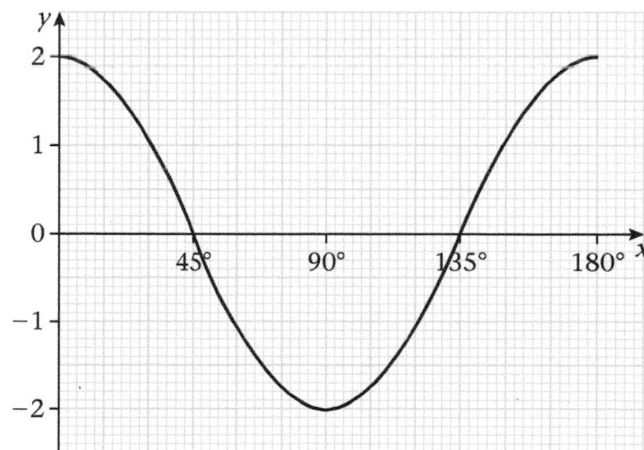

Fig. R29

5 For what value of x is the function $y = \frac{3x - 6}{x - 5}$ undefined?

A 5 B 3 C 2 D 1

Questions 6 and 7 refer to Fig. R30.

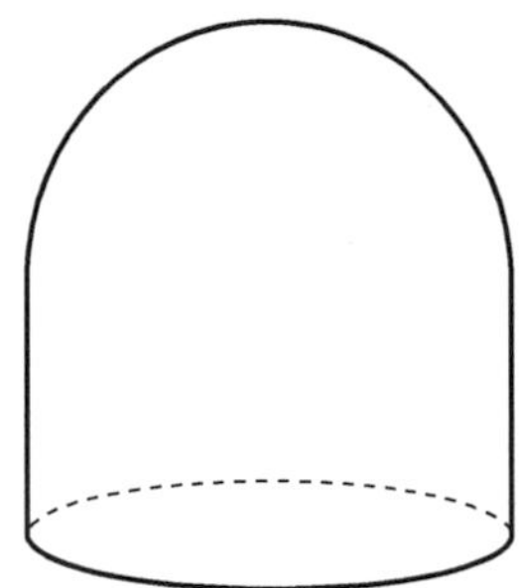

Fig. R30

The solid shown in Fig. R30 is a hemisphere of radius r joined to a cylinder of height r.

6 The total surface area of the solid is given by
A $4\pi r^2$ B $5\pi r^2$ C $6\pi r^2$ D $7\pi r^2$

7 The volume of the solid is equal to
A $\frac{5}{3}\pi r^3$ B $\frac{7}{3}\pi r^3$ C $\frac{8}{3}\pi r^3$ D $\frac{10}{3}\pi r^3$

8 The sketch graph in Fig. R31 contains four curved parts. Which of the parts show the relationship
$x \propto \frac{1}{y}$ for both negative and positive values of x?

A I, II, III and IV
B I and II only
C I and III only
D II and IV only

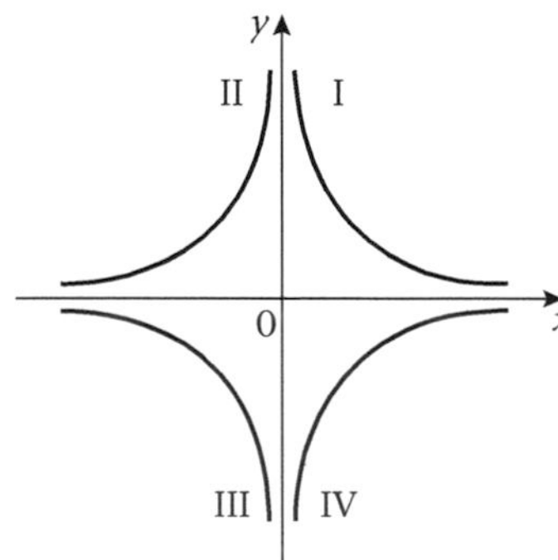

Fig. R31

9 Given $f: x \rightarrow x^2 - 2$ then $f^{-1}(x)$ is

A $x^2 + 2$
B $\sqrt{x^2 + 2}$
C $\sqrt{x + 2}$
D $\sqrt{x - 2}$

10 Fig. R32 is the speed–time graph of a motorist's journey.

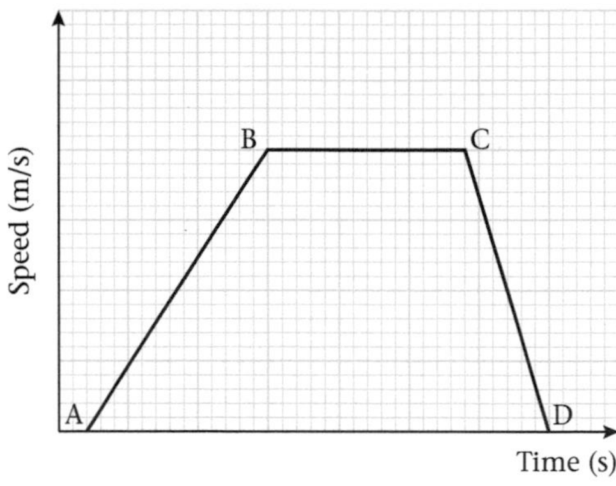

Fig. R32

Which of the following statements is not true for the graph?

A The motorist travelled at a constant speed from A to B.
B The motorist travelled at a constant speed from B to C.
C The motorist accelerated at a constant rate from A to B.
D The rate of deceleration from C to D was greater than the rate of acceleration from A to B.

11 (a) Simplify $\frac{2}{a + 2} + \frac{1}{a + 1} - \frac{2a + 3}{a^2 + 3a + 2}$.

(b) Find the value of a for which the simplified expression is not defined.

12 In a triangular plot of land ABC, AB = 14 m, AC = 13 m and $B\widehat{A}C = 130°$. Calculate, correct to one decimal place,

(a) the length of the side BC,
(b) the angle between the sides AB and BC,
(c) the shortest distance from A to the side BC,
(d) the area of the plot of land.

13 F varies directly as v^2 and inversely as r. If v is halved and r is doubled, find the ratio of the new value of F to the original value.

14 Two variables A and B are found experimentally to have the set of corresponding values as given in Table R3.

Table R3

A	1	3	6	10	15
B	6	16	34	55	85

Show graphically that there is direct variation between A and B in the form $B = kA$. Find (a) A when $B = 47.6$, (b) B when $A = 12$.

15 The height of a vertical flag-pole TF is 8 m. A point P is 15 m due South of the base F of the flag-pole. Another point R is on a bearing of 090° from F. The angle of elevation of T, the top of the flag-pole, from R is 53°.

(a) Sketch a diagram to represent the above information, showing clearly the North direction, the flag-pole TF, and the points P and R.

(b) Calculate, to one decimal place, (i) the angle of elevation of T from P, (ii) the distance FR, (iii) the distance PR.

16 (a) Copy and complete the following table for the function $f(x) = 3 \sin 2x - 1$, where $0° \leq x \leq 90°$.

Table R4

x	0	15.00	22.5	30.00	45	60	67.5	75	90
$f(x)$	−1	0.50		1.60			1.12		

(b) Using 1 cm to represent 5° on the x-axis and 4 cm to represent 1 unit on the $f(x)$-axis, draw the graph of $f(x)$ for the given domain.

(c) On the graph, draw in the axis of symmetry for $f(x)$ and state the value of x where this occurs.

(d) Use the graph to estimate for the given domain

(i) the maximum value of $f(x)$,

(ii) the solutions of the equation $3 \sin 2x = 1$,

(iii) the interval for which $f(x) < 0$.

17 The velocity v m/s of a particle after time t seconds is given by the equation $v = 7t - t^2$.

(a) Draw the graph of the equation of motion for values of t from 0 to 7 seconds.

(b) Using the graph, estimate (i) the maximum speed of the particle and the time when it occurs, (ii) the acceleration of the particle when $t = 2$ and when $t = 5\frac{1}{2}$, (iii) the distance travelled during the seven seconds.

18 Draw the graph of $y = 3x + 4 - 2x^2$ for $-1 \leq x \leq 3$. Using your graph, estimate

(a) the gradient of the curve at $x = 2$,

(b) the maximum value of $3x + 4 - 2x^2$.

19 A motorist travelled for 30 mins at 48 km/h and for another 30 minutes at 60 km/h to a town A where she stopped for 36 minutes. She then travelled at 80 km/h to another town B, arriving there 24 minutes after leaving town A.

(a) Using a scale of 2 cm for 30 minutes on the horizontal axis and 1 cm for 5 km on the vertical axis, draw the graph of the motorist's journey.

Using the graph:

(b) What was the distance travelled between towns A and B?

(c) If the motorist started out from home at 2:00 p.m., what is the slowest average speed at which she must ravel on the return journey to arrive back at home no later than 5:30 p.m.? Give your answer to the nearest whole number.

Chapter 13

Statistics and probability

Cumulative frequency

Pre-requisites

- frequency and group frequency; statistical averages – mean, mode and median; ratios; sets; Venn diagrams

Cumulative frequency

Table 13.1 is the frequency distribution of the masses of 40 pupils in a class.

Table 13.1

Mass (kg)	Number of pupils
41–45	3
46–50	7
51–55	12
56–60	10
61–65	6
66–70	2

Table 13.2 shows the running total of the frequencies of the classes. Note that $45\frac{1}{2}$, $50\frac{1}{2}$,... are the upper boundaries of the different classes. The running total is noted in the third column. The table shows that 3 pupils weigh less than $45\frac{1}{2}$ kg; 10 pupils (= 3 + 7) weigh less than $50\frac{1}{2}$ kg.

Table 13.2

Mass (kg) less than	Number weighing	Total
$45\frac{1}{2}$	3	3
$50\frac{1}{2}$	3 + 7	10
$55\frac{1}{2}$	10 + 12	22
$60\frac{1}{2}$	22 + 10	32
$65\frac{1}{2}$	32 + 6	38
$70\frac{1}{2}$	38 + 2	40

Table 13.3 is the frequency distribution of the marks for a class of 50 pupils. The table shows that 2 pupils scored 30 marks or less. 7 pupils (= 5 + 2) scored 40 marks or less, and so on. Each number in the third column is found by adding the number in the second column to the previous total. This progressive increase in the total is what is meant by *cumulative*.

Table 13.3

Class interval	Frequency	Cumulative frequency
21–30	2	2
31–40	5	5 + 2 = 7
41–50	7	7 + 7 = 14
51–60	9	9 + 14 = 23
61–70	11	11 + 23 = 34
71–80	8	8 + 34 = 42
81–90	5	5 + 42 = 47
91–100	3	3 + 47 = 50

It is customary to write only the result of the addition in the last column. The sum has been given in Table 13.3 to show the method.

From Table 13.2 it is easy to tell how many pupils weigh less than 45 kg, 50 kg and so on. From Table 13.3 it is easy to tell how many pupils gained up to 30 marks (30 or less), 40 and so on.

Cumulative frequency curve (ogive)

The graphs of the cumulative frequency distributions in Tables 13.2 and 13.3 are shown in Fig. 13.1 and Fig. 13.2. In Fig. 13.1 the cumulative frequencies are plotted against the corresponding upper boundaries. In Fig. 13.2 the cumulative frequencies are plotted against the corresponding upper limits of the class intervals. The points are joined by a smooth curve called an **ogive** (from a term used in architecture for this shape of curve).

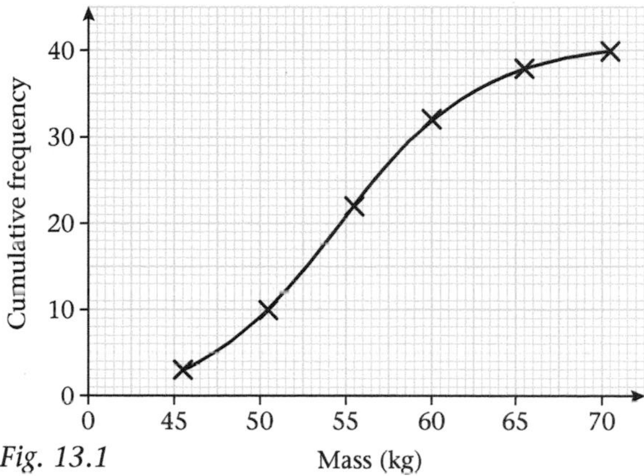

Fig. 13.1

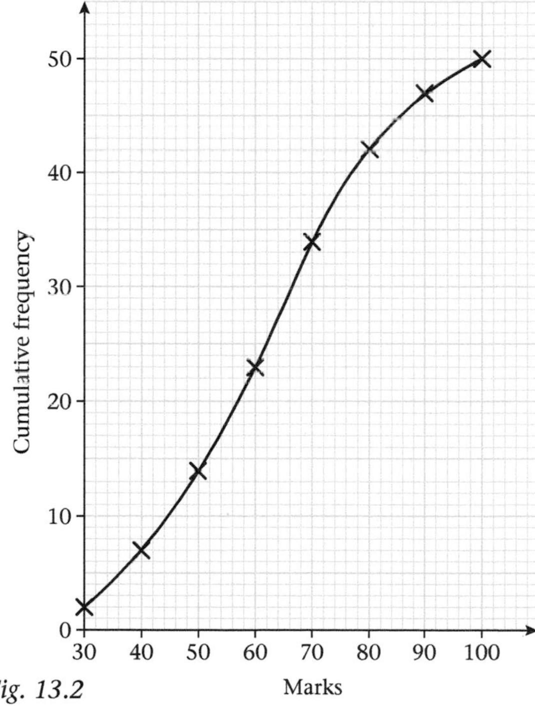

Fig. 13.2

Example 1

Using the cumulative frequency curve in Fig. 13.2 estimate (i) the upper quartile, (ii) the lower quartile (iii) the interquartile range (iv) the semi-interquartile range of the distribution.

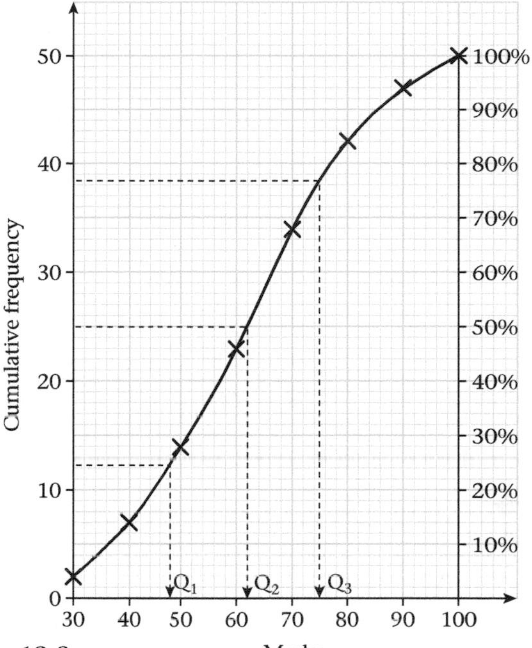

Fig. 13.3

In Fig. 13.3

(i) the lower quartile (Q_1) is one quarter the way up.
$Q_1 = 48$

(ii) The upper quartile (Q_3) is three-quarters the way up.
$Q_3 = 75$

(iii) The interquartile range
$Q_3 - Q_1 = 75 - 48 = 27$

(iv) The semi-interquartile range
$Q = \frac{Q_3 - Q_1}{2} = \frac{27}{2} = 13\frac{1}{2}$

From the graph the value of Q shows that half of the students scored within $13\frac{1}{2}$ marks of the median (Q_2). Hence the semi-interquartile range is a measure of the spread of the distribution.

Percentiles

The right-hand vertical axis of Fig. 13.3 shows the distribution divided into 10%, 20%, 30% ... These divisions are called **percentiles**. The percentiles divide the distribution into one hundred equal parts.

The lower quartile is the twenty-fifth percentile.
The median is the fiftieth percentile.
The upper quartile is the seventy-fifth percentile.

Example 2

Suppose that in Example 1 students who score over 45 marks pass the test. Use Fig. 13.2 to estimate the percentage of students that passed.

In Fig. 13.3, 45 marks is at the 21st percentile. This means that 21% of the students scored 45 marks or less.

Percentage scoring over 45 marks

$= 100\% - 21\% = 79\%$

Note: From Table 13.3 it can be seen that 39 students, or 78%, actually scored over 45 marks. Hence the estimate of 79% taken from the cumulative frequency curve is reasonably accurate.

Exercise 13a

1. In a test, the students are graded according to the marks scored as in Table 13.4.

Table 13.4

Marks scored	0–50	51–70	71–90	91–100
Grade	re-sit	pass	credit	distinction

Use Fig. 13.3 to estimate
(a) the percentage of students required to re-sit,
(b) the number of students who obtained a pass grade,
(c) the percentage of students awarded a distinction.

2. Table 13.5 shows the numbers of students who scored marks within 10-mark class intervals in a test.

Table 13.5

Marks (class intervals)	1–10	11–20	21–30	31–40	41–50	51–60	61–70	71–80	81–90	91–100
Number of students	2	7	9	11	13	16	16	15	8	3

(a) Make a cumulative frequency table and hence draw an ogive showing the mark distribution.
(b) Estimate the median and upper and lower quartiles.
(c) Calculate the semi-interquartile range for the test.
(d) If any mark over 45 is a pass, estimate the percentage of students that passed.

3. Table 13.6 gives the mark distribution in a test.

Table 13.6

Class interval	11–20	21–30	31–40	41–50	51–60	61–70	71–80	81–90	91–100
Frequency	3	17	60	48	27	20	13	8	4

(a) Draw a cumulative frequency curve for the test.
(b) Find its median and semi-interquartile range.
(c) Estimate the percentage of the candidates that obtained more than 56 marks.
(d) Which mark is at the 70th percentile?

4. Table 13.7 shows the lives in hours, to the nearest hour, of 50 electric light bulbs.
(a) Make a frequency distribution by grouping the values in Table 13.7 in 10-hour class intervals.

Table 13.7

563	608	607	632	590	621	614	576	602	582
599	624	580	595	582	581	605	584	596	562
599	598	596	626	596	617	615	589	556	603
594	589	617	560	610	630	571	592	610	597
616	594	622	597	576	595	601	600	592	638

(b) By drawing a cumulative frequency curve, estimate the median and 80th percentile of the distribution.

5 100 students participated in a fund-raising run. Their times were recorded and Table 13.8 shows the frequency distribution.

Table 13.8

Time (minutes)	2–4	5–7	8–10	11–13	14–16	17–19
Frequency	3	9	32	46	6	4

(a) Make a cumulative frequency table and draw a cumulative frequency curve for the distribution.
(b) Estimate the number of students who took more than 12 minutes to do the run.
(c) Estimate the interquartile range.

Probability

Probability is a numerical measure of the likelihood of an event happening or not happening. For example, if it has rained in Guyana in 9 out of the last 12 Septembers, then, statistically, the probability of rain falling in Guyana next September is $\frac{9}{12}$ (or $\frac{3}{4}$ or 0.75). This is an example of **experimental probability**. Since experimental probability uses numerical records of past events to predict the future, its predictions cannot be taken to be absolutely accurate. Alternatively, the probability of throwing a five on a fair six-sided die is $\frac{1}{6}$, since any one of the six faces is equally likely. This is an example of **theoretical probability**. Theoretical probabilities are exact values which can be calculated by considering the physical nature of the given situations.

Example 3

Trevor and Samuel have played each other at tennis 15 times this season. Trevor has won 12 of the matches. They play each other in a championship. What is the probability that (a) the match is drawn, (b) Trevor wins, (c) either Trevor or Samuel wins?

(a) Tennis matches are either won or lost. They are never drawn.
Probability of a draw = 0

(b) Trevor has won 12 of the last 15 matches. Experimental probability of Trevor winning the match
$= \frac{12}{15} = \frac{4}{5} = 0.8$

(c) Since one or other of Trevor or Samuel must win, probability of either person winning = 1.

If p is the probability of an event happening then p lies in the range $0 \leqslant p \leqslant 1$. The probability of an event *not* happening is p' where $p' = 1 - p$. For instance, in Example 4, the probability of Trevor *not* winning is $1 - 0.8$, i.e. 0.2.

Probability can also be described in set language. If p(R) is the probability of a required outcome happening, then

$$\text{p(R)} = \frac{\text{n(R)}}{\text{n(U)}}$$

where R = {required outcomes}
U = {all possible outcomes}

Example 4

A letter is chosen at random from the alphabet. Find the probability that it is (a) F, (b) F or T, (c) one of the letters of the word F R E Q U E N C Y, (d) not one of the letters of the word T A B L E.

In every case,

U = {A, B, C, ..., Z}

(a) Let A = {F}
then $\text{p(A)} = \frac{\text{n(A)}}{\text{n(U)}} = \frac{1}{26}$
The probability that F is chosen is $\frac{1}{26}$.

(b) Let B = {F, T}
then $\text{p(B)} = \frac{\text{n(B)}}{\text{n(U)}} = \frac{2}{26} = \frac{1}{13}$
There is a $\frac{1}{13}$ probability that F or T is chosen.

(c) Let C = {F, R, E, Q, U, E, N, C, Y}
then $\text{p(C)} = \frac{\text{n(C)}}{\text{n(U)}} = \frac{8}{26} = \frac{4}{13}$
The probability of choosing one of the letters of the word FREQUENCY is $\frac{4}{13}$.

(d) Let D = {T, A, B, L, E}
then $\text{p(D)} = \frac{\text{n(D)}}{\text{n(U)}} = \frac{5}{26}$
then $\text{p}'\text{(D)} = 1 - \text{p(D)} = 1 - \frac{5}{26} = \frac{21}{26}$
There is a $\frac{21}{26}$ chance that none of the letters is in the word TABLE.

Notes:
1 'At random' means 'in a free irregular way'.

2 The letter E in part (c) of Example 4 was *not* counted twice.

3 Part (d) of Example 4 is most conveniently solved using the method of subtraction as shown.

Exercise 13b (revision)

1. A statistical survey shows that 28% of all men take size 9 shoes. What is the probability that your friend's father takes size 9 shoes?

2. A school has 357 boys and 323 girls. If a student is chosen at random, what is the probability that a girl is chosen?

3. A State Lottery sells $1\frac{1}{2}$ million tickets of which 300 are prizewinners. What is the probability of getting a prize by buying just one ticket?

4. Statistics show that 92 out of every 100 adults are at least 150 cm tall. What is the probability that a person chosen at random from a large crowd is less than 150 cm tall?

5. Fig. 13.4 is a magic square.

16	2	3	13
5	11	10	8
9	7	6	12
4	14	15	1

Fig. 13.4

If a number is picked at random from Fig. 13.4, what is the probability that it is
(a) odd,
(b) prime,
(c) less than 10,
(d) exactly edivisible by 3.
(e) a perfect square,
(f) a perfect cube?

6. A bag contains 2 black balls, 3 green balls, 4 red balls. A ball is picked from the bag at random. What is the probability that it is
(a) black, (b) green,
(c) red, (d) yellow,
(e) not black, (f) either black or red?

7. A fair six-sided die is thrown. Find the probability of getting
(a) a 3, (b) a 4,
(c) a 9, (d) either 1, 2 or 3,
(e) a number divisible by 3,
(f) a number less than 5.

8. A letter is chosen at random from the alphabet. Find the probability that it is
(a) M,
(b) not A or Z,
(c) either P, Q, R or S,
(d) one of the letters of T R I N I D A D.

9. Table 13.9 gives the numbers of students in age groups in a school.

Table 13.9

Age	12	13	14	15	16	17	18
Number	42	130	125	131	110	84	53

Find the probability that a student chosen at random is (a) 14, (b) 14 or less.

10. A card is picked at random from a pack of playing cards*. Find the probability of picking
(a) the 5 of ♡, (b) the K of ♠.
(c) a 9, (d) a black Queen,
(e) a diamond,
(f) either a Jack, a 2 or an Ace,
(g) a red card,
(h) a red club.

*A pack of playing cards contains 52 cards in 4 suits: clubs (♣), diamonds (♢), hearts (♡), spades (♠). There are 13 cards in each suit: Ace (A), 2, 3, 4, 5, 6, 7, 8, 9, 10, Jack (J), Queen (Q), King (K). Clubs and spades are black, diamonds and hearts are red.

Outcome tables, tree diagrams

Example 5

Two dice are thrown at the same time. Find the probability of getting (a) at least one 5, (b) a total score divisible by 5.

Table 13.10 shows all the possible outcomes when two dice are thrown.

The number of possible outcomes in Table 13.10 is n(U) where n(U) = 36.

(a) Referring to the shaded column and row,
P = {outcomes with 5 on the first die}
Q = {outcomes with 5 on the second die}
$n(P \cup Q) = 11$

Probability of getting at least one five

$$= \frac{n(P \cup Q)}{n(U)} = \frac{11}{36}$$

(b) In Table 13.10, all the total scores which are exactly divisible by 5 have been ringed.
Number of outcomes divisible by five
= 7

Probability of getting a total score divisible by five $= \frac{7}{36}$

In Example 5, notice how the table helps to overcome the problem of finding the various numbers of outcomes.

Table 13.10

second die					P ↓		
	6	7	8	9	(10)	11	12
	5	6	7	8	9	(10)	11 ← Q
	4	(5)	6	7	8	9	(10)
	3	4	(5)	6	7	8	9
	2	3	4	(5)	6	7	8
	1	2	3	4	(5)	6	7
		1	2	3	4	5	6
		first die					

Example 6

A bag contains 3 black balls and 2 white balls.

(a) A ball is taken from the bag and then replaced. A second ball is chosen. What is the probability that (i) they are both black, (ii) one is black and one is white?

(b) Find out how those probabilities are affected if two balls are chosen without any replacement.

(a) The various possible ways of selecting the balls are shown in Fig. 13.5 on a **tree diagram.** In the diagram the branches of the tree show the different ways of choosing, with the related probabilities given as fractions.

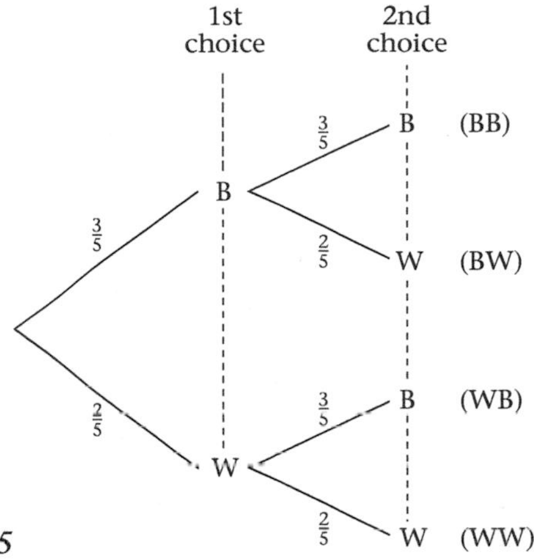

Fig. 13.5

(i) From the tree diagram, probability that both balls are black $= \frac{3}{5} \times \frac{3}{5} = \frac{9}{25}$.

(ii) Probability that the first is white and the second is black:

$$p(WB) = \frac{2}{5} \times \frac{3}{5} = \frac{6}{25}$$

p(BW) and p(WB) are probabilities of mutually exclusive events. Hence the probability of getting a black ball and a white ball when the order does not matter

$$= p(BW) + p(WB)$$

$$= \frac{6}{25} + \frac{6}{25} = \frac{12}{25}$$

(b) If there is no replacement, then there are only 4 balls left after the first is taken. Compare the probability fractions in Fig. 13.5 with those of Fig. 13.6.

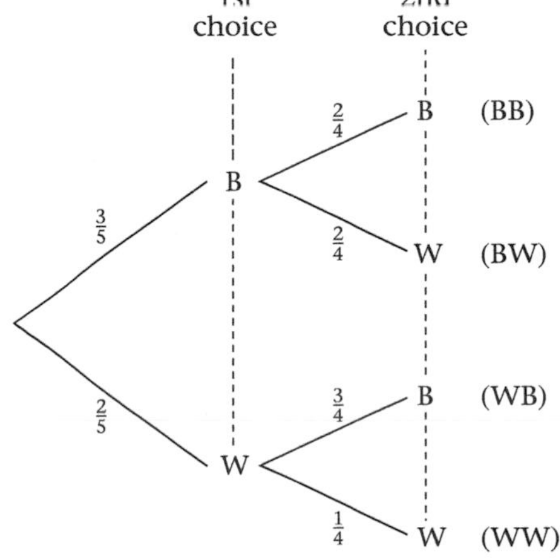

Fig. 13.6

From the tree diagram,

(i) $p(BB) = \frac{3}{5} \times \frac{2}{4} = \frac{3}{10}$

(ii) $p(BW) = \frac{3}{5} \times \frac{2}{4} = \frac{3}{10}$

(iii) $p(WB) = \frac{2}{5} \times \frac{3}{4} = \frac{3}{10}$

Probability of getting a black and white ball, regardless of order

$$= \frac{3}{10} + \frac{3}{10} = \frac{3}{5}$$

Notice in Example 6 that there are four possible outcomes: BB, BW, WB, WW and that in each case the sum of the probabilities of the outcomes is 1:

	(a)	(b)
BB	$\frac{9}{25}$	$\frac{3}{10}$
BW	$\frac{6}{25}$	$\frac{3}{10}$
WB	$\frac{6}{25}$	$\frac{3}{10}$
WW	$\frac{4}{25}$	$\frac{1}{10}$
Sum	1	1

This provides a useful check on calculations.

Example 7

If three cards are chosen from a pack without replacement, what is the probability of getting at least two spades?

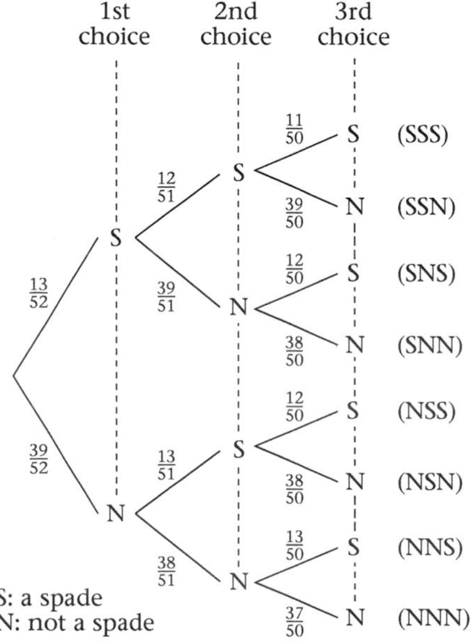

Fig. 13.7

Fig. 13.7 shows the various ways of choosing the three cards.

Probability of choosing 3 spades

$$= \frac{13}{52} \times \frac{12}{51} \times \frac{11}{50} = \frac{11}{850}$$

Probability of choosing 2 spades

$$= \frac{13}{52} \times \frac{12}{51} \times \frac{39}{50} + \frac{13}{52} \times \frac{39}{51} \times \frac{12}{50} + \frac{39}{52} \times \frac{13}{51} \times \frac{12}{50}$$

$$= \frac{3 \times 12 \times 13 \times 39}{52 \times 51 \times 50} = \frac{117}{850}$$

Probability of getting at least 2 spades (i.e. 2 spades or 3 spades)

$$= \frac{11}{850} + \frac{117}{850}$$

$$= \frac{128}{850}$$

$$= \frac{64}{425}$$

Exercise 13c

1. A pair of dice are thrown. What is the probability of getting (a) at least one six, (b) a total score of seven?

2. A game is played with a pentagonal spinner with sides marked 1 to 5. The score is on the side which comes to rest on the table. In two spins, what is the probability of getting
   (a) two 5s, (b) at least one 5, (c) a total score of 5, (d) a total score greater than 5?

3. When two dice are thrown what is the probability of the total score being a prime number?

4. In a school, 4 out of 5 students have pens. If 2 students are picked at random, what is the probability that (a) both will have a pen, (b) one has a pen and the other has not?

5. A number is chosen at random from the following: 1, 2, 1, 2, 1, 3, 1, 2.
   (a) State the probability of choosing a 1.
   (b) If two numbers are chosen find the probability of choosing
   (i) a total of 6, (ii) total of 4.

6. The probability of a seed germinating is $\frac{1}{3}$. If 3 of the seeds are planted, what is the probability that

(a) none germinate,
(b) at least one will germinate,
(c) only one will germinate?

7 A coin is tossed 3 times. What is the probability of getting
(a) 2 heads and 1 tail,
(b) at least 1 head?

8 In a car park there are 40 vehicles; 20 of them are cars, 13 of them are buses and 7 are lorries. If they are equally likely to leave, what is the probability of:
(a) a car leaving first?
(b) a bus leaving next if a car left first?

9 A bag contains 3 black balls, 4 white balls and 5 red balls. Three balls are removed without replacement. What is the probability of obtaining
(a) one of each colour,
(b) at least two red balls?

10 Two cards are chosen from a pack of playing cards without replacement. What is the probability of getting:
(a) a King or a 10,
(b) the King of Spades and a red card.

Summary

A **cumulative frequency curve** (ogive) is a smooth curve in which the cumulative frequencies are plotted against the corresponding upper limits/boundaries of the class intervals. The lower quartile, Q_1, median Q_2, and upper quartile, Q_3, mark the one-quarter, half and three-quarters points of the distribution.

Interquartile range $= Q_3 - Q_1$

Semi-interquartile range $= \dfrac{Q_3 - Q_1}{2}$

The **interquartile range** and **semi-interquartile range** are measures of spread of a set of observations.

Probability is a numerical measure of the likelihood of an event happening or not happening. A **tree diagram** may be used to represent the possible outcomes of a situation and to calculate the probability of a particular event.

Practice exercise P13.1

1 Using the cumulative frequency curve in Fig. 13.10, estimate
(a) the median,
(b) the upper quartile,
(c) the lower quartile,
(d) the interquartile range,
(e) the semi-interquartile range,
(f) the 60th percentile.

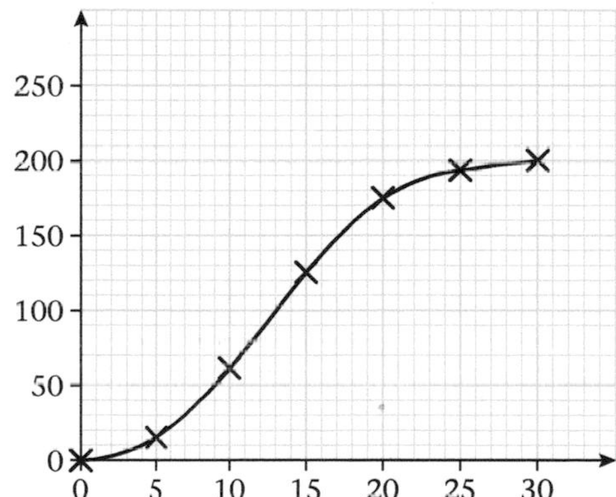

Fig. 13.10

2 Table 13.12 shows the frequency distribution of the marks of 80 students on a test.

Table 13.12

Marks	Frequency
11–20	2
21–30	4
31–40	9
41–50	19
51–60	28
61–70	13
71–80	2

(a) Construct a cumulative frequency table and hence draw a cumulative frequency curve for the distribution.

(b) Estimate the median and the semi-interquartile range.
(c) Given that 52 students passed the test, estimate the pass mark for the test.

3 Table 13.13 shows the frequency distribution of the heights of 600 plants.

Table 13.13

Height (cm)	Frequency
0–10	11
11–20	67
21–30	138
31–40	167
41–50	129
51–60	70
61–70	14
71–80	4

(a) Construct a cumulative frequency table and hence draw a cumulative frequency curve for the distribution.
(b) Estimate the median and the semi-interquartile range.
(c) Estimate the number of plants over 25 cm tall.

Practice exercise P13.2

1 A bag contains 5 red balls, 4 blue balls and 6 white balls.
A ball is taken from the bag and then replaced. A second ball is taken. What is the probability that
(a) both balls are red,
(b) neither ball is red,
(c) one ball is blue and one is white,
(d) one of the two balls is white?

2 Three cards are chosen from a pack of playing cards without replacement. What is the probability of getting
(a) an ace, a two and a three,
(b) three diamonds,
(c) three aces,
(d) three cards of the same suit?

3 The probability of being left-handed is 0.17. If two people are chosen at random, what is the probability that
(a) they are both left-handed,
(b) neither is left-handed,
(c) just one is left-handed?

4 The probability of a person being able to roll their tongue is $\frac{2}{5}$. If three people are chosen at random, what is the probability that
(a) they can all roll their tongues,
(b) only one can roll their tongue,
(c) at least two can roll their tongues?

5 In a bag there are 45 red beads, 17 white beads and 38 blue beads. What is the probability of picking
(a) a red bead,
(b) a white bead,
(c) a blue bead,
(d) two white beads if the first bead is replaced before the second is taken,
(e) two white beads if the first bead is not replaced?

Geometrical transformations and matrices

Combined transformations, matrices and transformations

Pre-requisites

- matrices; matrix multiplication; geometrical transformations

Combined transformations

Glide reflection

In Fig. 14.1, the triangle ABC is reflected in line KM to triangle A′B′C′ and then translated parallel to KM to triangle A″B″C″. Triangle ABC is said to be mapped onto triangle A″B″C″ by a **glide reflection**. The line KM is the **glide axis.**

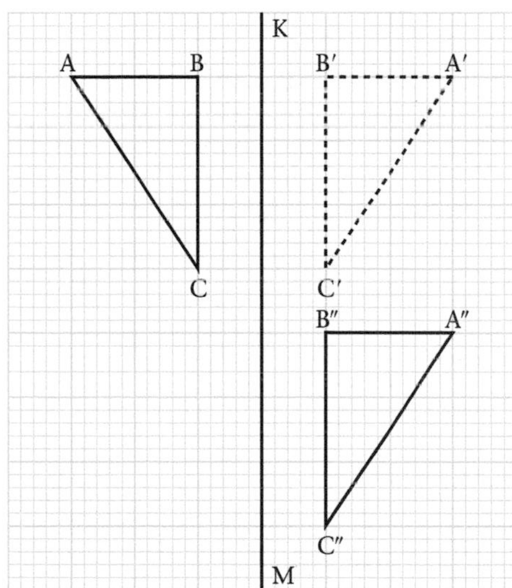

Fig. 14.1

Example 1

On graph paper draw the triangle A(1, 1), B(3, 1), C(3, 3). Translate triangle ABC to A′(1, 5), B′(3, 5), C′(3, 7), and then reflect triangle A′B′C′ in the y-axis. Draw the position of the final image A″B″C″, and describe the mapping of triangle ABC to triangle A″B″C″.

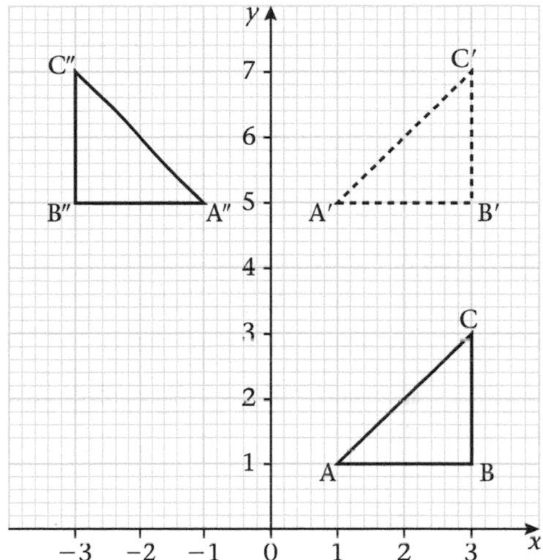

Fig. 14.2

Triangle A″B″C″ in Fig. 14.2 is the position of the final image. Triangle ABC is mapped to triangle A″B″C″ by means of a glide reflection, that is, a translation by the vector $\begin{pmatrix} 0 \\ 4 \end{pmatrix}$ followed by a reflection in the y-axis.

Example 1 shows that a glide reflection is equivalent to a translation followed by a reflection, or vice versa.

A common example of an object and its glide reflection is successive footprints (Fig. 14.3).

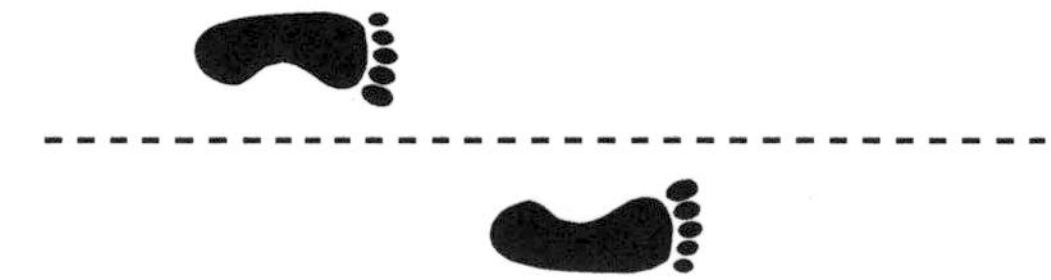

Fig. 14.3

Example 2

On squared paper draw the triangle A(1, 0), B(0, 2), C(2, 5) and the triangle P(5, 5), Q(6, 7), R(4, 10). Mark the mid-points of AP, BQ, CR and write down the equation of the line through these mid-points.

(a) Are triangles ABC and PQR congruent?

(b) What transformation maps $\triangle$ABC to $\triangle$PQR?

(c) Draw the position of the intermediate image.

(d) Name the line through the mid-points of AP, BQ, CR.

Equation of the line through the midpoints, that is, through M_1, M_2 and M_3 is $x = 3$.

(a) Triangles ABC, PQR in Fig. 14.4, are congruent.

(b) A glide reflection; that is, reflection in the line $x = 3$ followed by a translation by the vector $\begin{pmatrix}0\\5\end{pmatrix}$, [or, the translation followed by the reflection].

(c) Either K or L in Fig. 14.4

(d) The glide axis.

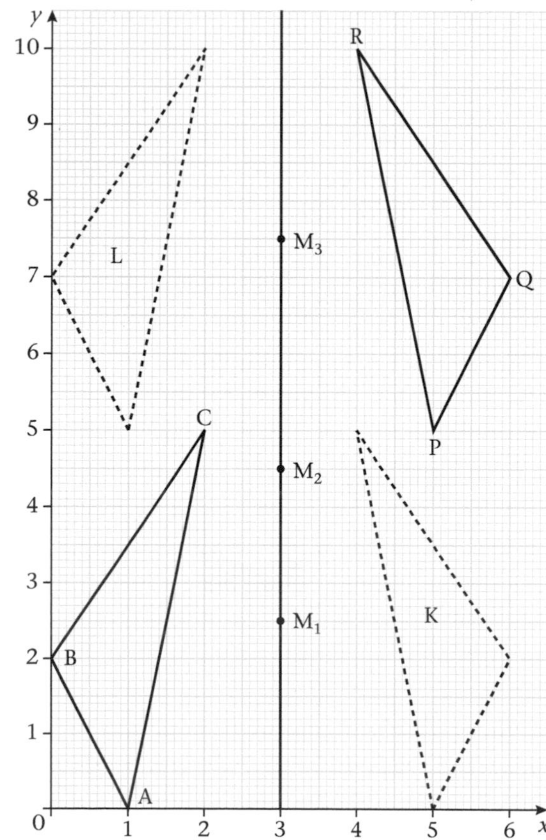

Fig. 14.4

Properties of glide reflections

1 A glide reflection, G, is completely defined by the glide axis and the magnitude and direction of the translation.

2 The mid-point of the lines joining corresponding points on the object and the image lie on the glide axis (reflecting line).

3 Under a glide reflection the object and the image are congruent. Thus a glide reflection is an isometry.

4 Under a glide reflection the object and the image have opposite orientation.

Exercise 14a (discussion)

1. Triangle A(1, 0), B(3, 2), C(−1, 2) is mapped by a glide reflection to triangle P(−1, 3), Q(−3, 5), R(1, 5). Draw a diagram to show the triangle and its image. By using mid-points, find the glide axis.

2. A triangle A(−6, 2), B(−2, 0), C(−1, 5) is translated through 5 units parallel to the y-axis and then mapped to triangle A″(8, 7), B″(4, 5), C″(3, 10). Draw a diagram to show triangles ABC, A″B″C″ and, by using mid-points, find the glide axis. Hence or otherwise draw the reflection of A″B″C″ in the glide axis.

Other combined transformations

Example 3

The points A(1, 3), B(1, 1) and C(4, 1) are the vertices of $\triangle$ABC. Plot $\triangle$ABC on graph paper. The triangle is transformed under an enlargement centre (0, 0) and scale factor 2. Plot $\triangle$A′B′C′, the image of $\triangle$ABC under the enlargement. State the coordinates of the points, A′, B′ and C′. The image, $\triangle$A′B′C′, is then reflected in the x-axis to form the image, $\triangle$A″B″C″. Plot $\triangle$A″B″C″ and state the coordinates of the points A″, B″ and C″.

The coordinates are A′(2, 6), B′(2, 2) and C′(8, 2); and A″(2, −6), B″(2, −2) and C″(8, −2). [Fig. 14.5].

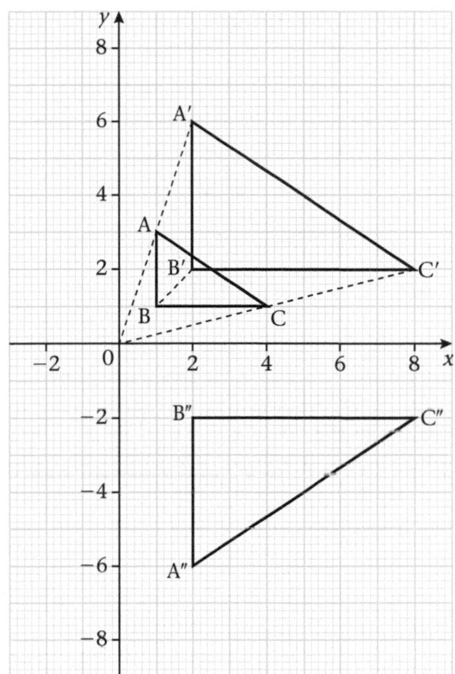

Fig. 14.5

Example 4

The triangle ABC in Example 3 is enlarged, centre (1, 1), scale factor 2 and then reflected in the line $y = x$. Plot the image under each of the transformations. State the coordinates of the points A′, B′ and C′, and of points, A″, B″ and C″.

The coordinates are A′(1, 5), B′(1, 1) and C′(7, 1); and A″(5, 1), B″(1, 1) and C″(1, 7). [Fig. 14.6].

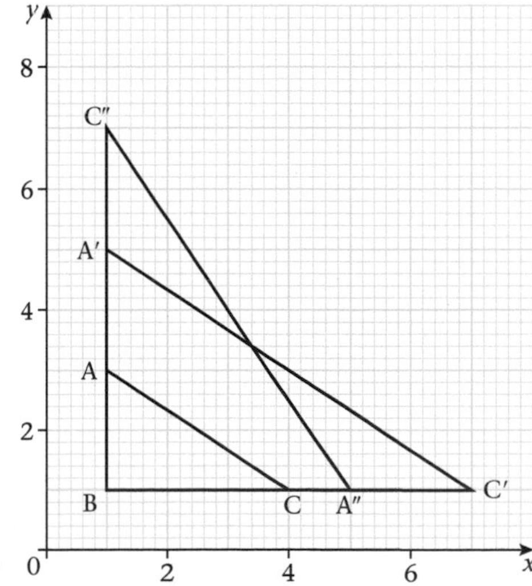

Fig. 14.6

Notice that B is an invariant point, that is, the position of B does not change under either transformation.

Exercise 14b

1. The parallelogram ABCD shown in Fig. 14.7 is reflected in the line $y = 1$. Plot the image A′B′C′D′ under the reflection and state the coordinates of each vertex.

   The image A′B′C′D′ is then translated by the vector $\begin{pmatrix} -3 \\ 0 \end{pmatrix}$. Plot the final image A″B″C″D″ under the translation, and state the coordinates of each vertex.

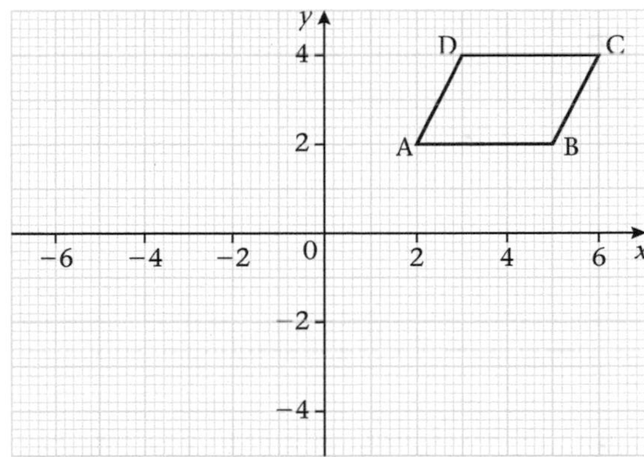

Fig. 14.7

2. The vertices of $\triangle$PQR are P(0, 1), Q(4, 1) and R(0, 2). Plot and state the coordinates of the image $\triangle$P′Q′R′ under an enlargement centre (0, 0) and scale factor 3.
   $\triangle$P′Q′R′ is then translated by the vector $\begin{pmatrix} 2 \\ -1 \end{pmatrix}$. Plot and state the coordinates of the image $\triangle$P″Q″R″.

3. The vertices of $\triangle$KMN are K(4, 7), M(2, 3) and N(6, 3). The triangle is rotated 90° clockwise about the origin. Plot and state the coordinates of the image $\triangle$K′M′N′. $\triangle$K′M′N′ is then reflected in the y-axis. Plot and state the coordinates of the image $\triangle$K″M″N″.

4. The vertices of a rectangle EFGH are at E(−1, 2), F(2, 2), G(2, 4) and H(−1, 4). The rectangle is reflected in the line $x = -2$. Plot the image E′F′G′H′ under the reflection and state the coordinates of each vertex.

The image E′F′G′H′ is then transformed under an enlargement, centre F′ and scale factor 2. Plot the final image E″F″G″H″ and state the coordinates of each vertex.

5 The triangle SVW, shown in Fig. 14.8, undergoes two successive transformations.

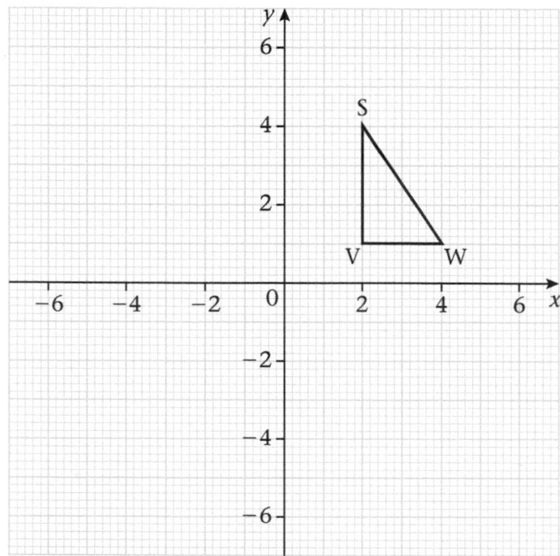

Fig. 14.8

(a) △SVW is enlarged, centre (0, 0) and scale factor −1. The image △S′V′W′ is then transformed under a rotation centre (0, 0) through 90° anticlockwise to the image △S″V″W″. Plot the images on a copy of the diagram.

(b) △SVW is first transformed under the rotation centre (0, 0) through 90° anticlockwise to an image $\triangle S_1V_1W_1$. The image $\triangle S_1V_1W_1$ is then enlarged, centre (0, 0) and scale factor −1, to form the image $\triangle S_2V_2W_2$. Plot the images on a copy of the diagram.

(c) Describe the relationship between the two final images, △S″V″W″ and $\triangle S_2V_2W_2$.

6 The points A(3, 2), B(7, 2) and C(3, 5) are vertices of triangle ABC. Triangle ABC is mapped onto triangle A′B′C′ by a glide reflection consisting of a reflection in the line $y = -x$ and a translation by the vector $\begin{pmatrix} -1 \\ 1 \end{pmatrix}$.

(a) Draw on graph paper triangles ABC and A′B′C′.

(b) A″B″C″ is the image of triangle A′B′C′ after reflection in the x-axis.
   (i) Construct the centre of rotation that maps triangle ABC onto triangle A″B″C″.
   (ii) State the coordinates of the centre of rotation.
   (iii) Describe fully the single transformation that maps triangle A″B″C″ onto triangle ABC.

[CXC Gen June '86] (15 marks)

7 Fig. 14.9 shows the rectangles OABC and PQRS. O is the point (0, 0) and B is the point (2, 1).

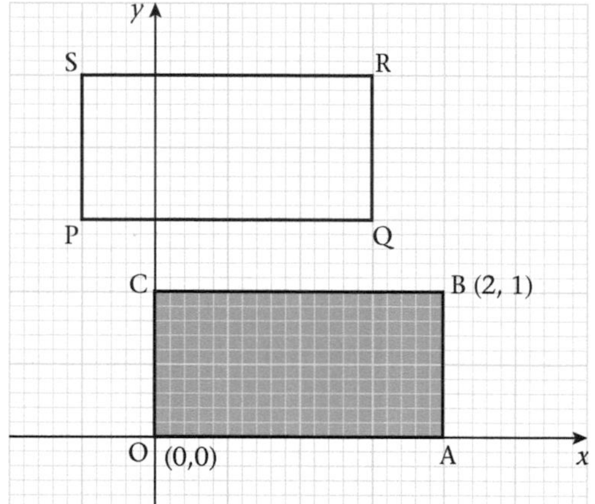

Fig. 14.9

(a) Use the diagram to determine the coordinates of A′, the image of A, after OABC is reflected in the line $y = 1\frac{1}{2}$. (2 marks)

(b) Describe fully the single transformation that maps OABC onto PQRS. (2 marks)

(c) **On graph paper**, draw
   (i) the position vector OB′ after an enlargement of OABC with centre C, scale factor 2,
   (ii) the image of OABC after a clockwise rotation about A through an angle of 270°.

[CXC Gen Jan '90] (4 marks)

Transformations and matrices

Translation

In Fig. 14.10, A is any point (p, q) on the Cartesian plane.

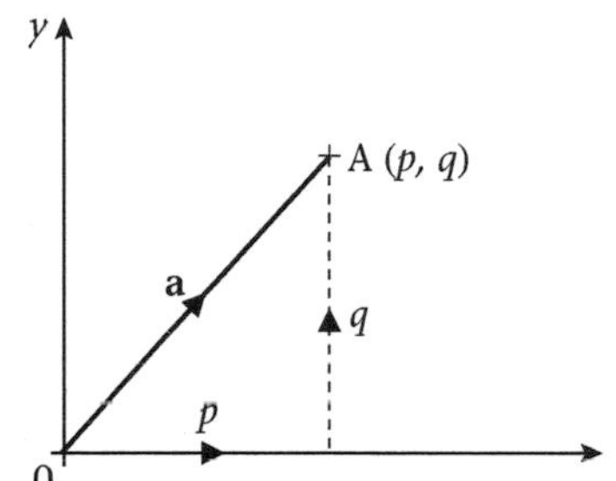

Fig. 14.10 0

$\overrightarrow{OA}$ or **a** is the **position vector** of A.

$\overrightarrow{OA} = \mathbf{a} = \begin{pmatrix} p \\ q \end{pmatrix}$ = the displacement of A from the origin. If A is the point (1, 2) and it is translated by vector $\begin{pmatrix} 4 \\ 2 \end{pmatrix}$, its final position B is shown in Fig. 14.11.

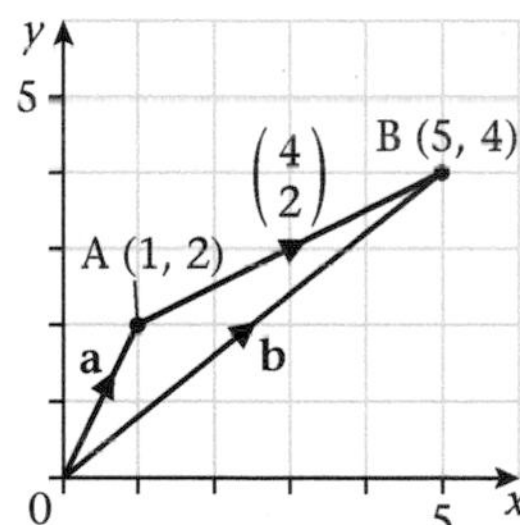

Fig. 14.11

The position of B is calculated by adding the translation vector to the position vector of A:

$$\overrightarrow{OB} = \mathbf{b} = \begin{pmatrix} 1 \\ 2 \end{pmatrix} + \begin{pmatrix} 4 \\ 2 \end{pmatrix} = \begin{pmatrix} 5 \\ 4 \end{pmatrix}$$

Translation is a geometrical transformation often represented by the letter T. Hence if T represents the translation $\begin{pmatrix} 4 \\ 2 \end{pmatrix}$ then $T(\mathbf{a}) = \mathbf{b}$.

Similarly,

$$T(\mathbf{b}) = \mathbf{b} + \begin{pmatrix} 4 \\ 2 \end{pmatrix} = \begin{pmatrix} 5 \\ 4 \end{pmatrix} + \begin{pmatrix} 4 \\ 2 \end{pmatrix} = \begin{pmatrix} 9 \\ 6 \end{pmatrix} = \mathbf{c}$$

Hence $T(T(\mathbf{a})) = \mathbf{c}$.
This is usually written as $T^2(\mathbf{a}) = \mathbf{c}$.

Note that if the Cartesian plane is given a translation T, then every point on the plane is translated through T.

Rotation

Fig. 14.12 shows successive rotations of a **unit square** through 90° clockwise about the origin. (*Note*: a unit square has vertices at (0, 0), (1, 0), (1, 1), (0, 1).)

In Fig. 14.12(a), the position vectors of the vertices A, B, C are **a**, **b**, **c** where

$$\mathbf{a} = \begin{pmatrix} 0 \\ 1 \end{pmatrix}, \quad \mathbf{b} = \begin{pmatrix} 1 \\ 1 \end{pmatrix}, \quad \mathbf{c} = \begin{pmatrix} 1 \\ 0 \end{pmatrix}$$

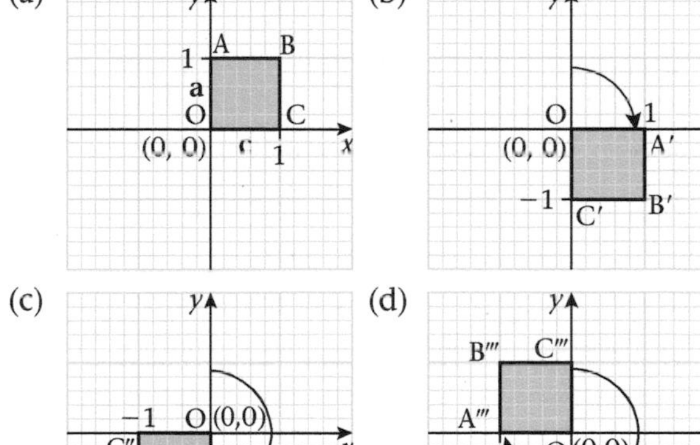

Fig. 14.12

If R represents a rotation through 90° clockwise about the origin, then from Fig. 14.12 parts (a) and (b):

$$R(\mathbf{a}) = \begin{pmatrix} 1 \\ 0 \end{pmatrix} \qquad (1)$$

$$R(\mathbf{c}) = \begin{pmatrix} 0 \\ -1 \end{pmatrix} \qquad (1)$$

It is possible to represent R by a matrix. Let this matrix be $\begin{pmatrix} p & q \\ r & s \end{pmatrix}$. From (1) and (2) above:

$$\begin{pmatrix} p & q \\ r & s \end{pmatrix} \begin{pmatrix} 0 \\ 1 \end{pmatrix} = \begin{pmatrix} 1 \\ 0 \end{pmatrix} \qquad (3)$$

and $$\begin{pmatrix} p & q \\ r & s \end{pmatrix} \begin{pmatrix} 1 \\ 0 \end{pmatrix} = \begin{pmatrix} 0 \\ -1 \end{pmatrix} \qquad (4)$$

By multiplying out the matrices in (3),

$0p + 1q = 1$ and $0r + 1s = 0$

$\Rightarrow q = 1$ $\qquad \Rightarrow s = 0$

Similarly, from (4),

$1p + 0q = 0$ and $1r + 0s = -1$

$\Rightarrow p = 1$ $\qquad \Rightarrow r = -1$

Hence, $R = \begin{pmatrix} 0 & 1 \\ -1 & 0 \end{pmatrix}$

In Fig. 14.12(c), R^2 represents R followed by R, i.e. a rotation of 180° about O.

$$R^2 = \begin{pmatrix} 0 & 1 \\ -1 & 0 \end{pmatrix}\begin{pmatrix} 0 & 1 \\ -1 & 0 \end{pmatrix} = \begin{pmatrix} -1 & 0 \\ 0 & -1 \end{pmatrix}$$

In Fig. 14.12(d), R^3 represents R followed by R followed by R, i.e. a clockwise rotation of 270° about O.

$$R^3 = R \times R^2 = \begin{pmatrix} 0 & 1 \\ -1 & 0 \end{pmatrix}\begin{pmatrix} -1 & 0 \\ 0 & -1 \end{pmatrix} = \begin{pmatrix} 0 & -1 \\ 1 & 0 \end{pmatrix}$$

R^4 represents represents four successive clockwise rotations of 90°, i.e. a rotation of 360°.

$$R^4 = R \times R^3 = \begin{pmatrix} 0 & 1 \\ -1 & 0 \end{pmatrix}\begin{pmatrix} 0 & -1 \\ 1 & 0 \end{pmatrix} = \begin{pmatrix} 1 & 0 \\ 0 & 1 \end{pmatrix}$$

or

$$R^4 = R^2 \times R^2 = \begin{pmatrix} -1 & 0 \\ 0 & -1 \end{pmatrix}\begin{pmatrix} -1 & 0 \\ 0 & -1 \end{pmatrix} = \begin{pmatrix} 1 & 0 \\ 0 & 1 \end{pmatrix}$$

in both cases.

Hence R^4 is equivalent to the identity matrix, I, as might be expected.

The matrices $= \begin{pmatrix} 0 & 1 \\ -1 & 0 \end{pmatrix}, \begin{pmatrix} -1 & 0 \\ 0 & -1 \end{pmatrix}, \begin{pmatrix} 0 & -1 \\ 1 & 0 \end{pmatrix}, \begin{pmatrix} 1 & 0 \\ 0 & 1 \end{pmatrix}$ represent **clockwise rotations** of 90°, 180°, 270°, 360° about the origin.

Note that if the Cartesian plane is given a rotation R then every point, except the centre of rotation, is rotated through R. The centre of rotation is said to be an **invariant** point. Remember that *invariant* means 'unchanging'.

Reflection

In Fig. 14.13, the unit square in part (a) is shown in (b) reflected in the x-axis.

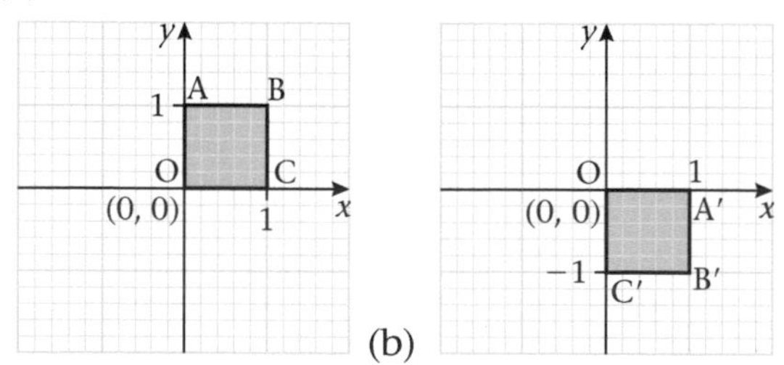

Fig. 14.13

If M represents the reflection in the x-axis, then by considering what happens to $\overrightarrow{OA}$ and $\overrightarrow{OC}$.

$$M(\overrightarrow{OA}) = \overrightarrow{OA}'$$

and $M(\overrightarrow{OC}) = \overrightarrow{OC}'$

If M is the matrix $\begin{pmatrix} p & q \\ r & s \end{pmatrix}$, then

$$\begin{pmatrix} p & q \\ r & s \end{pmatrix}\begin{pmatrix} 0 \\ 1 \end{pmatrix} = \begin{pmatrix} 0 \\ -1 \end{pmatrix} \quad (1)$$

and $$\begin{pmatrix} p & q \\ r & s \end{pmatrix}\begin{pmatrix} 1 \\ 0 \end{pmatrix} = \begin{pmatrix} 1 \\ 0 \end{pmatrix} \quad (2)$$

By multiplying out the matrices, $q = 0$, $s = -1$, $p = 1$ and $r = 0$.

Hence $M = \begin{pmatrix} 1 & 0 \\ 0 & -1 \end{pmatrix}$

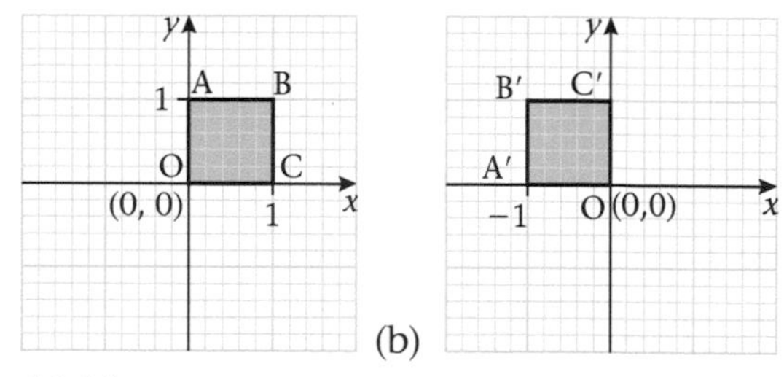

Fig. 14.14

Similarly, by considering Fig. 14.14(a) and (b), if N represents reflection in the y-axis, and N is the matrix $\begin{pmatrix} p & q \\ r & s \end{pmatrix}$

so that $$\begin{pmatrix} p & q \\ r & s \end{pmatrix}\begin{pmatrix} 0 \\ 1 \end{pmatrix} = \begin{pmatrix} 0 \\ 1 \end{pmatrix} \quad (1)$$

and $$\begin{pmatrix} p & q \\ r & s \end{pmatrix}\begin{pmatrix} 1 \\ 0 \end{pmatrix} = \begin{pmatrix} -1 \\ 0 \end{pmatrix}. \quad (2)$$

it can be shown that

$$N = \begin{pmatrix} -1 & 0 \\ 0 & 1 \end{pmatrix}$$

Notice that

$$M^2 = \begin{pmatrix} 1 & 0 \\ 0 & -1 \end{pmatrix}\begin{pmatrix} 1 & 0 \\ 0 & -1 \end{pmatrix} = \begin{pmatrix} 1 & 0 \\ 0 & 1 \end{pmatrix}$$

and

$$N^2 = \begin{pmatrix} -1 & 0 \\ 0 & 1 \end{pmatrix}\begin{pmatrix} -1 & 0 \\ 0 & 1 \end{pmatrix} = \begin{pmatrix} 1 & 0 \\ 0 & 1 \end{pmatrix}$$

Hence $M^2 = N^2 = I$ as might be expected.

Note that if the Cartesian plane is given a reflection M, then every point on the plane, except those on the line of reflection, is transformed. The line of reflection is an **invariant line**.

Example 5

The vertices of a triangle are A(1, 2), B(3, 1) and C(−2, 1). If △ABC is reflected in the x-axis, calculate the coordinates of the vertices of its image.

The matrix M represents reflection in the x-axis where

$$M = \begin{pmatrix} 1 & 0 \\ 0 & -1 \end{pmatrix}$$

If the image of $\triangle ABC$ is $\triangle A'B'C'$, then

$M(\overrightarrow{OA}) = \overrightarrow{OA'}$ (1)

$M(\overrightarrow{OB}) = \overrightarrow{OB'}$ (2)

$M(\overrightarrow{OC}) = \overrightarrow{OC'}$ (3)

From (1), $\overrightarrow{OA'} = \begin{pmatrix} 1 & 0 \\ 0 & -1 \end{pmatrix}\begin{pmatrix} 1 \\ 2 \end{pmatrix} = \begin{pmatrix} 1 \\ -2 \end{pmatrix}$

From (2), $\overrightarrow{OB'} = \begin{pmatrix} 1 & 0 \\ 0 & -1 \end{pmatrix}\begin{pmatrix} 3 \\ 1 \end{pmatrix} = \begin{pmatrix} 3 \\ -1 \end{pmatrix}$

From (3), $\overrightarrow{OC'} = \begin{pmatrix} 1 & 0 \\ 0 & -1 \end{pmatrix}\begin{pmatrix} -2 \\ 1 \end{pmatrix} = \begin{pmatrix} -2 \\ -1 \end{pmatrix}$

The vertices of the image of $\triangle ABC$ are A′(1, −2), B′(3, −1), C′(−2, −1).

Note: The working in Example 5 can be written more neatly by representing the vertices of $\triangle ABC$ by a single matrix,

$$\begin{pmatrix} 1 & 3 & -2 \\ 2 & 1 & 1 \end{pmatrix}$$

Then, by matrix multiplication,

$$\begin{pmatrix} 1 & 0 \\ 0 & -1 \end{pmatrix}\overset{\begin{matrix} A & B & C \end{matrix}}{\begin{pmatrix} 1 & 3 & -2 \\ 2 & 1 & 1 \end{pmatrix}} = \overset{\begin{matrix} A' & B' & C' \end{matrix}}{\begin{pmatrix} 1 & 3 & -2 \\ -2 & -1 & -1 \end{pmatrix}}$$

Note that matrix multiplication is possible because the number of columns in the first matrix equals the number of rows in the second matrix.

Example 6

The vertices of $\triangle HJK$ are H(1, 4), J(2, 1) and K(2, 7). Under a transformation Z, $\triangle HJK$ is mapped onto $\triangle H'J'K'$. The vertices of the image, $\triangle H'J'K'$, are H′(−4, 1), J′(−1, 2) and K′(−7, 2).

(a) Express the matrix of the transformation Z in the form $\begin{pmatrix} p & q \\ r & s \end{pmatrix}$

(b) Give a geometrical description of the transformation.

(c) Determine the matrix, V, that maps the image $\triangle H'J'K'$ back onto $\triangle HJK$.

(a) $\begin{pmatrix} p & q \\ r & s \end{pmatrix}\begin{pmatrix} 1 & 2 & 2 \\ 4 & 1 & 7 \end{pmatrix} = \begin{pmatrix} -4 & -1 & -7 \\ 1 & 2 & 2 \end{pmatrix}$

$p + 4q = -4$ and $r + 4s = 1$

$2p + q = -1$ $\quad$ $2r + s = 2$

Solving the equations, we get

$$p = 0, \quad q = -1, \quad r = 1, \quad s = 0$$

Hence,

$$Z = \begin{pmatrix} 0 & -1 \\ 1 & 0 \end{pmatrix}$$

(b) Z is a clockwise rotation about the origin through 270°.

(c) If $V = \begin{pmatrix} a & b \\ c & d \end{pmatrix}$,

then $\begin{pmatrix} a & b \\ c & d \end{pmatrix}\begin{pmatrix} -4 & -1 & -7 \\ 1 & 2 & 2 \end{pmatrix} = \begin{pmatrix} 1 & 2 & 2 \\ 4 & 1 & 7 \end{pmatrix}$

and $a = 0,\ b = 1,\ c = -1,$ and $d = 0$

$$V = \begin{pmatrix} 0 & 1 \\ -1 & 0 \end{pmatrix}$$

Note that

$$ZV = \begin{pmatrix} 0 & -1 \\ 1 & 0 \end{pmatrix}\begin{pmatrix} 0 & 1 \\ -1 & 0 \end{pmatrix} = \begin{pmatrix} 1 & 0 \\ 0 & 1 \end{pmatrix}$$
$$= I, \text{ the identity matrix.}$$

Hence $V = Z^{-1}$

Exercise 14c

1. Copy and complete Table 14.1.

Table 14.1

Transformation	Matrix
identity	$\begin{pmatrix} 1 & 0 \\ 0 & 1 \end{pmatrix}$
reflection in x-axis	
reflection in y-axis	
rotation of 180° about origin	

2. Find the matrices which are equivalent to anticlockwise rotations of (a) 90°, (b) 270° about the origin.

3. A triangle has vertices (1, 1), (2, 4) and (3, 7). Find the coordinates of the images of its vertices if it is rotated through 90° clockwise about the origin.

4. P′Q′ is the image of a line PQ after a translation $\begin{pmatrix} 7 \\ -4 \end{pmatrix}$. If the coordinates of P′ are (6, 1), calculate the coordinates of P.

5. A triangle has vertices A(0, 0), B(1, −1) and C(1, 1). Calculate the position of its vertices after a rotation of 180° about the origin.

6. Reflect the triangle of question 5 about the y-axis. Compare your answer with that of question 5.

7. A rhombus has vertices (2, 2), (1, −1), (−2, −2) and (−1, 1). Find the coordinates of its vertices when it is reflected in (a) the x-axis, (b) the y-axis.

8. In Fig. 14.15, there are three triangles, ABC, $A_1B_1C_1$ and $A_2B_2C_2$.

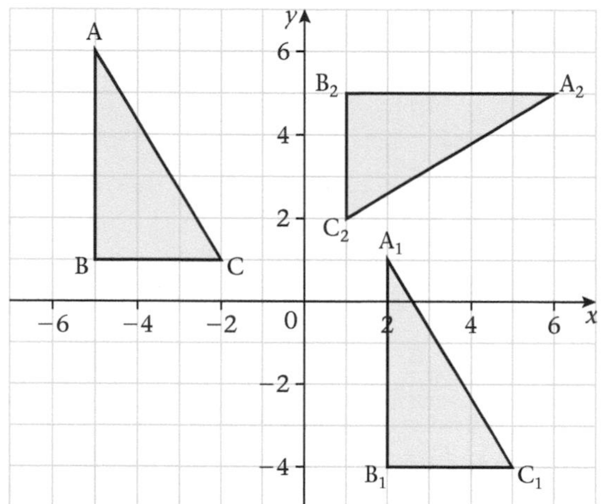

Fig. 14.15

(a) $A_1B_1C_1$ is the image of ABC under a single translation. Write down the column vector of this translation.

(b) $A_2B_2C_2$ is the image of ABC under an anticlockwise rotation about the origin. Write down (i) the angle of the rotation, (ii) the matrix of the rotation.

(c) $A_3B_3C_3$ (not shown in the diagram) is the image of ABC under a reflection represented by the matrix $\begin{pmatrix} 1 & 0 \\ 0 & -1 \end{pmatrix}$. Write down the equation of the straight line through B_3 and C_3.

Enlargement

In Fig. 14.16, $\triangle$AOB is enlarged to $\triangle$OPQ by scale factor 2 with the origin as the centre of enlargement.

From Fig. 14.16, $\mathbf{a} = \begin{pmatrix} 4 \\ 4 \end{pmatrix}$ and $\mathbf{b} = \begin{pmatrix} 3 \\ 1 \end{pmatrix}$

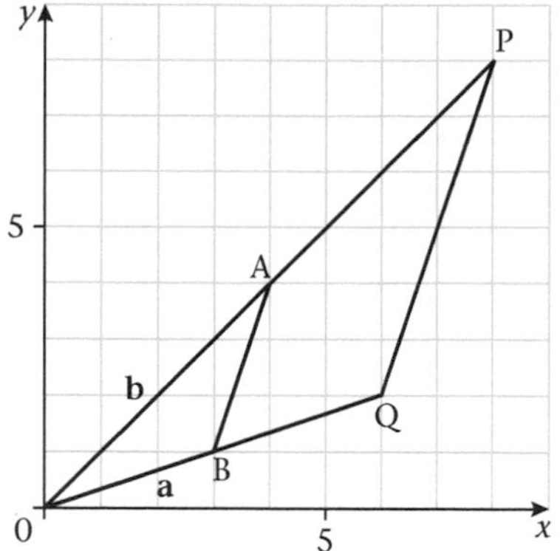

Fig. 14.16

$\overrightarrow{OP} = \begin{pmatrix} 8 \\ 8 \end{pmatrix}$ and $\overrightarrow{OQ} = \begin{pmatrix} 6 \\ 2 \end{pmatrix}$.

If operator E represents the enlargement, then

$$E(\mathbf{a}) = \overrightarrow{OP} \text{ and } (\mathbf{b}) = \overrightarrow{OQ}$$

Let E be the matrix $\begin{pmatrix} p & q \\ r & s \end{pmatrix}$, then

$$\begin{pmatrix} p & q \\ r & s \end{pmatrix}\begin{pmatrix} 4 \\ 4 \end{pmatrix} = \begin{pmatrix} 8 \\ 8 \end{pmatrix}$$

and $\begin{pmatrix} p & q \\ r & s \end{pmatrix}\begin{pmatrix} 3 \\ 1 \end{pmatrix} = \begin{pmatrix} 6 \\ 2 \end{pmatrix}$

from which $p = 2$, $q = 0$, $r = 0$, $s = 2$,

and $E = \begin{pmatrix} 2 & 0 \\ 0 & 2 \end{pmatrix} = 2\begin{pmatrix} 1 & 0 \\ 0 & 1 \end{pmatrix} = 2I$

In general, for any enlargement E with scale factor k and centre (0, 0),

$$E = \begin{pmatrix} k & 0 \\ 0 & k \end{pmatrix} = k\begin{pmatrix} 1 & 0 \\ 0 & 1 \end{pmatrix} = kI$$

Note that if the Cartesian plane is given an enlargement, there is only one invariant point: the centre of enlargement.

Example 7

Quadrilateral OABC has vertices O(0, 0), A(6, −1), B(−4, 2), C(9, 9). OABC is enlarged with scale factor $-\frac{1}{3}$ with the origin as centre. Find

(a) the coordinates of the vertices of the enlargement O′A′B′C′

(b) the matrix, V, that maps the image O′A′B′C′ back onto OABC.

(a) Enlargement matrix, Z

$$= -\tfrac{1}{3}I = -\tfrac{1}{3}\begin{pmatrix} 1 & 0 \\ 0 & 1 \end{pmatrix} = \begin{pmatrix} -\frac{1}{3} & 0 \\ 0 & -\frac{1}{3} \end{pmatrix}$$

$$\begin{pmatrix} -\frac{1}{3} & 0 \\ 0 & -\frac{1}{3} \end{pmatrix} \overset{\quad O \quad A \quad B \quad C}{\begin{pmatrix} 0 & 6 & -4 & 9 \\ 0 & -1 & 2 & 9 \end{pmatrix}}$$

$$= \overset{O' \quad A' \quad B' \quad C'}{\begin{pmatrix} 0 & -2 & 1\frac{1}{3} & -3 \\ 0 & \frac{1}{3} & -\frac{2}{3} & -3 \end{pmatrix}}$$

The enlargement has coordinates O′ (0, 0), A′ $(-2, \frac{1}{3})$, B′ $(1\frac{1}{3}, -\frac{2}{3})$ and C′ (−3, −3).

(b) If $V = \begin{pmatrix} a & b \\ c & d \end{pmatrix}$,

$$\begin{pmatrix} a & b \\ c & d \end{pmatrix} \begin{pmatrix} 0 & -2 & \frac{4}{3} & -3 \\ 0 & \frac{1}{3} & -\frac{2}{3} & -3 \end{pmatrix}$$

$$= \begin{pmatrix} 0 & 6 & -4 & 9 \\ 0 & -1 & 2 & 9 \end{pmatrix}$$

and $a = -3$, $b = 0$, $c = 0$, and $d = -3$

$$\therefore V = \begin{pmatrix} -3 & 0 \\ 0 & -3 \end{pmatrix}$$

Note that

$$ZV = \begin{pmatrix} -\frac{1}{3} & 0 \\ 0 & -\frac{1}{3} \end{pmatrix} \begin{pmatrix} -3 & 0 \\ 0 & -3 \end{pmatrix} = \begin{pmatrix} 1 & 0 \\ 0 & 1 \end{pmatrix}$$

= I, the identity matrix.

Hence, $V = Z^{-1}$

Notice in Examples 6 and 7 that if a shape is mapped onto an image by a matrix A, then the image can be mapped back onto the original shape by the inverse of A, sometimes written as A^{-1}. Remember that $AA^{-1} = I$.

Exercise 14d

1. Use the matrix $\begin{pmatrix} -3 & 0 \\ 0 & -3 \end{pmatrix}$ to enlarge △ABC in Fig. 14.17.

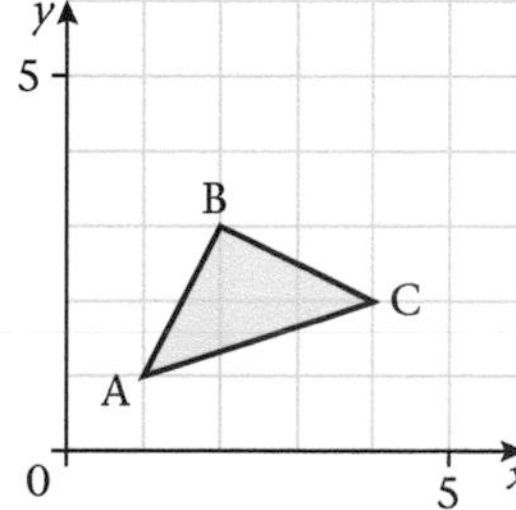

Fig. 14.17

2. Find the matrix E which has the effect of enlarging plane shapes by a scale factor $1\frac{1}{2}$ with the origin as centre of enlargement.

3. Use E of question 2 to enlarge the rectangle whose vertices are (0, 0), (3, 0), (0, 2), (3, 2).

4. A single transformation U maps △PQR of Fig. 14.18 onto $\triangle P_1Q_1R_1$ which has vertices $P_1(0, 16)$, $Q_1(12, 20)$, $R_1(8, 4)$.

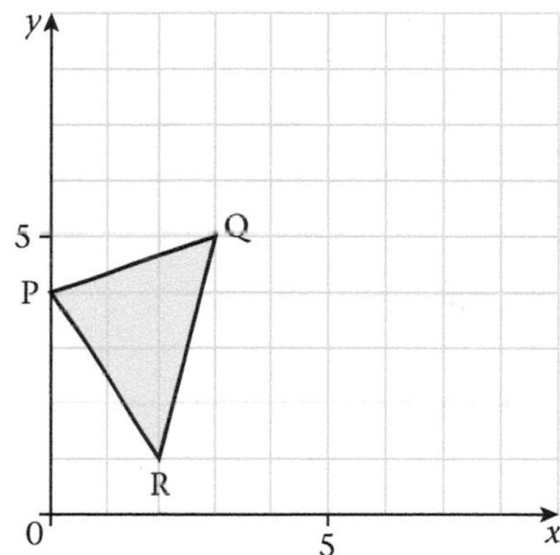

Fig. 14.18

Describe U, writing down the matrix which represents the transformation.

Combined transformations and matrices

Example 8

Triangle OAB has vertices at O(0, 0), A(2, 1), B(−1, 3). Find its new vertices if it is first enlarged by E, then translated through T where

$E = \begin{pmatrix} 2 & 0 \\ 0 & 2 \end{pmatrix}$ *and* $T = \begin{pmatrix} -3 \\ -5 \end{pmatrix}$.

First, enlargement by E:

$$\begin{pmatrix} 2 & 0 \\ 0 & 2 \end{pmatrix} \overset{O \quad A \quad B}{\begin{pmatrix} 0 & 2 & -1 \\ 2 & 1 & 3 \end{pmatrix}} = \overset{O' \quad A' \quad B'}{\begin{pmatrix} 0 & 4 & -2 \\ 0 & 2 & 6 \end{pmatrix}}$$

Second, translation of points O′, A′, B′ through T:

$$\begin{pmatrix} 0 \\ 0 \end{pmatrix} + \begin{pmatrix} -3 \\ -5 \end{pmatrix} = \begin{pmatrix} -3 \\ -5 \end{pmatrix}$$

$$\begin{pmatrix} 4 \\ 2 \end{pmatrix} + \begin{pmatrix} -3 \\ -5 \end{pmatrix} = \begin{pmatrix} 1 \\ -3 \end{pmatrix}$$

$$\begin{pmatrix} -2 \\ 6 \end{pmatrix} + \begin{pmatrix} -3 \\ -5 \end{pmatrix} = \begin{pmatrix} -5 \\ 1 \end{pmatrix}$$

The vertices of the transformed triangle are O″(−3, −5), A″(1, −3), and B″(−5, 1).

The combined transformation is shown in Fig. 14.19.

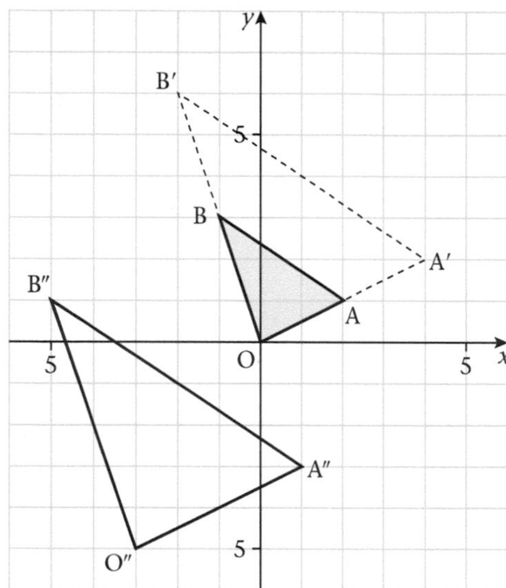

Fig. 14.19

The process in Example 8 can be written as TE(**a**) = **p**

where **p** is the final image of **a**
and TE means that E is done *before* T.

TE(**a**) = **p** ⇒ T[E(**a**)] = **p**

The order in which the operations are carried out is usually important. In general, TE ≠ ET. For example, with the data of Example 8:

$$\begin{aligned}\text{TE}\begin{pmatrix}2\\1\end{pmatrix} &= \text{T}\left[\text{E}\begin{pmatrix}2\\1\end{pmatrix}\right]\\ &= \text{T}\begin{pmatrix}4\\2\end{pmatrix}\\ &= \begin{pmatrix}1\\-3\end{pmatrix}\end{aligned}$$

$$\begin{aligned}\text{ET}\begin{pmatrix}2\\1\end{pmatrix} &= \text{E}\left[\text{T}\begin{pmatrix}2\\1\end{pmatrix}\right]\\ &= \text{E}\left[\begin{pmatrix}2\\1\end{pmatrix} + \begin{pmatrix}-3\\-5\end{pmatrix}\right]\\ &= \text{E}\begin{pmatrix}-1\\-4\end{pmatrix}\\ &= \begin{pmatrix}2&0\\0&2\end{pmatrix}\begin{pmatrix}-1\\-4\end{pmatrix}\\ &= \begin{pmatrix}-2\\-8\end{pmatrix}\end{aligned}$$

that is TE ≠ ET

Example 9

Triangle XYZ has vertices at X(3, 4), Y(5, −1), Z(−2, 2). Find the coordinates of the images of X, Y, Z if the triangle is first translated by vector $\begin{pmatrix}-2\\7\end{pmatrix}$ and then rotated through 180° about the origin.

Let △X′Y′Z′ be the image of △XYZ after translation $\begin{pmatrix}-2\\7\end{pmatrix}$.

$$\overrightarrow{OX'} = \begin{pmatrix}3\\4\end{pmatrix} + \begin{pmatrix}-2\\7\end{pmatrix} = \begin{pmatrix}1\\11\end{pmatrix}$$

$$\overrightarrow{OY'} = \begin{pmatrix}5\\-1\end{pmatrix} + \begin{pmatrix}-2\\7\end{pmatrix} = \begin{pmatrix}3\\6\end{pmatrix}$$

$$\overrightarrow{OZ'} = \begin{pmatrix}-2\\2\end{pmatrix} + \begin{pmatrix}-2\\7\end{pmatrix} = \begin{pmatrix}-4\\9\end{pmatrix}$$

Let △X″Y″Z″ be the image of △X′Y′Z′ after rotation through 180° about the origin.

$\begin{pmatrix}-1&0\\0&-1\end{pmatrix}$ is the matrix of rotation:

$$\begin{pmatrix}-1&0\\0&-1\end{pmatrix}\overset{\begin{matrix}X'&Y'&Z'\end{matrix}}{\begin{pmatrix}1&3&-4\\11&6&9\end{pmatrix}} = \overset{\begin{matrix}X''&Y''&Z''\end{matrix}}{\begin{pmatrix}-1&-3&4\\-11&-6&-9\end{pmatrix}}$$

The vertices of the final image of △XYZ are at X″(−1, −11), Y″(−3, −6) and Z″(4, −9).

We may use matrices to solve Example 3 on p. 170.

Example 3 (alternative method)

The points A(1, 3), B(1, 1) and C(4, 1) are the vertices of △ABC. The triangle is transformed under an enlargement centre (0, 0) and scale factor 2. △A′B′C′ is the image of △ABC under the enlargement. State the coordinates of the points A′, B′ and C′. The image, △A′B′C′, is then reflected in the *x*-axis to form the image, △A″B′C″. State the coordinates of the points A″, B″ and C″.

The matrix of an enlargement, scale factor 2, centre (0, 0) is $\begin{pmatrix}2&0\\0&2\end{pmatrix}$.

$$\text{Thus}\quad \begin{pmatrix}2&0\\0&2\end{pmatrix}\overset{\begin{matrix}A&B&C\end{matrix}}{\begin{pmatrix}1&1&4\\3&1&1\end{pmatrix}}$$

$$= \overset{\begin{matrix}A'&B'&C'\end{matrix}}{\begin{pmatrix}2&2&8\\6&2&2\end{pmatrix}}$$

The coordinates are A′(2, 6), B′(2, 2) and C′(8, 2).

The matrix of a reflection in the x-axis is $\begin{pmatrix}1 & 0\\ 0 & -1\end{pmatrix}$.

$$\text{Thus}\quad \begin{pmatrix}1 & 0\\ 0 & -1\end{pmatrix}\overset{\begin{matrix}A' & B' & C'\end{matrix}}{\begin{pmatrix}2 & 2 & 8\\ 6 & 2 & 2\end{pmatrix}}$$

$$= \overset{\begin{matrix}A'' & B'' & C''\end{matrix}}{\begin{pmatrix}2 & 2 & 8\\ -6 & -2 & -2\end{pmatrix}}$$

The coordinates are A″(2, −6), B″(2, −2) and C″(8, −2).

The combined transformations may be represented by

$$\begin{pmatrix}1 & 0\\ 0 & -1\end{pmatrix}\begin{pmatrix}2 & 0\\ 0 & 2\end{pmatrix} = \begin{pmatrix}2 & 0\\ 0 & -2\end{pmatrix}$$

$$\text{Thus}\quad \begin{pmatrix}2 & 0\\ 0 & -2\end{pmatrix}\overset{\begin{matrix}A & B & C\end{matrix}}{\begin{pmatrix}1 & 1 & 4\\ 3 & 1 & 1\end{pmatrix}}$$

$$= \overset{\begin{matrix}A'' & B'' & C''\end{matrix}}{\begin{pmatrix}2 & 2 & 8\\ -6 & -2 & -2\end{pmatrix}}$$

Exercise 14e

1. Triangle XYZ in Fig. 14.20 is first rotated through 90° anticlockwise about the origin and then reflected by the operator $\begin{pmatrix}1 & 0\\ 0 & -1\end{pmatrix}$. Calculate the coordinates of the vertices of its image.

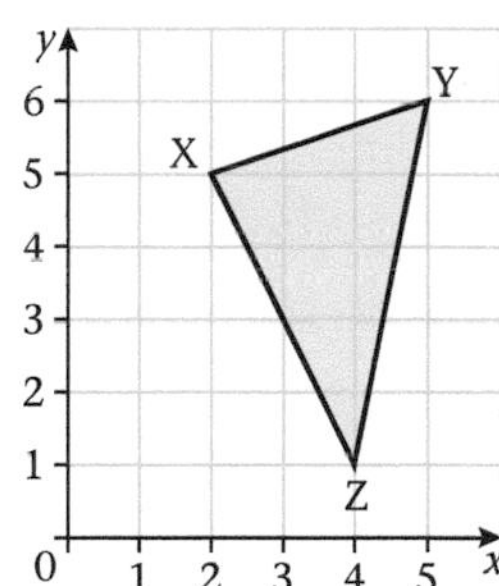

Fig. 14.20

2. Use the matrix $\begin{pmatrix}2 & 0\\ 0 & 2\end{pmatrix}$ to enlarge △XYZ shown in Fig. 14.20. Then reflect the result in the y-axis. Calculate the coordinates of its final image.

3. T is a translation $\begin{pmatrix}2\\ 8\end{pmatrix}$ and E is an enlargement represented by the matrix $\begin{pmatrix}3 & 0\\ 0 & 3\end{pmatrix}$. Calculate the image of A(3, 2) under the following transformations:
   (a) ET(A) (b) TE(A) (c) E^{-1}(A)

4. In Fig. 14.21, semicircle A can be mapped onto semicircle B by an anticlockwise rotation about the origin followed by a translation.
   (a) State the angle of rotation.
   (b) Find the matrix which represents the rotation.
   (c) Find the column vector of the translation.
   (d) Given that A can be mapped onto B by a single reflection in a line m, find the equation of m.

5. In Fig. 14.21 semicircle C is the image of semicircle A under a transformation given by TE(A) = C, where E is an enlargement with the origin as centre and T is a translation.
   (a) State the scale factor of E.
   (b) Write down the matrix which represents E.
   (c) Express T as a column vector.
   (d) The transformation can also be given by ET′(A) = C where T′ is a different translation and E is the same as before. Express T′ as a column vector.
   (e) A can be mapped onto C by a single enlargement with centre (h, k). State the values of h and k.

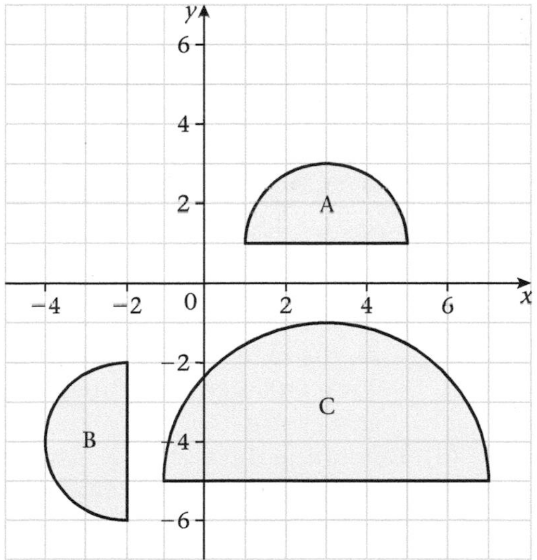

Fig. 14.21

6. The following equation represents a transformation

$$\begin{pmatrix}x'\\ y'\end{pmatrix} = \begin{pmatrix}4 & 0\\ 0 & 4\end{pmatrix}\begin{pmatrix}x\\ y\end{pmatrix} + \begin{pmatrix}-3\\ 1\end{pmatrix}$$

(a) Describe the transformation in words
(b) Find the image of the point $(-1, 4)$.

7 Answer this question on graph paper.
(a) Using a scale of 1 cm to 1 unit on each axis, draw x- and y-axes, taking values of x from -8 to 12 and values of y from -6 to 14. Draw and label the triangle X, with vertices (2, 4), (4, 4) and (4, 1).
(b) The single transformation U maps the triangle X onto the triangle U(X) which has vertices (6, 12), (12, 12) and (12, 3). Draw and label the triangle U(X) and describe fully the transformation U.
(c) The transformation R is a clockwise rotation of 90° about the origin. Draw and label the triangle R(X).
(d) The transformation T is the translation $\begin{pmatrix} -8 \\ 4 \end{pmatrix}$. Draw and label the triangle T(X) and the triangle RT(X).
(e) The single transformation V is represented by the matrix $\begin{pmatrix} 0 & -1 \\ -1 & 0 \end{pmatrix}$. Draw and label the triangle V(X) and describe fully the transformation V.

8 (a) (i) Write down the 2×2 matrix, R, corresponding to a reflection in the origin.
(ii) Determine the coordinates of the images, A′, B′, of the points, A(1, 3), B(2, 4) respectively under the transformation, R. (4 marks)
(b) (i) Write down the 2×2 matrices associated with
- M, the reflection in the x-axis
- T, an enlargement of magnitude 2 and centre (0, 0).

(ii) Determine the matrix, K, associated with a reflection M followed by an enlargement T.
(iii) Calculate the inverse of K.
(iv) Hence, calculate the coordinates (x, y) of the point whose image is $(3, -2)$ under the transformation K.
(11 marks)

[CXC (General) Jan 98] (15 marks)

Summary

A **glide reflection** is a combination of a reflection in a line, the glide axis, with a translation in a direction parallel to the glide axis. The order of the transformations is not important.

Matrices of transformation

Transformation	*Matrix*
identity	$\begin{pmatrix} 1 & 0 \\ 0 & 1 \end{pmatrix}$
reflection in the x-axis	$\begin{pmatrix} 1 & 0 \\ 0 & -1 \end{pmatrix}$
reflection in the y-axis	$\begin{pmatrix} -1 & 0 \\ 0 & 1 \end{pmatrix}$
rotation of 90° about origin clockwise	$\begin{pmatrix} 0 & 1 \\ -1 & 0 \end{pmatrix}$
rotation of 180° about origin	$\begin{pmatrix} -1 & 0 \\ 0 & -1 \end{pmatrix}$
rotation of 90° about origin anticlockwise	$\begin{pmatrix} 0 & -1 \\ 1 & 0 \end{pmatrix}$
enlargement scale factor k and centre (0, 0)	$\begin{pmatrix} k & 0 \\ 0 & k \end{pmatrix}$

Practice exercise P14.1

1 The vertices of △ABC are A(2, 3), B(3, 5) and C(6, 1). △ABC is translated by the vector $\begin{pmatrix} 3 \\ -1 \end{pmatrix}$ and then reflected in the line $x = 1$. Draw △ABC and the images, △A′B′C′ and △A″B″C″ on graph paper.

2 The vertices of △PQR are P(1, 1), Q(4, 3) and R(1, 3). △PQR is translated by the vector $\begin{pmatrix} -4 \\ 2 \end{pmatrix}$ and then reflected in the x-axis. Draw △PQR and the images, △P′Q′R′ and △P″Q″R″ on graph paper.

3 Triangle A(2, 1), B(2, 4), C(4, 3) is mapped by a glide reflection to triangle P(2, 6), Q(2, 3), R(4, 4).
(a) Draw a diagram to show the triangle and its image.
(b) Find the glide axis.

4 Triangle A(5, 1), B(6, 4), C(4, 3) is reflected in the line $x = 3$. The image, △A′B′C′, is then rotated through 90° clockwise about the point A′. Draw △ABC and the images, △A′B′C′ and △A″B″C″, on graph paper.

5. The vertices of △ABC are A(1, 1), B(4, 0) and C(0, 2). The triangle is translated by the vector $\begin{pmatrix}-1\\-4\end{pmatrix}$. Plot and state the coordinates of the image △A′B′C′. △A′B′C′ is then enlarged by a scale factor of 2, centre the origin. Plot and state the coordinates of the image △A″B″C″.

Practice exercise P14.2

1. T is the translation $\begin{pmatrix}-2\\3\end{pmatrix}$. Find the images under the translation T of the points A(3, –3), B(–3, 2) and C(–5, –5).

2. The vertices of △ABC are A(2, 1), B(3, 1) and C(2, 3). It is transformed by the matrix $\begin{pmatrix}0 & 1\\-1 & 0\end{pmatrix}$.
   (a) Give a geometrical description of the transformation.
   (b) Calculate the coordinates of the image △A′B′C′.

3. Find the images of the points A(2, 0) and B(0, 2) under an enlargement, centre the origin and scale factor 3. Hence write down the matrix representing the enlargement.

4. The single transformation T is represented by the matrix $\begin{pmatrix}1 & 0\\0 & -1\end{pmatrix}$.
   (a) Give a geometrical description of the transformation.
   (b) Calculate the coordinates of the image of the points A(2, 5) and B(−1, 3) under transformation T.
   (c) By finding the inverse of the matrix, find the coordinates of the point of which (−5, −10) is the image.

5. The vertices of △ABC are A(3, 2), B(2, −3) and C(4, 1). Under the transformation T, △ABC is mapped onto △A′B′C′. The vertices of the image △A′B′C′ are A′(3, 2), B′(4, −3) and C′(3, 2).
   (a) Express the matrix of the transformation T in the form $\begin{pmatrix}p & q\\r & s\end{pmatrix}$.
   (b) Draw and label △ABC and its image △A′B′C′.
   (c) Describe the transformation T.

Practice exercise P14.3

1. The following equation represents a transformation.
   $$\begin{pmatrix}x'\\y'\end{pmatrix} = \begin{pmatrix}3 & 0\\0 & 3\end{pmatrix}\begin{pmatrix}x\\y\end{pmatrix} + \begin{pmatrix}-2\\1\end{pmatrix}$$
   (a) Describe the transformation in words.
   (b) Find the image of the point (2, −5).

2. A point P(3, −2) is translated by the vector $\begin{pmatrix}2\\4\end{pmatrix}$ and is then reflected in the y-axis.
   (a) Express the transformation as a matrix equation.
   (b) Find the image of the point P.

3. (a) The transformation R is a rotation of 90° anticlockwise about the origin. Write down the matrix representing R.
   (b) The transformation M is a reflection in the x-axis. Write down the matrix representing M.
   (c) The point A has coordinates (4, 3). Calculate the coordinates of the image of A under the following transformations.
   (i) RM(A) (ii) MR(A) (iii) R^{-1}(A)

4. (a) The transformation R is a rotation of 180° about the origin. Write down the matrix representing R.
   (b) Calculate the coordinates of A(6, 2) and B(−2, −4) under the transformation R.
   (c) The transformation E is an enlargement of scale factor $\frac{1}{2}$, centre the origin. Write the matrix representing E.
   (d) Determine the matrix C which represents the rotation R followed by the enlargement E.
   (e) Find the inverse of the matrix C.
   (f) Hence calculate the coordinates of the point with an image (−5, 2) after undergoing the transformation C.

5. M = $\begin{pmatrix}2 & 3\\1 & 0\end{pmatrix}$ is a matrix of transformation. Find the images of the points P(3, 2), Q(2, −3) and R(4, 1) under the transformation M.

Chapter 15

Graphs (2)

Exponential, inverse and cubic functions

Pre-requisites

- linear and non-linear functions

Exponential functions

In Book 2 Chapter 21, it was seen that, in practice, when banks pay interest on savings or charge interest on loans, the calculated interest is called compound interest. By this method the amount calculated for a given fixed period of time is used as the principal for the next period; for example, the fixed period may be 1 month, 6 months, 1 year, and so on.

The amount A (i.e. principal P plus interest I) at the end of each period is given by

$$A = P + \frac{PRT}{100}, \text{ where } I = \frac{PRT}{100}$$

Write the principal and the amount for the first period as P_1 and A_1: for the second period as P_2 and A_2; and so on.

For each period, $T = 1$ and P is equal to the value of A for the previous period, so that

$$P_1 = P, P_2 = A_1, P_3 = A_2 \text{ and so on.}$$

Hence, $A_1 = P_1 + \dfrac{P_1R}{100}$

and $A_1 = P_2 = P_1\left(1 + \dfrac{R}{100}\right)$ (1)

Then $A_2 = P_2 + \dfrac{P_2R}{100}$

From (1)

$$\begin{aligned} A_2 &= P_1\left(1 + \frac{R}{100}\right) + P_1\left(1 + \frac{R}{100}\right)\frac{R}{100} \\ &= \left[P_1\left(1 + \frac{R}{100}\right)\right]\left(1 + \frac{R}{100}\right) \\ &= P_1\left(1 + \frac{R}{100}\right)^2 \end{aligned}$$

Similarly, it can be shown that

$$A_3 = P_1\left(1 + \frac{R}{100}\right)^3$$

and for x periods.

$$A_x = P_1\left(1 + \frac{R}{100}\right)^x$$

Thus, in general,

$$A_x = P\left[1 + \frac{R}{100}\right]^x \qquad (2)$$

Hence, for a given initial value of a loan or savings, that is, for a fixed value of P at a fixed rate of interest R, the value of the amount A_x depends on x, the number of periods.

Let A_x be a variable y, P be a constant k and $\left(1 + \dfrac{R}{100}\right)$ be a constant a.

Equation (2) becomes $y = ka^x$.

This is an example of **exponential growth**. A relation of this form is called an **exponential function** since the variable x is the exponent or index. The increase of world population and the growth of bacteria are shown approximately by this type of exponential function.

Fig. 15.1 shows the curves obtained by plotting corresponding values of x and y for different values of a when $k = 1$.

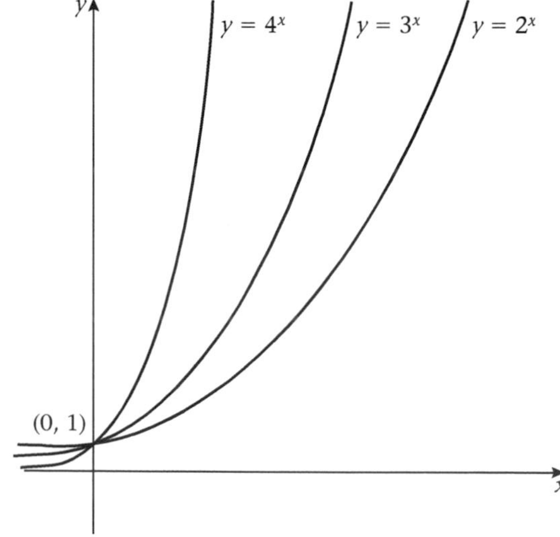

Fig. 15.1

In the equation $y = ka^x$, a is termed the growth factor, that is the factor by which the variable y changes for equal intervals in x, and k is the initial value of x when $x = 0$.

Note that $y > 0$ for all values of x, when $x \in R$; since for $x < 0$, $y = \frac{k}{a^z}$ where $z > 0$ and $|x| = z$.

Example 1

Draw the graph of the function $y = 2^x$ for values of x from -2 to 4.
Table 15.1 gives the corresponding values of x and y.

Table 15.1

x	-2	-1	0	1	2	3	4
y	$\frac{1}{4}$	$\frac{1}{2}$	1	2	4	8	16

Fig. 15.2 is the required graph of $y = 2^x$.

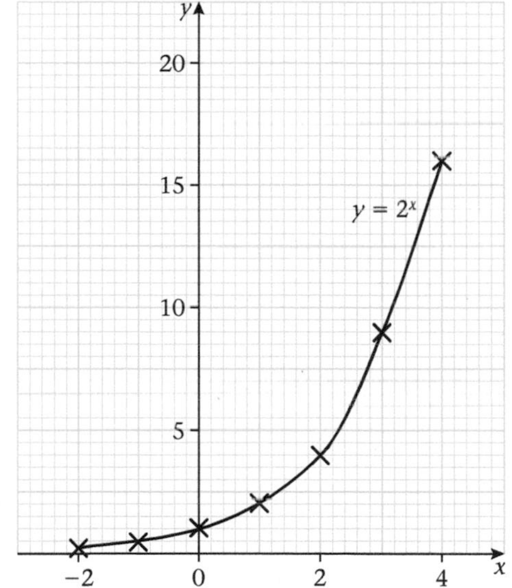

Fig. 15.2

Example 2

The amount \$$A$ earned on an investment of \$$P$ for x years is calculated using the relation $A = P\left(1 + \frac{R}{100}\right)^x$.

(a) Calculate the value of A at the end of each year for an investment of \$20 000 over a 5-year period at a rate of 12% per annum.

(b) Using a scale of 10 mm to represent 1 year on the horizontal axis and 1 cm to represent \$1000 on the vertical axis, draw a graph of these values to show the rate of growth of A.

(c) From the graph, estimate (i) the value of the investment (correct to 3 significant figures) at the end of $2\frac{1}{2}$ years, (ii) the time (to the nearest half-year) for which the money must be invested in order to earn \$10 000 interest.

(a) *Table 15.2*

x (years)	0	1	2	3	4	5
A (\$000)	20.0	22.4	25.1	28.1	31.5	35.3

(b) Fig. 15.3 is the required graph.

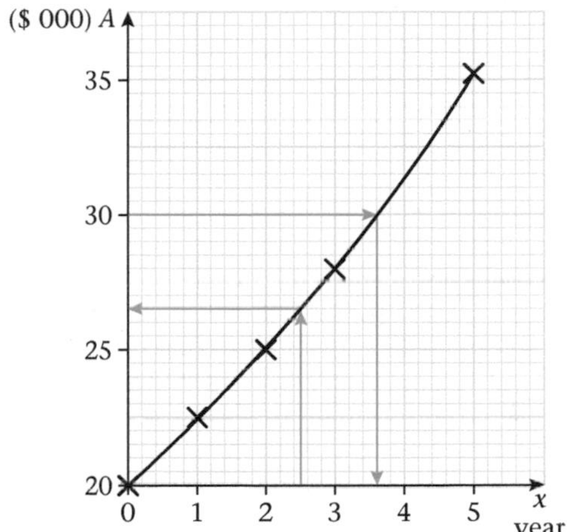

Fig. 15.3

(c) (i) \$26 000 (ii) $3\frac{1}{2}$ years

Exercise 15a

1. Draw the graph of the function $y = 3^x$ for values of x in the domain $-3 < x < 3$.

2. Table 15.3 shows the amounts due on a loan of \$1000 at the end of four one-month intervals.

Table 15.3

Time (month)	0	1	2	3	4
Amount (\$)	1000	1020	1040.40	1061.21	1082.43

Draw a graph to show this relation.

3 The population of a town increases at the rate of 3.5% every 5 years. In 1985 the population was 8000.

(a) Calculate the population (correct to 3 significant figures) at 5-year intervals for the years 1985–2005.

(b) Using a scale of 2 cm to represent 5 years on the horizontal axis and 1 cm to represent 100 people on the vertical axis, draw a graph to show the relation between the population and the year.

(c) From your graph, estimate the population, correct to 3 s.f., in the year 2010.

4 Table 15.4 shows the returns on two investments of \$1500 at different rates of interest.

Table 15.4

Year	0	1	2	3	4	5
Amount 1	1500	1590	1685	1787	1893	2007
Amount 2	1500	1680	1882	2107	2360	2644

(a) Calculate the growth factor, correct to 2 d.p. for each investment.

(b) Write an equation for the return on each investment in terms of the years for which the money is invested.

(c) Using a scale of 2 cm to represent 1 year on the horizontal axis and 2 cm to represent \$100 on the vertical axis, draw, on the same axes, graphs to show the relation between the amount and the number of years invested at each rate.

(d) From your curves, estimate, to the nearest half-year, the period before amount 1 will equal the value amount 2 has at the end of the fifth year.

Inverse functions

An **inverse** function of x is an expression in which x appears in the denominator of a fraction. For example, $\frac{6}{x}, \frac{2x^2}{1-3x}$ are inverse relations of x.

Note that in chapter 10, the inverse of a function defines $f^{-1}(x)$ as the inverse operation of $f(x)$.

Example 3

Draw the graph of the function $y = \frac{2}{x}$ for values of x between $x = -4$ and $x = +4$.

Table 15.5

x	-4	-3	-2	-1	0	1	2	3	4	$\pm 1\frac{1}{2}$
y	$-\frac{1}{2}$	$-\frac{2}{3}$	-1	-2		2	1	$\frac{2}{3}$	$\frac{1}{2}$	$\pm 1\frac{1}{3}$

Fig. 15.4 is the required graph.

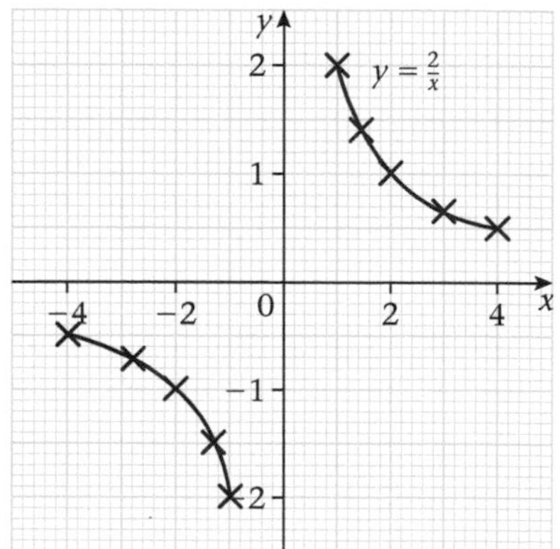

Fig. 15.4

Notice that there is a break in continuity of the graph in Fig. 15.4. It is in two branches which are separated by the axes. As x increases towards 0, y decreases in value; as x decreases towards 0, y increases in value. This kind of curve is called a **hyperbola**.

Example 4

Draw the graph of the function $y = \frac{1}{x-1}$ in the domain $-2 < x < 3$.

Table 15.6 and Fig. 15.5 give a table of values and the corresponding graph of the function $y = \frac{1}{x-1}$.

Table 15.6

x	-1	0	$\frac{1}{2}$	$\frac{3}{4}$	$\frac{9}{10}$	$1\frac{1}{10}$	$1\frac{1}{4}$	$1\frac{1}{2}$	2
$x-1$	-2	-1	$-\frac{1}{2}$	$-\frac{1}{4}$	$-\frac{1}{10}$	$\frac{1}{10}$	$\frac{1}{4}$	$\frac{1}{2}$	1
y	$-\frac{1}{2}$	-1	-2	-4	-10	10	4	2	1

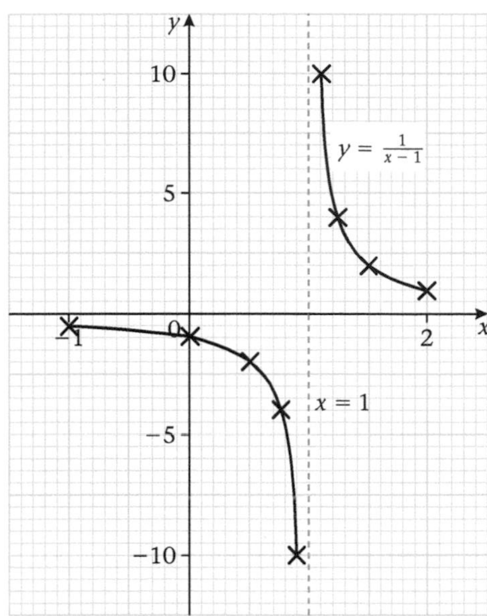

Fig. 15.5

Notice the following:

1 As the value of x approaches 1 from below, that is from values of x less than 1, the value of $\frac{1}{x-1}$ *decreases* rapidly. For example, when $x = 0.999$,
$\frac{1}{x-1} = \frac{1}{0.999-1} = \frac{1}{-0.001} = -1000$

2 As the value of x approaches 1 from above, that is from values of x greater than 1, the value of $\frac{1}{x-1}$ *increases* rapidly. For example, when $x = 1.001$,
$\frac{1}{x-1} = \frac{1}{1.001-1} = \frac{1}{0.001} = 1000$
When $x = 1$, it is impossible to say what the value of $\frac{1}{x-1}$ is; y is *undefined* when $x = 1$.

Example 5

Draw the graph of the function $y = \frac{3}{x^2}$ for values of x in the domain $-4 \leq x \leq 4$.

Table 15.7

x	−4	−3	−2	−1	0	1	2	3	4	$\pm\frac{1}{2}$
x^2	16	9	4	1	0	1	4	9	16	$\frac{1}{4}$
y	$\frac{3}{16}$	$\frac{1}{3}$	$\frac{3}{4}$	3		3	$\frac{3}{4}$	$\frac{1}{3}$	$\frac{3}{16}$	12

Fig. 15.6 is the required graph.

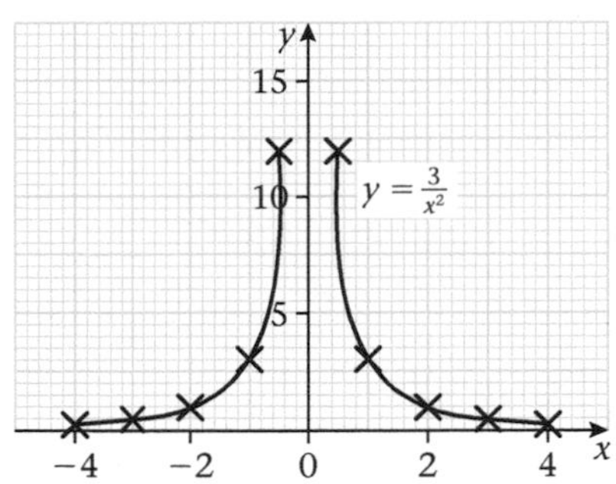

Fig. 15.6

Notice the following:

1 The graph is symmetrical about the y-axis.

2 There are no negative values of y.

3 The graph of the function $y = \frac{-3}{x^2}$ may be obtained by reflection of the graph in Fig. 15.6 in the x-axis.

4 It is possible to use a graph to determine, for a function, values that are not stated explicitly in the given domain and range, only if the plotted points lie on a continuous graph (straight line or smooth curve). This process is called *interpolation* when the point lies within the given domain, and *extrapolation* when the graph has to be extended beyond the limiting values, that is, the least or greatest values of the domain.

Example 6

(a) Draw the graph of $y = \frac{6}{x}$ for values of x from −4 to +2.

(b) Using the same scale and on the same axes, draw the graph of the straight line $y = 2x + 3$.

(c) Find the values of x at the point where the two graphs intersect.

(d) Of what equation in x are these values the roots?

(a) In Table 15.8 the values of y for integral values of x are first calculated. The extra values of y for $x = \pm 1\frac{1}{2}$ are then added.

Table 15.8

x	−4	−3	−2	−1	0	1	2	$\pm1\frac{1}{2}$
$y = \frac{6}{x}$	$-1\frac{1}{2}$	−2	−3	−6		6	3	±4

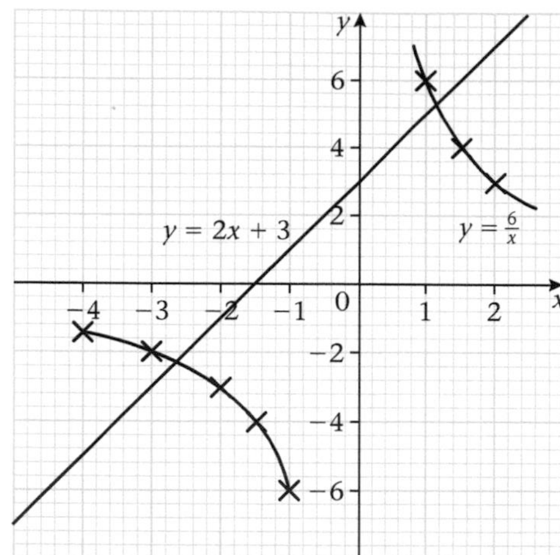

Fig. 15.7

Notice that when $x = 0$, $\frac{6}{x}$ is undefined.
Fig. 15.7 shows the graph of $y = \frac{6}{x}$.

(b) Draw the line $y = 2x + 3$ by plotting the points in Table 15.9.

Table 15.9

x	-4	0	2
y	-5	3	7

(c) The line cuts the curve where $x \simeq -2.6$ and $x \simeq 1.1$.

(d) The curve and the line intersect where y simultaneously equals $\frac{6}{x}$ and $2x + 3$.
Hence the values of x above are the roots of the equation:

$\frac{6}{x} = 2x \pm 3$, i.e. $2x^2 \pm 3x - 6 = 0$.

Exercise 15b

1. Draw the graphs of
   (a) $y = \frac{1}{x}$,
   (b) $y = \frac{1}{x^2}$,
   for values of x in the domain $-4 \leqslant x \leqslant 4$.

2. Draw the graphs of the functions
   (a) $y = \frac{6}{x^2}$,
   (b) $y = \frac{-6}{x^2}$
   for values of x in the domain $-4 \leqslant x \leqslant 4$.

3. Using the same scale and on the same axes draw the graphs of the functions $y = \frac{2}{x}$ and $2y = 3x - 1$ for values of x from -4 to 4. Read off the value(s) of x at the point(s) of intersection.
   Derive the equation in x for which these values are the roots.

4. Repeat question 3 for each of the following pairs of functions.
   (a) $y = \frac{2}{x + 1}$ and $y = -2x + 3$ for x from -2 to 3
   (b) $3y = x + 2$ and $y = \frac{2}{3x + 1}$ for x from -4 to 5
   (c) $y = \frac{5}{x}$ and $y = x(x - 1)$ from $x = -2$ to $x = 5$

Cubic functions

A **cubic function** of x is an expression in x in which 3 is the highest power of x. For example, $2x^3 + 5x^2 - x - 8$ is a cubic function of x.

Example 7

(a) Draw the graph of $y = x^3$ for values of x from -3 to $+3$.
(b) Hence solve the equation $x^3 + 20 = 0$.

(a) Table 15.10 gives the necessary table of values. Notice that additional values of y for $x = \pm\frac{1}{2}$ have been calculated. These will be helpful when drawing the graph.

Table 15.10

x	-3	-2	-1	0	1	2	3	$\pm\frac{1}{2}$
y	-27	-8	-1	0	1	8	27	$\pm\frac{1}{8}$

Fig. 15.8 shows the graph of $y = x^3$.

(b) If $x^3 + 20 = 0$, then $x^3 = -20$. $x^3 = -20$ when, in Fig. 15.8, $y = -20$. From Fig. 15.8, $y = -20$ at $x \simeq -2.7$; $x = -2.7$ is the approximate solution of $x^3 + 20 = 0$.

Remember that readings from graphs usually give approximate results only. A bigger scale in Fig. 15.8 would give more accurate results.

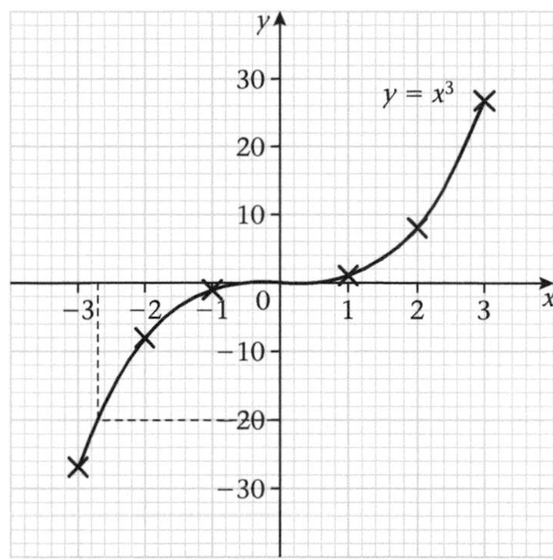

Fig. 15.8

Exercise 15c

1 (a) Draw the graph of $y = x^3$ for values of x from -4 to $+4$.
(b) On the same axes, draw the graph of $y = -x^3$.
(c) Use either graph to solve the equation $x^3 + 50 = 0$.

2 Solve the equation $x^3 = 5x + 2$ by drawing graphs of $y = x^3$ and $y = 5x + 2$ for values of x between -3 and $+3$.

3 Solve graphically the equation $x^3 = 5x - 3$.

Hint: Use the method of question 2, drawing the graphs of $y = x^3$ and $y = 5x - 3$, for values of x in the domain -3 to $+3$.

4 Solve the equation $x^3 + 3x - 7 = 0$ graphically. Use values of x in the domain 0 to 4.

5 Draw, on the same axes, the graphs of $y = x^3$ and $y = x(4x - 3)$ for values of x between 0 and 4. Use a scale of 2 cm to 1 unit on the x-axis and 2 cm to 10 units on the y-axis. From your graph,
(a) find the values of x satisfying the equation $x^3 - 4x^2 + 3x = 0$;
(b) find the range of values of x for which $x^3 < x(4x - 3)$;
(c) find the gradient of the curve (i) $y = x^3$ at the point $x = 1$, (ii) $y = x(4x - 3)$ at $x = 2$.

Sketch graphs

A sketch graph is a simple freehand drawing which shows the main features of a line or curve.

Linear functions

Example 8

Sketch the graph of $2x - 3y = 24$.

Method: Find the intercepts on the axes, i.e. the positions where $x = 0$ and $y = 0$.

$$2x - 3y = 24$$
When $x = 0$, $-3y = 24$
$$y = -8$$
When $y = 0$, $2x = 24$
$$x = 12$$

The sketch graph can now be drawn (Fig. 15.9).

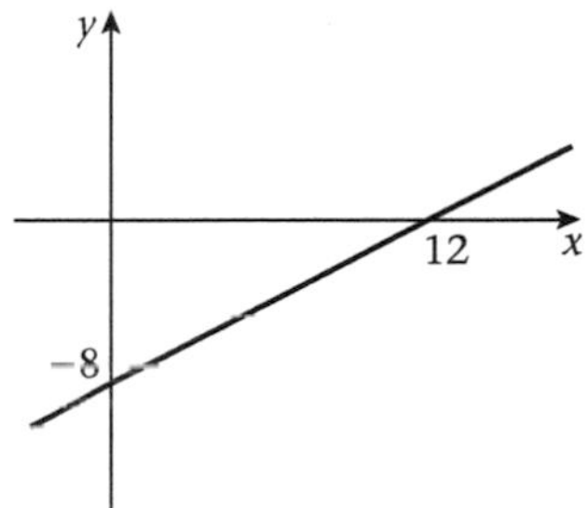

Fig. 15.9

Always label the axes and origin. If possible, show where the line crosses the axes.

Example 9

Find the equation of the line represented by the sketch graph in Fig. 15.10

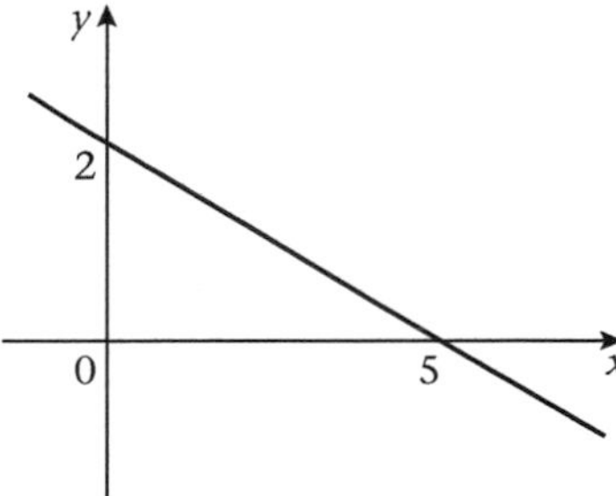

Fig. 15.10

Since the graph is a straight line, its equation is of the form $y = mx + c$, where m is the gradient of the line and c is the intercept on the y-axis.

From Fig. 15.10, $m = -\frac{2}{5}$*
$$c = +2$$

The equation is $y = -\frac{2}{5}x + 2$
i.e. $5y = -2x + 10$
or $2x + 5y = 10$
*As x increases by 5 units, y decreases by 2 units.

Quadratic functions

A quadratic function has an equation of the form $y = ax^2 + bx + c$, where a, b and c are positive or negative constants. When a, the coefficient of x^2, is positive, the graph is a cup-shaped parabola. When a is negative, the graph is a cap-shaped parabola (Fig. 15.11).

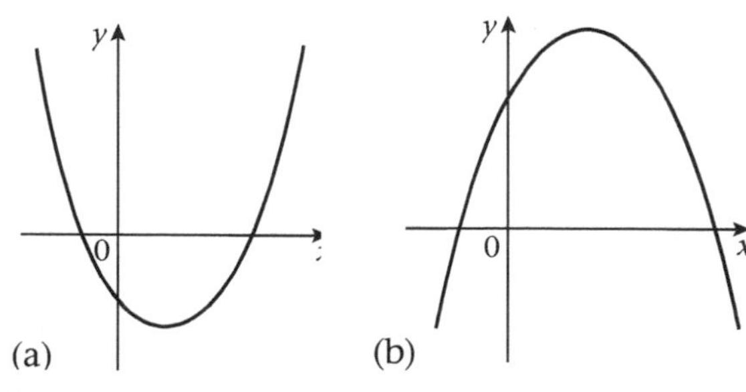

Fig. 15.11

Example 10

Sketch the graph of $y = x^2 - x - 12$, showing where the curve cuts the axes.

$y = x^2 - x - 12$

1 The curve cuts the y-axis when $x = 0$.
When $x = 0$, $y = -12$.

2 The curve cuts the x-axis when $y = 0$.
When $y = 0$, $x^2 - x - 12 = 0$
$(x - 4)(x + 3) = 0$
$x = 4$ or -3

3 The coefficient of x^2 is positive, hence the curve is a cup-shaped parabola.

The sketch in Fig. 15.12 is drawn using the data in 1, 2 and 3 above.

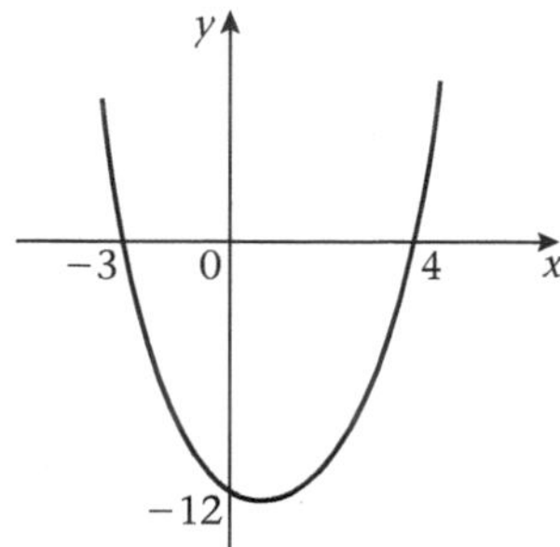

Fig. 15.12

Example 11

Find the equation of the curve represented by the sketch in Fig. 15.13

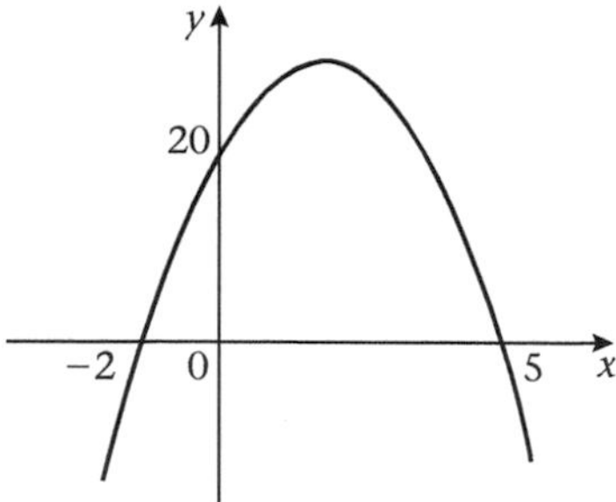

Fig. 15.13

Let the curve have the equation
$y = ax^2 + bx + c$.

1 c is the intercept on the y-axis.
Hence, from Fig. 15.13, $c = +20$.

2 When $y = 0$, $ax^2 + bx + c = 0$. Hence, from the intercepts on the x-axis in Fig. 15.13, the roots of the equation are -2 and $+5$.
$(x + 2)(x - 5) = 0$
$x^2 - 3x - 10 = 0$ (1)
Since $c = +20$, multiply each term in (1) by -2.
$-2x^2 + 6x + 20 = 0$

Hence $y = -2x^2 + 6x + 20$ is the required equation.

Notice, in Example 11, that the coefficient of x^2 is negative and that Fig. 15.13 shows a cap-shaped parabola.

Inverse functions

Fig. 15.14 shows the graphs of

(a) $y = \frac{1}{x + 2}$ and (b) $y = \frac{1}{x - 3}$.

In each graph, the continuity of the curve is broken at the line $x = k$, where k is the value of x for which the fraction is undefined. The intercept on the y-axis is found by substituting $x = 0$ in the given equation.

Neither curve cuts the x-axis.

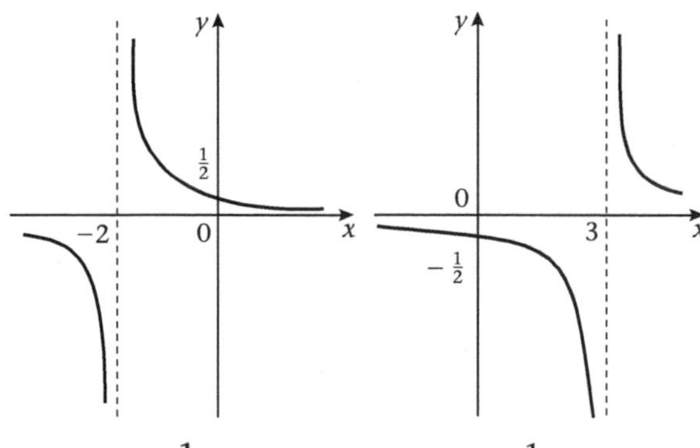

(a) $y = \dfrac{1}{x + 2}$ (b) $y = \dfrac{1}{x - 3}$

Fig. 15.14

Exercise 15d

Draw freehand graphs throughout this exercise. Whenever possible, show where the graph cuts the axes.

1. Sketch the graphs of the following.
   (a) $3x - 2y = 12$
   (b) $6x + 3y = 18$
   (c) $y - 2x = 7$
   (d) $y = 3x - 1$

2. Find the equations of the lines shown by the sketch graphs in Fig. 15.15.

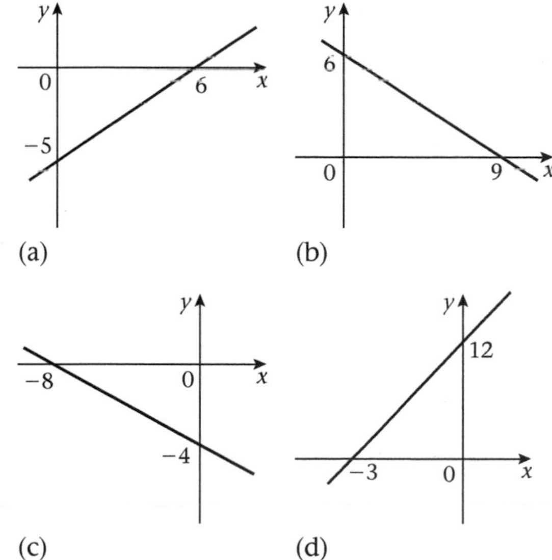

Fig. 15.15

3. What is the value of y at the point where the curve $y = x^2 + 3x - 11$ cuts the y-axis?

4. Sketch the graphs of the following, showing where the curve cuts the axes.
   (a) $y = x^2 - 5x + 4$
   (b) $y = 15 - 2x - x^2$
   (c) $y = x^2 - 12x + 36$
   (d) $y = 16 - x^2$
   (e) $y = 3x^2 + 3x - 6$
   (f) $y = 10 - 8x - 2x^2$

5. Fig. 15.16 is a sketch graph of $y = 6 + x - x^2$.
   (a) What is the value of x at M?
   (b) Find the tangent of the angle OMN.

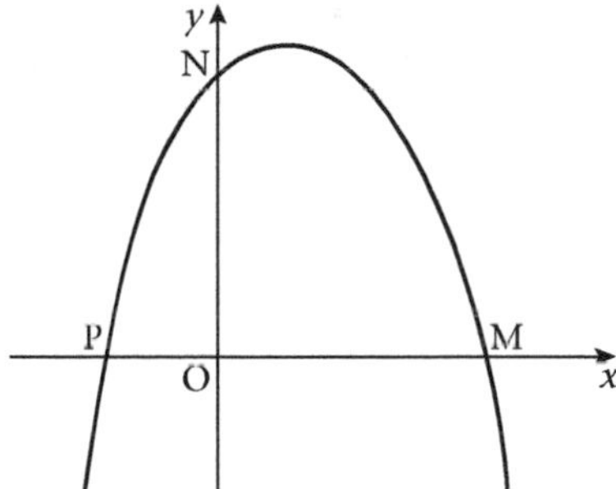

Fig. 15.16

6. What equation is represented by the sketch graph in Fig. 15.17?

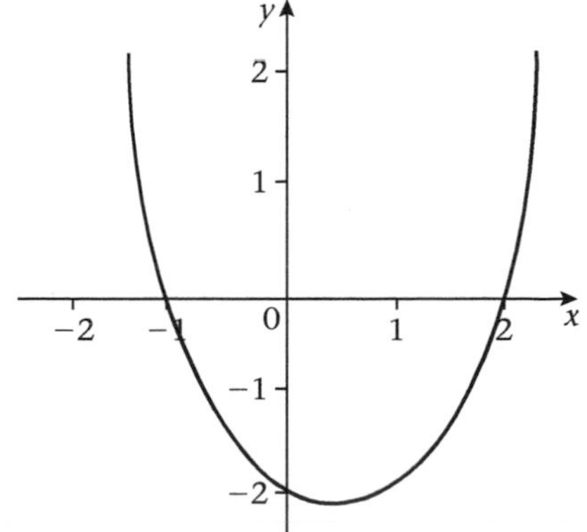

Fig. 15.17

Summary

Sketch graphs of various types of functions are illustrated below.

(a) **Linear functions**
$y = mx + c$

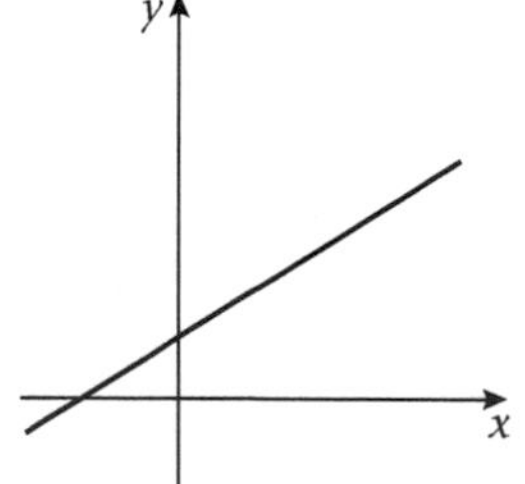

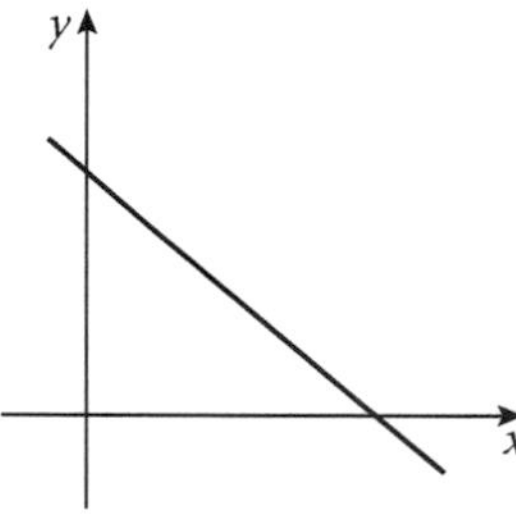

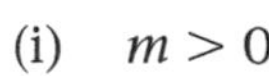
(i) $m > 0$

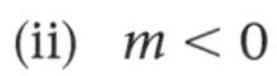
(ii) $m < 0$

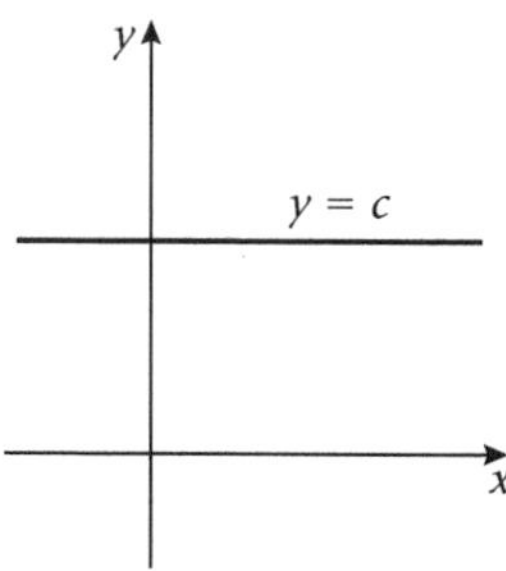

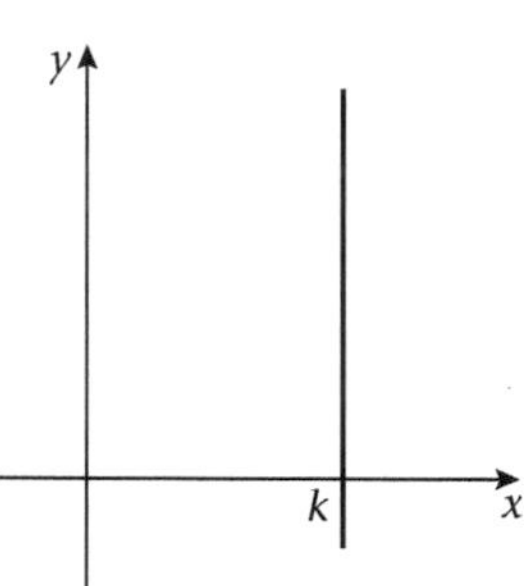

(iii) $m = 0$

(iv) $x = k$
m is undefined

Fig. 15.18

See page 192 for inverse functions.

(b) **Quadratic functions**
$y = ax^2 + bx + c$

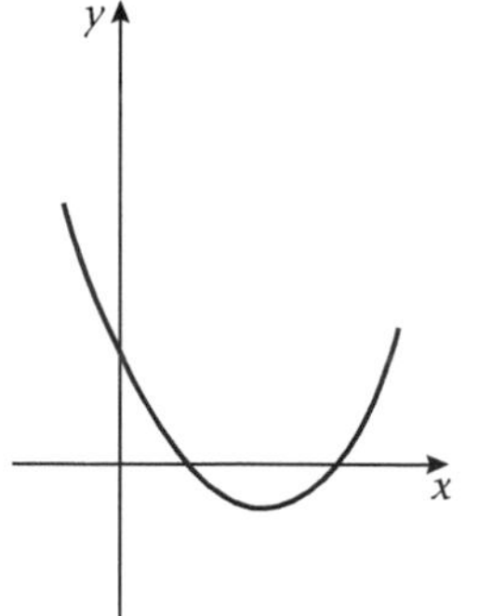

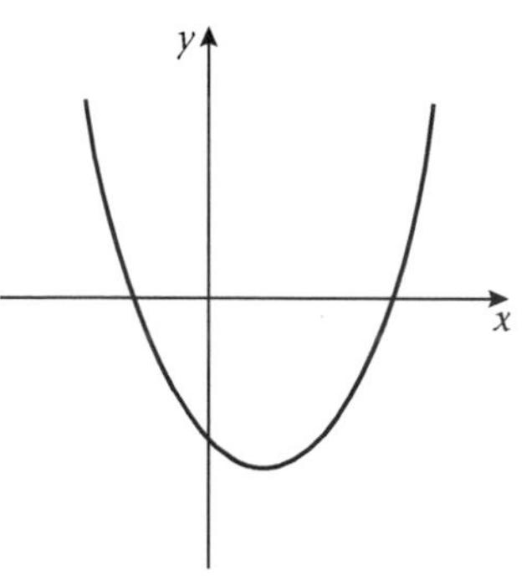

(i) $a > 0, c > 0$

(ii) $a > 0, c < 0$

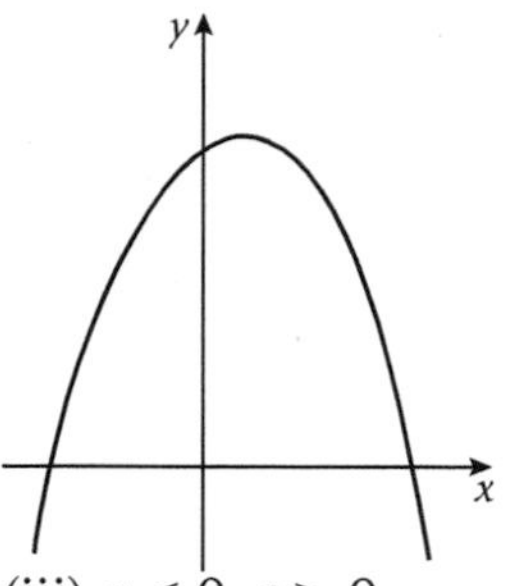

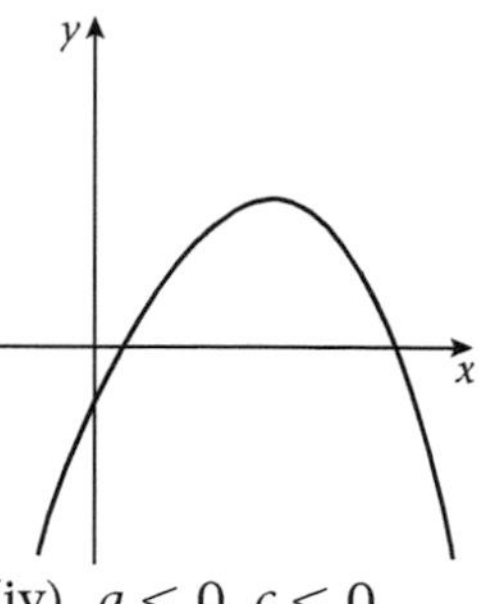

(iii) $a < 0, c > 0$

(iv) $a < 0, c < 0$

Fig. 15.19

(c) **Exponential functions**
$y = a^x$

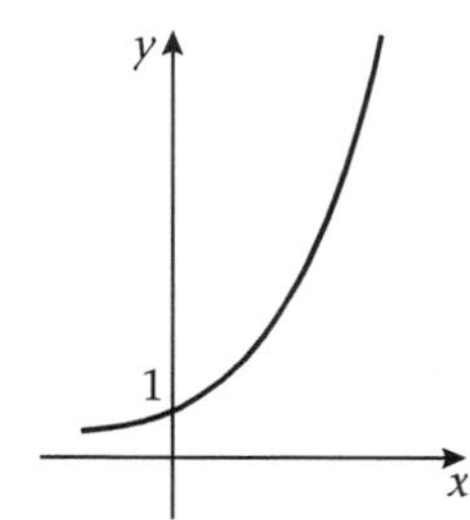

Fig. 15.20

(d) **Cubic functions**
$y = ax^3$

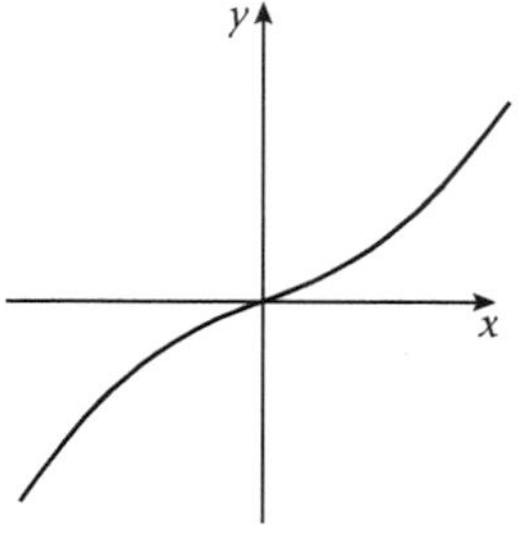

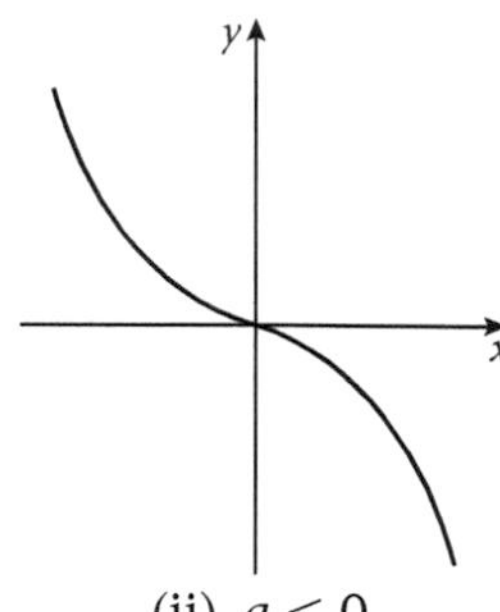

(i) $a > 0$

(ii) $a < 0$

Fig. 15.21

(e) **Inverse functions**
$y = \frac{a}{x^n}$

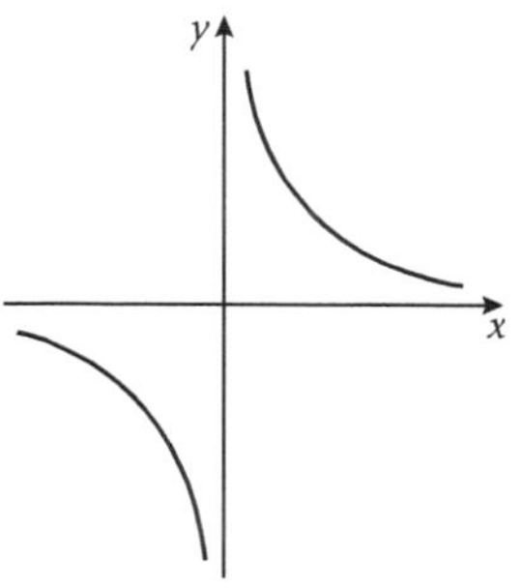

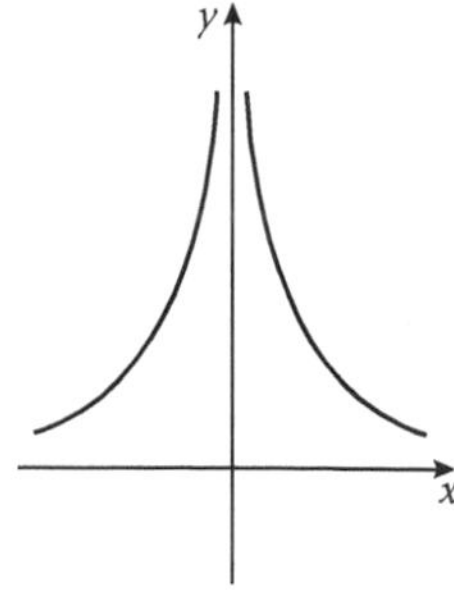

(i) $n = 1$

(ii) $n = 2$

Fig. 15.22

Practice exercise P15.1

For the graphs in Fig. 15.23 and Fig. 15.24,

(a) write the coordinates of points marked on the graph, that is, ordered pairs of the function.
(b) find the relation between the x-coordinate and the corresponding y-coordinate.
(c) write the relation in the form $y = f(x)$.

1

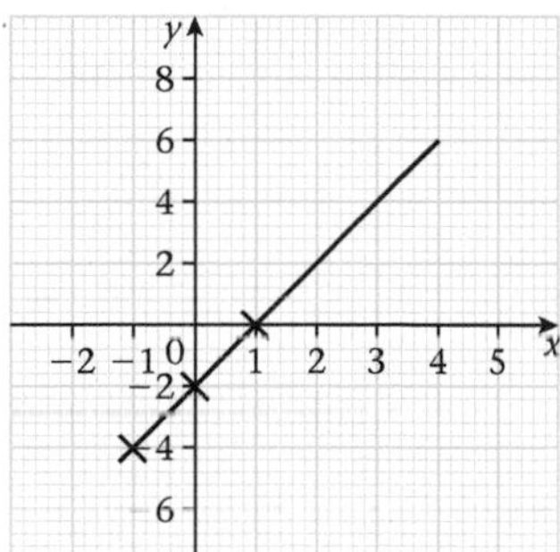

Fig. 15.23

2

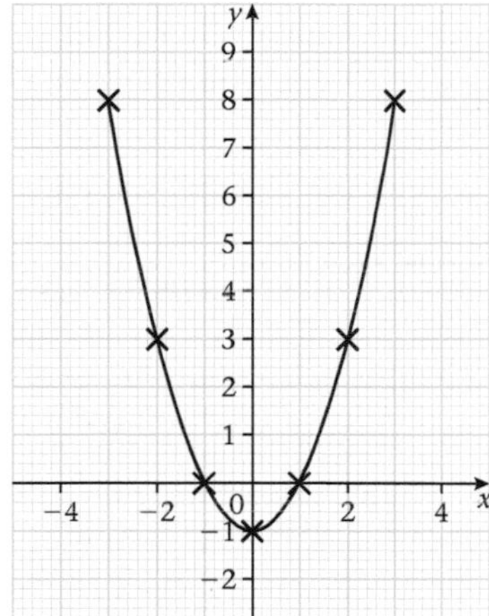

Fig. 15.24

Practice exercise P15.2

For each of the following functions
(a) draw up a table of values,
(b) plot the points and draw the graphs.

1. $f: x \rightarrow 2x - 3, x \in \{-2, -1, 0, 1, 2, 3\}$
2. $f(x) = x^2 + 1$ where $-3 < x < 3$
3. $y = x^2 - 2x + 4$, where $x = -3, -2, -1, 0, 1, 2, 3$

Practice exercise P15.3

1. (a) Given the function $y = 4^x$, draw up a table of corresponding values of x and y for $-3 \leqslant x \leqslant 4$.
   (b) Using a scale of 2 cm to represent 1 unit on the x-axis and 1 cm to represent 5 units on the y-axis, plot the points and draw the curve.
   (c) By drawing a suitable tangent, find the gradient of the curve at
   (i) $x = 0$, (ii) $x = 2$, (iii) $x = 3$.
2. The amount, to the nearest dollar, due on a loan of \$1000 at the end of four one-month periods is given in Table 15.11.

Table 15.11

x	0	1	2	3	4
A	1000	1020	1040	1061	1082

(a) Using a scale of 2 cm to represent 1 month on the x-axis and 1 cm to represent \$10 on the A-axis and taking values of A from \$1000 to \$1120, draw a graph to show this relation.
(b) Estimate the amount due at the end of 5 months.

Practice exercise P15.4

1. Draw up a table of values for the function $y = -\frac{2}{x^2}$ for values of x in the domain $-4 \leqslant x \leqslant 4$.
2. Draw the graph of $y = -\frac{2}{x^2}$ for values of x in the given domain.

Practice exercise P15.5

For each of the following functions,
(a) draw up a table of values for integral values of x,
(b) use suitable scales to plot the ordered pairs and draw the graph,
(c) draw the tangent to the curve at each of the following points,
(i) $x = -2$, (ii) $x = 0$, (iii) $x = 2$, (iv) $x = 3$,
(d) find the gradient of the curve at the given points.

1. $y = 2x^3$ where $-3 \leqslant x \leqslant 3$
2. $y = \frac{1}{2}x^3$ where $-6 \leqslant x \leqslant 6$

Practice exercise P15.6

Sketch the graphs of the following functions.

1. $y = \frac{1}{x}$
2. $y = -\frac{6}{x}$
3. $y = -\frac{4}{x^2}$
4. $y = -\frac{1}{x^2}$
5. $y = \frac{4}{x}$
6. $y = -\frac{4}{x}$
7. $y = -\frac{3}{x^2}$
8. $y = \frac{2}{x^2}$

Chapter 16

Inequalities (3)

Linear inequalities in two variables, linear programming

Pre-requisites

- linear equations; algebraic inequalities; Cartesian plane

Inequalities in two variables

(x, y) represents any point on the Cartesian plane which has coordinates x and y. In Fig. 16.1 the unshaded region represents the set of points given by $\{(x, y): x \geqslant 1 \text{ and } y < 2\}$.

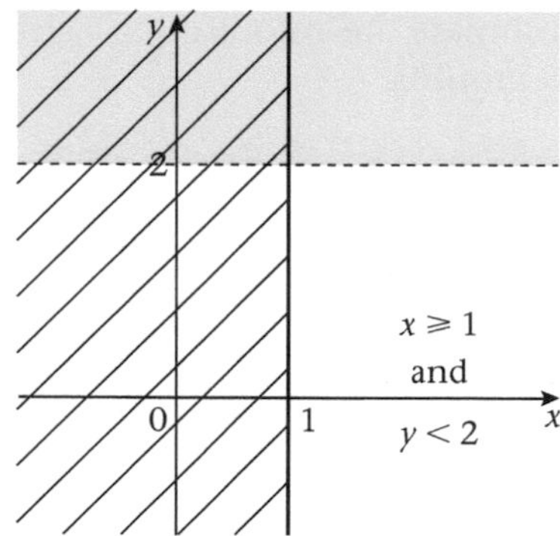

Fig. 16.1

In Fig. 16.1, note that

1. $x \geqslant 1$ is the set of all points to the right of the boundary line $x = 1$. The line $x = 1$ is continuous to show that the points on the line are included. The region to the left of the line is shaded to show that it is *not* required.
2. $y < 2$ is the set of all points below the boundary line $y = 2$. The line $y = 2$ is broken to show that the points on the line are *not* included. The region above the line is shaded to show that it is *not* required.

Note that generally the required region is not shaded so that the values of the variables may be easily read from the graph.

Example 1

Show on a graph the region which contains the set of points $\{(x, y): 2x + y < 3\}$.

Method 1

The line $2x + y = 3$ is the boundary between the required region and the set of points which are not required.

Given $2x + y = 3$
when $x = 0, y = 3$
when $y = 0, x = 1\frac{1}{2}$

Since the points on the line $2x + y = 3$ are not included, a broken line is drawn through the points $(0, 3)$ and $(1\frac{1}{2}, 0)$. (See Fig. 16.2.)

In order to determine the region which satisfies an inequality, we consider whether the origin of coordinates, that is, the point $(0, 0)$ satisfies the inequality.

If the coordinates of the origin satisfy the inequality, then the origin lies in the same region as the points which satisfy the inequality, that is, the origin is on the same side of the boundary line as the required points.

If the coordinates of the origin do not satisfy the inequality, then the required region is on the other side of the boundary line.

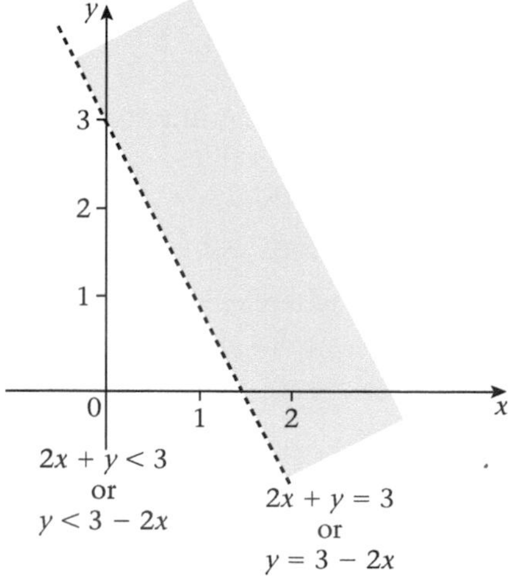

Fig. 16.2

Substituting $x = 0$ and $y = 0$ in the inequality, $2x + y < 3$, we get

$$2(0) + 0 < 3$$

that is, $0 < 3$

Since this is true, (0, 0) lies in the same region which contains the set of required points.

The region above the line is shaded to show that it is not the required region. (See Fig. 16.2.)

Method 2
First make y the subject of the given inequality:

$$y < 3 - 2x$$

Then the equation of the boundary line may be written as $y = 3 - 2x$
Therefore, in order to satisfy the inequality,

$$y < 3 - 2x$$

the points *below* the line $y = 3 - 2x$ are required.

In Fig. 16.2, the region above the line is shaded to show that it is not required.

Either method may be used to determine the required region; in some examples, it may be simpler to apply *Method 1*, in other examples, *Method 2* may work out more easily.

Note the following:

1 The boundary lines cross the axes at the points where $x = 0$ and $y = 0$. It is usually most convenient to draw boundary lines through such points since their coordinates are easily calculated.
2 A boundary line may be continuous or broken, depending on whether the inequality is included or not.
3 The region which is *not* required is shaded. However, note that this is the convention used in this book and care must be taken to ensure that in the problem you are asked to solve, it is given that the required region is not shaded.

Exercise 16a

1 Show on a graph the region which contains the set of points
(a) $\{(x, y): x \geqslant 3\}$
(b) $\{(x, y): x < -2\}$
(c) $\{(x, y): y < 4\}$
(d) $\{(x, y): 2x \leqslant 1\}$
(e) $\{(x, y): y \geqslant -2\}$
(f) $\{(x, y): 2y > 3\}$

2 Draw graphs to illustrate the set of points
(a) $\{(x, y): -3 \leqslant x < 2\}$
(b) $\{(x, y): 2 \geqslant y \geqslant -1\}$
(c) $\{(x, y): 1 > x \geqslant -3\}$
(d) $\{(x, y): 1 < y < 4\}$

3 Draw graphs to show the solution set of the following.
(a) $\{(x, y): x \geqslant 0 \text{ and } y > -1\}$
(b) $\{(x, y): x < 3 \text{ and } y \leqslant 2\}$
(c) $\{(x, y): y \geqslant 0 \text{ and } x > -2\}$
(d) $\{(x, y): 2y \leqslant 1 \text{ and } x \leqslant 4\}$

4 In each of the following diagrams
(i) find the equation of the boundary line,
(ii) if the unshaded area is the required region, state the solution sets of the inequalities.

(a)

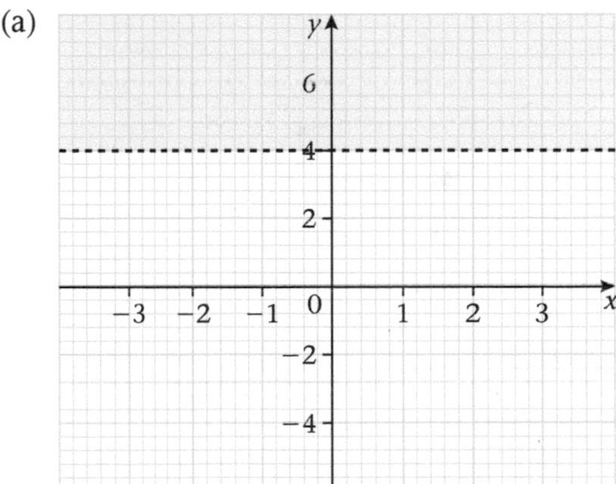

(b)

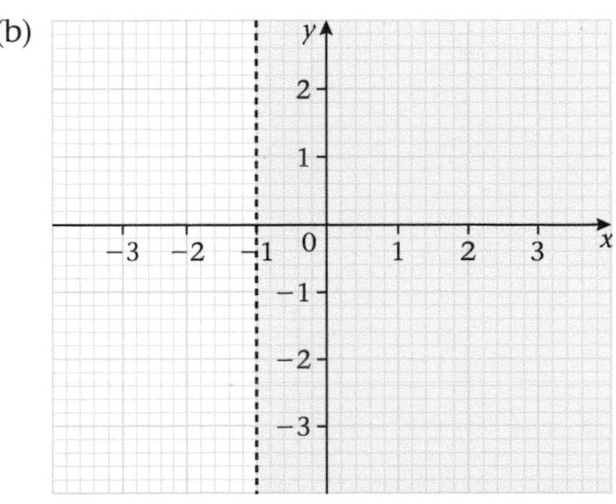

(c)

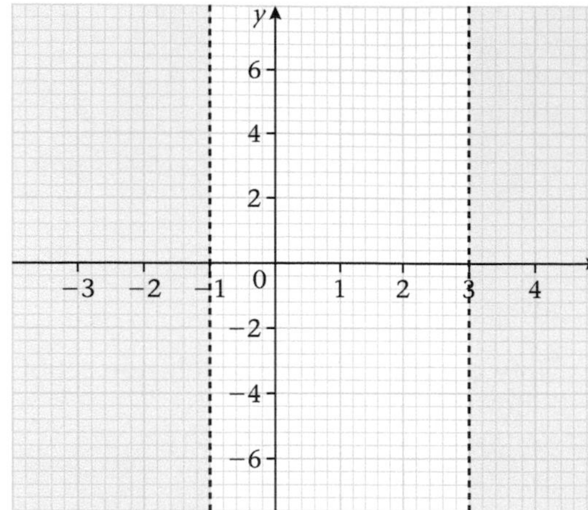

(d)

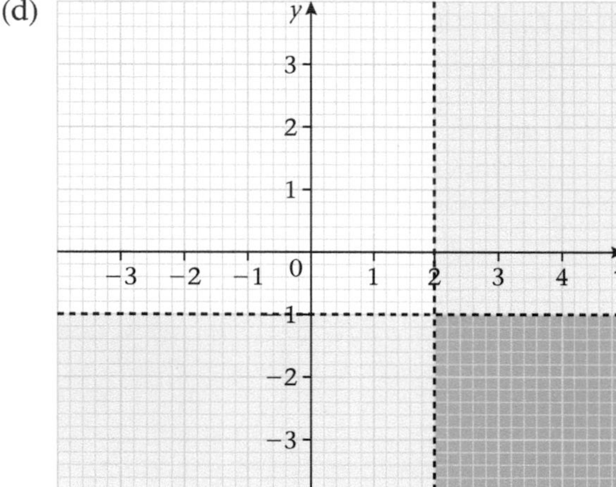

(e)

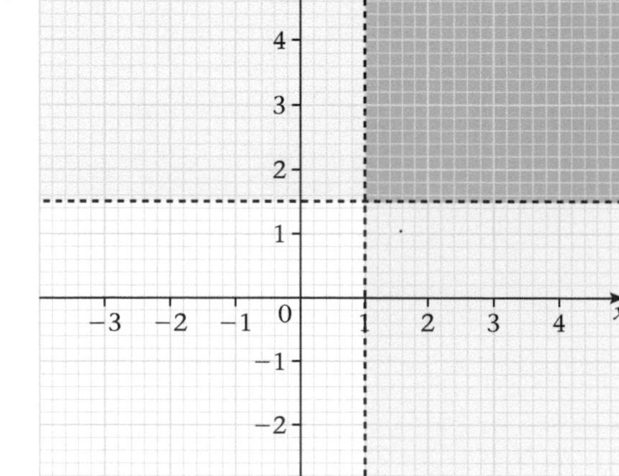

Fig. 16.3

5. Draw graphs to illustrate the following sets of points.
   (a) $\{(x, y): y > x - 3\}$
   (b) $\{(x, y): y > 2x - 1\}$
   (c) $\{(x, y): 2y - x < 2\}$
   (d) $\{(x, y): 3x - y > 6\}$
   (e) $\{(x, y): x + 2y > 2\}$
   (f) $\{(x, y): y + 2x < -2\}$

6. In each of the following diagrams
   (i) find the equation of the boundary line,
   (ii) if the unshaded area is the required region, state the solution sets of the inequalities.

(a)

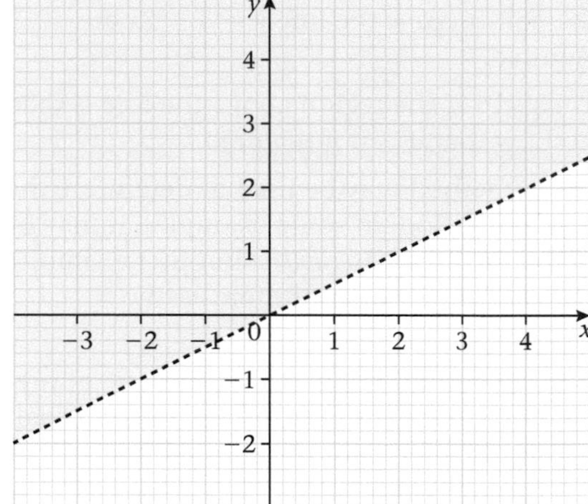

(b)

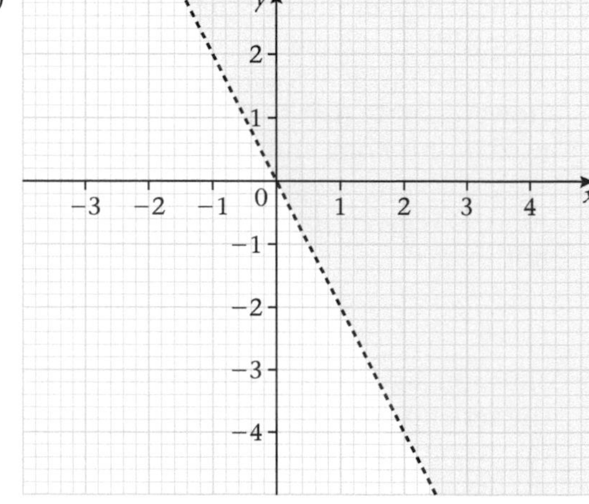

(c)

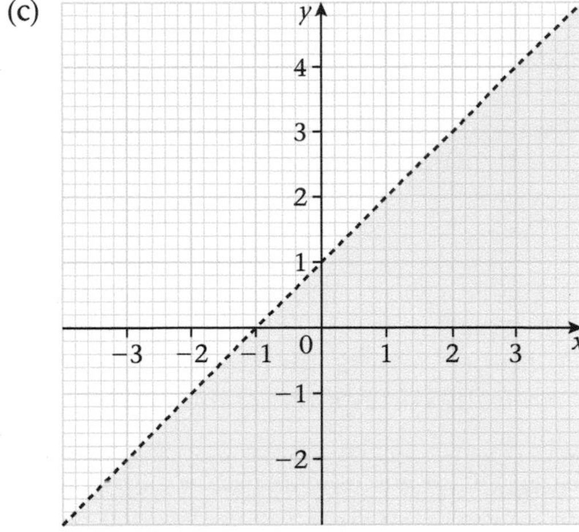

(d)

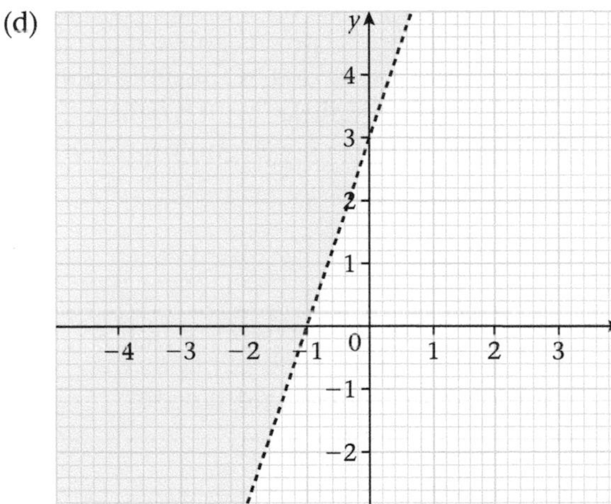

(e)

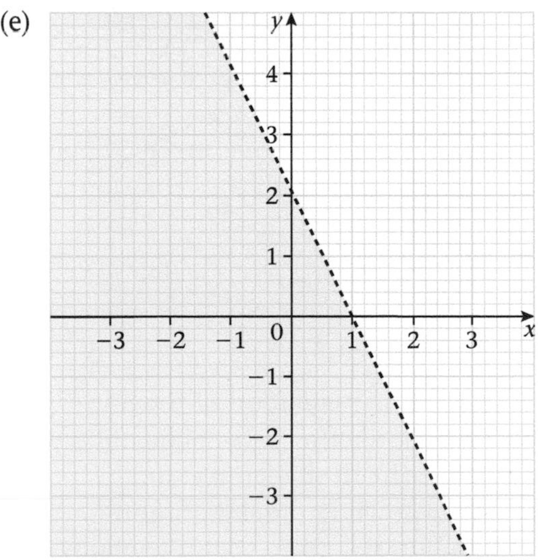

Fig. 16.4

Example 2

Solve graphically the simultaneous inequalities $4x + 3y < 12$, $y \geqslant 0$, $x > 0$ for integral values of x and y.

The equations of two of the boundary lines are

$4x + 3y = 12$

and $x = 0$ *(the y-axis)*

These must be shown as broken lines since the points on these boundary lines are *not* included in the required region.

The equation of the third boundary line is

$y = 0$ *(the x-axis)*

and is shown as a continuous line since the points on this boundary line are included in the required region.

Consider the equation of the boundary line

$4x + 3y = 12$

When $x = 0$, $y = 4$
$y = 0$, $x = 3$

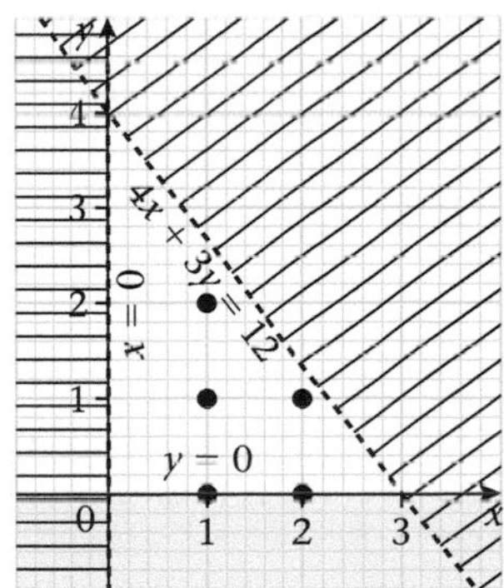

Fig. 16.5

In Fig. 16.5,

$4x + 3y = 12$	(broken)
$y = 0$	(continuous)
$x = 0$	(broken)

are the boundary lines. The solution set lies within the unshaded region and is shown by the five points marked by spots. The solution set is $\{(1, 0), (1, 1), (1, 2), (2, 0), (2, 1)\}$.

Example 3

Show on a graph the region which contains the solution set of the simultaneous inequalities $2x + 3y < 6$, $y - 2x \leqslant 2$, $y \geqslant 0$.

Consider the first inequality, $2x + 3y < 6$.

$2x + 3y = 6$ is a boundary line
when $x = 0$, $y = 2$
when $y = 0$, $x = 3$

Points below the broken line through (0, 2) and (3, 0) satisfy the inequality $2x + 3y < 6$.

Similarly, points on and below the continuous line through (0, 2) and (−1, 0) satisfy the inequality $y - 2x \leqslant 2$.

Likewise, points on and above the x-axis (i.e. the line $y = 0$) satisfy the inequality $y \geqslant 0$.

The solution set is the unshaded triangular region shown in Fig. 16.6.

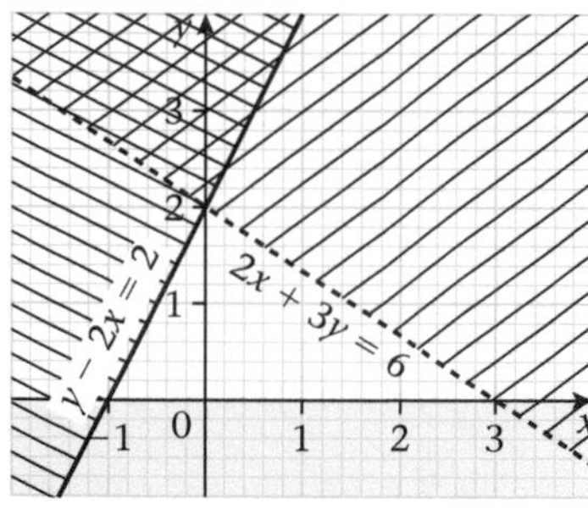

Fig. 16.6

Example 4

Write down the three inequalities which define the unshaded area labelled A in Fig. 16.7.

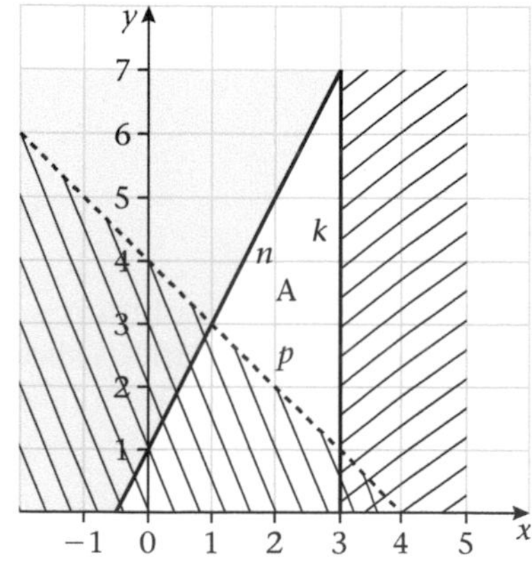

Fig. 16.7

The lines are labelled k, p, n for convenience.

Line k:
k is the line $x = 3$, k is continuous, showing that the points on the line are included in the required region; points to the right of k are not required. Hence the corresponding inequality is $x \leqslant 3$.

Line p:
p has a gradient of −1 and cuts the y-axis at (0, 4). Its equation is $y = -x + 4$. p is a broken line. Points below p are not required. Hence $y > 4 - x$ is the corresponding inequality.

Line n:
n has a gradient of 2 and cuts the y-axis at (0, 1). Its equation is $y = 2x + 1$. n is continuous showing that the points on the line are included in the required region: points above n are not required. Hence $y \leqslant 2x + 1$ is the corresponding inequality. The inequalities which define the region A are $x \leqslant 3$, $y > 4 - x$ and $y \leqslant 2x + 1$.

Exercise 16b

1. In Fig. 16.8 the lines p, $x + y = 2$ and $x + 2y = 5$ are the boundaries of the unshaded region which contains the solution set of three simultaneous inequalities.

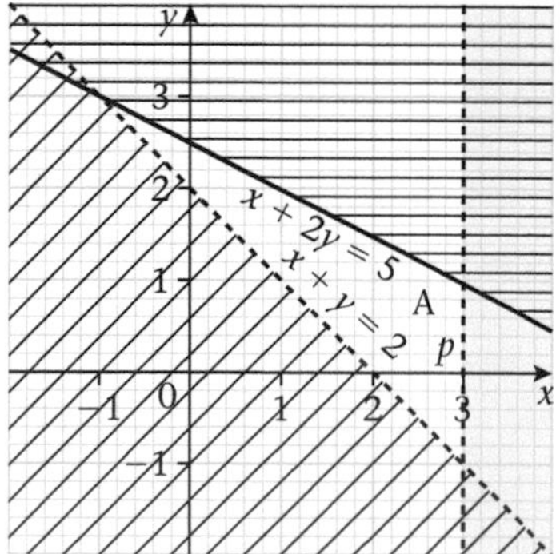

Fig. 16.8

(a) What is the equation of the line p?
(b) Write down the three inequalities which define the unshaded region, A.
(c) Write down the members of the solution set, given that it contains integral values of x and y only.

2 Using graph paper, draw the regions defined by each of the following. (Use solid and broken lines as explained earlier; leave each required region unshaded.)

(a) $y \geqslant 0, y < 3x, x + y \leqslant 4.$
(b) $x \geqslant -3, y \leqslant 2, x - y < 2$
(c) $y \leqslant 5, x - y \leqslant 1, 4x + 3y \geqslant 12$
(d) $x \geqslant 0, y \geqslant 0, x + y < 6, y - x < 2$
(e) $y < 3, x < 4, 2x + y + 2 \geqslant 0,$
$x - y - 2 \leqslant 0$

3 Solve each of the following graphically for integral values of x and y.

(a) $y \geqslant 0, x - y \geqslant 1, 3x + 4y < 12$
(b) $y \geqslant 1, y - x < 5, 2x + y \leqslant 0$
(c) $y > -2, x > 0, 2x + y < 4$
(d) $x + y \leqslant 2, x - y \leqslant 2, 2x + y \geqslant 2$
(e) $y \geqslant 0, y \leqslant 4, 4x + 3y > 0, 5x + 2y < 10$

4 Write down the three inequalities which define the unshaded area labelled A in Fig. 16.9.

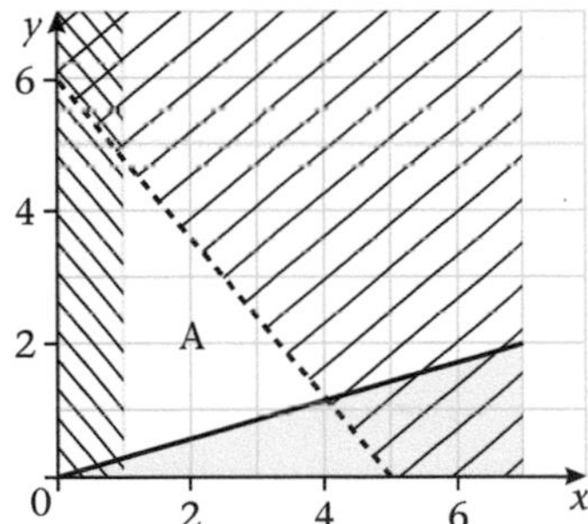

Fig. 16.9

5 Write down the three inequalities which define the unshaded region R in Fig. 16.10.

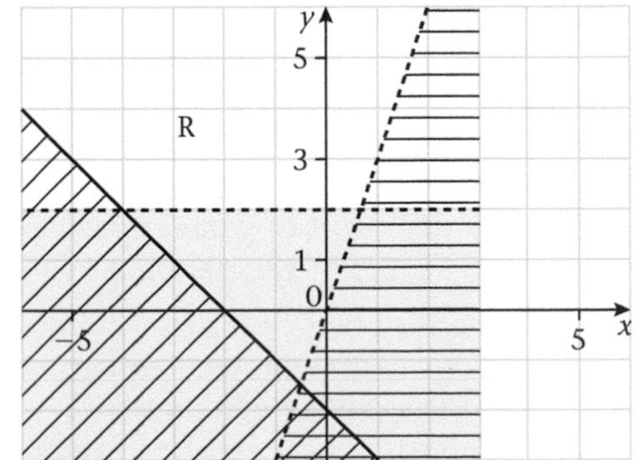

Fig. 16.10

6

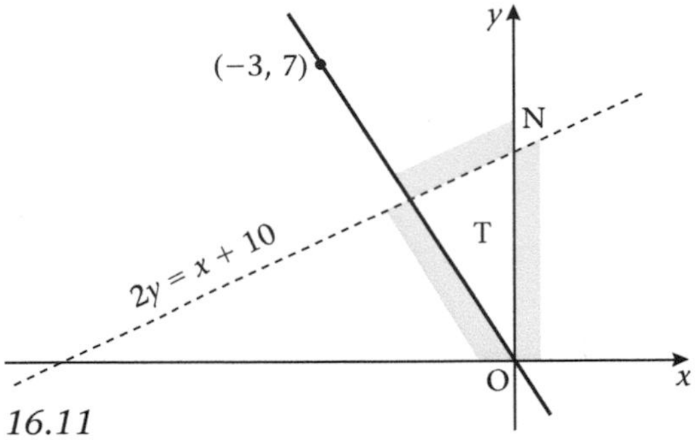

Fig. 16.11

In Fig. 16.11 find

(a) the coordinates of th point N where the line $2y = x + 10$ crosses the y-axis,
(b) the equation of the line which passes through the origin O and the point $(-3, 7)$,
(c) the three inequalities which define the triangular region T in the diagram.

Linear programming

The relationship between a number of quantities can often be expressed in mathematical form as an equation or an inequality. If there are only two quantities under varying conditions, each condition may be expressed as a linear inequality in two variables. By drawing, on the same axes, a straight line graph to represent the limiting values of each condition as the quantities vary, we are able to determine the values of the variables which satisfy the given conditions. This method of solving the problem is called **linear programming**.

It is also possible by this method to determine the maximum or minimum value of one quantity in relation to the other under different sets of conditions.

Example 5

A student has \$25. She buys ballpens at \$2.50 each and pencils at \$1 each. She gets at least five of each and the money spent on ballpens is over \$5 more than that spent on pencils.

Find (a) how many ways the money can be spent, (b) the greatest number of ballpens that can be bought, (c) the greatest number of pencils that can be bought.

Let the student buy x ballpens at $\$2\frac{1}{2}$ each and y pencils at $1 each.
Then, from the first two sentences.

$$\tfrac{5}{2}x + y \leqslant 25$$
$$\Rightarrow \quad 5x + 2y \leqslant 50$$

Since she gets at least 5 of each,

$$x \geqslant 5$$
and $y \geqslant 5$

Also, from the third sentence

$$\tfrac{5}{2}x - y > 5$$
$$\Rightarrow \quad 5x - 2y > 10$$

These inequalities are shown in Fig. 16.12.

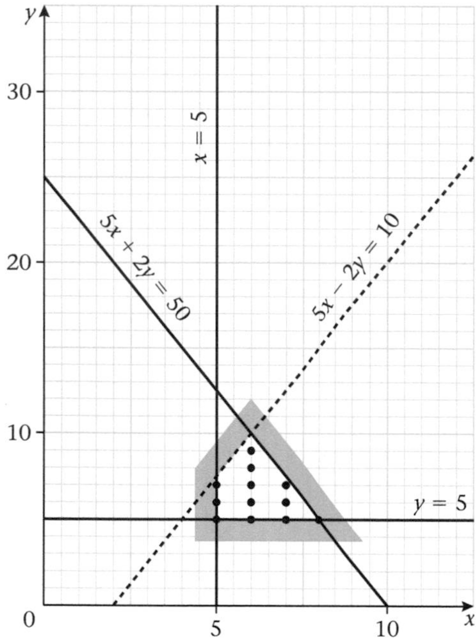

Fig. 16.12

(a) The solution set of the four inequalities is shown by the twelve points marked inside the unshaded region. For example the point (6, 8) shows that the student can buy 6 ballpens and 8 pencils. Hence there are 12 ways of spending the money.
(b) The greatest number of ballpens that can be bought is 8, corresponding to the point (8, 5).
(c) The greatest number of pencils is 9, corresponding to the point (6, 9).

The kind of problem solved in Example 5 involves making decisions in a situation in which there are conditions or restrictions. Each restriction, such as the limit on the amount of money available, can be represented by a linear inequality. Hence the solution to the problem can be found graphically.

Example 6

The student in Example 5 wants to buy as many items as possible. How many can she get and how much change will there be from the $25?

The number of items bought is $x + y$.
If the total is n, then $x + y = n$.

The problem is to determine the greatest value of n which satisfies the conditions given in Example 5.

From Fig. 16.12, the greatest possible value of n is 15 when $x = 6$ and $y = 9$, that is, at the point (6, 9). The 15 items are made up as follows:

6 ballpens at $2.50:	$15
9 pencils at $1.00:	$ 9
total cost:	$24

There will be $1 change from $25.

We may have to find the maximum or minimum value of some quantity that depends on the values of the variables when the variables satisfy given conditions. This maximum or minimum value of the quantity is usually a function of the variables for values at a point located at a vertex of the required region found graphically. If the value of the variables must be an integer, the point may be near to the vertex.

Example 7

To start a new bus company, a businessman needs at least 5 buses and 10 minibuses. He does not want to have more than 30 vehicles altogether. A bus takes up 3 units of garage space, a minibus takes up 1 unit of garage space and there are only 54 units of garage space available.

If x and y are the numbers of buses and minibuses respectively, (a) write down four inequalities which represent the restrictions on the businessman and (b) draw a graph which shows a region representing possible values of x and y.

(a) From the first sentence,
$x \geqslant 5$
$y \geqslant 10$
From the second sentence,
$x + y \leqslant 30$
From the third sentence,
$3x + y \leqslant 54$

(b) In Fig. 16.13, R is the region which contains the possible values of x and y.

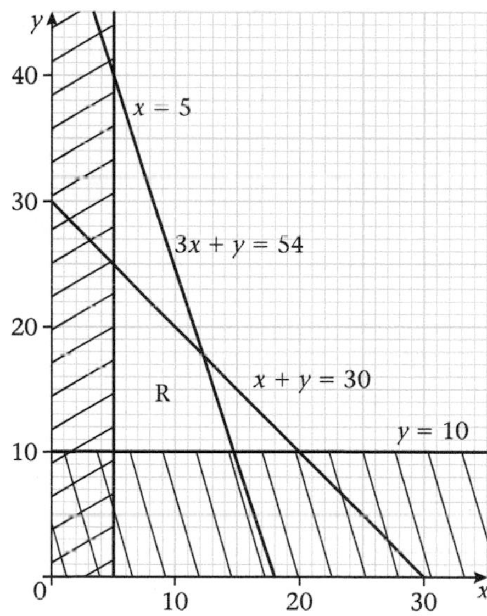

Fig. 16.13

Exercise 16c

1. Use the graph in Fig. 16.13 to answer the following. When the bus company is running at full efficiency, the daily profit on a bus is four times that on a minibus. Find the numbers of buses and minibuses the businessman should buy to maximise his profit.

2. Notebooks cost \$3 each and pencils \$1.80 each. A boy has \$18 to spend and needs at least 3 notebooks and 3 pencils. He decides to spend as much as possible of his \$18.
   (a) How many ways can he spend his money?
   (b) Do any of the ways give him change? If so, how much?

3. A car repair workshop uses large numbers of two types of space part, one costing \$3 and the other \$4. The workshop owner allows \$300 to buy spare parts and he needs twice as many cheap ones as expensive ones. There must be at least 50 cheap and 20 expensive parts.
   (a) What is the largest number of spare parts he can buy, and in what way?
   (b) If he decides to get as many of the expensive parts as conditions allow, how many of each type can he get?

4. A storeman fills a new warehouse with two types of goods, A and B. They both come in tall boxes which cannot be stacked. A box of A takes up $\frac{1}{2}$ m² of floor space and costs \$50. A box of B takes up $1\frac{1}{2}$ m² of floor space and costs \$300. The storeman has up to 100 m² of floor space available and can spend up to \$15 000 altogether. He wants to buy at least 50 boxes of A and and 20 boxes of B.
   (a) How many boxes of each should he buy in order to (i) spend all the money available and also to use as much space as possible? (ii) use all the space for the least cost?
   (b) What is the cost in the second case?

5. Following an illness, a patient is required to take pills containing minerals and vitamins. The contents and costs of two types of pill, Feelgood and Getbetta, together with the patient's daily requirement are shown in Table 16.1.

Table 16.1

	Mineral	Vitamin	Cost
Feelgood	160 mg	4 mg	\$2.00
Getbetta	40 mg	3 mg	\$1.00
Daily requirement	800 mg	30 mg	

A daily prescription contains x Feelgood pills and y Getbetta pills.
   (a) State the inequalities to be satisfied by x and y.
   (b) Use a graphical method to show the solution set of x and y.
   (c) Find the cheapest way of prescribing the pills and the cost.

6. While exploring for oil, it was necessary to carry at least 18 tonnes of supplies and 80 people into a desert region. There were two types of lorry available, Landmasters and Sandrovers. Each Landmaster could carry 900 kg of supplies and 6 people; each Sandrover could carry 1350 kg of supplies and 5 people. If there were only 12 of each type in good running order, find the smallest number of lorries necessary for the journey.

7. A shopkeeper orders packets of soap powder. The cost price of a large packet is \$9 and that of a small packet is \$4. She is prepared to spend up to \$200 altogether and needs twice as many small packets as large packets with a minimum of 10 large and 20 small packets.
   (a) What is the greatest number of packets she can buy?

   The profit is 80c on a large packet and 40c on a small packet.
   (b) Which arrangement gives the greatest profit?
   (c) What is that profit?

8. A manufacturer plans to buy new machines for his factory. Table 16.2 shows the cost, the necessary floor space and the output of each machine.

Table 16.2

Machine	Cost	Floor space	Output in components/ hour
Machine A	\$6000	$3\,m^2$	10 per hour
Machine B	\$8600	$2\frac{1}{2}\,m^2$	15 per hour

   He can spend \$72 000 altogether and he has $27\,m^2$ of floor space. Trade restrictions are such that he has to buy at least 3 of Machine A and 4 of Machine B.
   (a) What is the maximum number of machines he can buy?
   (b) What arrangement gives the biggest output?

Summary

The set of points satisfying an **inequality in two variables** can be represented by a region on a Cartesian plane.

When the inequality sign is replaced by the equals sign, the equation of a line is obtained. This line is the boundary between the required region and the rest of the plane.

When the inequality sign is $>$ or $<$, the line is broken to indicate that points on the line are not included.

When the sign is $\geqslant$ or $\leqslant$, the line is continuous indicating that points on the line are to be included in the required region.

By conventions, the regions which are *not* required are *shaded.*

Linear programming is a graphical method used to solve a problem represented by linear inequalities in two variables. For any linear inequality, the graph of the equation of the corresponding straight line represents the limiting values of the variables. The area bounded by these straight lines represents the region in which lie the solution sets of the variables satisfying the given conditions. Linear programming can be used to solve a variety of realistic problems.

Practice exercise P16.1

1. On squared paper, draw the line $y = x$. Take five points above the line and five points below the line. Compare the x- and y-values for each ordered pair.
   Shade the region of the points that do *not* satisfy the set $\{(x, y) : x < y\}$.
   Comment on the points on the line.

2. Repeat question 1 for the line $y = -x$ and the set of points $\{(x, y) : y \leqslant -x\}$.

3. Repeat question 1 for the line $y = 2x + 6$, comparing the y-values with $2x + 6$ for each point. Hence, leave unshaded the set of points where $\{(x, y): y \geqslant 2x + 6\}$.

Practice exercise P16.2

For each question, show on a graph the region which contains the set of points given. (Shade the region where the points do *not* satisfy the inequality.)

1 $\{(x, y) : x \geqslant -1, y < 1\}$

2 $\{(x, y) : x + 2y < 4\}$

For each of the following inequalities, draw a graph showing the region that contains the set of points.

3 $\{(x, y) : 2y > 5\}$

4 $\{(x, y) : x \leqslant -2\}$

5 $\{(x, y) : 2x \geqslant 1\}$

6 $\{(x, y) : y < 2\}$

7 $\{(x, y) : -2 < x < 2\}$

8 $\{(x, y) : 4 \geqslant y \geqslant 1\}$

9 $\{(x, y) : x > -2 \text{ and } y \geqslant 3\}$

10 $\{(x, y) : x < 0 \text{ and } y \leqslant -1\}$

Practice exercise P16.3

For each of the graphs in Fig. 16.14,
(a) find the equation of the boundary line(s),
(b) state the solution set of the inequalities, where the unshaded area is the required region.

1
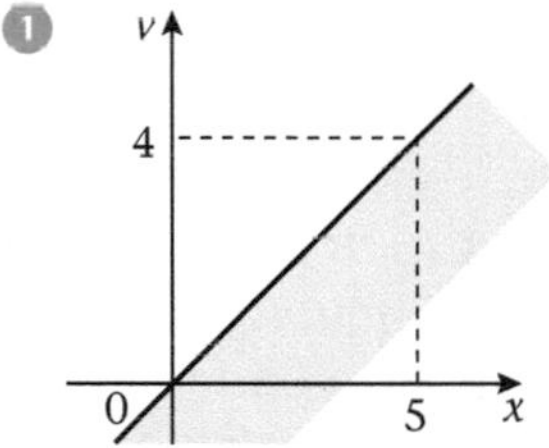

2
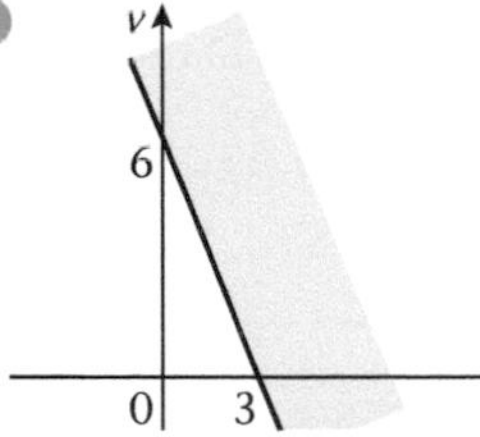

Fig. 16.14

Practice exercise P16.4

1 Indicate by leaving unshaded the region in which the solution sets lie for each of the following pairs of inequalities.
(a) $x + y > 5$, $2x - 3y < 3$
(b) $2x - y \leqslant 7$, $x + 3y > 5$

2 (a) On graph paper, draw the lines $x = 3$, $y = 1$ and $x + y = 6$.
(b) Indicate the set of points that satisfy the inequalities $x \geqslant 3$, $y \geqslant 1$ and $x + y \leqslant 6$ by shading the regions of the points that do *not* satisfy the inequalities.
(c) Write down all the ordered pairs of integers that satisfy these inequalities.
(d) If $u = 2x + y$, calculate the value of u at each of the points in part (c).
(e) At which point is the value of u greatest?

3 In Fig. 16.15, the equations of the straight lines are:
AB: $x + 2y = 12$
BC: $y = 2$
CA: $2x + y = 6$.

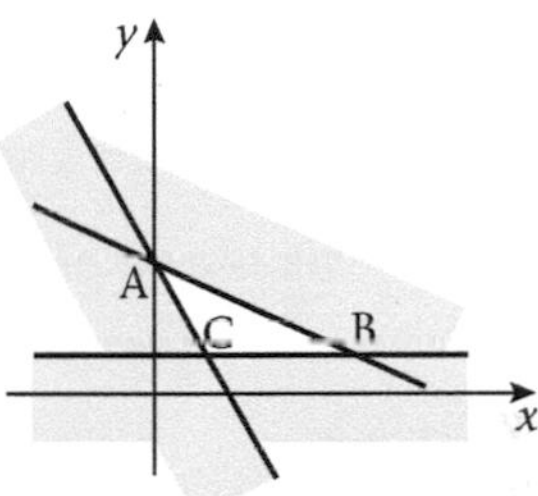

Fig. 16.15

(a) Write down three inequalities for the unshaded region.
(b) Find the coordinates of A, B and C.
(c) If $w = 2x + 3y$, find the value of w at A, B, and C.
(d) At which point is w greatest?

4 If $4x + 5y = k$, find the greatest value of k when (x, y) are in the set $P = \{(x, y) : 0 \leqslant x \leqslant 2\frac{1}{2}, 5y \leqslant 2x + 25\}$.

Practice exercise P16.5

1 The owner of a piece of land plans to divide it into not more than 36 plots and to build either a house or a shop on each plot. He decides that he will build at least 20 houses and that there will be at least twice as many houses as shops.
Taking h to represent the number of houses and s the number of shops, where $h \geqslant 0$ and $s \geqslant 0$,
(a) write the other three inequalities which satisfy the given conditions,
(b) deduce the equations of the lines that define the limiting values,
(c) use a scale of 1 cm to represent 1 house and 1 cm to represent 1 shop to draw a graph to show the required region.

2. Andrew has \$840 and wishes to buy cassettes at \$70 each and CDs at \$120 each. He decides to buy more than two but not more than five cassettes, and at least three CDs.
   (a) Using a scale of 1 cm to represent 1 unit on each axis, and c and d to represent the number of cassettes and CDs respectively, draw on the same axes a graph to show these inequalities.
   (b) Shade the areas that do not satisfy the inequalities and label the region S where the points satisfy all the given conditions.

3. A bakery makes cakes and pies so that there are
   (i) 15 or more cakes
   (ii) more than 25 pies
   (iii) at least 45 but fewer than 60 cakes and pies altogether.

   Using p to represent the number of pies and c to represent the number of cakes,
   (a) write the inequalities in p and c that satisfy these conditions,
   (b) show on a graph the region containing the solution set of p and c by shading the unwanted region.

4. A distribution agent can carry 400 cartons of juice in his van. He has a standing order for 150 cartons of orange juice and 50 cartons of grapefruit juice and he knows that he never sells more than four times as much orange juice as grapefruit juice. If the profit on one carton of orange juice is \$3 and on one carton of grapefruit juice is \$1.20, use a graphical method to find how many cartons of each type of juice he should take on each load for a maximum profit.

5. A school has to hire vehicles for an excursion. They need to carry at least 1.2 tonnes of supplies and 60 people. They plan to use two types of vehicles – minibuses and taxis. Each minibus can carry 120 kg of supplies and 6 people; each taxi can carry 90 kg and 5 people. There are only 12 of each type of vehicle available. Use a graphical method to find the smallest number of vehicles necessary for the trip.

6. A vendor sells tomatoes at \$30 per kg and carrots at \$24 per kg. A customer decides that she will buy at least 1 kg of each but she cannot carry more than 6 kg altogether and she must not spend more than \$160.
   (a) Represent these conditions graphically on the same axes.

   The vendor makes a profit of \$10 on 1 kg of tomatoes and \$6 on 1 kg of carrots.
   (b) Find how many kilograms of each vegetable the customer buys if the vendor makes the maximum profit.
   (c) Calculate the maximum profit .

7. Two factories, A and B make two products, X and Y. It takes $\frac{1}{2}$ an hour to make item X at factory A and 1 hour to make item X at factory B. It takes 1 hour to make item Y at both factories.
   The profit from item X is \$12 and from item Y is \$16. Factory A works for 8 hours a day and factory B works for 10 hours a day.
   (a) Use a graphical method to find the number of each item to be produced each day for a maximum profit.
   (b) Work out the maximum profit.

8. A cook blends two types of mixed herbs for a special dish. In mix X there are 4 units of A and 10 units of B per packet while in mix Y there are 5 units of A and 5 units of B per packet. She needs a mixture containing at least 20 units of A and 30 units of B in $(x + y)$ packets. Mix X costs \$5 per packet and mix Y costs \$4 per packet. Use a graphical method to find the cheapest way she can get the correct mix.

Latitude and longitude

Pre-requisites

- length of arc; trigonometric ratios

Great circles and small circles

Fig. 17.1(a) shows a spherical orange. The top has been cut off.

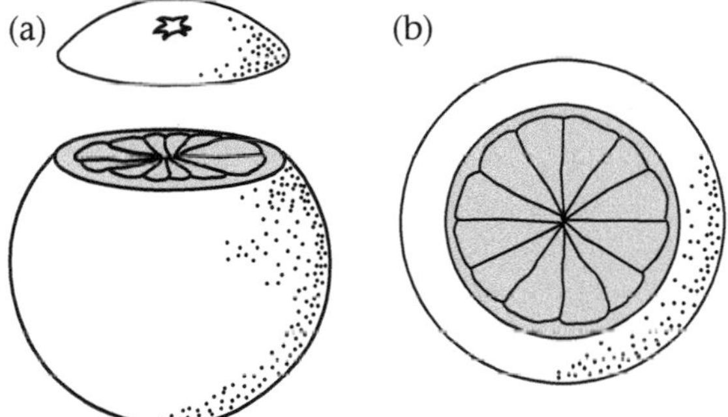

Fig. 17.1

The shape of the cross-section is a circle (Fig. 17.1(b)). In Fig. 17.2(a) another cut is made closer to the centre of the orange; the cross-section is a larger circle.

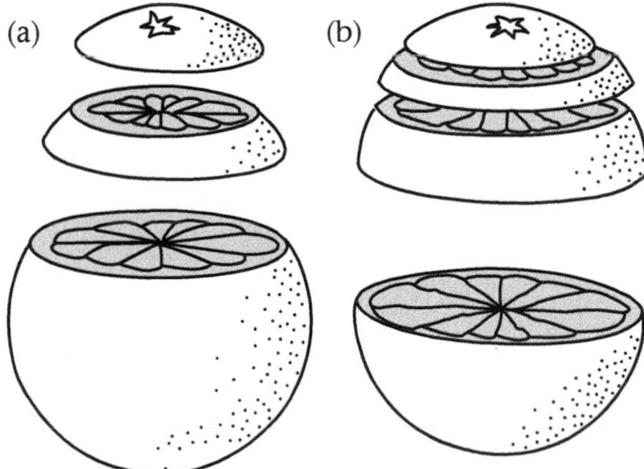

Fig. 17.2

In Fig. 17.2(b) a third cut is made through the *centre* of the orange. The cross-section gives the largest possible circle that can be obtained on the orange. This is known as a **great circle**. The radius of a great circle of a sphere is equal to the radius of the sphere. Since there is any number of ways of making a straight cut through the centre of a sphere, there are any number of great circles on a sphere.

The circles of the cross-sections in Figs. 17.1 and 17.2(a) are known as **small circles**. The radius of a small circle is less than the radius of the sphere.

Example 1

Fig. 17.3 shows a ball, centre O, resting on a circular hole in a piece of cardboard (shaded). If the radius of the ball is 3 cm and its centre is 2 cm above the plane of the cardboard, calculate the area of the circular hole.

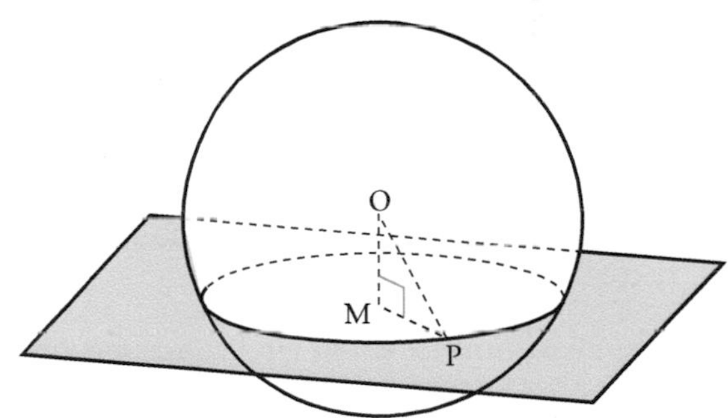

Fig. 17.3

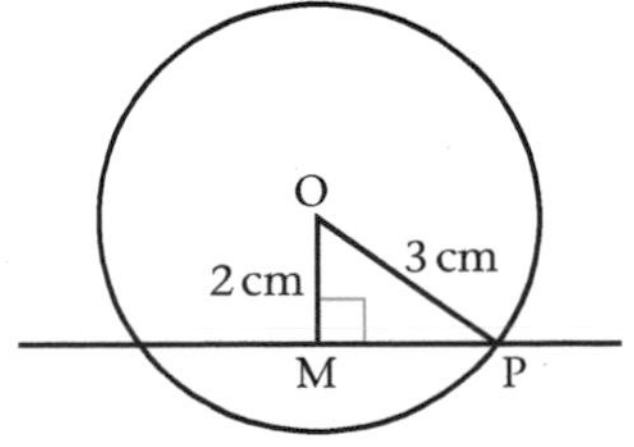

Fig. 17.4

Fig. 17.4 is a vertical section of the ball showing O, the centre of the ball, and M, the centre of the hole (a small circle).

In Figs. 17.3 and 17.4,

MP is a radius of the circular hole.

$$MP^2 = 3^2 - 2^2 \quad (\textit{Pythagoras})$$
$$= 9 - 4 = 5$$
$$\Rightarrow MP = \sqrt{5}\text{ cm}$$

Area of the hole $= \pi \times MP^2$
$= \pi \times (\sqrt{5})^2\text{ cm}^2$
$= 5\pi\text{ cm}^2$

Exercise 17a

1. Fig. 17.5(a) shows a spherical orange which has been peeled. Are the circles partly shown in Fig. 17.5(a) great circles or small circles?

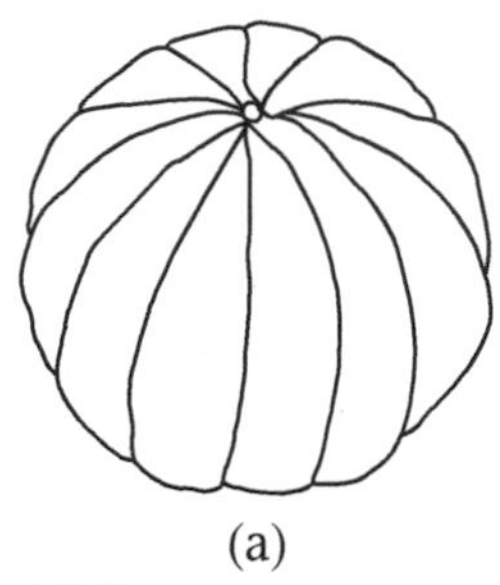

(a)

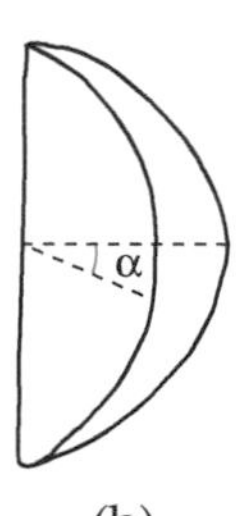

(b)

Fig. 17.5

2. The orange in Fig. 17.5(a) has 12 sectors, each assumed to be identical. What is the angle between the semicircular faces of each sector (i.e. α in Fig. 17.5(b))?

3. Fig. 17.6 shows a small circle and part of a great circle drawn on the surface of a sphere.

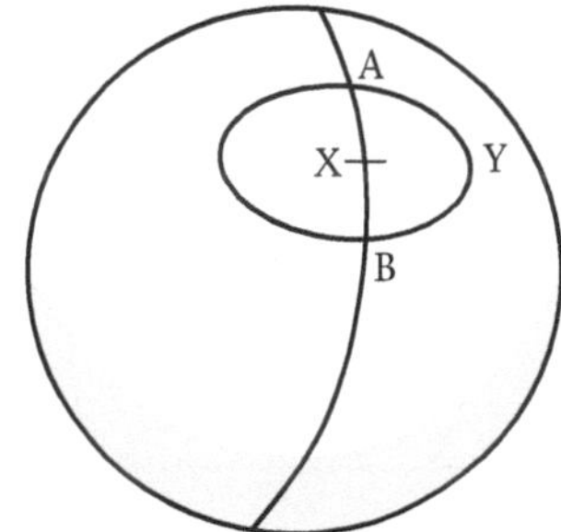

Fig. 17.6

Which length is smaller, arc AXB or arc AYB?

4. If the ball in Fig. 17.3 is of radius 10 cm and if O is 6 cm above the cardboard, calculate (a) the radius of the circular hole, (b) the area of the hole in terms of π.

5. Fig. 17.7 shows a vertical section of the orange in Fig. 17.1.

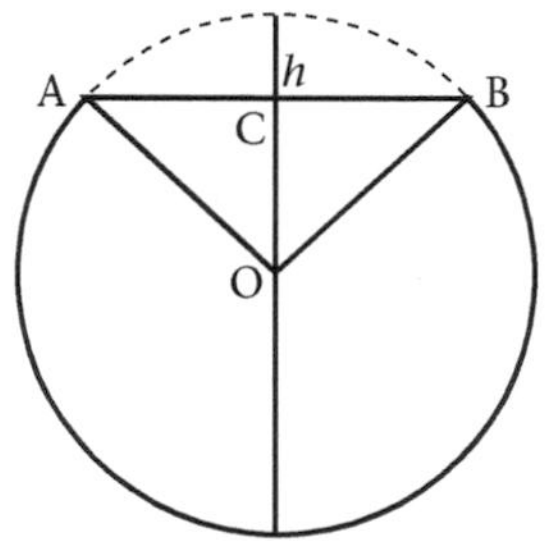

Fig.17.7

If the diameter of the small circle is 6 cm and the diameter of the orange is 10 cm:

(a) what is OA?
(b) what is CA?
(c) calculate OC.
(d) Hence find h, the height of the slice which has been cut off.

The earth, latitude and longitude

The earth is approximately spherical in shape, its radius varying between 6360 km and 6380 km. Fig. 17.8 is a view of a model of the earth.

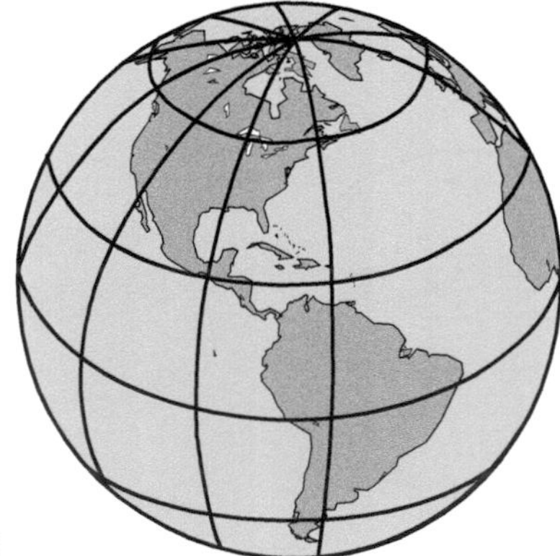

Fig. 17.8

Notice the circles drawn on the surface of the earth. These are imaginary lines which are used for giving the position of points on the earth's surface. They are called lines of **latitude** and **longitude**.

Longitude

The earth rotates about its **polar axis**. This is a straight line through the centre of the earth joining the North and South Poles. in Fig. 17.9, NOS is the polar axis.

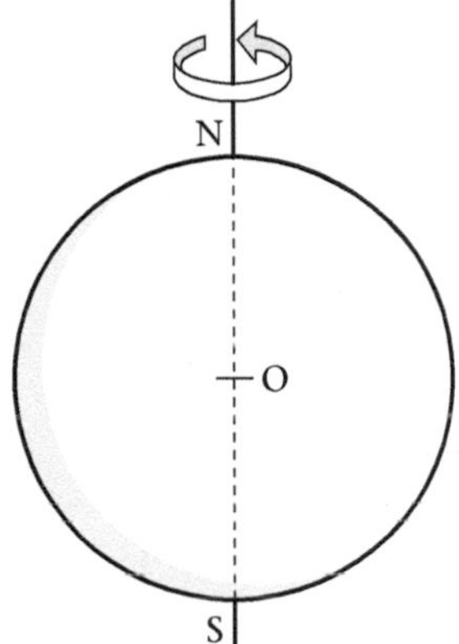

Fig. 17.9

If the earth is cut by planes through NS, the circles formed on the surface are great circles called **meridians**. Fig. 17.10(a) shows some meridians.

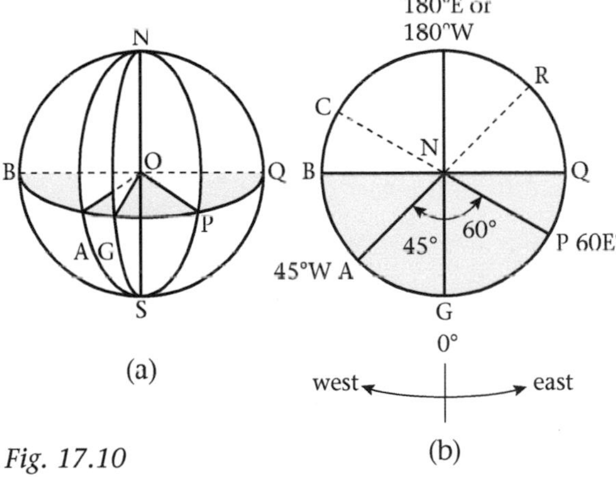

Fig. 17.10

In Fig. 17.10(a) NGS represents the meridian which passes through Greenwich in London. The **Greenwich Meridian** is used as the standard from which the positions of other meridians are measured in degrees east or west.

Fig. 17.10(b) is the plan view of the meridians in Fig. 17.10(a). The **longitude** of P is the angle that the half-circle NPS makes with the half-circle NGS, i.e. $P\widehat{O}G$, which is 60° in Fig. 17.10(b).

Hence NPS is the **line of longitude** 60°E. Similarly, NAS is the line of longitude 45°W. Longitudes vary from 180°E to 180°W.

Latitude

If the earth is cut by planes perpendicular to the polar axis, the circles formed on its surface are called **lines of latitude**. The line of latitude whose centre is also the centre of the earth is a great circle called the **equator**. In Fig. 17.11, O is the centre of the earth and the equator has been drawn.

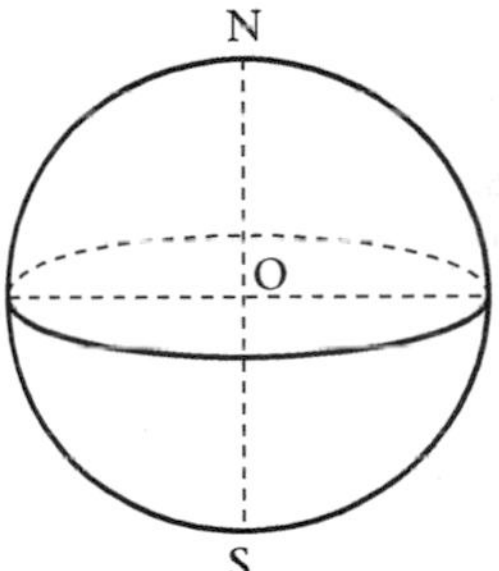

Fig. 17.11

All of the other lines of latitude are small circles, often called **parallels of latitude** since they are parallel to each other. Fig. 17.12(a) shows the equator and two parallels of latitude.

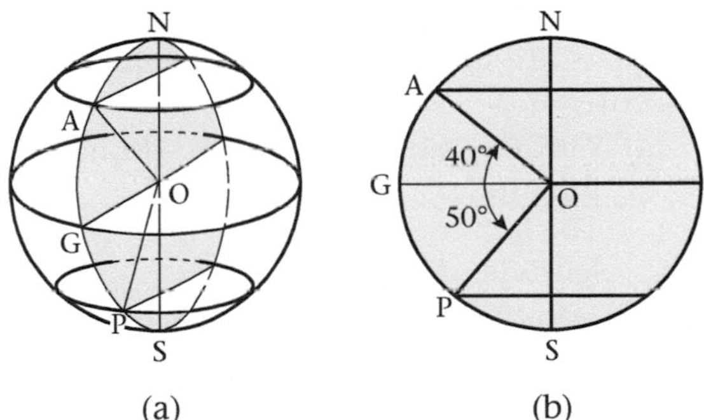

Fig. 17.12

The latitude of a point is measured in degrees north or south of the equator. Fig. 17.12(b) is a cross-section in the plane NAGPS. The latitude of A is the angle that OA makes with OG. Hence A is latitude 40°N. Similarly P is latitude 50°S. Latitudes vary from 90°N to 90°S.

The position of a point on the earth's surface is completely defined by its latitude and longitude.

Fig. 17.13 shows the point A (40°N, 60°E). There is only one such place on the surface of the earth.

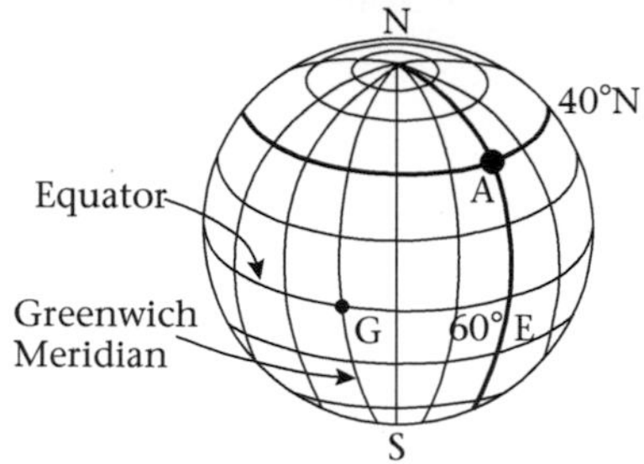

Fig. 17.13

Exercise 17b

1. In Fig. 17.10(b) state the longitude of (a) B, (b) Q, (c) R, (d) C.
2. In Fig. 17.10, what is the difference in longitude between A and P?
3. How many parallels of latitude have a radius of 2500 km?
4. In Fig. 17.12(b), what is the difference in latitude between (a) A and P, (b) A and S?
5. Use an atlas to find the country in which A is (40°N, 60°E), Fig. 17.13.
6. Use an atlas to find, to the nearest degree, the latitude and longitude of your town.
7. X and Y are points on the earth's surface at opposite ends of a diameter through the centre of the earth.
   (a) The longitude of X is 19°E; what is the longitude of Y?
   (b) The latitude of X is 52°N; what is the latitude of Y?

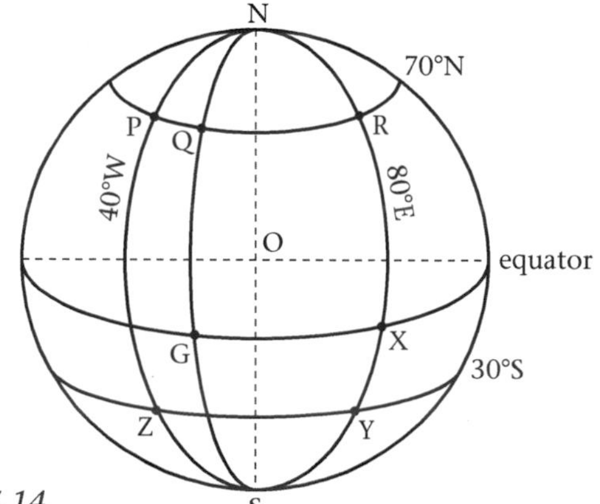

Fig. 17.14

8. In Fig. 17.14, G is the point where the Greenwich Meridian crosses the equator. Lines of latitude 70°N and 30°S, and longitudes 80°E and 40°W are given. State the latitude and longitude of the following points.

   (a) P (b) Q (c) R
   (d) X (e) Y (f) Z

9. Make a rough sketch of Fig. 17.15. On your sketch, mark the following points.

   (a) M(35°S, 55°W) (b) P(60°N, 55°W)
   (c) Q(60°N, 20°E) (d) R(35°S, 20°E)

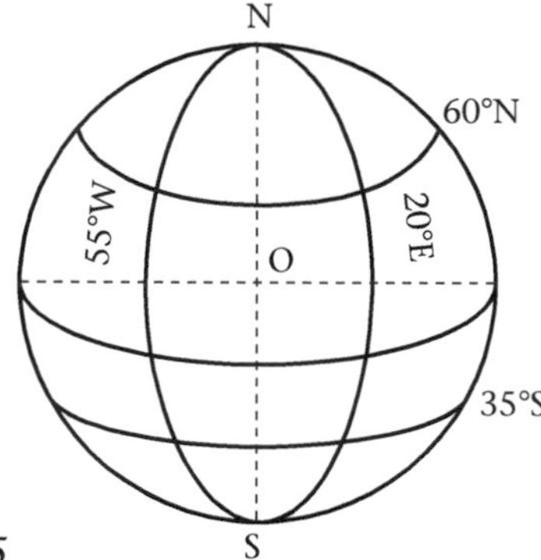

Fig. 17.15

10. In the 18th century the length of $\frac{1}{4}$ of the earth's circumference was taken to be 10 000 km. Use this value to calculate the radius of the earth to 3 s.f.

Distances along great circles

For the purposes of this course, only great-circle distances along either the equator or lines of longitude will be considered. Note, however, that it is possible to describe great circles on the earth's surface other than these.

Kilometres

The radius of the earth, R, is often given as 6400 km, which is correct to 2 s.f. The circumference of the earth, $2\pi R$, is 40 000 km, which is correct to 3 s.f. (See question 10 above.) Either of these values may be used in the remainder of the chaper, unless you are told otherwise.

Example 2

The positions of L and K to the nearest degree are (0°N, 9°E) and (0°N, 32°E) respectively. Calculate their distance apart along the equator.

Note that latitude 0°N is the equator. Fig. 17.16(a) shows L and K, on the equator. NGS is the Greenwich Meridian. Fig. 17.16(b) shows the cross-section through the equator.

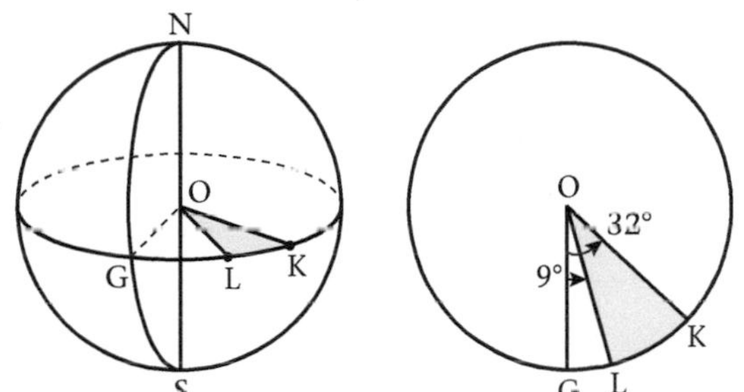

Fig. 17.16

From Fig. 17.16(b),

$L\hat{O}K = 32° - 9° = 23°$

$\text{arc LK} = \frac{23}{360} \times 2\pi R$

$= \frac{23 \times 2 \times \pi \times 6400}{360}\text{km}$

$= \frac{23 \times \pi \times 6400}{180}\text{km}$

$= 2570\text{km} = 2600\,\text{km to 2 s.f.}$

No	Log
23	1.362
π	0.497
6400	3.806
	5.665
180	2.255
2570	3.410

Since the given data are correct to 2 s.f., it is reasonable to round off the answer to 2 s.f. Notice that the answer assumes that the shortest distance between L and K is required. The distance along the equator could also be given as 37 km (40 000 − 2600); this is the distance going the long way round.

Example 3

B and G both lie on longitude 25.9°E. Their latitudes are 31.6°N and 24.8°S respectively. Calculate the shortest distance between the towns.

Fig. 17.17(a) shows the positions of B and G.

Fig. 17.17(b) shows the cross-section NBGS. OE represents the radius of the earth at the equator.

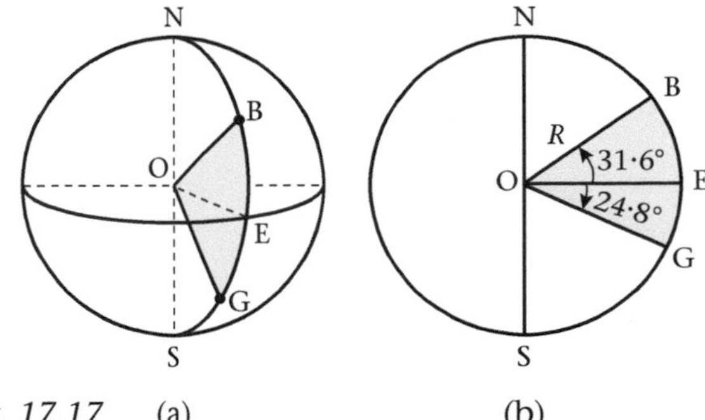

Fig. 17.17 (a) (b)

From Fig. 17.17(b),

$B\hat{O}G = 31.6° + 24.8° = 54°$

$\Rightarrow \text{arc BG} = \frac{56.4}{360} \times 2\pi R = \frac{56.4}{360} \times 40\,000\,\text{km}$

$= \frac{56.4 \times 1000}{9}\text{km} \simeq 6270\,\text{km}$

In this example, the data is given to 3 s.f. It is reasonable to round the answer off to 3 s.f.

Figs. 17.16 and 17.17 were of considerable help when solving Examples 2 and 3. Each contained a perspective view and a cross-section which showed the required arc clearly. Similar sketches should always be drawn when solving latitude and longitude problems.

Nautical miles

A **nautical mile** is the length of an arc of a great circle on the earth's surface which subtends an angle of 1 minute at the centre of the earth. From this definition, circumference of the earth

$= 360 \times 60$ nautical miles $= 21\,000$ n.mi.

Since nautical miles are defined in terms of angle only, they are often used by navigators in ships and aircraft to express distance on the earth's surface.

Example 4

Express the answers to Examples 2 and 3 in nautical miles.

Example 2

$L\hat{O}K = 23°$

$\text{arc LK} = 23 \times 60\,\text{n.mi.} = 1380\,\text{n.mi.}$

Example 3

$B\hat{O}G = 56.4°$

$\text{arc BG} = 56.4 \times 60\,\text{n.mi.} = 3384\,\text{n.mi.}$

Example 4 shows that if an arc of a great circle subtends an angle of $\alpha°$ at the centre of the earth, the length of the arc is 60α nautical miles.

Exercise 17c

Use the values of 6400 km and 40 000 km for the earth's radius and circumference as appropriate. Take log π to be 0.497.

1. Two places lie on the same meridian. Their latitudes are 15°N and 36°N. Find their distance apart (a) in km, (b) in nautical miles, measured along the meridian.
2. Two places have the same longitude. Their latitudes are 15°N and 36°S. Calculate their great-circle distance apart in km.
3. Two places lie on the equator and have longitudes of 92°E and 57°E. In nautical miles, find their distance apart, measured along the equator.
4. Two places, with longitudes 152°E and 171°W both lie on the equator. Find the shortest distance between them, in km, on the earth's surface.
5. The positions of A and B to the nearest degree are (9°N, 7°E) and (51°N, 7°E) respectively. Calculate their distance part in km to 2 s.f.
6. The Sao Tome and Galapagos Islands lie on the equator on longitudes 7°E and 90°W respectively. Find their distance apart in nautical miles to 3 s.f.
7. By roads, two towns are 560 km apart. Their positions are (21.7°S, 29.9°E) and (18.3°S, 29.9°E) respectively.
   (a) Calculate their great-circle distance apart in km.
   (b) Explain why your result differs from the road distance.
8. Find the distance, in nautical miles along a meridian, of the South Pole from any point on the parallel of latitude 65°S.
9. Show that the points (20°N, 20°E) and (60°N, 160°W) lie on the same great circle. Find their great-circle distance apart in nautical miles.
10. Two places are 264 km apart on the same line of longitude. Calculate the difference between their latitudes.

Distance along parallels of latitude

Radius of a parallel of latitude

In Fig. 17.18, the latitude of point P is θ. C is the centre of the circle of latitude on which P lies.

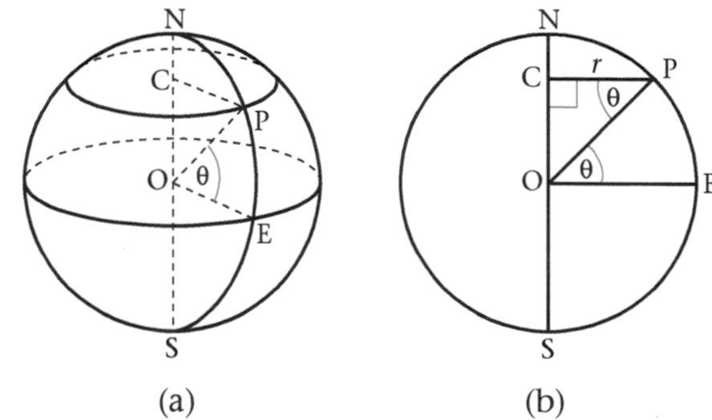

Fig. 17.18

Let r be the radius of the parallel of latitude through P. Then, in $\triangle$PCO,

$$\hat{CPO} = \theta \qquad \textit{(alt. angles, CP}\|\textit{OE)}$$
$$OP = R \qquad \textit{(radius of earth)}$$
$$\cos\theta = \frac{r}{R}$$
$$\Rightarrow r = R\cos\theta$$

The relationship $r = R\cos\theta$ is true for all parallels of latitude.

Example 5

Find the distance, measured along the parallel of latitude 56°N, between the two points (56°N, 23°E) and (56°N, 17°W). Give the answer (a) in km, (b) in n.mi.

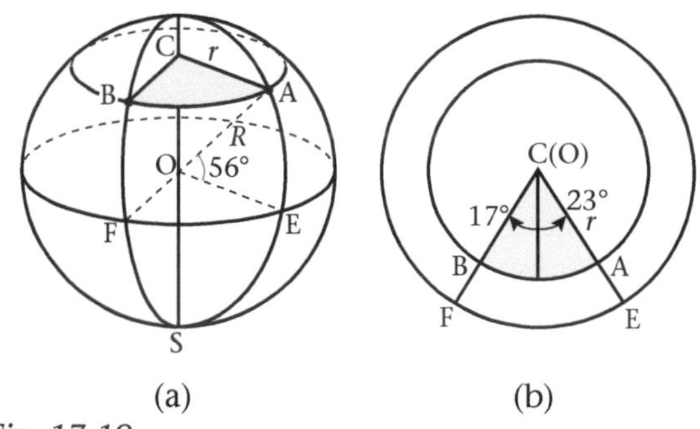

Fig. 17.19

In Fig. 17.19, A and B are the points (56°N, 23°E) and (56°N, 17°W) respectively. C is on the polar axis and is the centre of the parallel of latitude 56°N. Fig. 17.19(b) is a plan showing the circle of latitude and the equator.

(a) *in kilometres:*
In $\triangle$ACO, $r = \cos 56°$
In Fig. 17.19(b),

$$A\widehat{C}B = 23° + 17° = 40°$$

$$\text{arc } AB = \frac{40}{360} \times 2\pi r = \frac{40}{360} \times 2\pi R \cos 56°$$

But $2\pi R = 40\,000$ km

$$\text{arc } AB = \frac{40}{360} \times 40\,000 \times \cos 56° \text{ km}$$

$$= \frac{40\,000 \times 0.559}{9} \text{ km} \simeq 2480 \text{ km}$$

(b) *in nautical miles:*
From Fig. 17.19(b), sector ABC is similar to sector EFO.

$$\Rightarrow \frac{\text{arc } AB}{\text{arc } EF} = \frac{CA}{OE} = \frac{r}{R} = \frac{R \cos 56°}{R} = \cos 56°$$

$$\Rightarrow \text{arc } AB = \text{arc } EF \times \cos 56°$$

$$= (40 \times 60) \times \cos 56° \text{ n.mi.}$$

$$= 40 \times 60 \times 0.559 \text{ n.mi.}$$

$$= 1340 \text{ n.mi.}$$

Latitude and nautical miles

Example 5(b), above, demonstrates that the length of an arc of a circle of latitude θ°N or S which subtends an angle of α° at the centre of that circle is 60α cos θ nautical miles.

Example 6

Find the parallel of latitude in the northern hemisphere along which a journey of 640 km makes a change of 8° in longitude.

Let the required parallel of latitude be θ°N. Fig. 17.20 shows the data of the question.

In Fig. 17.20(a), $r = R \cos \theta$,
In Fig. 17.20(b),

$$640 = \frac{8}{360} \times 2\pi r$$

$$= \frac{8}{360} \times 2\pi R \cos \theta$$

$$= \frac{8}{360} \times 40\,000 \cos \theta$$

$$\cos \theta = \frac{640 \times 360}{8 \times 40\,000} = 0.72$$

$$\theta = 43.95°$$

$$\theta = 44° \text{ to the nearest degree.}$$

The parallel of latitude is 44°N.

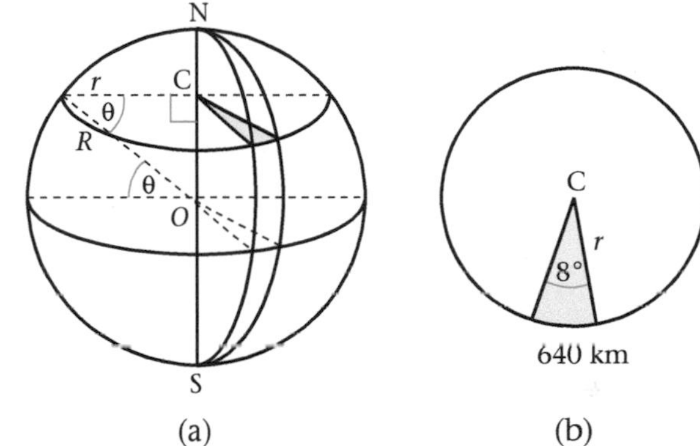

Fig. 17.20

Exercise 17d

Unless told otherwise, use the value 6400 km for R or 40 000 km for $2\pi R$. Take log p to be 0.497.

1. A geographical globe has a radius of 18 cm. Find the radius of the circle formed by the parallel of latitude 56°S.
2. On a globe of radius 60 cm, find the length in cm of the parallel of latitude 67°N.
3. Find the distance of a place with latitude 68°S from the polar axis of the earth.
4. Find the length of the parallel of latitude 42°N (a) in km, (b) in nautical miles.
5. Calculate in nautical miles:
   (a) the distance along a meridian between the Tropic of Cancer, $23\frac{1}{2}$°N, and the Tropic of Capricorn, $23\frac{1}{2}$°S;
   (b) the length of the Tropic of Capricorn.
6. Find the distance, measured along the parallel of latitude between two places with the same latitude, 18°S, and with longitudes 96°E and 57°E. (Answer in nautical miles.)
7. Two places are on the parallel of latitude 77°N. Their longitudes are 154°E and 142°W. Find their distance apart measured along the parallel of latitude. (Answer in km.)

8 Find the distance apart, measured along the parallel of latitude, of two places which both have latitude 63°N, and whose longitudes differ by 1°. (Answer in nautical miles.)

9 Find the distance in km, measured along the parallel of latitude, between M (6.3°N, 10.8°W) and B (6.3°N, 5.7°E).

10 Find the distance in nautical miles between H (17.8°S, 31.2°E) and T (17.8°S, 47.4°E) along the parallel of latitude.

Miscellaneous examples

Speed in km/h

Example 7

An aeroplane leaves a point P (21°N, 32°E) and flies due south at 850 km/h. Find its position Q after 4 hours. (Assume that the earth is a sphere of radius 6400 km.)

Distance travelled in 4 hours

$= 850 \times 4\,\text{km}$
$= 3400\,\text{km}$

With the lettering of Fig. 17.21,

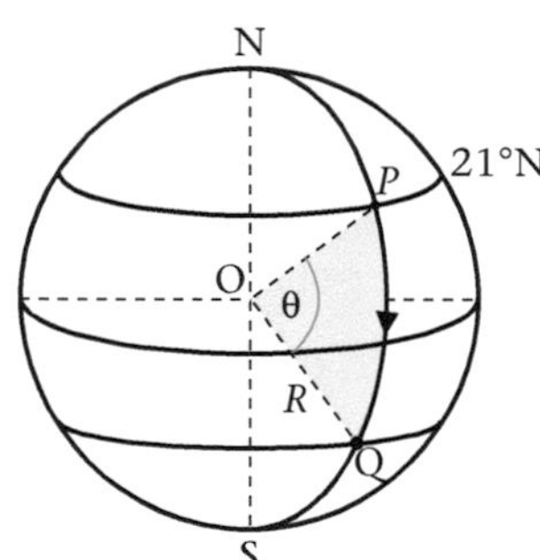

Fig. 17.21

$$3400 = \frac{\theta}{360} \times 2\pi R$$
$$= \frac{\theta \times 2 \times \pi \times 6400}{360}$$
$$\theta = \frac{3400 \times 360}{2 \times \pi \times 6400}$$
$$= \frac{1530}{16\pi}$$
$$= 30 \text{ to 2 s.f.}$$

No	Log
1530	3.185
16	1.204
π	0.497
	1.701
30.5	1.484

Latitude of Q = 21°N − 30°
= 9°S

Q is (9°S, 32°E).

Speed in knots

A speed of 1 nautical mile per hour is called a **knot**. The knot is a unit of speed commonly used by sailors and airmen.

Example 8

A cargo ship leaves New York (21°N, 73°W) and sails due east for 32 hours to a point 41°N, 57°W). Calculate its average speed in knots.

Distance travelled by ship

$= (73 - 57) \times 60 \times \cos 41°$ n.mi.
$= 16 \times 60 \times \cos 41°$ n.mi.

Time taken

$= 32\,\text{h}$

Average speed

$$= \frac{16 \times 60 \times \cos 41°}{32} \text{ knots}$$
$= 30 \times 0.755$ knots
$= 22.7$ knots to 3 s.f.

Shortest distance between points

Example 9

A and B are two points on latitude 70°N. Their longitudes are 62°W and 118°E respectively. Find the distance from A to B (a) along a great circle, (b) along a parallel of latitude.

(a) The longitudes of A and B differ by 180° (62° + 118° = 180°). Hence A and B lie on the same meridian, a great circle passing through the North Pole. This is shown in Fig. 17.22. In Fig. 17.22(b), O is the centre of the earth.

$$A\hat{O}B = 180° - 2 \times 70° = 40°$$

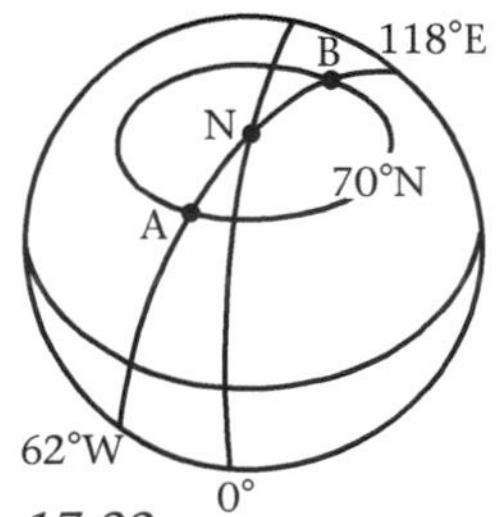

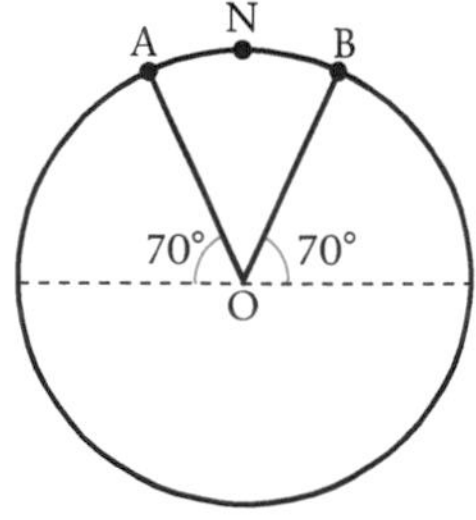

Fig. 17.22

$$\text{arc ANB} = \frac{40}{360} \times 2\pi R$$
$$= \frac{40 \times 40\,000}{360}\,\text{km} \simeq 4440\,\text{km}$$

(b) Distance from A to B along the parallel of latitude

$= \frac{1}{2} \times$ circumference of small circle, 70°N

$= \frac{1}{2} \times 2\pi R \cos 70°$

$= \frac{1}{2} \times 40\,000 \times 0.342$ km

$= 6840$ km

Notice that the great-circle distance between A and B is much shorter than the small-circle distance between A and B. **The great-circle distance between any two points on the earth's surface is the shortest possible distance between the points along the surface of the earth.**

Exercise 17e

Unless told otherwise, use the value 40 000 km for $2\pi R$. Take log π to be 0.497.

1. Find the distance between two places (55°S, 28°W) and (55°S, 152°E) (a) along latitude 55°S, (b) across the South Pole.
2. P is due north of Q and the distance PQ along the surface of the earth is 1120 km. If the latitude of P is 3.2°N, calculate the latitude of Q.
3. Two places A and B are 480 km apart and A is due south of B. If the latitude of A is 47.8°N, find the latitude of B.
4. A boat leaves V, latitude 15°45′N, and travels due south for 50 hours at 25 knots. What is its latitude then?
5. A aeroplane flies due south from C 30°N, to H, 18°S. If it takes 9 hours what is its average airspeed in knots?
6. An aeroplane leaves a point on latitude 52°S at 10 a.m. It flies due south at an average speed of 285 knots.
   (a) What is its latitude at mid-day?
   (b) At what time does it pass over the South Pole?
7. An aeroplane leaves the point X (43°N, 71°W) and flies due east at 500 km/h. After 3 hours (a) how far has it travelled? (b) what is its longitude?
8. A plane passes over a point P (20°E, 26°N) and flies due south for 8 hours at a rate of 540 knots until it is over a point Q (20°E, θ°S). The plane then flies east to a point R for 1 hour at the same speed. Calculate, to the nearest degree, (a) the value of θ, (b) the longitude of R.
9. X and Y are two points on latitude 65°N. Their longitudes differ by 180°. Calculate the distance in km from X to Y (a) along a great circle, (b) along the parallel of latitude.
10. A place X is at 70°S, 90°E and Y is at 70°S, 90°W. Find the distance from X to Y (a) along a great circle, (b) along the parallel of latitude. (Answers in nautical miles.)

Summary

The earth is approximately spherical in shape. The lines of **longitude** and the **equator** are all circles whose centre is the centre of the earth. These are all **great circles. Parallels of latitude** are parallel to the equator and are **small circles.**

The radius of a (small circle) parallel of latitude

$r = R \cos \theta$

where R is the radius of the earth and θ is the angle between the radius from the centre of the earth to the parallel of latitude and the radius of the equator.

A **nautical mile** is the length of an arc of a **great circle** which subtends an angle of 1 minute at the centre of the earth.

A speed of 1 nautical mile per hour is called a **knot**.

Practice exercise P17.1

Unless otherwise told, use the value 6 400 km for R or 40 000 km for $2\pi R$. Take log π to be 0.497.

1. The points P and Q lie on the surface of the earth at the positions P(60°N, 60°W), Q(30°N, 105°W). Calculate the distance along the surface of the earth between the points P and Q.

2. Calculate the shortest distance between the points P(42°N, 72°W) and Q(42°N, 108°E) measured along the surface of the earth.

3. The point K is 3 400 km due north of H(22°S, 95°E). Find the latitude and longitude of K.

4. Find the latitude and longitude of the point G which is 3600 km due west of the point H(22°S, 95°W).

5. A plane flew 4 800 km due north from a point D(13°S, 32°E) to a point E. Calculate the latitude and longitude of E.

6. P and Q are points on latitude 52°S. P is 15° due east of Q. Calculate the distance between P and Q, measured along the circle of latitude.

7. P and Q are two places on the same meridian. P has a latitude of 24°S and Q has a latitude of 80°S. Calculate the distance between the two places measured along the meridian.

8. Two places A and B both lie on longitude 120° E. Their latitudes differ by 36°. Calculate the shortest distance between A and B.

9. An aeroplane leaves the point X(21°S, 61°E) and flies due east at 560 km/h for $2\frac{1}{2}$ hours to a point Y. What is the longitude of Y?

10. (a) Show that the distance between two places A(25°N, 35°E) and B(35°S, 35°E) measured along the meridian, is $\frac{1}{3}\pi R$.
    (b) Calculate, in nautical miles, the distance between the two places.

Practice exercise P17.2

Unless otherwise told, use the value 6400 km for R or 40 000 km for $2\pi R$. Take log π to be 0.497.

1. Two places, A and B, on the same meridian are on an arc subtended by an angle of 60° at the centre of the earth. The position of A is (3°N, 50°W).
    (a) State two possible positions of B.
    (b) Calculate the distance between A and B.

2. Two towns, A and B, are on the circle of latitude 34°N. The longitude of town A is 110°E and the longitude of town B is 77°E. Calculate, correct to 2 significant figures,
    (a) the radius of the circle of latitude 34°,
    (b) the shortest distance between A and B, measured along the circle of latitude.

3. An aeroplane left from airport A (63°N, 20°W) and flies due south at 500 km/h for 3 hours to airport B.
    (a) Calculate the position of airport B.

    The aeroplane then flies due west for 1 hour and 12 minutes at 475 km/h to airport C.
    (b) Calculate the position of airport C.
    (c) Calculate the average air speed to the nearest kilometre per hour.

4. A and B are two points on the equator. Point A is on longitude 15°E and B is 255 nautical miles west of A. Calculate the position of point B.

5. (a) Calculate the circumference of a great circle in nautical miles.
    (b) Given that 1 knot is 1 nautical mile per hour and that 1 nautical mile equals approximately 1.852 kilometres, calculate the speed of a plane in kilometres per hour if the plane travels at an average speed of 150 knots.

Chapter 18
Vectors

Pre-requisites
- matrices; Cartesian coordinates; translation

Naming vectors (revision)

A vector is any quantity which has direction as well as size. Displacement (or translation), velocity, force, acceleration are all examples of vectors. Fig. 18.1 shows a vector which moves a point from position A to position B.

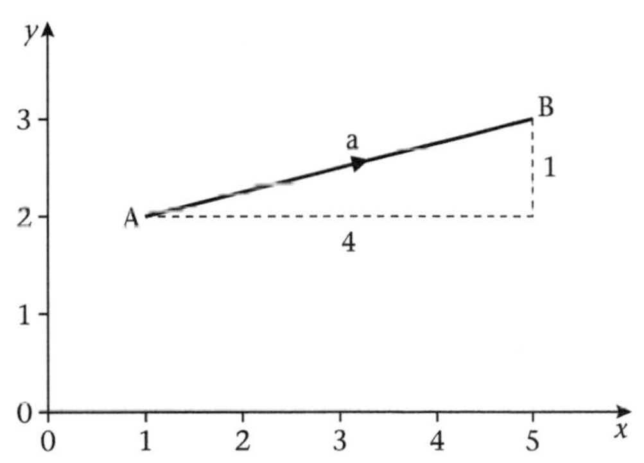

Fig. 18.1

The vector in Fig. 18.1 can be written in many ways:

either $\overline{AB}$, $\overrightarrow{AB}$, **AB**, $\underset{\sim}{AB}$

or $\bar{a}$, $\vec{a}$, **a**, $\underset{\sim}{a}$

Since the points are on a Cartesian plane, $\overrightarrow{AB}$ can also be written as a column matrix, or **column vector**:

$$\overrightarrow{AB} = \mathbf{a} = \begin{pmatrix} 4 \\ 1 \end{pmatrix}$$

Direction is important. $\overrightarrow{BA}$ is in the opposite direction to $\overrightarrow{AB}$, although they are parallel and have the same size:

$$\overrightarrow{BA} = -\overrightarrow{AB} = -\begin{pmatrix} 4 \\ 1 \end{pmatrix} = \begin{pmatrix} -4 \\ -1 \end{pmatrix}$$

Example 1

In Fig. 18.2, $\overrightarrow{AB}$, $\overrightarrow{BC}$, $\overrightarrow{CD}$, $\overrightarrow{DE}$ are vectors as shown.

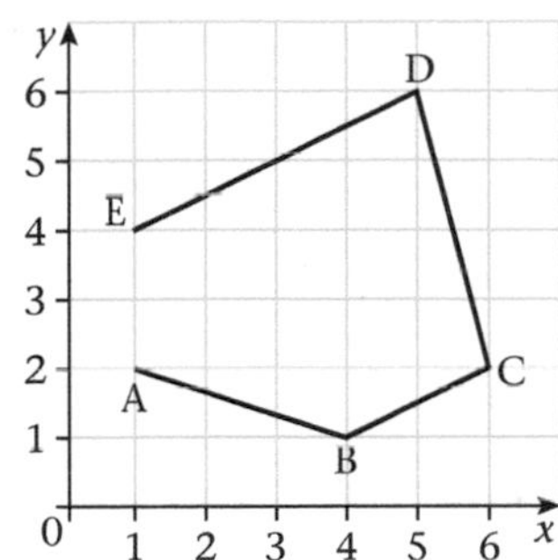

Fig. 18.2

Express each vector in the form $\begin{pmatrix} a \\ b \end{pmatrix}$.

$$\overrightarrow{AB} = \begin{pmatrix} 3 \\ -1 \end{pmatrix}, \overrightarrow{BC} = \begin{pmatrix} 2 \\ 1 \end{pmatrix},$$

$$\overrightarrow{CD} = \begin{pmatrix} -1 \\ 4 \end{pmatrix}, \overrightarrow{DE} = \begin{pmatrix} -4 \\ -2 \end{pmatrix}$$

Example 2

A point P(2, 3) is displaced by a translation vector to Q(2, 4). Write PQ as a column vector.

From Fig. 18.3, $\overrightarrow{PQ} = \begin{pmatrix} 0 \\ 1 \end{pmatrix}$

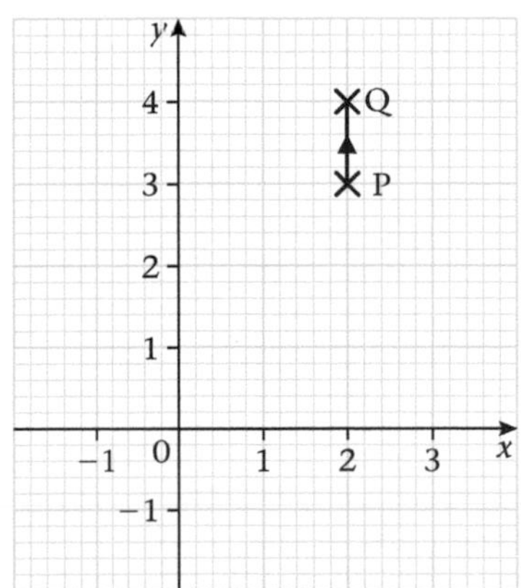

Fig. 18.3

Exercise 18a (revision)

1. If $\overrightarrow{AB} = \begin{pmatrix} 9 \\ -2 \end{pmatrix}$, write $\overrightarrow{BA}$ in the form $\begin{pmatrix} x \\ y \end{pmatrix}$.

2. Draw line segments to represent the following vectors.
   (a) $\overrightarrow{AB} = \begin{pmatrix} 4 \\ 1 \end{pmatrix}$ (b) $\overrightarrow{CD} = \begin{pmatrix} -6 \\ 2 \end{pmatrix}$
   (c) $\overrightarrow{EF} = \begin{pmatrix} 2 \\ -5 \end{pmatrix}$ (d) $\overrightarrow{GH} = \begin{pmatrix} -3 \\ -2 \end{pmatrix}$

3. Draw line segments to represent the following vectors.
   (a) $\overrightarrow{KL} = \begin{pmatrix} 4 \\ 0 \end{pmatrix}$ (b) $\overrightarrow{LM} = \begin{pmatrix} 0 \\ 4 \end{pmatrix}$
   (c) $\overrightarrow{MN} = \begin{pmatrix} -2 \\ 4 \end{pmatrix}$ (d) $\overrightarrow{NK} = \begin{pmatrix} -2 \\ -8 \end{pmatrix}$

4. (a) Draw line segments to represent the following vectors.
   $\overrightarrow{AB} = \begin{pmatrix} 3 \\ -1 \end{pmatrix}$, $\overrightarrow{BC} = \begin{pmatrix} 0 \\ -2 \end{pmatrix}$, $\overrightarrow{CD} = \begin{pmatrix} -5 \\ -3 \end{pmatrix}$
   (b) Hence give the vector $\overrightarrow{AD}$ in the form $\begin{pmatrix} a \\ b \end{pmatrix}$.
   (c) Write $\overrightarrow{DA}$ in the form $\begin{pmatrix} x \\ y \end{pmatrix}$.

5. (a) Draw the following vectors.
   $\overrightarrow{PQ} = \begin{pmatrix} 1 \\ 4 \end{pmatrix}$ $\overrightarrow{QR} = \begin{pmatrix} -7 \\ 5 \end{pmatrix}$
   (b) What is the vector $\overrightarrow{PR}$?
   (c) What is the vector $\overrightarrow{RP}$?

6. In Fig. 18.2 state the column vector which would displace E to A.

7. Express the following as positive vectors.
   (a) $-\begin{pmatrix} -3 \\ 1 \end{pmatrix}$ (b) $-\begin{pmatrix} 4 \\ -5 \end{pmatrix}$
   (c) $-\begin{pmatrix} -8 \\ -6 \end{pmatrix}$ (d) $-\begin{pmatrix} 2 \\ 7 \end{pmatrix}$

8. If $\overrightarrow{XY} = \begin{pmatrix} -8 \\ 5 \end{pmatrix}$ what is $\overrightarrow{YX}$?

9. A shape is translated through $\begin{pmatrix} 5 \\ -3 \end{pmatrix}$. It is then translated through $\begin{pmatrix} 2 \\ 8 \end{pmatrix}$.
   (a) What single translation is this equivalent to?
   (b) How far is the shape from its starting position?

10. $\triangle PQR$ is such that $\overrightarrow{PQ} = \begin{pmatrix} 6 \\ -2 \end{pmatrix}$ and $\overrightarrow{QR} = \begin{pmatrix} -9 \\ 9 \end{pmatrix}$.
    Sketch $\triangle PQR$ and hence or otherwise express $\overrightarrow{PR}$ and $\overrightarrow{RP}$ as column vectors.

Addition of vectors

In Fig. 18.4 it is clear that a translation $\overrightarrow{AB}$ followed by a translation $\overrightarrow{BC}$ is equivalent to the single translation $\overrightarrow{AC}$. We write this as the **vector sum**:

$$\overrightarrow{AB} + \overrightarrow{BC} = \overrightarrow{AC}$$

or, writing the vector sum with the small letters given in Fig. 18.4, **a** + **b** = **c**.

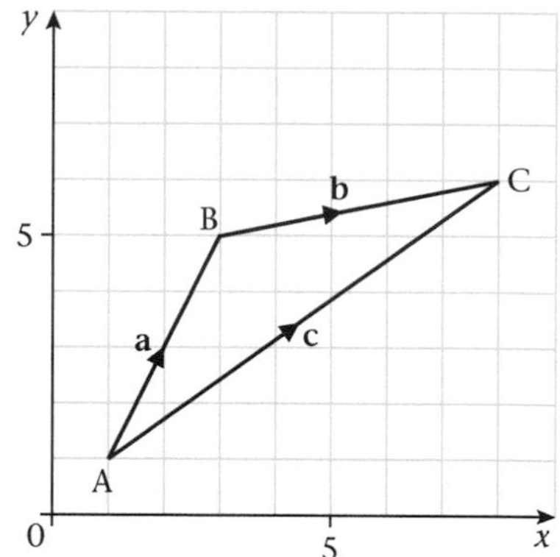

Fig. 18.4

By counting units in Fig. 18.4,

$$\mathbf{a} = \begin{pmatrix} 2 \\ 4 \end{pmatrix}, \mathbf{b} = \begin{pmatrix} 5 \\ 1 \end{pmatrix}, \mathbf{c} = \begin{pmatrix} 7 \\ 5 \end{pmatrix}$$

By matrix addition,

$$\mathbf{a} + \mathbf{b} = \begin{pmatrix} 2 \\ 4 \end{pmatrix} + \begin{pmatrix} 5 \\ 1 \end{pmatrix} = \begin{pmatrix} 2+5 \\ 4+1 \end{pmatrix} = \begin{pmatrix} 7 \\ 5 \end{pmatrix} = \mathbf{c}$$

In general, if $\mathbf{a} = \begin{pmatrix} x_1 \\ y_1 \end{pmatrix}$ and $\mathbf{b} = \begin{pmatrix} x_2 \\ y_2 \end{pmatrix}$

then $$\mathbf{a} + \mathbf{b} = \begin{pmatrix} x_1 \\ y_1 \end{pmatrix} + \begin{pmatrix} x_2 \\ y_2 \end{pmatrix} = \begin{pmatrix} x_1 + x_2 \\ y_1 + y_2 \end{pmatrix}$$

Note: There are many ways of writing vectors. The vector in Fig. 18.5 can be written

(a) using capital letters:
$\overrightarrow{AB}$ or **AB** or $\overline{AB}$

(b) using small letters:
$\vec{a}$ or **a** or $\bar{a}$

(c) using a column matrix:
$\overrightarrow{AB} = \mathbf{a} = \begin{pmatrix} p \\ q \end{pmatrix}$

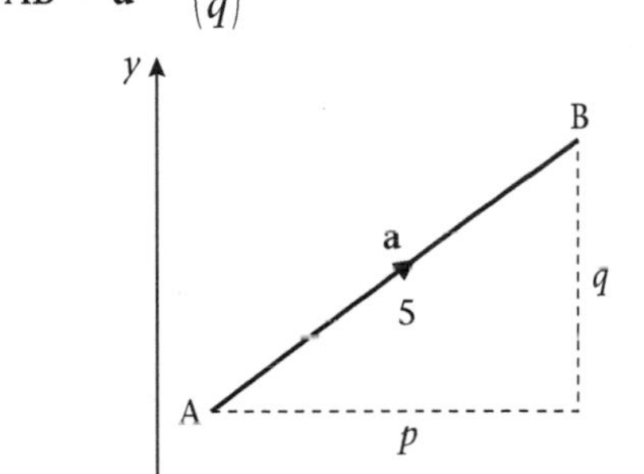

Fig. 18.5

Example 3

If $\mathbf{p} = \begin{pmatrix} 3 \\ 5 \end{pmatrix}$, $\mathbf{q} = \begin{pmatrix} -4 \\ 3 \end{pmatrix}$, $\mathbf{r} = \begin{pmatrix} 1 \\ -2 \end{pmatrix}$, find

(a) **p** + **q**, (b) **p** + **r**, (c) **q** + **r**, (d) **p** + **q** + **r**, showing the vector sum in part (d) on a diagram.

(a) $\mathbf{p} + \mathbf{q} = \begin{pmatrix} 3 \\ 5 \end{pmatrix} + \begin{pmatrix} -4 \\ 3 \end{pmatrix} = \begin{pmatrix} -1 \\ 8 \end{pmatrix}$

(b) $\mathbf{p} + \mathbf{r} = \begin{pmatrix} 3 \\ 5 \end{pmatrix} + \begin{pmatrix} 1 \\ -2 \end{pmatrix} = \begin{pmatrix} 4 \\ 3 \end{pmatrix}$

(c) $\mathbf{q} + \mathbf{r} = \begin{pmatrix} -4 \\ 3 \end{pmatrix} + \begin{pmatrix} 1 \\ -2 \end{pmatrix} = \begin{pmatrix} -3 \\ 1 \end{pmatrix}$

(d) $\mathbf{p} + \mathbf{q} + \mathbf{r} = \begin{pmatrix} 3 \\ 5 \end{pmatrix} + \begin{pmatrix} -4 \\ 3 \end{pmatrix} + \begin{pmatrix} 1 \\ -2 \end{pmatrix} = \begin{pmatrix} 0 \\ 6 \end{pmatrix}$

In Fig. 18.6, the broken line represents the vector **p** + **q** + **r**.

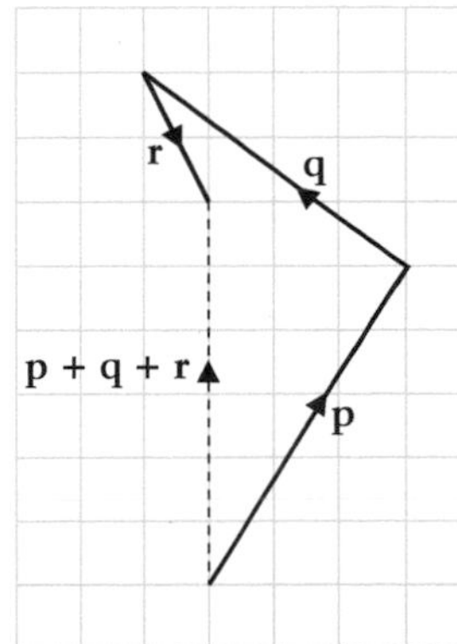

Fig. 18.6

This shows that $\mathbf{p} + \mathbf{q} + \mathbf{r} = \begin{pmatrix} 0 \\ 6 \end{pmatrix}$.

Exercise 18b

When using graph paper, take a scale of 1 cm to 1 unit.

1. In each part of Fig. 18.7, find **p** + **q**, giving the results in the form $\begin{pmatrix} a \\ b \end{pmatrix}$.

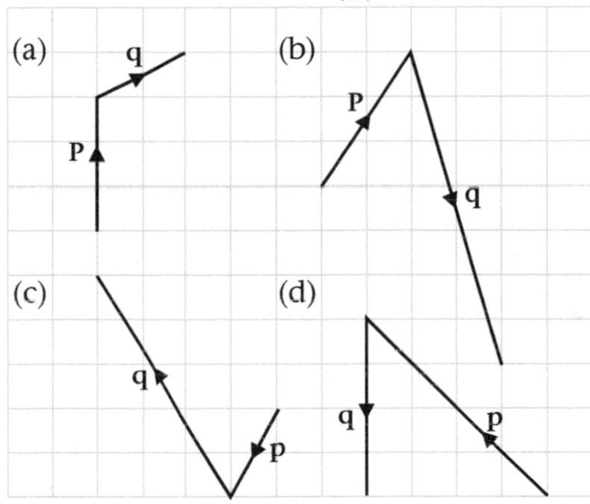

Fig. 18.7

2. If vectors **p**, **q**, **r**, **s** are as given in Fig. 18.8, express each of the following as a single vector in the form $\begin{pmatrix} a \\ b \end{pmatrix}$.

   (a) **p** + **q** (b) **p** + **q** + **r**
   (c) **q** + **r** (d) **q** + **r** + **s**
   (e) **r** + **s** (f) **p** + **q** + **r** + **s**

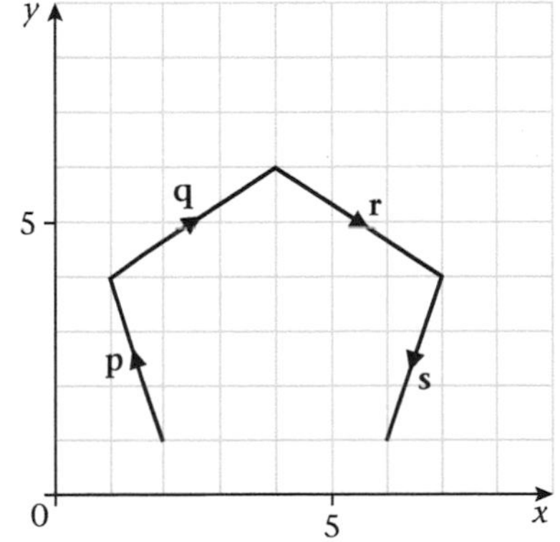

Fig. 18.8

3. (a) Draw vectors $\overrightarrow{PQ}$ and $\overrightarrow{QR}$ such that $\overrightarrow{PQ} = \begin{pmatrix} 2 \\ -1 \end{pmatrix}$ and $\overrightarrow{QR} = \begin{pmatrix} -5 \\ 3 \end{pmatrix}$.

   (b) Find $\overrightarrow{PR}$ (i) by drawing, (ii) by matrix addition.

4 If $\mathbf{p} = \begin{pmatrix} 0 \\ 4 \end{pmatrix}$, $\mathbf{q} = \begin{pmatrix} -2 \\ -3 \end{pmatrix}$, $\mathbf{r} = \begin{pmatrix} 4 \\ -2 \end{pmatrix}$, $\mathbf{s} = \begin{pmatrix} 2 \\ 0 \end{pmatrix}$, find

(a) $\mathbf{p} + \mathbf{q}$ (b) $\mathbf{p} + \mathbf{r}$
(c) $\mathbf{p} + \mathbf{s}$ (d) $\mathbf{q} + \mathbf{r}$
(e) $\mathbf{p} + \mathbf{s}$ (f) $\mathbf{r} + \mathbf{s}$
(g) $\mathbf{p} + \mathbf{q} + \mathbf{r}$ (h) $\mathbf{p} + \mathbf{q} + \mathbf{s}$
(i) $\mathbf{q} + \mathbf{r} + \mathbf{s}$ (j) $\mathbf{p} + \mathbf{q} + \mathbf{r} + \mathbf{s}$

5 A vector **b** is such that $\begin{pmatrix} 7 \\ 2 \end{pmatrix} + \mathbf{b} = \begin{pmatrix} 4 \\ -5 \end{pmatrix}$. Find **b**.

6 The Cartesian plane is translated through $\begin{pmatrix} 6 \\ -1 \end{pmatrix}$. It is then translated by $\begin{pmatrix} -2 \\ -2 \end{pmatrix}$.
(a) What single translation is this equivalent to?
(b) What are the coordinates of the final image of the point (−2, 7)?
(c) The final image of a point P is at (−2, 0). What are the coordinates of P?

Magnitude of a vector

In Fig. 18.9, $\overrightarrow{AB} = \begin{pmatrix} 3 \\ 4 \end{pmatrix}$ and $\overrightarrow{CD} = \begin{pmatrix} -4 \\ 3 \end{pmatrix}$.

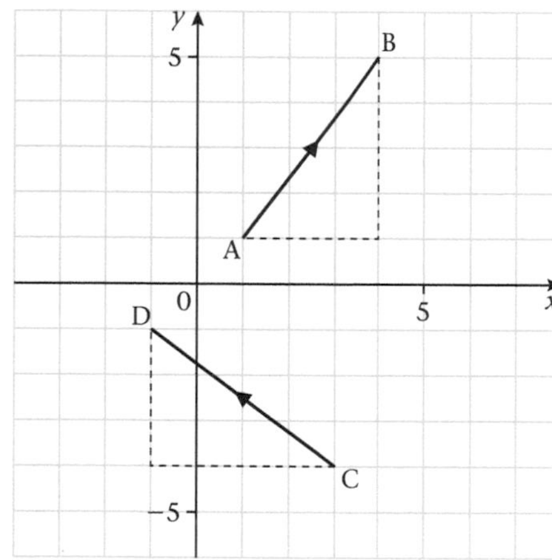

Fig. 18.9

The **magnitude**, or size, of $\overrightarrow{AB}$ is the length of the line segment $\overrightarrow{AB}$. This is often written as $|\overrightarrow{AB}|$ and is called the **modulus** of $\overrightarrow{AB}$.

$|\overrightarrow{AB}| = \sqrt{3^2 + 4^2}$ *(Pythagoras)*
$= 5$ units

Similarly,
$|\overrightarrow{CD}| = \sqrt{(-4)^2 + 3^2} = 5$ units
Hence different vectors may have the same magnitude.

In general, if $\mathbf{a} = \begin{pmatrix} x \\ y \end{pmatrix}$, then $|\mathbf{a}| = \sqrt{x^2 + y^2}$ where **a** is the magnitude of **a**. Notice that the magnitude of a vector is always given as a positive number of units.

A vector of unit length is called a **unit vector**.

Subtraction of vectors

In Fig. 18.10, $\mathbf{a} = \begin{pmatrix} 5 \\ 2 \end{pmatrix}$ and $\mathbf{b} = \begin{pmatrix} -5 \\ -2 \end{pmatrix}$.

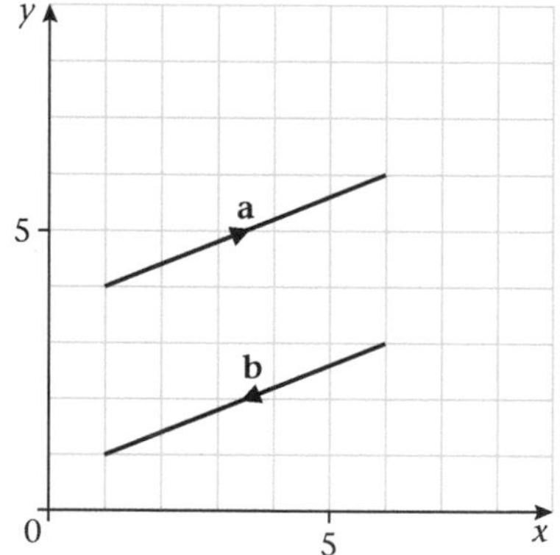

Fig. 18.10

Notice that **b** is a vector which has the same magnitude as **a** but which is in the opposite direction. We say that **b** = −**a**.

In general, if $\mathbf{a} = \begin{pmatrix} x \\ y \end{pmatrix}$, then $-\mathbf{a} = \begin{pmatrix} -x \\ -y \end{pmatrix}$.

Example 4

If $\mathbf{p} = \begin{pmatrix} 4 \\ 1 \end{pmatrix} - \begin{pmatrix} -2 \\ 7 \end{pmatrix}$, find **p** (a) by calculation. (b) by drawing, (c) Hence find the magnitude of **p**.

(a) Using matrix arithmetic:

$$\mathbf{p} = \begin{pmatrix} 4 \\ 1 \end{pmatrix} - \begin{pmatrix} -2 \\ 7 \end{pmatrix} = \begin{pmatrix} 4 - (-2) \\ 1 - \quad 7 \end{pmatrix} = \begin{pmatrix} 6 \\ -6 \end{pmatrix}$$

(b) If $-\begin{pmatrix} -2 \\ 7 \end{pmatrix} = +\begin{pmatrix} 2 \\ -7 \end{pmatrix}$

then $\begin{pmatrix} 4 \\ 1 \end{pmatrix} - \begin{pmatrix} -2 \\ 7 \end{pmatrix} = \begin{pmatrix} 4 \\ 1 \end{pmatrix} + \begin{pmatrix} 2 \\ -7 \end{pmatrix}$

Note: To subtract a vector is equivalent to adding a vector of the same size in the opposite direction. Fig. 18.11 shows the vector sum

$\begin{pmatrix} 4 \\ 1 \end{pmatrix} + \begin{pmatrix} 2 \\ -7 \end{pmatrix}$

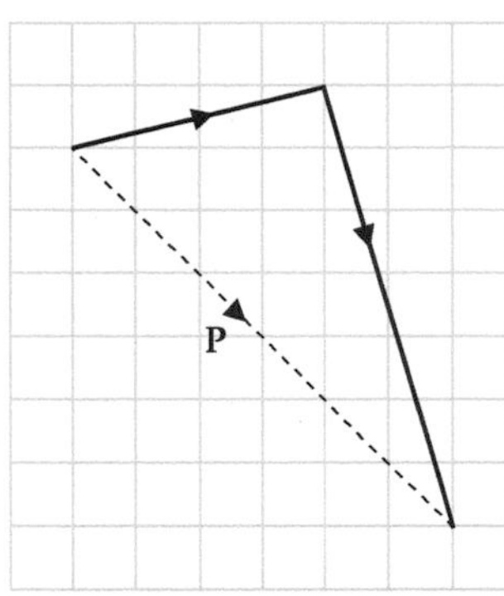

Fig. 18.11

In Fig. 18.11 the broken line represents the vector **p**.

$\mathbf{p} = \begin{pmatrix} 6 \\ -6 \end{pmatrix}$

(c) $|\mathbf{p}| = \sqrt{6^2 + (-6)^2}$

$= \sqrt{36 + 36} = \sqrt{72}$ units

Exercise 18c

1. Find the magnitudes of the following vectors.

   (a) $\begin{pmatrix} 4 \\ -3 \end{pmatrix}$ (b) $\begin{pmatrix} -5 \\ -12 \end{pmatrix}$ (c) $\begin{pmatrix} 8 \\ -6 \end{pmatrix}$

   (d) $\begin{pmatrix} 0 \\ 7 \end{pmatrix}$ (e) $\begin{pmatrix} -2 \\ 0 \end{pmatrix}$ (f) $\begin{pmatrix} -15 \\ 8 \end{pmatrix}$

2. Express the following as positive vectors.

   (a) $-\begin{pmatrix} 8 \\ -1 \end{pmatrix}$ (b) $-\begin{pmatrix} -2 \\ 3 \end{pmatrix}$ (c) $-\begin{pmatrix} 0 \\ 2 \end{pmatrix}$

   (d) $-\begin{pmatrix} -6 \\ -5 \end{pmatrix}$ (e) $-\begin{pmatrix} -4 \\ 0 \end{pmatrix}$ (f) $-\begin{pmatrix} 7 \\ 3 \end{pmatrix}$

3. If $\overrightarrow{AB} = \begin{pmatrix} 11 \\ 5 \end{pmatrix} + \begin{pmatrix} -2 \\ 7 \end{pmatrix}$, find $|\overrightarrow{AB}|$.

4. Find vector **p** such that $\begin{pmatrix} 5 \\ -6 \end{pmatrix} + \mathbf{p} = \begin{pmatrix} 1 \\ -2 \end{pmatrix}$.

5. Find vector **q** such that $\begin{pmatrix} -3 \\ 1 \end{pmatrix} - \mathbf{q} = \begin{pmatrix} -8 \\ 4 \end{pmatrix}$. Hence find $|\mathbf{q}|$.

6. Express each of the following as a single vector.

   (a) $\begin{pmatrix} 5 \\ 1 \end{pmatrix} - \begin{pmatrix} 4 \\ 2 \end{pmatrix}$ (b) $\begin{pmatrix} 8 \\ -3 \end{pmatrix} - \begin{pmatrix} -1 \\ 0 \end{pmatrix}$

   (c) $\begin{pmatrix} -1 \\ 4 \end{pmatrix} - \begin{pmatrix} -9 \\ -1 \end{pmatrix}$ (d) $\begin{pmatrix} -3 \\ -5 \end{pmatrix} - \begin{pmatrix} -4 \\ -9 \end{pmatrix}$

7. $\overrightarrow{AB} = \begin{pmatrix} 4 \\ -3 \end{pmatrix}$ and $\overrightarrow{BC} = \begin{pmatrix} -6 \\ 4 \end{pmatrix}$. Find $\overrightarrow{AB} - \overrightarrow{BC}$
   (a) using matrix arithmetic, (b) by drawing.

8. If $\mathbf{p} = \begin{pmatrix} 7 \\ -3 \end{pmatrix}$ and $\mathbf{q} = \begin{pmatrix} -6 \\ 2 \end{pmatrix}$ find

   (a) $\mathbf{p} - \mathbf{q}$ (b) $\mathbf{p} + \mathbf{q}$ (c) $\mathbf{q} - \mathbf{p}$

   (d) $|\mathbf{p} + \mathbf{q}|$ (e) $|\mathbf{p} - \mathbf{q}|$

9. If $\mathbf{p} = \begin{pmatrix} 6 \\ 3 \end{pmatrix}$, $\mathbf{q} = \begin{pmatrix} -1 \\ 5 \end{pmatrix}$, $\mathbf{r} = \begin{pmatrix} -4 \\ 2 \end{pmatrix}$,

   $\mathbf{s} = \begin{pmatrix} 0 \\ -3 \end{pmatrix}$, find

   (a) $\mathbf{p} - \mathbf{q}$ (b) $\mathbf{p} + \mathbf{q} - \mathbf{r}$

   (c) $\mathbf{p} - \mathbf{r}$ (d) $\mathbf{p} + \mathbf{r} - \mathbf{s}$

   (e) $\mathbf{s} - \mathbf{q}$ (f) $\mathbf{q} + \mathbf{r} - \mathbf{s}$

   (g) $(\mathbf{p} - \mathbf{q}) - \mathbf{r}$ (h) $\mathbf{p} - (\mathbf{q} - \mathbf{r})$

10. Draw $\overrightarrow{OP} = \begin{pmatrix} 3 \\ -5 \end{pmatrix}$ and $\overrightarrow{OQ} = \begin{pmatrix} 6 \\ 7 \end{pmatrix}$.

    (a) Use your drawing to find $\overrightarrow{PQ}$.
    (b) Use any method to find $\overrightarrow{OQ} - \overrightarrow{OP}$.
    (c) What do you notice?
    (d) Does this happen for any two vectors $\overrightarrow{OP}$ and $\overrightarrow{OQ}$?

Multiplication by a scalar

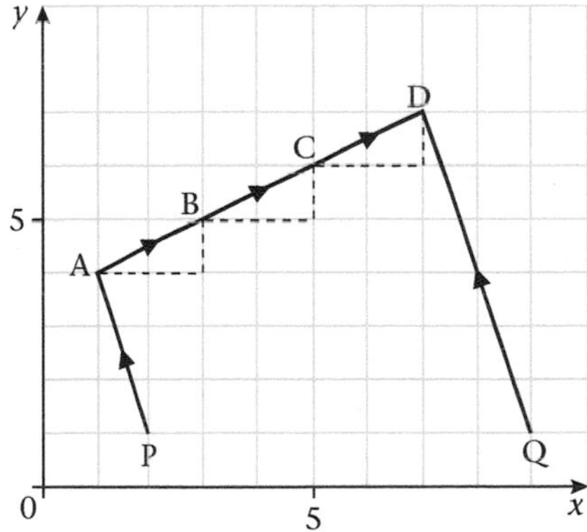

Fig. 18.12

In Fig. 18.12, $\overrightarrow{AB} = \overrightarrow{BC} = \overrightarrow{CD} = \begin{pmatrix} 2 \\ 1 \end{pmatrix}$

Also $\overrightarrow{AD} = \overrightarrow{AB} + \overrightarrow{BC} + \overrightarrow{CD}$

$= \begin{pmatrix} 2 \\ 1 \end{pmatrix} + \begin{pmatrix} 2 \\ 1 \end{pmatrix} + \begin{pmatrix} 2 \\ 1 \end{pmatrix}$

$= \begin{pmatrix} 6 \\ 3 \end{pmatrix} = 3\begin{pmatrix} 2 \\ 1 \end{pmatrix}$

Hence $\overrightarrow{AD} = 3\overrightarrow{AB}$ or $3\overrightarrow{BC}$ or $3\overrightarrow{CD}$.

If any of the vectors AB, BC or CD are multiplied by the **scalar** 3, the result is a vector 3 times as big.

Note: A scalar is simply a numerical multiplier.

Also in Fig. 18.12,

$$\overrightarrow{QD} = \begin{pmatrix} -2 \\ 6 \end{pmatrix}$$

and $\overrightarrow{PA} = \begin{pmatrix} -1 \\ 3 \end{pmatrix} = \frac{1}{2}\begin{pmatrix} -2 \\ 6 \end{pmatrix}$

Hence $\overrightarrow{PA} = \frac{1}{2}\overrightarrow{QD}$.

In the first case the vectors $\overrightarrow{AB}$, $\overrightarrow{BC}$, $\overrightarrow{CD}$ lie along the line of vector $\overrightarrow{AD}$. In the second case vectors $\overrightarrow{PA}$ and $\overrightarrow{QD}$ are *not* in the same straight line. In both cases the resultant vector is parallel to the original vector.

In general, if a vector **a** is multiplied by a scalar k, the result is a new vector k**a** which is in the same direction as **a** but which is k times as big. Vectors **a** and k**a** are parallel.

Example 5

Vectors $\overrightarrow{OA}$, $\overrightarrow{OB}$, $\overrightarrow{OC}$, $\overrightarrow{OP}$, $\overrightarrow{OQ}$, $\overrightarrow{OR}$ are such that:

$\overrightarrow{AB} = \begin{pmatrix} 2 \\ 1 \end{pmatrix}$, $\overrightarrow{OB} = \begin{pmatrix} 2 \\ 4 \end{pmatrix}$, $\overrightarrow{OC} = \begin{pmatrix} 3 \\ 1 \end{pmatrix}$ and

$\overrightarrow{OP} - 2\overrightarrow{OA}$, $\overrightarrow{OQ} = -2\overrightarrow{OB}$, $\overrightarrow{OR} = -2\overrightarrow{OC}$.

Take O as origin and draw the vectors on a Cartesian plane. Compare △PQR with △ABC.

The coordinates of A, B, C are (2, 1), (2, 4), (3, 1) respectively.

$$\overrightarrow{OP} = -2\overrightarrow{OA} = -2\begin{pmatrix} 2 \\ 1 \end{pmatrix} = \begin{pmatrix} -4 \\ -2 \end{pmatrix}$$

Similarly $\overrightarrow{OQ} = \begin{pmatrix} -4 \\ -8 \end{pmatrix}$ and $\overrightarrow{OR} = \begin{pmatrix} -6 \\ -2 \end{pmatrix}$

The coordinates of P, Q, R are (−4, −2), (−4, −8), (−6, −2) respectively.

Fig. 18.13 shows the vectors and △sABC and PQR.

Comparing the triangles,

1 △PQR is an enlargement of △ABC.
2 Each side of △PQR is twice as long as the corresponding side of △ABC.
3 △PQR is four times the area of △ABC.
4 △PQR appears to be rotated with respect to △ABC.

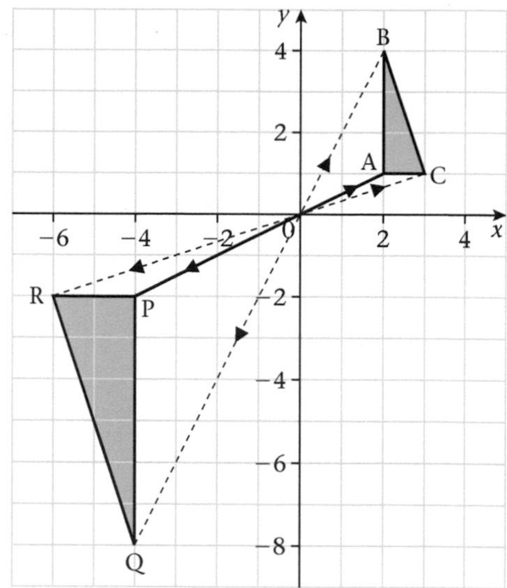

Fig. 18.13

Exercise 18d

Use Fig. 18.14 when answering questions 1 to 7.

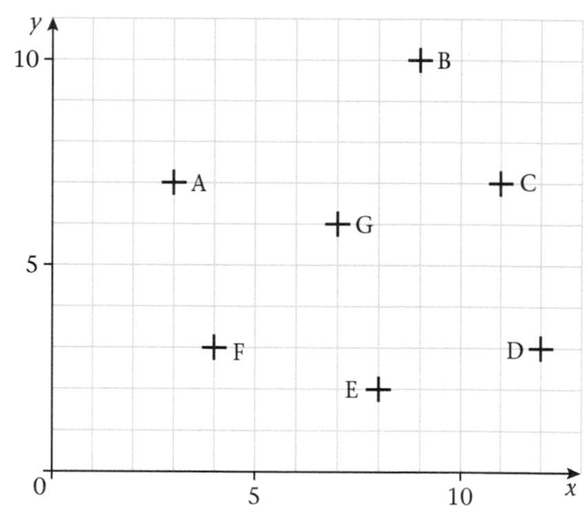

Fig. 18.14

1 Express each of the following as a column vector.

(a) $3\overrightarrow{OA}$ (b) $5\overrightarrow{ED}$ (c) $6\overrightarrow{CA}$
(d) $2\overrightarrow{CE}$ (e) $4\overrightarrow{AF}$ (f) $4\overrightarrow{FA}$
(g) $3\overrightarrow{GC}$ (h) $3\overrightarrow{ED}$

2 Express each of the following as a column vector.

(a) $\frac{1}{2}\overrightarrow{EO}$ (b) $\frac{1}{4}\overrightarrow{AC}$
(c) $\frac{1}{3}\overrightarrow{BA}$ (d) $\frac{1}{3}\overrightarrow{OD}$

3 Express each of the following as a column vector.

(a) $-2\overrightarrow{AG}$ (b) $-5\overrightarrow{BC}$

(c) $-\frac{3}{4}\overrightarrow{AC}$ (d) $-\frac{1}{2}\overrightarrow{EG}$

4 Calculate the following.

(a) $2\overrightarrow{OA} + \overrightarrow{BA}$ (b) $5\overrightarrow{ED} + 2\overrightarrow{DO}$

(c) $3\overrightarrow{OF} - 2\overrightarrow{AB}$ (d) $4\overrightarrow{EG} - \overrightarrow{ED}$

5 Calculate the following.

(a) $\frac{1}{4}\overrightarrow{OF} + \frac{3}{4}\overrightarrow{AG}$ (b) $\frac{2}{3}\overrightarrow{AB} + \frac{1}{2}\overrightarrow{EO}$

(c) $\frac{1}{2}\overrightarrow{BC} - 1\frac{1}{2}\overrightarrow{ED}$ (d) $\frac{1}{3}\overrightarrow{OD} - \frac{3}{4}\overrightarrow{CA}$

6 Calculate the following.

(a) $\frac{1}{6}(\overrightarrow{OA} + \overrightarrow{EC})$ (b) $\frac{1}{2}(\overrightarrow{OA} + \overrightarrow{AG} + \overrightarrow{GE})$

(c) $2(\overrightarrow{EC} - \overrightarrow{EG})$ (d) $\frac{1}{2}(\overrightarrow{AG} - \overrightarrow{GC})$

7 Calculate the following.

(a) $3(2\overrightarrow{OA} + \overrightarrow{GE})$ (b) $6\overrightarrow{OA} + 3\overrightarrow{GE}$

(c) $5(\overrightarrow{OE} - 3\overrightarrow{EC})$ (d) $5\overrightarrow{OE} - 15\overrightarrow{EC}$

(e) $\frac{1}{2}(2\overrightarrow{BC} - 3\overrightarrow{OE})$ (f) $\overrightarrow{BC} - 1\frac{1}{2}\overrightarrow{OE}$

(g) $-2(3\overrightarrow{ED}\ 1\ \overrightarrow{BC})$ (h) $-6\overrightarrow{ED} - 2\overrightarrow{BC}$

In questions 8 and 9 use a scale of 1 cm to 1 unit.

8 Vectors $\overrightarrow{OA}$, $\overrightarrow{OB}$, $\overrightarrow{OC}$, $\overrightarrow{OP}$, $\overrightarrow{OQ}$, $\overrightarrow{OR}$ are such that

$\overrightarrow{OA} = \begin{pmatrix} 3 \\ 4 \end{pmatrix}$, $\overrightarrow{OB} = \begin{pmatrix} 1 \\ -2 \end{pmatrix}$, $\overrightarrow{OC} = \begin{pmatrix} 4 \\ -3 \end{pmatrix}$.

and $\overrightarrow{OP} = 3\overrightarrow{OA}$, $\overrightarrow{OQ} = 3\overrightarrow{OB}$, $\overrightarrow{OR} = 3\overrightarrow{OC}$. Take O as origin and draw the vectors on a Cartesian plane.

Compare $\triangle$PQR with $\triangle$ABC.

9 Given the same vectors $\overrightarrow{OA}$, $\overrightarrow{OB}$, $\overrightarrow{OC}$ as in question 8, find, by drawing, the positions of K, L, M such that $\overrightarrow{OK} = -2\overrightarrow{OA}$, $\overrightarrow{OL} = -2\overrightarrow{OB}$, $\overrightarrow{OM} = -2\overrightarrow{OC}$. Compare $\triangle$KLM with $\triangle$ABC.

Resultant vectors

A translation $\overrightarrow{AB}$ followed by a translation $\overrightarrow{BC}$ is equivalent to the single translation $\overrightarrow{AC}$, i.e. $\overrightarrow{AB} + \overrightarrow{BC} = \overrightarrow{AC}$ as in Fig. 18.15.

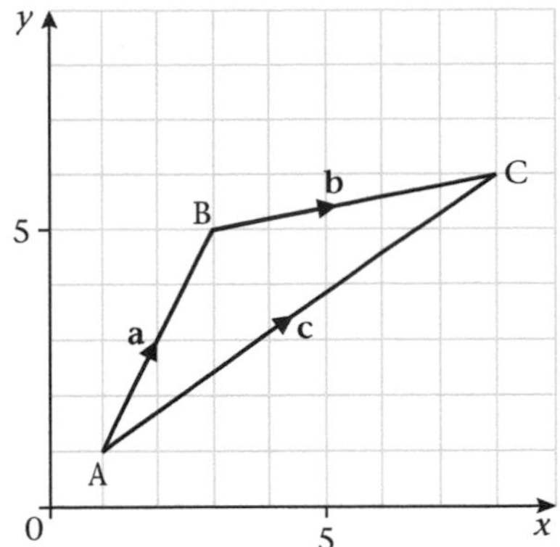

Fig. 18.15

$\overrightarrow{AC}$ is known as the **resultant vector**. By the **triangle law** the resultant of any two vectors is given by the length and direction of the line needed to complete the triangle.

Example 6

Find, by drawing, the resultant of the vectors $\overrightarrow{AB}$ and $\overrightarrow{AC}$ in Fig. 18.16(a).

(a)

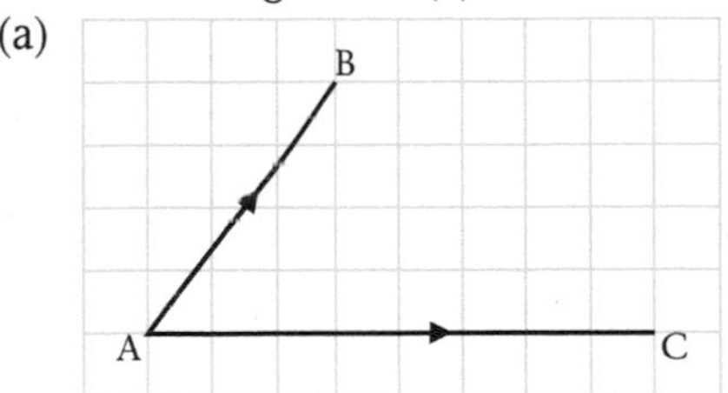

(b)

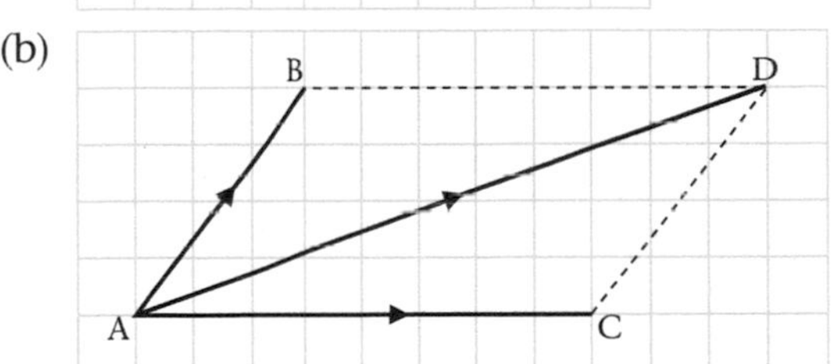

Fig. 18.16

$\overrightarrow{AD}$ in Fig. 18.16(b) is the required resultant vector.

Note that in Fig. 18.16(a) the arrows on the vectors $\overrightarrow{AB}$ and $\overrightarrow{AC}$ are pointing away from A. The triangle law could therefore not be used to find the resultant vector. The **parallelogram law** for the addition of the two vectors $\overrightarrow{AB}$ and $\overrightarrow{AC}$ with a common starting point A states that the resultant is given by the diagonal $\overrightarrow{AD}$ of the *completed parallelogram.*

Example 7

Find the resultant of the vectors in Fig. 18.17(a).

(a)

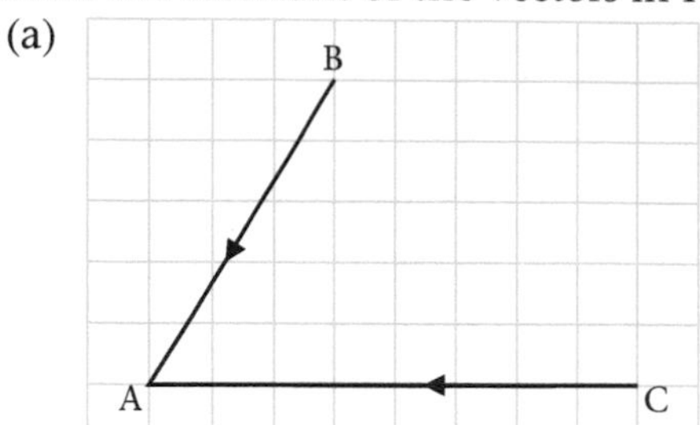

(b)

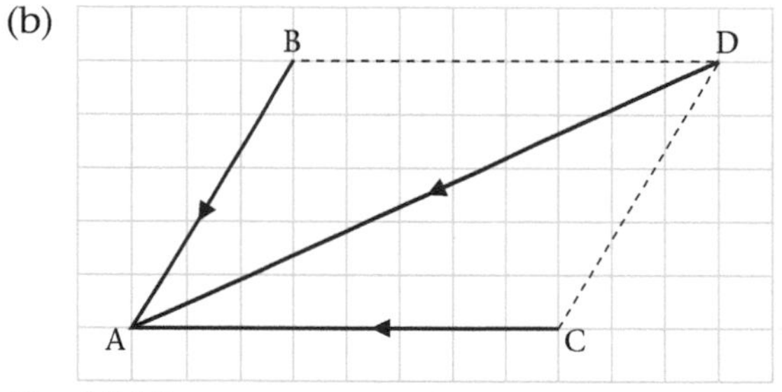

Fig. 18.17

$\overrightarrow{DA}$ in Fig. 18.17(b) is the required resultant vector.

Note the direction of the arrows in Examples 6 and 7. The arrows on the original vectors and on the resultant vector all point either away from or towards A.

Position vectors

In Fig. 18.18, P is a point (x, y) on the Cartesian plane, origin O.

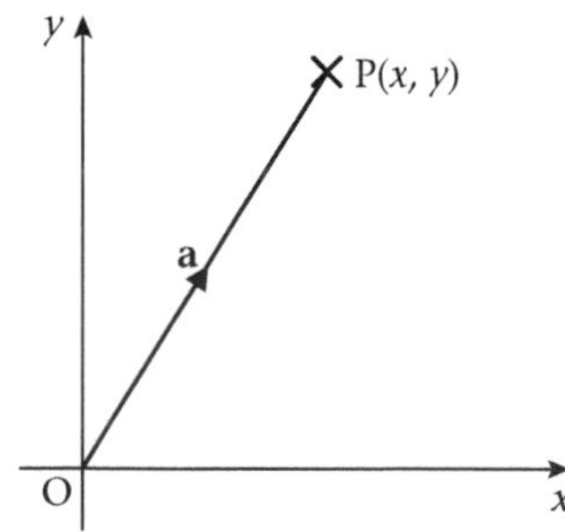

Fig. 18.18

Vector **a** is the displacement of P from O. Since this displacement gives the position of P relative to the origin, **a** is called the **position vector** of P.

In Fig. 18.18, $\mathbf{a} = \overrightarrow{OP} = \begin{pmatrix} x \\ y \end{pmatrix}$.

Hence if a point has coordinates (x, y), its position vector is $\begin{pmatrix} x \\ y \end{pmatrix}$.

Position vectors can be used to find displacements between points.

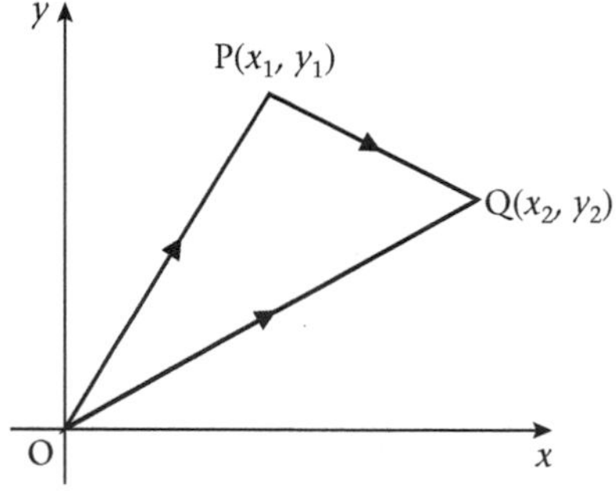

Fig. 18.19

In Fig. 18.19, by adding vectors,

$$\overrightarrow{OP} + \overrightarrow{PQ} = \overrightarrow{OQ}$$

$$\overrightarrow{PQ} = \overrightarrow{OQ} - \overrightarrow{OP}$$

$$\overrightarrow{PQ} = \begin{pmatrix} x_2 \\ y_2 \end{pmatrix} - \begin{pmatrix} x_1 \\ y_1 \end{pmatrix}$$

$$= \begin{pmatrix} x_2 - x_1 \\ y_2 - y_1 \end{pmatrix}$$

Also, by Pythagoras' theorem,

$$|\overrightarrow{PQ}| = \sqrt{(x_2 - x_1)^2 + (y_2 - y_1)^2}$$

The above results hold for any two general points $P(x_1, y_1)$ and $Q(x_2, y_2)$.

Example 8

If P and Q are the points (3, 7) and (11, 13) respectively, find $\overrightarrow{PQ}$ and $|\overrightarrow{PQ}|$.

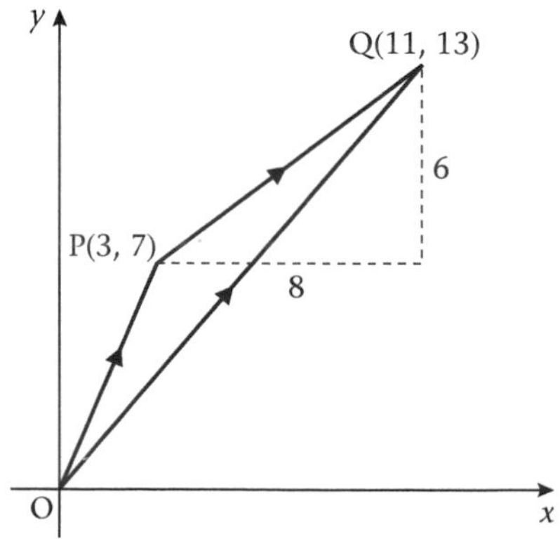

Fig. 18.20

$$\overrightarrow{OQ} = \overrightarrow{OP} + \overrightarrow{PQ} \qquad \textit{(triangle law)}$$
$$\Rightarrow \overrightarrow{PQ} = \overrightarrow{OQ} - \overrightarrow{OP}$$
$$= \begin{pmatrix}11\\13\end{pmatrix} - \begin{pmatrix}3\\7\end{pmatrix} = \begin{pmatrix}11-3\\13-7\end{pmatrix} = \begin{pmatrix}8\\6\end{pmatrix}$$
$$|\overrightarrow{PQ}| = \sqrt{(11-3)^2 + (13-7)^2}$$
$$= \sqrt{8^2 + 6^2} = \sqrt{100} = 10$$

Example 9

Quadrilateral OPQR is as shown in Fig. 18.21.
(a) Show that OPQR is a parallelogram and
(b) find the coordinates of the point of intersection of its diagonals.

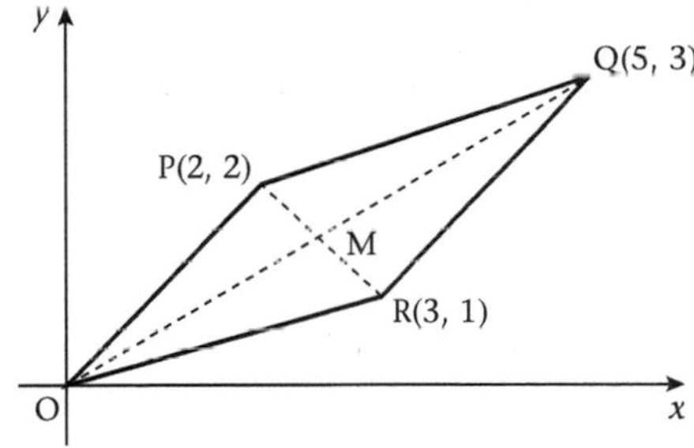

Fig. 18.21

(a) Using the position vectors of O, P, Q and R,

$\overrightarrow{OP} = \begin{pmatrix}2\\2\end{pmatrix}$ (position vector of P)

$\overrightarrow{PQ} = \begin{pmatrix}5\\3\end{pmatrix} - \begin{pmatrix}2\\2\end{pmatrix} = \begin{pmatrix}3\\1\end{pmatrix}$

$\overrightarrow{RQ} = \begin{pmatrix}5\\3\end{pmatrix} - \begin{pmatrix}3\\1\end{pmatrix} = \begin{pmatrix}2\\2\end{pmatrix}$

$\overrightarrow{OR} = \begin{pmatrix}3\\1\end{pmatrix}$ (position vector of R)

Hence $\overrightarrow{OP} = \overrightarrow{RQ}$
and $\overrightarrow{PQ} = \overrightarrow{OR}$

Considering the sides OP and RQ:
If $\overrightarrow{OP} = \overrightarrow{RQ}$
then $|\overrightarrow{OP}| = |\overrightarrow{RQ}|$ and $\overrightarrow{OP} \parallel \overrightarrow{RQ}$ since equal vectors have the same magnitude and direction.
Hence OPQR is a parallelogram since it has a pair of opposite sides equal and parallel.

(b) Let the diagonals intersect at M.

$\overrightarrow{OM} = \frac{1}{2}\overrightarrow{OQ}$ *(the diagonals of a parallelogram bisect each other)*

$\overrightarrow{OM} = \frac{1}{2}\begin{pmatrix}5\\3\end{pmatrix} = \begin{pmatrix}2\frac{1}{2}\\1\frac{1}{2}\end{pmatrix}$

The diagonals intersect at the point $(2\frac{1}{2}, 1\frac{1}{2})$.

Example 9 makes use of the following important result:

If $\mathbf{a} = \mathbf{b}$
then $|\mathbf{a}| = |\mathbf{b}|$ and $\mathbf{a} \parallel \mathbf{b}$.

Example 10

In Fig. 18.22 A(5, 6), B(1, 8), C(p, 4), D(q, r) are the vertices of a rhombus in the positive quadrant of the Cartesian plane.

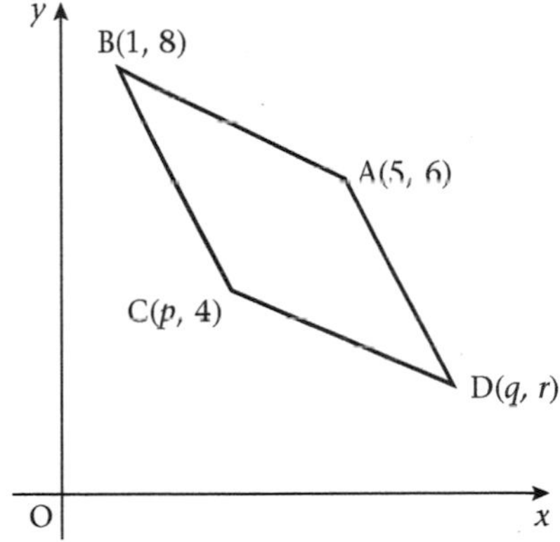

Fig. 18.22

Find p and hence find the coordinates of D.

If ABCD is a rhombus, then adjacent sides are equal:

$$|\overrightarrow{AB}| = |\overrightarrow{BC}| \qquad (1)$$

and opposite sides are equal and parallel:

$$\overrightarrow{AB} = \overrightarrow{DC} \qquad (2)$$

From (1),

$$\sqrt{(1-5)^2 + (8-6)^2} = \sqrt{(p-1)^2 + (4-8)^2}$$
$$\Rightarrow \quad (-4)^2 + (2)^2 = (p-1)^2 + (-4)^2$$
$$\Rightarrow \quad (p-1)^2 = (2)^2$$
$$\Rightarrow \quad p - 1 = 2, \text{ since } p > 0$$
$$\Rightarrow \quad p = 3$$

Let D have coordinates (q, r),
From (2),

$$\begin{pmatrix}1-5\\8-6\end{pmatrix} = \begin{pmatrix}p-q\\4-r\end{pmatrix}$$

Hence $\begin{pmatrix}-4\\2\end{pmatrix} = \begin{pmatrix}3-q\\4-r\end{pmatrix}$

$\Rightarrow \quad q = 7$
and $\quad r = 2$

$p = 3$ and D is the point (7, 2).

Exercise 18e

1. Given points A(7, 8) and B(2, −1), find (a) $\overrightarrow{AB}$, (b) $\overrightarrow{BA}$.

2. The points O, P, Q, R, S have coordinates (0, 0), (1, 5), (3, 8), (7, 10), (10, 3) respectively. Express each of the following as a column vector.
   (a) $\overrightarrow{OQ}$ (b) $\overrightarrow{OS}$
   (c) $\overrightarrow{PQ}$ (d) $\overrightarrow{QR}$
   (e) $\overrightarrow{QS}$ (f) $\overrightarrow{RP}$

3. 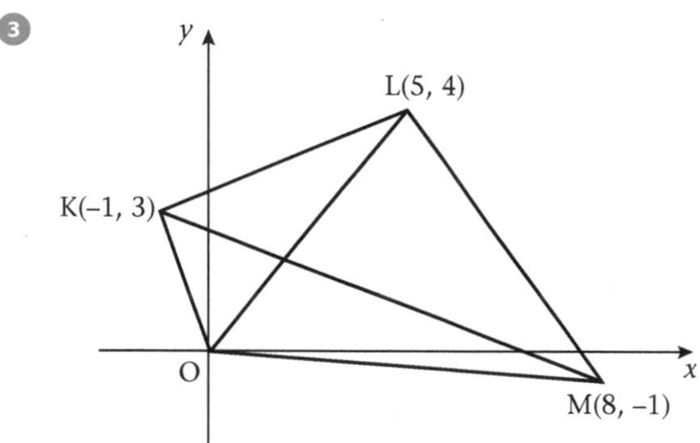

Fig. 18.23

   Given Fig. 18.23, express each of the following as a single column vector.
   (a) $\overrightarrow{OK}$ (b) $\overrightarrow{OM}$
   (c) $\overrightarrow{KL}$ (d) $\overrightarrow{LM}$
   (e) $\overrightarrow{OL}$ (f) $\overrightarrow{KM}$
   (g) $\overrightarrow{OK} + \overrightarrow{KL}$ (h) $\overrightarrow{OL} + \overrightarrow{LM}$
   (i) $\overrightarrow{MK} + \overrightarrow{KL}$ (j) $\overrightarrow{ML} + \overrightarrow{LO}$

4. OABC is a parallelogram where O is the origin, $\overrightarrow{OA} = \begin{pmatrix} 1 \\ 4 \end{pmatrix}$, $\overrightarrow{OC} = \begin{pmatrix} 5 \\ 2 \end{pmatrix}$.
   (a) On graph paper mark and clearly label the points A, B and C.
   (b) Express as a column vector (i) $\overrightarrow{OB}$, (ii) $\overrightarrow{CA}$.

5. Use vectors to show that the quadrilateral P(−3, 0), Q(−1, 6), R(3, 5), S(5, −2) is a trapezium.

6. Use vectors to show that the quadrilateral A(3, −5), B(8, 5), C(6, 16), D(1, 6) is a rhombus.

7. Prove that the quadrilateral O(0, 0), A(4, 0), B(7, 5), C(3, 5) is a parallelogram.

8. Show that P(3, 2), Q(9, 4), R(11, 8), S(5, 6) is a parallelogram. Use a vector method to find the coordinates of the point of intersection of its diagonals.

9. O(0, 0), P(4, 6), Q, R(8, 2) are vertices of a quadrilateral. Find the coordinates of Q such that OPQR is a parallelogram. Find the coordinates of the point of intersection of its diagonals.

10. Points M and N have position vectors **m** and **n** respectively relative to the origin O. If $\mathbf{m} = \begin{pmatrix} 1 \\ -4 \end{pmatrix}$ and $\overrightarrow{MN} = \begin{pmatrix} 7 \\ 10 \end{pmatrix}$, find
    (a) **n** (b) $|\mathbf{n}|$
    (c) the coordinates of a point P such that OM is the short diagonal of parallelogram MNOP.

Vector geometry

In the previous section, position vectors were restricted to the Cartesian plane. However, vector methods can be used in any geometrical situation. They are often used to discover and prove properties of shapes.

In Fig. 18.24, PQRS is a parallelogram as shown.

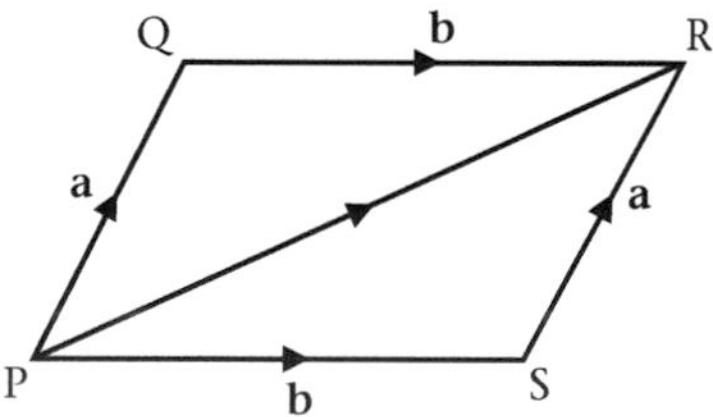

Fig. 18.24

$\overrightarrow{PR} = \overrightarrow{PQ} + \overrightarrow{QR}$ or $\overrightarrow{PS} + \overrightarrow{SR}$
$= \mathbf{a} + \mathbf{b}$ or $\mathbf{b} + \mathbf{a}$

Hence $\mathbf{a} + \mathbf{b} = \mathbf{b} + \mathbf{a}$

This result shows that the addition of vectors is not affected by the order in which they are taken.

In Fig. 18.25, ABCD is any quadrilateral with vectors **a**, **b**, **c**, **d** as shown.

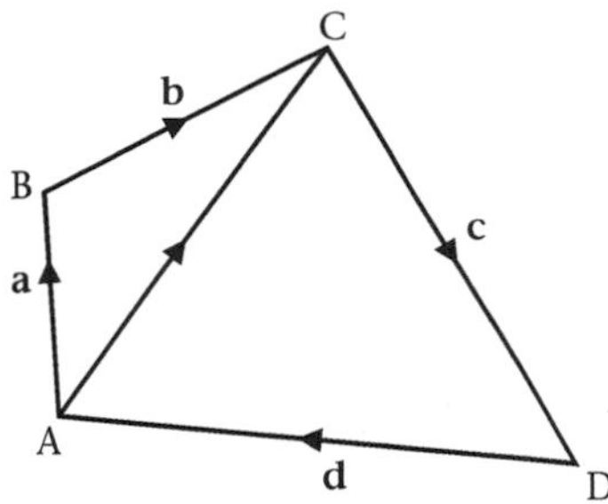

Fig. 18.25

$$\mathbf{a} + \mathbf{b} = \overrightarrow{AC}$$
$$\mathbf{c} + \mathbf{d} = \overrightarrow{CA}$$

adding,

$$\mathbf{a} + \mathbf{b} + \mathbf{c} + \mathbf{d} = \overrightarrow{AC} + \overrightarrow{CA}$$

but $\quad \overrightarrow{AC} + \overrightarrow{CA} = \mathbf{0}$

so $\quad \mathbf{a} + \mathbf{b} + \mathbf{c} + \mathbf{d} = \mathbf{0}$

If the vectors in Fig. 18.25 are taken to be displacements and the + sign is thought of as meaning 'followed by', the above result is hardly surprising. The total final displacement from the starting point, A, is zero when the vectors form the sides of a closed polygon.

Notice how the above results are used in the following examples.

Example 11

PQRS is any quadrilateral, A, B, C, D are the mid-points of PQ, QR, RS, SP respectively. Prove that ABCD is a parallelogram.

Let $\overrightarrow{PQ} = 2\mathbf{p}$, $\overrightarrow{QR} = 2\mathbf{q}$, $\overrightarrow{RS} = 2\mathbf{r}$, $\overrightarrow{PS} = 2\mathbf{s}$, as shown in Fig. 18.26.

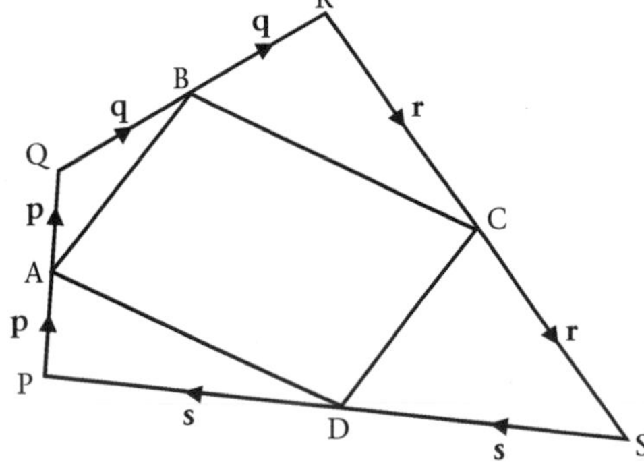

Fig. 18.26

Considering the opposite sides AB and CD of quadrilateral ABCD:

$$\overrightarrow{AB} = \mathbf{p} + \mathbf{q}$$
$$\overrightarrow{CD} = \mathbf{r} + \mathbf{s}$$

But $\quad 2\mathbf{p} + 2\mathbf{q} + 2\mathbf{r} + 2\mathbf{s} = \mathbf{0}$

$\Leftrightarrow \quad \mathbf{p} + \mathbf{q} + \mathbf{r} + \mathbf{s} = \mathbf{0}$

$\Leftrightarrow \quad \mathbf{p} + \mathbf{q} = -\mathbf{r} - \mathbf{s}$

$\Leftrightarrow \quad \mathbf{p} + \mathbf{q} = -(\mathbf{r} + \mathbf{s})$

Hence $\overrightarrow{AB} = \mathbf{p} + \mathbf{q} = -(\mathbf{r} + \mathbf{s}) = -\overrightarrow{CD}$

i.e. $\overrightarrow{AB} = \overrightarrow{DC}$

If $\overrightarrow{AB} = \overrightarrow{DC}$, then AB $\parallel$ DC and AB = DC. ABCD is a parallelogram since it has a pair of opposite sides which are parallel and equal.

Example 12

In Fig. 18.27, P divides the line AB in the ratio AP : PB = 7 : 3. If $\overrightarrow{OA} = \mathbf{a}$ and $\overrightarrow{OB} = \mathbf{b}$, express $\overrightarrow{OP}$ in terms of $\mathbf{a}$ and $\mathbf{b}$.

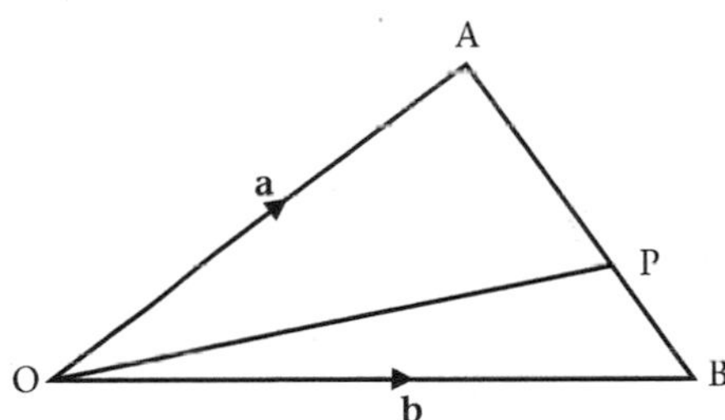

Fig. 18.27

In $\triangle$OAB,

$$\overrightarrow{OA} + \overrightarrow{AB} = \overrightarrow{OB}$$
$$\mathbf{a} + \overrightarrow{AB} = \mathbf{b}$$
$$\overrightarrow{AB} = \mathbf{b} - \mathbf{a}$$

Along AB,

$$\overrightarrow{AP} = \tfrac{7}{10}\overrightarrow{AB}$$
$$= \tfrac{7}{10}(\mathbf{b} - \mathbf{a})$$

In $\triangle$OAP

$$\overrightarrow{OP} = \overrightarrow{OA} + \overrightarrow{AP}$$
$$= \mathbf{a} + \tfrac{7}{10}(\mathbf{b} - \mathbf{a})$$
$$= \mathbf{a} + \tfrac{7}{10}\mathbf{b} - \tfrac{7}{10}\mathbf{a}$$
$$= \tfrac{3}{10}\mathbf{a} + \tfrac{7}{10}\mathbf{b}$$

Example 13

In Fig. 18.28, $\overrightarrow{OA} = \mathbf{a}$ and $\overrightarrow{OB} = \mathbf{b}$.

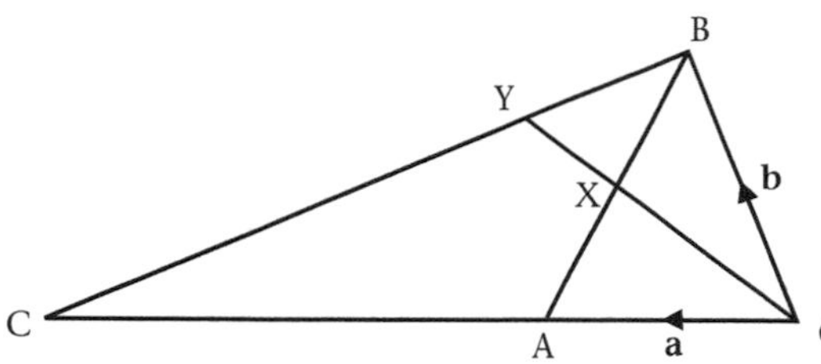

Fig. 18.28

(a) Express $\overrightarrow{BA}$ in terms of **a** and **b**.
(b) If X is the mid-point of BA, show that $\overrightarrow{OX} = \frac{1}{2}(\mathbf{a} + \mathbf{b})$.
(c) Given that $\overrightarrow{OC} = 3\mathbf{a}$, express $\overrightarrow{BC}$ in terms of **a** and **b**.

(a) In △OAP,
$$\overrightarrow{OB} + \overrightarrow{BA} = \overrightarrow{OA}$$
$$\mathbf{b} + \overrightarrow{BA} = \mathbf{a}$$
$$\overrightarrow{BA} = \mathbf{a} - \mathbf{b}$$

(b) In △OBX,
$$\overrightarrow{BX} = \tfrac{1}{2}\overrightarrow{BA}$$
$$= \tfrac{1}{2}(\mathbf{a} - \mathbf{b})$$
$$\overrightarrow{OX} = \overrightarrow{OB} + \overrightarrow{BX}$$
$$= \mathbf{b} + \tfrac{1}{2}(\mathbf{a} - \mathbf{b})$$
$$= \tfrac{1}{2}(\mathbf{a} + \mathbf{b})$$

(c) In △OBC,
$$\overrightarrow{BC} = \overrightarrow{BO} + \overrightarrow{OC}$$
$$= -\mathbf{b} + 3\mathbf{a}$$
$$= 3\mathbf{a} - \mathbf{b}$$

Exercise 18f

Make sketches where necessary.

1. Represent each of the following in Fig. 18.29 by a single vector.
   (a) $\overrightarrow{PQ} + \overrightarrow{QR}$
   (b) $\overrightarrow{PR} + \overrightarrow{RS}$
   (c) $\overrightarrow{PS} + \overrightarrow{ST}$
   (d) $\overrightarrow{PR} + \overrightarrow{RT}$
   (e) $\overrightarrow{PQ} + \overrightarrow{QR} + \overrightarrow{RS}$
   (f) $\overrightarrow{PQ} + \overrightarrow{QT} + \overrightarrow{TS}$
   (g) $\overrightarrow{PQ} + \overrightarrow{QR} + \overrightarrow{RS} + \overrightarrow{ST}$
   (h) $\overrightarrow{PQ} + \overrightarrow{QT} + \overrightarrow{TR} + \overrightarrow{RS}$

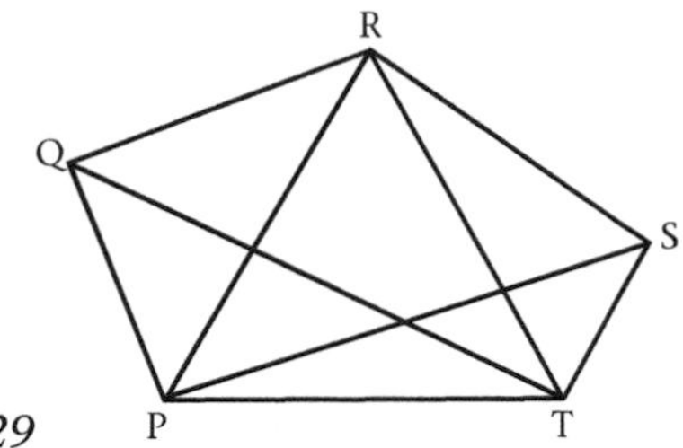

Fig. 18.29

2. In Fig. 18.30 $\overrightarrow{OA} = \mathbf{a}$, $\overrightarrow{OB} = \mathbf{b}$ and M is the mid-point of AB.

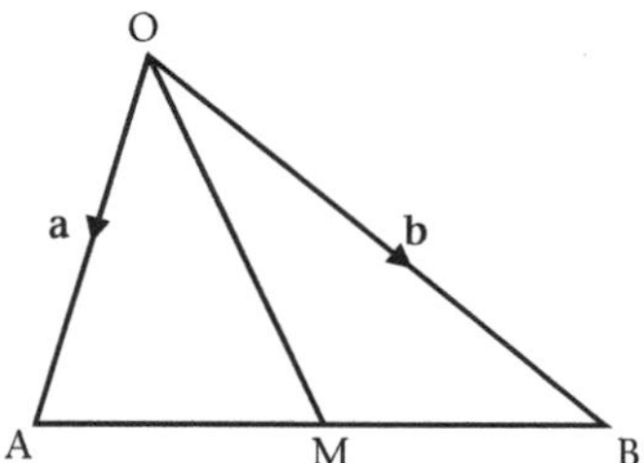

Fig. 18.30

Find $\overrightarrow{OM}$ in terms of **a** and **b**.

3. Given Fig. 18.31, express $\overrightarrow{XY}$, $\overrightarrow{YZ}$ and $\overrightarrow{ZX}$ in terms of **a**, **b** and **c**.

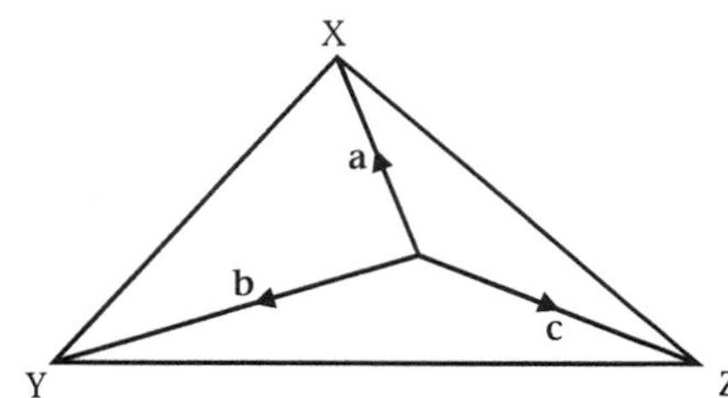

Fig. 18.31

4. In Fig. 18.32 $\overrightarrow{OP} = 2\mathbf{a}$, $\overrightarrow{PQ} = 2\mathbf{b} - 3\mathbf{a}$, $\overrightarrow{OR} = 3\mathbf{b}$.

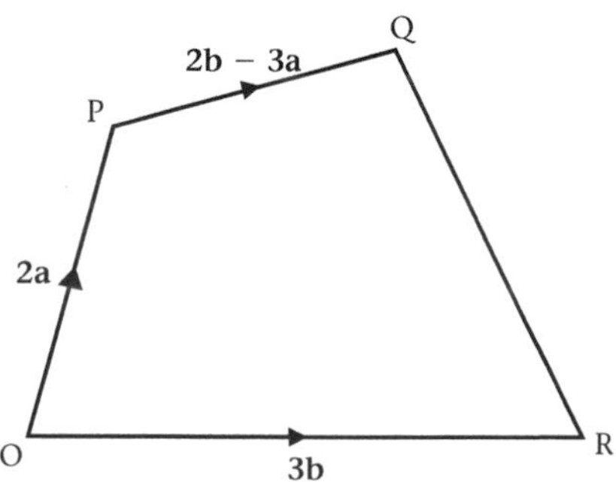

Fig. 18.32

Express (a) $\overrightarrow{OQ}$, (b) $\overrightarrow{QR}$ in terms of **a** and **b** as simply as possible.

5 In Fig. 18.33, ORST is a parallelogram, $\overrightarrow{OR} = \mathbf{r}$ and $\overrightarrow{OT} = \mathbf{t}$.

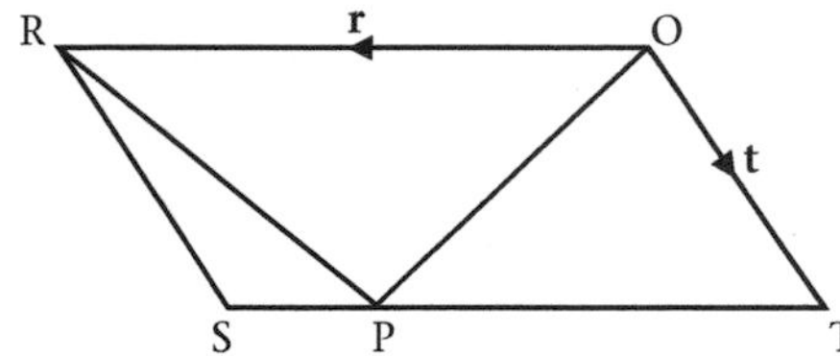

Fig. 18.33

If $\overrightarrow{ST} = 4\overrightarrow{SP}$, express the following in terms of **r** and/or **t**.

(a) $\overrightarrow{RS}$ (b) $\overrightarrow{ST}$ (c) $\overrightarrow{SP}$
(d) $\overrightarrow{RP}$ (e) $\overrightarrow{OP}$

6 In $\triangle PQR$, $\overrightarrow{PQ} = \mathbf{a}$, $\overrightarrow{PR} = \mathbf{b}$ and S is the mid-point of PR. Express the following in terms of **a** and/or **b**.

(a) $\overrightarrow{QR}$ (b) $\overrightarrow{PS}$ (c) $\overrightarrow{QS}$

7 PQRS is a trapezium in which PQ || SR, $\overrightarrow{PQ} = \mathbf{a}$ and $\overrightarrow{QR} = \mathbf{b}$. E is the mid-point of PS and PQ is half as long as SR. Express the following in terms of **a** and/or **b**.

(a) $\overrightarrow{SR}$ (b) $\overrightarrow{PS}$ (c) $\overrightarrow{EQ}$ (d) $\overrightarrow{ER}$

8 ABCDEF is a regular hexagon. If $\overrightarrow{AB} = \mathbf{x}$ and $\overrightarrow{AF} = \mathbf{y}$, express the following in terms of **x** and **y**.

(a) $\overrightarrow{FC}$ (b) $\overrightarrow{BC}$ (c) $\overrightarrow{FE}$
(d) $\overrightarrow{AE}$ (e) $\overrightarrow{AD}$ (f) $\overrightarrow{AC}$

9 In $\triangle PQR$, A is a point on PR such that $\overrightarrow{PA} = \frac{4}{5}\overrightarrow{PR}$ and B is the mid-point of QR. Point C lies on PQ produced so that $\overrightarrow{PC} = \frac{3}{2}\overrightarrow{PQ}$. If $\overrightarrow{PR} = \mathbf{x}$ and $\overrightarrow{PQ} = \mathbf{y}$ express the following in terms of **x** and **y**.

(a) $\overrightarrow{PA}$ (b) $\overrightarrow{PB}$ (c) $\overrightarrow{PC}$
(d) $\overrightarrow{AB}$ (e) $\overrightarrow{BC}$

10 Use vectors to show that if the diagonals of a quadrilateral bisect each other the quadrilateral is a parallelogram.

11 Use vectors to show that the diagonals of a parallelogram bisect each other.

12 In Fig. 18.34, OAB is any triangle, M and N are the mid-points of OA and OB respectively.

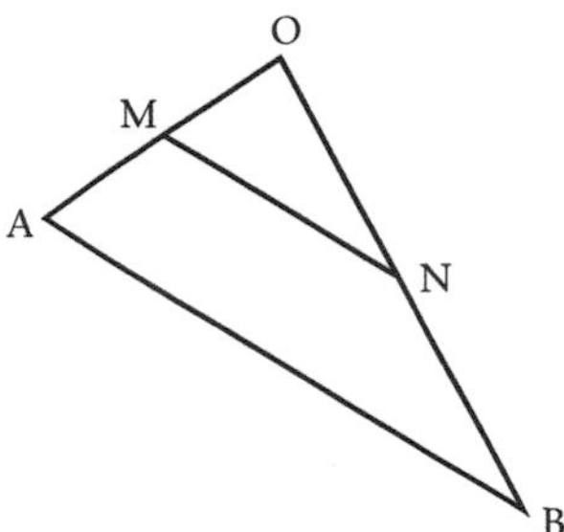

Fig. 18.34

If $\overrightarrow{OA} = \mathbf{a}$ and $\overrightarrow{OB} = \mathbf{b}$,

(a) express $\overrightarrow{AB}$, $\overrightarrow{OM}$, $\overrightarrow{ON}$, $\overrightarrow{MN}$ in terms of **a** and/or **b** and (b) hence describe any relationship between line segments MN and AB.

13 ABCD is a quadrilateral whose diagonals are equal in length. The mid-points of AB, BC, CD, DA are joined in order to form a quadrilateral. Use a vector method to show that the quadrilateral so formed is a rhombus.

14 ABCD is a kite. The mid-points of AB, BC, CD, DA are joined to form a quadrilateral. Show that the quadrilateral so formed is a rectangle.

15 In trapezium PQRS, $\overrightarrow{QP} = \mathbf{a}$, $\overrightarrow{RQ} = \mathbf{b}$, $\overrightarrow{RS} = 3\mathbf{a}$ and the diagonals intersect at X.

(a) Express $\overrightarrow{RP}$ and $\overrightarrow{QS}$ in terms of **a** and **b**.
(b) Show that PX :PR = QX : QS = 1 : 4.

16 In Fig. 18.35, P is a point on AB such that $\overrightarrow{BA} = 4\overrightarrow{BP}$ and Q is the mid-point of OA. OP and BQ intersect at X.

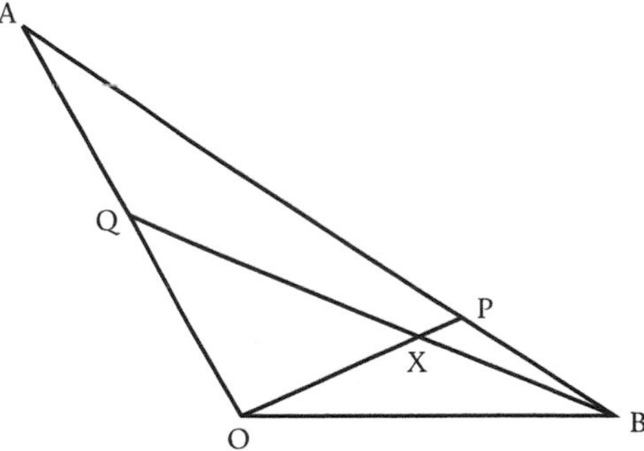

Fig. 18.35

Given $\overrightarrow{OA} = \mathbf{a}$ and $\overrightarrow{OB} = \mathbf{b}$:

(a) Express the following in terms of **a** and **b**.
(i) AB (ii) OP (ii) BQ

(b) If $\overrightarrow{BX} = h\overrightarrow{BQ}$, express $\overrightarrow{OX}$ in terms of **a**, **b** and h.
(c) If $\overrightarrow{OX} = k\overrightarrow{OP}$ use the previous result to find h and k.
(d) Hence express $\overrightarrow{OX}$ in terms of **a** and **b** only.

17 In Fig. 18.36, OABC is a parallelogram, M is the mid-point of $\overrightarrow{OA}$ and $\overrightarrow{AX} = \frac{2}{7}\overrightarrow{AC}$. $\overrightarrow{OA} = \mathbf{a}$ and $\overrightarrow{OC} = \mathbf{c}$.

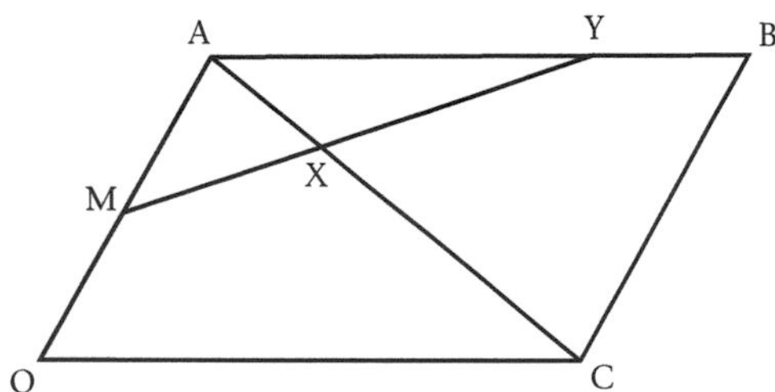

Fig. 18.36

(a) Express the following in terms of **a** and **c**.
(i) $\overrightarrow{MA}$ (ii) $\overrightarrow{AB}$ (iii) $\overrightarrow{AC}$ (iv) $\overrightarrow{AX}$
(b) Using $\triangle$MAX, express $\overrightarrow{MX}$ in terms of **a** and **c**.
(c) If $\overrightarrow{AY} = p\overrightarrow{AB}$, use $\triangle$MAY to express $\overrightarrow{MY}$ in terms of **a**, **c** and p.
(d) Also if $\overrightarrow{MY} = q\overrightarrow{MX}$, use the result in (b) to express $\overrightarrow{MY}$ in terms of **a**, **c** and q.
(e) Hence find p and q and the ratio AY : YB.

18 ABC is any triangle. M and N are the mid-points of BC and AC respectively and AM and BN intersect at G. $\overrightarrow{AB} = \mathbf{x}$ and $\overrightarrow{AC} = \mathbf{y}$.
(a) Express $\overrightarrow{AM}$ in terms of **x** and **y**.
(b) If $\overrightarrow{AG} = h\overrightarrow{AM}$, express $\overrightarrow{AG}$ in terms of **x**, **y** and h.
(c) Express $\overrightarrow{BN}$ in terms of **x** and **y**.
(d) If $\overrightarrow{BG}$ 5 k$\overrightarrow{BN}$, express $\overrightarrow{AG}$ in terms of **x**, **y** and k.
(e) Use the results of (b) and (d) to find h and k.
(f) What can you deduce about the three lines joining the mid-points of the sides of a triangle to the opposite vertices?

Velocity vectors

Velocities can be represented by vectors.

Example 14

Two trains are travelling at velocities of 50 km/h and 75 km/h respectively. Find the difference of their velocities if (a) they are travelling in the same direction, (b) they are travelling in opposite directions.

A line segment 2 cm long represents vel_A and a line segment 3 cm long represents vel_B. The arrows represent their direction. (Figs. 18.37 and 18.38)

(a) Velocity of train A ——→—— vel_A

Velocity of train B ———→——— vel_B

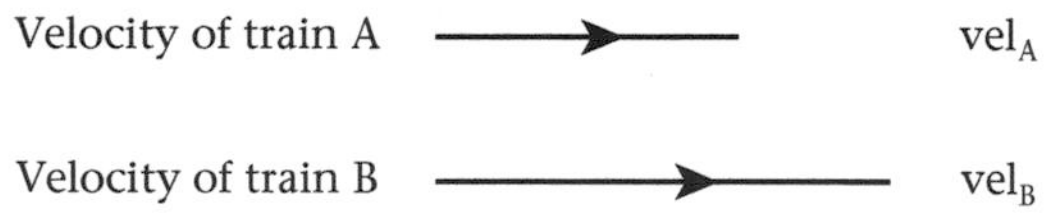

Fig. 18.37

$vel_B - vel_A = 75\text{ km/h} - 50\text{ km/h} = 25\text{ km/h}$

(b) Velocity of train A ——→—— vel_A

Velocity of train B ———←——— vel_B

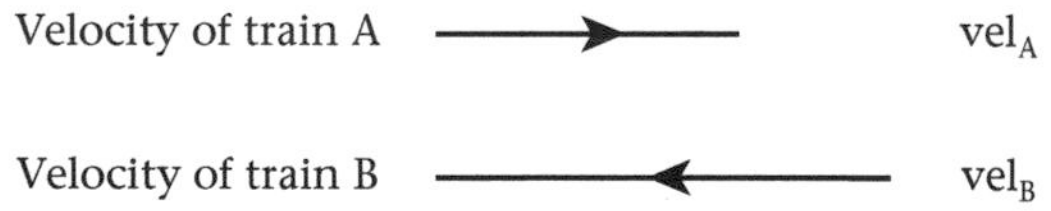

Fig. 18.38

Since the direction of train B is opposite to the direction of train A, $vel_B = -75$ km/h.
$vel_B - vel_A$
$= -75\text{ km/h} - 50\text{ km/h}$
$= -125\text{ km/h}$

$vel_B - vel_A$ is the velocity of B *relative to* A. Thus to a passenger sitting in train A, train B would appear to be travelling at 25 km/h if both trains are travelling in the same direction, or 125 km/h if the trains are travelling in opposite directions.

Example 15

An aircraft can fly at 500 km/h when there is no wind. At what speed will the aircraft travel over the ground if it is flying (a) with a wind of 60 km/h blowing from directly behind, (b) directly into a wind of 60 km/h?

The velocity of the aircraft will be the resultant of the velocity due to its engines and the velocity of the wind.

(a) Speed of aircraft over the ground $= (500 + 60)$ km/h $= 560$ km/h

(b) Speed of aircraft over the ground $= (500 - 60)$ km/h $= 440$ km/h

Vector diagrams

A **vector diagram** is an accurate scale drawing. Bearings are angles measured in a clockwise direction from the north. The wind direction is given by the direction *from* which the wind is blowing. The speed of an aircraft in still air is called its **airspeed.** its actual speed relative to the ground is its **groundspeed**. The direction in which the aircraft is steered is called its **course**. The actual direction of the aircraft over the ground is called its **track**.

Example 16

The airspeed of an aeroplane travelling due east is 500 km/h with a wind blowing from the north at 100 km/h. What is the resultant velocity of the aeroplane over the ground?

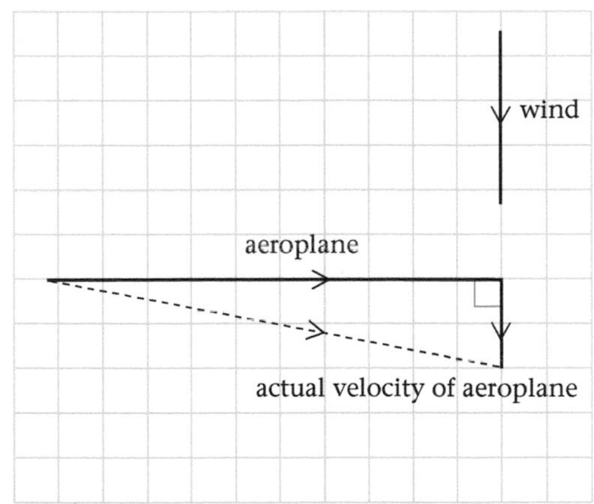

Fig. 18.39

Fig. 18.39 is a vector diagram of the information. The scale is 1 cm = 100 km/h. From the diagram, the resultant velocity of the aeroplane is 51 km/h in a direction 101° which gives (515 km/h, 101°).

Example 17

An aeroplane heads due north at 450 km/h. However, a wind of 135 km/h blows from a direction 315°. Draw a vector diagram to show the resultant velocity of the aeroplane.

The required vector diagram is Fig. 18.40. The aeroplane actually travels at (372 km/h, 016°).

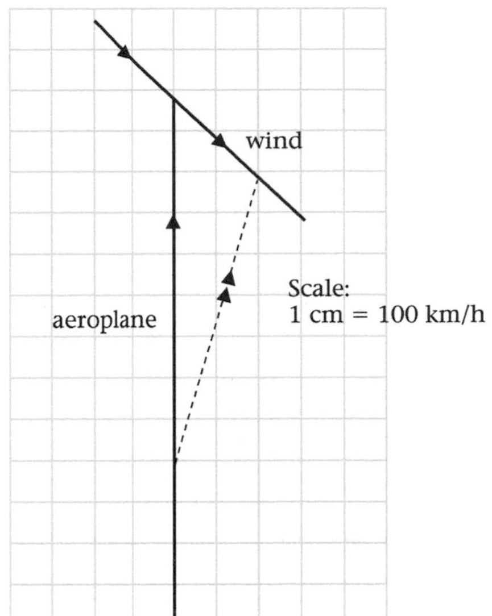

Fig. 18.40

Exercise 18g

1. Find, by drawing, the resultant of the two velocities: (80 km, N) and (50 km, E).
2. If the resultant of two velocities A and B is a velocity (132.5 km/h, 027°) and velocity A is (120 km/h, N), find the magnitude and size of velocity B.
3. A pilot steers and aircraft due east with an airspeed of 450 km/h. There is a wind blowing from the south at a speed of 60 km/h. Find, by drawing, the direction in which the aircraft travels and its speed over the ground.
4. An aeroplane has an airspeed of 500 km/h and is travelling on a bearing of 020°. The wind blowing from a bearing of 315° has a speed of 75 km/h. Find, by drawing, the track and groundspeed of the aeroplane.

Summary

A **vector** is any quantity which has direction as well as size. A **translation vector** describes both the horizontal and vertical movements of a point or points and is written as a single column matrix or column vector, e.g. $\overrightarrow{AB} = \begin{pmatrix} x \\ y \end{pmatrix}$

Using matrix arithmetic the operations addition, subtraction and multiplication by a **scalar** quantity can be performed on vectors.

Two vectors are parallel if one vector is a scalar multiple of the other.

The **magnitude** of a vector $\overrightarrow{AB}$ is the length of the line segment AB and is written $|\overrightarrow{AB}|$.

In general if $\overrightarrow{AB} = \begin{pmatrix} p \\ q \end{pmatrix}$

then $|\overrightarrow{AB}| = \sqrt{p^2 + q^2}$

A **unit vector** is a vector of unit length.

A vector can be multiplied by a scalar quantity.

Vectors can be added by the **triangle law** (Fig. 18.41) or by the **parallelogram law** (Fig. 18.42).

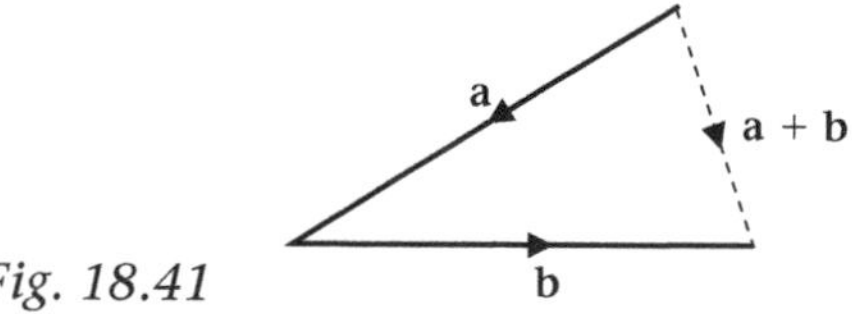

Fig. 18.41

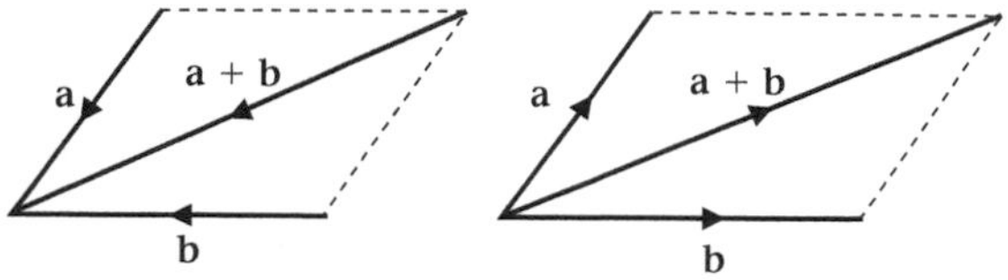

Fig. 18.42

If $\overrightarrow{AB} = \overrightarrow{DC}$ then AB ∥ DC and AB = DC.

Practice exercise P18.1

1. If $\mathbf{a} = \begin{pmatrix} 3 \\ 4 \end{pmatrix}$, $\mathbf{b} = \begin{pmatrix} -2 \\ 5 \end{pmatrix}$ and $\mathbf{c} = \begin{pmatrix} 5 \\ -3 \end{pmatrix}$, find, in the form $\begin{pmatrix} x \\ y \end{pmatrix}$,

   (a) $\mathbf{a} + \mathbf{b}$, (b) $2\mathbf{a} - \mathbf{c}$, (c) $3\mathbf{c} + \mathbf{b}$.

2. The position vectors of the points A, B and C relative to an origin O are $\overrightarrow{OA} = \begin{pmatrix} 4 \\ 9 \end{pmatrix}$, $\overrightarrow{OB} = \begin{pmatrix} 2 \\ 3 \end{pmatrix}$ and $\overrightarrow{OC} = \begin{pmatrix} 4 \\ 6 \end{pmatrix}$. Express, in the form $\begin{pmatrix} a \\ b \end{pmatrix}$, the vectors $\overrightarrow{AB}$ and $\overrightarrow{CB}$.

3. Draw on graph paper a diagram to show PQRS with P(−3, −2), Q(1, −1), R(5, 4) and S(1, 3). Express in the form $\begin{pmatrix} x \\ y \end{pmatrix}$ the vectors $\overrightarrow{PQ}$ and $\overrightarrow{RS}$.

4. $\overrightarrow{OP}$ is the position vector $\begin{pmatrix} 3 \\ 2 \end{pmatrix}$ and $\overrightarrow{OR}$ is the position vector $\begin{pmatrix} 3 \\ -2 \end{pmatrix}$.
   RQ = $\begin{pmatrix} 3 \\ 2 \end{pmatrix}$.

   (a) Draw on graph paper the diagram to show OPQR.
   (b) State the vector $\overrightarrow{OQ}$ in the form $\begin{pmatrix} x \\ y \end{pmatrix}$.
   (c) What shape is the quadrilateral OPQR?

5. Triangle ADE is an enlargement of triangle ABC by scale factor 2 and centre of enlargement A. Use vectors to show that the length of BC is $\frac{1}{2}$DE.

6. P, Q, R and S are points such that $\overrightarrow{PQ} = \begin{pmatrix} 2 \\ 5 \end{pmatrix}$, R has coordinates (−4, 3) and S has coordinates (0, 13). Show that PQ is parallel to RS.

7. A(3, 2), B(5, 3) and C(4, −1) are points on the Cartesian plane.

   (a) Write down the position vectors of A, B and C.
   (b) Calculate the lengths of the vectors $\overrightarrow{OA}$, $\overrightarrow{OB}$ and $\overrightarrow{OC}$.

8. ABC is an equilateral triangle. The vertices of △DEF are the midpoints of the sides of △ABC. Use vectors to show that △DEF is an equilateral triangle.

Sets

Sets were revised in Book 3. These exercises provide further revision.

Revision exercises

Revision exercise 19a: Symbols, definitions and operations

1 For each Venn diagram in Fig. 19.1, describe the members of
(i) the universal set,
(ii) the subset,
(iii) the complement of the subset.

(a)

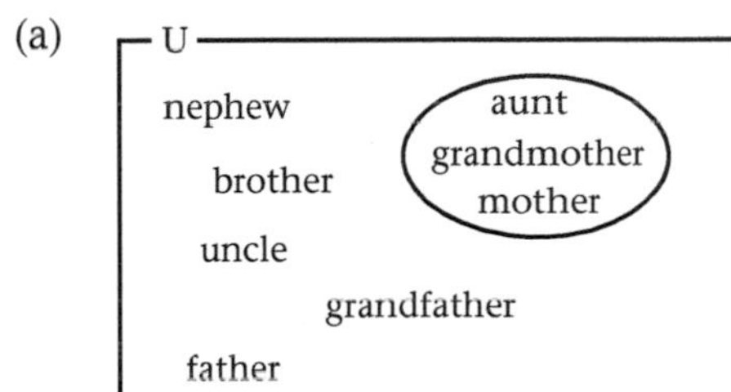

(b)

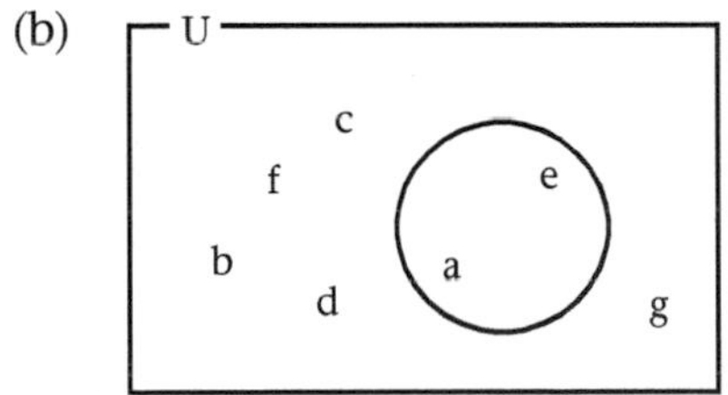

(c)

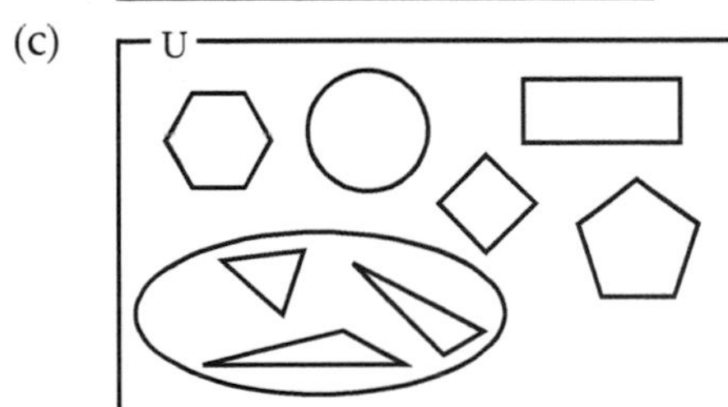

(d)

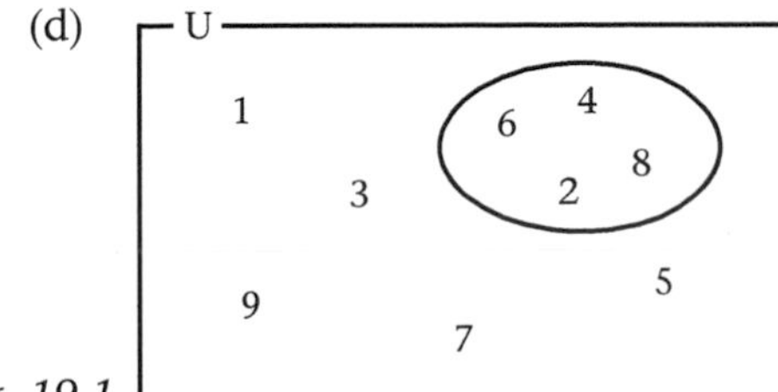

Fig. 19.1

2 For each of the following sets, state whether the set is infinite, finite or empty.
(a) {shoes in a shoe factory}
(b) {fractions that are equal to 0.2}
(c) {15, 16, 17, …, 20}
(d) {hours in 1 week}
(e) {prime numbers between 24 and 28}
(f) {letters in the word *Mississippi*}
(g) {teachers in the Caribbean}
(h) {cyclists that move at 40 m/s}
(i) {factors of 6 greater than 12}
(j) {integers less than 0}

3 For the universal sets given below,
(i) state the number of subsets,
(ii) write all the subsets.
(a) U = {0, 1, 2}
(b) U = {letters of the word *metre*}

4 U, Y, S and H are sets as defined below.
U = {all men},
Y = {young men},
S = {successful men}, and
H = {happy men}
(a) Write the symbols that represent each of the following.
(i) {old men}
(ii) {unsuccessful men}
(iii) {young happy men}
(iv) {unsuccessful old men}
(b) If $S' \subset H'$ and $H' \subset Y'$, write in words the conclusion you can draw from the two statements.

5 Use a Venn diagram to show that if $A \subset B$, then
(a) $B' \subset A'$ and vice versa,
(b) $A \cap B' = \emptyset$,
(c) $A' \cup B = U$.

Revision exercise 19b: Relations among sets

1. List the elements of the following sets.
   (a) $\{f : f \text{ is a factor of both 42 and 105}\}$
   (b) $\{p : p \text{ is a prime factor of 66}\}$
   (c) $\{\text{integer } x : x \geqslant -3\}$
   (d) $\{\text{integer } y : y < -4\}$
   (e) $\{x : -4 < x \leqslant 3, x \in \mathbb{Z}\}$
   (f) $\{x : 3x - 1 = 5\}$

2. For each of the following pairs of sets, state whether the sets are equivalent, equal or not equivalent.
   (a) (i) {factors of 10}
       (ii) {1, 2, 5, 10}
   (b) (i) {multiples of 2 less than 12}
       (ii) {vowels in the alphabet}
   (c) (i) {1, 2, 3}
       (ii) {4, 5, 6}
   (d) (i) {even numbers less than 7}
       (ii) {whole numbers less than 5}
   (e) (i) {letters of the word *rained*}
       (ii) {letters of the word *drainer*}
   (f) (i) {all shapes with four sides}
       (ii) {square, rectangle}

Revision exercise 19c: Intersecting sets

1. (a) Copy and complete the Venn diagram in Fig. 19.2 where X = {p, q, r, s}, Y = {r, s, t, w, a} and Z = {p, s, a, b}.

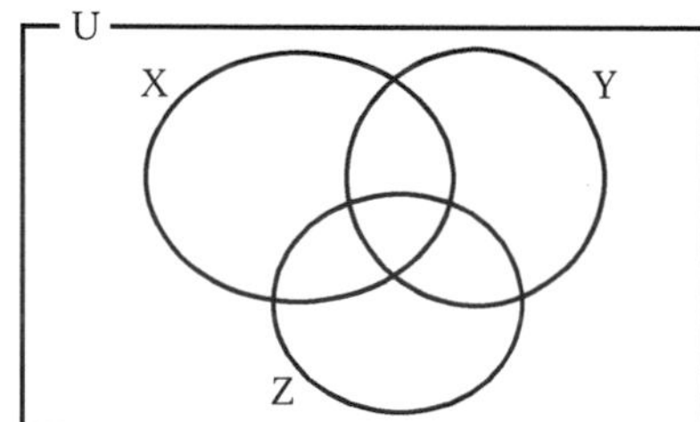

Fig. 19.2

   (b) Using your diagram or otherwise, list the members of
       (i) $(X \cap Y) \cup Z$
       (ii) $(Y \cup Z) \cap X$

2. Make seven copies of each of the Venn diagrams in Fig. 19.3 and shade these sets.
   (a) $P \cap R$ (b) $R \cup S$
   (c) $(S \cup P)'$ (d) $(P \cup R) \cap S$
   (e) $(P \cap R) \cup S$ (f) $P \cup (R \cap S')$
   (g) $P \cap (R \cup S')$

(i)

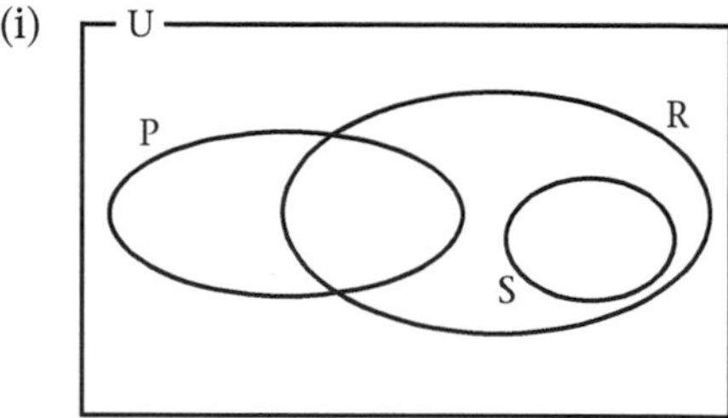

(ii)

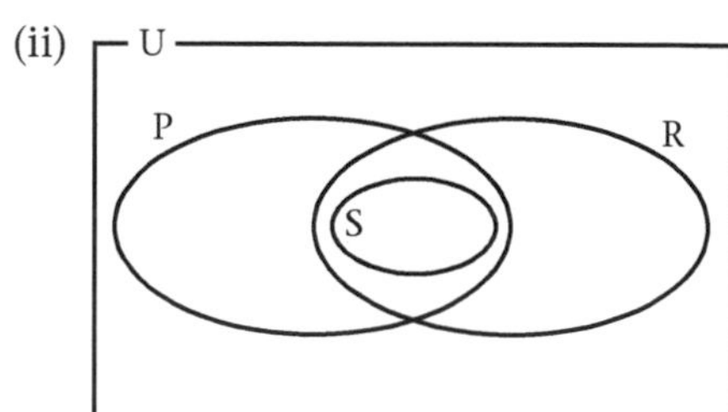

(iii)

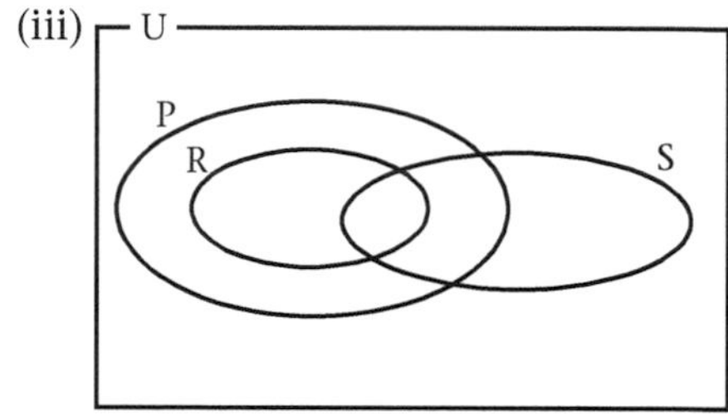

Fig. 19.3

3. $U = A \cup B \cup C$ and $n(U) = 37$. The number of the members in each of the subsets is shown in the Venn diagram in Fig. 19.4.

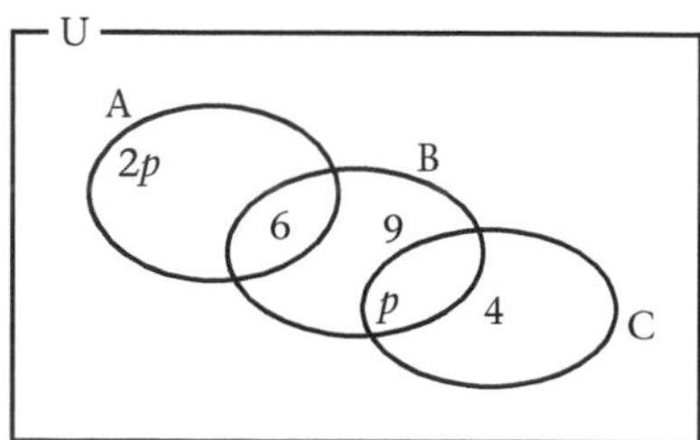

Fig. 19.4

   Find
   (a) p (b) $n(A \cup C)$
   (c) $n(A \cap C)$ (d) $n(C' \cap B)$

4. The members of subsets of the universal set, U, where U = $\{x : 3 \leqslant x < 30, x \in \mathbb{Z}\}$ are described below. In each of the following questions, list the members of
   (i) the universal set,
   (ii) each subset,
   (iii) each intersection of the subsets,
   (iv) the complement of each intersection,
   (v) each union of the sets,
   (vi) the complement of each union.
   Draw the Venn diagram.
   (a) A = {odd numbers greater than 15}
       B = {prime numbers}
   (b) F = (factors of 120}
       P = {prime factors of 210}
       M = {multiples of 5 less than 20}

5. Describe the shaded areas in the Venn diagrams in Fig. 19.5.
   (a)

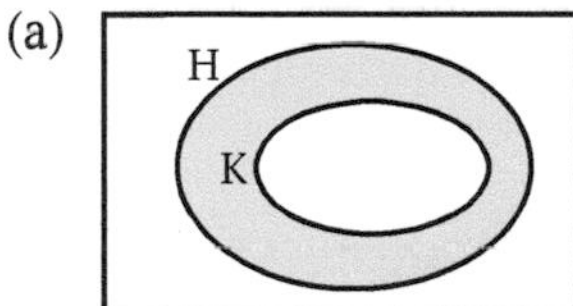

   (b)

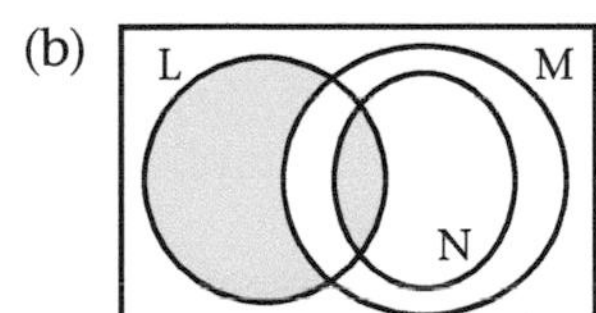

Fig. 19.5

6. For each of the Venn diagrams in Fig. 19.6,
   (i) describe the members of each set,
   and list the members of
   (ii) each intersection,
   (ii) the complement of the intersection,
   (iii) each union,
   (iv) the complement of the union.
   (a)

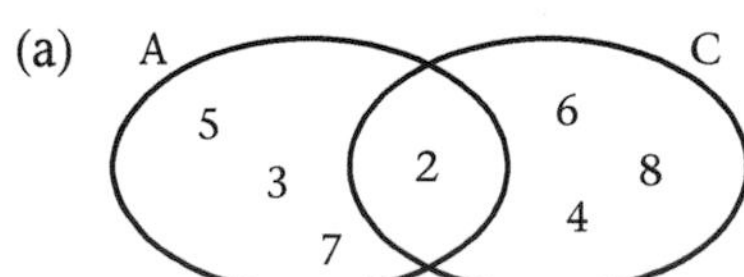

   (b)

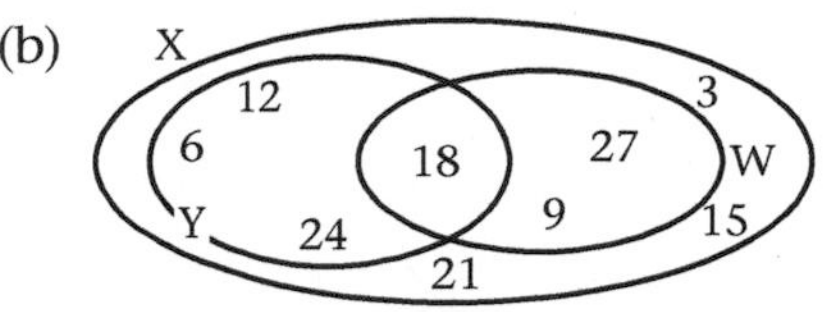

Fig. 19.6

7. In the Venn diagram in Fig. 19.7,
   U = {members of a youth club},
   A = {group A of the youth club} and
   B = {group B of the youth club},
   The letters m, r, x and y in the diagram represent the number of students in each subset.
   If n(U) = 100, n(A) = 40 and n(B) = 15,
   (a) express m in terms of x,
   (b) find the smallest possible value of y,
   (c) find the largest possible value of x,
   (d) find the value of x, r and y if m = 28.

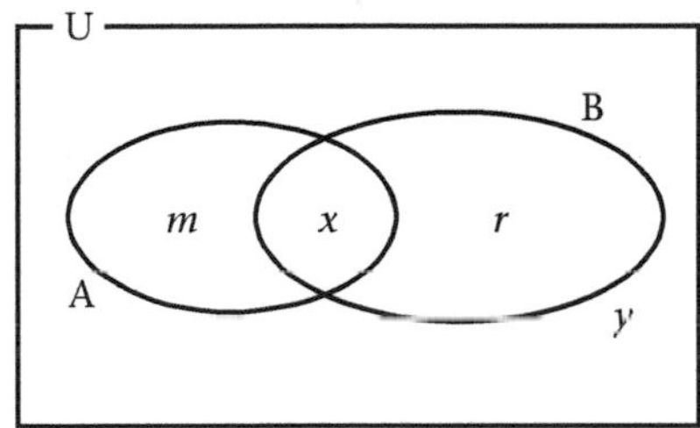

Fig. 19.7

8. *Some students use computers.*
   Some students are clever.
   All clever students read.
   If U = {students}, draw the Venn diagram that illustrates the statements above. Justify your arrangements of the sets.

9. For the Venn diagrams in Fig. 19.8,
   U = {books},
   M = {mathematics books},
   G = {Geometry Books} and
   P = {books with pictures}.
   For each of the Venn diagrams,
   (i) write down statements using the words 'all', 'some', 'not all' or 'not',
   (ii) write each statement in part (i) using set notation.

(a)

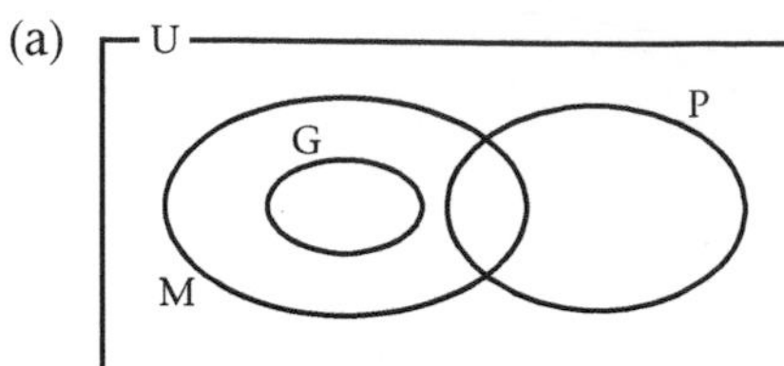

(b)

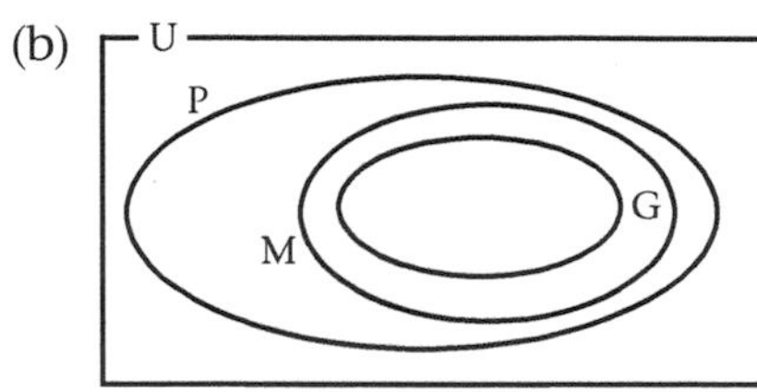

(c)

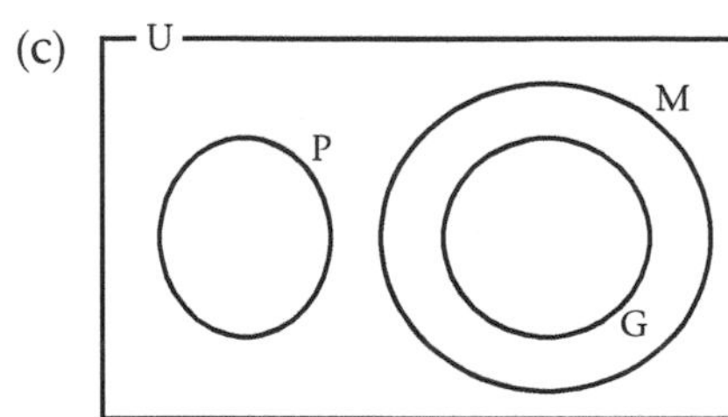

(d)

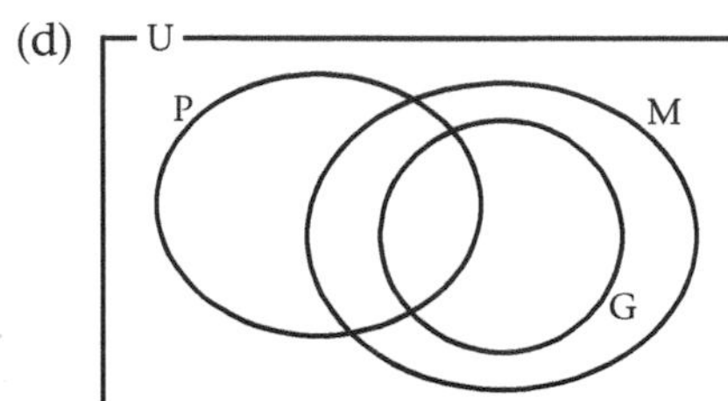

Fig. 19.8

10 In each of the following problems,
(i) use set-builder notation to describe the sets,
(ii) list the members of the sets,
(iii) solve the problem.
(a) Find the lowest common multiple of 6 and 16.
(b) Find the smallest perfect square that is greater than 15 and less than 40.

11 The subsets of a universal set U are X, Y and Z, where $X \cap Z = \{\}$. Given that $n(X) = 9$, $n(Y) = 16$, $n(X \cap Y) = 3$, $n(Y \cap Z) = 5$, $n(Y \cup Z) = 16$ and $n(X \cup Y \cup Z)' = 2$,
(a) draw and label clearly a Venn diagram to show the information above,
(b) use set-builder notation to write a statement representing the number of members in the universal set,
(c) calculate the value of n(U),
(d) state the relation between Y and Z.

12 For each of the following problems,
(i) use set-builder notation to describe all the sets and rewrite the statements,
(ii) draw the Venn diagram,
(iii) write the set relations that represent the statements in *italics*.
(a) All the boxes have straight edges.
Some boxes are made of glass.
Some boxes are made of wood.
All gift boxes are made of glass.
V is a gift box.
(b) All the students are fifth formers.
Some study Geography and Mathematics.
Some study Biology and Mathematics.
No student studies Geography, Biology and Mathematics.
Danielle studies Geography and Mathematics.
Tim studies neither Geography nor Biology.

Chapter 20
Algebra

Factorisation

The expression $(a + b)(c + d)$ can be expanded in two ways:

either: $(a + b)(c + d) = (a + b)\,c + (a + b)\,d$
$= ac + bc + ad + bd$

or: $(a + b)(c + d) = a(c + d) + b(c + d)$
$= ac + ad + bc + bd$
$= ac + bc + ad + bd$

The result is the same in both cases.
Then $(a + b)$ and $(c + d)$ are the factors of $ac + bc + ad + bd$.
Hence, factorisation is the reverse of expansion.

Example 1

Factorise $am + 3bn - an - 3bm$.

The terms am and an have a in common.

The terms $3bm$ and $3bn$ have $3b$ in common.

Grouping pairs in this way,

$am + 3bn - an - 3bm = am - an - 3bm + 3bn$
$= a(m - n) - 3b(m - n)$
$= (m - n)(a - 3b)$

(since $m - n$ is a common factor in the two terms, namely, $a(m - n)$ and $-3b(m - n)$.

Example 2 (revision)

Expand $(2n + 3)(3n - 7)$.

$(2n + 3)(3n - 7)$
$= (2n + 3)3n + (2n + 3)(-7)$
$= 6n^2 + 9n - 14n - 21$
$= 6n^2 - 5n - 21$

Note that $2n + 3$ and $3n - 7$ are the factors of the quadratic expression $6n^2 - 5n - 21$.

Example 3 (revision)

Expand $(2x - 3y)(5x - 2y)$.

$(2x - 3y)(5x - 2y)$
$= 2x(5x - 2y) - 3y(5x - 2y)$
$= 10x^2 - 4xy - 15xy + 6y^2$
$= 10x^2 - 19xy + 6y^2$

Again, $2x - 3y$ and $5x - 2y$ are the factors of $10x^2 - 19xy + 6y^2$.

Example 4

Factorise $10x^2 + 5x - 2x - 1$.

$10x^2 + 5x - 2x - 1$
$= 5x(2x + 1) - 1(2x + 1) = (2x + 1)(5x - 1)$

Generally, the x-terms in Example 4 will be combined and written as $+3x$, and the quadratic expression will be given as $10x^2 + 3x - 1$.

Example 5

Factorise $a^2 - 17a + 42$.

1st step: $a^2 - 17a + 42 = (a \quad)(a \quad)$

2nd step: Find two numbers such that their product is $+42$ and their sum is -17. Since the constant term is positive and the middle term is negative, consider negative factors only.

	factors of $+42$	*sum of factors*
(a)	-42 and -1	-43
(b)	-21 and -2	-23
(c)	-14 and -3	-17
(d)	-7 and -6	-13

Of these, only (c) gives the required result.
Hence $a^2 - 17a + 42 = (a - 14)(a - 3)$
Check: $(a - 14)(a - 3) = a^2 - 14a - 3a + 42$
$= a^2 - 17a + 42$

alternative method:
1st step: Find the product of the first and last terms:

$a^2 \times (+42) = +42a^2$

2nd step: Find two terms such that their product is $+42a^2$ and their sum is $-17a$ (the middle term).

	factors of $+42a^2$	*sum of factors*
(a)	$-42a$ and $-a$	$-43a$
(b)	$-21a$ and $-2a$	$-23a$
(c)	$-14a$ and $-3a$	$-17a$
(d)	$-7a$ and $-6a$	$-13a$

Of these, only (c) gives the required result.

3rd step: Replace $-17a$ with $-14a -3a$ in the given expression.

Factorise by grouping:

$$\begin{aligned} a^2 - 17a + 42 &= a^2 - 14a - 3a + 42 \\ &= a(a - 14) - 3(a - 14) \\ &= (a - 14)(a - 3) \end{aligned}$$

Example 6

Factorise $3x^2 - 13x - 10$.

1st step: Find the product of the first and last terms.

$$3x^2 \times (-10) = -30x^2$$

2nd step: Find two terms such that their product is $-30x^2$ and their sum is $-13x$ (the middle term). Since the middle term is negative, consider only those factors in which the negative term is numerically greater than the positive term.

	factors of $-30x^2$	*sum of factors*
(a)	$-30x$ and $+x$	$-29x$
(b)	$-15x$ and $+2x$	$-13x$
(c)	$-10x$ and $+3x$	$-7x$
(d)	$-6x$ and $+5x$	$-x$

Of these, only (b) gives the required result.

3rd step: Replace $-13x$ in the given expression by $-15x + 2x$. Factorise by grouping.

$$\begin{aligned} 3x^2 - 13x - 10 &= 3x^2 - 15x + 2x - 10 \\ &= 3x(x - 5) + 2(x - 5) \\ &= (x - 5)(3x + 2) \end{aligned}$$

Notice in Example 6 that when the required result was found in line (b), it was not really necessary to do lines (c) and (d).

Example 7

Factorise $6 - 15x + 9x^2$.

3 is a common factor of all the terms. Take this out first.

$$\begin{aligned} 6 - 15x + 9x^2 &= 3(2 - 5x + 3x^2) \\ &= 3(2 - 2x - 3x + 3x^2)^* \\ &= 3[2(1 - x) - 3x(1 - x)] \\ &= 3(1 - x)(2 - 3x) \end{aligned}$$

*The reason why $-2x -3x$ was substituted for $-5x$ is left as an exercise. See Example 6.

Perfect squares

By expanding brackets and collecting terms it can be shown that:

$$\boldsymbol{(a + b)^2 = a^2 + 2ab + b^2}$$
$$\boldsymbol{(a - b)^2 = a^2 - 2ab + b^2}$$
$$\boldsymbol{(a + b)(a - b) = a^2 - b^2}$$

Example 8 (revision)

Expand

(a) $(4p + q)^2$, (b) $(5x - 3y)^2$, (c) $(2r - s)(2r + s)$.

(a) $(4p + q)^2 = (4p)^2 + 2(4p)(q) + (q)^2 = 16p^2 + 8pq + q^2$

(b) $(5x - 3y)^2 = (5x)^2 - 2(5x)(3y) + (3y)^2 = 25x^2 - 30xy + 9y^2$

(c) $(2r - s)(2r + s) = 4r^2 - 2rs + 2rs - s^2 = 4r^2 - s^2$

Note that $a^2 + 2ab + b^2$ and $a^2 - 2ab + b^2$ are perfect squares so that the factors of

$$\boldsymbol{a^2 + 2ab + b^2 = (a + b)^2}$$
$$\boldsymbol{a^2 - 2ab + b^2 = (a - b)^2}$$

and $\boldsymbol{a^2 - b^2 = (a + b)(a - b)}$

but $\boldsymbol{a^2 + b^2}$ **has no factors**

Example 9

Factorise $d^2 - 10dm + 25m^2$.

d^2 is the square of d.
$25m^2$ is the square of $5m$.
$10dm$ is twice the product of d and $5m$.
$d^2 - 10dm + 25m^2 = (d - 5m)^2$

Example 10

Factorise $16a^2 - 25b^2$.
$16a^2 - 25b^2 = (4a)^2 - (5b)^2 = (4a + 5b)(4a - 5b)$

Example 11

Factorise $5m^2 - 80$.
$5m^2 - 80 = 5(m^2 - 16) = 5(m + 4)(m - 4)$

Exercise 20a

Factorise:

1. $(3c - d)(m - n) + (3c - d)(2m - 3n)$
2. $ac + bc + 2ad + 2bd$

3. $x^2 - 16a^2$
4. $4c^2 - 25d^2$
5. $a^2 - 3a - 10$
6. $a^2 - 3ab - 10b^2$
7. $a^2b^2 - 3ab - 10$
8. $5m^2 - 45n^2$
9. $hm - 2km - 2hn + 4kn$
10. $(2a + b)^2 - (2a + b)(a - 3b)$
11. $3m^2 - 10m + 3$
12. $c^2d^2 - 81$
13. $16x^2 - 9a^2m^2$
14. $6n^2 + 13n + 6$
15. $4a^2 + 20a + 25$
16. $9h^2 - 36k^2$
17. $(5a - 3b)(a - 2b) - (4a + b)(a - 2b)$
18. $25a^2b^2c^2 - 9d^2$
19. $\frac{m^2}{9} - \frac{n^2}{4}$
20. $3su + tu - 6sv + 2tv$
21. $9x^2 - 12xy + 4y^2$
22. $12d^2 + 5d - 2$
23. $x^4 - y^2$
24. $10m^2n^2 - 7mn - 12$
25. $16 - n^4$
26. $(a + b)^2 - c^2$
27. $x^2 - (m - n)^2$
28. $(m + 2n)(3a - b) - (a - 3b)(m + 2n)$
29. $2 - h - 15h^2$
30. $2am - bm + 3bn - 6an$
31. $a^2 - 15ab + 54b^2$
32. $m^2 - 15mn - 54n^2$
33. $(c - 2d)^2 - 9e^2$
34. $12x^2 + 35xy + 18y^2$
35. $(h - k)(2h - 3k) + (h - k)^2$
36. $6acx - 8ady + 4acy - 12adx$
37. $6a^2 - 19ax - 36x^2$
38. $25a^2 - 4(m + 2n)^2$
39. $25a^2 - 4(a - 2b)^2$
40. Factorise completely $x^2 - y^2 - 4x + 4y$ [CXC (General) June 89]

Factorisation can often be used to simplify calculations.

Example 12

Evaluate $17.9^2 - 12.1^2$ by using factorisation.

$$17.9^2 - 12.1^2 = (17.9 + 12.1)(17.9 - 12.1) = 30 \times 5.8 = 174$$

Exercise 20b

Calculate:

1. $106^2 - 94^2$
2. $8.78^2 - 1.22^2$
3. $5 \times 9.2^2 - 5 \times 4.8^2$
4. $\pi R^2h - \pi r^2h$, where $\pi = 3\frac{1}{7}$, $R = 18$ cm, $r = 10$ cm and $h = 15$ cm.

Example 13

Simplify $\frac{6 - x - x^2}{x^2 - 4}$.

$$\frac{6 - x - x^2}{x^2 - 4} = \frac{(3 + x)(2 - x)}{(x + 2)(x - 2)} = -\frac{(3 + x)(x - 2)}{(x + 2)(x - 2)} = -\frac{3 + x}{x + 2}$$

Example 14

Simplify $\frac{3a^3}{3a^2 - 6ab} + \frac{4b^3}{2b^2 - ab}$.

$$\frac{3a^3}{3a^2 - 6ab} + \frac{4b^3}{2b^2 - ab}$$

$$= \frac{3a^3}{3a(a - 2b)} + \frac{4b^3}{b(2b - a)}$$

$$= \frac{a^2}{a - 2b} + \frac{4b^2}{2b - a}$$

$$= \frac{a^2}{a - 2b} - \frac{4b^2}{a - 2b}$$

$$= \frac{a^2 - 4b^2}{a - 2b} = \frac{(a + 2b)(a - 2b)}{(a - 2b)}$$

$$= a + 2b$$

Exercise 20c

Simplify:

1. $\frac{a^2 - b^2}{a^2 - ab}$
2. $\frac{m^2 - 2mn + n^2}{m^2 - n^2}$
3. $\frac{2x^2 - x - 1}{x - 1}$
4. $\frac{x^2 - 5x + 6}{x^2 - 9}$
5. $\frac{15 - 2x - x^2}{x^2 - 9}$
6. $\frac{9 - a^2}{a^2 + 6a + 9}$
7. $\frac{2a + b}{a^2 - ab} - \frac{2b + a}{ab - b^2}$
8. $\frac{2}{x + 2} - \frac{x - 6}{x^2 - 4}$

Indices

The following **laws of indices** are true for all non-zero values of a, b and x.

1. $x^a \times x^b = x^{a+b}$
2. $x^a \div x^b = x^{a-b}$
3. $x^0 = 1$
4. $x^{-a} = \dfrac{1}{x^a}$
5. $(x^a)^b = x^{ab}$
6. $x^{\frac{1}{a}} = \sqrt[a]{x}$
7. $x^{\frac{a}{b}} = \sqrt[b]{x^a}$ or $\left(\sqrt[b]{x}\right)^a$

Example 15

Table 20.1

Simplify	Working	Result
(a) $25^{\frac{1}{2}}$	$\sqrt{25}$	$= \pm 5$
(b) $9^{\frac{1}{6}} \times 9^{\frac{1}{3}}$	$= 9^{\frac{1}{6}+\frac{1}{3}} = 9^{\frac{1}{2}} = \sqrt{9}$	$= \pm 3$
(c) $27^{-\frac{2}{3}}$	$= \dfrac{1}{27^{\frac{2}{3}}} = \dfrac{1}{\sqrt[3]{27}^2} = \dfrac{1}{3^2}$	$= \frac{1}{9}$
(d) $\left(\frac{81}{16}\right)^{-\frac{3}{4}}$	$= \left(\sqrt[4]{\frac{16}{81}}^3\right) = \left(\pm\frac{2}{3}^3\right)$	$= \pm\frac{8}{27}$

Example 16

Simplify: (a) $3d^3 \times 2d^2$ (b) $(5c^{-4})^2$ (c) $\left(\dfrac{3x}{2}\right)^{-2}$ (d) $(32n)^{\frac{1}{5}}$

(a) $3d^3 \times 2d^2 = (3 \times 2)d^{3+2}$
$= 6d^5$

(b) $(5c^{-4})^2 = 5^2c^{-4\times 2}$
$= 25c^{-8}$

(c) $\left(\dfrac{3x}{2}\right)^{-2} = \left(\dfrac{2}{3x}\right)^2$
$= \dfrac{4}{9x^2}$

(d) $(32n)^{\frac{1}{5}} = 32^{\frac{1}{5}} \times n^{\frac{1}{5}}$
$= \sqrt[5]{32} \times \sqrt[5]{n}$
$= 2 \times \sqrt[5]{n}$

Exercise 20d

Rewrite the following using positive indices only $\left(\text{e.g. } ab^{-2} = \dfrac{a}{b^2}\right)$.

1. x^{-3}
2. xy^{-1}
3. $(xy)^{-1}$
4. $a^{-2}b^3$
5. $(ab^{-3})^2$
6. $3x^{-\frac{1}{2}}$

Exercise 20e

Simplify the following expressions.

1. $\dfrac{1}{3^{-2}}$
2. 2×2^{-3}
3. $\dfrac{a^6}{(-a)^4}$
4. $\dfrac{-(d^2)^3}{d^4 \times (-d)}$
5. $5^{-2} \times 16^{\frac{3}{4}}$
6. $(32)^{\frac{2}{5}}$
7. $8^{\frac{2}{3}}$
8. $4^{-\frac{3}{2}}$
9. $\sqrt{1\frac{9}{16}}$
10. $\left(\frac{1}{9}\right)^{-\frac{1}{2}}$
11. $2^{\frac{2}{3}} \times 2^{\frac{4}{3}}$
12. $0.09^{\frac{1}{2}}$
13. $5^x \times 5^{-x}$
14. $\sqrt[3]{4^{1.5}}$
15. $\left(\frac{27}{48}\right)^{-\frac{3}{2}}$
16. $0.216^{-\frac{2}{3}}$
17. $\sqrt[3]{8a^{-6}}$
18. $3^{n-1} \times 3^{1-n}$
19. $64^{-\frac{5}{6}}$
20. $2x^{\frac{1}{2}} \times 3x^{-\frac{5}{2}}$
21. $\left(\frac{16}{9}\right)^{-\frac{3}{2}}$
22. $\sqrt{16a^{-12}}$
23. $\left(\frac{18}{32}\right)^{-\frac{3}{2}}$
24. $4a^3b \times (3ab)^{-2}$
25. $\dfrac{75a^2b^{-2}}{5a^3b^{-3}}$

Exercise 20f

Solve the following equations.

1. $x^{\frac{1}{2}} = 3$
2. $x^{\frac{1}{3}} = 2$
3. $x^{-2} = 16$
4. $3x^3 = 24$
5. $x^{-\frac{2}{3}} = 9$
6. $5x = 80x^{-\frac{1}{3}}$

Variation

Direct variation

'y varies directly as x' is written as $y \propto x$.
$y \propto x$ means that $y = kx$, where k is a constant.

Example 17

If $P \propto R$ and $P = 10$ when $R = 6$, find the relationship between P and R. Hence find R when $P = 12.5$.

Let $P = kR$

$10 = k(6)$ (from 1st sentence)

$k = \dfrac{10}{6} = \dfrac{5}{3}$

$\therefore P = \frac{5}{3}R$ is the required relationship between P and R.

When $P = 12.5$, $12.5 = \frac{5}{3}R$

$R = \frac{5}{3} \times 12.5 = 7.5$

Fig. 20.1 is a graph of the relationship between R and P.

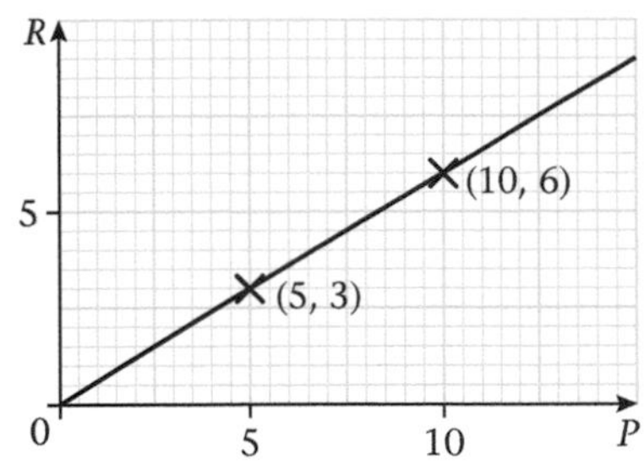

Fig. 19.1

Inverse variation

If y varies inversely as x, then $y \propto \frac{1}{x}$ or $y = \frac{k}{x}$, where k is a constant. Notice that if $y \propto \frac{1}{x}$, then $x \propto \frac{1}{y}$.

Similarly, if $x \propto y^3$, then $y \propto \sqrt[3]{x}$, and so on.

Example 18

Given that N varies inversely as D^3 and that $N = 240$ when $D = 3$, find N when $D = 2$.

Let $N = \frac{k}{D^3}$

then $240 = \frac{k}{3^3}$

$k = 240 \times 27$

When $D = 2$,

$$N = \frac{240 \times 27}{2^3} = \frac{240 \times 27}{8} = 810$$

Example 19

$R \propto \sqrt{M}$ and $R = 6$ when $M = 16$. Find the law connecting R and M. Find R when $M = 6\frac{1}{4}$ and find M when $R = 15$.

Let $R = k\sqrt{M}$

then $6 = k\sqrt{16} = 4k$

$k = \frac{3}{2}$

Hence $R = \frac{3}{2}\sqrt{M}$, which is the required law.

When $M = 6\frac{1}{4}$,

$$R = \frac{3}{2}\sqrt{6\tfrac{1}{4}} = \frac{3}{2}\sqrt{\frac{25}{4}} = \frac{3}{2} \times \frac{5}{2} = \frac{15}{4} = 3\tfrac{3}{4}$$

When $R = 15$, $15 = \frac{3}{2}\sqrt{M}$

$\sqrt{M} = 15 \times \frac{2}{3} = 10$

$M = 10^2 = 100$

Notice that if $R = \frac{3}{2}\sqrt{M}$, then $R^2 = \frac{9}{4}M$ or $M = \frac{4}{9}R^2$. It is often convenient to use an alternative form of the original.

Joint variation

Joint variation involves three or more variables. The relationships between them can be of many forms, for example,

$$M \propto R^2T,\ F \propto \frac{Mm}{d^2},\ W = kDS,\ t = k\frac{mV^2}{r}$$

Example 20

The mass of a solid metal ball varies jointly as its specific gravity and the cube of its diameter. When the diameter is 6 cm and the specific gravity 7.5, the mass is 850 g. Find the mass of a ball of specific gravity 10.5 and diameter 8 cm.

Let M = mass in grams
D = diameter in cm
S = specific gravity

Then $M = kSD^3$

$\therefore\ 850 = k \times 7.5 \times 6^3$ (1)

and $M = k \times 10.5 \times 8^3$ (2)

Dividing (2) by (1),

$$\frac{M}{850} = \frac{10.5}{7.5} \times \left(\frac{8}{6}\right)^3 = \frac{7 \times 64}{5 \times 27}$$

$$M = 850 \times \frac{7}{5} \times \frac{64}{27} \doteq 2821$$

The mass is approximately 2821 g.

Example 20 shows a method that may be used when the value of k is not required.

Exercise 20g

1. Express the following as relationships using the given letters and *either* the symbol $\propto$ *or* any constants which may be necessary.
   (a) The length, l, of a rectangle of constant area, varies inversely as its breadth, b.
   (b) The resonance frequency, r, of a series circuit varies inversely as the square root of its capacitance, c.
   (c) The gravitational attraction, G, between two particles of mass m_1 and m_2, varies jointly as the product of their masses and the inverse of the square of their distance apart, d.
2. If $y \propto x$ and $y = 10$ when $x = \frac{1}{2}$ find the law of the variation. Find x if $y = 35$.

3. $A \propto M$ and $A = 8$ when $M = 20$. Find A when $M = 15$ and M when $A = 7$.
4. $P \propto Q$ and $P = 14$ when $Q = 8$. Find P when $Q = 6$ and Q when $P = 28$.
5. $D \propto V$ and $D = 108$ when $V = 3$. Find D when $V = 3.75$ and V when $D = 189$.
6. $P \propto Q^2$. $P = 27$ when $Q = 6$. Find (a) the law, (b) P when $Q = 10$, (c) Q when $P = 18\frac{3}{4}$.
7. $x \propto \frac{1}{y}$. $x = 7\frac{1}{2}$ when $y = 4$. Find (a) the law, (b) x when $y = 12$, (c) y when $x = 20$.
8. $M \propto R^3$. $M = 40$ when $R = 4$. Find (a) the law, (b) M when $R = 10$, (c) R when $M = 2.56$.
9. $\sqrt{Y} \propto Z$. $Y = 4$ when $Z = 3$. Find (a) the law, (b) Y when $Z = 15$, (c) Z when $Y = 16$.
10. $A \propto BC$. When $B = 6$ and $C = 3$, $A = 7\frac{1}{2}$. Find A when $B = 8$ and $C = 9$; also B if $A = 25$ and $C = 8$.
11. $P \propto \frac{Q}{R^2}$. When $Q = 5$ and $R = 3$, $P = 20$. Find P when $Q = 6$ and $R = 4$; also R when $P = 21.6$ and $Q = 15$.
12. $x \propto y$ and $y \propto z^2$. How does x vary with z?
13. $x \propto y^2$ and $y \propto \frac{1}{z}$. How does x vary with z?
14. A car takes 6 hours to travel from X to Y at a constant speed. How long does that same journey take for (a) a lorry travelling at half the speed of the car? (b) a helicopter travelling at 3 times the speed of the car? (c) an aeroplane travelling at 6 times the speed of the car?
15. If y is inversely proportional to x, complete Table 20.2.

Table 20.2

x	10		20	25	30	
y		$\frac{1}{3}$	$\frac{1}{4}$			$\frac{1}{7}$

Formulae

A formula gives a relationship or rule between two or more variables which are represented by different letters. One of the variables is stated as a function of the other variables. That variable (letter) is called the subject of the formula.

Example 21

The sum of the squares of the first n integers is given by

$$S_n = \frac{n(n+1)(2n+1)}{6}$$

Calculate (a) S_{20}, (b) the sum of the squares from 21 to 40 inclusive.

(a) S_{20} means the value of S_n when $n = 20$.

$$\begin{aligned} S_{20} &= \frac{20(20+1)(2\times 20+1)}{6} \\ &= \frac{20\times 21\times 41}{6} \\ &= 10\times 7\times 41 \\ &= 2870 \end{aligned}$$

(b) The sum of squares from 21 to 40
= sum of squares from 1 to 40
− sum of squares from 1 to 20
$= S_{40} - S_{20}$

$$\begin{aligned} S_{40} &= \frac{40(40+1)(2\times 40+1)}{6} \\ &= \frac{40\times 41\times 81}{6} \\ &= 20\times 41\times 27 \\ &= 540\times 41 \\ &= 22\,140 \end{aligned}$$

$$\begin{aligned} S_{40} - S_{20} &= 22\,140 - 2870 \\ &= 19\,270 \end{aligned}$$

Exercise 20h

1. The formula $d = \sqrt{(l^2 + b^2 + h^2)}$ gives the length d of the longest diagonal in a cuboid of length l, breadth b and height h. (d, l, b and h are in the same units.) Find the length of the diagonal of a cuboid which is 6 cm long, 2 cm wide and 3 cm high.
2. Find the value of $2\pi\sqrt{\frac{l}{g}}$ when $\pi = 3\frac{1}{7}$, $l = 98$ and $g = 32$.
3. The formula $A = \pi r(r + s)$ gives the surface area, A, of a cone of base radius r cm and slant height s cm. Find the surface area of a cone of base radius 6 cm and slant height 22 cm, using the value $\frac{22}{7}$ for π.
4. Given that $S_n = \frac{4n^3 - 3n^2 + 6}{n}$, evaluate $S_{20} - S_{10}$.

5 Given that $c = 2\pi r$,
(a) find c when $\pi = 3.142$ and $r = 50$,
(b) find r when $c = 286$ and $\pi = 3\frac{1}{7}$.

6 In a certain country, the cost, c cents, of sending a telegram of 12 words or over is given by the formula $c = 3(w - 2)$ where w is the number of words in the telegram.
(a) Find the cost of sending a telegram of 35 words.
(b) If it costs \$1.41 to send a telegram, how many words does it contain?

7 The formula $b = 40 + \frac{7W}{50}$ is used to work out the electricity bill, b dollars, for a month in which W kilowatt-hours of electricity are used.
(a) Find the bill for a month in which 705 kilowatt-hours are used.
(b) Find the number of kilowatt-hours used by a consumer who receives a bill for \$61.14.

8 The formula $d \simeq 5\sqrt{\frac{h}{2}}$ gives the approximate distance, d km, of the horizon which can be seen from a point h m above ground level.
(a) Find the approximate distance of the horizon from the top of a building 72 m high.
(b) From the top of a tower, a man can see for about 35 km. How high is the man above ground level?

Change of subject of a formula

It may be useful to rearrange the letters so that another letter becomes the subject – to do this is **to change the subject of the formula**.

To change the subject of a formula:

1 treat the formula as an algebraic equation;
2 solve the equation for the letter which is to be the subject of the formula.

Example 22

Make P the subject of the formula $R = \dfrac{Q^2 - PR}{Q + P}$.

$$R = \frac{Q^2 - PR}{Q + P}$$

Multiply both sides by $(Q + P)$ to clear fractions:

$$R(Q + P) = Q^2 - PR$$

Clear brackets:

$$RQ + PR = Q^2 - PR$$

Collect terms in P on the LHS of the equation:

$$2PR = Q^2 - RQ$$

Divide both sides by $2R$:

$$P = \frac{Q^2 - RQ}{2R} = \frac{Q(Q - R)}{2R}$$

Example 23

Make r the subject of the formula $b = \sqrt{\dfrac{4\pi r^3}{3h}}$.

$$b = \sqrt{\frac{4\pi r^3}{3h}}.$$

Square both sides:

$$b^2 = \frac{4\pi r^3}{3h}$$

Multiply both sides by $\frac{3h}{4\pi}$:

$$\frac{3hb^2}{4\pi} = r^3 \text{ or } r^3 = \frac{3hb^2}{4\pi}$$

Take the cube root of both sides:

$$r = \sqrt[3]{\frac{3hb^2}{4\pi}}$$

Exercise 20i

In questions 1 to 17 a formula is given. A letter is printed in heavy type after it. Make that letter the subject of the formula. If more than one letter is given, make each letter the subject in turn.

1 $I = \dfrac{PRT}{100}$ **T**

2 $V = \frac{1}{4}\pi d^2 h$ **h, d**

3 $F = \dfrac{mv - mu}{t}$ **t**

4 $R = \dfrac{N - M}{D}$ **M**

5 $T = a + bN^2$ **N**

6 $V = \frac{4}{3}\pi r^3$ **r**

7 $S = 2\pi rh + 2\pi r^2$ **h**

8 $x^2 = a^2(1 - w^2)$ **a, w**

9 $\dfrac{1}{f} = \dfrac{1}{u} + \dfrac{1}{v}$ **f, u**

10 $s = \frac{n}{2}\{a + (n - 1)d\}$ **a**

11 $\frac{P + 3Q}{Q - 3P} = \frac{x}{y}$ **Q**

12 $v^2 = u^2 - 2as$ **s, u**

13 $E = \frac{m}{2g}(v^2 - u^2)$ **u**

14 $e = \frac{P - p}{PT - pt}$ **p**

15 $T - W = \frac{Wv^2}{gx}$ **W**

16 $t = 2\pi\sqrt{\frac{l}{g}}$ **g**

17 $V = \pi r^2 (h + \frac{2}{3}r)$ **h**

18 Given that $x = \frac{y - 2}{y - 3}$, express y in terms of x. [CXC (General) June 89]

19 When a sum of money, $\$P$, is invested at $R\%$ per annum simple interest, its value $\$A$ after T years is given by $A = P\left(1 + \frac{RT}{100}\right)$
(a) Make T the subject of this formula.
(b) A man invests \$2500 at 9% per annum. After how many years will its value be \$4300?

20 $C = 2\pi r$ and $V = \pi r^2 h$. (a) Express r in terms of C and π. (b) Hence find a formula with V as subject, eliminating r.

Equations and identities

The conditions which relate different quantities may be expressed by an algebraic statement in terms of variables, such as $a, b, x, y, m, \ldots$ When there are specific values of the variable(s) which satisfy the given conditions, the statement is called an **equation**. Then the left-hand side of the equation is equal to the right-hand side for these specific values *only*. These values are termed the **solutions** of the equation.

For example, in an algebraic statement such as

$$2x + 3 = 5, \quad \text{then} \quad x = 1.$$

This is the *only* value of x that satisfies the statement. This statement is a **linear equation**.

In a statement such as

$$y^2 - 3y + 2 = 0, \quad \text{then} \quad y = 1, \text{ or } y = 2.$$

These are the *only* two values of y that satisfy the statement. This statement is a **quadratic equation.**

If the left-hand side of the statement is equal to the right-hand side for all values of the variable(s), the statement is called an **identity**. For example, in $y + 3(y - 1) \equiv 2(2y + 1) - 5$ and in
$x^2 - 3x + 2 \equiv 2(x^2 - x - 1) - (x^2 + x - 4)$,
the two sides of each statement are equal for *all* values of the variables. These statements are identities.

Since identities are true for *all* values of a variable, we cannot calculate specific value(s) of the variable for which the identity is true, that is, we cannot '*solve*' an identity. However, we can show in an identity that

(i) the terms on both sides of the identity can be rearranged and written in a common format;

(ii) the terms on one side of the identity can be rearranged and written in the same format as the other side of the identity.

Example 24

Show that $y + 3(y - 1) \equiv 2(2y + 1) - 5$

$$y + 3(y - 1) \equiv 2(2y + 1) - 5$$

LHS: $y + 3(y - 1) \equiv y + 3y - 3$
$\equiv 4y - 3$

RHS: $2(2y + 1) - 5 \equiv 4y + 2 - 5$
$\equiv 4y - 3$
$\equiv$ LHS

Example 25

Show that
$x^2 - 3x + 2 \equiv 2(x^2 - x - 1) - (x^2 + x - 4)$

$$x^2 - 3x + 2 \equiv 2(x^2 - x - 1) - (x^2 + x - 4)$$

RHS: $2(x^2 - x - 1) - (x^2 + x - 4)$
$\equiv 2x^2 - 2x - 2 - x^2 - x + 4$
$\equiv x^2 - 3x + 2$
$\equiv$ LHS

Exercise 20j

Show that the following statements are identities.

1. $3(x - 2) - x - 2 \equiv 2(x - 4)$
2. $2x^2 + 6x + 1 \equiv 2(x + \frac{3}{2})^2 - \frac{5}{4}$
3. $9 + 4y - y^2 \equiv 13 - (y - 2)^2$
4. $3 - \dfrac{2}{m + 1} \equiv \dfrac{3m + 1}{m + 1}$
5. $\dfrac{2a^2 + a - 27}{a^2 - 9} \equiv \dfrac{2}{a + 3} + \dfrac{2 - 1}{a - 3}$
6. $\sqrt{(x - 3)^4} \equiv x^2 - 6x + 9$
7. $(x - 2)^{\frac{1}{2}}(x + 2)^{\frac{1}{2}} \equiv \sqrt{(x^2 - 4)}$
8. $\dfrac{a^4}{5(a^4/16)^{\frac{3}{4}}} \equiv \dfrac{8a}{5}$

Linear inequalities in two variables

In a problem with two variable quantities, a linear inequality in the two variables divides a Cartesian plane into the required region where the inequality is satisfied and a region in which the points do not meet the required condition. The equation of the corresponding linear equation is the equation of the boundary line between the required region and the region in which the points are not included.

Note also that:

(i) the boundary line is drawn continuous to show that the set of points on the line is included in the required region;

(ii) the boundary line is drawn using broken lines to show that the points on the line are *not* included in the required region;

(iii) regions outside the boundaries are shaded to show that they are not required.

Example 26

$y - x \leq 1$, $2x < 5$, $5y > -4x$ are simultaneous inequalities. (a) Show on a graph the region which contains the solution set of the inequalities. (b) If the solution set contains only integral values of x and y, list its members.

(a) The boundary lines of the region are
p: $y - x = 1$
q: $2x = 5$
r: $5y = -4x$
The unshaded region in Fig. 20.2 gives the solution set of all points (x, y) which satisfy the three inequalities.

(b) In Fig. 20.2 the solution set is shown by heavy points, indicating that the values of x and y are integral. The solution set is as follows:
$\{(0, 1), (1, 0), (1, 1), (1, 2), (2, -1), (2, 0), (2, 1), (2, 2), (2, 3)\}$

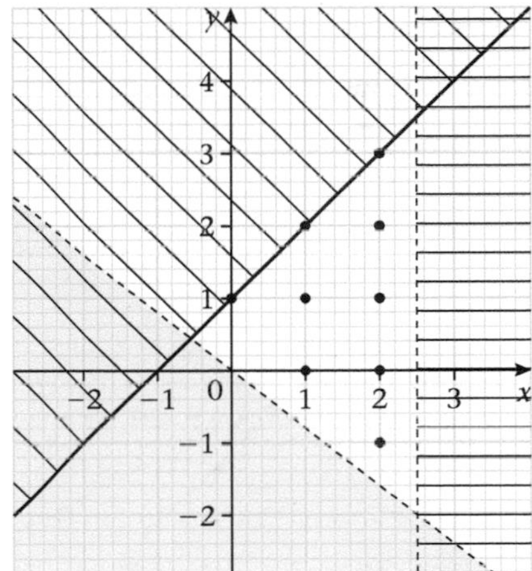

Fig. 20.2

Exercise 20k

1.

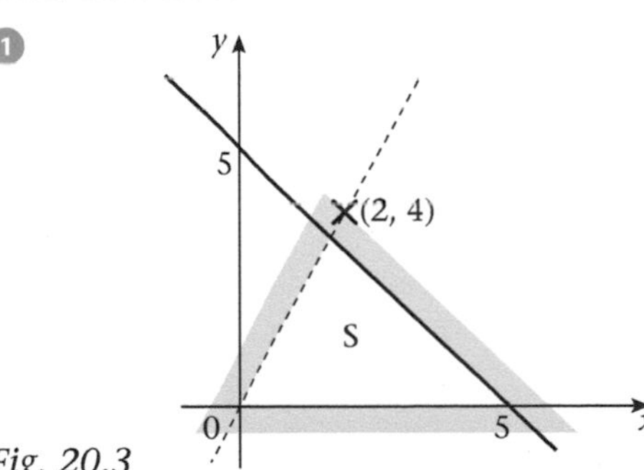

Fig. 20.3

Given Fig. 20.3, find

(a) the equation of the line through the origin and the point (2, 4),

(b) the equation of the line through (0, 5) and (5, 0),

(c) the inequalities which define the triangular shape S.

2 Write down the three inequalities which define the unshaded triangular area A in Fig. 20.4.

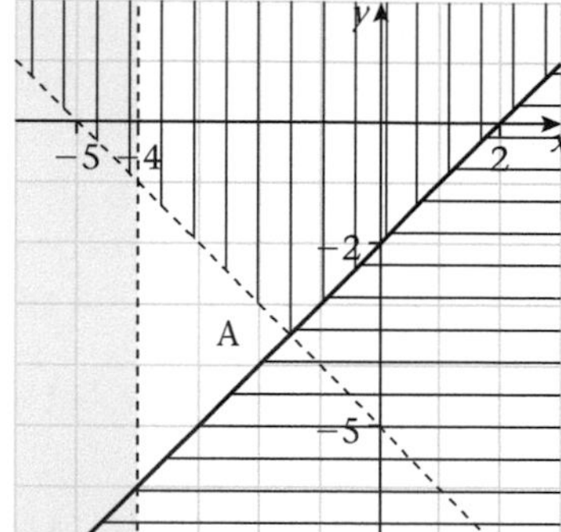

Fig. 20.4

3 Write down the four inequalities which define the unshaded region R in Fig. 20.5.

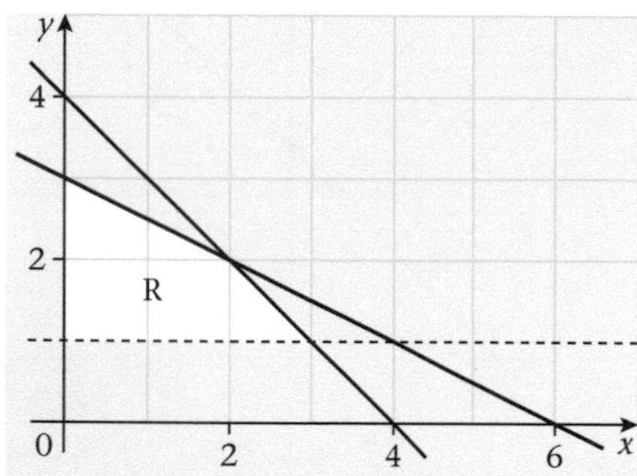

Fig. 20.5

4 Using graph paper, show the regions defined by each of the following. (Use solid and broken lines where appropriate and leave each required region unshaded.)
(a) $4y - x < 4$, $x - y < 3$, $x \geqslant -2$
(b) $x - y > -2$, $x + y < 4$, $x \geqslant -1$, $y > 0$
(c) $x < 4$, $y - 2x \leqslant 2$, $2 < y < 4$
(d) $6 \leqslant 2x + 3y \leqslant 12$, $x - 2y < 8$, $y < 3$

5 Solve each of the following graphically for integral values of x and y.
(a) $y > x$, $y \leqslant 3x$, $y + 2x < 8$
(b) $3x + 4y \leqslant 12$, $y - x \leqslant 2$, $y > 1$

6 K is the set of points (x, y) which satisfies the four inequalities $y - x \leqslant 1$, $2x \leqslant 5$, $5y > -4x$, $y \leqslant 2$. Show on a graph the region which represents K. Use your graph to find the greatest value of $(x + y)$.

7 It takes 2 m of cloth to make a shirt and 3 m to make a dress. A tailor has 36 m of cloth and he needs to make at least 6 of each.
(a) Using x to represent the number of dresses and y the number of shirts, show this information on a graph.
(b) If the profit on a shirt is the same as that on a dress, what arrangement of shirts and dresses gives the greatest profit?

8 A shopkeeper orders two sizes of notebooks, large at 90c each and small at 40c each. She needs at least twice as many small ones as large ones, with minimum quantities of 10 large and 20 small.
(a) Show this information graphically.
(b) If she spends up to $20, what is the maximum number she can buy?
(c) If the profit is 10c on a large notebook and 5c on a small one, (i) what arrangement gives the greatest profit, (ii) how much is that profit?

9 A company plans to spend $500 000 on new machines. Table 20.3 shows the cost and necessary floor space for the two types of machine to be bought.

Table 20.3

Machine	Cost	Floor space
A	$20 000	6 m^2
B	$25 000	4 m^2

More of machine A than of machine B are needed and there are only 120 m^2 of factory floor space available.
(a) Represent this information on a graph.
(b) Which purchase arrangement
(i) uses the maximum space available,
(ii) is more expensive?

10 A farmer wishes to buy some goats and cows. He has pasture for only 50 animals. Goats cost $300 each and cows $600 each. He can expect to make a profit of $250 on each goat and $350 on each cow.
(a) If he has $24 000 to spend and he buys g goats and c cows, obtain *two* inequalities connecting g and c.

(b) Illustrate these inequalities on the same graph.

(c) Determine how many of *each* type of animal he should buy to obtain a maximum profit.

[CXC (General) June 86]

Quadratic equations

If the product of two numbers is 0, then one of the numbers (or possibly both of them) must be zero. For example,

$3 \times 0 = 0$, $0 \times -5 = 0$ and $0 \times 0 = 0$

In general, if $a \times b = 0$

then either $a = 0$

or $b = 0$

Example 27

Solve the equation $(x - 2)(x + 7) = 0$.

If $(x - 2)(x + 7) = 0$

then either $x - 2 = 0$ or $x + 7 = 0$

$x = 2$ or $x = -7$

Example 28

Solve $a(a + 3) = 0$.

If $a(a + 3) = 0$

then either $a = 0$ or $a + 3 = 0$

$a = 0$ or $a = -3$

The solutions of the equations in Examples 27 and 28 are called the **roots** of the equations.

Example 29

Solve the equation $(m - 5)^2 = 0$.

If $(m - 5)^2 = 0$

then $(m - 5)(m - 5) = 0$

Hence $m - 5 = 0$ (twice)

$m = 5$ (twice)

Notice that Example 29 contains a repeated root.

Exercise 20l

Solve the following equations.

1. $(a - 3)(a + 5) = 0$
2. $(b - 2)(b - 1) = 0$
3. $(x + 2)(x + 6) = 0$
4. $(y - 5)y = 0$
5. $(m + 3)(m - 4) = 0$
6. $(n - 5)^2 = 0$
7. $u(u + 1) = 0$
8. $(a + 3)(5a + 2) = 0$
9. $(4x + 3)^2 = 0$
10. $(2y - 7)(y + 2) = 0$
11. $(4b - 12)(b - 5) = 0$
12. $(4h - 1)(2h + 3) = 0$
13. $(5 - d)(5 - 2d) = 0$
14. $(5 + 3m)(2 - 5m) = 0$
15. $(3n + 7)(4n - 1) = 0$

Example 30

Solve the equation $y^2 - 4y = 0$.

$y^2 - 4y = 0$

$y(y - 4) = 0$

either $y = 0$ or $y - 4 = 0$

$y = 0$ or $y = 4$

Example 31

Solve the equation $3x^2 + 5x - 28 = 0$.

$3x^2 + 5x - 28 = 0$

Factorise the quadratic expression on the LHS of the equation:

$(x + 4)(3x - 7) = 0$

either $x + 4 = 0$ or $3x - 7 = 0$

$x = -4$ or $3x = 7$

i.e. $x = -4$ or $x = 2\frac{1}{3}$

Exercise 20m

Use factorisation to solve the following equations.

1. $x^2 - 10x + 21 = 0$
2. $m^2 + 3m + 2 = 0$
3. $a^2 + a - 6 = 0$
4. $n^2 - 3n - 10 = 0$
5. $x^2 + x - 2 = 0$
6. $y^2 + 3y = 0$
7. $2d^2 - 7d + 6 = 0$
8. $b^2 - 6b + 9 = 0$
9. $4e^2 + 11e + 6 = 0$
10. $3k^2 - 17k + 10 = 0$
11. $a^2 - 4 = 0$
12. $4m^2 - 12m + 9 = 0$

13. $a^2 - 4a = 0$
14. $12d^2 - 19d - 18 = 0$
15. $2t^2 + 7t + 5 = 0$
16. $2x^2 + 20x + 50 = 0$
17. $3u^2 - 10u = 8$
18. $8w^2 - 18w + 9 = 0$
19. $12d^2 + 36d + 27 = 0$
20. $8n^2 + 2n = 21$

When the roots of a quadratic equation are non-rational, the equation may be solved by completing the square.

The general quadratic equation

$$ax^2 + bx + c = 0$$

where a, b and c are constants, is written as

$$\left(x^2 + \frac{bx}{a}\right) + \frac{c}{a} = 0$$

Then, completing the square,

$$\left(x^2 + \frac{bx}{a} + \left(\frac{b}{2a}\right)^2\right) + \frac{c}{a} - \left(\frac{b}{2a}\right)^2 = 0$$

$$\left(x + \frac{b}{2a}\right)^2 + \frac{(4ac - b^2)}{4a^2} = 0$$

$$\left(x + \frac{b}{2a}\right)^2 + \frac{(b^2 - 4ac)}{4a^2}$$

Taking the square root of both sides of the equation,

$$x + \frac{b}{2a} = \frac{\pm\sqrt{(b^2 - 4ac)}}{2a}$$

$$x + \frac{-b}{2a} = \frac{\sqrt{(b^2 - 4ac)}}{2a}$$

The formula for solving a quadratic equation is derived from this result, namely

$$x = \frac{-b \pm\sqrt{(b^2 - 4ac)}}{2a}$$

Exercise 20n

1. Add the term to make each of the following expressions into a perfect square and write the result as the square of a bracketed expression:
   (a) $v^2 + \frac{1}{4}v$ (b) $c^2 - 1\frac{1}{2}c$ (c) $w^2 - \frac{3}{4}w$
2. Solve the following equations by finding the square root of both sides of the equation.
   (a) $(x + 3)^2 = 49$ (b) $(x - 1)^2 = 7$
3. Use the formula to solve the following equations.
   (a) $u^2 - 14u - 3 = 0$
   (b) $x^2 + 3x - 2 = 0$
   (c) $3m^2 - 7m + 1 = 0$
   (d) $2v^2 + 10v + 9 = 0$

Exercise 20o

Solve the following equations by factorising if possible, but otherwise by completing the square, or by using the formula. If the roots involve decimals, give them correct to two places. Square root tables may be used.

1. $m^2 - 2m - 2 = 0$
2. $n^2 - 2n - 1 = 0$
3. $x^2 - 2x - 4 = 0$
4. $y^2 + 2y - 1 = 0$
5. $z^2 + 2z - 3 = 0$
6. $b^2 = 11b + 26$
7. $b^2 - 4b + 1 = 0$
8. $c^2 - 8c + 13 = 0$
9. $d^2 + 6d + 7 = 0$
10. $e^2 - 6e + 4 = 0$
11. $h^2 + 4h = 11$
12. $k^2 + 4k + 2 = 0$
13. $x^2 - 3x + 2 = 0$
14. $y^2 - 3y - 11 = 0$
15. $m^2 + 5m = 1$
16. $n^2 = n + 5$
17. $2d^2 - 4d + 1 = 0$
18. $2e^2 - e - 1 = 0$
19. $3m^2 = 6m + 2$
20. $p^2 - 10p + 5 = 0$
21. $y^2 + y = 8$
22. $x^2 = x + 6$
23. $a^2 - 6a - 3 = 0$
24. $2b^2 = 8b + 11$
25. $2c^2 - 3c - 9 = 0$
26. $x^2 - 10x + 23 = 0$
27. $h^2 = 3h + 40$
28. $4c^2 - 8c + 1 = 0$
29. $d^2 = 12d - 35$
30. $e^2 = 3e + 11$
31. $x^2 + 5x = 15$
32. $3m^2 - 5m + 2 = 0$
33. $4h^2 = 8h + 3$
34. $3k^2 = 9k - 2$
35. $5f^2 = 20f + 28$

Graphical solution of quadratic equations

Example 32

Solve $x^2 - 2x - 4 = 0$ graphically.

Let $y = x^2 - 2x - 4$. Make a table of corresponding values of x and y (Table 20.4).

Table 20.4 $y = x^2 - 2x - 4$

x	−2	−1	0	1	2	3	4
x^2	4	1	0	1	4	9	16
$-2x$	4	2	0	−2	−4	−6	−8
-4	−4	−4	−4	−4	−4	−4	−4
y	4	−1	−4	−5	−4	−1	4

Fig. 20.6 is the graph of $y = x^2 - 2x - 4$.

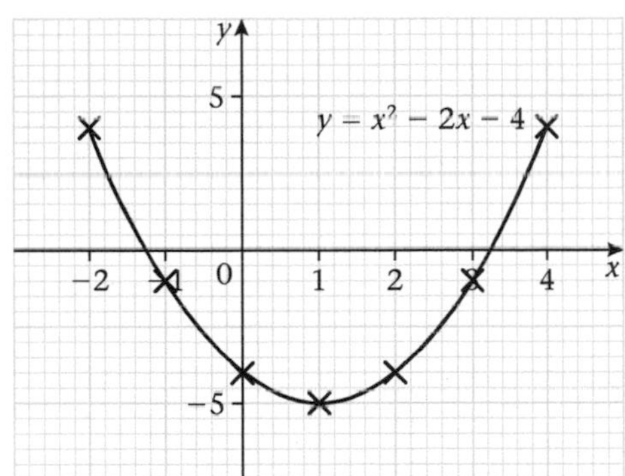

Fig. 20.6

$x^2 - 2x - 4 = 0$ when $y = 0$, i.e. where the curve intersects the x-axis. The curve cuts the x-axis at $x = 3.2$ and $x = -1.2$. These are the approximate solutions to the equation $x^2 - 2x - 4 = 0$.

Example 33

(a) Draw the graph of $y = 2x^2 + 3x - 6$ for values of x from −4 to 2.

(b) Use the graph to solve the equations (i) $2x^2 + 3x - 6 = 0$, (ii) $2x^2 + 3x - 3 = 0$.

(c) By drawing the line $y = 2x + 1$ on the same axes, solve the equation $2x^2 + x - 7 = 0$.

(a) The values in Table 20.5 are used to draw the curve in Fig. 20.7.

Table 20.5 $y = 2x^2 + 3x - 6$

x	−4	−3	−2	−1	0	1	2
x^2	32	18	8	2	0	2	8
$+3x$	−12	−9	−6	−3	0	3	6
-6	−6	−6	−6	−6	−6	−6	−6
y	14	3	−6	−7	−6	−1	−8

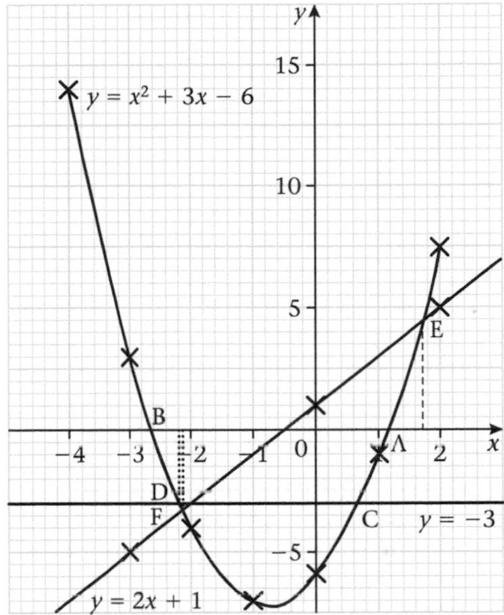

Fig. 20.7

(b) (i) $2x^2 + 3x - 6 = 0$ when $y = 0$, i.e. along the x-axis, at the points A and B in Fig. 20.7. Hence $x = 1.1$ and $x = -2.6$ are the approximate solutions.

(ii) $2x^2 + 3x - 3 = 0$

$2x^2 + 3x = 3$

$2x^2 + 3x - 6 = 3 - 6$

i.e. $2x^2 + 3x - 6 = -3$

This is true at the points where the line $y = -3$ cuts the curve; i.e. at the points C and D in Fig. 20.7. Hence $x = 0.7$ and $x = -2.2$ are the approximate solutions.

(c) Table 20.6 is used to draw the straight line $y = 2x + 1$ in Fig. 20.7.

Table 20.6

x	−3	0	2
y	−5	1	5

The curve and the straight line intersect at the points E and F in Fig. 20.7. At these points

$y = 2x^2 + 3x - 6$ and $y = 2x + 1$

$2x^2 + 3x - 6 = 2x + 1$

$2x^2 + x - 7 = 0$

Hence $x = 1.6$ and $x = -2.1$ are the approximate solutions of $2x^2 + x - 7 = 0$.

Exercise 20p

1. Copy and complete Table 20.7 for the relation $y = x^2 - 4x + 2$.

   Table 20.7

x	-2	-1	0	1	2	3	4	5
y	14			-1	-2		2	

   Use a scale of cm to 1 unit on both axes and draw a graph of the relationship. Use your graph to find
   (a) the solutions of $x^2 - 4x + 2 = 0$,
   (b) the least value of $x^2 - 4x + 2$.

2. Copy and complete Table 20.8 for the relation $y = 3x^2 - 6x + 1$.

   Table 20.8

x	-2	-1	0	1	2	3	4	5
y	25			-2			10	

   Using scales of 2 cm to 1 unit on the x-axis and 2 cm to 5 units on the y-axis, draw a graph of the relationship. Use the graph to solve $3x^2 - 6x + 1 = 0$.

3. Draw the graph of $y = x^2 - 3x - 2$, taking values of x from -1 to 4. Use the graph to read off the roots of the equation
   (a) $x^2 - 3x - 2 = 0$, (b) $x^2 - 3x + 1 = 0$,
   (c) $x^2 - 3x - 4 = 0$.

4. Draw the graph of $y = 3x^2 + 2x - 1$, taking values of x from -3 to 2. Use the same graph to read off the roots of the equations:
   (a) $3x^2 + 2x = 0$ (b) $3x^2 + 2x - 7 = 0$
   (c) $3x^2 + 2x = 3$ (d) $3x^2 + 2x - 12 = 0$

5. Draw the graph of $y = 3x^2 - 5x + 3$ and use it to find the roots of (a) $3x^2 - 5x + 3 = 0$, (b) $3x^2 - 5x = 0$, (c) $3x^2 - 5x - 1 = 0$.

6. Draw the graph of $y = 2x$ to cut the curve drawn in question 5. Hence find the solution of the equation $3x^2 - 7x + 3 = 0$.

7. Draw the graph of $y = x(x - 1)$ for values of x from -2 to 3. Read off the values of x at the points where the curve cuts the line $y = 3 - x$. Of what equation in x are these values the roots?

8. Draw the graphs of $y = 2x^2 - 3x - 6$ and $y = 1 - 3x$ on the same axes for values of x from -2 to 3. Use the graph to find the roots of the equation $2x^2 - 7 = 0$.

9. Copy and complete Table 20.9 for the relation $y = 2 + x - x^2$.

   Table 20.9

x	-2	-1.5	-1	-0.5	0	0.5	1	1.5	2	2.5	3
y			0	1.25	2		2	1.25			-4

   (a) Draw the graph of the relation using a scale of 2 cm to 1 unit on each axis.
   (b) From your graph, find the greatest value of y and the value of x at which this occurs.
   (c) Using the same axes, draw the graph of $y = 1 - x$
   (d) From your graphs, determine the roots of the equation $1 + 2x - x^2 = 0$.

10. Solve the equation $x^2 = \frac{1}{2}x + 4$ by drawing the graphs of $y = x^2$ and $y = \frac{1}{2}x + 4$ on the same axes for values of x from -3 to 3. Check your result by drawing a separate graph of $y = x^2 - \frac{1}{2}x - 4$ for the same range of values of x.

Simultaneous linear and quadratic equations

Example 34

Solve the simultaneous equations $2x - 5y = 1$, $4x^2 + 25y^2 = 41$.

$$2x - 5y = 1 \quad (1)$$
$$4x^2 + 25y^2 = 41 \quad (2)$$

From (1): $x = \dfrac{1 + 5y}{2}$ (3)

Substitute $\frac{1}{2}(1 + 5y)$ for x in (2):

$$4 \times [\tfrac{1}{2}(1 + 5y)]^2 + 25y^2 = 41$$
$$4 \times \tfrac{1}{4}(1 + 10y + 25y^2) + 25y^2 = 41$$
$$50y^2 + 10y - 40 = 0$$
$$5y^2 + y - 4 = 0$$
$$(5y - 4)(y + 1) = 0$$
$$y = \tfrac{4}{5} \text{ or } -1$$

Substitute for y in (3):

When $y = \frac{4}{5}, x = \frac{1+4}{2} = 2\frac{1}{2}$

When $y = -1, x = \frac{1-5}{2} = -2$

The solutions may be given in the form of ordered pairs: $(2\frac{1}{2}, \frac{4}{5})$ and $(-2, -1)$

Notice that part (c) of Example 33 gives a graphical method of solving simultaneous linear and quadratic equations. Questions 6–10 of Exercise 20p provide practice in the use of graphical methods.

Exercise 20q

Solve the following pairs of equations.

1. $2x + y = 5$, $x^2 + y^2 = 25$
2. $4x - y = 7$, $xy = 15$
3. $x + y = 3$, $x^2 - y^2 = -3$
4. $4x^2 + y^2 = 61$, $2x + y = 1$
5. $2x - y = 5$, $4x^2 - y^2 = 15$
6. $2x^2 - y^2 = -2$, $3x + y = 1$
7. $2x + 3y = 1$, $4x^2 - 9y^2 = -17$
8. $x + 2y = 2$, $x^2 + 2xy = 8$
9. $x^2 + 2y^2 = 3$, $x - 3y = 2$
10. $xy = 30$, $3x + y = 21$
11. $25x^2 - 4y^2 = 36$, $5x - 2y = 2$
12. $x - 3y = 1$, $x^2 - 2xy - y^2 = 7$
13. $3xy - y^2 = 2$, $2x - 3y = -4$
14. $25x^2 - 7y^2 = 29$, $5x + 7y + 1 = 0$
15. $9x^2 + 16y^2 = 52$, $3x - 4y = 2$

Word problems

Example 35

Find two numbers whose difference is 4 and whose product is 192.

Let the smaller number be x.
Then the larger number is $x + 4$.
Their product is $x(x + 4)$.

Hence $x(x + 4) = 192$

$$x^2 + 4x - 192 = 0$$
$$(x - 12)(x + 16) = 0$$
$$x = 12 \text{ or } -16$$

The other number is 4 more, i.e. $12 + 4$ or $-16 + 4$, i.e. 16 or -12.
The two numbers are 12 and 16, or -16 and -12.
Check: $12 \times 16 = 192$ and $-16 \times -12 = 192$.

Compare Example 35 with Example 36 which follows. Notice the use of units and the elimination of one root because it is not a sensible result.

Example 36

The length of a rectangular hall is 5 m more than the width. Its area is 500 m². Find the width and length of the hall.

Let the width be x m. Then, from the 1st sentence, the length is $(x + 5)$ m. The area is $x(x + 5)$ m².

From the second sentence:

$$x(x + 5) = 500$$
$$x^2 + 5x - 500 = 0$$
$$(x - 20)(x + 25) = 0$$
$$\Rightarrow x = 20 \text{ or } -25$$

An answer of -25 m is clearly not sensible for the width of a hall.
Hence the width is 20 m and the length, 5 m more, is 25 m.
Check: 20 m × 25 m = 500 m²

Exercise 20r

1. A girl bought some pencils for \$10.80. If she had paid 12c less for each pencil she could have bought 3 more pencils. How many pencils did she buy?
2. Divide 27 into two parts such that their product is 180.
3. The area of a rectangle is 360 cm² and its length is 2 cm more than its width. Find the width.
4. The perimeter of a rectangle is 42 cm and its area is 68 cm². Find its length and breadth.
5. Two brothers are 11 and 8 years of age. In how many years' time will the product of their ages be 378?
6. A rectangular garden measures 12 m by 5 m. A path of uniform width runs along one side and one end. If the total area of the garden and path is 98 m², find the width of the path.

7. A number is subtracted from 20 and from 17. The product of the numbers so obtained is 180. Find the original number.

8. A rectangular piece of cardboard measures 21 cm by 16 cm. When strips of equal width are cut off one side and one end, the area of the remaining piece is 234 cm^2. Find the width of the strips.

9. A woman is 35 years old and her son is 12 years old. How many years ago was the product of their ages 174?

10. Two rectangles have the same area of 24 cm^2. The second rectangle is 4 cm shorter and 1 cm wider than the first. What is the length and breadth of the first rectangle?

Exercise 20s

1. (a) Solve the equation
   $\frac{x + 5}{5} - \frac{2x + 3}{3} = 7$
   (b) (i) Make R the subject of the formula
   $2\pi Rh + A = 60$
   (ii) Find the value of R when $h = 5$, $A = 16$, $\pi = \frac{22}{7}$

2. (a) Write as a single fraction
   $\frac{3}{y + 1} + \frac{1}{y - 1} - \frac{3}{2}$
   (b) Given that $a \# b = a^2 - \frac{2b}{3}$, find the value of
   (i) $5 \# 6$ (ii) $3 \# (5 \# 6)$

3. (a) If w is proportional to v^2, and $w = 81$ when $v = 3$,
   (i) determine w as a function of v
   (ii) find the value of w when $v = 6$.
   (b) y is inversely proportional to $x + p$, where p is a constant. If $y = 1$ when $x = 2$, and $y = -1$ when $x = 4$, determine
   (i) the value of p
   (ii) y as a function of x.

4. (a) Solve the equation
   $\frac{2}{x} + \frac{x - 1}{3} = 2$
   (b) (i) Determine the values of x for which
   $3(x + 1) - 2(3x + 2) < 5$
   (ii) Show the values on a number line.

5. A manufacturer produces two types of ballpoint pens: Type L and Type M. There are at least 50 of Type L and at least 25 of Type M pens.
   The manufacturer, however, does not produce more than 80 of Type L or more than 60 of Type M or more than 120 of both Type L and Type M taken together.
   (a) Using x to represent the number of Type L pens produced and y to represent the number of Type M pens produced, write **three** inequalities (not including $x \geqslant 0$ and $y \geqslant 0$) which represent the above conditions.
   (b) Using a scale of 1 cm to represent 10 pens on **each** axis, draw the graph of the inequalities.
   Identify the region which satisfies the inequalities.
   (c) The manufacturer makes a profit of \$1.50 on each Type L pen and \$1.10 on each Type M pen.
   (i) Write an expression to represent his total profit.
   (ii) Use the graph to determine the values of x and y which give a maximum profit, and hence determine the maximum profit.
   (15 marks)
   [CXC (General) June 92]

6. For exercise, Ken must swim and walk. Each exercise session must last for 1 hour. He must swim for at least 3 hours but not more than 10 hours each week. He must walk for more than 5 hours but for less than 12 hours weekly. He must exercise for a total of at least 18 hours each week.
   (a) Using x to represent the number of sessions that he swims, and y the number of sessions that he walks, write three inequalities to represent the above conditions.
   (b) Use a scale of 1 cm to represent 1 session on each axis, and draw the graphs of the inequalities.
   (c) Label the region K that satisfies the inequalities.

(d) Mark on the diagram the possible combinations of walking and swimming sessions that satisfy the conditions.
(e) State these combinations as a set of ordered pairs.

7 Seeds of Type A and Type B are sold in packets. Each packet must contain
- both Type A and Type B seeds
- at least twice the number of Type B seeds as there are of Type A seeds
- no more than 12 seeds.

(a) State the minimum number in each packet of Type A and of Type B seeds. (2 marks)
(b) If there are x Type A and y Type B seeds in each packet, write four inequalities to represent the above conditions. (4 marks)
(c) (i) Using a scale of 1 cm for each unit on both axes, draw graphs on the same axes to represent the inequalities.
(ii) Label the region S which satisfies all four inequalities. (6 marks)
(d) If a profit of 30¢ is made on the sale of each Type A seed, and 60¢ on each Type B seed, determine the packet of seeds on which maximum profit will be made. (3 marks)

[CXC (General) Jan 97]

8 (a) Calculate the values of x for which
$$7 - 4x > 15$$
Show your solution on a number line. (3 marks)
(b) Factorise completely:
(i) $x^2 + 3x + 2$
(ii) $(x + 3)(x - 2) + ax + 3a$ (4 marks)
(c) Find the value of p, if 3 is a root of $5x^2 - px - 18 = 0$. (3 marks)

[CXC (General) Jan 98]

9 Fig. 20.8 shows the graph of $f(x) = x^2 + x - c$. From the graph, state
(a) the domain of x, using inequality notation,
(b) the roots of the equation $x^2 + x - c = 0$,

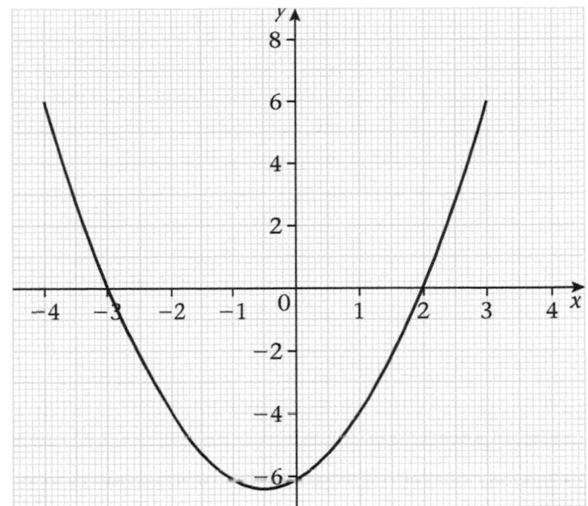

Fig. 20.8

(c) the values of x, to 2 significant figures, for which $x^2 + x - c = 2$. (7 marks)

[CXC (General) Jan 97]

10

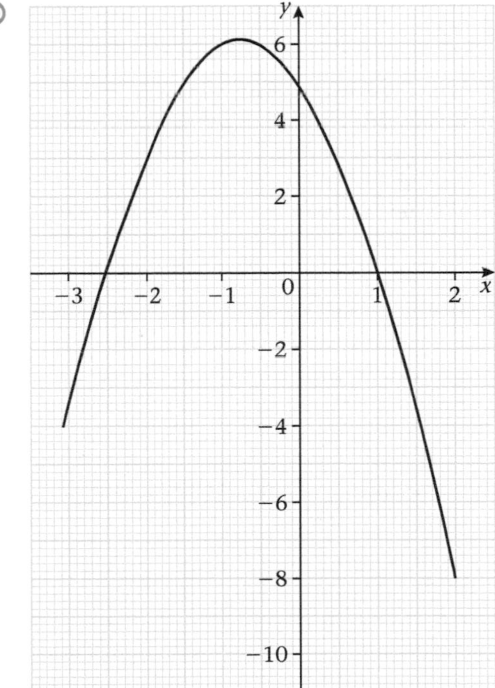

Fig. 20.9

(a) Use the graph in Fig. 20.9 to solve the equation: $2x^2 + 3x = 5$
(b) (i) Draw on the same axes the graph of $y = 3$

(ii) Hence, solve the equation:
$2x^2 + 3x = 2$

(c) (i) Draw on the same axes the graph of
$y = x + 2$

(ii) Hence solve the equation:
$2x^2 + 4x = 3$

11 Express $9x^2 - 30x + 21$ in the form $(ax - b)^2 - c$, where a, b and c are integers.
[CXC (General) June 86]

12 Prove the identity
$(x + y + z)^2 - (x + y)^2 \equiv 2\,(x + y)z + z^2$
for all real x, y, z. (3 marks)
[CXC (General) Jan 90]

13 (a) Simplify $\dfrac{9x^2 - y^2}{9x + 3y}$. (3 marks)

(b) Solve $3a - \frac{1}{2}b = 4$
$9a + 2b = -2$. (5 marks)
[CXC (General) Jan 91]

14 Factorise completely
(a) $3r^2 + 6rR$
(b) $km - mn - kn + n^2$
(c) $2a^2 - 18$
(d) $5y^2 + y - 4$

15 (a) Write $f(x) = 3x^2 - x + 1$ in the form: $a + b(x + c)^2$, where a, b and c are constants.

(b) Hence, state whether the function has a maximum or a minimum value, explaining the reason for your answer.

(c) State this value of the function and the x-value at which it occurs.

16 Solve the following pairs of simultaneous equations:
(a) $3v + w = -4$
$2v +- 3w = -10$
(b) $2y^2 + z^2 = 6$
$3y + 2z = 1$

17 (a) (i) Make t the subject of the equation $S = \frac{1}{2}gt^2$

(ii) Find the positive value of t when $S = 50$ and $g = 10$, writing your answer correct to 2 significant figures.

(b) (i) Write the expression $2x^2 + 5x - 1$ in the form $p(x + q)^2 + r$, where p, q and r are constants.

(ii) Hence find the solution of $2x^2 + 5x - 1 = 0$, correct to 2 decimal places.

18 (a) Simplify the following expressions, writing each answer in its lowest terms:

(i) $\dfrac{3x^3y^4z^2}{15xy^3z^6}$

(ii) $\dfrac{3b^2 - 12c^2}{b^2 + 2bc} \times \dfrac{2b^3 + 3b^2}{5b^2 - 10bc}$

(iii) $\dfrac{1}{h^2 + 4h + 3} - \dfrac{1}{3h^2 + 8h - 3}$

(b) Show that, when $p = 2$, $q = -1$, $r = -2$, the value of
$\dfrac{12}{p^2qr^3} - \dfrac{15}{3pq^2r} + \dfrac{4}{p^3q^2r^2}$
is $1\frac{3}{4}$

(c) Given that
$(x + m)^2 + (y + n)^2 \equiv (x + y + 3)(x - y + 2)$
is an identity, find the values of m and n

Revision test (Chapter 20)

1 Which of the following graphs represents the solution set of $x \leqslant 2$?

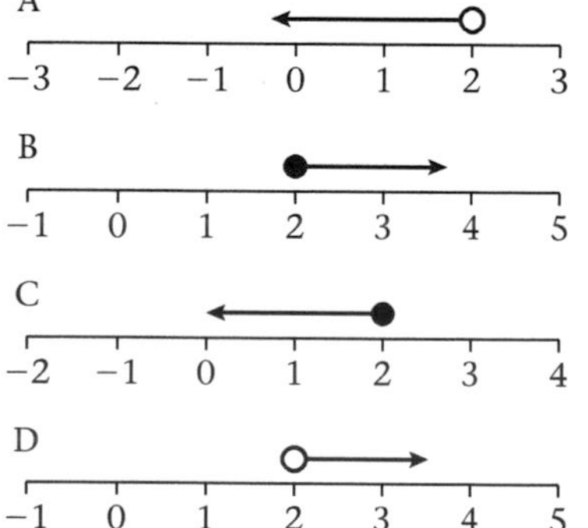

Fig. 20.10

2 One factor of $25x^2 - 9$ is $5x - 3$. The other is

A $-5x - 3$ B $-5x + 3$
C $5x - 3$ D $5x + 3$

3 If $A = 2\pi rh$, then $r =$

A $\frac{2\pi h}{A}$ B $\frac{A}{2\pi h}$

C $-\frac{A}{2\pi h}$ D $A - 2\pi h$

4 Given that $3 : x = 8 : 2$, $x =$

A $\frac{3}{4}$ B $\frac{4}{3}$ C 6 D 12

5 $(-2)^2\,(2)^{-3} =$

A -2 B 2 C $\frac{1}{2}$ D $\frac{1}{32}$

6 Which one of the following ordered pairs is *not* a solution set of $x - 2y \leqslant 6$?

A $(2, -2)$ B $(5, 1)$

C $(0, 0)$ D $(1, -6)$

7 Which one of the following has the same solution set as $5x + 10 = 55$?

A $x = 45$ B $x + 10 = 11$

C $5x = 65$ D $5x + 15 = 60$

8 Given that n is a positive number, which of these symbolic statements means 'the square of a positive number decreased by 10 equals 6'?

A $10 - 2n = 6$

B $10 - n^2 = 6$

C $n^2 - 10 = 6$

D $2n - 10 = 6$

9 Which of the following inequalities is represented by the graph in Fig. 20.11?

A $-2 \leqslant x \leqslant 3$

B $-2 \leqslant x < 3$

C $-2 \geqslant x < 3$

D $-2 < x \leqslant 3$

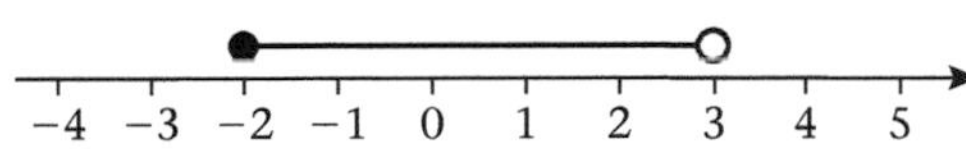

Fig. 20.11

10 Given that $\frac{2}{x} + \frac{3}{y} = 16$

and $\frac{1}{x} - \frac{2}{y} = 1$,

$\{x, y\} =$

A $\{\frac{1}{5}, \frac{1}{2}\}$ B $\{-5, -2\}$

C $\{5, 2\}$ D $\{-\frac{1}{5}, -\frac{1}{2}\}$

11 If $x^2 - 5x - 6 = 0$, then $x =$

A $\{6, -1\}$ B $\{3, -2\}$

C $\{3, 2\}$ D $\{-6, 1\}$

12 If $\frac{x}{x-2} - \frac{3}{x+2} = 1$, then $x =$

A -2 B $\frac{4}{3}$ C $\frac{8}{3}$ D 10

13 If y varies inversely as x and $y = 3$ when $x = 5$, then, in terms of x, $y =$

A $\frac{15}{x}$ B $\frac{x}{15}$ C $\frac{3x}{5}$ D $15x$

14 If $2^{r-3} = 2\sqrt{2}$, then $r =$

A $-\frac{9}{2}$ B $-\frac{3}{2}$ C $+\frac{3}{2}$ D $+\frac{9}{2}$

15 $\frac{y+3}{(y-1)(y-2)}$ is undefined when

a $y = -3$ **b** $y = 1$

c $y = 2$ **d** $y = 3$

A **a** only B **b** and **c** only

C **a**, **b** and **c** only D **b**, **c** and **d** only

16

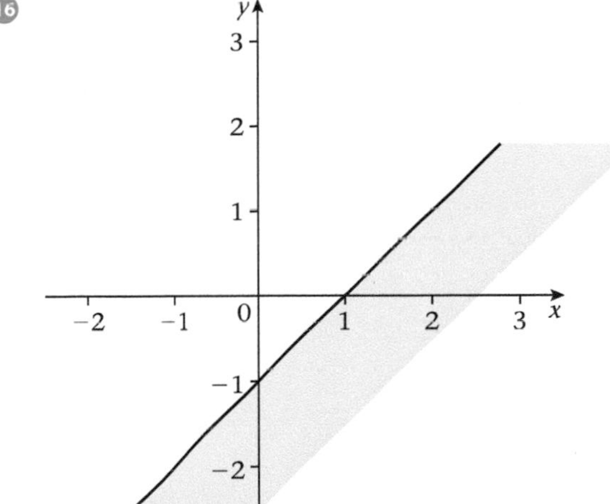

Fig. 20.12

In Fig. 20.12, the unshaded region represents the graph of the inequality

A $y > x - 1$ B $y \geqslant x - 1$

C $y < x - 1$ D $y \leqslant x - 1$

17 Which of the following statements is an identity?

a $\frac{(x-3)}{2} = \frac{(x+1)}{3}$

b $\frac{3}{y-1} + \frac{2}{y+1} = \frac{5}{(y-1)(y+1)}$

c $x^{\frac{3}{2}} + 1 = \sqrt{(x+1)^3}$

d $(x-1)^2 + 2 = x^2 - 2x + 3$

A **b** only B **d** only

C **b** and **c** only D **a**, **c** and **d** only

18 If $x^2 + xy = 2$
and $y = 3x - 2$,
the solution set of these simultaneous equations is

A $\{(\frac{1}{2}, -\frac{1}{2}), (-1, -5)\}$
B $\{(-\frac{1}{2}, -\frac{7}{2}), (1, 1)\}$
C $\{(\frac{1}{2}, \frac{7}{2}), (-1, 1)\}$
D $\{(-\frac{1}{2}, \frac{7}{2}), (1, -1)\}$

19 Chairs cost \$55 each and desks cost \$72 each. At least three times as many chairs as desks must be bought. The total spent must be less than \$1 200. If c chairs and d desks are bought, the two inequalities to be satisfied are

A $55c + 72d < 1\,200$ and $3c \geqslant d$
B $55c + 72d < 1\,200$ and $3c \leqslant d$
C $55c + 72d < 1\,200$ and $c \geqslant 3d$
D $55c + 72d < 1\,200$ and $c \leqslant 3d$

20

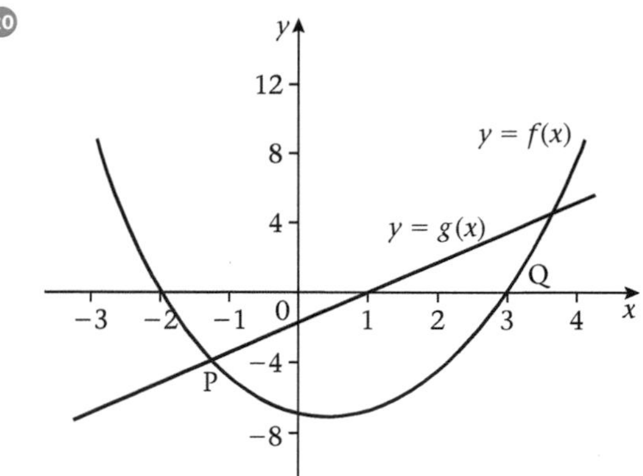

Fig. 20.13

In Fig. 20.13 the points of intersection P and Q of $f(x) = x^2 - x - 6$ and $g(x) = 2x - 2$ are the solutions of

A $2x - 2 = 0$ B $x^2 - x - 6 = 0$
C $x^2 + x - 8 = 0$ D $x^2 - 3x - 4 = 0$

Chapter 21
Consumer arithmetic

Consumer arithmetic was revised in Book 3. These exercises provide further revision.

Revision exercises

Revision exercise 21a: Profit and loss

1. *Home Store* buys 100 tables for $50 each. It costs $3000 to store all the tables.
   (a) Find the cost price of each table.
   They sell the tables for $100 each.
   (b) Calculate the percentage profit on each table.
2. *Drinkies* coffee shop is open seven days a week and just sells cups of coffee. The coffee beans cost $780 per day, the milk $420 per day, water $270 per day and labour costs $500 per day. The shop is rented for $1 200 per week.
   (a) How much does it cost to run *Drinkies* for a week?
   (b) A cup of coffee costs $2.50. Find the number of cups they must sell to cover the costs of running *Drinkies* for a week.
   (c) Last week they sold 7 125 cups of coffee. Calculate the profit they made.
3. Hyacinth wants to sell a CD for $10.80. This would give her a 20% profit. However, she only sells it for $10.35. Calculate the percentage profit she makes.
4. Carl bought a bike for $300. He sold it to Aaron for a 15% profit. Aaron then sold the bike to Lyris for a 5% profit.
   (a) How much more did Lyris pay for the bike than Carl?
   (b) Write the increase as a percentage of the original cost.

Revision exercise 21b: Discount

1. *Sam's Spades* is giving a 40% discount on all spades. The marked price of a spade is $45.
   (a) How much, before VAT, does the buyer pay for a spade?
   (b) The VAT on each spade is 17.5%. How much is the VAT?
2. Kirsten pays $15.85 for a book that would have cost $17.50.
   (a) Calculate the percentage discount on the book.
   (b) The sales tax on the book is 8%. If the shop bought the book for $14.20, calculate the percentage profit they make on the sale.
3. *Comfy Clothes* sells shoes at a 25% discount. The selling price is $34.50. They make a 20% profit on the selling price. What percentage profit would they have made on the marked price?

Revision exercise 21c: Mortgages and hire purchase

1. Linda is buying a car on hire purchase. The cash price is $15 950. She pays a ten percent deposit, then $270 a month for 60 months.
   (a) How much does she pay for the car?
   (b) How much interest does she pay on the outstanding balance?
2. *Homely Houses* has two mortgage options, as shown in Table 21.1.

   Table 21.1

	Deposit	Instalments
Mortgage A	7%	300 monthly instalments of 0.5%
Mortgage B	3%	250 monthly instalments of 0.7%

   Terence and Alaina want to buy a house costing $150 000.
   (a) Which mortgage costs less?
   (b) What percentage profit do *Homely Homes* make on each mortgage?

Revision exercise 21d: Bills and invoices

1 Calculate the amount owed to the electricity board in each of the cases shown in Table 21.2.

Table 21.2

	Previous reading	Present reading	Rate ($ per kWh)
(a)	532 981	533 187	0.435
(b)	000 177	000 208	0.66

2 Table 21.3 shows how *Sparky Electricity* calculate their bills.

Table 21.3

Fixed rate	$11.75
First 100 kWh	$0.44 per kWh
Next 250 kWh	$0.39 per kWh
Over 350 kWh	$0.30 per kWh
VAT	17.5%

Calculate the amount owed in the following cases.

(a) Previous reading: 004 195
Present reading: 004 473

(b) Previous reading: 984 560
Present reading: 986 355

3 Pamela is changing her telephone supplier. Her new supplier has three tariffs, as shown in Table 21.4.

Table 21.4

	Fixed charge	Daytime rate	Evening rate
Tariff A	$10 per month	7¢ per minute	10¢ per minute
Tariff B	$5 per month	9¢ per minute	11¢ per minute
Tariff C	$15 per month	Free	12¢ per minute

Pamela expects to spend an average of 10 minutes per day on the phone during the daytime and about 25 minutes in the evening. Which tariff will cost her least money? (Assume a 30-day month.)

Revision exercise 21e: Investments and depreciation

1 Kenny borrows $550 from *Easy Money* bank. They charge 6.8% simple interest and Kenny pays his loan back over 3 years.

(a) How much does Kenny repay?

Sunita also takes a loan from *Easy Money* at the same rate of interest. She pays $275.40 interest and pays the loan back over four-and-a-half years.

(b) How much was her loan?

2 Vivian invests $1 200 in a savings account. The rate paid is 4.5% simple interest per annum.

(a) Calculate the amount in the account after three years.

(b) After three years she adds another $900. Calculate the amount in the account after a further two years.

3 Vincent borrowed $9 000 from his bank at a rate of 12% p.a. compound interest. If he doesn't make any repayments, how much does he owe after three years?

4 Candice owns a vintage table that her grandmother gave her. She takes it to an antiques shop for an evaluation. The expert says the table is worth $325 and will appreciate at 35% per annum.

(a) How much will it be worth in three years?

(b) She accidentally chips the surface on the way out of the antiques shop. The expert says it is now worth only $190 and will appreciate at 20% per annum. How much less will the table be worth after three years as a result of the damage?

5 Mr Griffith bought a vintage car for $17 999. One year later it is worth $24 298.65.

(a) By what percentage did the car appreciate in the year?

(b) Assuming the same rate of appreciation, how much is the car worth two years after he bought it?

6 Chantal borrows $2 400 from the bank. The bank charges 12% interest per annum. He

makes monthly repayments to the bank. At the end of the first year the bank charges him $72 interest.
Calculate
(a) how much he owed at the end of the first year,
(b) his monthly payments,
(c) after how many months of the second year he pays off the loan.

Revision exercise 21f: Wages, salary and income tax

1. Mr Boon works $7\frac{1}{2}$ hours a day for 22 days a month. He has a six-month contract and gets paid $1 221 every month. He occasionally has to work overtime for which he gets paid triple his usual rate. Calculate his wages during his contract if
   (a) he just works his normal hours,
   (b) he does a total of 59 hours overtime
   (c) he takes four days unpaid holiday but works 102 hours overtime.

2. Use Table 21.5 to calculate the income tax to be paid by Alvin and Lorna Best who are married. They have one child at university. Alvin is currently out of work. Lorna earns $23 000 per year. They have mortgage interest payments of $1 350 per year and pay a total of $660 per year for national insurance.

Table 21.5

Tax-free allowances (per year)	Tax rates (per year)
Personal allowance = $8 000 Spousal allowance = $5 000 (if not working) Per child (at school) = $900 Per child (at university) = $1 300 Mortgage interest = all National insurance contributions = first $1 500	10% on first $3 000 25% on next $8 000 40% on next $24 000 50% on the remaining

Revision exercise 21g: Exchange rates

Use the rates in Table 21.6 for this exercise.

Table 21.6

	Bds$	EC$	Ja$	TT$	Pesos	US$	UK£
Bds$ 1.00	1	1.35	30.87	3.10	5.72	0.50	0.28
EC$ 1.00	0.74	1	22.85	2.30	4.23	0.37	0.20
Ja$ 1.00	0.03	0.04	1	0.10	0.19	0.02	0.01
TT$ 1.00	0.32	0.44	9.96	1	1.85	0.16	0.09
Pesos 1.00	5.72	0.24	5.40	0.54	1	0.09	0.05
US$ 1.00	2.00	2.70	61.59	6.19	11.41	1	0.55
UK£ 1.00	3.63	4.91	112.20	11.27	20.79	1.82	1

1. Devon wants to convert 1 350 Trinidad & Tobago dollars (TT$1 350). He has to pay 10% commission. Use the table to calculate how much of the following currencies he would get.
   (a) US dollars (US$)
   (b) East Caribbean dollars (EC$)
   (c) Jamaican dollars (Ja$)
   (d) Barbados dollars (Bds$)

2. Roger is travelling from Miami to St Kitts. He exchanges US$1 750 into East Caribbean dollars and pays 7% commission.
   (a) How many EC$ does he receive?
   (b) He doesn't spend any money in St Kitts before travelling to Barbados. He changes his East Caribbean dollars to Barbados dollars and pays 8% commission. How many Bds$ does he get?

3. Andy leaves Trinidad with TT$5 600. He travels to Jamaica where he spends Ja$15 000. He then converts the remainder into Mexican Pesos. Assuming he doesn't pay commission on any exchange, how many Pesos does he get?

Chapter 22

Relations, functions and graphs

Relations and functions

For more detailed study of the topic see Book 3, Chapter 20 and Book 4, Chapters 10 and 11.

Exercise 22a

1. Given the relation $x \to 3x + 2$.
   (a) write the ordered pairs for integral values in the domain $\{x : -1 \leqslant x \leqslant 2\}$;
   (b) plot the ordered pairs and draw the graph of the relation;
   (c) use the graph to state
      (i) the values of p, q, r in the following ordered pairs: $(-2, p)$, $(q, 11)$, $(r, 0)$;
      (ii) if the relation is a function, giving reasons.

2. Given that $f: x \to 7 - \frac{2}{x}$,
   (a) derive the expression for f^{-1};
   (b) evaluate (i) $f(4)$, (ii) $f^{-1}(3)$;
   (c) find x when (i) $f(x) = \frac{1}{3}$,
   (ii) $f^{-1}(x) = -2$.

3. If $f(x) = 3x + 1$ and $g(x) = \frac{1}{3}x - 1$ derive the expressions for
   (a) $fg(x)$, (b) $gf(x)$.
   (c) Find the value of (i) $fg(-1)$,
   (ii) $gf(\frac{8}{3})$.
   (d) Find the value of x when (i) $fg(x) = -1$,
   (ii) $(fg)^{-1}(x) = -\frac{7}{3}$.

4. (a) Given that $f(x) = 3x^2 + 3x - 2$ and x is an integer, write down the range of values of $f(x)$ for $-2 \leqslant x \leqslant 2$.
   (b) The function $h(x) = \frac{x+2}{x-2}$.
      (i) Calculate $h(\frac{1}{2})$
      (ii) Find the values of x for which $h(x) = 0$, $h(x)$ is undefined.

Graphs

Straight-line graphs

Exercise 22b

1. 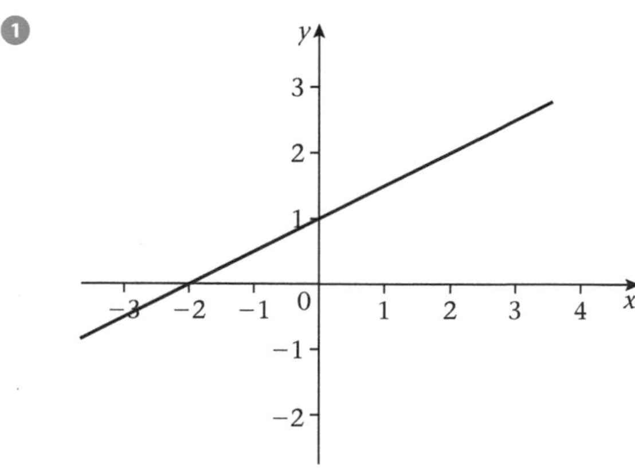

Fig. 22.1

Derive the equation of the line shown in Fig. 22.1.

2. Find the equation of the straight line that passes through the points $(-3, 3)$ and $(2, -1)$.

3. The straight line $3y = 2x + 6$ cuts the y-axis at C and the x-axis at D. Find (a) the gradient of the line, (b) the coordinates of the mid-point of CD, (c) the length of CD.

4. A straight line passes through the point $(4, 3)$ and has a gradient of $\frac{3}{2}$. Find (a) the equation of the straight line, (b) the intercepts on the axes.

5 Find the equation of (a) the straight line shown in Fig. 22.2, (b) the straight line parallel to the line in Fig. 22.2 that passes through (i) (0, 0), (ii) (5, 0).

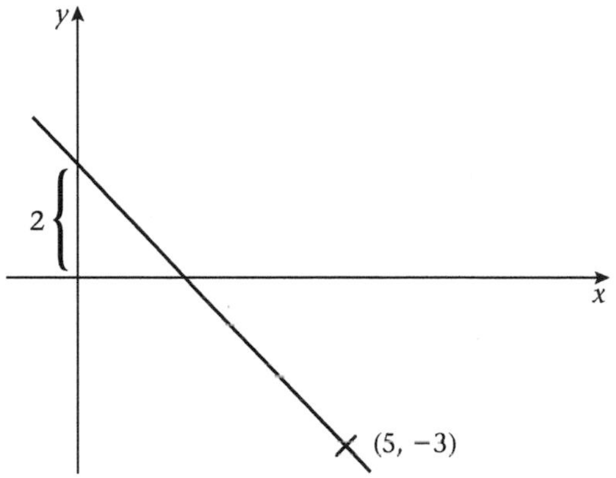

Fig. 22.2

Example 1

Draw the graph of $y = x^2 + 3x - 4$ for values of x from −5 to +2. Find (a) y when $x = 1.5$, (b) y when $x = -2.5$, (c) x when $y = 4$, (d) the least value of $x^2 + 3x - 4$, (e) the value of x for which y is least.

Obtain the values of y by adding the values of x^2, $3x$ and −4 for each value of x. Table 22.1 shows the method.

Table 22.1

x	−5	−4	−3	−2	−1	0	1	2
x^2	25	16	9	4	1	0	1	4
$+3x$	−15	−12	−9	−6	−3	0	3	6
−4	−4	−4	−4	−4	−4	−4	−4	−4
y	6	0	4	6	6	4	0	6

Fig. 22.3 is the graph of $y = x^2 + 3x - 4$. Note that scales of 2 cm to 1 unit on the x-axis and 1 cm to 1 unit on the y-axis would give a larger, clearer graph.

The dotted construction lines (a), (b), (c), (d) and (e) in Fig. 22.3 are used to obtain the required results:

(a) $y = 2.8$ (b) $y = -5.2$
(c) $x = 1.7$ or -4.7 (d) -6.2
(e) $x = -1.5$

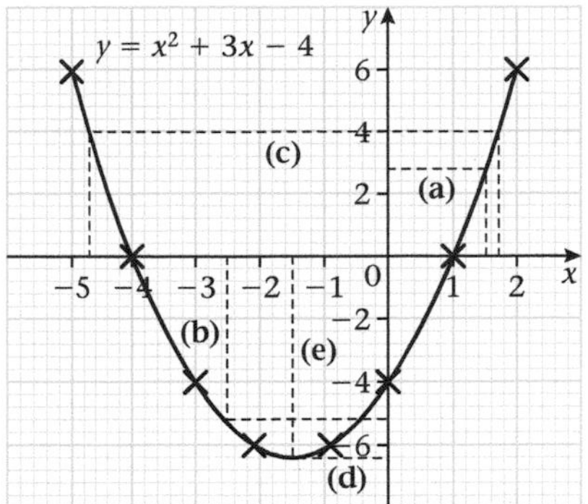

Fig. 22.3

Example 2

On the same axes, draw the graphs of $y = \frac{3}{x \quad 2}$ and $y = 2x - 3$ for values of x from −2 to +4. Find their points of intersection, if any.

Tables 22.2 and 22.3 give corresponding values of x and y for each graph.

Table 22.2 $y = \frac{3}{x - 2}$

x	−2	−1	0	1	2	3	4	1.5	2.5
y	−0.75	−1	−1.5	−3	*udf*	3	1.5	6	6

Notice that no value of y is given for $x = 2$; y is undefined (*udf*) when $x = 2$. Values of y when $x = 1.5$ and $x = 2.5$ are found instead.

Table 22.3 $y = 2x - 3$

x	0	2	4
y	−3	1	5

Since $y = 2x - 3$ is a linear relation, three points are sufficient.

Fig. 22.4 shows the graphs drawn on the same axes.

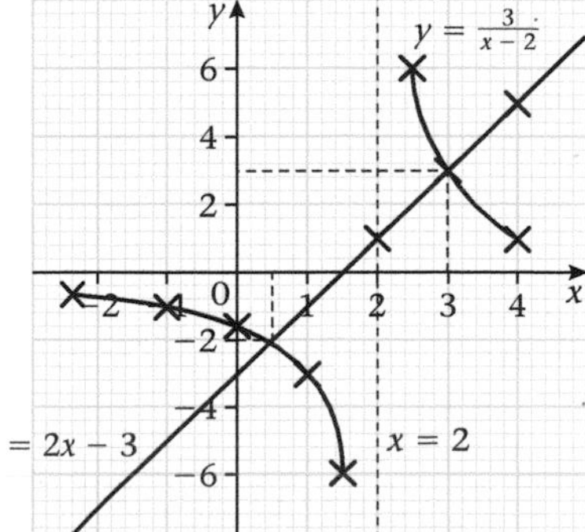

Fig. 22.4

The graphs intersect at the points (3, 3) and $(\frac{1}{2}, -2)$

Exercise 22c

1 Copy and complete Table 22.4 giving values for the relation $y = 3x^2 - 5x + 3$.

Table 22.4

x	−2	−1	0	1	2	3	4
y	25					15	31

(a) Draw a graph of the relation using scales of 2 cm to 1 unit on the x-axis and 2 cm to 5 units on the y-axis.
(b) What is the minimum value of $3x^2 - 5x + 3$?
(c) Write down the equation of the line of symmetry of the curve.
(d) Find the values of x when $y = 5$.

2 Draw the graph of $y = x^2 - 3x - 4$ for values of x from -2 to $+5$. Use scales of 2 cm to 1 unit on the x-axis and 1 cm to 1 unit on the y-axis.

(a) Read off the value of y when $x =$ (i) 4.5, (ii) 2.5.
(b) Find the values of x when $y =$ (i) 4, (ii) 0.
(c) For what value of x is y a minimum?

3 Draw the graph of $y = 4x - x^2$ for values of x from -1 to $+5$.

(a) Read off the value of y at the point where the line $x = -0.5$ cuts the curve.
(b) Find the points of intersection of the curve with the line $y = -3$.
(c) For what value of x does $4x - x^2$ have its greatest value, and what is this greatest value?

4 Choose a suitable scale and draw the graph of $y = x^2 + 4x + 1$ for values of x from -5 to $+1$. Find (a) the range of values of x for which y increases as x increases, and (b) the coordinates of the point at which y has its least value.

5 Draw the graph of $y = (x + 1)(x - 2)$ for values of x from -3 to $+4$. Find the range of values of x for which (a) y decreases as x increases, (b) y is negative.

6 The coordinates of two points P and Q are P(2, 1), Q(6, 3) and S is the mid-point of PQ.

(a) Calculate
  (i) the length of PQ,
  (ii) the gradient of PQ,
  (iii) the coordinates of S.
(b) Determine the equation of the line through S perpendicular to PQ.

7 Choose suitable scales and then, for each given domain of x, draw the graph of

(a) $y = \frac{3}{x+3}$ from -6 to 0
(b) $y = \frac{12}{x-2}$ from -1 to $+5$
(c) $y = x(x - 1)(x - 3)$ from -2 to $+5$
(d) $y = x^3 - 4x - 1$ from -3 to $+3$

8 *Without drawing the graph* of the relation $y = (4 - x)(2 + x)$, write down the coordinates of the points where the curve cuts (a) the x-axis, (b) the y-axis.
(c) State whether the curve has a maximum or minimum value of y and
(d) write down the value of x at this point.
(e) Hence calculate the corresponding value of y.

9 On the same axes, draw the graphs of $y = x^2 - 5x$ and $y = 5x - x^2$ for values of x from -1 to $+6$.

(a) At what points do the curves intersect?
(b) Write down the equations of *two* lines of symmetry of the completed curves.

10 On the same axes, draw the graphs of $y = \frac{2}{x}$ and $y = \frac{2}{3}x - 1$, for values of x from -3 to $+3$.
Find the values of x and y at the points of intersection of the graphs.

11 On the same axes, draw the graphs of $y = \frac{-1}{x+2}$ and $x + y + 2 = 0$ for values of x from -5 to $+2$. Read off the values of x and y at the points of intersection of the graphs.

Travel graphs

Distance–time graphs

Fig. 22.5 is a typical distance–time graph of a journey which is in three stages.

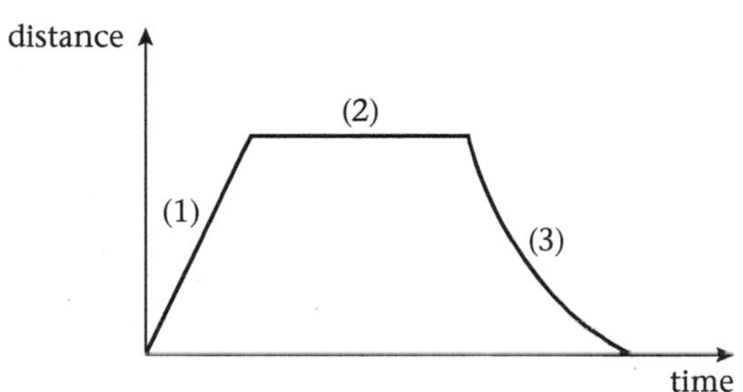

Fig. 22.5

Stage 1: Distance is changing uniformly with time. The **gradient** of the line is a measure of the **speed**, or rate of change of distance with time.
Stage 2: Distance does not change with time. The object is at rest.
Stage 3: The gradient of the tangent at any point on the curve gives the speed at that point. During this stage the speeds are negative. This shows that the object is travelling in a direction opposite to the original direction.

Example 3

At 08:00 hrs a cyclist left home and cycled at 13 km/h to a village 20 km away. She stayed for 2 hours at the village before cycling home at 9 km/h. Use a graphical method to find the time when she arrived home.

Fig. 22.6 is a graph of the cyclist's journey.

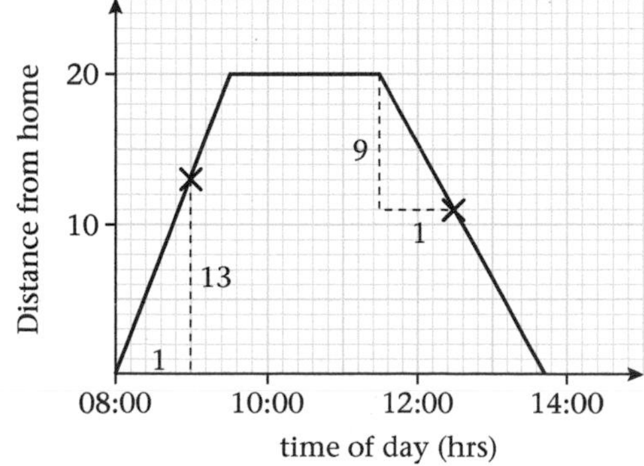

Fig. 22.6

From the graph, the time of arrival is found at E, approximately 13.45 hrs.

Method
Choose suitable scales with time on the horizontal axis. Then:

(a) In 1 hour the cyclist covers 13 km. Plot the point A (09:00, 13).
(b) Produce the line through OA to B, 20 km from the start.
(c) On the horizontal through B, mark C, 2 hours after B. C represents the starting point of the journey home.
(d) Going home the cyclist covers 9 km in 1 hour, Plot D, 1 hour and 9 km from C.
(e) Produce the line through CD to cut the time axis at E.

Speed–time graphs

Fig. 22.7 is a typical speed–time graph.

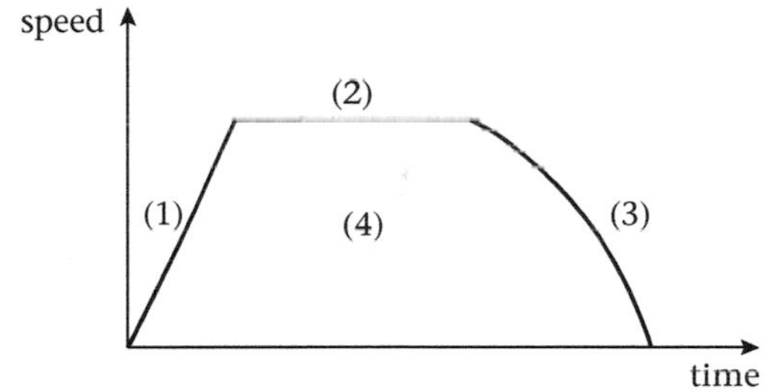

Fig. 22.7

Stage 1: Speed is changing uniformly with time. The **gradient** of this line is a measure of the **acceleration**, or rate of change of speed with time.
Stage 2: Speed does not change with time. The horizontal line represents a period of constant speed.
Stage 3: Speed is decreasing with time. This gives a **negative acceleration** or **deceleration**. The gradient of the tangent at any point on the curve gives the acceleration at that point.
Region 4: The **area** under the graph is a measure of the **distance** travelled.

Chapter 10 gives a full explanation of how to estimate the gradient at a point on a curve.

Example 4

Fig. 22.8 is the speed–time graph of a train journey. If the total distance travelled in the 80 seconds is 920 m, calculate (a) V, (b) the acceleration of the train during the first 15 seconds, (c) the distance travelled in the final 40 seconds.

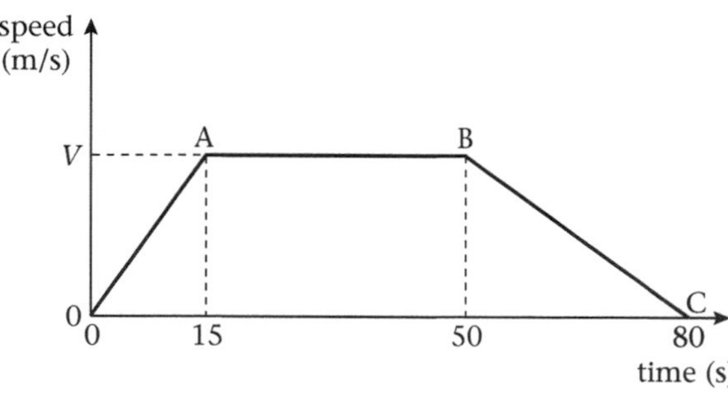

Fig. 22.8

(a) Distance travelled

$$= \text{area of trapezium OABC}$$
$$= \tfrac{1}{2}(\text{AB} + \text{OC})V$$
$$= \tfrac{1}{2}(35 + 80)V$$

Hence, $920 = \frac{1}{2}(35 + 80)V$

$$1\,840 = 115V$$
$$V = \frac{1\,840}{115} = 16$$

(b) Acceleration during first 15 seconds

$$= \text{gradient of OA}$$
$$= \frac{V}{15} = \frac{16}{15}\,\text{m/s}^2 = 1\tfrac{1}{15}\,\text{m/s}^2$$

(c) Distance travelled during the last 40 seconds

$$= \text{Area under DBC in Fig. 22.9}$$
$$= \tfrac{1}{2}(\text{DB} + 40)\ 16\,\text{m}$$
$$= \tfrac{1}{2}(10 + 40)\ 16\,\text{m} = 400\,\text{m}$$

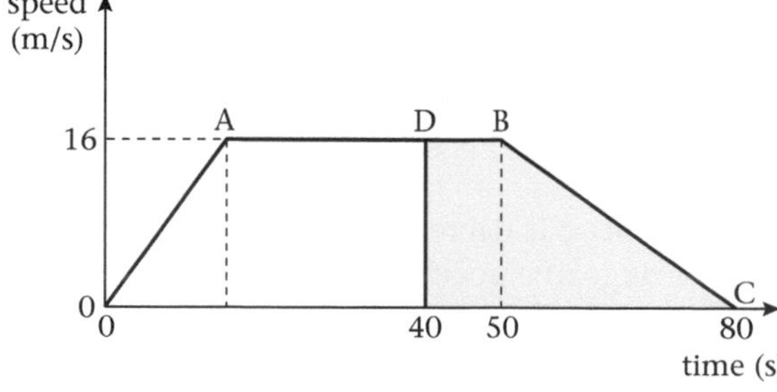

Fig. 22.9

Example 5

A particle moves in a straight line so that after t seconds its velocity v m/s is given by $v = 5t^2 - 12t + 8$. Some corresponding values of v and t are given in Table 22.5.

Table 22.5

t	0	$\frac{1}{2}$	1	$1\frac{1}{2}$	2	$2\frac{1}{2}$	3	$3\frac{1}{2}$	4
v	8	$3\frac{1}{4}$	p	$1\frac{1}{4}$	4	q	17	$27\frac{1}{4}$	40

(a) Calculate p and q.
(b) Draw the graph of $v = 5t^2 - 12t + 8$ for the range $0 \leqslant t \leqslant 4$.
(c) Use the graph to estimate (i) the time at which the acceleration is zero, (ii) the acceleration when $t = 3$.

(a) When $t = 1$.

$$p = 5(1)^2 - 12(1) + 8 = 5 - 12 + 8$$
$$= 1$$

When $t = 2\frac{1}{2}$,

$$q = 5(2\tfrac{1}{2})^2 - 12(2\tfrac{1}{2}) + 8$$
$$= 31\tfrac{1}{4} - 30 + 8 = 9\tfrac{1}{4}$$

(b) Fig. 22.10 is the required graph.

(c) (i) The acceleration is zero when the tangent to the curve is horizontal, i.e. at the lowest point of the curve in Fig. 22.10. At this point, $t = 1.2$ seconds.

(ii) In Fig. 22.10, a tangent has been drawn at the point where $t = 3$.
Gradient of tangent

$$= \frac{18\,\text{m/s}}{1\,\text{s}} = 18\,\text{m/s}^2$$

When $t = 3$, the acceleration is $18\,\text{m/s}^2$.

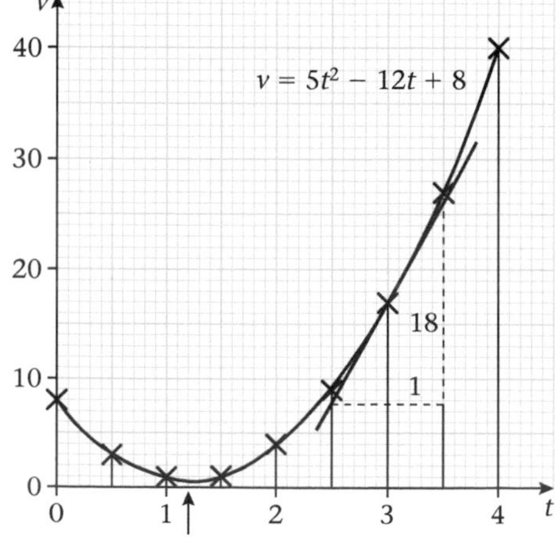

Fig. 22.10

Exercise 22d

1 Fig. 22.11 shows the journey of a car from Hiroad to Kinville and of a bus from Kinville to Hiroad on the same road.

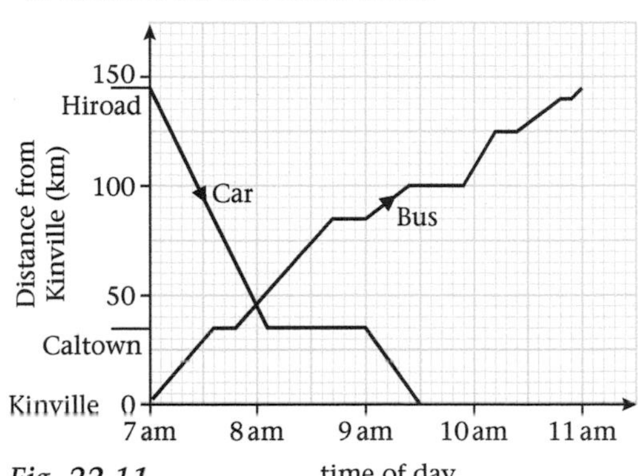

Fig. 22.11

(a) How many times did the bus stop between Kinville and Hiroad?
(b) What is the distance between Kinville and Hiroad?
(c) What was the car's average speed for the whole journey?
(d) What was the bus's average speed to the nearest km/h for the whole journey?
(e) What was the car's average speed between Hiroad and Caltown?
(f) What was the car's average speed between Caltown and Kinville?
(g) How far were they both from Caltown when they passed each other on the road?

2 Fig. 22.12 shows how Sara walked from school to the Post Office and how Gina walked from the Post Office to the school.

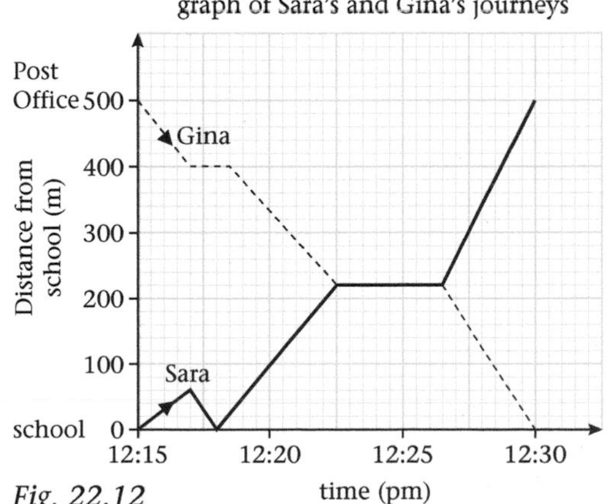

Fig. 22.12

(a) How far were they from the Post Office when they met?
(b) How long did they stand talking to each other?
(c) After leaving school, Sara suddenly remembered a letter she had to post. She returned to school for the letter. At what time did she remember the letter?
(d) Gina stopped for $1\frac{1}{2}$ min to buy some bread. How far is the bakery from the school?
(e) After leaving Gina what was Sara's walking speed in m/min?
(f) How much further did Sara walk than Gina?

3 Three cars, A, B, C, start one after the other in alphabetical order at 5-minute intervals. They travel at 80, 100, 120 km/h respectively. Given that A leaves at midday, draw a distance–time graph which enables you to find when (a) B passes A, (b) C passes A, (c) C passes B.

4 An aircraft travels at an average speed of 800 km/h. It left airport A at 12:30 hrs and arrived at airport B at 13:45 hrs. After stopping at airport B for 45 min it left for airport C, arriving there at 19:00 hrs.

Draw a distance–time graph of its journey using a scale of 2 cm to 1 hour on the time axis and 1 cm to 400 km on the distance axis. Use your graph to find (a) the distance from A to C, (b) the distance of the plane from B at 15:00 hrs.

5 A cyclist sets out at 07:05 hrs to reach her office 13 km away at 08:00 hrs. After 5 min she finds she has forgotten her briefcase. She turns round and rides home at 16 km/h and then takes 2 min to find the briefcase. Use a graphical method to answer the following.
(a) How fast must she ride to get to work on time?
(b) If her top speed is 17 km/h, how many minutes will she be late for work?

6 During a journey a train accelerates uniformly for 6 min. Its speed, v km/h, is given in 1-min intervals in Table 22.6

Table 22.6

Time (min)	0	1	2	3	4	5	6
v (km/h)	7	16	25	34	43	52	61

By drawing a speed–time graph find,
(a) the train's acceleration in km/h per min,
(b) the distance travelled in km during the whole 6 min.

7 Fig. 22.13 is the speed–time graph of a sprinter.

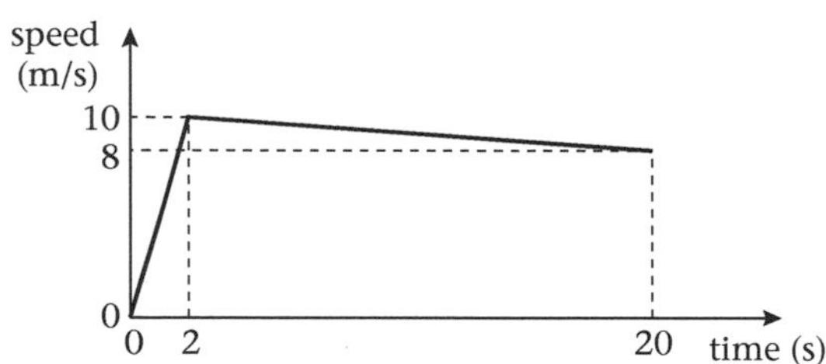

Fig. 22.13

Calculate
(a) the acceleration of the sprinter over the first 2 seconds.
(b) the deceleration of the sprinter over the next 18 seconds.
(c) the distance covered in the 20 seconds.

8 Fig. 22.14 is the speed–time graph of a journey.

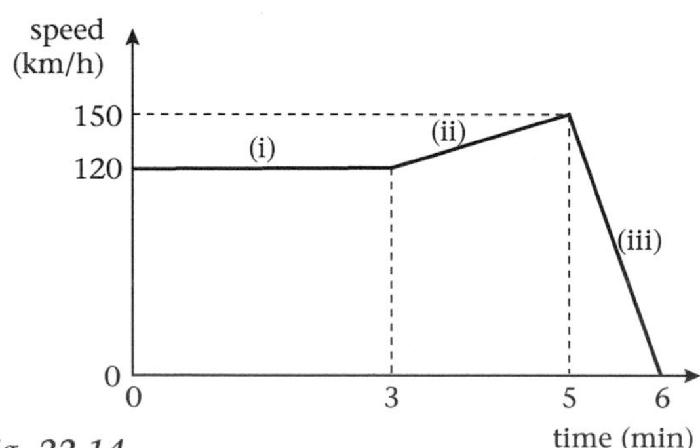

Fig. 22.14

(a) Describe the three parts of the journey in your own words.
(b) Calculate the accelerations represented by parts (ii) and (iii) of the graph (answers in km/h per minute).
(c) Calculate the total distance travelled (answer in km).

9 Fig. 22.15 is the speed–time graph of a car journey.

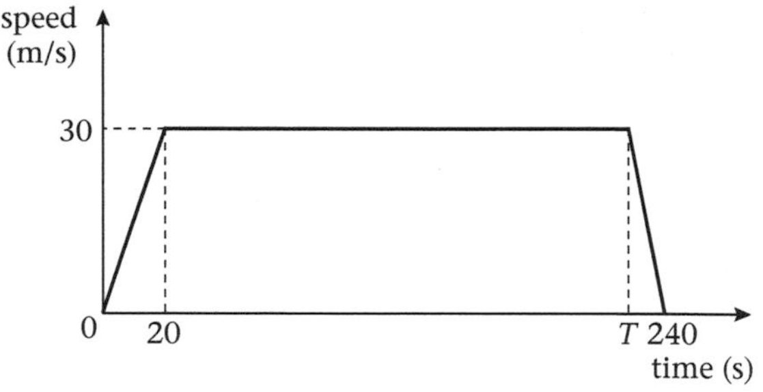

Fig. 22.15

(a) Calculate the acceleration of the car during the first 20 seconds.
(b) Given that the final deceleration is $2\,m/s^2$, calculate (i) T, (ii) the total distance travelled by the car.

10 Fig. 22.16 is the speed–time graph of the last ten seconds of a car journey. It travels at a constant speed of 31 m/s for 4 seconds and slows down uniformly first to 16 m/s then to rest after a further 5 seconds and 1 second respectively.

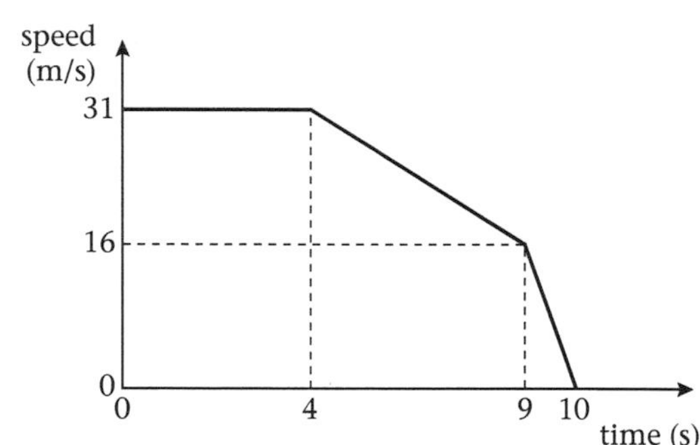

Fig. 22.16

Calculate (a) the speed of the car 2 seconds before it stopped, (b) the average speed of the car during the whole 10 seconds.

11 An aeroplane accelerates from rest at 40 km/h per minute until its speed is 840 km/h. It then travels at this speed for $2\frac{1}{2}$ h before decelerating at an average of 15 km/h per minute until it comes to rest.
(a) Sketch the journey on a speed (km/h)–time (min) graph.

(b) Calculate the total time taken for the journey in minutes.
(c) Calculate the distance travelled in km.

12 Fig. 22.17 is a speed–time curve of part of a taxi journey.

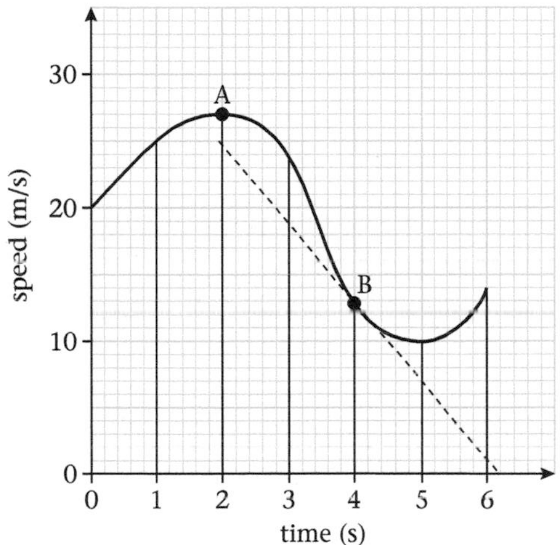

Fig. 22.17

(a) What is the acceleration of the taxi at A?
(b) Use the tangent drawn at B to estimate the acceleration of the taxi at that point.

13 Table 22.17 gives the speed, v m/s, of an object taken at 1-second intervals.

Table 22.7

t (s)	0	1	2	3	4	5	6	7	8
v (m/s)	32	35	36	35	32	27	20	11	0

(a) Draw a v–t graph of the motion.
(b) By constructing a suitable tangent, estimate the acceleration of the object after 5 seconds.

14 The velocity, v m/s, of an object after t seconds is given by the equation $v = 4t^2 - 12t + 11$. Table 22.8 contains some corresponding values of v and t.

Table 22.8

t (s)	0	$\frac{1}{2}$	1	$1\frac{1}{2}$	2	$2\frac{1}{2}$	3	$3\frac{1}{2}$	4
v (m/s)	11	6	3	m	32	6	11	n	27

(a) Calculate m and n.
(b) Taking 2 cm to represent 1 second on the horizontal axis and 2 cm to represent 5 m/s on the vertical axis, draw the graph of $v = 4t^2 - 12t + 11$ for the range $0 \leqslant t \leqslant 4$.
(c) From your graph, find the times when the velocity is 8 m/s.
(d) By drawing a tangent, find the acceleration of the object after 3 seconds.

15 The velocity, v m/s, of a car after t seconds is given by $v = 7 + 6t - t^2$. Table 22.9 contains some corresponding values of v and t.

Table 22.9

t	0	1	2	3	4	5	6	7
v	7	12	15	a	15	12	7	b

(a) Calculate a and b.
(b) Draw the graph of $v = 7 + 6t - t^2$ for the range $0 \leqslant t \leqslant 7$ using scales of 2 cm to 1 second horizontally and 1 cm to 1 m/s vertically.
(c) Use your graph to find the speed of the car after 4.3 seconds.
(d) By drawing suitable tangents, find the acceleration of the car after (i) 1 second, (ii) 6 seconds.

Revision test (Chapter 22)

Select the correct answer and write down your choice.

1 In Fig. 22.18, which of the labelled sections of the plane satisfy *both* of the conditions listed below?

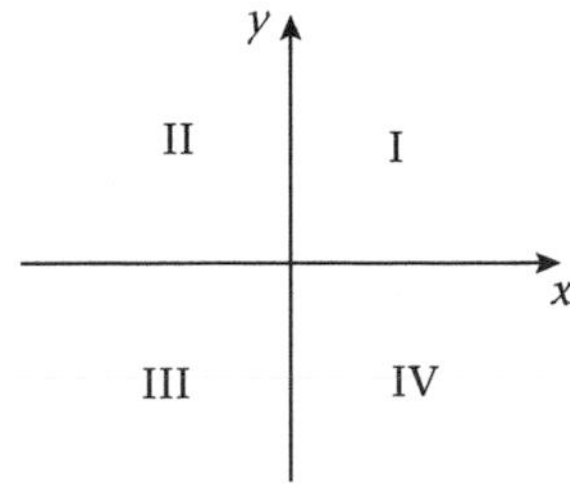

Fig. 22.18

(i) y is always positive when x is negative
(ii) y is always negative when x is positive

A II and III only B I and IV only
C II and IV only D I and III only

2 In Fig. 22.19 the arrow diagram that represents a *function* is

A

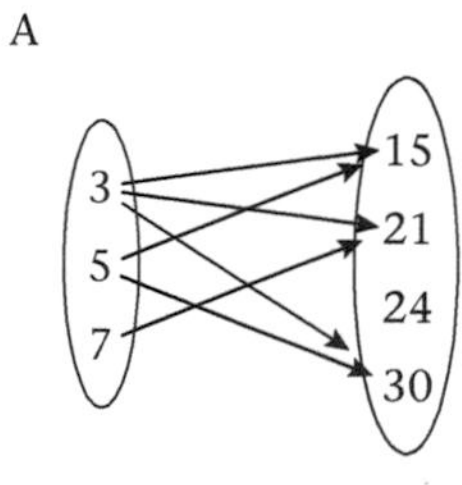

B

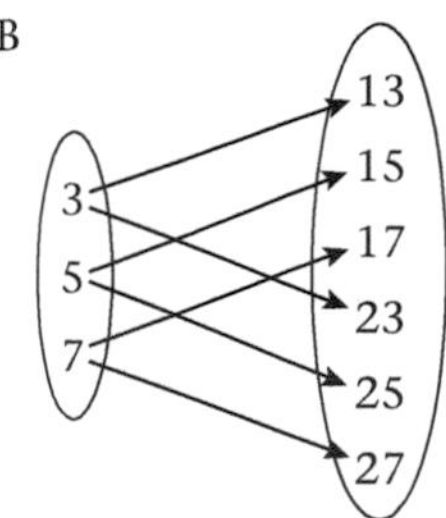

C

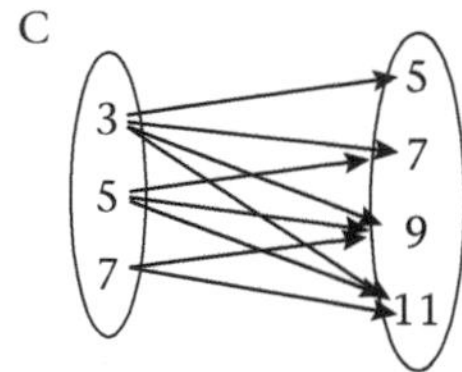

D

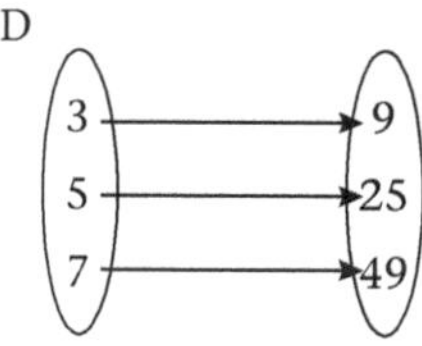

Fig. 22.19

3 The domain of the mapping $g : x \rightarrow x^2 - 3$ is $\{-1, 0, 1\}$. The range is

A $\{-2, -3, -2\}$ B $\{-5, -3, -1\}$
C $\{-1, 0, -1\}$ D $\{-4, -3, -2\}$

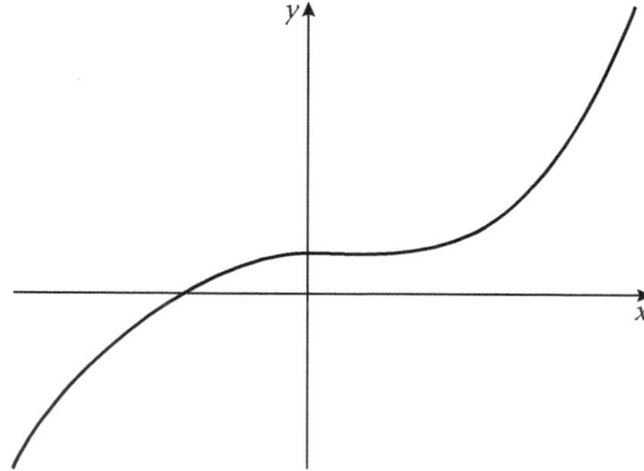

Fig. 22.20

Which of the following sets of ordered pairs lies on the graph shown in Fig. 22.20?

A $\{(-2, -9), (-1, -2), (0, 1), (1, 2), (2, 7)\}$
B $\{(-2, -8), (-1, -1), (0, 0), (1, 1), (2, 8)\}$
C $\{(-2, -7), (-1, 0), (0, 1), (1, 2), (2, 9)\}$
D $\{(-2, -9), (-1, -8), (0, 1), (1, 0), (2, 7)\}$

5 Which of the following points does *not* lie on the graph of the equation $2x + y = 3$?

A $(-1, -1)$ B $(-1, 5)$
C $(0, 3)$ D $(\frac{1}{2}, 2)$

6 The coordinates of the point of intersection of $x = 1$ and $y = 4$ are

A $(0, 0)$ B $(0, 4)$
C $(1, 4)$ D $(1, 0)$

7 If $g(x) = x(x + 1)^2$, then $g(-2) =$

A -18 B -2 C 4 D 18

8 Given that $f(x)$ and $g(x)$ are functions of x, then $(gf)^{-1} =$

A gf^{-1} B $g^{-1}f^{-1}$
C $f^{-1}g^{-1}$ D $f^{-1}g$

9 The graph of the equation $y = -2$ on the coordinate plane is a line parallel to

A the x-axis and 2 units below the axis
B the x-axis and 2 units above the axis
C the y-axis and 2 units to the right of the axis
D the y-axis and 2 units to the left of the axis

10 The gradient of the line joining the points $(-3, 5)$ and $(1, -2)$ is

A $-\frac{4}{7}$ B $-\frac{2}{3}$ C $-\frac{3}{4}$ D $-\frac{7}{4}$

11 The graph that represents the equation

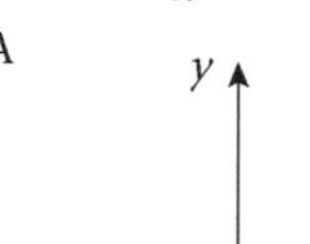

$y = \dfrac{3}{x^2}$ is

A

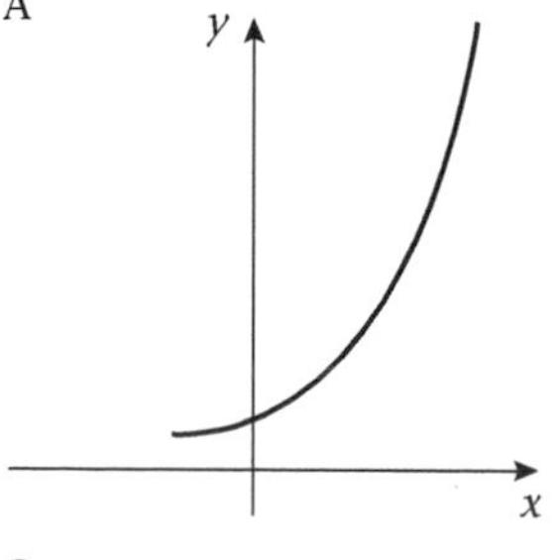

B

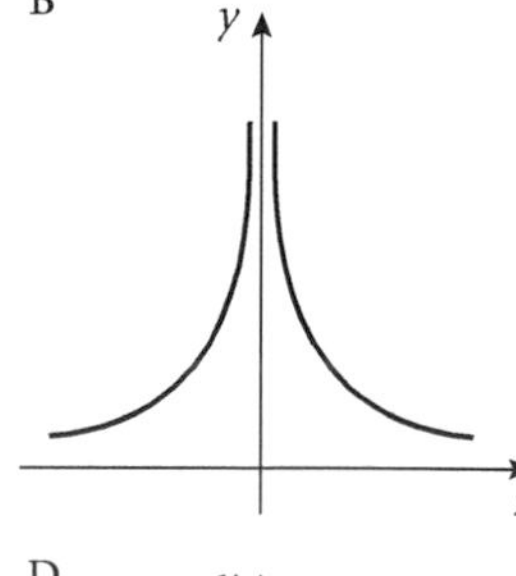

C

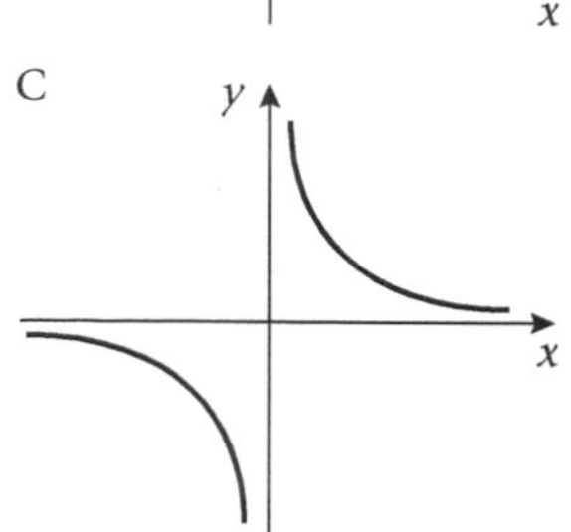

D

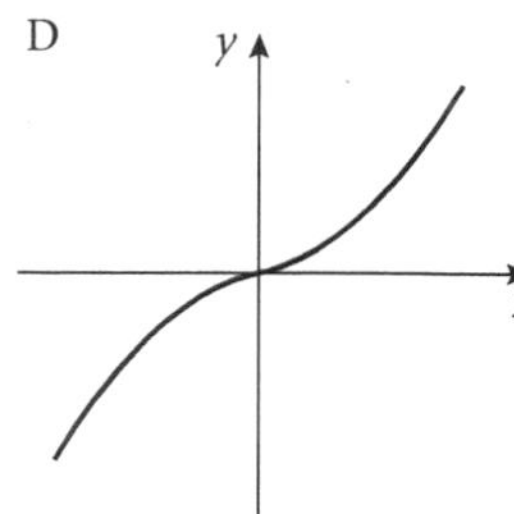

Fig. 22.21

12 Which of the following expressions is undefined when $x = 2$?

A $x - 2$ B $\frac{x}{2}$
C $\frac{1}{x - 2}$ D x^{-2}

13

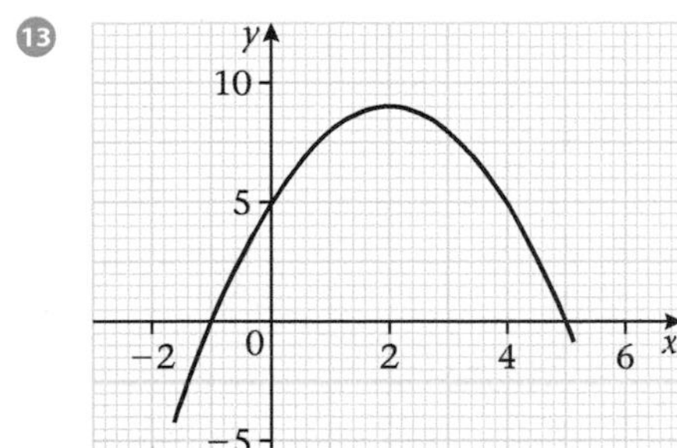

Fig. 22.22

In Fig. 22.22 the maximum value of the function is given by
A $y = 5$ B $y = 9$
C $x = 2$ D $x = 5$

14

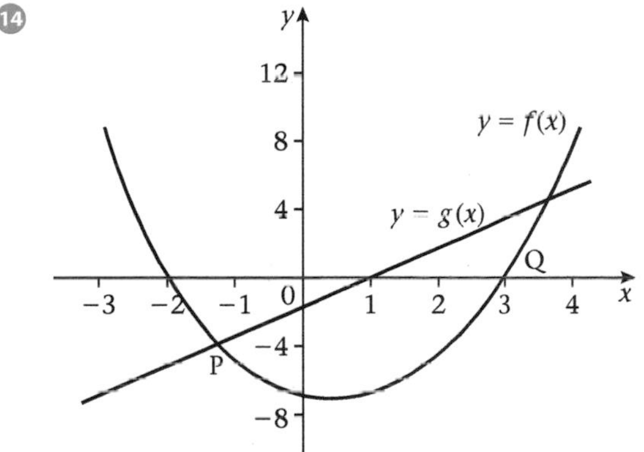

Fig. 22.23

Fig. 22.23 shows the points of intersection P and Q of $f(x) = x^2 - x - 6$ and $g(x) = 2x - 2$. The length of PQ is
A $\sqrt{13}$ units B 5 units
C $5\sqrt{5}$ units D 15 units

15 The gradient of a straight line is −2 and the intercept on the y-axis is 3 units. The equation of the line is
A $y + 2x = 3$
B $y = 3x - 2$
C $y = 3x + 2$
D $y - 2x = 3$

16 Which one of the following mappings gives (−5, 5) as the image of *both* (5, 5) and (4, 3)?
A $g : (x, y) \to (-x, y)$
B $g : (x, y) \to (y - 2x, y)$
C $g : (x, y) \to (y - 2x, 2x - y)$
D $g : (x, y) \to (y - 3y, 2x - y)$

17 If $f : x \to \frac{2(x + 1)}{3}$, $f^{-1} =$

A $\frac{2(1 - x)}{3}$ B $\frac{3 - 2x}{2}$
C $\frac{3(1 - x)}{2}$ D $\frac{3x - 2}{2}$

18 Given that $f(x) = \frac{x - 1}{2}$ and $g(x) = \frac{3x^2}{2}$ then $gf(x) =$

A $\frac{3x^2 - 1}{4}$ B $\frac{3x^2 - 2}{4}$
C $\frac{3(x - 1)^2}{4}$ D $\frac{3(x - 1)^2}{8}$

Fig. 22.24 represents four stages in the first four hours of a journey. Use Fig. 22.24 to answer questions 19 and 20.

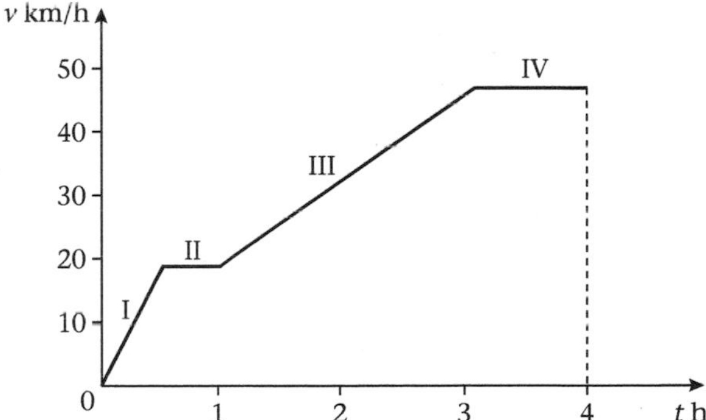

Fig. 22.24

19 The average speed during this time in km/h is
A $31\frac{1}{4}$
B $33\frac{3}{4}$
C 35
D $43\frac{3}{4}$

20 The average acceleration is a maximum during stage
A I
B II
C III
D IV

Chapter 23

Geometry

Cyclic quadrilaterals

Angles in opposite segments are supplementary.

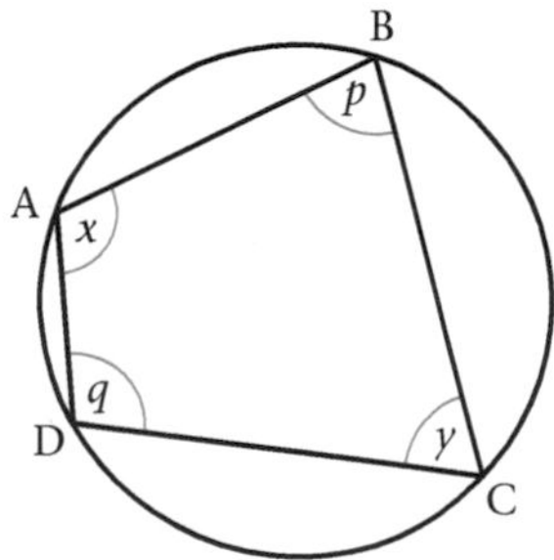

Fig. 23.1

In Fig. 23.1, $x + y = p + q = 180°$. ABCD is known as a **cyclic quadrilateral**. Hence, opposite angles of a cyclic quadrilateral are supplementary, It follows that the exterior angle of a cyclic quadrilateral is equal to the interior opposite angle (Fig. 23.2).

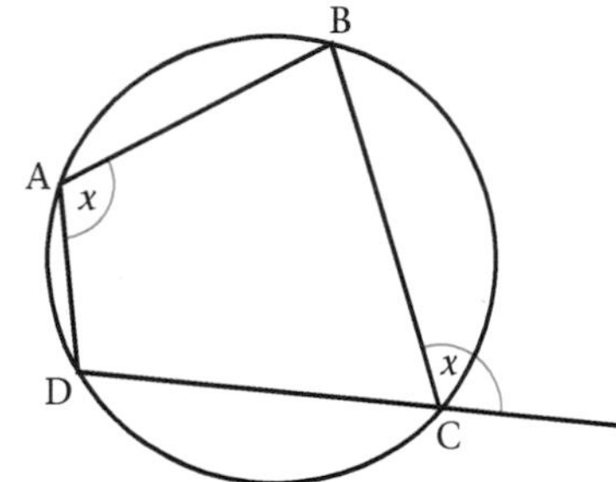

Fig. 23.2

Tangents to circles

A **tangent** to a circle is a line which meets the circle in one point only.

1 A tangent is perpendicular to the radius at the point of contact.

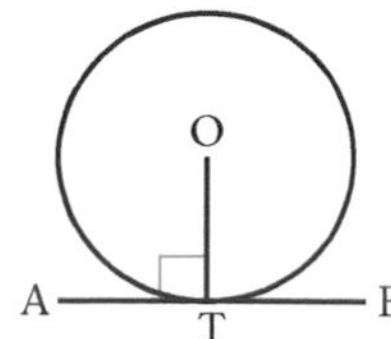

Fig. 23.3

In Fig. 23.3 OT $\perp$ ATB.

2 The tangents to a circle from an external point are equal in length.

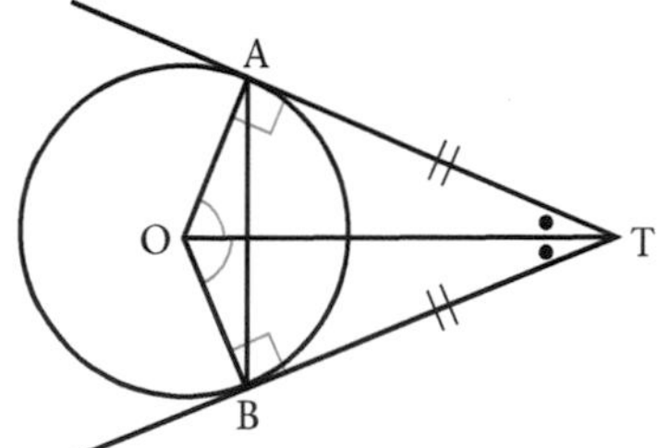

Fig. 23.4

In Fig. 23.4, TA = TB. Notice also that $A\hat{T}O = B\hat{T}O$, $A\hat{O}T = B\hat{O}T$ and OT bisects AB at right angles.

3 The angle between a tangent to a circle and a chord through the point of contact is equal to the angle in the alternate segment.

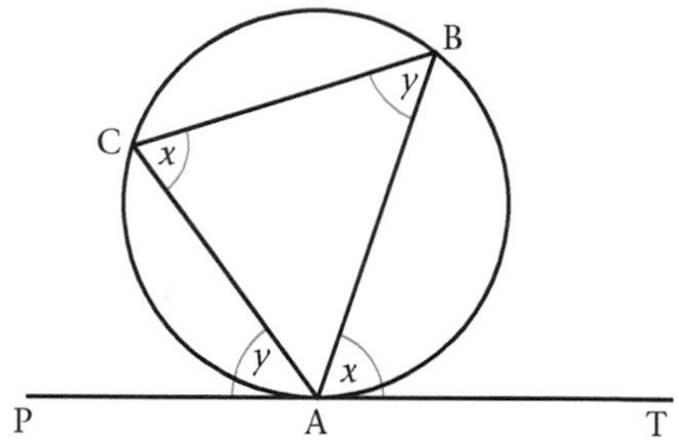

Fig. 23.5

In Fig. 23.5, $T\hat{A}B = A\hat{C}B$ and $P\hat{A}C = A\hat{B}C$. For other angle properties of the circle see Book 3, Chapter 23.

Exercise 23a

1. Redraw the figures in Fig. 23.6 and write in the sizes of the marked angles. (Some construction lines may be necessary).

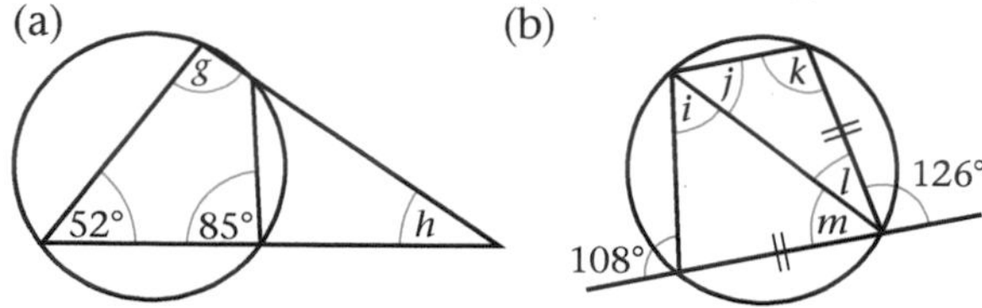

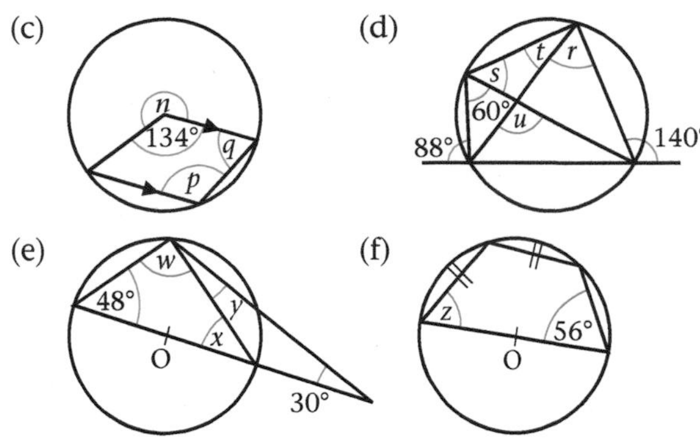

Fig. 23.6

2. Three circles, centres A, B and C with radii 4 cm, 3 cm and 2 cm respectively, touch one another externally. Calculate the lengths of the sides of △ABC.

3. Three circles touch one another externally. Their centres form a triangle with sides 10 cm, 9 cm and 7 cm. Find the radii of the circles.

4. In Fig. 23.7, the circle touches the sides of △ABC at X, Y, Z. If BC = 11 cm, CA = 10 cm and AB = 9 cm, find AY and BX.

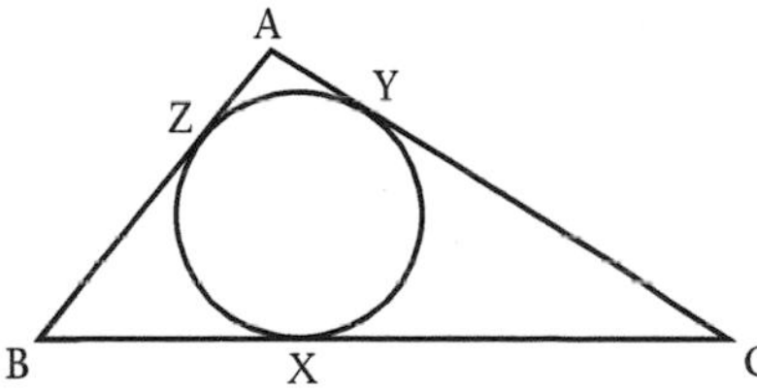

Fig. 23.7

5. In Fig. 23.8 (a)–(f), lines drawn through T are tangents and O is the centre of any given circle. Find the lettered angles.

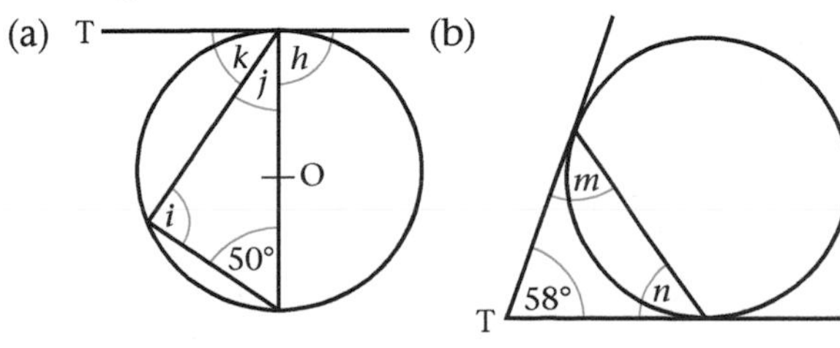

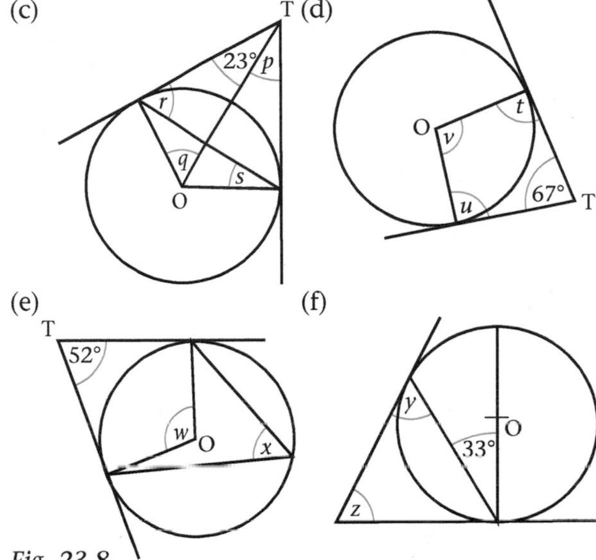

Fig. 23.8

6. A circle is drawn inside a triangle ABC to touch the sides BC, CA and AB at P, Q and R respectively. If $\widehat{A} = 56°$ and $\widehat{B} = 68°$, find the angles of △PQR.

7. In each part of Fig. 23.9, find the value of x, In part (d), O is the centre of the circle.

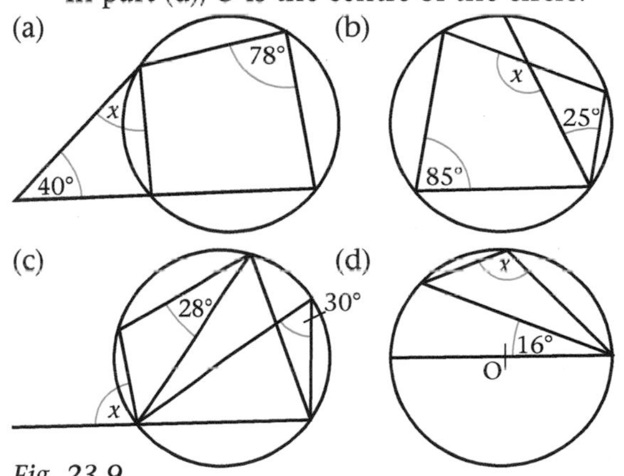

Fig. 23.9

8. In Fig. 23.10, find the sizes of the lettered angles.

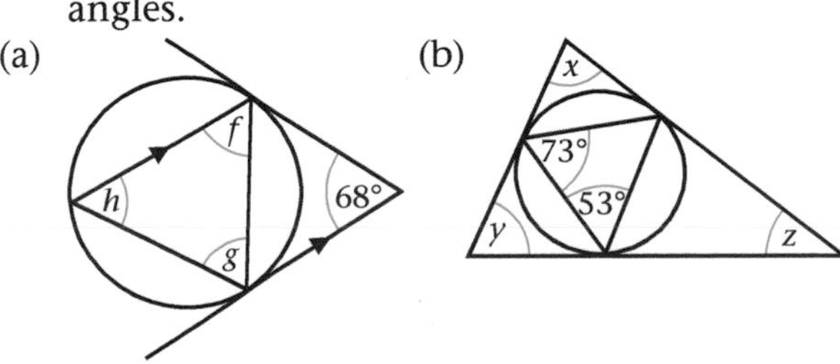

(c)

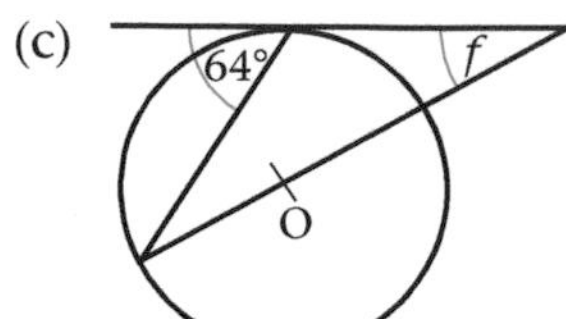

(d)

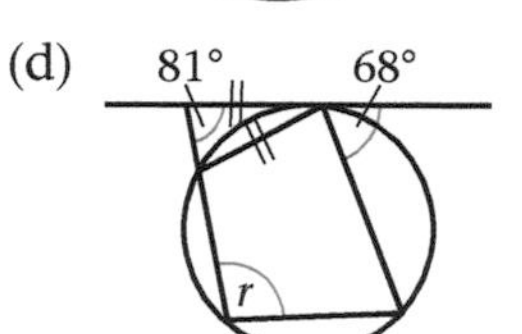

(e)

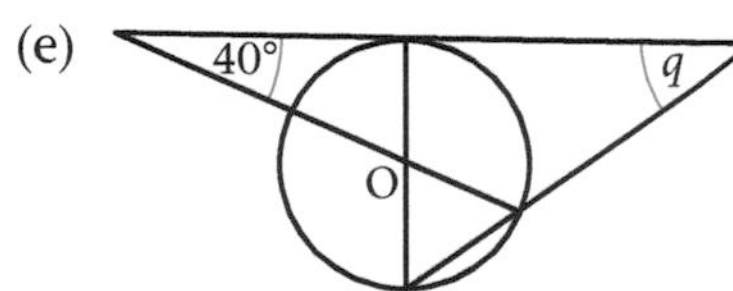

Fig. 23.10

9 In Fig. 23.11, express y in terms of x.

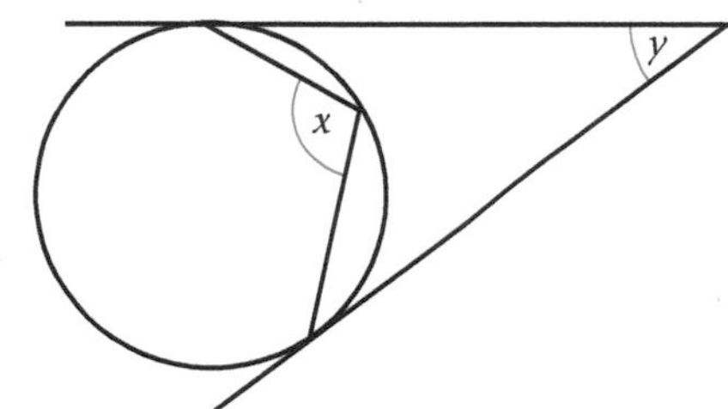

Fig. 23.11

10 In each part of Fig. 23.12, find the value of x.

(a) (b)

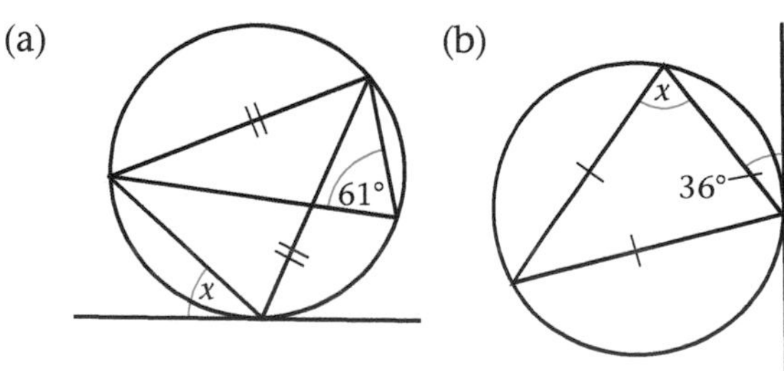

(c) (d)

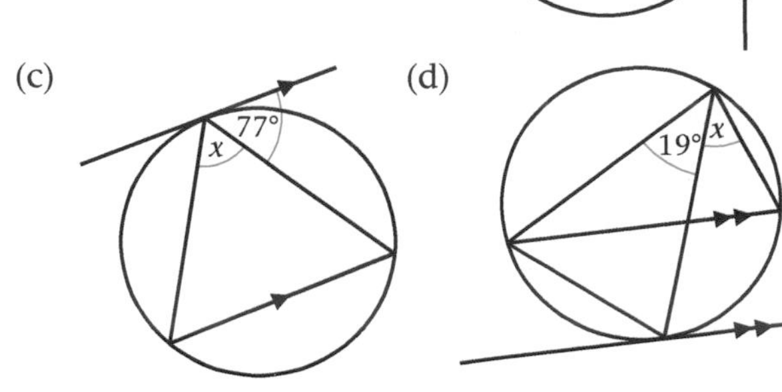

Fig. 23.12

Revision test (Chapter 23)

1 In Fig. 23.13, O is the centre of the circle.

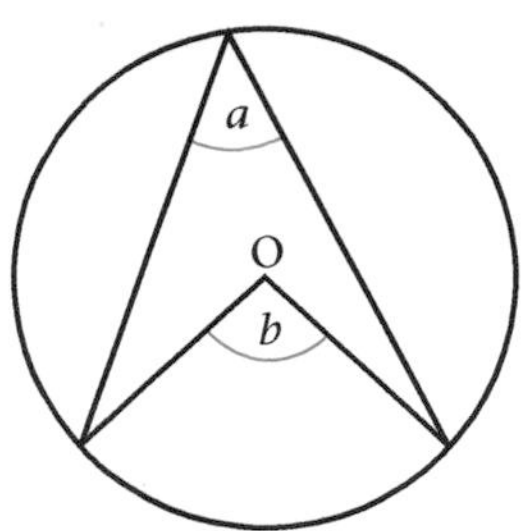

Fig. 23.13

In terms of b, a =

A $\frac{b}{2}$ B $2b$

C $\frac{360 - b}{2}$ D $\frac{180 - b}{2}$

2 In Fig. 23.14, O is the centre of the circle and $\widehat{AOB}$ is 62°.

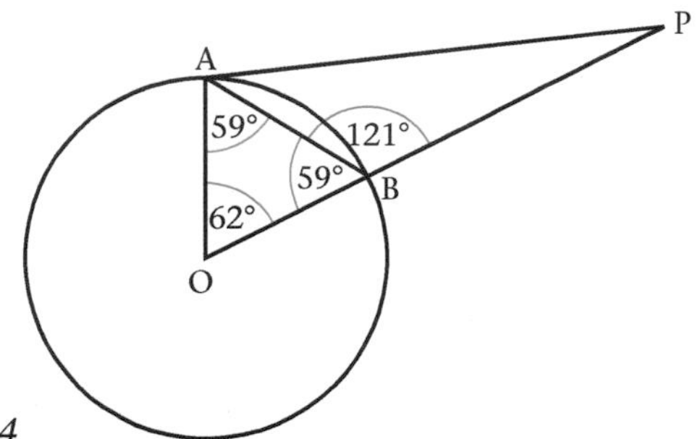

Fig. 23.14

The size of angle $\widehat{APB}$ is

A 28° B 31° C 59° D 62°

3 In Fig. 23.15, TR is a tangent to the circle, centre O

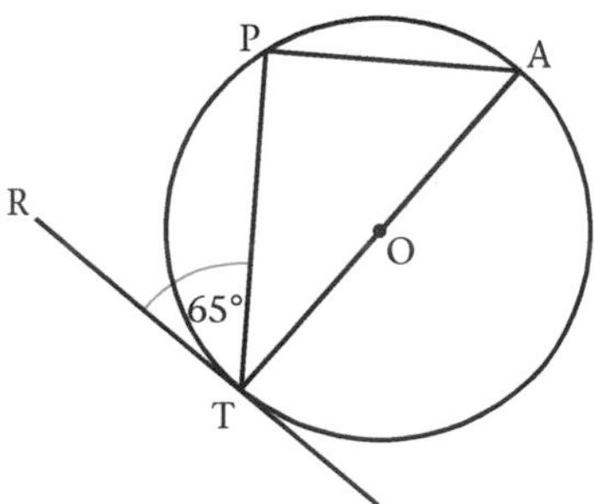

Fig. 23.15

$\widehat{TAP}$ =

A 25° B 55° C 65° D 90°

Use Fig. 23.16 to answer questions 4 and 5.

In Fig. 23.16, TP is a tangent to the circle.

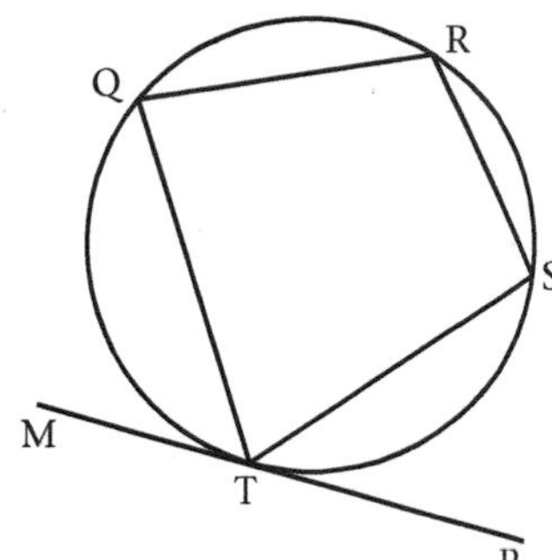

Fig. 23.16

4 $S\hat{T}P$ =
A $R\hat{Q}T$ B $Q\hat{R}S$ C $R\hat{T}S$ D $T\hat{R}S$

5 Which of the following is not necessarily true for Fig. 23.16?
A $R\hat{T}P$ is a right angle
B $Q\hat{T}M$ are $Q\hat{R}T$ are equal
C $R\hat{S}T$ and $R\hat{Q}T$ are supplementary
D $Q\hat{T}S$ and $Q\hat{R}S$ are supplementary

Use Fig. 23.17 to answer questions 6, 7 and 8.

In Fig. 23.17 SR, and PQ are produced to meet at O.

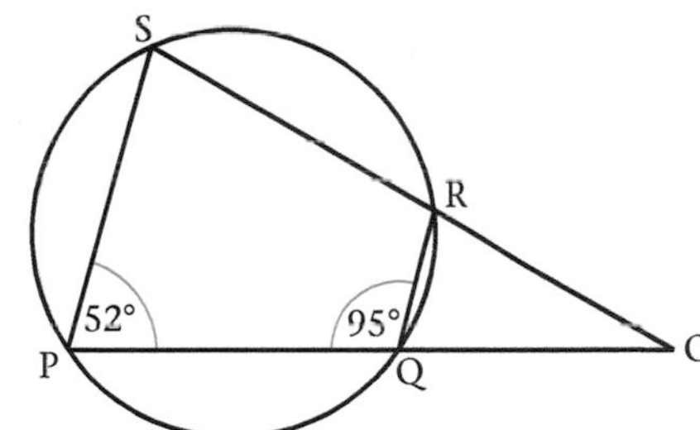

Fig. 23.17

6 $Q\hat{R}O$ =
A 95° B 90° C 85° D 52°

7 $P\hat{S}R$ =
A 95° B 90° C 85° D 52°

8 $Q\hat{O}R$ =
A 10° B 43° C 76° D 85°

Use Fig. 23.18 to answer questions 9 and 10.

In Fig. 23.18, OA and OB are tangents to the circle, and OA ∥ BC,

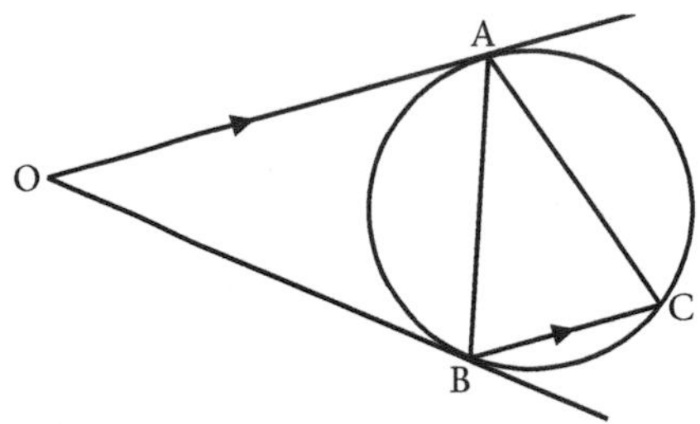

Fig. 23.18

9 Which of the following statements is *not* true for Fig. 23.18?
A $O\hat{B}A = A\hat{C}B$ B $O\hat{A}B = A\hat{B}C$
C $O\hat{B}A = O\hat{A}B$ D $C\hat{A}B = C\hat{B}A$

10 Given that $A\hat{C}B = 68°$, then $A\hat{O}B$ =
A 44° B 68° C 72° D 112°

Use Fig. 23.19 to answer questions 11 and 12.

In Fig. 23.19, TP is a tangent to the circle, centre O and $S\hat{T}P = 62°$.

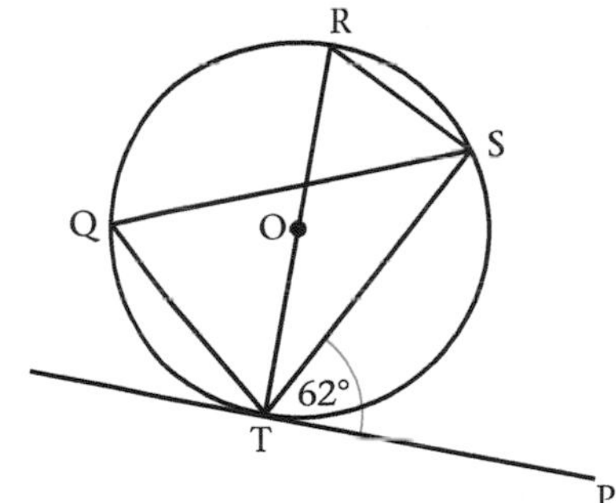

Fig. 23.19

11 $S\hat{O}T$ =
A 62° B 118° C 124° D 152°

12 Given that $R\hat{S}Q = 25°$, then $Q\hat{R}T$ =
A 90° B 65° C 62° D 28°

Chapter 24
Trigonometry

Sine, cosine, tangent

The trigonometric ratios of an acute angle, sine, cosine and tangent, may be defined in terms of the hypotenuse, opposite and adjacent sides of a right-angled triangle as follows:

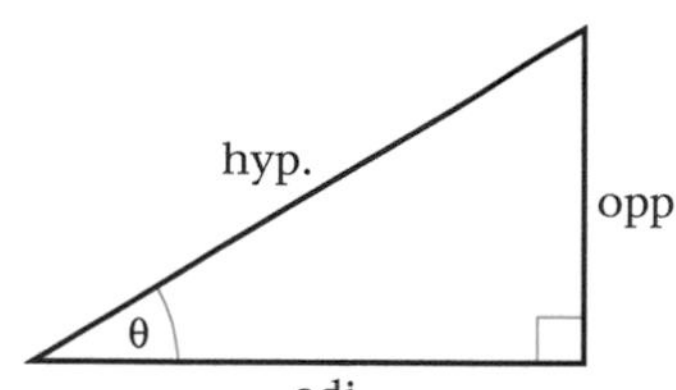

Fig. 24.1

$$\text{sine } \theta = \frac{\text{opp.}}{\text{hyp.}} \qquad \text{cosine } \theta = \frac{\text{adj.}}{\text{hyp.}}$$

$$\text{tangent } \theta = \frac{\text{opp.}}{\text{adj.}}$$

Exercise 24a

Make a sketch of the given information where necessary.

1. In Fig. 24.2, write down the ratios (a) tan α, (b) sin α, (c) cos α in as many ways as possible in terms of p, q, x, y, z.

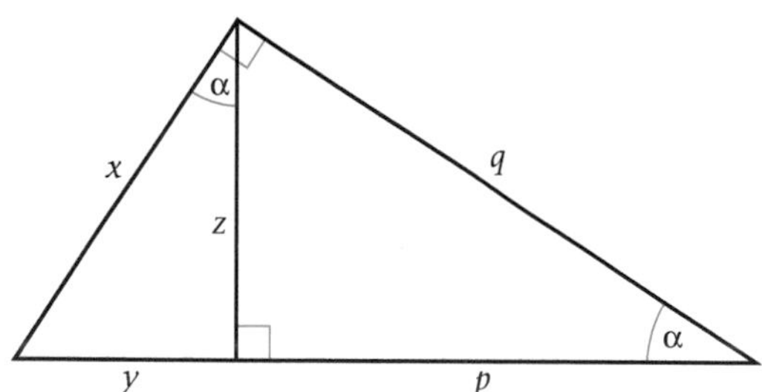

Fig. 24.2

2. Calculate, to the nearest 0.1°, the sizes of α, β, γ, δ, in Fig. 24.3.

(a)

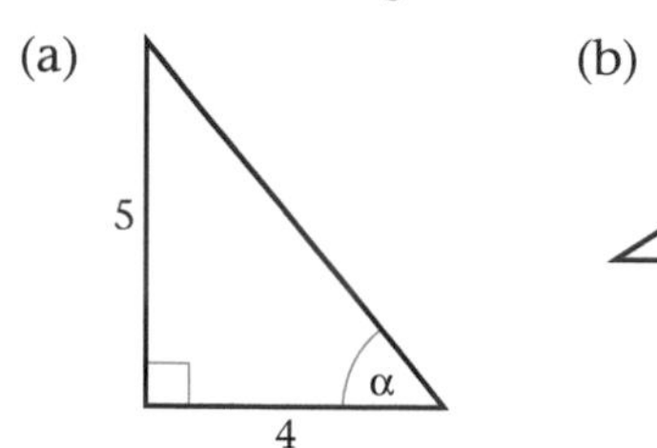

(b)

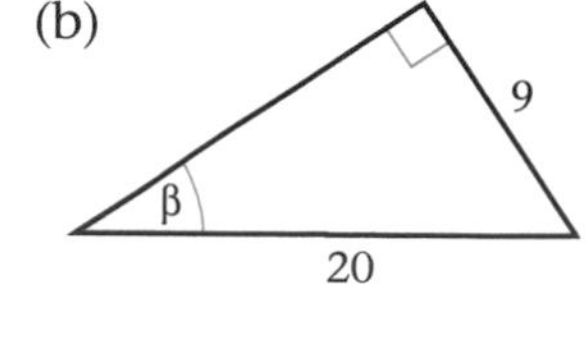

(c)

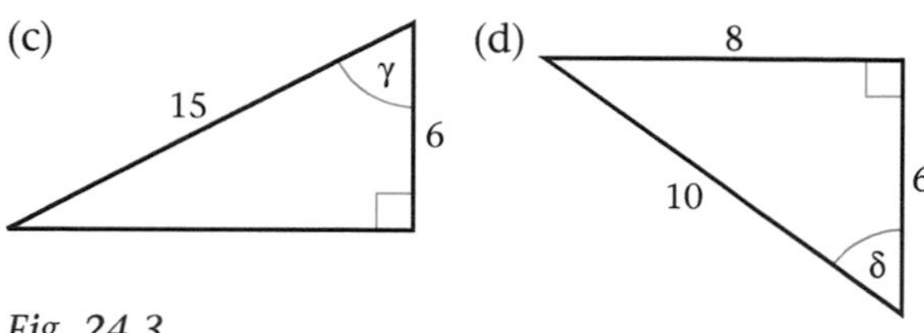

Fig. 24.3

3. In Fig. 24.4, M is the mid-point of XZ.
   (a) Calculate MZ.
   (b) Hence write down the value of XZ and calculate θ.

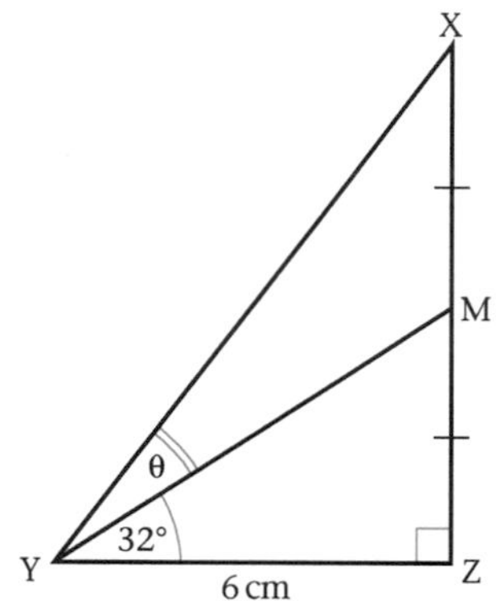

Fig. 24.4

4. In Fig. 24.5, calculate (a) $A\hat{O}B$, (b) AB.

5. In Fig. 24.6, LK ⊥ MN. Calculate sin $M\hat{N}L$.

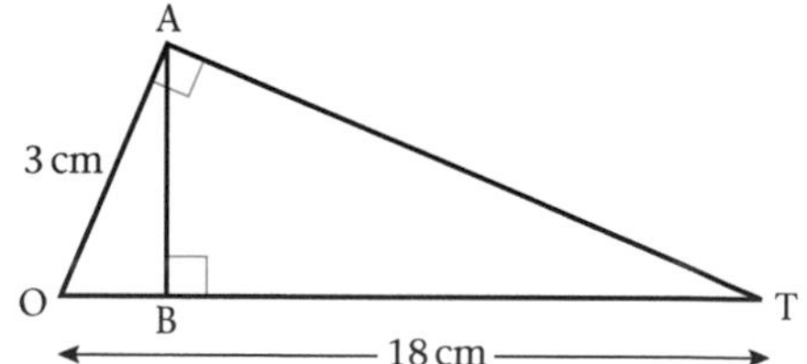

Fig. 24.5

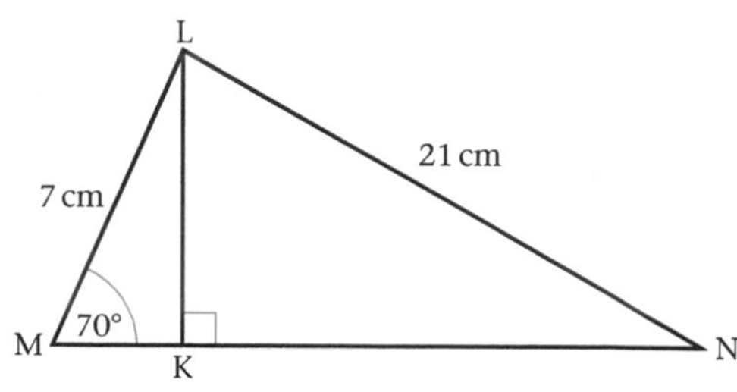

Fig. 24.6

6 A triangle has sides 8 cm and 5 cm and an angle of 90° between them. Calculate the smallest angle of the triangle.

7 A rhombus has sides 11 cm long. The shorter diagonal of the rhombus is 8 cm long. Find the size of one of the smaller angles of the rhombus correct to the nearest degree.

8 XYZ is an isosceles triangle.
$XY = XZ = 6$ cm and $Y\hat{X}Z = 100°$
Calculate YZ correct to 2 s.f.

Trigonometric ratios of angles of 45°, 60° and 30°

Fig. 24.7 is an isosceles right-angled triangle.

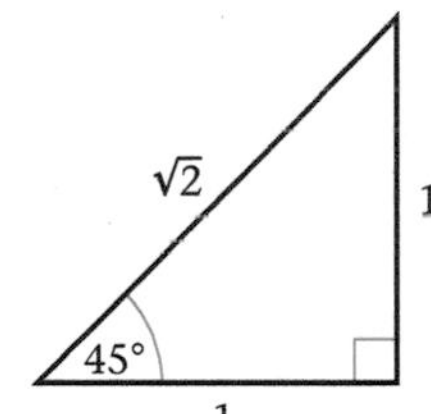

Fig. 24.7

From Fig. 24.7,

$$\sin 45° = \frac{1}{\sqrt{2}} = \frac{\sqrt{2}}{2}$$

$$\cos 45° = \frac{1}{\sqrt{2}} = \frac{\sqrt{2}}{2}$$

$$\tan 45° = \frac{1}{1} = 1$$

Fig. 24.8 is an equilateral triangle.

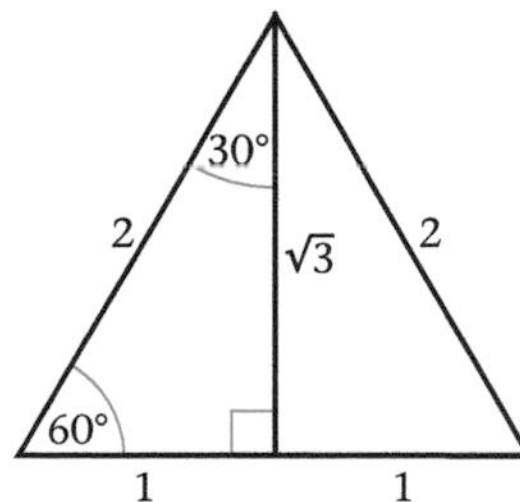

Fig. 24.8

From Fig. 24.8.

$$\sin 60° = \frac{\sqrt{3}}{2} \qquad \sin 30° = \frac{1}{2}$$

$$\cos 60° = \frac{1}{2} \qquad \cos 30° = \frac{\sqrt{3}}{2}$$

$$\tan 60° = \frac{\sqrt{3}}{1} = \sqrt{3} \qquad \tan 30° = \frac{1}{\sqrt{3}} = \frac{\sqrt{3}}{3}$$

Example 1

Given Fig. 24.9 calculate RQ and RS.

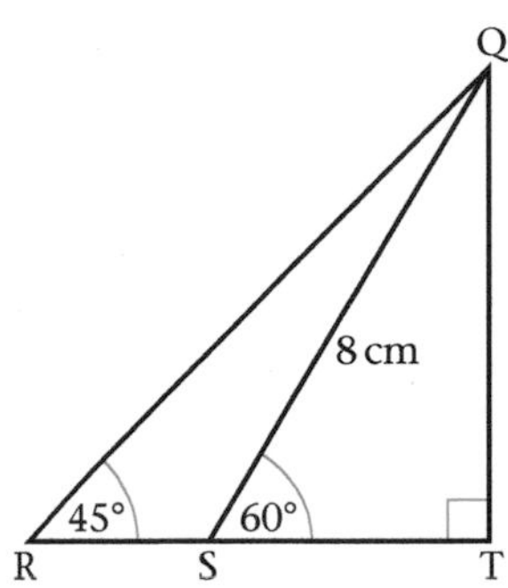

Fig. 24.9

In $\triangle QST$, the sides are in the ratio $1 : 2 : \sqrt{3}$ (a 30°, 60°, 90° $\triangle$).

$ST = \frac{1}{2} \times SQ = \frac{1}{2} \times 8\text{ cm} = 4\text{ cm}$
and $TQ = \sqrt{3} \times ST = 4\sqrt{3}$ cm

In $\triangle QRT$, the sides are in the ratio $1 : 1 : \sqrt{2}$ (a 45°, 45°, 90° $\triangle$).

$RQ = \sqrt{2} \times TQ$
$= \sqrt{2} \times 4\sqrt{3}$ cm
$= 4\sqrt{6}$ cm
$RT = TQ = 4\sqrt{3}$ cm
$RS = RT - ST$
$= (4\sqrt{3} - 4)$ cm
$= 4(\sqrt{3} - 1)$ cm

Exercise 24b

1 In each part of Fig. 24.10(a)–(h) the length of one side of a triangle is given in cm. Find the lengths of the other two sides, giving the answers in surd form with rational denominators where necessary.

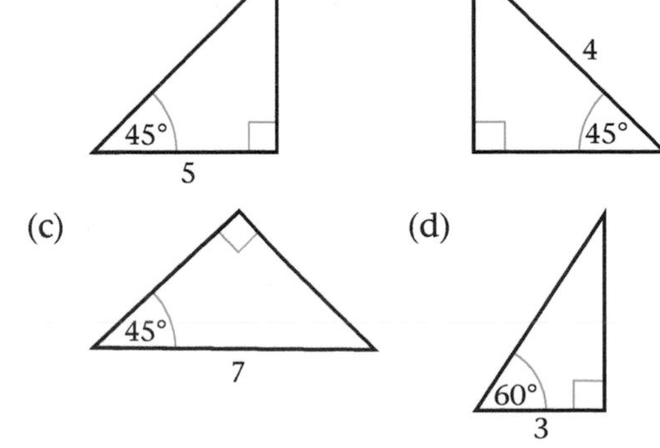

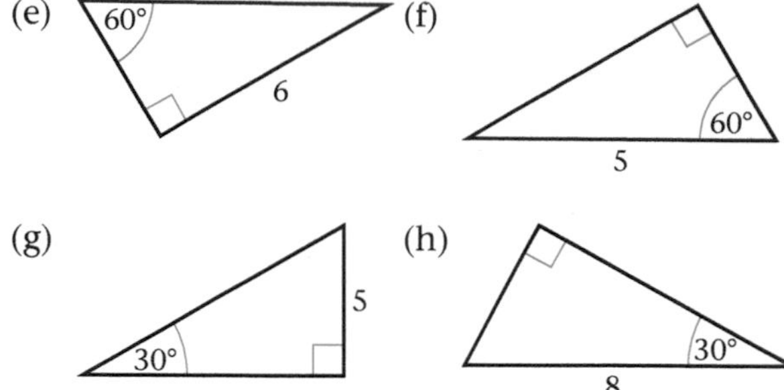

Fig. 24.10

2 In Fig. 24.11 the given lengths are in cm. In each part, find the length marked x, giving the answers in surd form with rational denominators where necessary.

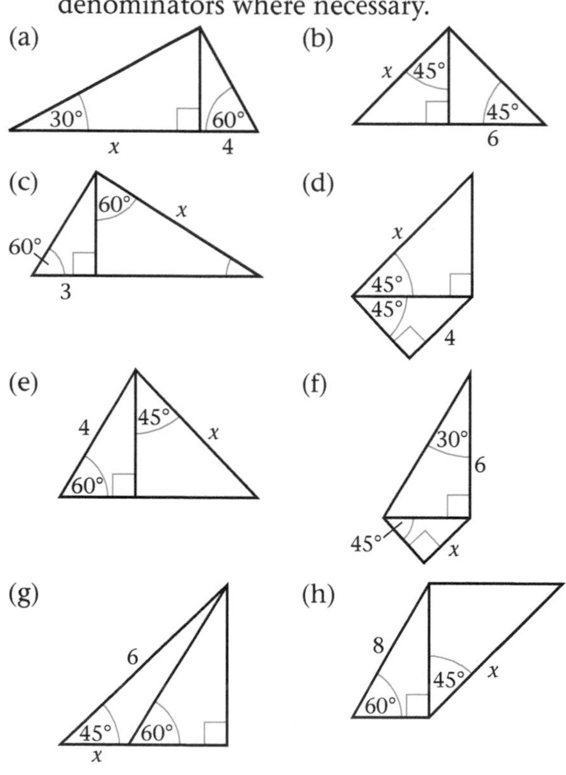

Fig. 24.11

Trigonometric ratios of angles of 0°, 90°, 180°, 270° and 360°

Using a line rotating anticlockwise and starting from the positive x-axis, the trigonometric ratios of angles of 0°, and 90° made with the positive x-axis may be calculated.

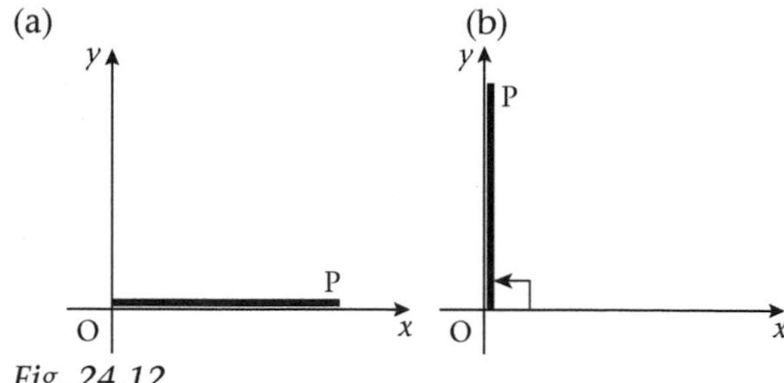

Fig. 24.12

From Fig. 24.12, the trigonometric ratios of 0° (see Fig. 24.12(a)), and 90° (see Fig. 24.12(b)) are as follows:

$\sin 0° = 0$ $\qquad$ $\sin 90° = 1$
$\cos 0° = 1$ $\qquad$ $\cos 90° = 0$
$\tan 0° = 0$ $\qquad$ $\tan 90°$ is undefined

Similarly, the trigonometric ratios of angles of 180°, 270° and 360° made with the x-axis may be calculated. Hence,

$\sin 180° = -1$ $\qquad$ $\sin 270° = -1$
$\cos 180° = 0$ $\qquad$ $\cos 270° = 0$
$\tan 180° = 0$ $\qquad$ $\tan 270°$ is undefined

$\sin 360° = 0$
$\cos 360° = 1$
$\tan 360° = 0$

Trigonometric ratios of angles between −360° and 360°

The trigonometric ratios of an angle between −360° and 360° can be expressed in terms of the trigonometric ratios of an angle between 0° and 90°, that is, of an acute angle, θ.

For the trigonometric ratios of obtuse angles

$\sin(180° - \theta) = \sin\theta$
$\cos(180° - \theta) = -\cos\theta$
$\tan(180° - \theta) = -\tan\theta$

For the trigonometric ratios of reflex angles

$\sin(180° + \theta) = -\sin\theta$
$\cos(180° + \theta) = -\cos\theta$
$\tan(180° + \theta) = \tan\theta$

and

$\sin(360° - \theta) = -\sin\theta$
$\cos(360° - \theta) = \cos\theta$
$\tan(360° - \theta) = -\tan\theta$

For the trigonometric ratios of negative angles

$\sin (2\theta) = -\sin \theta$
$\cos (-\theta) = \cos \theta$
$\tan (-\theta) = -\tan \theta$

Example 2

$\sin 125° = \sin (180 - 55)° = \sin 55°$
$\cos 225° = \cos (180 + 45)° = -\cos 45°$
$\tan 325° = \tan (360 - 35)° = -\tan 35°$

$\sin (-50°) = -\sin 50°$
$\cos (-140°) = -\cos 40°$
$\tan (-125°) = \tan 55°$

Exercise 24c

Use tables or a calculator to find the values, correct to 3 significant figures, of the following:

1. sin 235°
2. cos 235°
3. tan 235°
4. cos 110°
5. sin 125°
6. tan 320°
6. sin 93.4°
8. cos −80° 36′
9. cos 290.6°
10. tan −94.2°
11. tan 140° 129′
12. sin 42° 429

Find the values of θ between −360° and 360° in each of the following, giving your answer in degrees to 1 decimal place.

13. $\sin \theta = -0.550$
14. $\cos \theta = 0.234$
15. $\tan \theta = 5.204$
16. $\sin \theta = -0.474$
17. $\sin \theta = 0.631$
18. $\tan \theta = -0.598$
19. $\tan \theta = -0.965$
20. $\cos \theta = -0.862$
21. $\cos \theta = -0.728$
22. $\tan \theta = 2.942$
23. $\sin \theta = 0.504$
24. $\cos \theta = -0.609$

Solving right-angled triangles

In order to solve a right-angled triangle, that is, to find the unknown length of its sides or the unknown value of its angles, we may use
(i) trigonometric ratios (see Example 3)
(ii) Pythagoras' theorem.

Pythagoras' theorem

In a right-angled triangle the square on the hypotenuse is equal to the sum of the squares on the other two sides.

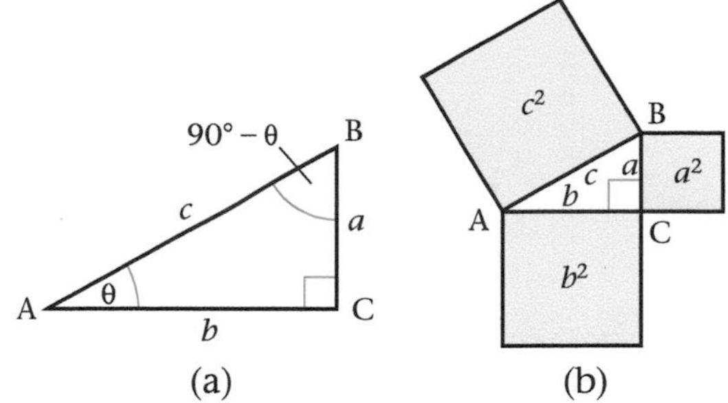

Fig. 24.13

In Fig. 24.13,

$AB^2 = BC^2 + AC^2$
or $c^2 = a^2 + b^2$

Notice also that

$a^2 = c^2 - b^2$
and $b^2 = c^2 - a^2$

Fig. 24.13(b) gives a geometrical interpretation of Pythagoras' theorem.
Note that, in the right-angled △ABC in Fig. 24.13(a),

(i) $\mathbf{\sin \theta = \cos (90° - \theta)}$
and $\mathbf{\cos \theta = \sin (90° - \theta)}$

(ii) $\cos \theta = \frac{b}{c} \Rightarrow \cos^2 \theta = \frac{b^2}{c^2}$

(iii) $\sin \theta = \frac{a}{c} \Rightarrow \sin^2 \theta = \frac{a^2}{c^2}$

$$\cos^2 \theta + \sin^2 \theta = \frac{b^2 + a^2}{c^2}$$
$$= \frac{c^2}{c^2} \text{ (\textit{Pythagoras' theorem})}$$

$\mathbf{\cos^2 \theta + \sin^2 \theta = 1}$ $(0° \leqslant \theta \leqslant 90°)$
is true for any right-angled triangle.

Example 3

Calculate (a) PR, (b) RS in Fig. 24.14. Give the answers correct to 2 s.f.

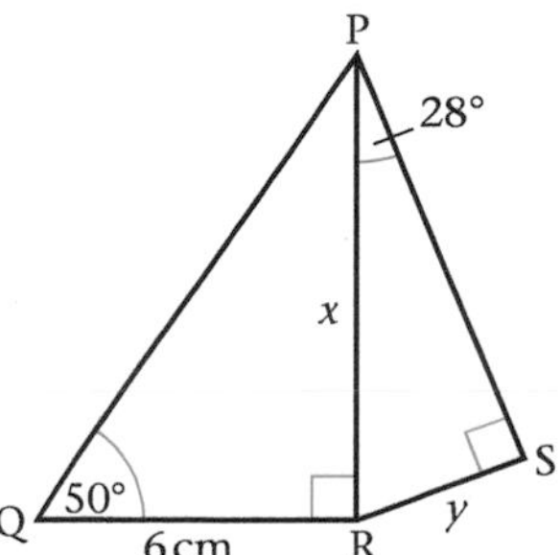

Fig. 24.14

(a) In $\triangle$PQR,

$$\tan 50° = \frac{x}{6}$$
$$x = 6 \times \tan 50°$$
$$= 6 \times 1.192$$
$$= 7.152$$
$$PR = 7.2\text{ cm to 2 s.f.}$$

(b) In $\triangle$PRS,

$$\sin 28° = \frac{y}{x}$$
$$y = x \sin 28°$$
$$= 7.152 \times \sin 28°$$
$$= 3.36$$
$$RS = 3.4\text{ cm to 2 s.f.}$$

working:

No	Log
7.152	0.854
sin 28°	$\bar{1}$.672
3.36	0.526

Notice in Example 3 that the value of x found in part (a) is used in part (b). In such cases, always use the full 4-figure value, *not* the rounded value. Table working should be set at the side so as not to get in the way of the main solution. The logarithm of sin 28° is found in the Logarithms of sines table (page 338).

Example 4

The ratio of the sides of an isosceles triangle is 7 : 6 : 7. Find the base angle to the nearest degree.

Let the sides of the triangle be $7n$, $6n$, $7n$ units.

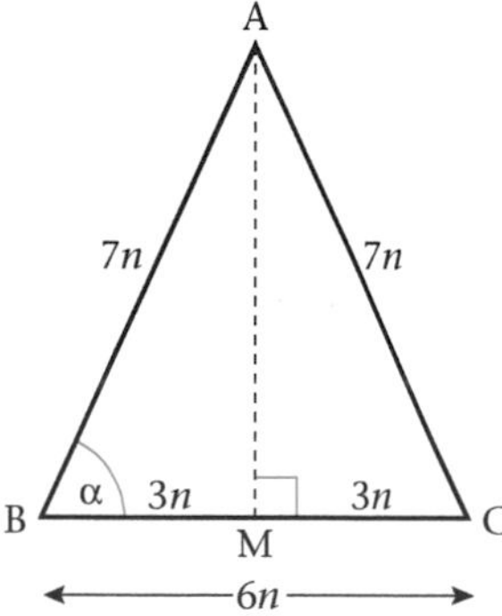

Fig. 24.15

Construct the line of symmetry as in Fig. 24.15.
In$\triangle$ABM,

$$\cos \alpha = \frac{3n}{7n} = \frac{3}{7} = 0.429$$
$$\alpha = 64.6°$$

The base angle is 65° to the nearest degree.

Example 5

In Fig. 24.16 if AC = 12 cm, BC = 5 cm, CD = 11 cm, find AD.

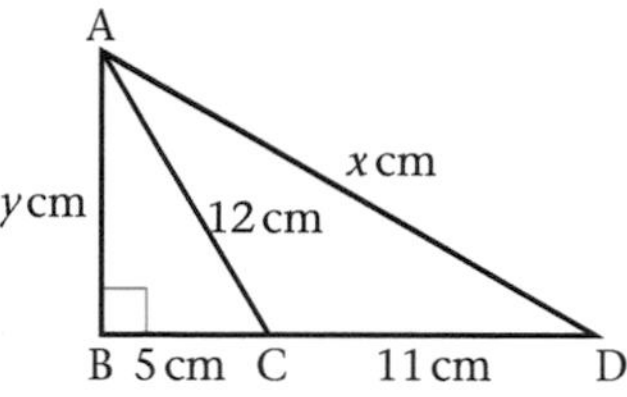

Fig. 24.16

Let AD = x cm and AB = y cm.

In $\triangle$ABC,

$$y^2 = 12^2 - 5^2$$
$$= 144 - 25$$
$$= 119$$

In $\triangle$ABD,

$$x^2 = y^2 + 16^2$$
$$= 119 + 256$$
$$= 375$$
$$\therefore \quad x = \sqrt{375} \simeq 19.4$$
$$AD = 19.4\text{ cm to 3 s.f.}$$

Notice that y represented an *intermediate* length. When y^2 was found to be 119 there was no need to find the value of y, since it was the value of y^2 that was needed in the subsequent working.

Exercise 24d

1. In $\triangle$ABC, AC = 2 m, BC = 3 m and $\widehat{C}$ is obtuse. The perpendicular from A to BC produced is AD. If CD = 1 m, calculate AB.

2. ABCD is a rectangle in which AB = 2.8 cm and AD = 3.3 cm. E is a point on DC produced such that $\triangle$AED and rectangle ABCD are equal in area. Calculate (a) DE, (b) AE.

3. Fig. 24.17 shows a prism such that M and N are the mid-points of AD and FE respectively.
   (a) Calculate the lengths of
   (i) CA (ii) CM.
   (b) Hence, find the angles
   (i) ACF (ii) MCN.

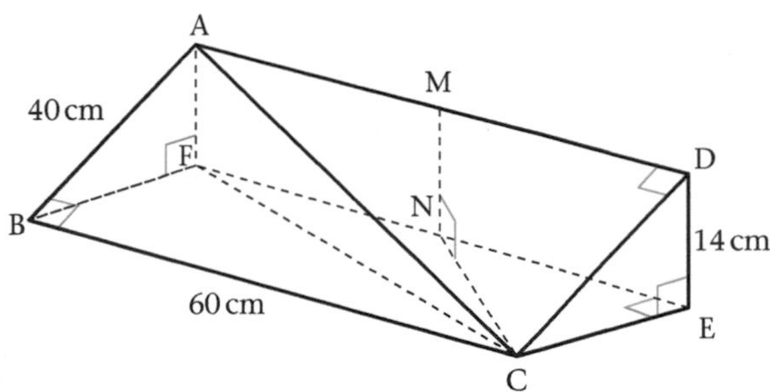

Fig. 24.17

4 A man walks 5 km from A to B on a bearing of 035°. He then walks 6 km from B to C on a bearing of 125°. Calculate (a) the distance, (b) the bearing of C from A.

5 A girl 160 cm tall stands 150 m from the foot of a building. She finds that the angle of elevation of the top of the building is 17°. Calculate the height of the building to the nearest $\frac{1}{2}$-metre.

6 A point Y is 400 m north-west of X. A tree is north-east of Y and on a bearing 015° from X. Find the distance of the tree from (a) X, (b) Y.

7 A, B and C are three points on the banks of a river with B and C on one side of the river and A on the opposite bank. Assume that the banks are parallel. BC = 20 m, angle $A\hat{B}C = 30°$, angle $A\hat{C}B = 45°$ and the width AD of the river is w metres.

(a) Express the lengths of DC and BD in terms of w.

(b) Hence, determine the width of the river correct to 2 significant figures.

[CXC (General) June 86]

8

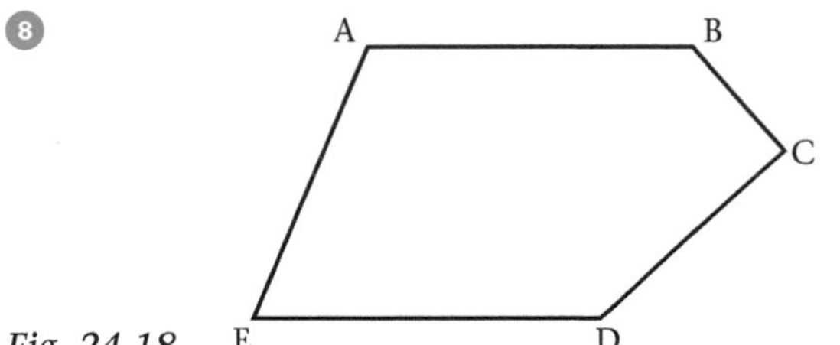

Fig. 24.18

In the figure ABCDE, not drawn to scale, AB || ED, AB = AE = ED = 12 cm, angle BCD = 90°, angle EAB = 120°, and angle EDC = 140°.

(a) Calculate, to 2 significant figures
  (i) angle BDC,
  (ii) the length of BC,
  (iii) the length of DC.

(b) Given that AF is the perpendicular from A to ED, calculate, giving your answer correct to 1 decimal place
  (i) the length of AF,
  (ii) the area of the figure ABCDE.

9

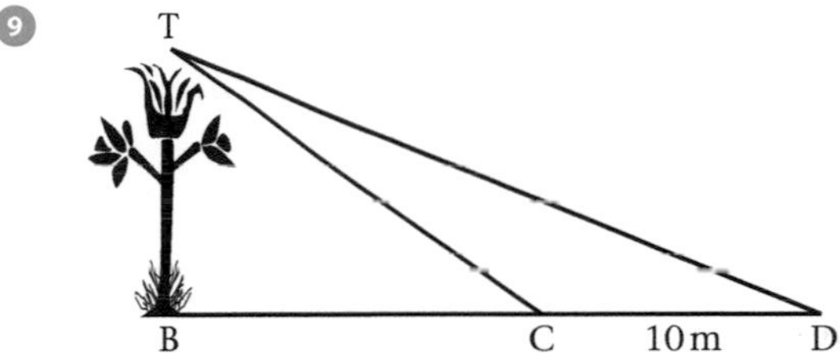

Fig. 24.19

The foot B of a tree T and points C and D are on the same horizontal level. The angle of elevation of T, the top of the tree, from C is 38°, and from D is 28°. The distance between C and D is 10 m.
Calculate the height of the tree, giving your answer correct to 2 significant figures.

10 PQ is a vertical tower such that R is a point due West of Q, the base of the tower, and T is due East of Q. The angles of depression of the points R and T from the top of the tower P are each equal to 40°.
If RT = 24 m, calculate, to the nearest whole number, the height of the tower.

11 A town F is 8 km due South of another town G. A man runs from G to F and then runs a further 6 km to H on a bearing S 73° W from F.

(a) Given that D is due West of F and due North of H, calculate the distances
  (i) FD (ii) DH

(b) Calculate the bearing of H from G.

12

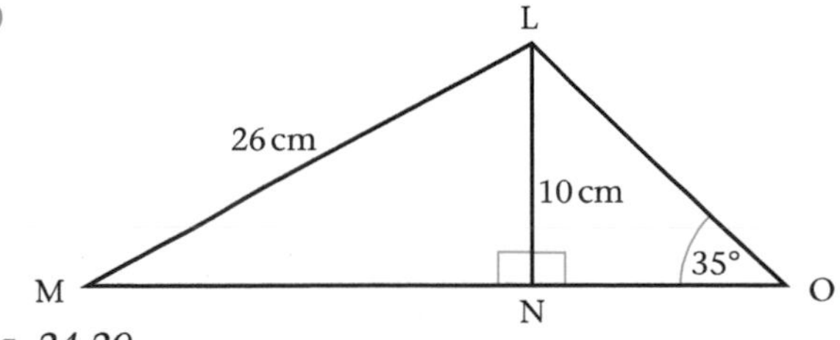

Fig. 24.20

Fig. 24.20, **not drawn to scale**, shows triangle LMO, with height LN = 10 cm. LM = 26 cm and angle LON = 35°. Calculate, in cm,

(a) MN, (b) MO. (5 marks)

[CXC (General) Jan 98]

Three-dimensional problems

It should be noted that in three-dimensional figures, the angle between a line and a plane is the angle between the line and its projection in the plane.

Example 6

In Fig. 24.21, ABCD and XYCD are rectangular planes such that XYCD is horizontal and B is 10 cm above CY. EF ∥ CB and EC = 16 cm, CP = 12 cm. Calculate the angle that (a) EF, (b) EB makes with the horizontal.

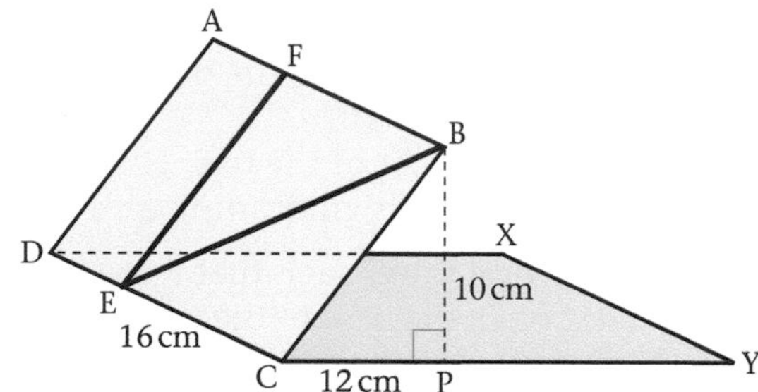

Fig. 24.21

(a) EF ∥ CB so $B\widehat{C}P$ equals the required angle.

In △BCP, $\tan B\widehat{C}P = \frac{10}{12} = \frac{5}{6} = 0.833$

$\Rightarrow B\widehat{C}P = 39.8°$

EF makes an angle of 39.8° with the horizontal.

(b) PE is the projection of BE on plane XYCD. Hence $B\widehat{E}P$ is the required angle. Fig. 24.22 shows the triangles used to find $B\widehat{E}P$.

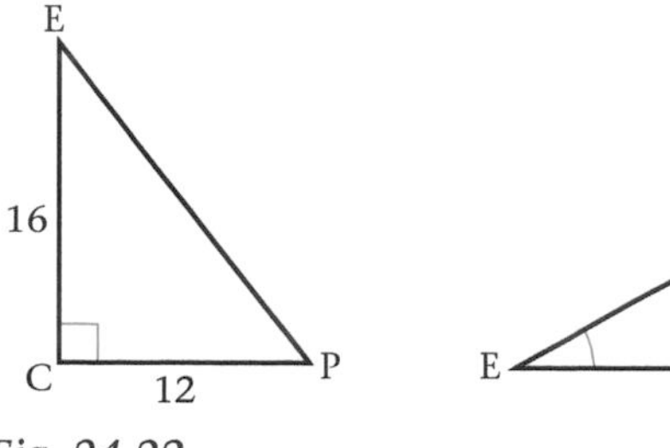

Fig. 24.22

In △ECP,

$EP^2 = 16^2 + 12^2$ (*Pythagoras*)

$= 256 + 144 = 400$

$EP = \sqrt{400} = 20\text{ cm}$

(*Note:* △ECP *is a* 3 : 4 : 5 *triangle*)

In △BEP,

$\tan B\widehat{E}P = \frac{BP}{EP} = \frac{10}{20} = 0.5$

$\Rightarrow B\widehat{E}P = 26.6°$ to the nearest 0.1 degree

EB makes an angle of 26.6° with the horizontal.

Example 7

A rectangular lid 25 cm by 20 cm is kept open at an angle of 65° to the horizontal, the hinges being on one of the long edges. Calculate the slope of a diagonal of the lid.

Fig. 24.23 shows the lid ABCD. θ is the required angle.

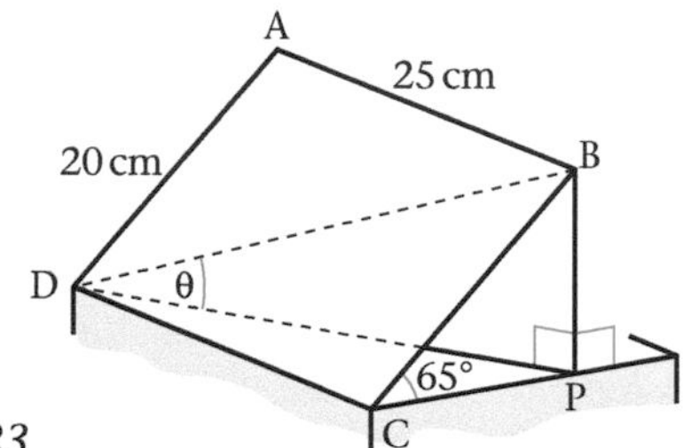

Fig. 24.23

Since θ is in the right-angled triangle BPD, it is necessary to find two sides of this triangle. The triangles in Fig. 24.24 each contain a side of △BPD.

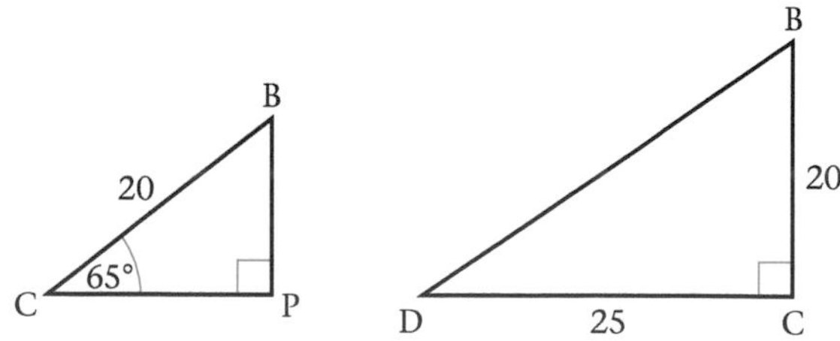

Fig. 24.24

In △BPC, $\sin 65° = \frac{BP}{20}$

$\Rightarrow BP = 20 \sin 65°$

In △BCD,

$BD^2 = 20^2 + 25^2$

$= 400 + 625 = 1\,025$

$BD = \sqrt{1\,025} \simeq 32.1\text{ cm}$

In △BPD,

$$\sin\theta = \frac{BP}{BD} = \frac{20 \sin 65°}{32.1}$$

$$\theta = 34.3°$$

working:

No	Log
20	1.301
sin 65°	$\bar{1}$.957
	1.258
32.1	1.507
sin 34.3°	$\bar{1}$.751

The slope of the diagonal is 34.3°

Notice in Example 7 that tables of log-sines were used. This is quicker than using natural sines first, then using logarithms.

Example 8

A pyramid with a vertex O and edges OA, OB, OC, OD each 10 cm long stands on a square base ABCD of side 8 cm. Calculate (a) the height OP of the pyramid. (b) the angle between an edge and the base.

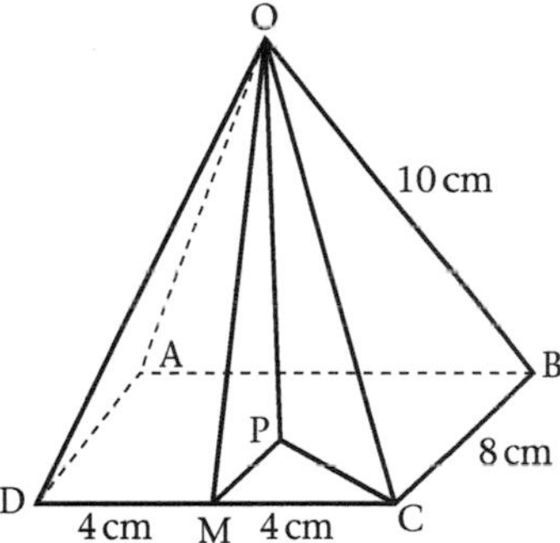

Fig. 24.25

In Fig. 24.25, M is the mid-point of edge DC.

(a)

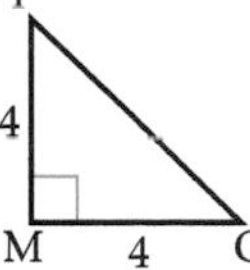

Fig. 24.26

In △PMC (Fig. 24.26),

$PC^2 = 4^2 + 4^2$ (*Pythagoras*)

$= 16 + 16 = 32$

Fig. 24.27

In △OPC (Fig. 24.27),

$OP^2 = OC^2 - PC^2$ (*Pythagoras*)

$= 100 - 32 = 68$

$OP = \sqrt{68}$ cm $= 8.25$ cm

(b) $O\hat{C}P$ is the angle between an edge OC and the base ABCD.

In △OPC,

$$\cos O\hat{C}P = \frac{\sqrt{32}}{10} = 0.566$$

$$O\hat{C}P = 55.5°$$

Notice that Examples 6, 7 and 8 were answered by solving the appropriate right-angled triangle. These examples show the value of sketching the various triangles used.

Exercise 24e

1. A cube has edges of 4 cm length. Calculate
   (a) the length of a diagonal of the cube,
   (b) the angle between the diagonal of the cube and its base.

2. The vertex of a pyramid on a square base of side 12 cm is 7 cm above the base. Calculate (a) the length of each sloping edge, (b) the angle between each sloping edge and the base.

3. A triangular prism like that of Fig. 24.28 has BE = 4 cm, AB = 20 cm and AD = 16 cm. Make a suitable sketch and calculate the slope of (a) BC, (b) BD.

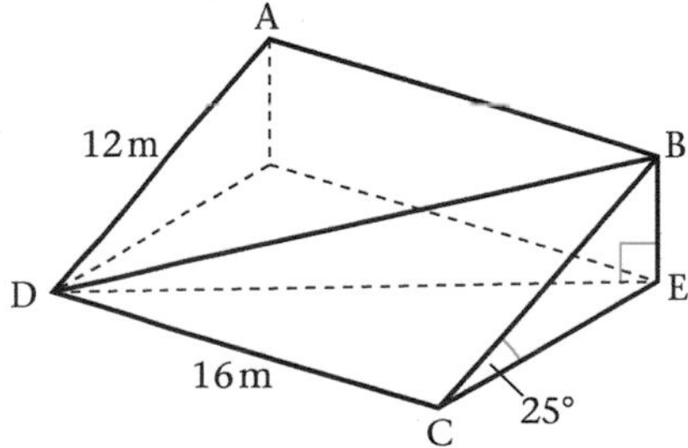

Fig. 24.28

4. A room in the shape of a cuboid has a floor which measures 5 m by 4 m. The longest diagonal of the room makes an angle of 35° with the floor. Calculate the height of the room.

5. A prism like that of Fig. 24.28 has AB = 24 cm, AD = 7 cm and BE = 3 cm. X is a point on AB such that $A\widehat{D}X = 45°$. Calculate the slope of BD and of DX.

6. A pole is resting in the corner of a room. The top of the pole is 2.5 m above the floor and the bottom is 1 m from each wall. Find the length of the pole and the angle which it makes with the floor.

7. A mast 45 m high is supported by four equal straight wires attached to the top of the mast and to the corners of a square of side 60 m on the level ground. Calculate the inclination of each wire to the horizontal.

8. Fig. 24.29 represents an open door. Assume that ABCD and ABEF are rectangles each measuring 2 m by 1.5 m.

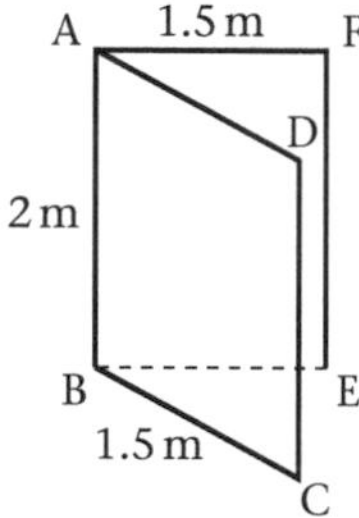

Fig. 24.29

If $D\widehat{A}F = 60°$ calculate the slope of DE.

9. Fig. 24.30 shows an open rectangular box 7 cm long. 6 cm wide and 6 cm high. A rod 12 cm long rests with its lower end in one bottom corner and is supported by the opposite corner.

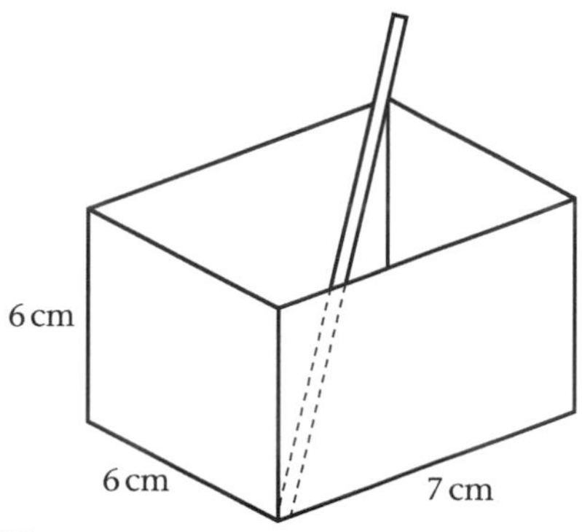

Fig. 24.30

Calculate (a) the inclination of the rod to the horizontal and (b) the height of its top end above the level of the base of the box.

10. The angles of depression from the top of a tower T to R and S are 32° and 22° respectively. The points R and S and the foot of the tower are on the same horizontal plane. The height of the tower TX is 52 m. The bearings of R and S from X are 270° and 220° respectively.
   (a) Draw a sketch to represent the information given above.
   (b) Hence or otherwise, calculate
      (i) the distance RS to one decimal place
      (ii) the bearing of S from R.

(13 marks)

[CXC (General) June 92]

11. A flagpole TF is 12 m tall. The foot of the pole F and the points A and B are in the same horizontal plane. The bearing of A from F is 035°, and of B from F is 115°. FA = AB = 8 m.
   (a) Draw a diagram to represent the given information. [*Hint: Show the North direction and all the measurements given*]
   (b) Calculate
      (i) the distance FB
      (ii) the angle of elevation of T from B.

Solving non-right-angled triangles

The sine rule

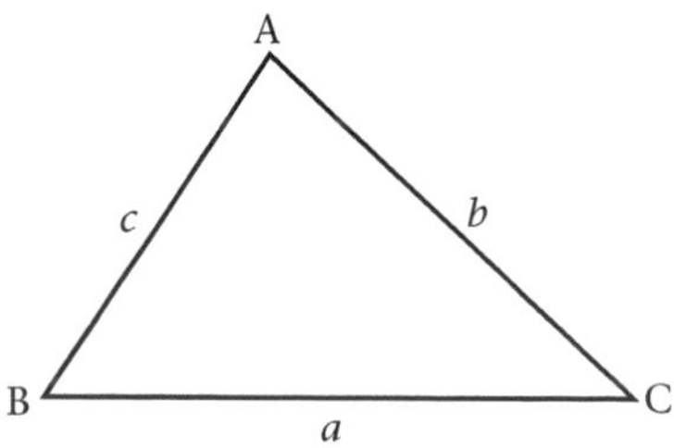

Fig. 24.31

The sine rule states that

$$\frac{a}{\sin A} = \frac{b}{\sin B} = \frac{c}{\sin C}$$

or

$$\frac{\sin A}{a} = \frac{\sin B}{b} = \frac{\sin C}{c}$$

for any triangle ABC in which A, B, C are the angles of the triangle and a, b, c are the lengths of the sides opposite these angles. In obtuse-angled triangles, $\sin \theta = \sin (180° - \theta)$.

Example 9

Solve completely the $\triangle$ABC in which $a = 12.4$ cm, $c = 14.7$ cm and $C = 72°4'$.

To 'solve completely' means to find all the unknown lengths and angles.

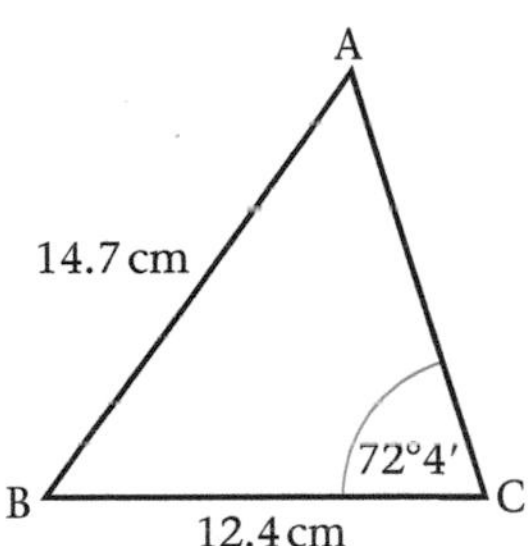

Fig. 24.32

In Fig. 24.32

$$\frac{\sin A}{a} = \frac{\sin C}{c}$$

$$\frac{\sin A}{12.4} = \frac{\sin 72°\,4'}{14.7}$$

$$\Rightarrow \quad \sin A = \frac{12.4 \sin 72°\,4'}{14.7}$$

$$A = 53°18' \text{ or } 126°\,42'$$

working:

No	Log
12.4	1.093
sin 72°4′	$\bar{1}.978$
	1.071
14.7	1.167
sin 53°18′	$\bar{1}.904$

But $a < c$, therefore $A < C$.

$$A = 53°18'$$

and $\quad B = 54°38' \qquad$ (angles of $\triangle$)

$$\frac{b}{\sin B} = \frac{c}{\sin C}$$

$$\frac{b}{\sin 54°\,38'} = \frac{14.7}{\sin 72°\,4'}$$

$$b = \frac{14.7 \sin 54°38'}{\sin 72°\,4'}$$

$$\simeq 12.6 \text{ cm}$$

working:

No	Log
14.7	1.167
sin 54°38′	$\bar{1}.911$
	1.078
sin 72°4′	$\bar{1}.978$
12.6	1.100

Hence $A = 53°18'$, $B = 54°38'$ and $b = 12.6$ cm.

The cosine rule

This is used for solving triangles in which two sides and the included angle are given.

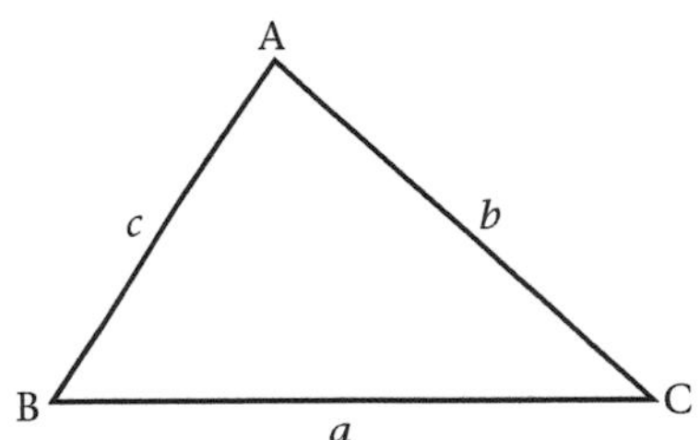

Fig. 24.33

In $\triangle$ABC,

$$\mathbf{a^2 = b^2 + c^2 - 2bc \cos A}$$
$$\mathbf{b^2 = a^2 + c^2 - 2ac \cos B}$$
$$\mathbf{c^2 = a^2 + b^2 - 2ab \cos C}$$

These formulae are true for both acute and obtuse angles. In obtuse-angled triangles, $\cos \theta = -\cos (180° - \theta)$.

Example 10

Find x in Fig. 24.34.

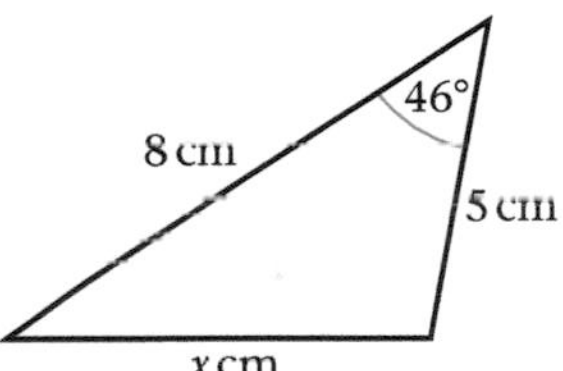

Fig. 24.34

$$x^2 = 8^2 + 5^2 - 2 \times 8 \times 5 \times \cos 46°$$
$$= 64 + 25 - 80 \times 0.695 = 89 - 55.6 = 33.4$$
$$x = \sqrt{33.4} = 5.78$$

Example 11

Find y in Fig. 24.35.

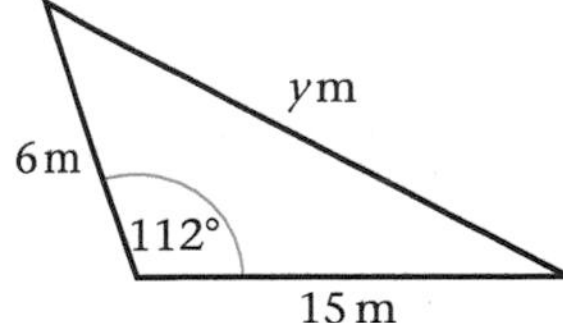

Fig. 24.35

$$y^2 = 6^2 + 15^2 - 2 \times 6 \times 15 \times \cos 112°$$
$$= 36 + 225 - 180 \times (-\cos 68°)$$
$$= 261 + 180 \times 0.375 = 261 + 67.5 = 328.5$$
$$y = \sqrt{328.5} = 18.1$$

When all three sides of a triangle are known, the angles can be calculated by rearranging the basic formulae as follows:

$$\cos A = \frac{b^2 + c^2 - a^2}{2bc}$$

$$\cos B = \frac{a^2 + c^2 - b^2}{2bc}$$

$$\cos C = \frac{a^2 + b^2 - c^2}{2bc}$$

Example 12

Calculate the angles of a triangle which has sides 5 cm, 8 cm and 11 cm.

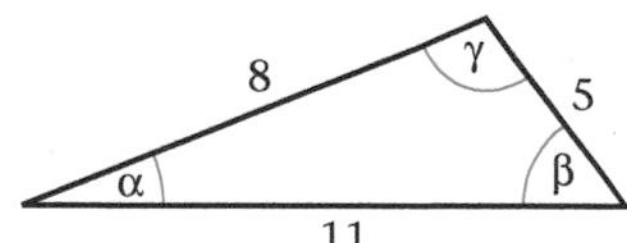

Fig. 24.36

Lettering the angles of the △ as in Fig. 24.36,

$$\cos \alpha = \frac{8^2 + 11^2 - 5^2}{2 \times 8 \times 11}$$

$$= \frac{160}{16 \times 11} = \frac{10}{11} = 0.909$$

$$\alpha = 24.6°$$

$$\cos \beta = \frac{5^2 + 11^2 - 8^2}{2 \times 5 \times 11}$$

$$= \frac{82}{110}$$

$$= 0.745$$

$$\beta = 41.8°$$

$$\cos \gamma = \frac{8^2 + 5^2 - 11^2}{2 \times 8 \times 5}$$

$$= \frac{-32}{80}$$

$$= -0.4$$

$$\therefore \quad \gamma = 180° - 66.4° = 113.6°$$

Check: $\alpha + \beta + \gamma = 180°$

Example 13

A village P is 10 km from a point X on a bearing 025° from X. Another village, Q, is 6 km from X on a bearing of 162°. Calculate the distance and bearing of P from Q.

Enter the details on a sketch such as Fig. 24.37.

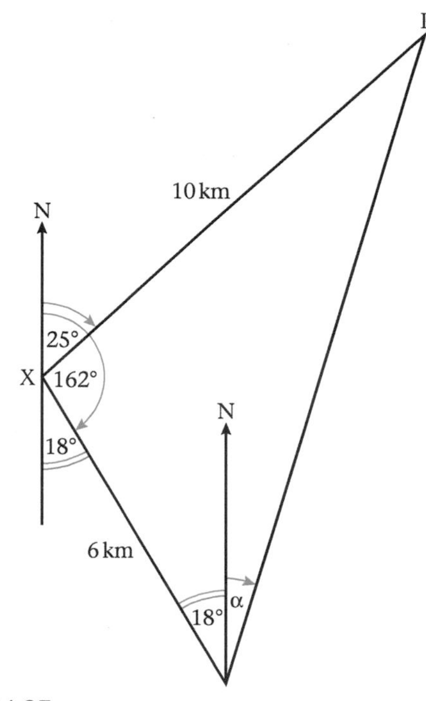

Fig. 24.37

$$P\widehat{X}Q = 162° - 25° = 137°$$

$$PQ^2 = 10^2 + 6^2 - 2 \times 10 \times 6 \times \cos 137°$$

$$= 100 + 36 - 120 \times (-\cos 43°)$$

$$= 136 + 120 \times 0.731$$

$$= 136 + 87.72$$

$$= 223.72$$

$$PQ = \sqrt{224}\text{ km}$$

$$= 15.0\text{ km to 3 s.f.}$$

$$\frac{\sin Q}{10} = \frac{\sin 137°}{15}$$

$$\sin Q = \frac{10 \sin 43°}{15}$$

$$Q = 27°3'$$

working:

No	Log
10	1.000
sin 43°	$\bar{1}$.834
	0.834
15	1.176
sin 27°3′	$\bar{1}$.658

From Fig. 24.37,

α = bearing of P from Q

$= 27°\ 3' - 18°$

$= 009°\ 3'$

Exercise 24f

1. In △ABC, $\widehat{B} = 38°$, $\widehat{C} = 48°$, $c = 18.8$ cm. Find b.
2. In △ABC, $\widehat{A} = 98°$, $\hat{C} = 36°$, $a = 34.4$ cm. Find c.

3 In $\triangle ABC$, $\hat{A} = 96°139$, $a = 39.4$ cm, $b = 11.2$ cm. Find $\hat{B}$.

4 In $\triangle ABC$, $\hat{A} = 60°$, $b = 5$ cm, $c = 3$ cm. Find a.

5 In $\triangle ABC$, $\hat{C} = 123°$, $a = 3$ m, $b = 2$ m. Find c.

6 In$\triangle ABC$, $\hat{B} = 143°$, $a = 25$ cm. $c = 40$ cm. Find b.

7 In$\triangle ABC$, $a = 5$ m, $b = 6$ m, $c = 7$ m. Find $\hat{B}$ and $\hat{C}$.

8 In $\triangle ABC$, $a : b : c = 15 : 21 : 24$. Find $\hat{B}$.

9 In $\triangle ABC$, $a = 9$ m, $b = 6$ m, $c = 4.8$ m. Find $\hat{A}$.

10 A girl starts from a point X and walks 220 m on a bearing 063°. She then walks to a point Y on a bearing 156°. If Y is due east of X, calculate XY.

11 Three villages X, Y and Z are connected by straight level roads. XY = 5 km, YZ = 4 km and $X\hat{Y}Z = 160°$. What distance is saved by walking from X to Z directly instead of through Y?

12 In Fig. 24.38 the lengths are in cm. Calculate d and α.

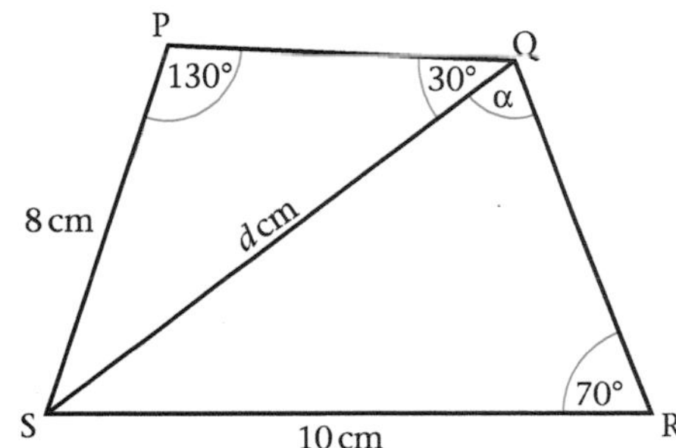

Fig. 24.38

13 In Fig. 24.39 the lengths are in m. Find x and θ.

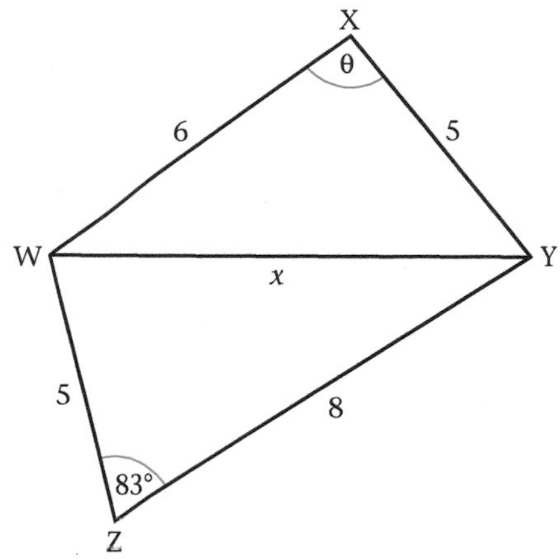

Fig. 24.39

14 In the quadrilateral ABCD, AD || BC, AB = 15 cm, DC = 14 cm, $D\hat{A}B = 60°$ and $A\hat{D}B = 40°$. Calculate (a) BD, to 2 s.f., (b) $B\hat{C}D$, correct to the nearest degree.

15 ABCD is a cyclic quadrilateral in which AB = 5 cm, BC = 4 cm, CD = 7 cm, DA = 6 cm. Calculate $A\hat{B}C$.

16 An aeroplane leaves an airport and flies due north for $1\frac{1}{2}$ hours at 500 km/h. It then flies 400 km on a bearing 053°. Calculate its final distance and bearing from the airport.

17 In Fig. 24.40, not drawn to scale, CA, AB and BW are three sides of a parallelogram.

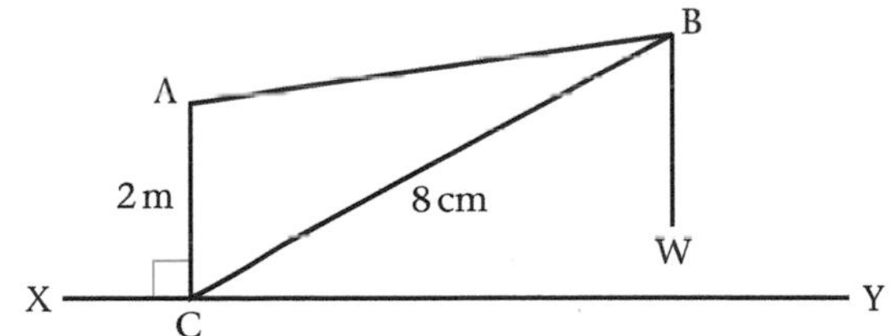

Fig. 24.40

CA = 2 m, CB = 8 m, angle ACX = 90° and angle BCY = 36.9°.
Calculate, to one decimal place,

(a) the length of AB
(b) the shortest distance between W and XY
(c) the angle which CW makes with XY.

(9 marks)

[CXC (General) Jan 91]

18

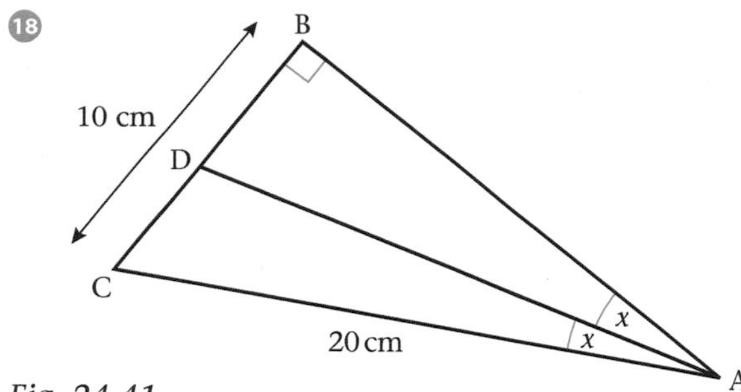

Fig. 24.41

Fig. 24.41, **not drawn to scale**, shows a triangle ABC right-angled at B. AC = 20 cm, BC = 10 cm and DA bisects angle CAB.

(a) Show that angle BDA = 75° (4 marks)

(b) Use the sine rule to show that CD = AC tan 15°. (4 marks)

(c) Given that tan 15° = 0.27, calculate to two significant figures, the length of

(i) BD

(ii) BA. (4 marks)

[CXC (General) Jan 90]

Trigonometric functions : the sine, cosine and tangent functions

Trigonometric functions are periodic functions in which the value of the function is repeated over a constant fixed interval as the size of the angle varies. For the sine and cosine functions, the value of the function varies between −1 and +1, and the period is 360°. However, the value of the cosine function of any angle θ is equal to the sine function of (θ + 90°) as shown in Fig 24.42.

The tangent function is also a periodic function but the period is 180°. The tangent function is undefined after intervals of 180° at ±90°, ±270°, etc., that is, the function is *not* a continuous function.

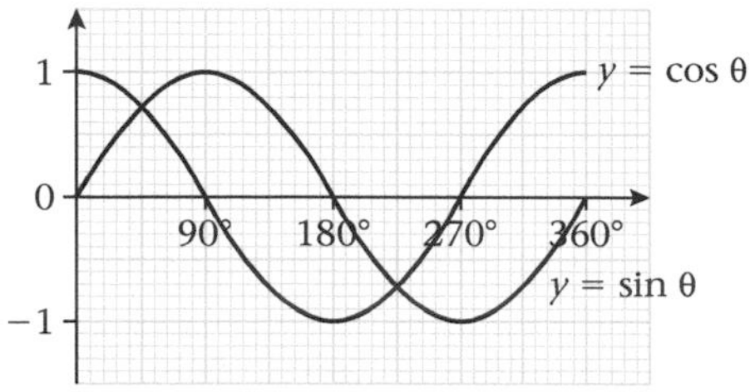

Fig. 24.42

Example 14

Using a scale of 2 cm to 1 unit on the vertical axis and 1 cm to 30° on the horizontal axis draw the graph of $y = 2\sin x$ for $-90° \leqslant x \leqslant 90°$. On the same axes draw the graph of $y = 1 + \cos x$, $-90° \leqslant x \leqslant 90°$. Use your drawing to find the solution of

$$2\sin x = 1 + \cos x$$

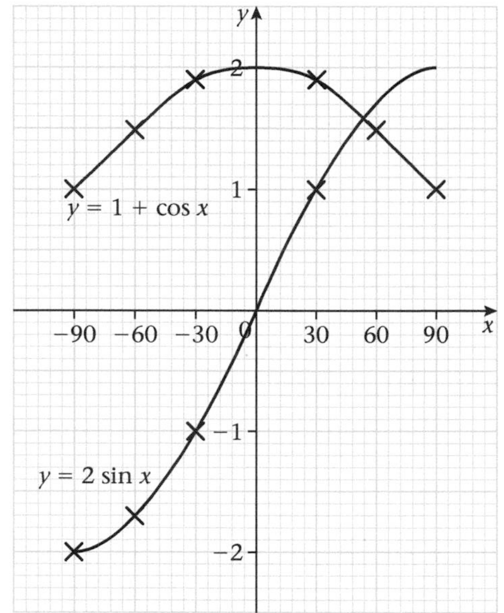

Fig. 24.43

Solution at $x \simeq 52°$

Exercise 24g

1. Given that sin 25° = 0.423, write down all the angles θ between −360° and 360° for which sin θ = 0.423.

2. Given that cos 36° = 0.809, write down all the angles θ between −360° and 360° for which cos θ = −0.809.

3. Given that tan x = 0.577, write down all the possible values of x for which $-360° \leqslant x \leqslant 360°$.

4. In a right-angled triangle, $\cos^2\theta = \frac{a^2}{x^2}$. Given that $x = 2a$, show that θ = 60°.

5. Find the values of θ in the range $-360° \leqslant \theta \leqslant 360°$, which satisfy the equation $2\sin\theta - 1 = \frac{1}{3}$.

6 Using a suitable scale draw the graph of $y = \frac{1}{2}\tan x$. On the same axes draw the graph of $y = \frac{1}{2}\sin x + 1$. Using your graph, find the solution of $\tan x - \sin x = 2$.

7 Table 24.1 shows values of
$f : x \to \tan x$ for $0° \leqslant x° \leqslant 80°$

Table 24.1

$x°$	0	10	20	30	40	50	60	70	80
$f(x)$	0	0.18	0.36	0.58	0.84	1.19	1.73	2.75	5.67

(a) Using a scale of 1 cm to represent 5° on the horizontal axis, and 2 cm to represent 1 unit on the vertical axis, draw the graph of

$y = f(x)$ for $0° \leqslant x° \leqslant 80°$

(b) Copy and complete Table 24.2 for the function

$g : x \to 3 \cos x$ for $0° \leqslant x° \leqslant 80°$

Table 24.2

$x°$	0	10	20	30	40	50	60	70	80	90
$g(x)$	3			2.60			1.5			0

(c) Using the same axes and scale as in (a), draw the graph of

$y = g(x)$ for $0° \leqslant x° \leqslant 80°$

(d) From your graph, determine the value of x for which $3 \cos x = \tan x$, giving your answer
(i) in degrees, to the nearest degree
(ii) in radians, to 2 significant figures.

(e) Use your answer to part (d) to deduce the other solutions of $3 \cos x = \tan x$

for $-360° \leqslant x \leqslant 360°$.

8 (a) Copy and complete Table 24.3 for $y = F(x)$, to 2 decimal places, where

$F(x) = 3 \sin x - 2 \cos x$

Table 24.3

$x°$	−60	−30	0	30	60	90	120	150
$F(x)$			−2	−0.23		3	3.60	

(b) Using a scale of 2 cm to represent 30° on the x-axis, and 2 cm to represent 1 unit on the y-axis, draw the graph of $F(x)$ for

$-60° \leqslant x° \leqslant 150°$

(c) Draw on the same axes the graph of

$y = 2$

(d) Using the graphs, estimate to the nearest degree
(i) the value of x for which

$3 \sin x - 2 \cos x = 2$

(ii) the value of x in the given range for which $3 \sin x - 2 \cos x$ has a maximum value

(e) Calculate, in radians, to 2 significant figures, the values of x which were determined in (d).

Longitude and latitude

Consider the earth to be a sphere of radius R, where R km = 6400 km (or of circumference $2\pi R$ km = 40 000 km). The distance between any two positions on the earth's surface is therefore measured along the arc of a circle. If the radius of the circle is R, then the circle is a **great circle**.

Distances from North to South are measured along the semi-circle of a great circle from the North Pole to the South Pole. These semi-circles are called **meridians** or **lines of longitude**. The longitude of a point varies between 180°E and 180°W. The distance, d, between two points A and B which are on the same line of longitude is given by

$$d = \frac{\alpha}{360} 2\pi R,$$

where α is the difference in latitude between the points A and B, that is, B is $\alpha°$ south of A (or vice versa).

Distances from East to West are measured along **parallels of latitude**. The radius of a parallel of latitude 0° varies with its distance from the **equator**, and is given by

$$r = R \cos \theta,$$

where θ is the angle shown in Fig. 24.44.

Note that at the equator, $\theta = 0°$ and $R \cos 0° = R$, so that the only parallel of latitude which is a 'great circle', that is with a radius equal to R, is the equator.

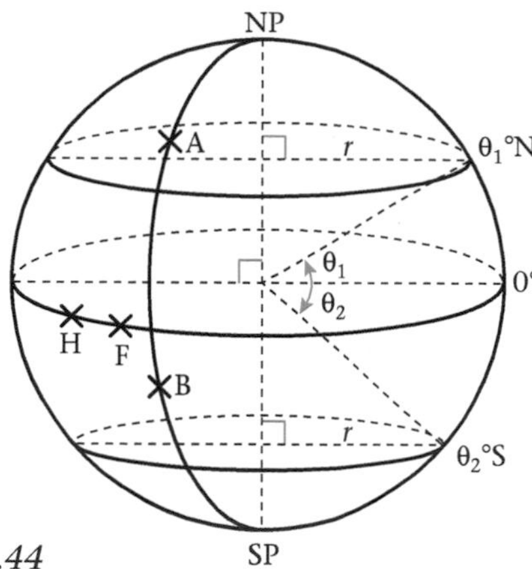

Fig. 24.44

The latitude of a point varies between 0° and 90°N, and 0° and 90°S. The distance, d, between two points F and H which are on the same parallel of latitude is given by

$$d = \frac{\beta}{360} 2\pi r.$$

where β is the difference in longitude between the points F and H, that is, F is β° east of H (or vice versa).

Distances between points on the earth's surface are given either in km or in nautical miles (n.mi) 1 n.mi. is the length of arc on a circumference of the earth, that is, on a 'great circle' which subtends an angle of 1 minute or $\frac{1}{60}$° at the centre of the earth.

Example 15

P and Q lie on the same line of longitude in latitudes 30°N and 12°N respectively. Find the great-circle distance between P and Q (a) in km, (b) in nautical miles.

Fig. 24.45 shows the points P and Q. O is the centre of the earth.

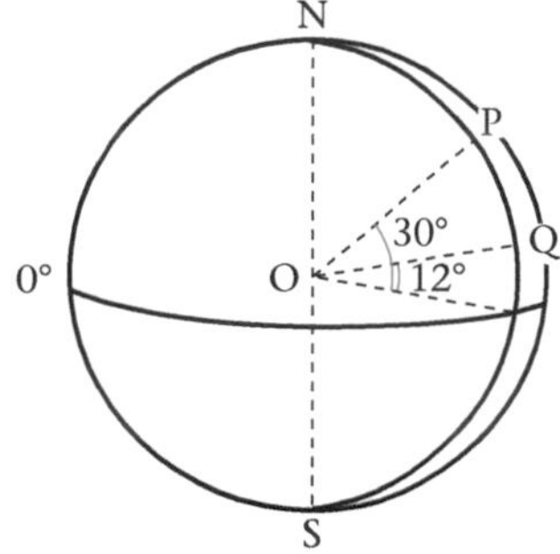

Fig. 24.45

(a) Taking the earth to be a sphere of radius 6 400 km,

$$\text{arc PQ} = \frac{(30° - 12°)}{360°} \times 2\pi \times 6\,400\,\text{km}$$
$$= \frac{18}{360} \times 2\pi \times 6\,400\,\text{km}$$
$$= 640\pi\,\text{km}$$
$$= 2\,000\,\text{km to 2 s.f.}$$

working:

No	Log
640	2.806
π	0.497
2010	3.303

(b) In nautical miles,

$$\text{arc PQ} = (30 - 12) \times 60\,\text{n.mi.}$$
$$= 18 \times 60\,\text{n.mi.}$$
$$= 1\,080\,\text{n.mi.}$$

Example 16

Calculate the distance, measured along the parallel of latitude, between two places in latitude 42°N, whose longitudes are 23°W and 17°E.

Fig. 24.46 shows the places, A and B, on the circle of latitude, centre C. X and Y lie on the equator and O is the centre of the earth.

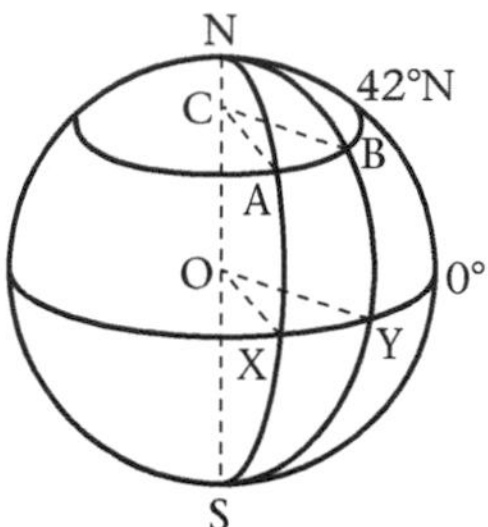

Fig. 24.46

Fig. 24.47 shows (a) the cross-section through N, A, X, S, and (b) the plan showing the arc AB.

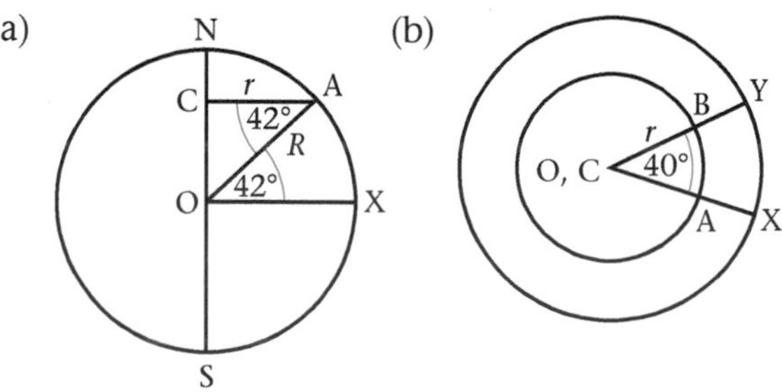

Fig. 24.47

In △OCA.

$$r = R \cos 42°$$

In Fig. 24.47(b),

$$\hat{ACB} = 23° + 17° = 40°$$

Then, arc AB $= \frac{40}{360}$ of $2\pi r$

$= \frac{1}{9} \times 2\pi R \cos 42°$

$= \frac{1}{9} \times 40\,000 \times 0.743$ km

$= \frac{29\,720}{9}$ km

$\simeq 3\,300$ km

Example 17

A (50°S, 20°E) and B (50°S, 160°W) are two points on the earth's surface. Calculate, in nautical miles (n.mi), (a) the distance between A and B along the parallel of latitude, (b) the distance between A and B along a great circle.

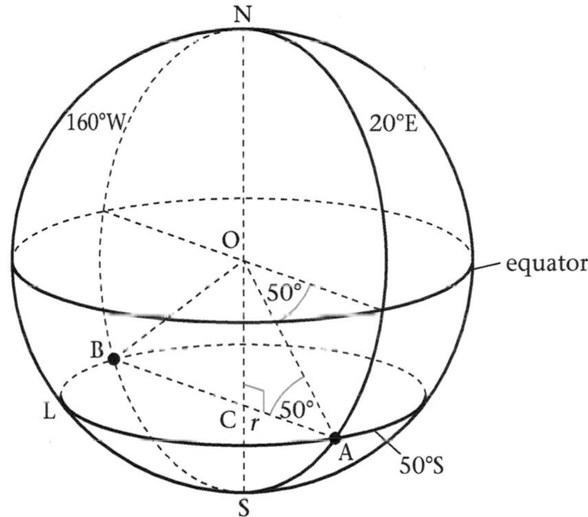

Fig. 24.48

(a) In Fig. 24.49(a),

arc AB $= (160 + 20) \times 60 \times \cos 50°$ n.mi.

$= 180 \times 60 \times 0.643$

$= 6\,940$ n.mi to 3 s.f.

(a)

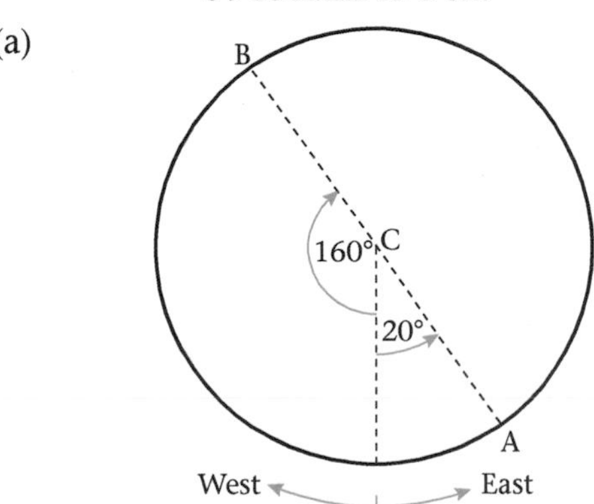

(b)

Fig. 24.49

(b) Since the longitudes of A and B differ by 180°, the two points lie on the same meridian. In Fig. 24.49(b),

$A\hat{O}B = 180° - 2 \times 50° = 80°$

Hence, arc ASB $= 80 \times 60$ n.mi.

$= 4\,800$ n.mi.

The great-circle distance between two points is the shortest possible distance between them on the surface of the earth.

Exercise 24h

Take $R = 6\,400$ km or $2\pi R = 40\,000$ km and $\pi = 3.14$ or $\log \pi = 0.497$ as appropriate. Give all answers to 3 s.f.

1. Two places on the same meridian have latitudes 23°S and 41°S. What is their distance apart measured along the meridian? (Answer in km.)
2. As question 1, but with latitudes 23°S and 41°N.
3. Two places on the equator have longitudes 63°E and 132°E. How far apart are they, measured along the equator? (Answer in n.mi.)
4. As question 3, but the places have longitudes 132°E and 126°W.
5. Two places on the same line of longitude are 3 940 km apart. If one of them is on latitude 18°N and the other is in the southern hemisphere, what is the latitude of the other?
6. What is the length of the Arctic Circle (66.5°N)? How far from the North Pole is any point on this circle? (Answer in km).

7 Calculate the shortest distance, in km, between two places, A (60°N, 72°W) and B (60°N, 108°E).

8 Two places are on latitude 37°S and their longitudes are 34°W and 29°E respectively. What is their distance apart measured along the parallel of latitude? (Answer in km.).

9 (In this question, use $\pi = \frac{22}{7}$ and take the circumference of the earth to be 40 000 km.)

(a) From an airport A, situated at latitude 42°N, longitude 15°W, an aeroplane travels due east to a landing strip B located at longitude 30°E.
Calculate the distance AB, measured along the surface of the earth, giving your answer to the nearest kilometre.

(b) The aeroplane continues its journey from B by flying due south for a distance of 1 557 km to a location C. Calculate
  (i) the latitude of C, giving your answer to the nearest degree;
  (ii) the shortest distance along the surface of the earth from C to the South Pole.

[CXC (General) Jan 89]

10 Find the distance in n.mi. between New York (42°N, 74°W) and Rome (42°N, 12°E) measured along a parallel of latitude.

11 A and B are points on the parallel of latitude 58.4°N, their longitudes being 148°W and 32°E respectively. What is their distance apart measured along (a) the parallel of latitude, (b) a great circle? (Answer in n.mi.)

12

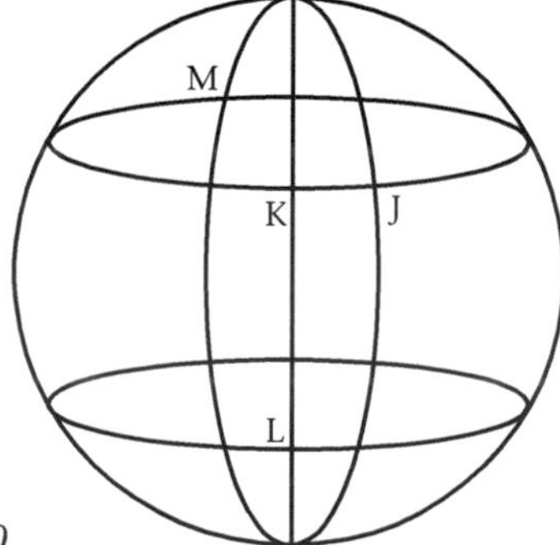

Fig. 24.50

[Take the radius of the earth to be 6 370 km and π to be 3.142]
The positions of two towns J and K are (15°N, 20°E) and (15°N, 70°W) respectively.

(a) Calculate, in kilometres, how far K is due West of J.

L is 2 560 km due South of K.

(b) Calculate the coordinates of L.

The position of another town M is (15°N, 110°E).

(c) Calculate the shortest distance between J and M, giving your answer in standard form.

13 In this question take the radius of the earth to be 6 400 km and π to be 3.142.

(a) The coordinates of the points P and Q on the earth's surface are (26°S, 25°W) and (60°N, 25°W) respectively.

Calculate
  (i) the shortest distance from P to Q.
  (ii) the circumference of the circle of latitude 60°N. (7 marks)

(b) Two tracking stations X and Y are both situated on latitude 60° N. Station X is situated at (60° N, 10° E) and station Y is situated west of X. The distance between X and Y along the latitude 60° N is 1 800 km. Calculate the position of the tracking station Y. (8 marks)

[CXC (General) June 92]

14

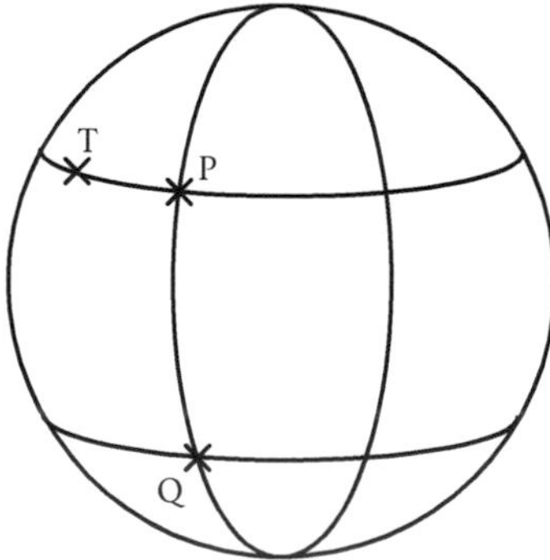

Fig. 24.51

(In this problem, use radius of earth $R = 6\,370$ km, $\pi = 3.14$.)
Fig. 24.51, **not drawn to scale**, shows the position of three towns, P, Q and T. P and Q are on the same circle of longitude. P and T are on the same circle of latitude.

P is located at (38°N, 35°W), and T at (38°N, 80°W).

The distance between P and Q, measured along the circle of longitude, is approximately 6890 km.
(a) Show that Q must be a town in the southern hemisphere.
(b) State the position of Q.
(c) Calculate the distance between P and T
[CXC (General) Jan 97] (9 marks)

15

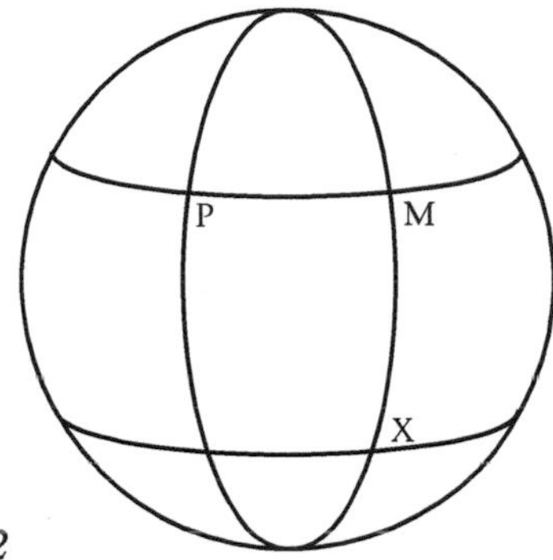

Fig. 24.52

M (40°N, 60°W) is due East of P and X is due South of M. Given that MX = 3 400 km and MP = MX, calculate to the nearest degree
(a) the position of X
(b) the position of P.
[Take the radius of the earth to be 6 400 km and π to be 3.142]

Revision test (Chapter 24)

Select the correct answer and write down your choice.

1. In a right-angled triangle, $\sin\alpha = \frac{15}{17}$. Then $\cos\alpha =$
   A $\frac{8}{17}$ B $\frac{8}{15}$ C $\frac{17}{15}$ D $\frac{15}{8}$
2. Given that $\sin 60° = \frac{\sqrt{3}}{2}$ and $\cos 60° = \frac{1}{2}$ then $\tan 60° =$
   A $\frac{1}{\sqrt{3}}$ B $\frac{\sqrt{3}}{4}$ C $\sqrt{3}$ D $\frac{\sqrt{3}}{2}$
3. In a triangle, $\sin^2\theta = \frac{16}{25}$. $\cos\theta =$
   A $\frac{9}{25}$ B $\frac{1}{5}$ C $\frac{3}{5}$ D $\frac{4}{5}$
4. 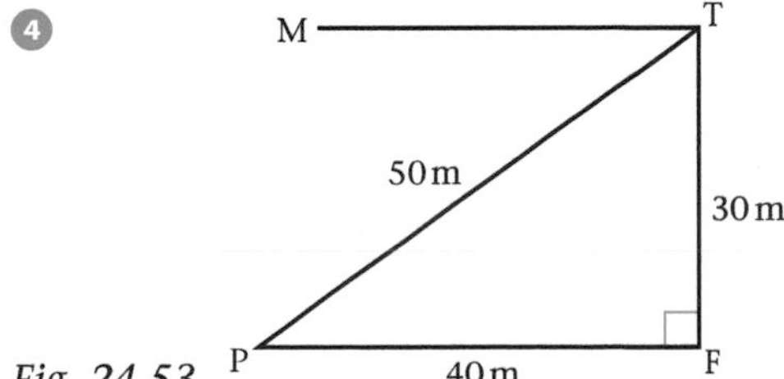

Fig. 24.53

In Fig. 24.53, the angle of elevation, E, from P is given by $\tan\widehat{E} =$
A $\frac{3}{5}$ B $\frac{3}{4}$ C $\frac{5}{4}$ D $\frac{4}{3}$

5. 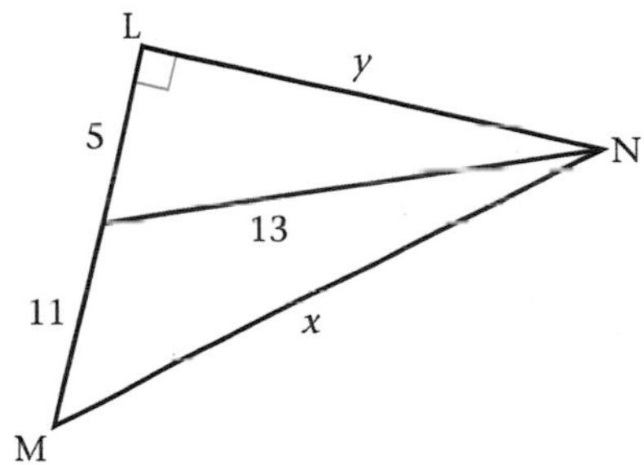

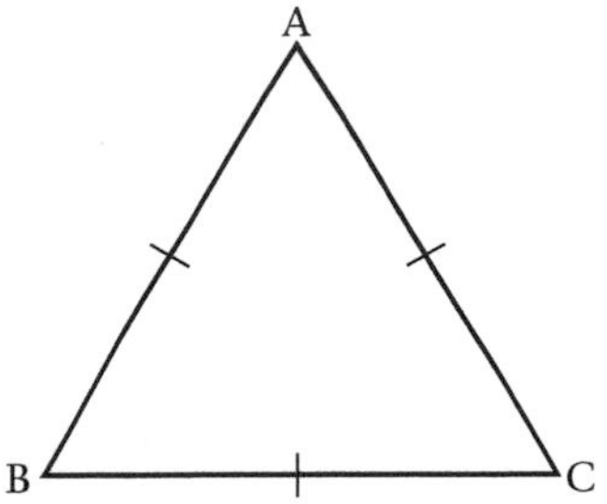

Fig. 24.54

In Fig. 24.54, given that RS = RT = ST = 6 cm, then the area of $\triangle$RST is
A 36 cm² B 18 cm²
C $18\sqrt{3}$ cm² D $9\sqrt{3}$ cm²

Questions 6 and 7 refer to Fig. 24.55.

Fig. 24.55

6. The value of y in Fig. 24.54. is
   A 8 B 9 C 12 D 18
7. The value of x in Fig. 24.54. is
   A $\sqrt{48}$ B $\sqrt{290}$ C 20 D 24
8. sin 230° =
   A −sin 40° B −sin 50°
   C sin 40° D sin 50°
9. 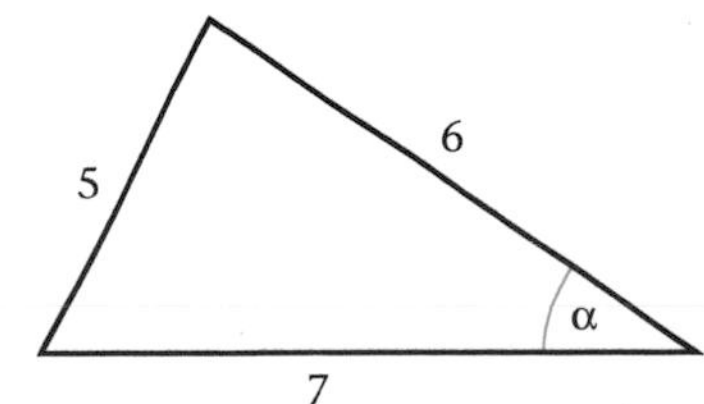

Fig. 24.56

In Fig. 24.56, the value of the angle α may be calculated from the expression given by

A $\dfrac{5^2 + 6^2 - 7^2}{2(5)(6)}$ B $\dfrac{5^2 + 6^2 + 7^2}{2(5)(7)}$

C $\dfrac{6^2 + 7^2 - 5^2}{2(6)(7)}$ D $\dfrac{5^2 + 6^2 + 7^2}{2(6)(7)}$

10 $\cos(260°) =$

A $-\sin 60°$ B $-\cos 60°$

C $\sin 60°$ D $\cos 60°$

11 The sides of a right-angled triangle are in the ratio $1 : 1 : \sqrt{2}$. Which of the following statements is *not* true?

A The triangle is an isosceles right-angled triangle.

B The sine of one of its acute angles is $\dfrac{1}{\sqrt{2}}$.

C The tangent of one of its acute angles is 1.

D The tangent of one of its acute angles is $\sqrt{2}$.

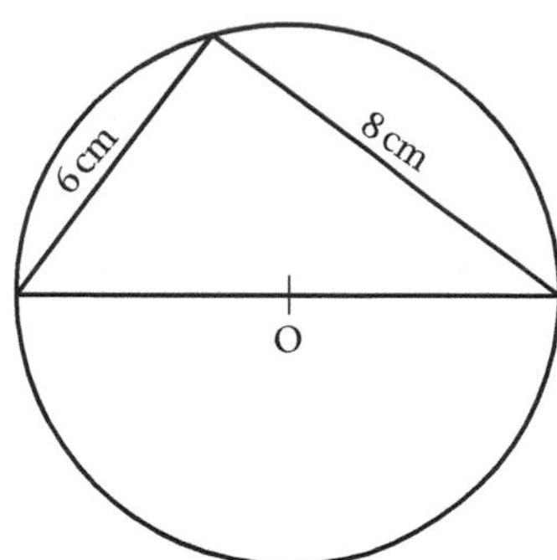

Fig. 24.57

In Fig. 24.57, O is the centre of the circle. The radius of the circle is

A $2\sqrt{7}$ cm B $\sqrt{7}$ cm

C 5 cm D 10 cm

13 A cone, on a base of radius 5 cm, has a slant height of 15 cm. The height of the cone is

A $\sqrt{8}$ cm B 10 cm

C $10\sqrt{2}$ cm D $5\sqrt{10}$ cm

14 M is on a bearing 026° from K, and N is on a bearing 126° from M. If KM = MN, then the bearing of N from K is

A 076° B 100° C 104° D 150°

15 X and Y lie on the same line of longitude in latitudes 25°N and 35°S respectively. The great-circle distance between X and Y in nautical miles is

A 600 B 3 600

C $\dfrac{20\,000}{3}$ D $\dfrac{10\,000}{9}$

16 A cube has edges of length 6 cm. The length of a diagonal of the cube is

A $\sqrt{72}$ B $\sqrt{108}$ C $\sqrt{180}$ D $\sqrt{216}$

17 Given that $\sin 25° = 0.423$, which of the following has the value 2 0.423?

A $\sin 155°$ B $\cos 155°$

C $\sin 115°$ D $\cos 115°$

18 Given that the circumference of the earth is 40 000 km, then the circumference of the parallel of latitude 60° is

A 20 000 km B $40\,000\sqrt{3}$ km

C $20\,000\sqrt{3}$ km C $80\,000\sqrt{3}$ km

Questions 19 and 20 refer to Fig. 24.58.
In Fig. 24.58, FT is a flag-pole. TJ and TK are light wires.

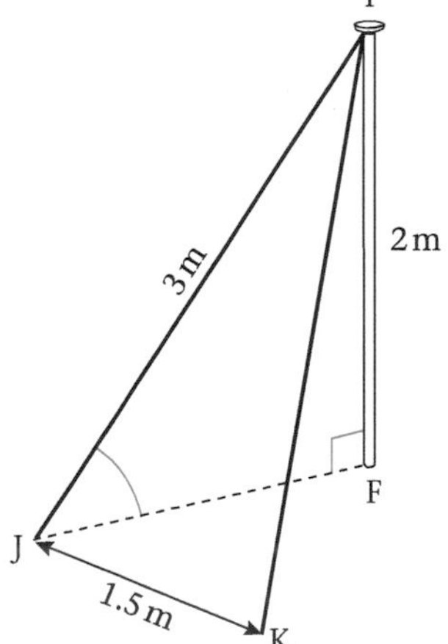

Fig. 24.58

19 Given that TF = 2 m and TK = 4 m then the sine of the angle of elevation of T from K is

A $\frac{1}{2}$ B $\frac{2}{3}$ C $\frac{3}{4}$ D $\frac{2}{5}$

20 Let θ be the angle between TJ and TK in Fig. 24.58, then $\cos\theta =$

A $\dfrac{22.75}{24}$ C $\dfrac{2}{3}$

B $\dfrac{-22.75}{24}$ D $\dfrac{1.5}{3}$

Chapter 25

Matrices and vectors

Matrices

A **matrix** is a set of elements, usually numbers, arranged in a rectangular pattern.

The **order of the matrix** is given in terms of the number of rows and columns. If there are r rows and c columns, the order of the matrix is r by c. Note that the number of rows is always given first. For example,

$\begin{pmatrix} 3 & 2 \\ 1 & 5 \\ 2 & 0 \end{pmatrix}$ is a **3 × 2 matrix** since it has 3 rows and 2 columns.

A matrix may be used as
(i) a store of information
(ii) an operator.

Example 1

The number of persons in four families is as follows: there were 5 in the Jones family, 3 in the Smith family, 4 in the Brown family and 2 in the Henry family. The average money budgeted per person on the following four items over three weeks is:

Food	\$150,	\$140,	\$160
Transport	\$50,	\$50,	\$50
Clothing	\$30,	\$40,	\$25
Recreation	\$25,	\$20,	\$30

Store each set of data in a matrix.

To represent the data on family members, we may write the information either as
(i) a row matrix, P, that is, a 1 × 4 matrix

$$\begin{matrix} & \text{J} & \text{S} & \text{B} & \text{H} \\ \text{P} = & (5 & 3 & 4 & 2) \end{matrix} \qquad \textit{(no. in family)}$$

or
(ii) a column matrix, Q, that is, a 4 × 1 matrix

no in family

$$\text{Q} = \begin{pmatrix} 5 \\ 3 \\ 4 \\ 2 \end{pmatrix} \begin{matrix} \text{(J)} \\ \text{(S)} \\ \text{(B)} \\ \text{(H)} \end{matrix}$$

The data on the money spent may be written either as
(i) a 4 × 3 matrix. Name this matrix S.

$$\begin{matrix} & \text{wk1} & \text{wk2} & \text{wk3} \end{matrix}$$
$$\text{S} = \begin{pmatrix} 150 & 140 & 160 \\ 50 & 50 & 50 \\ 30 & 40 & 25 \\ 25 & 20 & 30 \end{pmatrix} \begin{matrix} \text{(F)} \\ \text{(T)} \\ \text{(C)} \\ \text{(R)} \end{matrix}$$

or
(ii) a 3 × 4 matrix. Name this matrix W.

$$\begin{matrix} & \text{F} & \text{T} & \text{C} & \text{R} \end{matrix}$$
$$\text{W} = \begin{pmatrix} 150 & 50 & 30 & 25 \\ 140 & 50 & 40 & 20 \\ 160 & 50 & 25 & 30 \end{pmatrix} \begin{matrix} \text{(wk 1)} \\ \text{(wk 2)} \\ \text{(wk 3)} \end{matrix}$$

Addition and subtraction

Matrices can be added and subtracted only if they have the same number of rows and columns.

Matrices of the same order can be added or subtracted by combining corresponding elements.

Example 2

$$\begin{pmatrix} 5 \\ 9 \end{pmatrix} + \begin{pmatrix} 2 \\ -1 \end{pmatrix} = \begin{pmatrix} 5 \\ 9 \end{pmatrix} + \begin{pmatrix} 2 \\ (-1) \end{pmatrix} = \begin{pmatrix} 7 \\ 8 \end{pmatrix}$$

$$\begin{pmatrix} 3 & -4 \\ 0 & 2 \end{pmatrix} - \begin{pmatrix} 1 & 5 \\ -3 & 2 \end{pmatrix} = \begin{pmatrix} 3 - \ \ 1 & -4 - 5 \\ 0 - (-3) & 2 - 2 \end{pmatrix}$$
$$= \begin{pmatrix} 2 & -9 \\ 3 & 0 \end{pmatrix}$$

Example 3

Simplify

$$\begin{pmatrix} 2 \\ -1 \end{pmatrix} - \begin{pmatrix} -3 & 2 \\ 0 & -5 \end{pmatrix} + \begin{pmatrix} -1 & 8 \\ -5 & 7 \end{pmatrix}$$

$$\begin{pmatrix} 2 \\ -1 \end{pmatrix} - \begin{pmatrix} -3 & 2 \\ 0 & -5 \end{pmatrix} + \begin{pmatrix} -1 & 8 \\ -5 & 7 \end{pmatrix}$$
$$= \begin{pmatrix} 2 - (-3) + (-1) & 6 - 2 + 8 \\ 1 - 0 + (-5) & -4 - (-5) + 7 \end{pmatrix}$$
$$= \begin{pmatrix} 4 & 12 \\ -4 & 8 \end{pmatrix} = 4\begin{pmatrix} 1 & 2 \\ -1 & 2 \end{pmatrix}$$

In Example 3, notice that 4 is a factor of each element in the resultant matrix. The 4 can be taken out as a **scalar** factor.

Example 4

If $A = \begin{pmatrix} 3 & -2 \\ 1 & 0 \\ 0 & 4 \end{pmatrix}$ and $B = \begin{pmatrix} -5 & 2 \\ 2 & 3 \\ -1 & 0 \end{pmatrix}$

find

(a) A + B, (b) B + A, (c) A − B, (d) B − A, (e) 3A.

(a) $$A + B = \begin{pmatrix} 3 & -2 \\ 1 & 0 \\ 0 & 4 \end{pmatrix} + \begin{pmatrix} -5 & 2 \\ 2 & 3 \\ -1 & 0 \end{pmatrix} = \begin{pmatrix} 3+(-5) & (-2)+2 \\ 1+2 & 0+3 \\ 0+(-1) & 4+0 \end{pmatrix} = \begin{pmatrix} -2 & 0 \\ 3 & 3 \\ -1 & 4 \end{pmatrix}$$

(b) $$B + A = \begin{pmatrix} -5 & 2 \\ 2 & 3 \\ -1 & 0 \end{pmatrix} + \begin{pmatrix} 3 & -2 \\ 1 & 0 \\ 0 & 4 \end{pmatrix} = \begin{pmatrix} -5+3 & 2+(-2) \\ -2+1 & 3+0 \\ -1+0 & 0+4 \end{pmatrix} = \begin{pmatrix} -2 & 0 \\ 3 & 3 \\ -1 & 4 \end{pmatrix}$$

(c) $$A - B = \begin{pmatrix} 3 & -2 \\ 1 & 0 \\ 0 & 4 \end{pmatrix} - \begin{pmatrix} -5 & 2 \\ 2 & 3 \\ -1 & 0 \end{pmatrix} = \begin{pmatrix} 3-(-5) & (-2)-2 \\ 1-2 & 0-3 \\ 0-(-1) & 4-0 \end{pmatrix} = \begin{pmatrix} 8 & -4 \\ -1 & -3 \\ 1 & 4 \end{pmatrix}$$

(d) $$B - A = \begin{pmatrix} -5 & 2 \\ 2 & 3 \\ -1 & 0 \end{pmatrix} - \begin{pmatrix} 3 & -2 \\ 1 & 0 \\ 0 & 4 \end{pmatrix} = \begin{pmatrix} -5-3 & 2-(-2) \\ 2-1 & 3-0 \\ -1-0 & 0-4 \end{pmatrix} = \begin{pmatrix} -8 & 4 \\ 1 & 3 \\ -1 & 4 \end{pmatrix}$$

(c) $$3A = 3\begin{pmatrix} 3 & -2 \\ 1 & 0 \\ 0 & 4 \end{pmatrix} = \begin{pmatrix} 3\times 3 & 3\times(-2) \\ 3\times 1 & 3\times 0 \\ 3\times 0 & 3\times 4 \end{pmatrix} = \begin{pmatrix} 9 & -6 \\ 3 & 0 \\ 0 & 12 \end{pmatrix}$$

Revision notes:

1 Matrices A and B are of the same *order*. They are both 3 × 2 matrices, i.e. 3 rows by 2 columns.

2 Matrices can be added or subtracted only if they are of the same order, in which case corresponding elements are added or subtracted as in parts (a)–(d) of Example 4.

3 Addition of matrices is commutative so that A + B = B + A, but subtraction of matrices is *not* commutative so that, in general, A − B ≠ B − A.

4 When a matrix is multiplied by a *scalar*, as in part (e), the scalar multiplies every element of the matrix.

Multiplication

Matrices can be multiplied only if the number of columns in the first or **pre-multiplying** matrix is the same as the number of rows in the second or **post-multiplying** matrix. A $p \times q$ matrix will multiply a $q \times r$ matrix to give a $p \times r$ product. In general AB ≠ BA, where A and B are matrices.

Table 25.1

Order of matrices in product	General case	Numerical example
(1 × 2) × (2 × 1)	$(a\ b)\begin{pmatrix} p \\ q \end{pmatrix} = (ap + bq)$	$(-2\ 3)\begin{pmatrix} 5 \\ 1 \end{pmatrix} = (-2 \times 5 + 3 \times 1) = (-7)$
(2 × 1) × (1 × 2)	$\begin{pmatrix} a \\ b \end{pmatrix}(p\ q) = \begin{pmatrix} ap & aq \\ bp & bq \end{pmatrix}$	$\begin{pmatrix} -4 \\ 7 \end{pmatrix}(-1\ 2) = \begin{pmatrix} 4 & -8 \\ -7 & 14 \end{pmatrix}$
(2 × 2) × (2 × 2)	$\begin{pmatrix} a & b \\ c & d \end{pmatrix}\begin{pmatrix} p & q \\ r & s \end{pmatrix} =$	$\begin{pmatrix} 5 & 1 \\ 2 & 7 \end{pmatrix}\begin{pmatrix} -2 & 0 \\ 3 & 1 \end{pmatrix}$
	$\begin{pmatrix} ap + br & aq + bs \\ cp + dr & cq + ds \end{pmatrix}$	$= \begin{pmatrix} -7 & 1 \\ 17 & 7 \end{pmatrix}$

To multiply matrices, find the sum of the products of corresponding elements in each row of the pre-multiplying matrix and each column of the post-multiplying matrix. The examples in Table 25.1 demonstrate the method of matrix multiplication.

Example 5

If $M = \begin{pmatrix} -2 & 4 \\ 3 & 5 \end{pmatrix}$ and $N = \begin{pmatrix} 6 & 0 \\ -1 & 2 \end{pmatrix}$, find the matrix products (a) MN (b) NM (c) M^2.

(a) $$MN = \begin{pmatrix} -2 & 4 \\ 3 & 5 \end{pmatrix}\begin{pmatrix} 6 & 0 \\ -1 & 2 \end{pmatrix}$$
$$= \begin{pmatrix} (-2)\times 6 + 4\times(-1) & (-2)\times 0 + 4\times 2 \\ 3\times 6 + 5\times(-1) & 3\times 0 + 5\times 2 \end{pmatrix}$$
$$= \begin{pmatrix} -12 + (-4) & 0 + 8 \\ 18 + (-5) & 0 + 10 \end{pmatrix} = \begin{pmatrix} -16 & 8 \\ 13 & 10 \end{pmatrix}$$

(b) $$NM = \begin{pmatrix} 6 & 0 \\ -1 & 2 \end{pmatrix}\begin{pmatrix} -2 & 4 \\ 3 & 5 \end{pmatrix}$$
$$= \begin{pmatrix} -12 + 0 & 24 + 0 \\ 2 + 6 & -4 + 10 \end{pmatrix} = \begin{pmatrix} -12 & 24 \\ 8 & 6 \end{pmatrix}$$

(c) $$M^2 = \begin{pmatrix} -2 & 4 \\ 3 & 5 \end{pmatrix}\begin{pmatrix} -2 & 4 \\ 3 & 5 \end{pmatrix}$$
$$= \begin{pmatrix} 4 + 12 & -8 + 20 \\ -6 + 15 & 12 + 25 \end{pmatrix} = \begin{pmatrix} 16 & 12 \\ 9 & 37 \end{pmatrix}$$

Revision notes:

1. In Example 5(a) M *pre-multiplies* N.
   In Example 5(b) M *post-multiplies* N.
2. Matrices can be multiplied only if there are as many columns in the first matrix as there are rows in the second matrix.
3. A $p \times q$ matrix will multiply a $q \times r$ matrix to give a $p \times r$ product:
   $(p \times \boxed{q) \times (q} \times r) \rightarrow (p \times r)$
4. In general $AB \neq BA$ where A and B are matrices. Multiplication of matrices is *not* commutative.

Example 6

Using the data given in Example 1, calculate the money budgeted for each item by each family for week 1.

In order to do the calculation, we need to write the family data as given in the column matrix Q, and consider the first row of matrix W as a 1×4 matrix. Then the budgeted amounts are given by the product $Q \times W$ (row 1): thus

$$\begin{matrix} (J) \\ (S) \\ (B) \\ (H) \end{matrix}\begin{pmatrix} 5 \\ 3 \\ 4 \\ 2 \end{pmatrix}\ \overset{\begin{matrix} F & T & C & R \end{matrix}}{\begin{pmatrix} 150 & 50 & 30 & 25 \end{pmatrix}}\ (\text{wk } 1)$$

which gives the result

$$\begin{matrix} \\ (J) \\ (S) \\ (B) \\ (H) \end{matrix}\begin{matrix} \begin{matrix} F & T & C & R \end{matrix} \\ \begin{pmatrix} 750 & 250 & 150 & 125 \\ 450 & 150 & 90 & 75 \\ 600 & 200 & 120 & 100 \\ 300 & 100 & 60 & 50 \end{pmatrix} \end{matrix}$$

Note that $(4 \times 1) \times (1 \times 4) \Rightarrow (4 \times 4)$

Exercise 25a

1. Simplify the following, giving each result as a single matrix.
   (a) $\begin{pmatrix} 3 & 9 \\ 2 & 1 \end{pmatrix} + \begin{pmatrix} 4 & 0 \\ -6 & 2 \end{pmatrix}$
   (b) $\begin{pmatrix} 2 \\ 6 \end{pmatrix} + \begin{pmatrix} 10 \\ 3 \end{pmatrix}$
   (c) $\begin{pmatrix} 5 & -1 \\ 1 & -1 \\ -2 & 7 \end{pmatrix} - \begin{pmatrix} 2 & 6 \\ 1 & -7 \\ 3 & 0 \end{pmatrix}$
   (d) $\begin{pmatrix} -4 & 2 \\ 1 & -2 \end{pmatrix} - \begin{pmatrix} 6 & 9 \\ -2 & -3 \end{pmatrix}$
   (e) $\begin{pmatrix} 3 \\ 9 \end{pmatrix} - \begin{pmatrix} 0 \\ 5 \end{pmatrix} + \begin{pmatrix} 2 \\ 6 \end{pmatrix}$
   (f) $\begin{pmatrix} 3 & -2 \\ 1 & 5 \end{pmatrix} + \begin{pmatrix} -1 & 3 \\ -2 & -1 \end{pmatrix} - \begin{pmatrix} 2 & 4 \\ 7 & -2 \end{pmatrix}$
2. If $A = \begin{pmatrix} -1 & 5 \\ 2 & 3 \end{pmatrix}$ and $B = \begin{pmatrix} 6 & 0 \\ 4 & -8 \end{pmatrix}$ find
   (a) 3A (b) 2B (c) −2A
   (d) $\frac{1}{2}$B (e) 2A + B (f) A − 3B
3. Express each of the following products as a single-element matrix.
   (a) $(4\ \ 5)\begin{pmatrix} 6 \\ 1 \end{pmatrix}$ (b) $(2\ \ 3\ \ -1)\begin{pmatrix} 2\frac{1}{2} \\ -2\frac{1}{2} \\ 1\frac{1}{2} \end{pmatrix}$
   (c) $(3\ \ -1)\begin{pmatrix} x \\ y \end{pmatrix}$ (d) $(4\ \ 5\ \ 6)\begin{pmatrix} x \\ y \\ z \end{pmatrix}$
4. Find x and y if
   $$4\begin{pmatrix} 5 & x \\ -1 & 1 \end{pmatrix} - 3\begin{pmatrix} 6 & x \\ 2 & 2 \end{pmatrix} = 2\begin{pmatrix} 1 & 4 \\ y & -1 \end{pmatrix}$$

5 Express each of the following products as a single matrix.

(a) $\begin{pmatrix} 5 & 1 \\ 8 & 2 \end{pmatrix}\begin{pmatrix} -2 \\ 3 \end{pmatrix}$

(b) $(2\ \ 5)\begin{pmatrix} 6 & 2 & 3 \\ 1 & 0 & -1 \end{pmatrix}$

(c) $\begin{pmatrix} 4 & 9 \\ -2 & 5 \end{pmatrix}\begin{pmatrix} 1 & 3 \\ 0 & -1 \end{pmatrix}$

(d) $\begin{pmatrix} 1 & 3 \\ 0 & -1 \end{pmatrix}\begin{pmatrix} 4 & 9 \\ -2 & 5 \end{pmatrix}$

(e) $\begin{pmatrix} 6 & 3 \\ -1 & 2 \end{pmatrix}^2$

(f) $\begin{pmatrix} 1 & 0 \\ 0 & 1 \end{pmatrix}\begin{pmatrix} 2 & 3 & 0 \\ 5 & -1 & 6 \end{pmatrix}$

6 $A = \begin{pmatrix} 3 & 5 \\ 5 & 0 \end{pmatrix}$ and $B = \begin{pmatrix} 6 & 4 \\ -1 & -2 \end{pmatrix}$.

(a) Calculate the matrix product AB.

(b) If $C = \begin{pmatrix} x & -8 \\ 4 & y \end{pmatrix}$ and $2A + C = AB$, calculate the values of x and y.

(8 marks)

[CXC (General) June 90]

7 Given that $M = \begin{pmatrix} 1 & -2 \\ 2 & 3 \end{pmatrix}$, $N = \begin{pmatrix} a & 0 \\ -3 & a+b \end{pmatrix}$, and $R = \begin{pmatrix} -1 & -2 \\ -1 & 2 \end{pmatrix}$, and that $M + N = R$, calculate the values of a and b.

8 Let $A = \begin{pmatrix} 1 & 0 \\ -2 & 3 \end{pmatrix}$, $B = \begin{pmatrix} 3 & -5 \\ -1 & 2 \end{pmatrix}$.

Calculate

(a) A + B (1 mark)

(b) BA (2 marks)

[CXC (General) Jan 96]

9 $X = \begin{pmatrix} x & -1 \\ 0 & 2 \end{pmatrix}$, and $Y = \begin{pmatrix} 1 & 0 \\ y & 3 \end{pmatrix}$

Given that $XY = \begin{pmatrix} 5 & -3 \\ -4 & 6 \end{pmatrix}$, find the values of x and y.

10 The bills for utilities per month are calculated according to the rates and the number of units used. In January and February, a family uses the following chargeable units – for Electricity: 45, 30; and for Water: 40, 38 respectively; the rates charged per unit for the utilities are: Electricity: \$6; and Water: \$5.

(a) Write a matrix to represent the chargeable units used for the two months, labelling the rows and the columns.

(b) Writing either a row matrix or a column matrix to represent the rates charged, construct a matrix product to calculate the amount paid for each utility for the two months.

(c) State the amount paid for each utility.

Identity and inverse

I is the 2×2 **identity matrix** where

$$I = \begin{pmatrix} 1 & 0 \\ 0 & 1 \end{pmatrix}$$

Any 2×2 matrix is left unchanged when pre- or post-multiplied by I.

In most cases a matrix M has an **inverse**, M^{-1}, such that $M \times M^{-1} = M^{-1} \times M = I$.

If $M = \begin{pmatrix} a & b \\ c & d \end{pmatrix}$

then $\mathbf{M^{-1}} = \dfrac{\mathbf{1}}{\boldsymbol{ad - bc}}\begin{pmatrix} \boldsymbol{d} & \boldsymbol{-b} \\ \boldsymbol{-c} & \boldsymbol{a} \end{pmatrix}$

The cross-product difference, $ad - bc$, is the **determinant** of the given matrix.

Example 7

Find the inverse of the following where possible.

(a) $\begin{pmatrix} 5 & -2 \\ 9 & -3 \end{pmatrix}$ (b) $\begin{pmatrix} 10 & 3 \\ 7 & 2 \end{pmatrix}$ (c) $\begin{pmatrix} 6 & 9 \\ 4 & 6 \end{pmatrix}$

(a) The determinant of the given matrix
$= 5 \times (-3) - 9 \times (-2)$
$= -15 + 18 = 3$
Its inverse is $\frac{1}{3}\begin{pmatrix} -3 & 2 \\ -9 & 5 \end{pmatrix}$.

(b) The determinant of the given matrix
$= 10 \times 2 - 7 \times 3$
$= 20 - 21 = -1$
Its inverse is $-1\begin{pmatrix} 2 & -3 \\ -7 & 10 \end{pmatrix}$
or $\begin{pmatrix} -2 & 3 \\ 7 & -10 \end{pmatrix}$.

(c) The determinant of the given matrix
$= 6 \times 6 - 9 \times 4$
$= 36 - 36 = 0$

The inverse of the given matrix would contain the undefined fraction $\frac{1}{0}$. This is an example of a **singular** matrix; such a matrix has *no* inverse.

Matrices may be used as the operators in solving simultaneous linear equations. The simultaneous linear equations are written as a matrix equation so

that $\quad ax + by = p$

and $\quad cx + dy = q$

become $\begin{pmatrix} a & b \\ c & d \end{pmatrix}\begin{pmatrix} x \\ y \end{pmatrix} = \begin{pmatrix} p \\ q \end{pmatrix}$

Multiplying both sides of the matrix equation by the inverse matrix $\frac{1}{ad - bc}\begin{pmatrix} d & -b \\ -c & a \end{pmatrix}$ the matrix equation becomes

$$\begin{pmatrix} 1 & 0 \\ 0 & 1 \end{pmatrix}\begin{pmatrix} x \\ y \end{pmatrix} = \frac{1}{ad - bc}\begin{pmatrix} d & -b \\ -c & a \end{pmatrix}\begin{pmatrix} p \\ q \end{pmatrix}$$

$$\begin{pmatrix} x \\ y \end{pmatrix} = \frac{1}{ad - bc}\begin{pmatrix} d & -b \\ -c & a \end{pmatrix}\begin{pmatrix} p \\ q \end{pmatrix}$$

Example 8

Solve the following simultaneous equations using a matrix method.

$$3x - 2y = 8$$
$$x + y = 1$$

$$\begin{pmatrix} 3 & -2 \\ 1 & 1 \end{pmatrix}\begin{pmatrix} x \\ y \end{pmatrix} = \begin{pmatrix} 8 \\ 1 \end{pmatrix}$$

$$\begin{pmatrix} x \\ y \end{pmatrix} = \frac{1}{5}\begin{pmatrix} 1 & 2 \\ -1 & 3 \end{pmatrix}\begin{pmatrix} 8 \\ 1 \end{pmatrix}$$

$$\begin{pmatrix} x \\ y \end{pmatrix} = \frac{1}{5}\begin{pmatrix} 10 \\ -5 \end{pmatrix}$$

$$= \begin{pmatrix} 2 \\ -1 \end{pmatrix}$$

Hence $x = 2$, $y = -1$

Exercise 25b

1. Find the determinant of the matrix $\begin{pmatrix} -5 & 2 \\ 4 & -3 \end{pmatrix}$. Hence write down the inverse of the matrix.

2. Find the value of x for which the matrix $\begin{pmatrix} 5x - 3 & 7 \\ 2x + 4 & 2 \end{pmatrix}$ has no inverse.

3. $A = \begin{pmatrix} 2 & 0 \\ 3 & 1 \end{pmatrix}$ and $B = \begin{pmatrix} 1 & 2 \\ -1 & 3 \end{pmatrix}$.
   (a) Find $A + 2B$.
   (b) Given that $A\begin{pmatrix} x \\ 2 \end{pmatrix} = \begin{pmatrix} 8 \\ 2y \end{pmatrix}$, find the value of x and y.

4. Give that $A = \begin{pmatrix} 3 & -1 \\ 2 & 0 \end{pmatrix}$, $B = \begin{pmatrix} 0 & \frac{1}{2} \\ -1 & m \end{pmatrix}$, $C = \begin{pmatrix} -9 & 4 \\ 4 & n \end{pmatrix}$, find
   (a) A^2, (b) m if $B = A^{-1}$, (c) n if A and C have equal determinants.

5. Given that the value of the determinant of the matrix $\begin{pmatrix} a & 4 \\ -1 & 3 \end{pmatrix}$ is 10, find the value of a. Hence write down the inverse of the matrix.

6. Find a, b, c such that
   $\begin{pmatrix} a & b \\ 0 & 2 \end{pmatrix}\begin{pmatrix} 0 & 3 \\ 1 & -1 \end{pmatrix} = \begin{pmatrix} 1 & 9 \\ 5 & 0 \end{pmatrix} - \begin{pmatrix} 4 & -6 \\ 3 & 2c \end{pmatrix}$.

7. (a) Given that $P = \begin{pmatrix} 2 & 3 \\ 0 & 1 \end{pmatrix}\begin{pmatrix} 1 & -4 \\ 3 & 0 \end{pmatrix}$, find the matrix P.
   (b) Find the inverse of the matrix $\begin{pmatrix} 3 & 5 \\ 2 & 4 \end{pmatrix}$.
   (c) Given that R is a 2×2 matrix such that $R + \begin{pmatrix} 2 & 0 \\ 0 & 2 \end{pmatrix} R = \begin{pmatrix} 1 & 2 \\ 3 & 4 \end{pmatrix}$, find R.

8. Find two values of k such that $\begin{pmatrix} 2k + 2 & k \\ 4k - 3 & k + 3 \end{pmatrix}$ is a singular matrix.

9. (a) Write down the inverse of the matrix $\begin{pmatrix} -1 & 4 \\ 1 & 3 \end{pmatrix}$.
   (b) Hence or otherwise find x and y if $\begin{pmatrix} -1 & 4 \\ 1 & 3 \end{pmatrix}\begin{pmatrix} x \\ y \end{pmatrix} = \begin{pmatrix} 13 \\ 1 \end{pmatrix}$.

10. Express the simultaneous equations
    $2x + 5y = 7$
    $x - y = 7$
    as a single matrix equation
    $M\begin{pmatrix} x \\ y \end{pmatrix} = \begin{pmatrix} 7 \\ 7 \end{pmatrix}$,
    where M is a 2×2 matrix. Pre-multiply both sides of the matrix equation by M^{-1} to find the values of x and y.

11. (a) Calculate the determinant of the matrix $\begin{pmatrix} 4 & 3 \\ -1 & \frac{1}{2} \end{pmatrix}$
    (b) Find the inverse of the matrix $\begin{pmatrix} 4 & 3 \\ -1 & \frac{1}{2} \end{pmatrix}$.
    (c) Hence, use a matrix method to solve simultaneously the pair of equations
    $8x + 6y = 28$
    $-2x + y = 8$
    [CXC (General) June 88]

12 Given $V = \begin{pmatrix} 3 & -2 \\ 5 & -3 \end{pmatrix}$ and $W = \begin{pmatrix} x \\ y \end{pmatrix}$,

(a) solve the matrix equation $VW = \begin{pmatrix} 2 \\ 5 \end{pmatrix}$ for x and y.

(b) write the algebraic equations that are equivalent to the matrix equation.

13 (a) Given the matrix $W = \begin{pmatrix} x & 1 \\ 0 & y \end{pmatrix}$, determine the values of x and y for which the matrix has no inverse.

(b) If $M = \begin{pmatrix} 2 & 3 \\ -1 & 0 \end{pmatrix}$ and $N = \begin{pmatrix} 1 & 6 \\ -2 & -3 \end{pmatrix}$,

(i) calculate the value of the determinant of M

(ii) show that $M^2 = N$.

14 3 books and 2 tapes cost \$23,
and 2 books and 3 tapes cost \$22.
Given that \$$b$ and \$$t$ are the costs of 1 book and 1 tape respectively,

(a) write two algebraic equations to represent this information

(b) represent these equations as a matrix equation $V \begin{pmatrix} b \\ t \end{pmatrix} = M$

(c) Determine V^{-1}, and hence find the values of b and t.

Vectors

A **vector** is any quantity which has direction as well as size and is written as a single column matrix or column vector, e.g. $\overrightarrow{AB} = \begin{pmatrix} x \\ y \end{pmatrix}$

Magnitude

The **magnitude** or size of $\overrightarrow{AB}$ is represented by the length of the line segment AB. This is written as $|\overrightarrow{AB}|$ and is called the **modulus** of $\overrightarrow{AB}$. In Fig. 18.1 (p. 233),

$$\overrightarrow{AB} = \begin{pmatrix} 4 \\ 1 \end{pmatrix}$$

$$|\overrightarrow{AB}| = \sqrt{4^2 + 1^2} \qquad \textit{(Pythagoras)}$$

$$= \sqrt{17}$$

The modulus of a vector is always positive.
A vector of unit length is called a **unit vector**.
Using matrix arithmetic, the operations addition, subtraction and multiplication by a **scalar** quantity can be performed on vectors.

Addition and subtraction

Vectors may be added and subtracted. For example,

if $\mathbf{p} = \begin{pmatrix} -9 \\ 0 \end{pmatrix}$ and $\mathbf{q} = \begin{pmatrix} 6 \\ -2 \end{pmatrix}$

$$\text{then } \mathbf{p} + \mathbf{q} = \begin{pmatrix} -9 \\ 0 \end{pmatrix} + \begin{pmatrix} 6 \\ -2 \end{pmatrix}$$

$$= \begin{pmatrix} -9 + 6 \\ 0 + (-2) \end{pmatrix} = \begin{pmatrix} -3 \\ -2 \end{pmatrix}$$

$$\text{and } \mathbf{p} - \mathbf{q} = \begin{pmatrix} -9 \\ 0 \end{pmatrix} - \begin{pmatrix} 6 \\ -2 \end{pmatrix}$$

$$= \begin{pmatrix} -9 \\ 0 \end{pmatrix} + \left[- \begin{pmatrix} 6 \\ -2 \end{pmatrix} \right]$$

$$= \begin{pmatrix} -9 + (-6) \\ 0 + (+2) \end{pmatrix} = \begin{pmatrix} -15 \\ 2 \end{pmatrix}$$

Scalar multiplication

If a vector $\overrightarrow{AB}$ is multiplied by a **scalar** k, where k is any number, the result is a vector k times as big as $\overrightarrow{AB}$.
Two vectors are parallel if one vector is a scalar multiple of the other.

If $\overrightarrow{AB} = \begin{pmatrix} 4 \\ 1 \end{pmatrix}$

then $3\overrightarrow{AB} = \begin{pmatrix} 4 \\ 1 \end{pmatrix} = \begin{pmatrix} 12 \\ 3 \end{pmatrix}$

and $2\frac{1}{2}\overrightarrow{AB} = \begin{pmatrix} 4 \\ 1 \end{pmatrix} = \begin{pmatrix} -2 \\ -\frac{1}{2} \end{pmatrix}$

The effect of scalars can be summarised as follows:

1 If $\mathbf{a} = k\mathbf{b}$ then $\mathbf{a}$ is k times as big as $\mathbf{b}$ and parallel to it.

2 If $h\mathbf{a} = k\mathbf{b}$ then $\mathbf{a} \parallel \mathbf{b}$ or $h = 0$ and $k = 0$.

Exercise 25c

1 If $\mathbf{p} = \begin{pmatrix} 3 \\ 4 \end{pmatrix}$, $\mathbf{q} = \begin{pmatrix} 3 \\ -1 \end{pmatrix}$, $\mathbf{r} = \begin{pmatrix} -4 \\ 0 \end{pmatrix}$, express each of the following as a single column vector.

(a) $5\mathbf{p}$ (b) $-3\mathbf{q}$
(c) $\frac{1}{2}\mathbf{r}$ (d) $\mathbf{p} + \mathbf{q}$
(e) $\mathbf{r} - \mathbf{p}$ (f) $\mathbf{p} - \mathbf{r}$
(g) $3\mathbf{p} + \mathbf{r}$ (h) $\mathbf{p} - 2\mathbf{q}$
(i) $5\mathbf{p} - 4\mathbf{q} + \mathbf{r}$ (j) $3\mathbf{p} + \mathbf{q} - 6\mathbf{r}$

2 Given $\mathbf{p}$, $\mathbf{q}$, $\mathbf{r}$ of question 1, evaluate the following, leaving the answers in surd from where necessary.

(a) $|\mathbf{p}|$ (b) $|\mathbf{q}|$ (c) $|\mathbf{r}|$
(d) $|\mathbf{p} + \mathbf{r}|$ (e) $|\mathbf{q} + \mathbf{r}|$ (f) $|\mathbf{p} - \mathbf{q}|$

3 Find the length of $\overrightarrow{AB}$ and say whether it is a unit vector when A and B are

(a) (3, 5), (3, 6) (b) (3, 2), (4, 2)
(c) (5, 1), (4, 2) (d) (5, 7), (7, 6)
(e) (1, 4), (0, 4) (f) (8, 4), (7, 4)

4 (a) Find vector **q** such that $\begin{pmatrix}1\\3\end{pmatrix} - \mathbf{q} = \begin{pmatrix}9\\-3\end{pmatrix}$.

(b) Hence find $|\mathbf{q}|$.

5 Fig. 25.1 shows two vectors **a** and **b**.

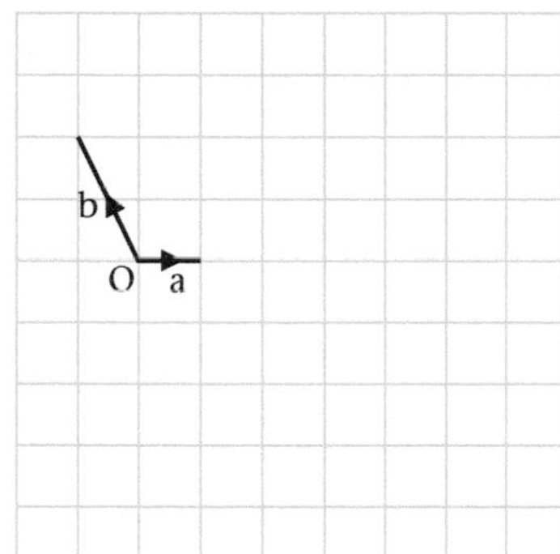

Fig. 25.1

Copy Fig. 25.1 on squared paper. Draw and clearly label the lines OP, OQ, OR, OS such that

(a) $\overrightarrow{OP} = 4\mathbf{a}$, (b) $\overrightarrow{OQ} = -2\mathbf{b}$,
(c) $\overrightarrow{OR} = 3\mathbf{a} + 2\mathbf{b}$, (d) $\overrightarrow{OS} = 5\mathbf{a} - 3\mathbf{b}$.

6 PQRS is a rhombus. $\overrightarrow{PQ}$ is shown in Fig. 25.2 and $\overrightarrow{OR}$ is the column vector $\begin{pmatrix}4\\1\end{pmatrix}$.

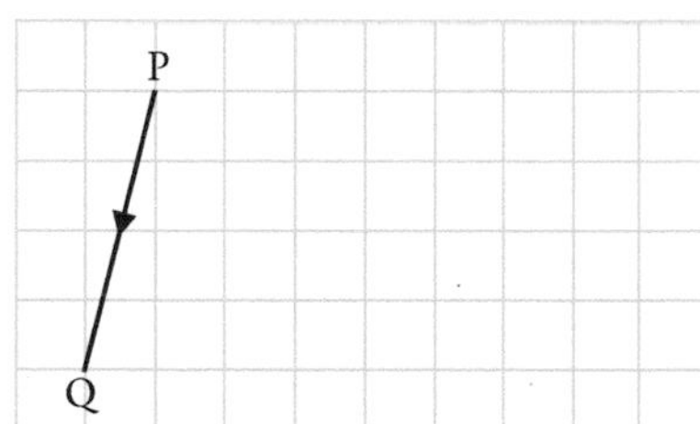

Fig. 25.2

(a) On a copy of Fig. 25.2, mark and label the points R and S.
(b) Express the following as column vectors.
(i) $\overrightarrow{SR}$ (ii) $\overrightarrow{PR}$ (iii) $\overrightarrow{QS}$

7 OABC is a trapezium such that O is the origin,

$\overrightarrow{OA} = \begin{pmatrix}2\\5\end{pmatrix}$, $\overrightarrow{OC} = 2\overrightarrow{AB} = \begin{pmatrix}6\\2\end{pmatrix}$.

(a) On squared paper, mark and label the points O, A, B and C.
(b) Express the following as column vectors.
(i) $\overrightarrow{OB}$ (ii) $\overrightarrow{BC}$ (iii) $\overrightarrow{CA}$

8 Given that **a** and **b** are two vectors in the same plane,

$\overrightarrow{OP} = 3\mathbf{a} + 2\mathbf{b}$
$\overrightarrow{OQ} = 5\mathbf{a} - 3\mathbf{b}$
$\overrightarrow{OR} = \mathbf{a} + 7\mathbf{b}$

(a) express the vector $\overrightarrow{PQ}$ in terms of **a** and **b**,
(b) show that the points P, Q and R lie on a straight line and indicate, on a diagram, their relative positions.

[CXC (General) June 88]

9 $\overrightarrow{OA} = \begin{pmatrix}-4\\9\end{pmatrix}$, $\overrightarrow{OB} = \begin{pmatrix}6\\-3\end{pmatrix}$ and M is the mid-point of AB. Express the following as column vectors: (a) $\overrightarrow{AB}$ (b) $\overrightarrow{AM}$ (c) $\overrightarrow{OM}$

10 In Fig. 25.3, C, D and F are the mid-points of BE, CE and AE respectively. $\overrightarrow{AB} = \mathbf{a}$ and $\overrightarrow{AC} = \mathbf{a} + \mathbf{b}$.

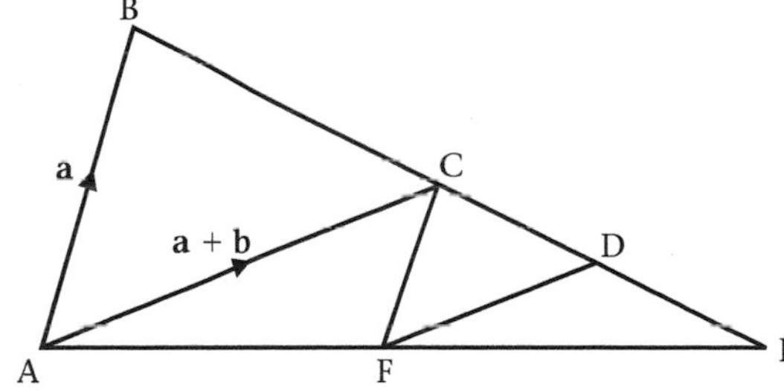

Fig. 25.3

Write down as simply as possible in terms of **a** and **b**, expressions for the following.

(a) $\overrightarrow{BC}$ (b) $\overrightarrow{CF}$ (c) $\overrightarrow{FD}$
(d) $\overrightarrow{DE}$ (e) $\overrightarrow{FE}$ (f) $\overrightarrow{BF}$

11 $\overrightarrow{OP} = \begin{pmatrix}-3\\4\end{pmatrix}$ and $\overrightarrow{OQ} = \begin{pmatrix}1\\q\end{pmatrix}$. Find

(a) $|\overrightarrow{OP}|$,
(b) q if O, P and Q are vertices of a square OPQR.

12 It is given that $\mathbf{u} = 4\mathbf{a} + 3\mathbf{b}$, $\mathbf{v} = 5\mathbf{a} - \mathbf{b}$ and $\mathbf{w} = h\mathbf{a} + (h + k)\mathbf{b}$, where h and k are constants. If $\mathbf{w} = 3\mathbf{u} - 2\mathbf{v}$, calculate the values of h and k.

13 If $\overrightarrow{OA} = 3\mathbf{p} - 2\mathbf{q}$, $\overrightarrow{OB} = \mathbf{p} + 7\mathbf{q}$ and $\overrightarrow{AB} = 2m\mathbf{p} + (m - n)\,\mathbf{q}$, find the values of m and n.

14 Triangle ABC has coordinates A(1, 0), B(0, 2), C(1, 3). X is the point (4, 4). $\triangle$ABC is displaced by vector $\overrightarrow{AX}$. Find
(a) the coordinates of the image of $\triangle$ABC,
(b) the modulus of $\overrightarrow{AX}$.

15 In Fig. 25.4, OABC is a square and X, Y, Z are the mid-points of OC, CB, BA respectively. $\overrightarrow{OA} = \mathbf{a}$ and $\overrightarrow{OC} = \mathbf{c}$.
(a) Express $\overrightarrow{XY}$ and $\overrightarrow{CZ}$ in terms of **a** and **c**.
(b) If $\overrightarrow{XP} = h\overrightarrow{XY}$, express $\overrightarrow{XP}$ in terms of **a**, **c** and h.
(c) If $\overrightarrow{CP} = k\overrightarrow{CZ}$. express $\overrightarrow{CP}$ in terms of **a**, **c** and k.
(d) Use the fact that $\overrightarrow{XP} = \overrightarrow{XC} + \overrightarrow{CP}$ to evaluate h and k.

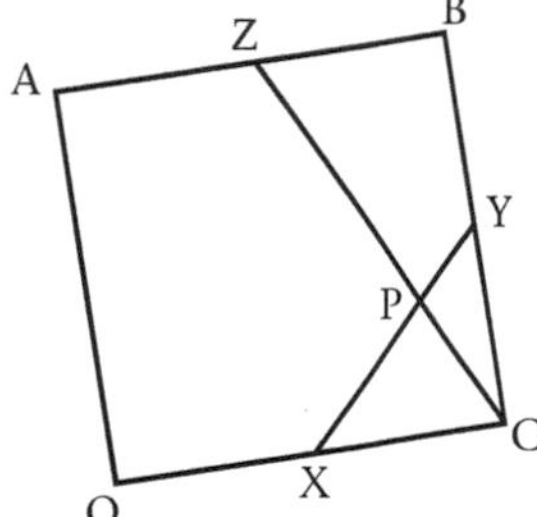

Fig. 25.4

16 The position vectors of the points A, B and C are $\begin{pmatrix}-2\\3\end{pmatrix}$, $\begin{pmatrix}3\\3\end{pmatrix}$ and $\begin{pmatrix}6\\7\end{pmatrix}$.
(a) Determine the vectors
(i) $\overrightarrow{AB}$ (ii) $\overrightarrow{BC}$
(b) Given that $\overrightarrow{CD} = \overrightarrow{BA}$, determine the position vector of D.
(c) Show by a vector method that ABCD is a rhombus.

17 (a) Given that $\overrightarrow{AB} = \begin{pmatrix}4\\-2\end{pmatrix}$ and $\overrightarrow{CB} = \begin{pmatrix}3\\2\end{pmatrix}$, determine the magnitude of the vector $\overrightarrow{AC}$.
(b) Given that $\overrightarrow{OA} = \begin{pmatrix}3\\0\end{pmatrix}$, calculate the tangent of the angle which $\overrightarrow{AC}$ makes with the line $x = 3$. (5 marks)
[CXC (General) Jan 91]

18 The coordinates of the points K and L are (1,5), and (5,3) respectively.
(a) Write the position vector of
(i) K (ii) L.
M is the mid-point of KL.
(b) Determine the position vector of M.

19 F is the point (2,1) and H is the point (−1,5). Given that u is a unit vector in the x-direction and v is a unit vector in the y-direction,
(a) express the position vectors of F and H in terms of u and v.
(b) determine the vector $\overrightarrow{HF}$
(c) calculate the modulus of $\overrightarrow{HF}$

20 The position vectors of the points P and R relative to an origin O are $\begin{pmatrix}-2\\1\end{pmatrix}$ and $\begin{pmatrix}1\\3\end{pmatrix}$ respectively.
(a) Draw a diagram to show the positions of P and R.
(b) Determine the vector $\overrightarrow{PR}$.
(c) Given that $\overrightarrow{OS} = \begin{pmatrix}a\\b\end{pmatrix}$ and $2\overrightarrow{PR} = \overrightarrow{RS}$, calculate the values of a and b.

21 (a) Given that $\overrightarrow{OK} = \begin{pmatrix}16\\2\end{pmatrix}$, $\overrightarrow{OL} = \begin{pmatrix}4\\-3\end{pmatrix}$ and that M and N are the mid-points of OK and OL respectively, (i) express $\overrightarrow{MN}$ as a column vector, (ii) find the value of $|\overrightarrow{KL}|$.
(b)

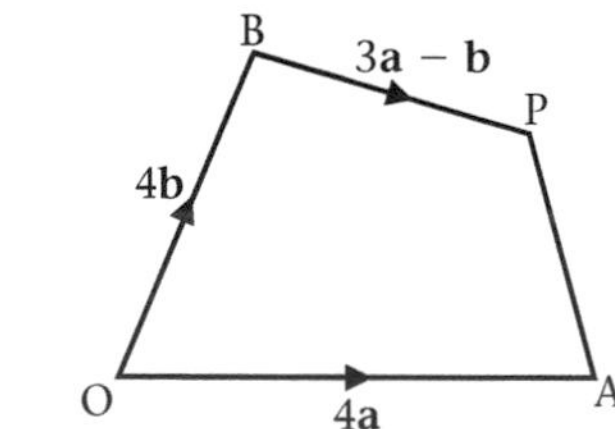

Fig. 25.5

In Fig. 25.5, given that $\overrightarrow{OA} = 4\mathbf{a}$, $\overrightarrow{OB} = 4\mathbf{b}$ and $\overrightarrow{BP} = 3\mathbf{a} - \mathbf{b}$, express as simply as possible, in terms of **a** and **b**, (i) $\overrightarrow{OP}$, (ii) $\overrightarrow{AP}$.
(c) The lines OA produced and BP produced meet at Q. Given that $\overrightarrow{BQ} = m\overrightarrow{BP}$ and $\overrightarrow{OQ} = n\overrightarrow{OA}$, form an equation connecting m, n, **a** and **b**.
(d) Hence deduce the values of m and n.

22

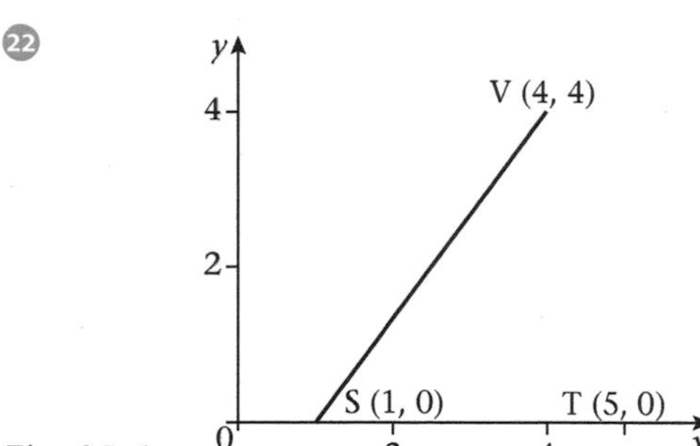

Fig. 25.6

(a) Using Fig. 25.6, not drawn to scale, write the position vectors of S, T, and V relative to the origin, O.

(b) Given that $\overrightarrow{TW} = 3\overrightarrow{SV}$, determine the position vector of W.

23 (a) The position vectors of the points A, B and C relative to an origin O are

$\overrightarrow{OA} = \begin{pmatrix} 3 \\ 2 \end{pmatrix}$, $\overrightarrow{OB} = \begin{pmatrix} -3 \\ -2 \end{pmatrix}$ and $\overrightarrow{OC} = \begin{pmatrix} 4 \\ -2 \end{pmatrix}$

respectively.

(i) Express the following vectors in the form $\begin{pmatrix} a \\ b \end{pmatrix}$:

$\overrightarrow{AC}$, $\overrightarrow{CB}$ and $\overrightarrow{AB}$. (3 marks)

(ii) Prove that A, O and B are on the same straight line. (3 marks)

(iii) Calculate the length of vectors $\overrightarrow{AC}$, $\overrightarrow{CB}$ and $\overrightarrow{AB}$. (2 marks)

(iv) Hence or otherwise, show by calculation that angle ACB is a right angle. (2 marks)

(b) In Fig. 25.7, **not drawn to scale**, the vectors $\overrightarrow{OP}$ and $\overrightarrow{OQ}$ are **p** and **q** respectively. R and S are the mid-points of OP and OQ respectively.

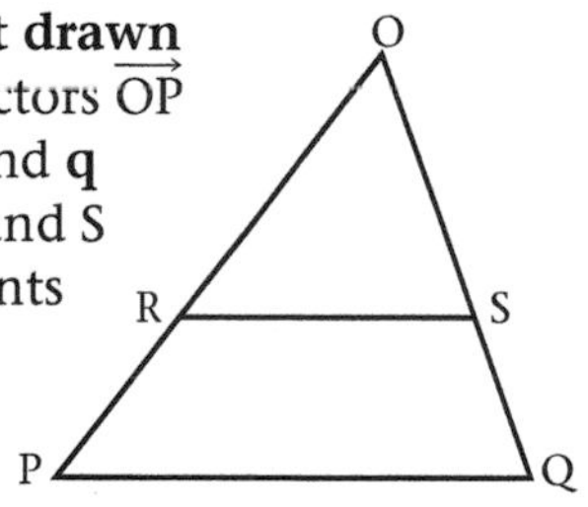

Fig. 25.7

(i) Express $\overrightarrow{PQ}$ in terms of **p** and **q**. (2 marks)

(ii) Prove that RS is parallel to PQ and half its length. (3 marks)

[CXC (General) Jan 96]

Geometrical transformations

A figure is **transformed** when its position and/or shape changes. The **image** of a shape is the figure obtained after a transformation.

If the image has the same dimensions as the original shape, the transformation is called a **congruency** or **isometry**. Fig. 25.8 shows a shaded triangle ABC and its images after a typical **translation**, T, **rotation**, R, **reflection**, M.

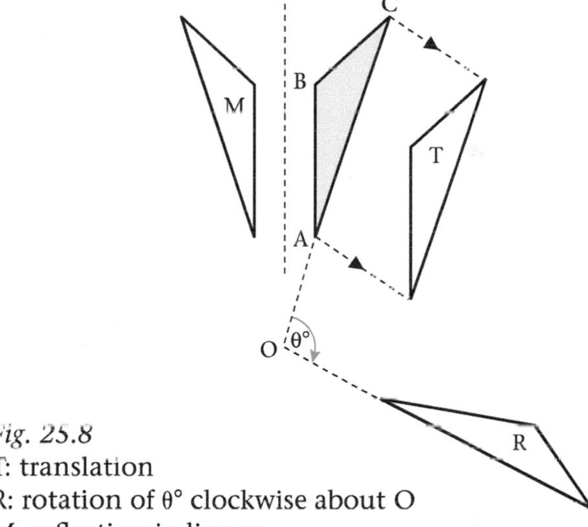

Fig. 25.8

T: translation

R: rotation of θ° clockwise about O

M: reflection in line m

A **glide reflection** is a composite of a reflection and a translation. Fig. 25.9 shows a triangle ABC and its final image triangle $A_2B_2C_2$ after a glide reflection. The object and its image are congruent but have opposite orientation.

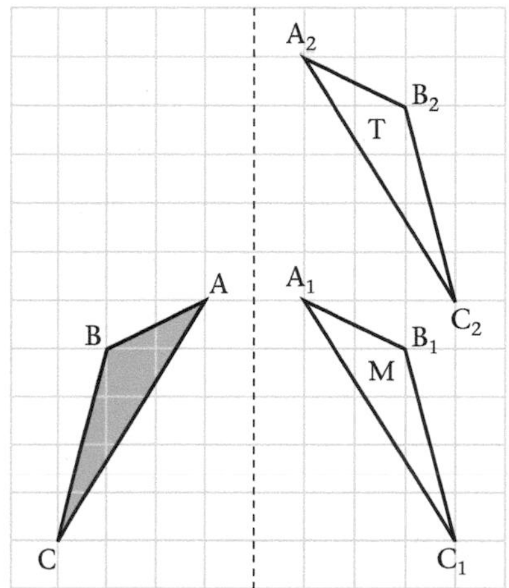

Fig. 25.9

Isometric (congruent) shapes have corresponding lengths and angles equal.

Fig. 25.10 shows triangle ABC and its images after **enlargements** E and E′.

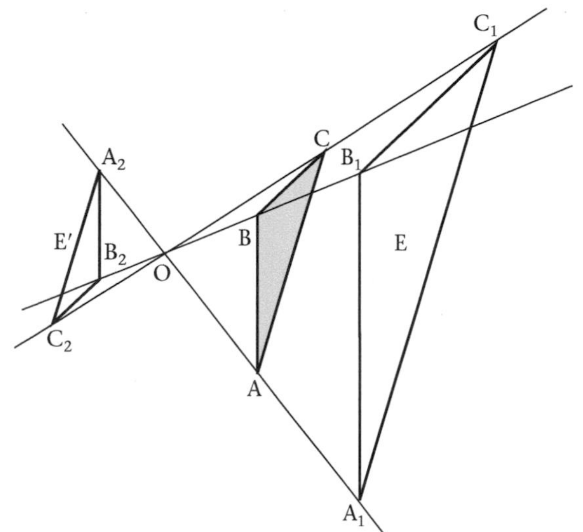

Fig. 25.10

E: enlargement of factor k

$$\text{where } k = \frac{OA_1}{OA} = \frac{OB_1}{OB} = \frac{OC_1}{OC}$$

E′: enlargement of factor K

$$\text{where } k = \frac{OA_2}{OA} = \frac{OB_2}{OB} = \frac{OC_2}{OC}$$

Notice the following:

(a) Enlarged shapes are geometrically similar and have corresponding angles equal.

(b) When $k > 1$, or $k < -1$, the size of the image is larger than the object;
when $-1 < k < 0$, or $0 < k < 1$, the image is smaller than the object.

(c) If the 'enlargement' factor is k, then the original area will be 'enlarged' by factor k^2.

(d) In Fig. 25.10, k is positive and greater than 1; in Fig. 25.10, K is negative and fractional.

Matrices can be used as operators which transform shapes drawn on the Cartesian plane. Table 25.2 gives a summary of the most common transformations of the Cartesian plane and their related matrices.

Table 25.2

Transformation	Sketch	Matrix
Identity		$\begin{pmatrix} 1 & 0 \\ 0 & 1 \end{pmatrix}$
Translation		$\begin{pmatrix} a \\ b \end{pmatrix}$
Reflection in x-axis		$\begin{pmatrix} 1 & 0 \\ 0 & -1 \end{pmatrix}$
Reflection in y-axis		$\begin{pmatrix} -1 & 0 \\ 0 & 1 \end{pmatrix}$
Rotation of 180° about (0, 0)		$\begin{pmatrix} -1 & 0 \\ 0 & -1 \end{pmatrix}$
Enlargement centre (0, 0)		$\begin{pmatrix} k & 0 \\ 0 & k \end{pmatrix}$

Example 9

△PQR has coordinates P(−3, 1), Q(−2, 4), R(0, 4).

(a) If △PQR is translated by a vector $\begin{pmatrix} 4 \\ 0 \end{pmatrix}$, find by drawing or calculation the coordinates of its image, △P′Q′R′.

(b) If △P′Q′R′ is given a transformation represented by the matrix $\begin{pmatrix} 1 & 0 \\ 0 & -1 \end{pmatrix}$ find the coordinates of the new image △P″Q″R″.

(a) *Either* by drawing as in Fig. 25.11:

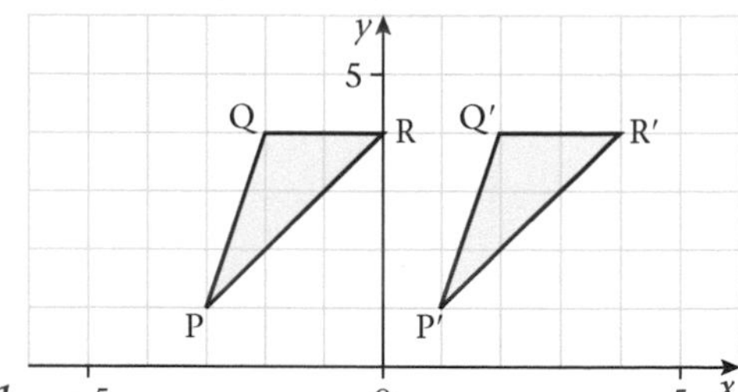

Fig. 25.11

ΔP′Q′R′ as coordinates P′(1, 1), Q′(2, 4), R′(4, 4).

or by calculation:

$$\begin{pmatrix}-3\\1\end{pmatrix} + \begin{pmatrix}4\\0\end{pmatrix} = \begin{pmatrix}1\\1\end{pmatrix}$$

$$\begin{pmatrix}-2\\4\end{pmatrix} + \begin{pmatrix}4\\0\end{pmatrix} = \begin{pmatrix}2\\4\end{pmatrix}$$

$$\begin{pmatrix}0\\4\end{pmatrix} + \begin{pmatrix}4\\0\end{pmatrix} = \begin{pmatrix}4\\4\end{pmatrix}$$

ΔP′Q′R′ has coordinates P′(1, 1), Q′(2, 4), R′(4, 4), as above

(b) P′ Q′ R′ P″ Q″ R″

$$\begin{pmatrix}1 & 0\\0 & -1\end{pmatrix}\begin{pmatrix}1 & 2 & 4\\1 & 4 & 4\end{pmatrix} = \begin{pmatrix}1 & 2 & 4\\1 & 4 & 4\end{pmatrix}$$

ΔP″Q″R″ has coordinates P″(1, −1), Q″(2, −4), R″(4, −4).

Exercise 25d

1. Describe completely the single transformation which maps ΔPQR onto ΔKLM in Fig. 25.12.

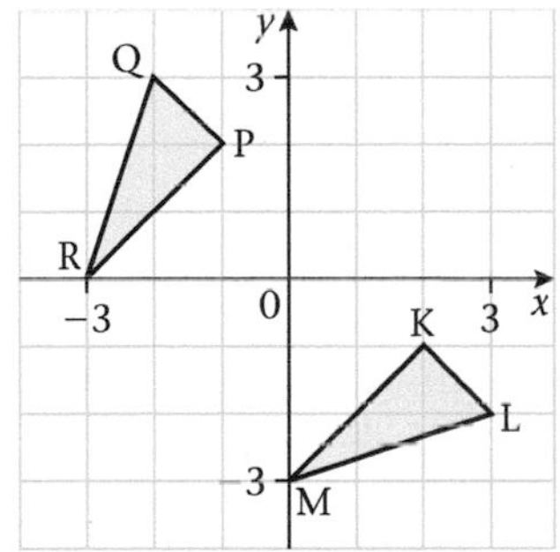

Fig. 25.12

2. T is the translation $\begin{pmatrix}3\\2\end{pmatrix}$ and R is an anticlockwise rotation of 90° about the origin. A is the point (2, −5), B is (−1, 4) and C is (−4, 4), Find the coordinates of (a) T(A), (b) R(B), (c) the point D if RT(D) = C.

3. F is a transformation of matrix $\begin{pmatrix}k & 0\\0 & k\end{pmatrix}$. The images of A(2, 2), B(−2, 4), C(0, 6) under F are A′(5, 5), B′(−5, 10), C′(0, 15).
   (a) Using a suitable scale, draw triangles ABC and A′B′C′ on the same graph.
   (b) Describe fully the transformation F and write down the value of k.
   (c) Find the coordinates of the vertices of the image of ABC after rotation of 270° clockwise about the point (3, 2).

4. P′(0, 0), Q′(3, 13), R′(−2, −11) are the images of P(0, 0), Q(3, 1), R(−2, −3) under a transformation represented by a matrix of the form $\begin{pmatrix}a & b\\c & d\end{pmatrix}$.
   (a) Find the transformation matrix.
   (b) Find the matrix which will transform ΔP′Q′R′ back to ΔPQR.

5. P is the point (−1, 7) on the shape given in Fig. 25.13.
   M is a reflection in the line 2 − y = 0.
   N is a reflection in the line $x + y = 0$.
   (a) Find the image of P under
      (i) MN(P),
      (ii) NM(P).
   (b) Describe fully the single transformation K such that K[MN(P)] = P.

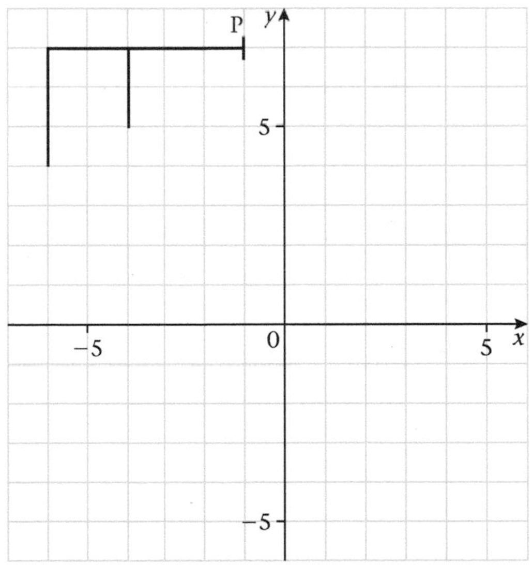

Fig. 25.13

6. Answer the whole of this question on a sheet of graph paper.
   The triangle PQR has vertices P(4, 1), Q(7, 3), R(1, 6). The triangle P′Q′R′ has vertices P′(−1, −4), Q′(−3, −7),

R′(−6, −1). The triangle P″Q″R″ has vertices P″(3, −4), Q″(5, −7), R″(8, −1). Draw the triangles on graph paper using a scale of 1 cm to 1 unit on each axis, and label the vertices.

(a) Describe fully the transformation that maps △PQR onto △P′Q′R′.
(b) State the matrix which represents this transformation.
(c) Describe fully the transformation that maps △P′Q′R′ onto △P″Q″R″.
(d) △PQR can be mapped onto △P″Q″R″ by a clockwise rotation. Find the centre of rotation and describe fully the rotation.

7 (a) Given that the matrix $L = \begin{pmatrix} 0 & 1 \\ 1 & 0 \end{pmatrix}$ and the matrix $M = \begin{pmatrix} 1 & 1 \\ 0 & 1 \end{pmatrix}$,
(i) show that L represents a reflection in the line $y = x$.
(ii) calculate the matrix product LM and show that the combination of transformations represented by L and M is *not* commutative.

(b) The position vectors of the points A, B, R, and S are as follows:
$\overrightarrow{OA} = \begin{pmatrix} 4 \\ 0 \end{pmatrix}$, $\overrightarrow{OA} = \begin{pmatrix} 4 \\ 4 \end{pmatrix}$ $\overrightarrow{OR} = \begin{pmatrix} 8 \\ 0 \end{pmatrix}$ and
$\overrightarrow{OS} = \begin{pmatrix} 8 \\ 2 \end{pmatrix}$ where O is the origin.
$\overrightarrow{OA}$ and $\overrightarrow{OB}$ are transformed into the position vectors $\overrightarrow{OC}$ and $\overrightarrow{OD}$ respectively by LM.
(i) Calculate $\overrightarrow{OC}$ and $\overrightarrow{OD}$.
(ii) Find the matrix that maps △OAB onto △ORS,

[CXC (General) June 87] (adapted)

8 (a) Given the matrices, $V = \begin{pmatrix} 3 & -1 \\ -5 & 2 \end{pmatrix}$ and $W = \begin{pmatrix} 2 & 1 \\ 5 & 3 \end{pmatrix}$ determine the image of
(i) the point A(2, 1) under the transformation represented by V,
(ii) the point B(5, −8) under the transformation represented by the matrix W.

(b) State the relationship between the matrices V and W, explaining the reason for your statement.

9 (a) Using a scale of 1 cm to represent 1 unit on each axis, plot points $P\begin{pmatrix} 2 \\ 2 \end{pmatrix}$, $Q\begin{pmatrix} 7 \\ 2 \end{pmatrix}$, and $R\begin{pmatrix} 5 \\ 5 \end{pmatrix}$ and draw △PQR.
(b) △P′Q′R′ is the image of △PQR under the transformation $\begin{pmatrix} 1 & 0 \\ 0 & -1 \end{pmatrix}$. Determine the position vectors of the vertices P′, Q′ and R′ and draw △P′Q′R′.
(c) △P″Q″R″ is the image of △P′Q′R′ under the transformation $\begin{pmatrix} 0 & -1 \\ 1 & 0 \end{pmatrix}$. Determine the position vectors of the vertices P″, Q″ and R″, and draw △P″Q″R″.
(d) Determine the single matrix, V, that will transform △PQR onto △″P″Q″R″.
(e) Describe the transformation under, V, geometrically.

10 (a) The vertices of triangle ABC have coordinates A (1, 1), B (2, 1) and C (1, 2). Matrix T transforms triangle ABC into triangle A′B′C′. The coordinates of the vertices of triangle A′B′C′ are A′(3, 3), B′(6, 3) and C′(3, 6).
(i) Express the transformation matrix, T, in the form $\begin{pmatrix} s & t \\ u & v \end{pmatrix}$. (4 marks)
(ii) Write a complete geometrical description of the transformation T. (2 marks)
(iii) State the ratio of the areas of triangle ABC to A′B′C′. (1 mark)

(b) A triangle XYZ, with coordinates X(4, 5), Y(−3, 2) and Z(−1, 4), is mapped onto triangle X′Y′Z′ by a transformation $M = \begin{pmatrix} -1 & 0 \\ 0 & 1 \end{pmatrix}$.
(i) Calculate the coordinates of the vertices of triangle X′Y′Z′. (3 marks)
(ii) A matrix $N = \begin{pmatrix} 0 & -1 \\ 1 & 0 \end{pmatrix}$ maps triangle X′Y′Z′ onto triangle X″Y″Z″. Determine the 2 × 2 matrix, Q, which maps triangle XYZ onto X″Y″Z″ (2 marks)
(iii) Show that the matrix which maps triangle X″Y″Z″ back onto triangle XYZ is equal to Q. (3 marks)

[CXC (General) Jan 96]

11 The coordinates of the vertices of a shape OKLN are (0, 0), (2, 0), (2, 2) and (0, 2) respectively.

(a) Write the positions of K, L, and N as column vectors.

The shape OKLN is transformed by the matrix $Z = \begin{pmatrix} 3 & 0 \\ 0 & 3 \end{pmatrix}$.

(b) Determine the position vectors of the images of the vertices under the transformation.

(c) Describe the transformation geometrically.

(d) Sketch the shape OKLN and its image.

(e) Write the equation of the line of symmetry of the figure formed by OKLN and its image.

12 The coordinates of △ABC are A(2, 0), B(6, 3) and C(6, 0).

(a) The matrix $P = \begin{pmatrix} -1 & 0 \\ 0 & 1 \end{pmatrix}$ describes a transformation on △ABC onto △A′B′C′.

(i) Determine the coordinates of A′, B′, C′.

(ii) Describe the transformation under P geometrically.

(b) The matrix $Q = \begin{pmatrix} -1 & 0 \\ 0 & -1 \end{pmatrix}$ describes a transformation on △A′B′C′ onto △A″B″C″.

(i) Determine the coordinates of A″, B″, C″.

(ii) Describe the transformation under Q geometrically.

(c) Given that △ABC is mapped directly onto △A″B″C″ by a transformation matrix R,

(i) determine the matrix R.

(ii) describe the transformation under R geometrically.

Revision test (Chapter 25)

Select the correct answer and write down your choice.

1 The order of the matrix $\begin{pmatrix} 1 & 1 & 0 \\ 0 & 1 & 1 \end{pmatrix}$ is

A 0×3 B 2×3

C 0×2 D 3×2

2 Given that $\overrightarrow{DA} = \mathbf{a}$ and $\overrightarrow{DB} = \mathbf{b}$, then $\overrightarrow{AB} =$

A $\mathbf{a} + \mathbf{b}$

B $\mathbf{a} - \mathbf{b}$

C $\mathbf{b} - \mathbf{a}$

D $\frac{1}{2}(\mathbf{a} + \mathbf{b})$

3 If the 2×2 matrix, V, is singular, then

A the value of the determinant is zero

B the product of the diagonal elements is zero

C all the elements are equal

D $V = V^{-1}$

4 Given that $\overrightarrow{AB} = \frac{1}{2}\overrightarrow{PQ}$ and $\overrightarrow{AB} = \begin{pmatrix} 6 \\ 4 \end{pmatrix}$ then $\overrightarrow{PQ} =$

A $\begin{pmatrix} 3 \\ 2 \end{pmatrix}$ B $\begin{pmatrix} 6 \\ 4 \end{pmatrix}$ C $\begin{pmatrix} 12 \\ 8 \end{pmatrix}$ D $\begin{pmatrix} 36 \\ 16 \end{pmatrix}$

5 Given that J and K are 2×2 matrices and that $J \times K = J$ then $K =$

A $\begin{pmatrix} 1 & 1 \\ 1 & 1 \end{pmatrix}$ B $\begin{pmatrix} 0 & 1 \\ 1 & 0 \end{pmatrix}$

C $\begin{pmatrix} 1 & 0 \\ 0 & 1 \end{pmatrix}$ D $\begin{pmatrix} 1 & 0 \\ 0 & 0 \end{pmatrix}$

6 If $X = \begin{pmatrix} 2 & 1 \\ -1 & 3 \end{pmatrix}$, $3X =$

A $\begin{pmatrix} 6 & 3 \\ -1 & 3 \end{pmatrix}$ B $\begin{pmatrix} 6 & 1 \\ -3 & 3 \end{pmatrix}$

C $\begin{pmatrix} 6 & 1 \\ -3 & 9 \end{pmatrix}$ D $\begin{pmatrix} 6 & 3 \\ -3 & 9 \end{pmatrix}$

7 Given that $\overrightarrow{OP} = \begin{pmatrix} 4 \\ 3 \end{pmatrix}$ and $\overrightarrow{PQ} = \begin{pmatrix} 1 \\ -1 \end{pmatrix}$, where O is the origin, the position vector of Q is

A $\begin{pmatrix} 5 \\ 2 \end{pmatrix}$ B $\begin{pmatrix} -3 \\ -4 \end{pmatrix}$

C $\begin{pmatrix} 5 \\ -2 \end{pmatrix}$ D $\begin{pmatrix} 3 \\ 4 \end{pmatrix}$

8 Given that $\begin{pmatrix} 3 & 1 \\ 2 & 0 \end{pmatrix} - K = \begin{pmatrix} 2 & 0 \\ 1 & -2 \end{pmatrix}$, $K =$

A $\begin{pmatrix} 1 & 1 \\ -1 & 2 \end{pmatrix}$ B $\begin{pmatrix} -1 & -1 \\ 1 & 2 \end{pmatrix}$

C $\begin{pmatrix} -1 & -1 \\ -1 & -2 \end{pmatrix}$ D $\begin{pmatrix} 1 & 1 \\ 1 & 2 \end{pmatrix}$

9 Given that $\begin{pmatrix} 5 & b \\ 4 & -2 \end{pmatrix}$ is singular, then the value of b is

A -6 B 0

C $2\frac{1}{2}$ D $-2\frac{1}{2}$

10 The matrix product $\begin{pmatrix}-1 & -2\\ 2 & 1\end{pmatrix}\begin{pmatrix}3 & 1 & 0\\ 0 & -3 & 2\end{pmatrix} =$

A $\begin{pmatrix}-5\\ -4\end{pmatrix}$ B $\begin{pmatrix}-3 & -2\\ 0 & -3\end{pmatrix}$

C $\begin{pmatrix}-3 & 5 & -4\\ 6 & -1 & 2\end{pmatrix}$ D $\begin{pmatrix}-3 & -2 & 0\\ 0 & -6 & 2\end{pmatrix}$

11 Given $\overrightarrow{LM} = \mathbf{p} + 3\mathbf{q}$ and $\overrightarrow{LN} = 3\mathbf{p} - 2\mathbf{q}$, $\overrightarrow{ML} + \overrightarrow{LN} =$

A $2\mathbf{p} - \mathbf{q}$ B $2\mathbf{p} - 5\mathbf{q}$
C $4\mathbf{p} + \mathbf{q}$ D $2\mathbf{p} + 5\mathbf{q}$

12 Under the enlargement represented by the matrix $\begin{pmatrix}3 & 0\\ 0 & 3\end{pmatrix}$, the point which remains invariant is

A (0, 3) B (0, 0) C (3, 0) D (3, 3)

13 If the matrix $\begin{pmatrix}1 & 0\\ 3 & 1\end{pmatrix}$ transforms △PQR to △P′Q′R′, then the matrix which transforms △P′Q′R′ to △PQR is

A $\begin{pmatrix}1 & -3\\ 0 & 1\end{pmatrix}$ B $\begin{pmatrix}1 & 0\\ -3 & 1\end{pmatrix}$

C $\begin{pmatrix}1 & 3\\ 0 & 1\end{pmatrix}$ D $\begin{pmatrix}-1 & 3\\ 0 & -1\end{pmatrix}$

14 M is a transformation represented by $\begin{pmatrix}-1 & 0\\ 1 & 0\end{pmatrix}$. If R is a transformation represented by $\begin{pmatrix}1 & 2\\ 0 & 1\end{pmatrix}$, then MR is the single transformation represented by

A $\begin{pmatrix}-1 & -2\\ 1 & 2\end{pmatrix}$ B $\begin{pmatrix}1 & 2\\ 0 & 1\end{pmatrix}$

C $\begin{pmatrix}1 & 1\\ -1 & -2\end{pmatrix}$ D $\begin{pmatrix}1 & 0\\ 1 & 0\end{pmatrix}$

15 Given that a clockwise rotation of 270° about the origin, O, of the coordinate axes is represented by the matrix $\begin{pmatrix}0 & -1\\ 1 & 0\end{pmatrix}$, then an anti-clockwise rotation of 90° about the origin O, is represented by

A $\begin{pmatrix}0 & 1\\ -1 & 0\end{pmatrix}$ B $\begin{pmatrix}-1 & 0\\ -0 & 1\end{pmatrix}$

C $\begin{pmatrix}0 & -1\\ 1 & 0\end{pmatrix}$ D $\begin{pmatrix}1 & 0\\ 0 & 1\end{pmatrix}$

16 P′(2, 12) is the image of P(2, −6) under a transformation of matrix $\begin{pmatrix}1 & 0\\ 0 & -2\end{pmatrix}$. Which of the following matrices represents the transformation under which P is the image of P′?

A $\begin{pmatrix}2 & 0\\ 0 & 1\end{pmatrix}$ B $\frac{1}{2}\begin{pmatrix}2 & 0\\ 0 & -1\end{pmatrix}$

C $\begin{pmatrix}2 & 0\\ 0 & -1\end{pmatrix}$ D $\frac{1}{2}\begin{pmatrix}-2 & 0\\ 0 & 1\end{pmatrix}$

17 The matrix which will produce a reflection in the line $y = x$ is

A $\begin{pmatrix}0 & 1\\ 1 & 0\end{pmatrix}$ B $\begin{pmatrix}-1 & 0\\ 0 & -1\end{pmatrix}$

C $\begin{pmatrix}1 & 0\\ 0 & -1\end{pmatrix}$ D $\begin{pmatrix}1 & 0\\ 0 & 1\end{pmatrix}$

18

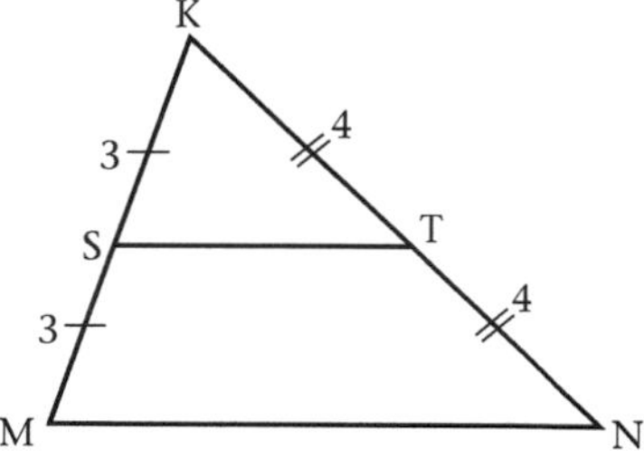

Fig. 25.15

In Fig. 25.15, KS = SM = 3 units and KT = TN = 4 units.
If ST = v units, then MN =

A $v + 3$ B $\frac{v + 7}{2}$

C $v + 4$ D $2v$

Chapter 26
Statistics and probability

See Book 3, Chapter 24, for revision of bar charts, pie charts, mean, median, mode, histograms, spread, probability.

Mean, **mode** and **median** are statistical averages used to represent a set of data. **Frequency** tells us how often an observation occurs; a **frequency distribution** can be shown in a **frequency table**. **Bar charts**, **frequency polygons**, **pie charts** and **histograms** are representations of statistical data. A **cumulative frequency curve** shows the cumulative frequency of observations. **Dispersion** is the spread of data. It can be measured using the **range**, **interquartile range** or **semi-interquartile range**.

Example 1

The heights of 60 plants are given in Table 26.1.

Table 26.1

Height (cm)	Frequency
10–19	4
20–29	11
30-39	12
40–49	18
50–59	12
60–69	3

(a) Calculate the mean height of the plants, to 2 significant figures.
(b) Estimate the interquartile range of the distribution.
(c) Draw a histogram of the distribution.
(d) What is the probability that a plant taken at random will be of a height greater than 50 cm?

(a) Mean height of the plants =

$$\frac{14.5 \times 4 + 24.5 \times 11 + 34.5 \times 12 + 44.5 \times 18 + 54.5 \times 12 + 64.5 \times 3}{60}$$

= 40 cm to 2 s.f

(b) The interquartile range is $49\frac{1}{2} - 29\frac{1}{2} = 20$

(c)

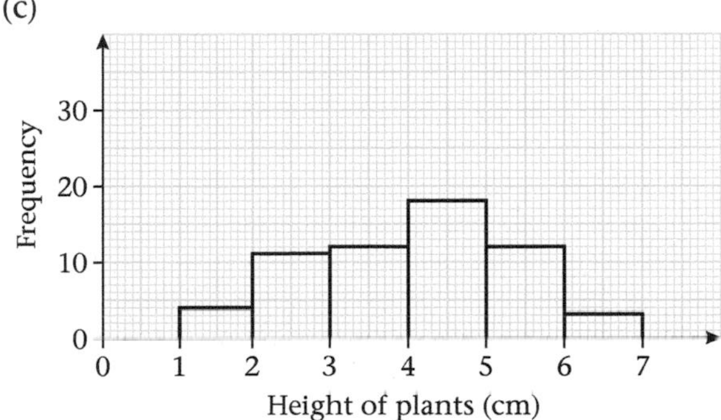

Fig. 26.1

(d) $p = \frac{15}{60} = \frac{1}{4}$

Cumulative frequency curve

Example 2

Table 26.2 is the frequency distribution of the masses of 40 pupils in a class.

Table 26.2

Mass (kg)	Number of pupils
41–45	3
46–50	7
51–55	12
56–60	10
61–65	6
66–70	2

(a) State the modal class of the distribution.
(b) Calculate the mean mass of the pupils.
(c) Draw a cumulative frequency curve of the distribution.
(d) Hence estimate (i) the median, (ii) the semi-interquartile range.

(a) The modal class is 51–55. This class has the highest frequency.
(b) The mid-values of each class can be taken to be representative of that class as a whole. Table 26.3 shows the mid-values and the corresponding deviations from a working mean of 53 kg.

Table 26.3

Mass (kg)	Frequency f	Mid-value	Deviation d	$f \times d$
41–45	3	43	−10	−30
46–50	7	48	−5	−35
51–55	12	53	0	0
56–60	10	58	+5	+50
61–65	6	63	+10	+60
66–70	2	68	+15	+30
			total deviation =	+75

Either, using the mid-values only:

mean mass

$$= \frac{3 \times 43 + 7 \times 48 + 12 \times 53 + 10 \times 58 + 6 \times 63 + 2 \times 68}{3 + 7 + 12 + 10 + 6 + 2}$$

$$= \frac{2\,195}{40}\text{ kg} = 54.875\text{ kg}$$

or, using a working mean of 53 kg, from Table 26.3

$$\text{mean deviation} = \frac{+75}{40}\text{ kg} = 1.875\text{ kg}$$

$$\text{mean mass} = 53 + (+\,1.875)\text{ kg} = 54.875\text{ kg}$$

(c) Table 26.4 is used to draw the cumulative frequency curve in Fig. 26.2.

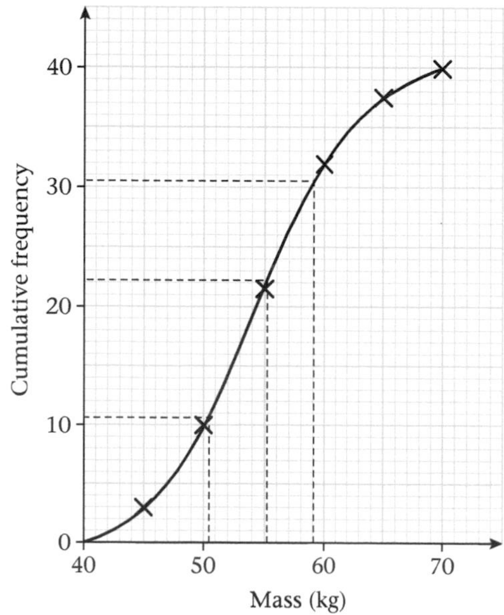

Fig. 26.2

Table 26.4

Mass (kg)	Frequency	Cumulative frequency
41–45	3	3
46–50	7	10
51–55	12	22
56–60	10	32
61–65	6	38
66–70	2	40

(d) From Fig. 26.2,

(i) median $Q_2 = 54\frac{1}{2}$ kg

(ii) semi-interquartile range

$$\frac{Q_3 - Q_1}{2}\text{ kg} = \frac{59\frac{1}{4} - 50\frac{1}{4}}{2}\text{ kg} = 4\frac{1}{2}\text{ kg}$$

Exercise 26

1 The following marks were the marks earned by two students in twelve tests.

Dwight: 8, 12, 13, 16, 17, 19, 17, 14, 15, 18, 20, 20

Vashan: 9, 11, 13, 14, 17, 19, 12, 15, 16, 18, 18, 11

For each student, calculate

(a) the mean mark,
(b) the median mark,
(c) the range,
(d) the probability that they earned more than 15 marks on a test picked at random.

2 The examination marks of 50 students are as follows:

65	58	51	36	23	40	53	59	70	51
46	59	50	67	46	39	61	62	73	60
71	51	47	32	48	40	40	51	58	67
60	69	43	52	37	26	38	50	59	40
44	54	42	47	68	74	45	39	48	55

(a) Make a frequency distribution using class intervals of 21–30, 31–40, ...
(b) Draw a cumulative frequency curve.
(c) Hence estimate (i) the median, (ii) the semi-interquartile range.
(d) Find the percentage of students that got more than 45 marks.

3 The daily earnings of 50 men are given in Table 26.5.

Table 26.5

Earnings per day ($)	Number of men (frequency)
51–75	1
76–100	4
101–125	17
126–150	15
151–175	11
176–200	2

(a) Construct a cumulative frequency table and draw the cumulative frequency curve. Use this to estimate the median daily earnings.

(b) Two men are chosen at random from the 50. Find the probability that
   (i) both are in the \$126–\$150 group,
   (ii) one is in that group and the other is not.

4 Table 26.6 is the frequency distribution of the heights of 40 pupils.

Table 26.6

Height (cm)	Number of pupils
131–140	2
141–150	11
151–160	14
161–170	10
171–180	3

(a) Draw a cumulative frequency curve of the distribution.

(b) Hence estimate the median height of the pupils.

(c) Estimate the probability that a student chosen at random will be taller than 156 cm.

5 The scores obtained by 100 applicants on an aptitude test for selection into a programme are shown in the frequency table (Table 26.7).

Table 26.7

Score	Frequency	Cumulative frequency
0–9	8	
10–19	13	
20–29	17	
30–39	20	
40–49	19	
50–59	14	
60–69	6	
70–79	3	

(a) Copy Table 26.7 and complete the cumulative frequency column.

(b) Draw the cumulative frequency curve using 2 cm to represent each class interval and 2 cm to represent 10 applicants.

(c) Using your curve, answer the following.
   (i) A score of 45 was considered as acceptable for the programme. Estimate the number of applicants who qualified for the programme.
   (ii) Assuming there were places for only 15 applicants, estimate the lowest score that would be needed to select them.
   (iii) Calculate the probability that an applicant chosen at random obtained a score of at least 45.

[CXC (General) June 88]

6 120 doctors took part in a run in a heart health awareness week. Table 26.8 shows the frequency distribution of the times taken by the runners.

Table 26.8

Time (minutes)	Frequency
81–90	3
91–100	18
101–110	36
111–120	49
121–130	11
131–140	3

(a) Construct a cumulative frequency table.

(b) Draw the cumulative frequency curve.

(c) Estimate the median time and the interquartile range.
(d) Estimate how many runners took less than 115 minutes.

7 Table 26.9 shows the frequency distribution of the heights of plants in a nursery.

Table 26.9

Height (cm)	1–20	21–30	31–40	41–50
Frequency	3	15	17	5

(a) How many plants were measured?
(b) What is the modal class of the distribution?
(c) State (i) the class limits (ii) the upper boundary of the 31–40 class.
(d) Draw a histogram to represent the information.
(e) What is the probability that a plant taken at random will measure between 30 cm and 40 cm?

8 The marks gained by a group of students in mathematics are shown below.

11	20	23	27	29	32
13	22	25	28	30	35
17	23	26	27	32	37
19	23	25	25	33	36

(a) Construct a frequency table to show the distribution of the marks using the classes 10–14, 15–19, 20–24, 25–29, 30–34 and 35–39.
(b) Draw a histogram to represent the information.
(c) Estimate the probability that a student chosen at random scored at most 32 marks.

9 The heights, in centimetres, of a sample of blades of grass were recorded and grouped as shown in Table 26.10.

Table 26.10

Height (cm)	Number of blades
3–7	5
8–12	16
13–17	23
18–22	12
23–27	4

(a) Calculate
(i) the *total* number of blades in the sample,
(ii) an estimate of the mean height of the blades in the sample.
(b) Using a scale of 2 cm to represent a height of 5 cm on the x-axis and 2 cm to represent 5 blades on the y-axis, draw on graph paper the frequency polygon to represent the data given in Table 26.10.
(c) Calculate the probability that a blade, selected at random, measures at most 12 cm in height.

Revision test (Chapter 26)

Select the correct answer and write down your choice.

1 Given a set of data, which of the following must represent an observation?
A mean B mode
C median D range

2 The median of the set of scores 5, 12, 19, 0, 7, 12, 0, 27, 11, 8 is
A 9.5 B 11 C 11.5 D 12

Use the following information to answer questions 3 and 4. In a pie chart representing type of transport an angle of 50° represents 285 'private motoring'.

3 The size of the sample of types of transport is
A 570 B 1710 C 2072 D 2720

4 The angle representing a group of 1425 would be
A 10° B 28.5° C 200° D 250°

5 If the smallest value in a set of observations is 15 and the largest is 67, then it is correct to say that
A the mean is 41
B the semi-interquartile range is 26
C the range is 52
D the interquartile range is 52

Use the information below for answering questions 7, 8, 9.

The following shows the heights in cm, of twelve students:

485	437	501	489	422	487
511	548	468	441	527	563

6. The median is
   A 487 B 488
   C 489 D 495

7. The range is
   A 422 to 563 B 487 to 563
   C 86 D 141

8. The mean height in cm is
   A 454 B 455
   C 454.8 D 454.75

9. The probability that the height of a student chosen at random is in the range 450 to 500 cm is
   A $\frac{1}{4}$ B $\frac{1}{3}$ C $\frac{2}{3}$ D $\frac{3}{4}$

10. Fig. 26.3 shows the number of novels read by a group of students over a given period.

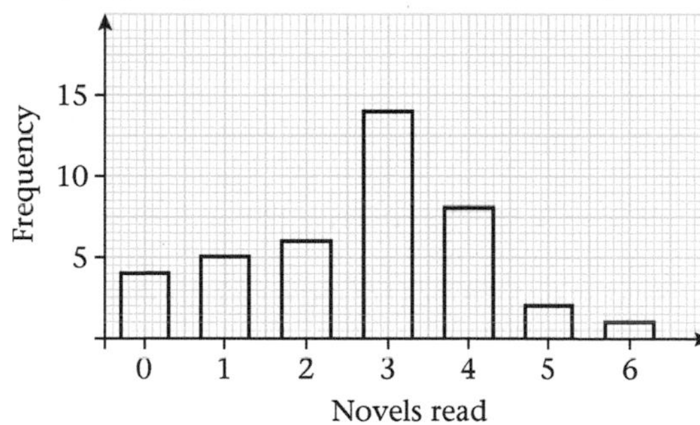

Fig. 26.3

   The total number of students in the group is
   A 21 B 25 C 36 D 40

11. A debating team of two students is to be chosen from 3 boys and 4 girls. The probability that one boy **and** one girl will be chosen is
   A $\frac{1}{12}$ B $\frac{2}{7}$ C $\frac{1}{2}$ D $\frac{3}{4}$

12. A set of measures has a mean of 42. If each measure is halved and 1 subtracted, the new mean will be
   A $42 \div 2 - 1$ B $(42 - 1) \div 2$
   C $(42 + 1) \div 2$ D $42 \div 2 + 1$

13. Which one of the following statements is **always** true for a set of observations?
   A The cumulative frequency curve is the same as the frequency polygon.
   B The frequency polygon must be imposed on the histogram.
   C The mean, mode and median are actual observations.
   D The area of the columns in a histogram is proportional to the frequency represented.

14. The mean of 17 numbers is 16. When one number is deleted, the mean is still 16. The number deleted is
   A 17 B 16 C 15 D 0

15. A box contains 6 red buttons and some black buttons. If the probability of choosing one red button **and** one black button at random, is $\frac{1}{18}$, the number of black buttons in the bag is
   A 12 B 9 C 6 D 3

16. The number of students writing examinations is distributed among four subjects as follows: Technical Drawing 24, Auto Mechanics 35, Mathematics 45, Art n. In a pie chart representing this information, the number doing Art is represented by an angle of size 72°. The number doing Art is
   A 16 B 26 C 35 D 45

17. A student rolled a die 30 times and recorded the number that turned up on each roll. The table below shows the result.

Number	1	2	3	4	5	6
Frequency	3	7	6	8	4	2

   What is the median number?
   A 3 B 3.5 C 4 D 7

18. If the student in question 19 needed to choose a 5 or 6 on any roll of the die, what was the probability of this event?
   A $\frac{1}{6}$ B $\frac{1}{5}$ C $\frac{1}{3}$ D $\frac{11}{30}$

Practice examination I

Paper 1 ($1\frac{1}{2}$ hours)

Attempt **all** *questions. Circle the letter of your choice.*

1 $3 + 17 \times 10 - 5 =$
A 88 B 100 C 168 D 195

2 The value of 6.056 7 to 3 s.f. is
A 6.05 B 6.056
C 6.057 D 6.06

3 The positive square root of 0.003 6 is
A 0.6 B 0.06
C 0.006 D 0.0006

4 In the set $\{\sqrt{2}, 24, \frac{22}{7}, 0.35\}$ the irrational number is
A $\sqrt{2}$ B -4 C $\frac{22}{7}$ D 0.35

5 P and Q denote two sets. Using set notation, the statement 'P does not include Q' is written
A $P \not\cup Q$ B $P \not\cap Q$
C $P \not\supset Q$ D $P \notin Q$

6

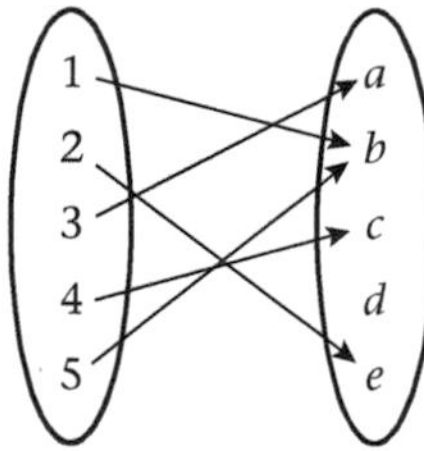

Fig. P1

The range of the function shown in Fig. P1 is given by the set
A $\{1, 2, 3, 4, 5\}$ B $\{a, c, e\}$
C $\{a, b, c, e\}$ D $\{a, b, c, d, e\}$

7 $(4x - 3)(2x + 5) =$
A $8x^2 - 26x - 15$ B $8x^2 - 2x - 15$
C $8x^2 + 7x - 15$ D $8x^2 + 14x - 15$

8 $\dfrac{3.75 \times 10^{-2} + 4.50 \times 10^{-1}}{3.75 \times 10} =$
A 1.30×10^{-2} B 2.20×10^{-2}
C 3.50×10^{-1} D 5.50×10^{-1}

9 A 16-metre long tube is cut into lengths in the ratio 1 : 3 : 4. The length of the shortest piece is
A 1 m B 2 m C 3 m D 4 m

10 The point P (2, −3) is reflected in the y-axis. The coordinates of the image of P are
A $(-2, 3)$ B $(2, 3)$
C $(-2, -3)$ D $(-3, 2)$

11 Given that $m * n = m^2 - 4n$, the value of $(3*2) * 1$ is
A -3 B 0 C 8 D 9

12 A farmer has 350 ha of land. After selling some land he is left with 329 ha. The percentage of land sold is
A 6% B 47% C 53% D 94%

13 A quadrilateral in which the diagonals meet at right angles but do *not* bisect each other is *best* called a
A rectangle
B square
C parallelogram
D kite

14 If V = {a, e, i, o, u}, the total number of subsets of V is
A 5 B 6 C 25 D 32

15 A circle of radius r is inscribed in a square as shown in Fig. P2. The area of the shaded region is
A $r^2(4 - \pi)$ B $r^2(2 - \pi)$
C $\frac{r^2}{4}(4 - \pi)$ D $\frac{r^2}{4}(2 - \pi)$

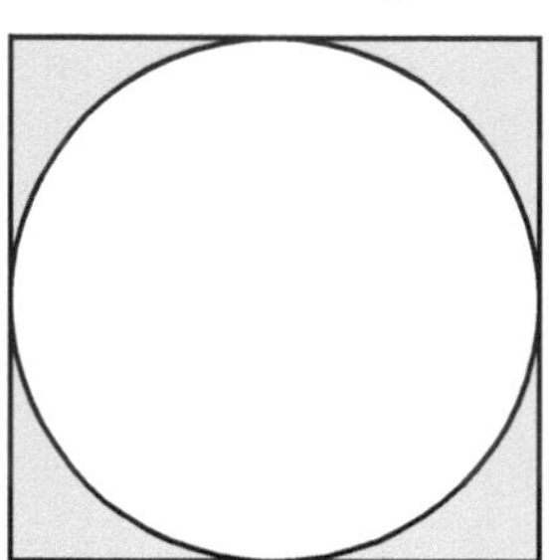

Fig. P2

16 $\frac{2}{x+2} - \frac{5}{3x-1} =$

A $\frac{-3}{2x-3}$ B $\frac{-4}{(x+1)(3x-1)}$

C $\frac{x-12}{(x+2)(3x-1)}$ D $\frac{x+1}{(x+2)(3x-1)}$

17 Given that $x = -3$, then $x^2 - x + 3 =$

A -9 B 3 C 9 D 15

18 Which one of the following defines a function?

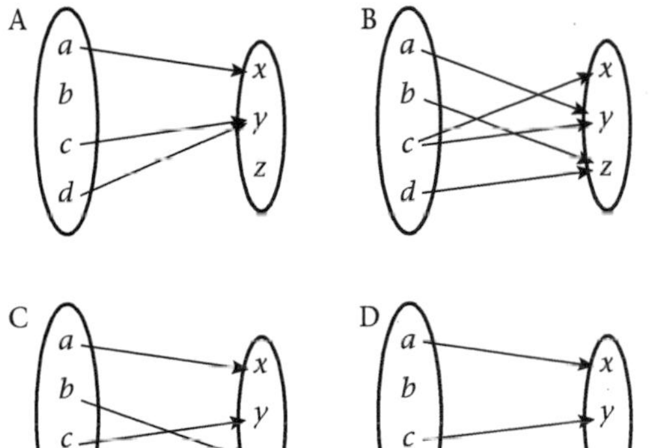

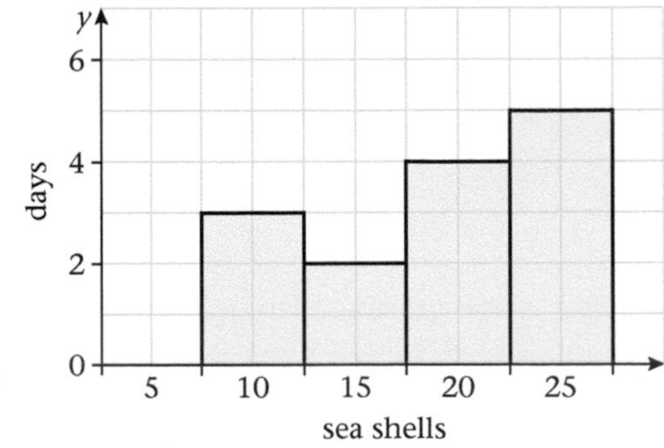

Fig. P3

19 A map is drawn to a scale of 1 : 50 000. The distance between two towns on the map is 2.5 cm. The actual distance, in km, between the two towns is

A 1.25 B 2.5 C 12.5 D 25

20 Given that n represents an *odd* integer, the next larger odd integer is

A $n - 1$ B $2n$

C $n + 1$ D $n + 2$

Fig. P4 shows the number of sea shells that a child collected over a period of time. Use Fig. P4 to answer questions 21 and 22.

Fig. P4

21 The total number of sea shells collected is

A 14 B 70 C 250 D 265

22 The number of days on which shells were collected is

A 4 B 14 C 50 D 70

23 If J\$1 = EC\$0.36 and EC\$1 = TT\$2.13, then, to the nearest dollar, TT\$150 =

A J\$25 B J\$115

C J\$196 D J\$391

24

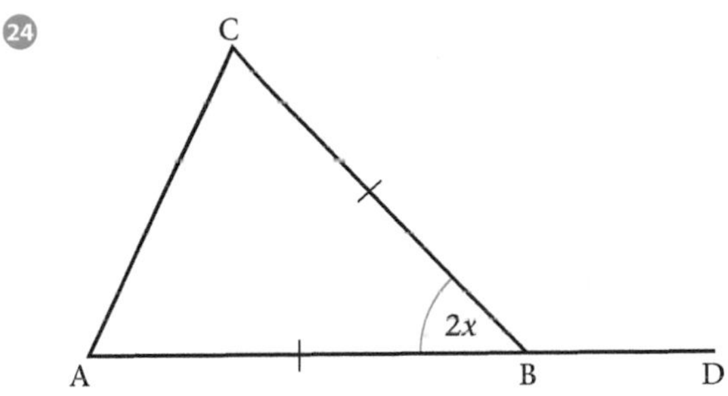

Fig. P5

In the triangle ABC (Fig. P5) BC = BA and $A\widehat{B}C = 2x°$. $A\widehat{C}B =$

A $2x°$ B $90° - 2x°$

C $90° - x°$ D $180° - 2x°$

25 If \$$P$ is invested at R% per annum for 3 years, then the simple interest earned is

A \$$\frac{100\,PR}{3}$ B \$$\frac{3\,PR}{100}$

C \$$\frac{3\,P}{100\,R}$ D \$$\frac{3 \times 100}{PR}$

26

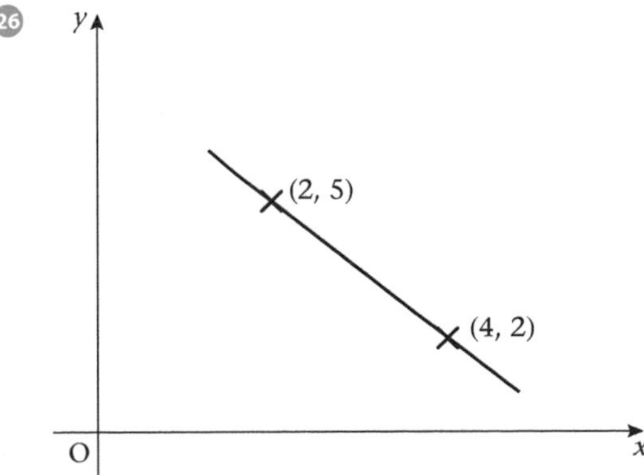

Fig. P6

The gradient of the line in Fig. P6 is

A $-\frac{3}{2}$ B $-\frac{2}{3}$ C $\frac{2}{3}$ D $\frac{3}{2}$

27 Which one of the following sets is *not* the null set?

A $\{x: x \text{ is a letter after } z \text{ in the alphabet}\}$
B $\{x: x^2 = 16 \text{ and } 3x = 9\}$
C $\{x: x + 5 = 5\}$
D $\{x: x \text{ is even and } x = 5^2\}$

28 If $\frac{8}{d+2} = \frac{5}{d+4}$, then $d =$

A $-\frac{2}{3}$ B -2 C $-7\frac{1}{3}$ D -14

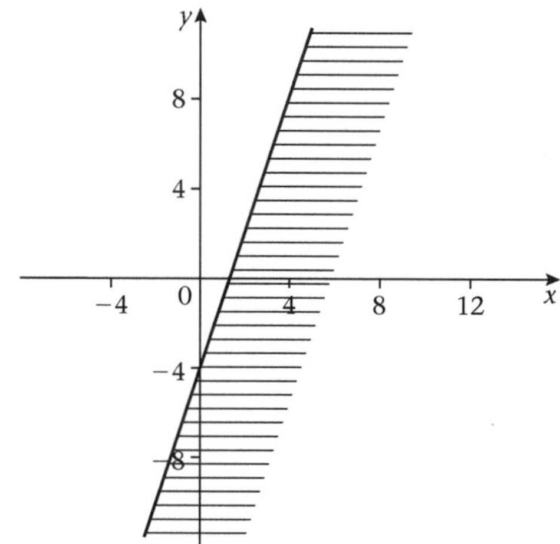

Fig. P7

The unshaded area in the graph in Fig. P7 represents the inequality

A $y < 2x - 4$
B $y \leqslant 2x - 4$
C $y > 2x - 4$
D $y \geqslant 2x - 4$

30 $0.04^{-1\frac{1}{2}} =$

A -0.008 B -0.06
C 37.5 D 125

Use the following information to answer questions 31, 32 and 33.
The heights in cm of twelve trees are

251	257	268	264	253	276
254	268	275	232	278	280

31 The range is

A 48 B 232 C 262 D 280

32 The median height is

A 263 B 266 C 268 D 280

33 The probability that a tree chosen at random is of height greater than 260 cm is

A $\frac{1}{3}$ B $\frac{1}{2}$ C $\frac{7}{12}$ D $\frac{2}{3}$

34 The total purchase price of a television set including a sales tax of 10% is \$1 100. The actual selling price of the set before tax is

A \$990 B 1 000
C \$1090 D \$1 210

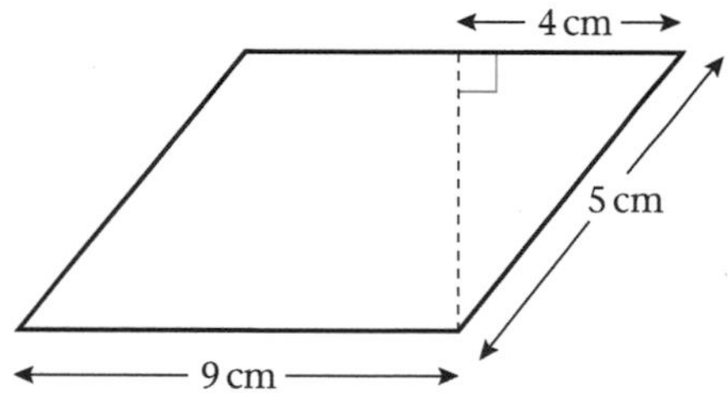

Fig. P8

The area, in cm², of the parallelogram in Fig. P8 (not drawn to scale) is

A 15 B 27 C 36 D 45

36 $b^2 - 5by + 6y^2 =$

A $(b + y)(b - 6y)$
B $(b - y)(b + 6y)$
C $(b - 2y)(b - 3y)$
D $(b + 2y)(b - 3y)$

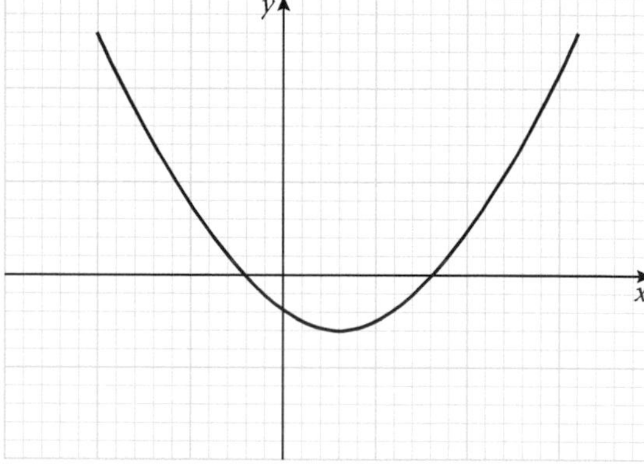

Fig. P9

Fig. P9 is the graph of the function

A $y = 2x^2 - 3x - 4$
B $y = x^2 - 3x - 4$
C $y = 2x^2 + 3x + 4$
D $y = x^2 - 3x + 4$

38

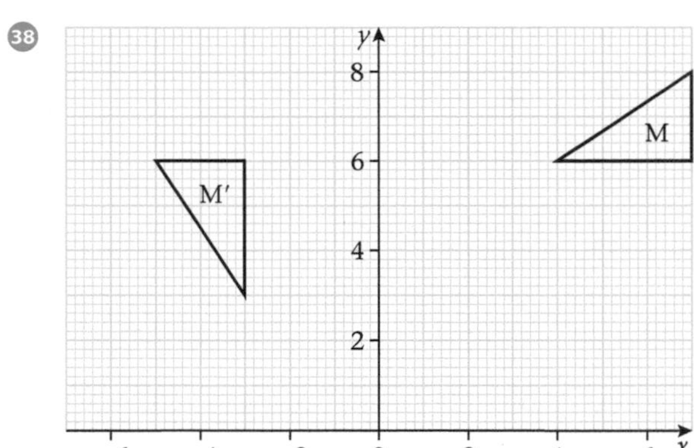

Fig. P10

In Fig. P10, M′ is the image of M under a rotation of 90°. The centre of rotation is
A (1,1) B (1,2) C (2,1) D (2,2)

39 Given that $\cos 30° = \frac{\sqrt{3}}{2}$, then $\tan 30° =$

A $\sqrt{3}$ B $\frac{1}{\sqrt{3}}$ C $\frac{2}{\sqrt{3}}$ D $\frac{\sqrt{3}+1}{2}$

40 Two geometrically similar tins have heights of 7 cm and 21 cm. If the smaller tin holds 250 g of powder, the capacity in kg of the larger tin is
A 0.75 B 2.25 C 5.25 D 6.75

41 Given that x is an even integer, the solution set which satisfies both $x \leqslant 4$ and $3x + 7 > 10$ is
A { } B {2} C {2, 3} D {2, 4}

42 If $\begin{pmatrix} 2 & a \\ 2a & 3 \end{pmatrix} + \begin{pmatrix} 4 & b \\ b & 2 \end{pmatrix} = \begin{pmatrix} 6 & 0 \\ 3 & 5 \end{pmatrix}$, then the value of b is
A -3 B $-\frac{3}{2}$ C $\frac{3}{2}$ D 3

Fig. P11 shows a circle, radius r, with a segment removed, Use Fig. P11 to answer questions 43 and 44.

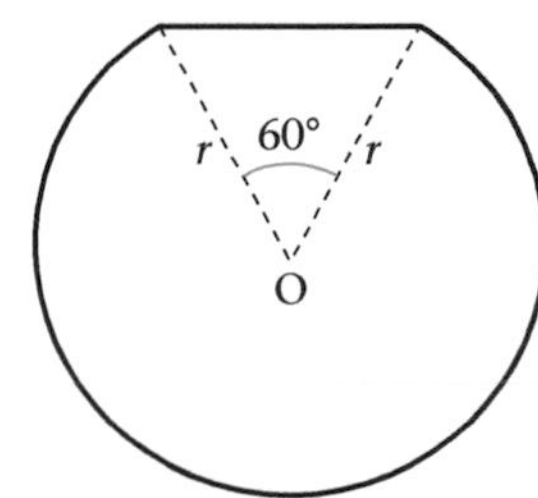

Fig. P11

43 The perimeter of the region shown is

A $\frac{5\pi + 1}{3}r$ B $\frac{5\pi + 3}{3}r$

C $\frac{7\pi}{3}r$ D $\frac{8\pi}{3}r$

44 The area of the segment removed is

A $\frac{\pi r^2}{6}$ B $r^2\left(\frac{\pi}{6} - \frac{1}{4}\right)$

C $r^2\left(\frac{2\pi}{3} - \frac{1}{4}\right)$ D $r^2\left(\frac{\pi}{6} - \frac{\sqrt{3}}{4}\right)$

45 The line AB passes through the points A(4, 4) and B(1, −2). The equation of the line AB is
A $y - x = 0$ B $y + 2x = 0$
C $y = 2x - 4$ D $y = 2x + 4$

46 The mean age of six pupils is 12 yr 5 mo. When a seventh pupil is added, the mean age becomes 12 yr 9 mo. The age of the seventh pupil is
A 12 yr 6 mo B 12 yr 7 mo
C 14 yr 7 mo D 14 yr 9 mo

47

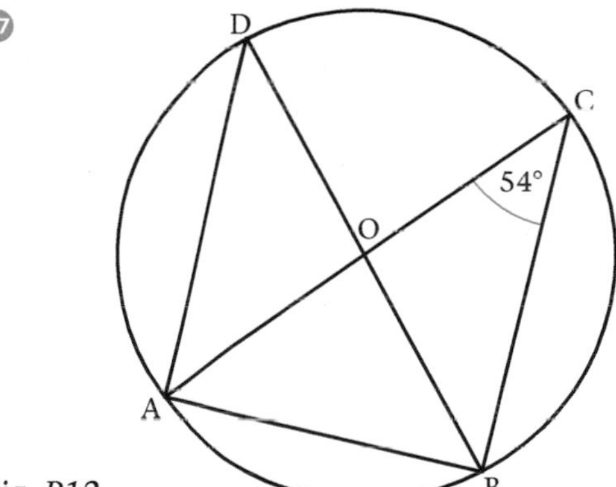

Fig. P12

In Fig. P12 O is the centre of the circle. $A\hat{B}D =$
A 72° B 63° C 54° D 36°

48 For which one of the following does an inverse function exist?

A

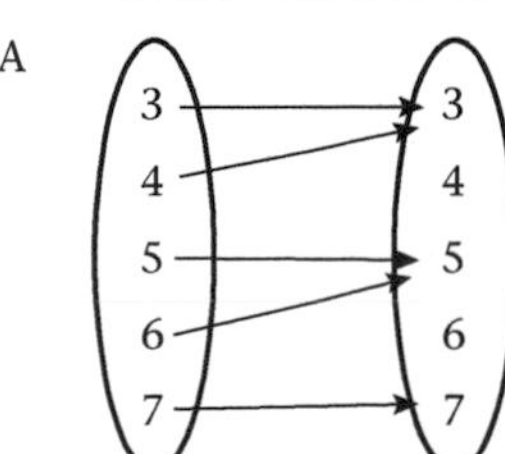

B

3, 4, 5, 6, 7 → 3, 4, 5, 6, 7

C

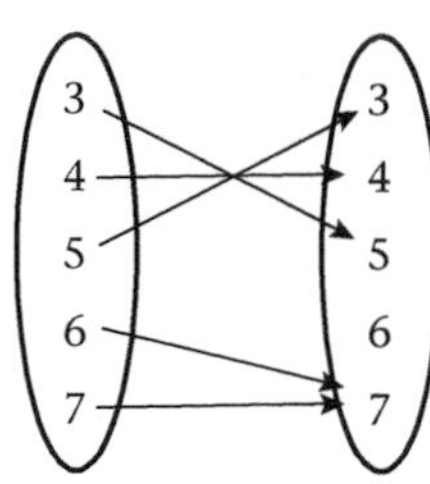

D

Fig. P13

49 If $V = \sqrt{\frac{2GM}{r}}$, then $G =$

A $\frac{Vr^2}{4M^2}$ B $Vr^2 - 4M^2$
C $V^2r - 2M$ D $\frac{V^2r}{2M}$

50 Given $\mathbf{p} = \mathbf{a} + 2\mathbf{b}$, $\mathbf{r} = 3\mathbf{a} - \mathbf{b}$, then $\mathbf{r} - \mathbf{p} =$
A $2\mathbf{a} - \mathbf{b}$ B $2\mathbf{a} + \mathbf{b}$
C $3\mathbf{b} - 2\mathbf{a}$ D $2\mathbf{a} - 3\mathbf{b}$

51 \$3 000 was invested and at the end of 6 months this sum of money had increased to \$3 090. The rate per cent per annum at which simple interest was added was
A $\frac{1}{2}$ B 1 C 6 D 18

52 Given that $\widehat{A}$ is acute and $\tan \widehat{A} = \frac{8}{15}$, then $\cos A =$
A $\frac{8}{17}$ B $\frac{15}{17}$ C $\frac{7}{15}$ D $\frac{15}{8}$

53 Each of the small circles in Fig. P14 is of radius r.

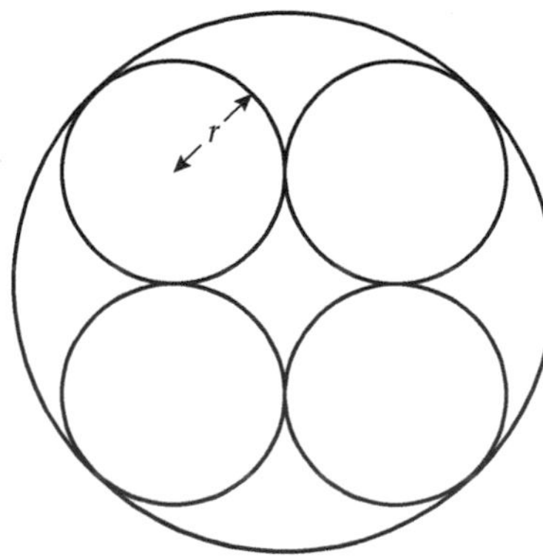

Fig. P14

The radius of the large circle in terms of r is given by
A $\sqrt{2}r$ B $2r$
C $r(1 + \sqrt{2})$ D $3r$

54 If $\begin{pmatrix} 4 & 0 \\ -2 & 2 \end{pmatrix} \begin{pmatrix} x \\ y \end{pmatrix} = \begin{pmatrix} -2 \\ 7 \end{pmatrix}$, then $\begin{pmatrix} x \\ y \end{pmatrix} =$

A $\begin{pmatrix} -\frac{1}{4} \\ -\frac{7}{2} \end{pmatrix}$ B $\begin{pmatrix} -\frac{1}{2} \\ 0 \end{pmatrix}$

C $\begin{pmatrix} -\frac{1}{2} \\ 3 \end{pmatrix}$ D $\begin{pmatrix} -\frac{1}{4} \\ -\frac{3}{2} \end{pmatrix}$

55 Given that $f(x) = 3x - 4$ and $g(x) = x^2 - 2$, $fg(-1) =$
A -7 B -1 C 1 D 5

56 X sold an article to Y at a profit of 20%. Y then sold it to Z at a loss of 20% of what it cost her. The ratio *final price*: *original price* in its simplest form is given by
A 3 : 2 B 1 : 1
C 24 : 25 D 4 : 5

57

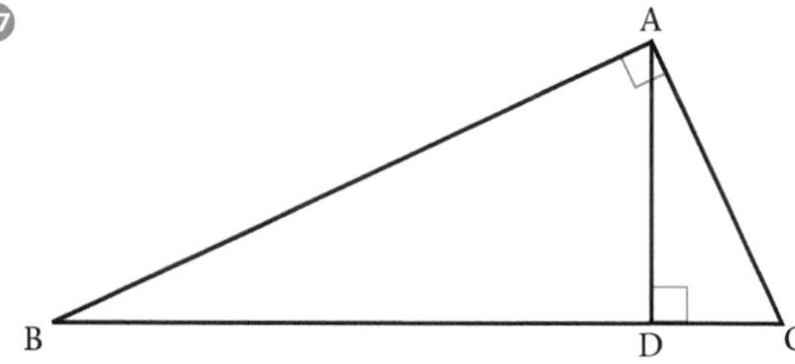

Fig. P15

In Fig. P15, $\frac{AB}{BC}$ is equivalent to

A $\frac{AD}{AC}$ B $\frac{DC}{CB}$ C $\frac{AD}{DC}$ D $\frac{AD}{DB}$

58 An investment was increased by 20%. The rate of interest was changed from 5% per annum to 10% per annum. The percentage increase in the simple interest earned at the end of 1 year was
A 140 B 70 C 40 D 25

59

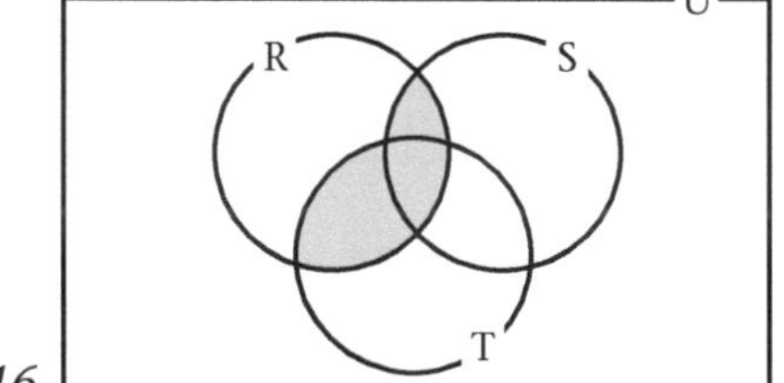

Fig. P16

In the Venn diagram (Fig. P16), the shaded area represents
A $R \cup (S \cap T)$ B $R \cap (S \cap T)$
C $(R \cap S) \cup (S \cap T)$ D $(R \cap S) \cup (R \cap T)$

60

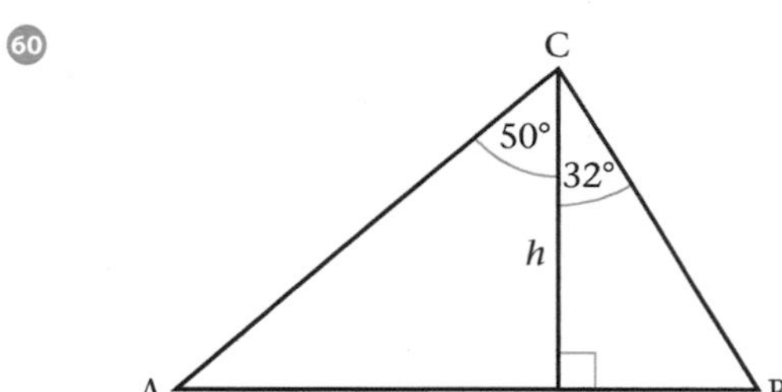

Fig. P17

The length of AB in the triangle ABC (Fig. P17) is

A $h(\tan 32° + \tan 50°)$
B $h \tan 82°$
C $h(\sin 32° + \sin 50°)$
D $h \sin 82°$

Paper 2 (2 hours 40 minutes)

1 *Answer* **all** *questions in Section I and any* **two** *in Section II.*
2 *Begin the answer for each question on a new page.*
3 *Full marks may not be awarded unless full working or explanation is shown with the answer.*
4 *Mathematical tables, formulae and graph paper are provided.*
5 *Silent electronic calculators may be used for this paper, except where instructions expressly forbid their use.*
6 *You are advised to use the first* 10 *minutes of the examination time to read through this paper. Writing may begin during this* 10 *minute period.*

Section I

Answer **all** *the questions in this section.* **All** *working must be clearly shown.*

1 (a) Calculate the exact value of $\dfrac{0.6 + 0.75}{0.6 \times 0.75}$ (3 marks)

(b) Find the value of $\dfrac{2\sqrt{1.21 \times 10^{-4}}}{4.4 \times 10^{-6}}$ giving the answer in standard form. [3 marks]

(c) A jug of water has a mass of 3.9 kg when full and 1.5 kg when quarter full. Calculate the mass of the jug when empty. [4 marks]

2 Given that U = {1, 2, 3, ..., 9, 10}
S = {perfect squares}
T = {1, 3, 6, 10}

(a) mark the members of these sets on a copy of the Venn diagram in Fig. P18.

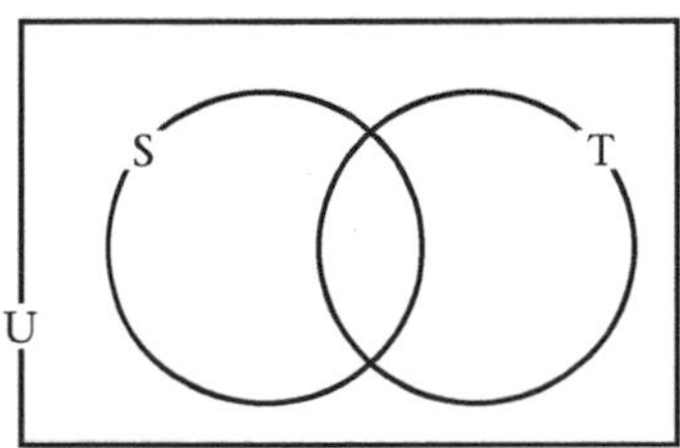

Fig. P18

(b) Using your diagram or otherwise,
(i) list the members of $S' \cap T'$,
(ii) find $n(T' \cup S)$. [7 marks]

3 A straight line passes through the points (5, 0) and (−4, 3). Find (a) the gradient of the line, (b) its equation, (c) the coordinates of the point where it cuts the line $x = 7$. [7 marks]

4 (a) If $c = ab - \frac{b}{a}$, find c when $a = \frac{2}{3}$ and $b = -6$ [3 marks]

(b) y varies inversely as the positive square root of x and $y = 12$ when $x = \frac{4}{9}$. Express y in terms of x. [3 marks]

(c) Factorise $16a^2 - 9$. Hence or otherwise express 1 591 as the product of two prime numbers. [4 marks]

5 The matrix $\begin{pmatrix} -3 & 0 \\ 0 & -3 \end{pmatrix}$ represents a transformation C.

(a) Find the image of (2, 7) under C.
(b) If (6, −9) is the image of point (r, s) under C, find r and s.
(c) Describe C as completely as possible. [9 marks]

6 (a) Sugar costing \$810/tonne is mixed with sugar costing \$920/tonne in the ratio 3 : 8. Calculate
(i) the cost of 1 tonne of the mixture.
(ii) the percentage profit if the mixture is sold at \$1157/tonne. [6 marks]

(b) The internal radius and height of a cylindrical oil drum are 30 cm and

90 cm respectively. The drum contains oil to a depth of 70 cm. Use the value 3.14 for π to calculate, correct to 2 significant figures (i) the area of the cylinder which is in contact with the oil, (ii) the extra volume of oil required to fill the drum to the top. [7 marks]

7 (a) The vertex, V, of a pyramid is vertically above the centre, O, of its rectangular base PQRS. If PQ = 9 cm, QR = 12 cm and VO = 6 cm, calculate, correct to 1 decimal place,

(i) the length of a diagonal of the base, PQRS,
(ii) the length of the sloping edge, VP,
(iii) the angle between VP and the base. [7 marks]

(b) *Answer this question on a sheet of plain paper.*

(i) Using ruler and compasses only, construct triangle ABC so that BC = 10 cm, $A\hat{B}C = 60°$ and $B\hat{C}A = 45°$. On the diagram write the length of AC.
(ii) Construct the perpendicular bisector of AB.
(iii) Draw the circumcircle of triangle ABC.
(iv) Hence mark a point P such that $A\hat{P}B$ 45° *and* AP = PB.
(v) Mark a point Q such that $A\hat{Q}B = 45°$ and AB = AQ.
(vi) Measure PQ.
(Give answers correct to nearest 0.1 cm.) [12 marks]

8 (a) A coffee farmer kept a record of his crop (Table P1).

Table P1

Days	No. of boxes
1–5	113
6–10	125
11–15	148
16–20	156
21–25	142
26–30	97

(i) Construct a cumulative frequency table to represent the data.
(ii) Using a scale of 1 cm to represent 5 days on the horizontal axis and 1 cm to represent 50 boxes on the vertical axis, draw a cumulative frequency curve.

Using the curve estimate

(iii) the number of boxes of coffee reaped in the first 16 days,
(iv) the time at which a quota of 500 boxes was reached. [11 marks]

(b) In a competition, the probability of hitting the target is $\frac{3}{4}$ and the probability of winning the door prize is $\frac{2}{3}$. Calculate the probability that someone will hit the target, or win the door prize or do both. [4 marks]

Section II

Answer any two *questions in this section.* [15 marks each]

Algebra and relations, functions and graphs

9 (a) Draw the graph of the function $y = 7 - x - x^2$ for values of x from -4 to 3, using a scale of 2cm to represent 1 unit on the x-axis and 1 cm for 1 unit on the y-axis.

(b) Estimate *from your graph* the greatest value of $7 - x - x^2$.

(c) Using the graph, find, correct to 1 decimal place, the solution of the equations

(i) $7 - x - x^2 = 0$,
(ii) $2 - 2x - x^2 = 0$.

10 A factory makes radios and cassette players. The manufacturing details are as follows:

	Radio	Cassette
Cost of making 1 item	\$9	\$18
Time to make 1 item	6 min	4 min
Profit/item	\$10	\$15

The factory works not more than 40 h per week and has up to \$7 200 available to cover weekly production costs.

(a) Using r and p to represent the numbers of radios and cassette players made in a week, write down two inequalities which satisfy the given imformation.

(b) The point (r, p) represents r radios and p cassette players. Using a scale of 2 cm to 100 items on each axis, construct the region in which all points (r, p) must lie. (Indicate the region by shading the *unwanted* region.)

(c) Use your diagram to find (i) the number of radios and cassette players that should be made to obtain the greatest profit, (ii) the greatest profit.

Geometry and trigonometry

11 Fig. P19. represents a field ABCD.

(a) Use the dimensions given in Fig. P19 to calculate (i) BD, (ii) BC, (iii) the area of the field in m².

(b) If D is due east of B, find the bearing of (i) B from C, (ii) C from B, (iii) A from B.

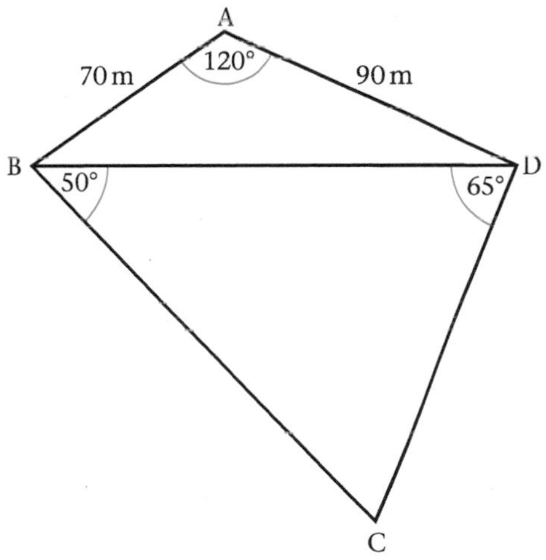

Fig. P19

12 P (44°W) and Q (31°E) are points on latitude 61°30′S. Calculate (a) the distance in nautical miles of P from the South Pole, (b) the distance in km from P to Q along the parallel of latitude, (c) the latitude of a place 870 km due north of Q. (*Note*: consider that the earth is a sphere of circumference 40 000 km.)

Vectors and matrices

13 (a)

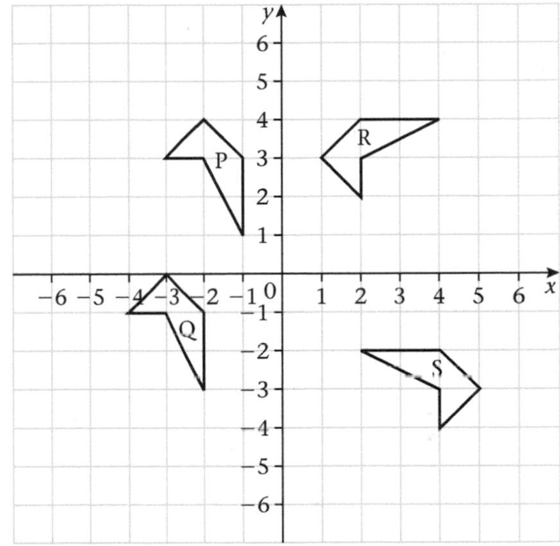

Fig. P20

Shapes P, Q, R and S are as given in Fig. P20.

(i) Write down the column vector which represents the single translation which maps P onto Q.

(ii) R is the image of P under a clockwise rotation. Find the angle of rotation and the coordinates of the centre of rotation.

(iii) S is a reflection of P in a line l. Find the equation of l.

(b) The transformation J is represented by the matrix $\begin{pmatrix} 2 & 1 \\ 0 & 3 \end{pmatrix}$, and the transformation K by $\begin{pmatrix} 1 & -1 \\ -1 & 0 \end{pmatrix}$.

(i) Find the image of the point D(2, 3) under the transformation J followed by the transformation K.

(ii) Show that the combination of transformations represented by J and K is *not* commutative.

14 (a) In Fig. P21 $\overrightarrow{AC}$ represents the column vector $\begin{pmatrix} 2 \\ -4 \end{pmatrix}$.

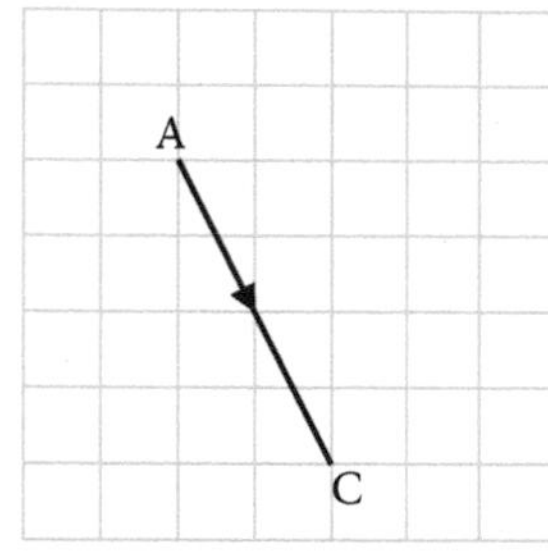

Fig. P21

AC is a diagonal of a square ABCD such that $\overrightarrow{AD} = \begin{pmatrix} p \\ q \end{pmatrix}$ where p and q are both negative.

(i) Sketch the square ABCD on a copy of Fig. P21.

(ii) Write down the values of p and q.

(iii) Write down $\overrightarrow{AB}$ in column vector form.

(iv) Write down $\overrightarrow{DB}$ in column vector form.

(b) $\begin{pmatrix} 4x & x-5 \\ 1-3x & x \end{pmatrix}$ is the inverse matrix of $\begin{pmatrix} x & 5-x \\ 3x-1 & 4x \end{pmatrix}$

Find two values of x for which this is true.

Practice examination II

Paper 1 ($1\frac{1}{2}$ hours)

Attempt **all** *the questions. Circle the letter of your choice.*

1. In decimal form, $\frac{3}{8} =$
   A 0.375 B 0.38 C 3.75 D 3.8

2. The L.C.M. of 3, 6, 8, 18 is
   A 24 B 36 C 72 D 96

3. $(-5)(-2) - (3)(-3) =$
   A -19 B -1 C 1 D 19

4. The number of lines of symmetry in a rhombus is
   A 4 B 3 C 2 D 1

5. $14.17 \div 0.13 =$
   A 1.09 B 10.9
   C 19.0 D 109

6. The subsets of the universal set U = {f, g} are
   A { },{f},{g}, {f, g} B { },{ f},{g}
   C { }, {f, g} D {f}, {g}

7. 1 litre =
   A 102 cm^2 B 103 cm^3
   C 103 cm^2 D 102 cm^3

8. The prime factors of 30 are
   A 1, 2, 3, 5 B 2, 3, 5, 6, 15
   C 2, 3, 5 D 1, 2, 3, 5, 6, 15, 30

9. 0.0235, written in standard form, is
   A 2.35×10^{-1} B 2.35×10^{-2}
   C 2.35×10^{-3} D 2.35×10^{-4}

10. The fractions are arranged in descending order of magnitude in
    A $\frac{3}{4}, \frac{5}{7}, \frac{2}{3}, \frac{3}{5}$ B $\frac{2}{3}, \frac{3}{4}, \frac{3}{5}, \frac{5}{7}$
    C $\frac{5}{7}, \frac{3}{5}, \frac{3}{4}, \frac{2}{3}$ D $\frac{3}{4}, \frac{2}{3}, \frac{3}{5}, \frac{5}{7}$

11. A trip began at 22:25 hrs and ended at 03:15 hrs. The time taken for the trip was
    A 4 h 50 min B 5 h 30 min
    C 7 h 10 min D 19 h 10 min

12. 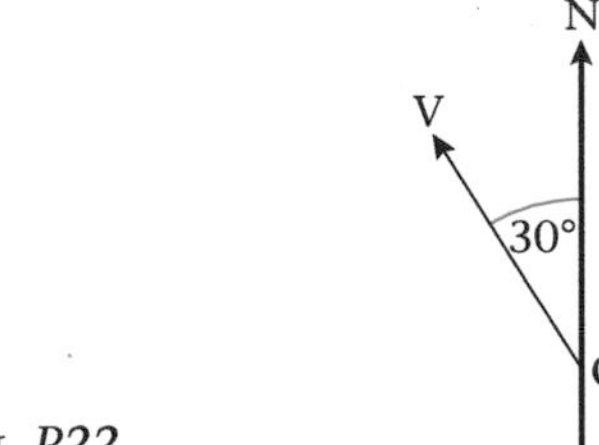

Fig. P22

In Fig, P22, **not drawn to scale**, the bearing of the point V from O is
A 030° B 060° C 150° D 330°

13. The total surface area in cm^2 of a cuboid with sides of length 3 cm, 4 cm and 5 cm is
    A 47 B 60 C 94 D 120

14. 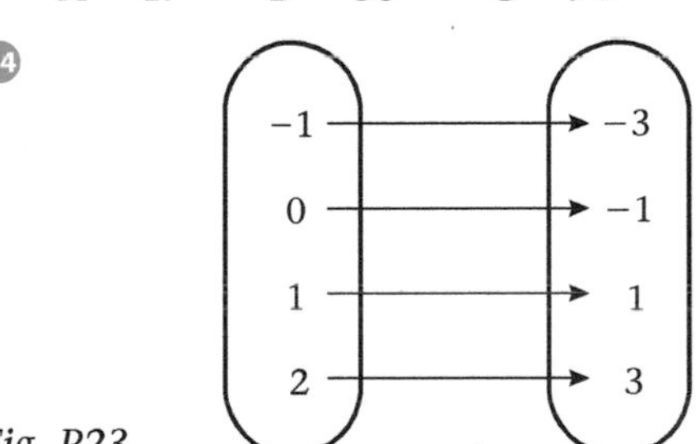

Fig. P23

The arrow diagram in Fig P23 shows the mapping *f*, where
A $f: x \to x - 2$ B $f: x \to -2x^2 - 1$
C $f: x \to 2x - 1$ D $f: x \to x^2 - 1$

15. The number that is 20% more than 40 is
    A 42 B 44 C 46 D 48

16. Given that $3x - y = 5$
    and $x + y = 21$,
    $(x, y) =$
    A $(-1, 2)$ B $(1, -2)$
    C $(2, -1)$ D $(-2, 1)$

17. Given that
    P = {a, b, c}
    Q = {1, 2, 3}
    R = {the first three letters of the alphabet}
    S = {c, a, b}
    the equal sets are
    A P and R B P and S
    C P, R and S D P, Q, R and S

18 On a map, 1 cm represents 10 km. The scale used is

A 1 : 10
B 1 : 1 000
C 1 : 10 000
D 1 : 1 000 000

19 $\frac{2}{5a} + \frac{3}{a-3} =$

A $\frac{1}{a(a-3)}$ B $\frac{2}{5(a-3)}$
C $\frac{17a-6}{5a(a-3)}$ D $\frac{17a-3}{5a(a-3)}$

20

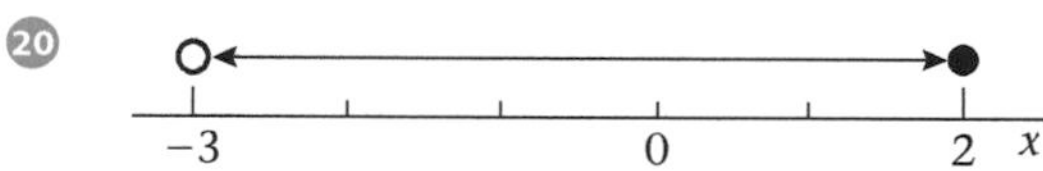

Fig. P24

Fig. P24 shows the inequality

A $-3 \leqslant x < 2$
B $-3 < x \leqslant 2$
C $-3 \geqslant x < 2$
D $-3 > x \leqslant 2$

21 The selling price of a book is \$126. If the cost price is \$105, the percentage profit on the cost price is

A $16\frac{2}{3}$ B 20 C 80 D $83\frac{1}{3}$

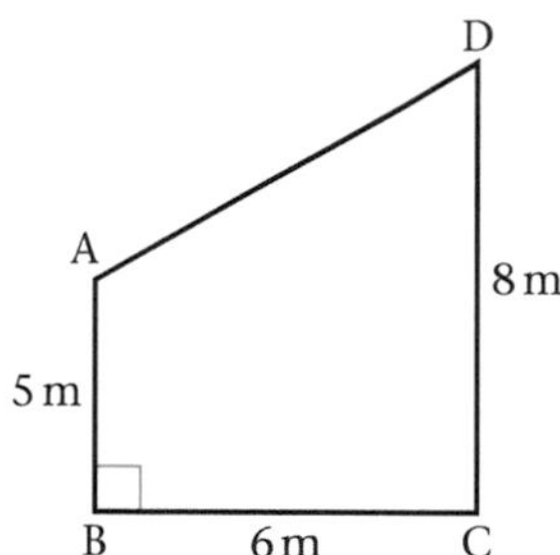

Fig. P25

Fig. P25, **not drawn to scale**, shows the floor of a room. AB is perpendicular to BC, and AB is parallel to DC. AB = 5 m, BC = 6 m, and DC = 8 m. The area in m² of ABCD is

A 38 B 39 C 42 D 48

23 The simple interest earned on \$1500 for 3 years was \$270. The rate percent per year was

A 1.8 B 2.0 C 5.4 D 6.0

24 Using the Venn diagram in Fig. P26, which of the following statements is **not** true?

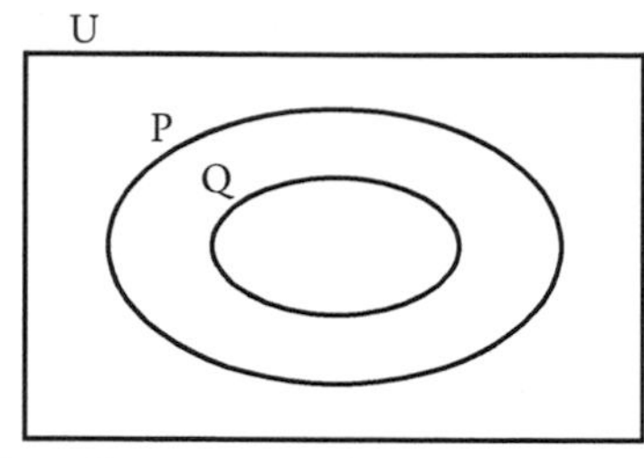

Fig. P26

A $P \subset Q$ B $P \cup Q = P$
C $Q \subset P$ D $P \cup Q = Q$

25 $(x^2y)^3 \div \frac{1}{xy^2} =$

A x^4y B x^5y C x^4y^5 D x^7y^5

26 When \$27.00 is shared in the ratio 2 : 3 : 4, the largest share is

A \$4.00 B \$9.00
C \$12.00 D \$22.00

27 $2\,347_8 + 461_8 =$

A $2\,808_8$ B $2\,828_8$
C $3\,010_8$ D $3\,030_8$

28 Which set of ordered pairs defines a function?

A {(2, −1), (2, 0), (2, 1)}
B {(2, −1), (3, −1), (4, −1)}
C {(2, −1), (3, 0), (3, 1)}
D {(2, −1), (3, 0), (2, 1)}

29 When $a = 2$, $b = 3$, $c = -2$, $\frac{3a^2b}{5a^2b - 3c^2} =$

A $-\frac{1}{4}$ B $\frac{1}{9}$ C $\frac{1}{2}$ D $\frac{3}{4}$

Questions 30 and 31 refer to Fig. P27. The bar graph shows the number of stamps collected by a group of students.

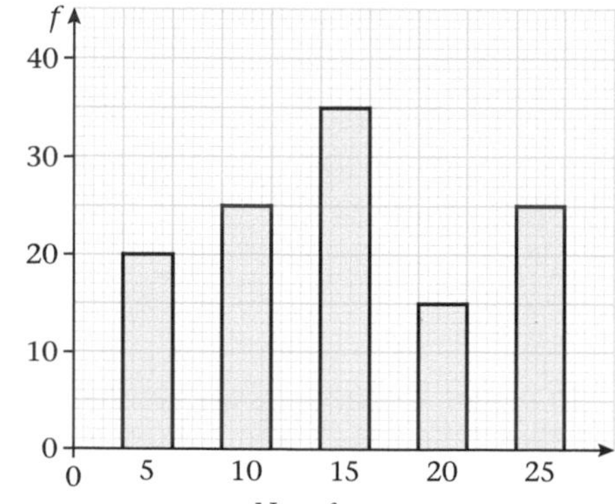

Fig. P27

30 The total number of students who collected stamps is

A 75 B 85 C 100 D 120

31 The mode of the distribution is

A 15 B 24 C 25 D 35

32 A discount of 15% is given for full payment by cash. The discount on a table costing $1 515 is

A $101.50 B $110.00
C $227.25 D $247.50

33 Given that $f(3) = -2$, then $f^{-1}(-2) =$

A $-\frac{1}{2}$ B 3 C $\frac{1}{3}$ D $\frac{1}{2}$

34 The heights in cm of ten students are 140, 142, 146, 156, 166, 148, 176, 158, 165, 162 The range is

A 162 B 157 C 36 D 18

35 If $\begin{pmatrix}-5\\6\end{pmatrix} + \mathbf{p} = \begin{pmatrix}2\\-3\end{pmatrix}$, then $\mathbf{p} =$

A $\begin{pmatrix}-3\\3\end{pmatrix}$ B $\begin{pmatrix}7\\-9\end{pmatrix}$

C $\begin{pmatrix}-7\\9\end{pmatrix}$ D $\begin{pmatrix}7\\-3\end{pmatrix}$

36 Given that $4c + 7 \geqslant 2c + 3$

A $c \leqslant -2$ B $c \leqslant -5$
C $c \geqslant -2$ D $c \geqslant -5$

37 In Fig. P28, O is the centre of the circle, radius 21 cm. The angle of the sector XOYW is 60°. The area in cm² of the shaded sector is

A 1 155 B 462 C 231 D 110

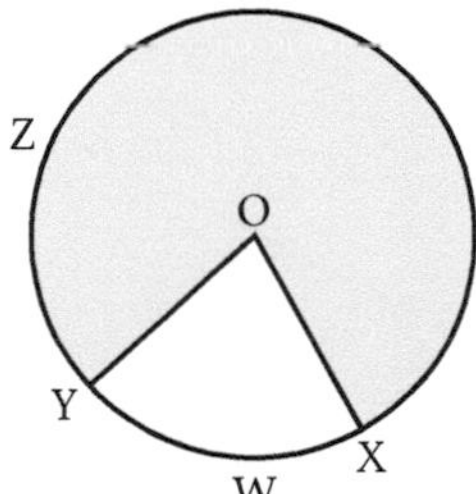

Fig. P28

38 Given that $p * q = \frac{p}{3} - \frac{4}{q}$,

$3 * (3 * 2) =$

A 5 B $\frac{11}{3}$ C $\frac{-1}{3}$ D -3

39 The volume of a cylinder is 250π m³. If the height of the cylinder is equal to the diameter of the base, the diameter in m of the base is

A 2 B 4 C 5 D10

40 If x varies directly as y^2, and $x = 9$ when $y = 2$, then

A $9x = 4y^2$ B $4x = 9y^2$
C $x = 36y^2$ D $36x = y^2$

41 The distribution of funds in a cash budget of a business is shown in the table below:

Salaries, wages and fees	60%
Rental charges: office	10%
: equipment	5%
General expenses	17%
Utilities	5%
Financial charges	3%

In a pie chart, the angle representing the total rental charges is

A 15° B 27° C 30° D 54°

42 Given the set of letters in the word ACCEPT, what is the probability of choosing the letter C?

A $\frac{1}{6}$ B $\frac{1}{5}$ C $\frac{1}{3}$ D $\frac{2}{3}$

43 'Twice a number is 5 less than 13' written in symbols is

A $13 - 2n = 5$ B $2n - 13 < 5$
C $13 - 2n < 5$ D $2n - 5 = 13$

Questions 44 and 45 use the following information. The cash price of a computer is $9 000. On hire purchase, the down-payment is $4 800 and equal monthly payments are made for 1 year. The total amount paid on hire purchase is $12 000.

44 Each monthly payment is equal to

A $400 B $600 C $750 D $1 000

45 The percentage increase of the hire purchase price over the cash price is

A $62\frac{1}{2}$% B 40% C $33\frac{1}{3}$% D 25%

46 The coordinates of the points J and K are (2, 3) and (6, 6) respectively. The length of the straight line joining J and K is d units, where $d =$

A 7 B 5 C 4 D 3

47 The next term in the sequence 1, 3, 7, 15, 31, ... is

A 39 B 43 C 47 D 63

48 If C\$2.09 ≡ G\$1.00
and B\$0.41 ≡ C\$1.00,
then B\$1.00 ≡ G\$$x$, where x =

A 5.10 B 1.17 C 0.86 D 0.20

49

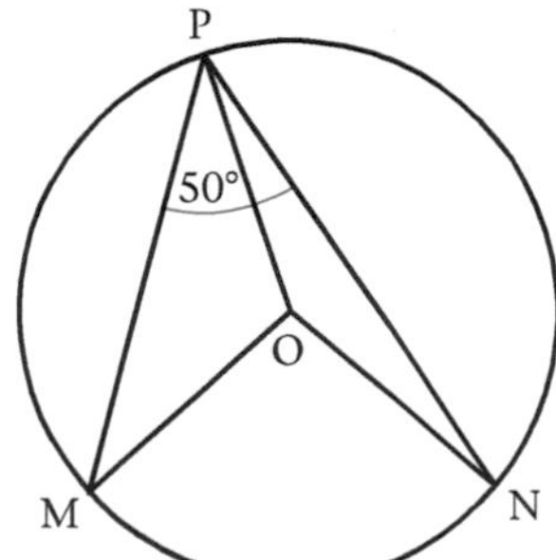

Fig. P29

In Fig P29, **not drawn to scale**, O is the centre of the circle and $M\widehat{P}N = 50°$.
$M\widehat{O}P + P\widehat{O}N =$

A 80° B 100° C 260° D 310°

50 The mean of a set of five numbers is 17.4. The number 15 is added to the set. The new mean is

A 20.4 B 17.0 C 16.2 D 14.0

51 A length is measured as 7.4 cm to the nearest 0.1 cm. Of the following values, the smallest value possible in cm of the length is

A 7.35 B 7.36 C 7.39 D 7.40

52 An article costing \$20 was sold for \$35. During a sale, the article was marked at \$25. The profit percentage was *reduced* by

A 25% B 40% C 50% D 60%

53

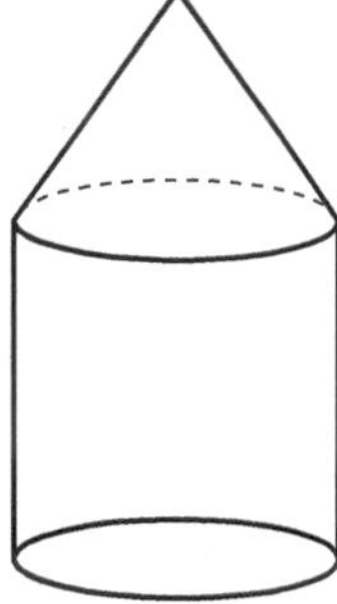

Fig. P30

The plan of the solid shown in Fig. P30 is

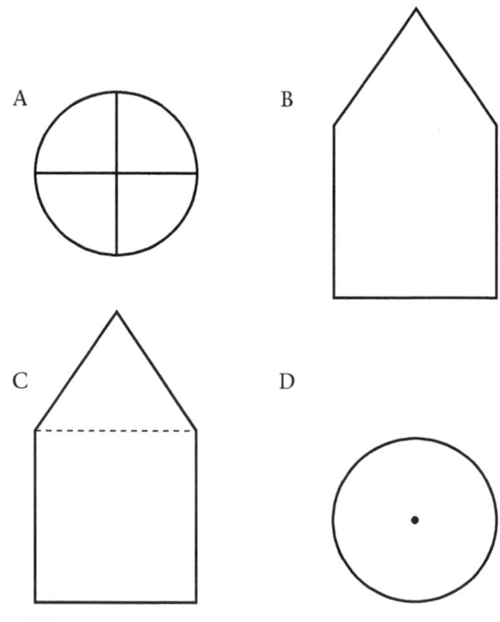

Fig. P31

54 A point P (3, 1) is reflected in the x-axis and 24 then translated by a vector $\begin{pmatrix}-4\\-2\end{pmatrix}$. The coordinates of the image P′ are

A (−1, 3) B (−1,−3)
C (−7,−1) D (−7,−3)

55 The distance x km moved in t min is shown in the graph in Fig. P32. The average speed is

A 1 km/h B 20 km/h
C 40 km/h D 60 km/h

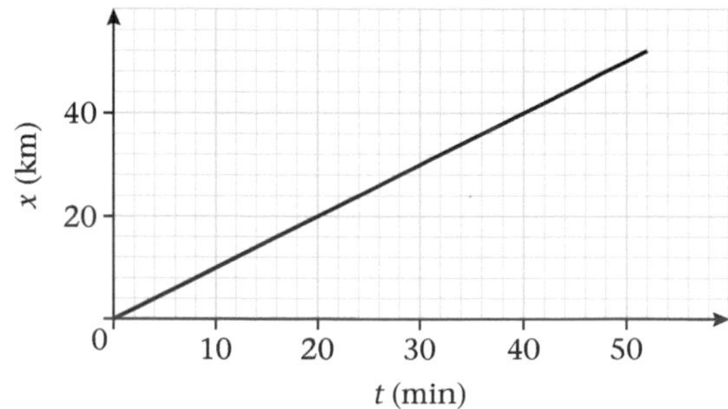

Fig. P32

56 If sin 20° = 0.342, and cos 20° = 0.940, then sin 110° =

A 0.940 B 0.342
C −0.342 D −0.940

57 Given that U = {animals}
D = {dogs}
B = {big dogs}

The statement '*Toro is a big dog*', where T represents Toro, is shown in

A

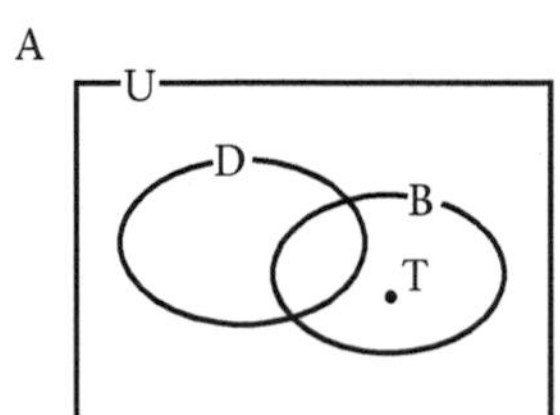

B

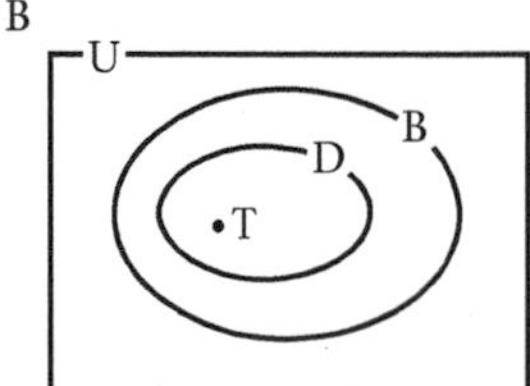

C

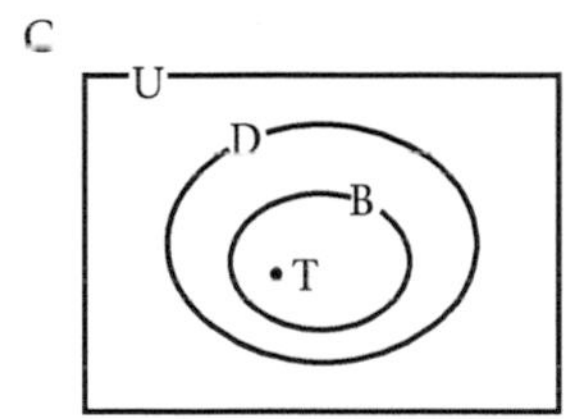

D

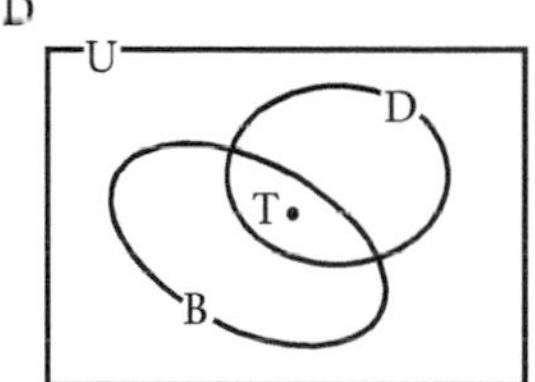

Fig. P33

58 The value of a machine depreciates each year by 10% of its value at the beginning of the year. The value of the machine at the beginning of 1998 was \$5 000. The value of the machine during the year 2000 is

A \$4 100
B \$4 050
C \$4 000
D \$3 500

59 In Fig. P34, △ABC and △DAC are right-angled triangles. In △ABC, $B\hat{A}C = 90°$, and in △DAC, $A\hat{D}C = 90°$. Given that AB = 4 cm, BC = 5 cm and AC = 3 cm, AD = h cm, where h =

A $\frac{12}{5}$ B $\frac{9}{5}$ C $\frac{9}{4}$ D $\frac{15}{4}$

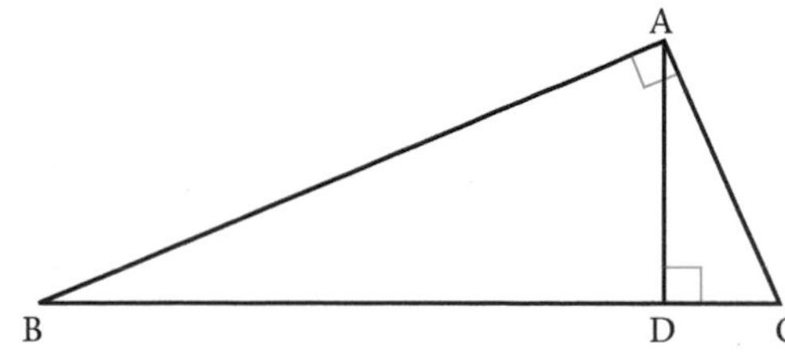

Fig. P34

60 The inequality $y \leq x + 1$ is represented graphically by the unshaded area in

A

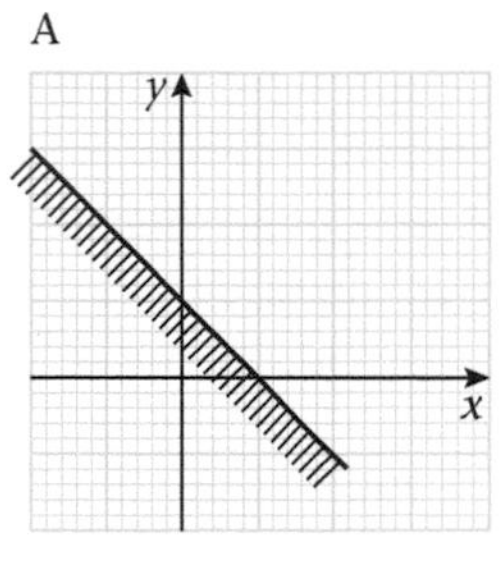

B

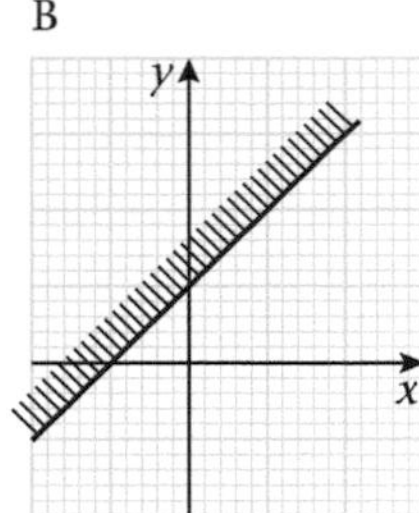

C

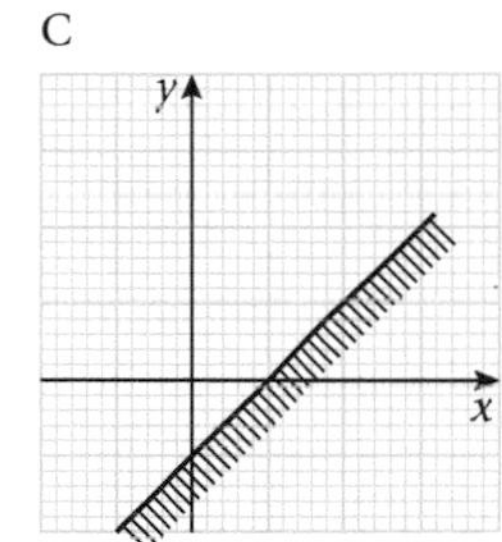

D

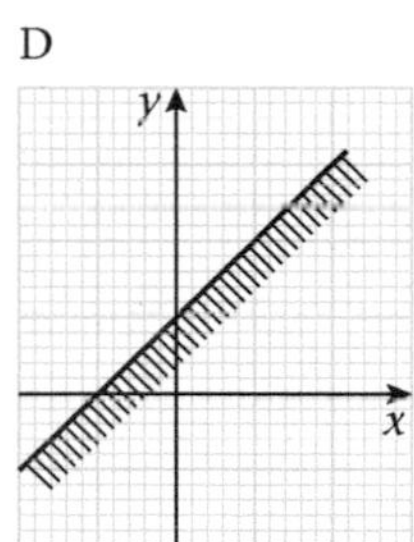

Fig. P35

Paper 2 (2 hours 40 minutes)

1 *Answer* **all** *questions in Section I and any* **two** *in Section II.*
2 *Begin the answer for each question on a new page.*
3 *Full marks may not be awarded unless full working or explanation is shown with the answer.*
4 *Mathematical tables, formulae and graph paper are provided.*
5 *Silent electronic calculators may be used for this paper, except where instructions expressly forbid their use.*
6 *You are advised to use the first 10 minutes of the examination time to read through this paper. Writing may begin during this 10 minute period.*

Section 1

Answer **all** *the questions in this section.*
All working must be clearly shown.

1 (a) Simplify

$$\frac{2.146 \times 10^{-3} + 1.34 \times 10^{-4}}{1.14 \times 10^{-4}}$$

[3 marks]

(b) The tax based on the value of a property is charged at an annual rate of \$4.50 per \$10 000.
   (i) Calculate the annual property tax when the value is \$200 000.

   In addition, there is a government tax for utilities of 3% per annum.
   (ii) Calculate the annual tax for utilities.
   (iii) Find the total monthly payment due for taxes. [7 marks]

Total 10 marks

2 Solve the following equations:

(a) $3(w - 5) = 2(w - 4)$

(b) $\frac{y}{2} - \frac{2y}{7} = 3$

(c) $\frac{x+4}{4x} - \frac{3}{x+3} = \frac{1}{2x}$

[10 marks]

3 (a) Using ruler and compasses only, construct a trapezium WXYZ so that WX is parallel to ZY. WX = 11 cm, XY = 5 cm, $Z\widehat{W}X = 60°$, and $W\widehat{X}Y = 45°$.

(b) Measure and state the length of WZ.

(c) Show, *by calculation*, that the perpendicular distance between WX and ZY is 3.5 cm, correct to 1 decimal place.

[10 marks]

4 Fig. P36, **not drawn to scale**, represents ABCDE, the cross-section of the model of a container. ABCD is a square and AED is a semicircle. The maximum height of the model is 10.5 cm, and the length of the model is 8 cm.

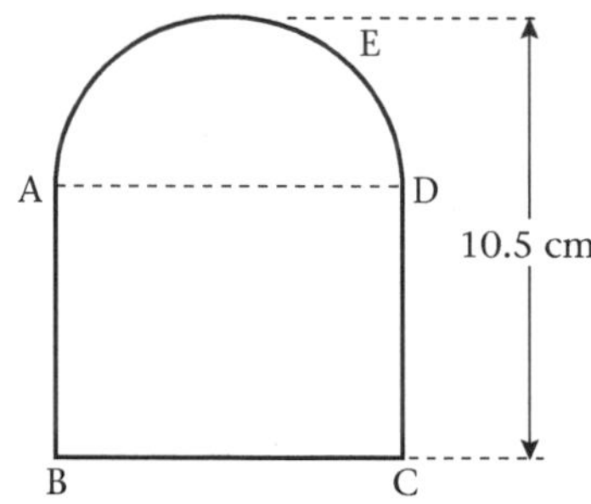

Fig. P36

(a) Taking $\pi = \frac{22}{7}$, calculate
   (i) BC, the maximum width of the cross-section of the model,
   (ii) the area of the cross-section, ABCDE,
   (iii) the capacity of the model.

(b) The scale used for the model is 1 : 25. Calculate the capacity in m^3 of the actual container, giving your answer correct to 2 decimal places.

[10 marks]

5 The table below shows the frequency distribution of the weights of a group of students.

Weight (kg)	Frequency (f)
40–44	3
45–49	5
50–54	9
55–59	12
60–64	10
65–69	7
70–74	3
75–79	1

(a) Draw a histogram of the distribution.

(b) Estimate, to the nearest whole number, the median weight of the distribution.

(c) Calculate, correct to 1 decimal place, the mean weight of the students.

[10 marks]

6 J is the point (−1, 2), and K the point (3, 5). Find, *by calculation*,

(a) the coordinates of the mid-point of JK,

(b) the slope of the line JK,

(c) the equation of the perpendicular bisector of JK. [10 marks]

7 (a) In a sports club, there are 60 members who play basketball [B], volleyball [V], and tennis [T]. All members play at least one game, and all the members who play volleyball also play basketball.
   (i) Write in set notation, the relation between V and B.

(ii) Given also that
25 members play volleyball
36 members play basketball
40 members play tennis
24 members play tennis only
12 members play all three games
show the information given on a Venn diagram. [6 marks]

(b) The triangle PQR has vertices P(3, 8), Q(2, 5) and R(5, 6).
(i) Find, by drawing or otherwise, the vectors $\overrightarrow{PS}$ and $\overrightarrow{PT}$, such that $\overrightarrow{PS} = 2\overrightarrow{PR}$ and $\overrightarrow{PT} = 2\overrightarrow{PQ}$.
(ii) Calculate the vector TS.
(iii) Show that $\triangle$PTS is an isosceles triangle. [9 marks]

Total 15 marks

8 (a) Given that $f(x) = \dfrac{x+1}{x-1}$ and $g(x) = x^2 - 2$,
(i) calculate $g(3)$, $fg(3)$,
(ii) show that $f^{-1}(x) = f(x)$. [8 marks]

(b) There are m apples and n oranges in a basket. The ratio of the number of apples to the number of oranges is 1 : 3.
(i) Write an equation to show the relation betwen m and n.
The apples are sold at \$2.00 each and the oranges at \$1.50 each.
(ii) Write an expression in m and n for the total amout of money received.
(iii) If the total amount received was \$65, write an equation in m and n to show this information.
(iv) Use the equations in (i) and (iii) to find the value of m and of n.
[7 marks]

Total 15 marks

Section II

Answer any **two** *questions in this section.*

Algebra and relations, functions and graphs

9 (a) Write $f(x) = 6 + 5x - 2x^2$ in the form $f(x) = a(x + b)^2 + c$ where a, b and c are constants.

(b) State
(i) the maximum value of $f(x)$,
(ii) the value of x at which the maximum value of $f(x)$ occurs.

(c) Find, correct to 1 decimal place, the values of x which satisfy the equation $6 + 5x - 2x^2 = 0$

(d) Sketch the graph of $f(x)$, indicating the maximum value of the function and the coordinates of the points of intersection with the axes. [15 marks]

10 Fig. P37 shows the speed–time graph of a motorcar during part of a jounery.

(a) Describe fully the information given by each of the parts, (i), (ii), and (iii), of the graph in Fig. P37.

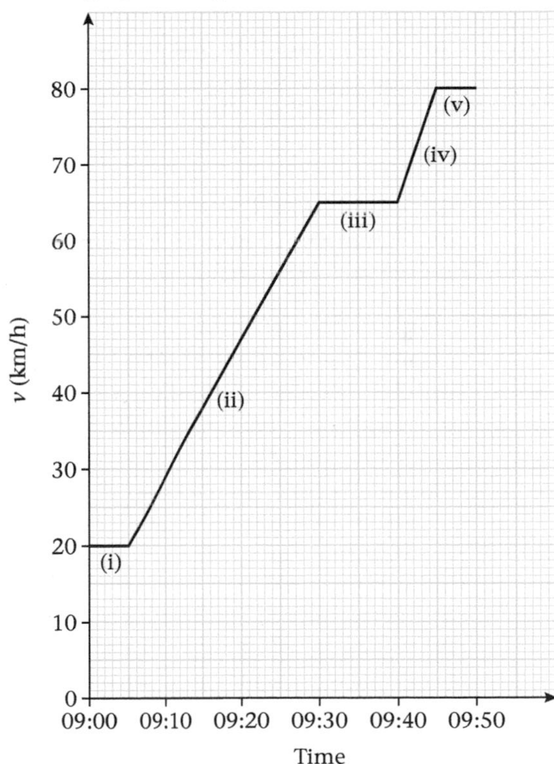

Fig. P37

(b) Calculate the acceleration in km/h per min during (iv) and (v).

(c) Calculate the distance travelled in the first half-hour.

(d) Find the average speed to the nearest km/h during the time given. [15 marks]

Geometry and trigonometry

11

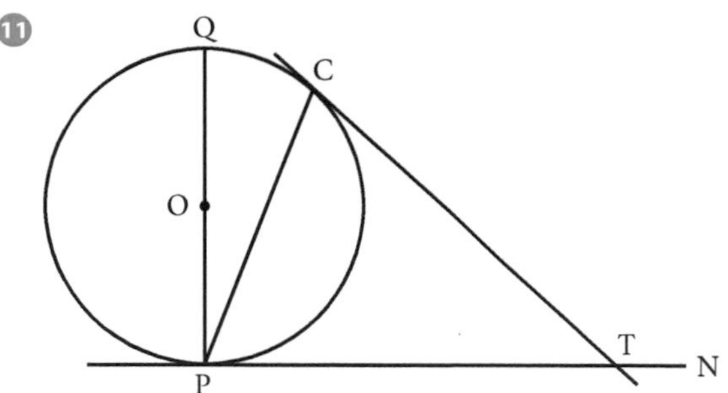

Fig. P38

Fig. P38, **not drawn to scale**, is a circle, centre O and radius 10.5 cm. PN is a tangent at P. CP is a chord such that $C\widehat{P}N$ is 80°.

(a) Calculate, correct to 1 decimal place, giving a reason for each statement,
 (i) the length of the chord, CP,
 (ii) the shortest distance from C to PN.

(b) The tangent at C intersects PN at T.
 (i) Show that CT = PT.
 (ii) Calculate, correct to 1 decimal place, the length of PT.

[15 marks]

12 (a) A point D is 30 m south of a tree, HF. Another point G is 24 m on a bearing of 060° from the tree. The angle of elevation of H, the top of the tree, from D is 27°.
 (i) Draw a sketch to show this information.

Calculate
 (ii) the height of the tree,
 (iii) the angle of elevation of H from G,
 (iv) the distance DG, correct to the nearest metre. [8 marks]

(b) (i) In Fig. P39, describe fully the single transformation which maps quadrilateral K onto quadrilateral K′.

If K′ = RE(K) where E is an enlargement and R is a rotation,
 (ii) state the scale factor of E,
 (iii) describe R fully. [7 marks]

Total 15 marks

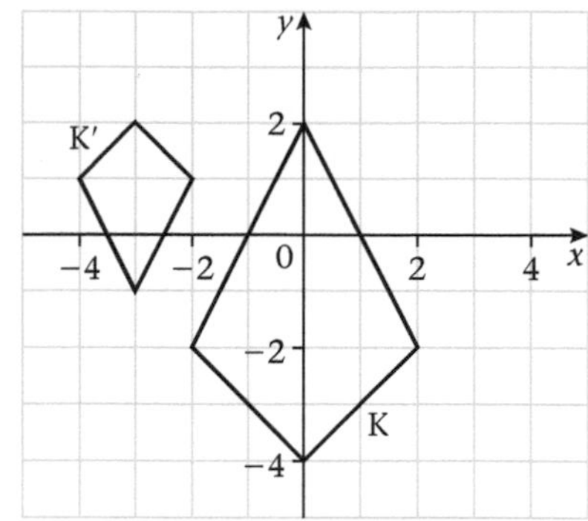

Fig. P39

Vectors and matrices

13 Fig. P40 shows quadrilaterals GHJK and G′H′J′K′.

(a) G′H′J′K′ is the image of GHJK under a transformation T.

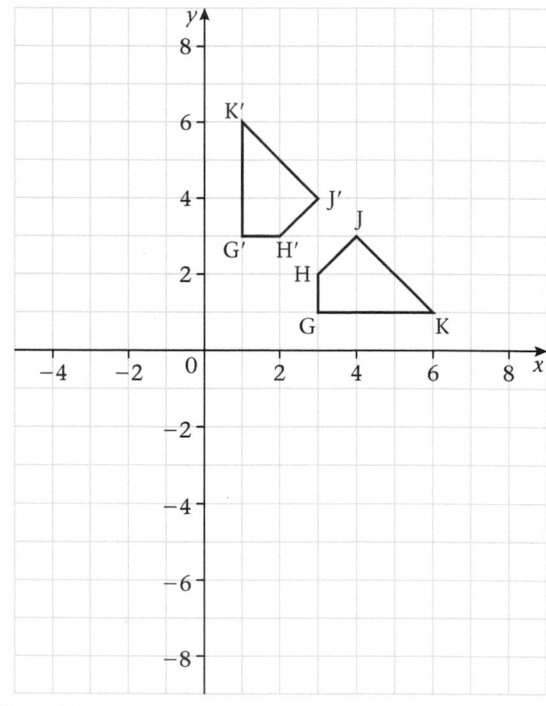

Fig. P40

 (i) State the coordinates of GHJK and of G′H′J′K′.
 (ii) Determine the matrix representing the transformation T.
 (iii) Determine the matrix that transforms G′H′J′K′ onto GHJK.

(b) G″H″J″K″ is the image of GHJK under the transformation V represented by the matrix $\begin{pmatrix} 1 & 0 \\ 0 & -1 \end{pmatrix}$.
   (i) Find the coordinates G″, H″, J″, K″.
   (ii) Draw on a copy of Fig. P40, the quadrilateral G″H″J″K″.
   (iii) Describe fully the transformation V.

[15 marks]

14 (a) The coordinates of the points A, B, and C are (3, 0), (5, 6) and (1, 3) respectively.
   (i) Using a scale of 1 cm to represent 1 unit on each axis, plot on graph paper the points A, B and C.
   (ii) Write down the position vectors of A, B and C.
   (iii) Using a **vector method**, describe the relation between $\overrightarrow{AB}$ and $\overrightarrow{OC}$, and between $|\overrightarrow{AB}|$ and $|\overrightarrow{OC}|$.
   (iv) State the name of the figure OABC.

[8 marks]

(b) (i) If $M = \begin{pmatrix} 2 & -1 \\ 1 & 3 \end{pmatrix}$, find M^{-1}.
   (ii) Hence, solve the equations

$$2x - y = 5$$
$$\text{and } x + 3y = -1$$

[7 marks]

Total 15 marks

Mensuration tables and formulae, three-figure tables

SI units

Mass

The **gram** is the basic unit of mass.

unit	abbreviation	basic unit
1 kilogram	1 kg	1000 g
1 hectogram	1 hg	100 g
1 decagram	1 dag	10 g
1 gram	1 g	1 g
1 decigram	1 dg	0.1 g
1 centigram	1 cg	0.01 g
1 milligram	1 mg	0.001 g

The **tonne** (t) is used for large masses. The most common measures of mass are the milligram, the gram, the kilogram and the tonne.

1 g = 1 000 mg
1 kg = 1 000 g = 1 000 000 mg
1 t = 1 000 kg = 1 000 000 g

Length

The **metre** is the basic unit of length.

unit	abbreviation	basic unit
1 kilometre	1 km	1000 m
1 hectometre	1 hm	100 m
1 decametre	1 dam	10 m
1 metre	1 m	1 m
1 decimetre	1 dm	0.1 m
1 centimetre	1 cm	0.01 m
1 millimetre	1 mm	0.001 m

The most common measures are the millimetre, the metre and the kilometre.

1 m = 1 000 mm
1 km = 1000 m = 1 000 000 mm

Time

The **second** is the basic unit of time.

unit	abbreviation	basic unit
1 second	1 s	1 s
1 minute	1 min	60 s
1 hour	1 h	3600 s

Area

The **square metre** is the basic unit of area. Units of area are derived from units of length.

unit	abbreviation	relation to other units of area
square millimetre	mm^2	
square centimetre	cm^2	$1\ cm^2 = 100\ mm^2$
square metre	m^2	$1\ m^2 = 10\,000\ cm^2$
square kilometre	km^2	$1\ km^2 = 1\,000\,000\ m^2$
hectare (for land measure)	ha	$1\ ha = 10\,000\ m^2$

Volume

The **cubic metre** is the basic unit of volume. Units of volume are derived from units of length.

unit	abbreviation	relation to other units of volume
cubic millimetre	mm^3	
cubic centimetre	cm^3	$1\ cm^3 = 1000\ mm^3$
cubic metre	m^3	$1\ m^3 = 1\,000\,000\ cm^3$

Capacity

The **litre** is the basic unit of capacity. 1 litre takes up the same space as 1 000 cm^3.

Unit	Abbreviation	Relation to other units of capacity	Relation to units of volume
millilitre	$m\ell$		$1\,m\ell = 1\,cm^3$
litre	ℓ	$1\,\ell = 1000\,m\ell$	$1\,\ell = 1000\,cm^3$
kilolitre	$k\ell$	$1\,k\ell = 1000\,\ell$	$1\,k\ell = 1\,m^3$

Mensuration formulae

Plane shapes	Perimeter	Area
Square side s	$4s$	s^2
Rectangle length l, breadth b	$2\ (l + b)$	lb
Triangle base b, height h		$\frac{1}{2}bh$
Parallelogram base b, height h		bh
Trapezium height h, parallels a and b		$\frac{1}{2}(a + b)\ h$
Circle radius r	$2\pi r$	πr^2
Sector of circle radius r, angle $\theta°$	$2r + \frac{\theta}{360}\ 2\pi r$	$\frac{\theta}{360}\ \pi r^2$

Solid shapes	Surface area	Volume
Cube edge s	$6s^2$	s^3
Cuboid length l, breadth b, height h	$2(lb + bh + lh)$	lbh
Prism height h, base area A		Ah
Cylinder radius r, height h	$2\pi rh + 2\pi r^2$	πr^2h
Cone radius r, slant height l, height h	$\pi rl + \pi r^2$	$\frac{1}{3}\pi r^2h$
Sphere radius r	$4\pi r^2$	$\frac{4}{3}\pi r^3$

Divisibility tests

Any whole number is exactly divisible by

2	if its last digit is even
3	if the sum of its digits is divisible by 3
4	if its last two digits form a number divisible by 4
5	if its last digit is 5 or 0
6	if its last digit is even and the sum of its digits is divisible by 3
8	if its last three digits form a number divisible by 8
9	if the sum of its digits is divisible by 9
10	if its last digit is 0

Trigonometric formulae

Right-angled triangles

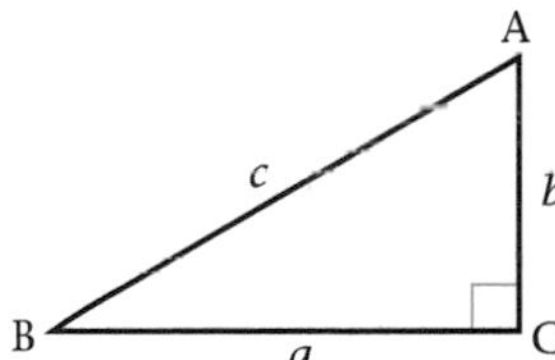

Fig. T1

In the right-angled triangle shown in Fig. T1,

$c^2 = a^2 + b^2$ (*Pythagoras' theorem*)

$\tan B = \frac{b}{a}$ $\tan A = \frac{a}{b}$

$\sin B = \frac{b}{c}$ $\sin A = \frac{a}{c}$

$\cos B = \frac{a}{c}$ $\cos A = \frac{b}{c}$

Obtuse angles

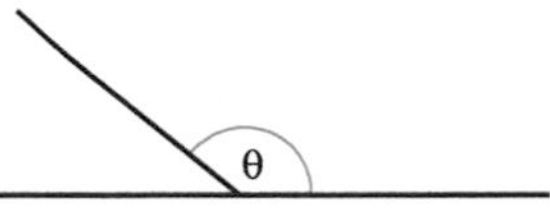

Fig. T2

$\sin\theta = \sin(180° - \theta)$
$\cos\theta = -\cos(180° - \theta)$
$\tan\theta = -\tan(180° - \theta)$

Any triangle

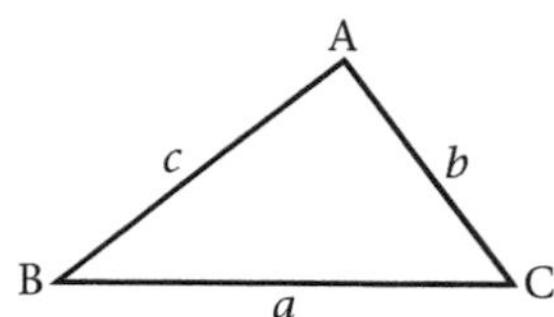

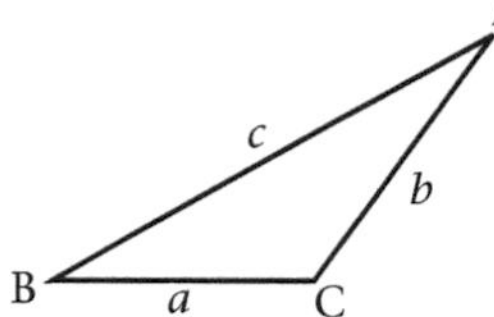

Fig. T3

In both triangles in Fig. T3,

$$\frac{a}{\sin A} = \frac{b}{\sin B} = \frac{c}{\sin C} \quad \textit{(sine rule)}$$

$$c^2 = a^2 + b^2 - 2ab \cos C \quad \textit{(cosine rule)}$$

$$b^2 = a^2 + c^2 - 2ac \cos B$$

$$a^2 = b^2 + c^2 - 2bc \cos A$$

and

$$\cos A = \frac{b^2 + c^2 - a^2}{2bc}$$

$$\cos B = \frac{c^2 + a^2 - b^2}{2ca}$$

$$\cos C = \frac{a^2 + b^2 - c^2}{2ab}$$

Symbols

Symbol	Meaning
$=$	is equal to
$\neq$	is not equal to
$\simeq$	is approximately equal to
$\equiv$	is identical or congruent to
$\Rightarrow$	leads to
$\therefore$	therefore
$\propto$	is proportional to
$>$	is greater than
$<$	is less than
$\geqslant$	is greater than or equal to

Symbol	Meaning
$\leqslant$	is less than or equal to
$^\circ$	degree (angle)
$^\circ$C	degrees Celsius (temperature)
A, B, C	points (geometry)
AB	the line through points A and B, *or* the distance between points A and B
$\triangle$ABC	triangle ABC
$\|^{gm}$ ABCD	parallelogram ABCD
$\frown$ or $\angle$	angle symbol
A$\widehat{\text{B}}$C	angle ABC
$\perp$	is perpendicular to
$\|$	is parallel to
π	pi
%	per cent
A = $\{p, q, r\}$	A is the set p, q, r
B = $\{1, 2, 3, \ldots\}$	B is the infinite set 1, 2, 3 and so on
C = $\{x : x$ is an integer$\}$	Set builder notation. C is the set of numbers x such that x is an integer
n(A)	number of elements in set A
$\in$	is an element of
$\notin$	is not an element of
A$'$	complement of A
{ } or $\varnothing$	the empty set
U	the universal set
A $\supset$ B	A is a subset of B
A $\subset$ B	A contains B
$\not\subset$, $\not\supset$	negations of $\subset$ and $\supset$
A $\cup$ B	union of A and B
A $\cap$ B	intersection of A and B
a or $\vec{a}$ or $\underline{a}$	vector **a**
AB or $\overrightarrow{AB}$	vector $\overrightarrow{AB}$
$\lvert\overrightarrow{AB}\rvert$	modulus of $\overrightarrow{AB}$
Σ	the sum of

Logarithms

$x \rightarrow \log x$

x	0	1	2	3	4	5	6	7	8	9
10	.000	004	009	013	017	021	025	029	033	037
11	.041	045	049	053	057	061	064	068	072	076
12	.079	083	086	090	093	097	100	104	107	111
13	.114	117	121	124	127	130	134	137	140	143
14	.146	149	152	155	158	161	164	167	170	173
15	.176	179	182	185	188	190	193	196	199	201
16	.204	207	210	212	215	217	220	223	225	228
17	.230	233	236	238	241	243	246	248	250	253
18	.255	258	260	262	265	267	270	272	274	276
19	.279	281	283	286	288	290	292	294	297	299
20	.301	303	305	307	310	312	314	316	318	320
21	.322	324	326	328	330	332	334	336	338	340
22	.342	344	346	348	350	352	354	356	358	360
23	.362	364	365	367	369	371	373	375	377	378
24	.380	382	384	386	387	389	391	393	394	396
25	.398	400	401	403	405	407	408	410	412	413
26	.415	417	418	420	422	423	425	427	428	430
27	.431	433	435	436	438	439	441	442	444	446
28	.447	449	450	452	453	455	456	458	459	461
29	.462	464	465	467	468	470	471	473	474	476
30	.477	479	480	481	483	484	486	487	489	490
31	.491	493	494	496	497	498	500	501	502	504
32	.505	507	508	509	511	512	513	515	516	517
33	.519	520	521	522	524	525	526	528	529	530
34	.531	533	534	535	537	538	539	540	542	543
35	.544	545	547	548	549	550	551	553	554	555
36	.556	558	559	560	561	562	563	565	566	567
37	.568	569	571	572	573	574	575	576	577	579
38	.580	581	582	583	584	585	587	588	589	590
39	.591	592	593	594	595	597	598	599	600	601
40	.602	603	604	605	606	607	609	610	611	612
41	.613	614	615	616	617	618	619	620	621	622
42	.623	624	625	626	627	628	629	630	631	632
43	.633	634	635	636	637	638	639	640	641	642
44	.643	644	645	646	647	648	649	650	651	652
45	.653	654	655	656	657	658	659	660	661	662
46	.663	664	665	666	667	667	668	669	670	671
47	.672	673	674	675	676	677	678	679	679	680
48	.681	682	683	684	685	686	687	688	688	589
49	.690	691	692	693	694	695	695	696	697	698
50	.699	700	701	702	702	703	704	705	706	707
51	.708	708	709	710	711	712	713	713	714	715
52	.716	717	718	719	719	720	721	722	723	723
53	.724	725	726	727	728	728	729	730	731	732
54	.732	733	734	735	736	736	737	738	739	740

x	0	1	2	3	4	5	6	7	8	9
55	.740	741	742	743	744	744	745	746	747	747
56	.748	749	750	751	751	752	753	754	754	755
57	.756	757	757	758	759	760	760	761	762	763
58	.763	764	765	766	766	767	768	769	769	770
59	.771	772	772	773	774	775	775	776	777	777
60	.778	779	780	780	781	782	782	783	784	785
61	.785	786	787	787	788	789	790	790	791	792
62	.792	793	794	794	795	796	797	797	798	799
63	.799	800	801	801	802	803	803	804	805	806
64	.806	807	808	808	809	810	810	811	812	812
65	.813	814	814	815	816	816	817	818	818	819
66	.820	820	821	822	822	823	823	824	825	825
67	.826	827	827	828	829	829	830	831	831	832
68	.833	833	834	834	835	836	836	837	838	838
69	.839	839	840	841	841	842	843	843	844	844
70	.845	846	846	847	848	848	849	849	850	851
71	.851	852	852	853	854	854	855	856	856	857
72	.857	858	859	859	860	860	861	862	862	863
73	.863	864	865	865	866	866	867	867	868	869
74	.869	870	870	871	872	872	873	873	874	874
75	.875	876	876	877	877	878	879	879	880	880
76	.881	881	882	883	883	884	884	885	885	886
77	.886	887	888	888	889	889	890	890	891	892
78	.892	893	893	894	894	895	895	896	897	897
79	.898	898	899	899	900	900	901	901	902	903
80	.903	904	904	905	905	906	906	907	907	908
81	.908	909	910	910	911	911	912	912	913	913
82	.914	914	915	915	916	916	917	918	918	919
83	.919	920	920	921	921	922	922	923	923	924
84	.924	925	925	926	926	927	927	928	928	929
85	.929	930	930	931	931	932	932	933	933	934
86	.934	935	936	936	937	937	938	938	939	939
87	.940	940	941	941	942	942	943	943	943	944
88	.944	945	945	946	946	947	947	948	948	949
89	.949	950	950	951	951	952	952	953	953	954
90	.954	955	955	956	956	957	957	958	958	959
91	.959	960	960	960	961	961	962	962	963	963
92	.964	964	965	965	966	966	967	967	968	968
93	.968	969	969	970	970	971	971	972	972	973
94	.973	974	974	975	975	975	976	976	977	977
95	.978	978	979	979	980	980	980	981	981	982
96	.982	983	983	984	984	985	985	985	986	986
97	.987	987	988	988	989	989	989	990	990	991
98	.991	992	992	993	993	993	994	994	995	995
99	.996	996	997	997	997	998	998	999	999	1.000

Antilogarithms

$x \rightarrow 10^x$

x	0	1	2	3	4	5	6	7	8	9
.00	100	100	100	101	101	101	101	102	102	102
.01	102	103	103	103	103	104	104	104	104	104
.02	105	105	105	105	106	106	106	106	107	107
.03	107	107	108	108	108	108	109	109	109	109
.04	110	110	110	110	111	111	111	111	112	112
.05	112	112	113	113	113	114	114	114	114	115
.06	115	115	115	116	116	116	116	117	117	117
.07	117	118	118	118	119	119	119	119	120	120
.08	120	121	121	121	121	122	122	122	122	123
.09	123	123	124	124	124	124	125	125	125	126
.10	126	126	126	127	127	127	128	128	128	129
.11	129	129	129	130	130	130	131	131	131	132
.12	132	132	132	133	133	133	134	134	134	135
.13	135	135	136	136	136	136	137	137	137	138
.14	138	138	139	139	139	140	140	140	141	141
.15	141	142	142	142	143	143	143	144	144	144
.16	145	145	145	146	146	146	147	147	147	148
.17	148	148	149	149	149	150	150	150	151	151
.18	151	152	152	152	153	153	153	154	154	155
.19	155	155	156	156	156	157	157	157	158	158
.20	158	159	159	160	160	160	161	161	161	162
.21	162	163	163	163	164	164	164	165	165	166
.22	166	166	167	167	167	168	168	169	169	169
.23	170	170	171	171	171	172	172	173	173	173
.24	174	174	175	175	175	176	176	177	177	177
.25	178	178	179	179	179	180	180	181	181	182
.26	182	182	183	183	184	184	185	185	185	186
.27	186	187	187	188	188	188	189	189	190	190
.28	191	191	191	192	192	193	193	194	194	195
.29	195	195	196	196	197	197	198	198	199	199
.30	200	200	200	201	201	202	202	203	203	204
.31	204	205	205	206	206	207	207	207	208	208
.32	209	209	210	210	211	211	212	212	213	213
.33	214	214	215	215	216	216	217	217	218	218
.34	219	219	220	220	221	221	222	222	223	223
.35	224	224	225	225	226	226	227	228	228	229
.36	229	230	230	231	231	232	232	233	233	234
.37	234	235	236	236	237	237	238	238	239	239
.38	240	240	241	242	242	243	243	244	244	245
.39	245	246	247	247	248	248	249	249	250	251
.40	251	252	252	253	254	254	255	255	256	256
.41	257	258	258	259	259	260	261	261	262	262
.42	263	264	264	265	265	266	267	267	268	269
.43	269	270	270	271	272	272	273	274	274	275
.44	275	276	277	277	278	279	279	280	281	281
.45	282	282	283	284	284	285	286	286	287	288
.46	288	289	290	290	291	292	292	293	294	294
.47	295	296	296	297	298	299	299	300	301	301
.48	302	303	303	304	305	305	306	307	308	308
.49	309	310	310	311	312	313	313	314	315	316

x	0	1	2	3	4	5	6	7	8	9
.50	316	317	318	318	319	320	321	321	322	323
.51	324	324	325	326	327	327	328	329	330	330
.52	331	332	333	333	334	335	336	337	337	338
.53	339	340	340	341	342	343	344	344	345	346
.54	347	348	348	349	350	351	352	352	353	354
.55	355	356	356	357	358	359	360	361	361	362
.56	363	364	365	366	366	367	368	369	370	371
.57	372	372	373	374	375	376	377	378	378	379
.58	380	381	382	383	384	385	385	386	387	388
.59	389	390	391	392	393	394	394	395	396	397
.60	398	399	400	401	402	403	404	405	406	406
.61	407	408	409	410	411	412	413	414	415	416
.62	417	418	419	420	421	422	423	424	425	426
.63	427	428	429	430	431	432	433	434	435	436
.64	437	438	439	440	441	442	443	444	445	446
.65	447	448	449	450	451	452	453	454	455	456
.66	457	458	459	460	461	462	463	465	466	467
.67	468	469	470	471	472	473	474	475	476	478
.68	479	480	481	482	483	484	485	486	488	489
.69	490	491	492	493	494	495	497	498	499	500
.70	501	502	504	505	506	507	508	509	511	512
.71	513	514	515	516	518	519	520	521	522	524
.72	525	526	527	528	530	531	532	533	535	536
.73	537	538	540	541	542	543	545	546	547	548
.74	550	551	552	553	555	556	557	558	560	561
.75	562	564	565	566	568	569	570	571	573	574
.76	575	577	578	579	581	582	583	585	586	587
.77	589	590	592	593	594	596	597	598	600	601
.78	603	604	605	607	608	610	611	612	614	615
.79	617	618	619	621	622	624	625	627	628	630
.80	631	632	634	635	637	638	640	641	643	644
.81	646	647	649	650	652	653	655	656	658	659
.82	661	662	664	665	667	668	670	671	673	675
.83	676	678	679	681	682	684	685	687	689	690
.84	692	693	695	697	698	700	701	703	705	706
.85	708	710	711	713	714	716	718	719	721	723
.86	724	726	728	729	731	733	735	736	738	740
.87	741	743	745	746	748	750	752	753	755	757
.88	759	760	762	764	766	767	769	771	773	774
.89	776	778	780	782	783	785	787	789	791	793
.90	794	796	798	800	802	804	805	807	809	811
.91	813	815	817	818	820	822	824	826	828	830
.92	832	834	836	838	839	841	843	845	847	849
.93	851	853	855	857	859	861	863	865	867	869
.94	871	873	875	877	879	881	883	885	887	889
.95	891	893	895	897	899	902	904	906	908	910
.96	912	914	916	918	920	923	925	927	929	931
.97	933	935	938	940	942	944	946	948	951	953
.98	955	957	959	962	964	966	968	971	973	975
.99	977	979	982	984	986	989	991	993	995	998

Sines of angles

$\theta \rightarrow \sin \theta$

θ	.0	.1	.2	.3	.4	.5	.6	.7	.8	.9
0	0.000	002	003	005	007	009	010	012	014	016
1	.017	019	021	023	024	026	028	030	031	033
2	.035	037	038	040	042	044	045	047	049	051
3	.052	054	056	058	059	061	063	065	066	068
4	.070	071	073	075	077	078	080	082	084	085
5	0.087	089	091	092	094	096	098	099	101	103
6	.105	106	108	110	111	113	115	117	118	120
7	.122	124	125	127	129	131	132	134	136	137
8	.139	141	143	144	146	148	150	151	153	155
9	.156	158	160	162	163	165	167	168	170	172
10	0.174	175	177	179	181	182	184	186	187	189
11	.191	193	194	196	198	199	201	203	204	206
12	.208	210	211	213	215	216	218	220	222	223
13	.225	227	228	230	232	233	235	237	239	240
14	.242	244	245	247	249	250	252	254	255	257
15	0.259	261	262	264	266	267	269	271	272	274
16	.276	277	279	281	282	284	286	287	289	291
17	.292	294	296	297	299	301	302	304	306	307
18	.309	311	312	314	316	317	319	321	322	324
19	.326	327	329	331	332	334	335	337	339	340
20	0.342	344	345	347	349	350	352	353	355	357
21	.358	360	362	363	365	367	368	370	371	373
22	.375	376	378	379	381	383	384	386	388	389
23	.391	392	394	396	397	399	400	402	404	405
24	.407	408	410	412	413	415	416	418	419	421
25	0.423	424	426	427	429	431	432	434	435	437
26	.438	440	442	443	445	446	448	449	451	452
27	.454	456	457	459	460	462	463	465	466	468
28	.469	471	473	474	476	477	479	480	482	483
29	.485	486	488	489	491	492	494	495	497	498
30	0.500	502	503	505	506	508	509	511	512	514
31	.515	517	518	520	521	522	524	525	527	528
32	.530	531	533	534	536	537	539	540	542	543
33	.545	546	548	549	550	552	553	555	556	558
34	.559	561	562	564	565	566	568	569	571	572
35	0.574	575	576	578	579	581	582	584	585	586
36	.588	589	591	592	593	595	596	598	599	600
37	.602	603	605	606	607	609	610	612	613	614
38	.616	617	618	620	621	623	624	625	627	628
39	.629	631	632	633	635	636	637	639	640	641
40	0.643	644	645	647	648	649	651	652	653	655
41	.656	657	659	660	661	663	664	665	667	668
42	.669	670	672	673	674	676	677	678	679	681
43	.682	683	685	686	687	688	699	691	692	693
44	.695	696	697	698	700	701	702	703	705	706

θ	.0	.1	.2	.3	.4	.5	.6	.7	.8	.9
45	0.707	708	710	711	712	713	714	716	717	718
46	.719	721	722	723	724	725	727	728	729	730
47	.731	733	734	735	736	737	738	740	741	742
48	.743	744	745	747	748	749	750	751	752	754
49	.755	756	757	758	759	760	762	763	764	765
50	0.766	767	768	769	771	772	773	774	775	776
51	.777	778	779	780	782	783	784	785	786	787
52	.788	789	790	791	792	793	794	795	797	798
53	.799	800	801	802	803	804	805	806	807	808
54	.809	810	811	812	813	814	815	816	817	818
55	0.819	820	821	822	823	824	825	826	827	828
56	.829	830	831	832	833	834	835	836	837	838
57	.839	840	841	842	842	843	844	845	846	847
58	.848	849	850	851	852	853	854	854	855	856
59	.857	858	859	860	861	862	863	863	864	865
60	0.866	867	868	869	869	870	871	872	873	874
61	.875	875	876	877	878	879	880	880	881	882
62	.883	884	885	885	886	887	888	889	889	890
63	.891	892	893	893	894	895	896	896	897	898
64	.899	900	900	901	902	903	903	904	905	906
65	0.906	907	908	909	909	910	911	911	912	913
66	.914	914	915	916	916	917	918	918	919	920
67	.921	921	922	923	923	924	925	925	926	927
68	.927	928	928	929	930	930	931	932	932	933
69	.934	934	935	935	936	937	937	938	938	939
70	0.940	940	941	941	942	943	943	944	944	945
71	.946	946	947	947	948	948	949	949	950	951
72	.951	952	952	953	953	954	954	955	955	956
73	.956	957	957	958	958	959	959	960	960	961
74	.961	962	962	963	963	964	964	965	965	965
75	0.966	966	967	967	968	968	969	969	969	970
76	.970	971	971	972	972	972	973	973	974	974
77	.974	975	975	976	976	976	977	977	977	978
78	.978	979	979	979	980	980	980	981	981	981
79	.982	982	982	983	983	983	984	984	984	985
80	0.985	985	985	986	986	986	987	987	987	987
81	.988	988	988	988	989	989	989	990	990	990
82	.990	991	991	991	991	991	992	992	992	992
83	.993	993	993	993	993	994	994	994	994	994
84	.995	995	995	995	995	995	996	996	996	996
85	0.996	996	996	997	997	997	997	997	997	997
86	.998	998	998	998	998	998	998	998	998	999
87	.999	999	999	999	999	999	999	999	999	999
88	0.999	999	1.000	1.000	1.000	1.000	1.000	1.000	1.000	1.000
89	1.000	1.000	1.000	1.000	1.000	1.000	1.000	1.000	1.000	1.000
90	1.000									

Cosines of angles

$\theta \to \cos\theta$

θ	.0	.1	.2	.3	.4	.5	.6	.7	.8	.9
0	1.000	000	000	000	000	000	000	000	000	000
1	1.000	000	000	000	000	000	000	000	000	0.999
2	0.999	999	999	999	999	999	999	999	999	999
3	.999	999	998	998	998	998	998	998	998	998
4	.998	997	997	997	997	997	997	997	996	996
5	0.996	996	996	996	996	995	995	995	995	995
6	.995	994	994	994	994	994	993	993	993	993
7	.993	992	992	992	992	991	991	991	991	991
8	.990	990	990	990	989	989	989	988	988	988
9	.988	987	987	987	987	986	986	986	985	985
10	0.985	985	984	984	984	983	983	983	982	982
11	.982	981	981	981	980	980	980	979	979	979
12	.978	978	977	977	977	976	976	976	975	975
13	.974	974	974	973	973	972	972	972	971	971
14	.970	970	969	969	969	968	968	967	967	966
15	0.966	965	965	965	964	964	963	963	962	962
16	.961	961	960	960	959	959	958	958	957	957
17	.956	956	955	955	954	954	953	953	952	952
18	.951	951	950	949	949	948	948	947	947	946
19	.946	945	944	944	943	943	942	941	941	940
20	0.940	939	938	938	937	937	936	935	935	934
21	.934	933	932	932	931	930	930	929	928	928
22	.927	927	926	925	925	924	923	923	922	921
23	.921	920	919	918	918	917	916	916	915	914
24	.914	913	912	911	911	910	909	909	908	907
25	0.906	906	905	904	903	903	902	901	900	900
26	.899	898	897	896	896	895	894	893	893	892
27	.891	890	889	889	888	887	886	885	885	884
28	.883	882	881	880	880	879	878	877	876	875
29	.875	874	873	872	871	870	869	869	868	867
30	0.866	865	864	863	863	862	861	860	859	858
31	.857	856	855	854	854	853	852	851	850	849
32	.848	847	846	845	844	843	842	842	841	840
33	.839	838	837	836	835	834	833	832	831	830
34	.829	828	827	826	825	824	823	822	821	820
35	0.819	818	817	816	815	814	813	812	811	810
36	.809	808	807	806	805	804	803	802	801	800
37	.799	798	797	795	794	793	792	791	790	789
38	.788	787	786	785	784	783	782	780	779	778
39	.777	776	775	774	773	772	771	769	768	767
40	0.766	765	764	763	762	760	759	758	757	756
41	.755	754	752	751	750	749	748	747	745	744
42	.743	742	741	740	738	737	736	735	734	733
43	.731	730	729	728	727	725	724	723	722	721
44	.719	718	717	716	714	713	712	711	710	708

θ	.0	.1	.2	.3	.4	.5	.6	.7	.8	.9
45	0.707	706	705	703	702	701	700	698	697	696
46	.695	693	692	691	690	688	687	686	685	683
47	.682	681	679	678	677	676	674	673	672	670
48	.669	668	667	665	664	663	661	660	659	657
49	.656	655	653	652	651	649	648	647	645	644
50	0.643	641	640	639	637	636	635	633	632	631
51	.629	628	627	625	624	623	621	620	618	617
52	.616	614	613	612	610	609	607	606	605	603
53	.602	600	599	598	596	595	593	592	591	589
54	.588	586	585	584	582	581	579	578	576	575
55	0.574	572	571	569	568	566	565	564	562	561
56	.559	558	556	555	553	552	550	549	548	546
57	.545	543	542	540	539	537	536	534	533	531
58	.530	528	527	525	524	522	521	520	518	517
59	.515	514	512	511	509	508	506	505	503	502
60	0.500	498	497	495	494	492	491	489	488	486
61	.485	483	482	480	479	477	476	474	473	471
62	.469	468	466	465	463	462	460	459	457	456
63	.454	452	451	449	448	446	445	443	442	440
64	.438	437	435	434	432	431	429	427	426	424
65	0.423	421	419	418	416	415	413	412	410	408
66	.407	405	404	402	400	399	397	396	394	392
67	.391	389	388	386	384	383	381	379	378	376
68	.375	373	371	370	368	367	365	363	362	360
69	.358	357	355	353	352	350	349	347	345	344
70	0.342	340	339	337	335	334	332	331	329	327
71	.326	324	322	321	319	317	316	314	312	311
72	.309	307	306	304	302	301	299	297	296	294
73	.292	291	289	287	286	284	282	281	279	277
74	.276	274	272	271	269	267	266	264	262	261
75	0.259	257	255	254	252	250	249	247	245	244
76	.242	240	239	237	235	233	232	230	228	227
77	.225	223	222	220	218	216	215	213	211	210
78	.208	206	204	203	201	199	198	196	194	193
79	.191	189	187	186	184	182	181	179	177	175
80	0.174	172	170	168	167	165	163	162	160	158
81	.156	155	153	151	150	148	146	144	143	141
82	.139	137	136	134	132	131	129	127	125	124
83	.122	120	118	117	115	113	111	110	108	106
84	.105	103	101	099	098	096	094	092	091	089
85	0.087	085	084	082	080	078	077	075	073	071
86	.070	068	066	065	063	061	059	058	056	054
87	.052	051	049	047	045	044	042	040	038	037
88	.035	033	031	030	028	026	024	023	021	019
89	.017	016	014	012	010	009	007	005	003	002
90	0.000									

Tangents of angles

$\theta \rightarrow \tan \theta$

θ	.0	.1	.2	.3	.4	.5	.6	.7	.8	.9
0	0.000	002	003	005	007	009	010	012	014	016
1	.017	019	021	023	024	026	028	030	031	033
2	.035	037	038	040	042	044	045	047	049	051
3	.052	054	056	058	059	061	063	065	066	068
4	.070	072	073	075	077	079	080	082	084	086
5	0.087	089	091	093	095	096	098	100	102	103
6	.105	107	109	110	112	114	116	117	119	121
7	.123	125	126	128	130	132	133	135	137	139
8	.141	142	144	146	148	149	151	153	155	157
9	.158	160	162	164	166	167	169	171	173	175
10	0.176	178	180	182	184	185	187	189	191	193
11	.194	196	198	200	202	203	205	207	209	211
12	.213	214	216	218	220	222	224	225	227	229
13	.231	233	235	236	238	240	242	244	246	247
14	.249	251	253	255	257	259	260	262	264	266
15	0.268	270	272	274	275	277	279	281	283	285
16	.287	289	291	292	294	296	298	300	302	304
17	.306	308	310	311	313	315	317	319	321	323
18	.325	327	329	331	333	335	337	338	340	342
19	.344	346	348	350	352	354	356	358	360	362
20	0.364	366	368	370	372	374	376	378	380	382
21	.384	386	388	390	392	394	396	398	400	402
22	.404	406	408	410	412	414	416	418	420	422
23	.424	427	429	431	433	435	437	439	441	443
24	.445	447	449	452	454	456	458	460	462	464
25	0.466	468	471	473	475	477	479	481	483	486
26	.488	490	492	494	496	499	501	503	505	507
27	.510	512	514	516	518	521	523	525	527	529
28	.532	534	536	538	541	543	545	547	550	552
29	.554	557	559	561	563	566	568	570	573	575
30	0.577	580	582	584	587	589	591	594	596	598
31	.601	603	606	608	610	613	615	618	620	622
32	.625	627	630	632	635	637	640	642	644	647
33	.649	652	654	657	659	662	664	667	669	672
34	.675	677	680	682	685	687	690	692	695	698
35	0.700	703	705	708	711	713	716	719	721	724
36	.727	729	732	735	737	740	743	745	748	751
37	.754	756	759	762	765	767	770	773	776	778
38	.781	784	787	790	793	795	798	801	804	807
39	.810	813	816	818	821	824	827	830	833	836
40	0.839	842	845	848	851	854	857	860	863	866
41	.869	872	875	879	882	885	888	891	894	897
42	.900	904	907	910	913	916	920	923	926	929
43	.933	936	939	942	946	949	952	956	959	962
44	.966	969	972	976	979	983	986	990	993	997

θ	.0	.1	.2	.3	.4	.5	.6	.7	.8	.9
45	1.000	003	007	011	014	018	021	025	028	032
46	.036	039	043	046	050	054	057	061	065	069
47	.072	076	080	084	087	091	095	099	103	107
48	.111	115	118	122	126	130	134	138	142	146
49	.150	154	159	163	167	171	175	179	183	188
50	1.192	196	200	205	209	213	217	222	226	230
51	.235	239	244	248	253	257	262	266	271	275
52	.280	285	289	294	299	303	308	313	317	322
53	.327	332	337	342	347	351	356	361	366	371
54	.376	381	387	392	397	402	407	412	418	423
55	1.428	433	439	444	450	455	460	466	471	477
56	.483	488	494	499	505	511	517	522	528	534
57	.540	546	552	558	564	570	576	582	588	594
58	.600	607	613	619	625	632	638	645	651	658
59	.664	671	678	684	691	698	704	711	718	725
60	1.732	739	746	753	760	767	775	782	789	797
61	.804	811	819	827	834	842	849	857	865	873
62	.881	889	897	905	913	921	929	937	946	954
63	1.963	971	980	988	997	2.006	2.014	2.023	2.032	2.041
64	2.050	059	069	078	087	097	106	116	125	135
65	2.145	154	164	174	184	194	204	215	225	236
66	.246	257	267	278	289	300	311	322	333	344
67	.356	367	379	391	402	414	426	438	450	463
68	.475	488	500	513	526	539	552	565	578	592
69	.605	619	633	546	660	675	689	703	718	733
70	2.747	762	778	793	808	824	840	856	872	888
71	2.904	921	937	954	971	989	3.006	3.024	3.042	3.060
72	3.078	096	115	133	152	172	191	211	230	251
73	.271	291	312	333	354	376	398	420	442	465
74	.487	511	534	558	582	606	630	655	681	706
75	3.732	758	785	812	839	867	895	923	952	981
76	4.011	041	071	102	134	165	198	230	264	297
77	4.331	366	402	437	474	511	548	586	625	665
78	4.705	745	787	829	872	915	959	5.005	5.050	5.097
79	5.145	193	242	292	343	396	449	503	558	614
80	5.671	730	789	850	912	976	6.041	6.107	6.174	6.243
81	6.314	386	460	535	612	691	772	855	940	7.026
82	7.115	207	300	396	495	596	700	806	916	8.028
83	8.144	264	386	513	643	777	915	9.058	9.205	9.357
84	9.514	9.677	9.845	10.02	10.20	10.39	10.58	10.78	10.99	11.20
85	11.43	11.66	11.91	12.16	12.43	12.71	13.00	13.30	13.62	13.95
86	14.30	14.67	15.06	15.46	15.89	16.35	16.83	17.34	17.89	18.46
87	19.08	19.74	20.45	21.20	22.02	22.90	23.86	24.90	26.03	27.27
88	28.64	30.14	31.82	33.69	35.80	38.19	40.92	44.07	47.74	52.08
89	57.29	63.66	71.62	81.85	95.49	114.6	143.2	191.0	286.5	573.0

Logarithms of sines

$\theta \rightarrow \log \sin \theta$

θ	.0	.1	.2	.3	.4	.5	.6	.7	.8	.9
0		$\bar{3}$.242	$\bar{3}$.543	$\bar{3}$.719	$\bar{3}$.844	$\bar{3}$.941	$\bar{2}$.020	$\bar{2}$.087	$\bar{2}$.145	$\bar{2}$.196
1	$\bar{2}$.242	283	321	356	388	418	446	472	497	521
2	.543	564	584	603	622	640	657	673	689	704
3	.719	733	747	760	773	786	798	810	821	833
4	.844	854	865	875	885	895	904	913	923	932
5	$\bar{2}$.940	949	957	966	974	982	989	997	$\bar{1}$.005	$\bar{1}$.012
6	$\bar{1}$.019	026	033	040	047	054	060	067	073	080
7	.086	092	098	104	110	116	121	127	133	138
8	.144	149	154	159	165	170	175	180	185	190
9	.194	199	204	208	213	218	222	227	231	235
10	$\bar{1}$.240	244	248	252	257	261	265	269	273	277
11	.281	284	288	292	296	300	303	307	311	314
12	.318	321	325	328	332	335	339	342	345	349
13	.352	355	359	362	365	368	371	374	378	381
14	.384	387	390	393	396	399	402	404	407	410
15	$\bar{1}$.413	416	419	421	424	427	430	432	435	438
16	.440	443	446	448	451	453	456	458	461	463
17	.466	468	471	473	476	478	481	483	485	488
18	.490	492	495	497	499	501	504	506	508	510
19	.513	515	517	519	521	523	526	528	530	532
20	$\bar{1}$.534	536	538	540	542	544	546	548	550	552
21	.554	556	558	560	562	564	566	568	570	572
22	.574	575	577	579	581	583	585	586	588	590
23	.592	594	595	597	599	601	602	604	606	608
24	.609	611	613	614	616	618	619	621	623	624
25	$\bar{1}$.626	628	629	631	632	634	636	637	639	640
26	.642	643	645	646	648	650	651	653	654	656
27	.657	659	660	661	663	664	666	667	669	670
28	.672	673	674	676	677	679	680	681	683	684
29	.686	687	688	690	691	692	694	695	696	698
30	$\bar{1}$.699	700	702	703	704	705	707	708	709	711
31	.712	713	714	716	717	718	719	721	722	723
32	.724	725	727	728	729	730	731	733	734	735
33	.736	737	738	740	741	742	743	744	745	746
34	.748	749	750	751	752	753	754	755	756	758
35	$\bar{1}$.759	760	761	762	763	764	765	766	767	768
36	.769	770	771	772	773	774	775	776	777	778
37	.779	780	781	782	783	784	785	786	787	788
38	.789	790	791	792	793	794	795	796	797	798
39	.799	800	801	802	803	804	804	805	806	807
40	$\bar{1}$.808	809	810	811	812	813	813	814	815	816
41	.817	818	819	820	820	821	822	823	824	825
42	.826	826	827	828	829	830	831	831	832	833
43	.834	835	835	836	837	838	839	839	840	841
44	.842	843	843	844	845	846	846	847	848	849

θ	.0	.1	..2	.3	.4	.5	.6	.7	.8	.9
45	$\bar{1}$.849	850	851	852	852	853	854	855	855	856
46	.857	858	858	859	860	861	861	862	863	863
47	.864	865	866	866	867	868	868	869	870	870
48	.871	872	872	873	874	874	875	876	876	877
49	.878	878	879	880	880	881	882	882	883	884
50	$\bar{1}$.884	885	886	886	887	887	888	889	889	890
51	.891	891	892	892	893	894	894	895	895	896
52	.897	897	898	898	899	899	900	901	901	902
53	.902	903	903	904	905	905	906	906	907	907
54	.908	909	909	910	910	911	911	912	912	913
55	$\bar{1}$.913	914	914	915	915	916	917	917	918	918
56	.919	919	920	920	921	921	922	922	923	923
57	.924	924	925	925	926	926	927	927	927	928
58	.928	929	929	930	930	931	931	932	932	933
59	.933	934	934	934	935	935	936	936	937	937
60	$\bar{1}$.938	938	938	939	939	940	940	941	941	941
61	.942	942	943	943	943	944	944	945	945	946
62	.946	946	947	947	948	948	948	949	949	949
63	.950	950	951	951	951	952	952	953	953	953
64	.954	954	954	955	955	955	956	956	957	957
65	$\bar{1}$.957	958	958	958	959	959	959	960	960	960
66	.961	961	961	962	962	962	963	963	963	964
67	.964	964	965	965	965	966	966	966	967	967
68	.967	967	968	968	968	969	969	969	970	970
69	.970	970	971	971	971	972	972	972	972	973
70	$\bar{1}$.973	973	974	974	974	974	975	975	975	975
71	.976	976	976	976	977	977	977	977	978	978
72	.978	978	979	979	979	979	980	980	980	980
73	.981	981	981	981	982	982	982	982	982	983
74	.983	983	983	983	984	984	984	984	985	985
75	$\bar{1}$.985	985	985	986	986	986	986	986	987	987
76	.987	987	987	987	988	988	988	988	988	989
77	.989	989	989	989	989	990	990	990	990	990
78	.990	991	991	991	991	991	991	991	992	992
79	.992	992	992	992	993	993	993	993	993	993
80	$\bar{1}$.993	993	994	994	994	994	994	994	994	994
81	.995	995	995	995	995	995	995	995	996	996
82	.996	996	996	996	996	996	996	996	997	997
83	.997	997	997	997	997	997	997	997	997	998
84	.998	998	998	998	998	998	998	998	998	998
85	$\bar{1}$.998	998	998	999	999	999	999	999	999	999
86	.999	999	999	999	999	999	999	999	999	999
87	$\bar{1}$.999	999	999	0.000	0.000	0.000	0.000	0.000	0.000	0.000
88	0.000	0.000	0.000	0.000	0.000	0.000	0.000	0.000	0.000	0.000
89	0.000	0.000	0.000	0.000	0.000	0.000	0.000	0.000	0.000	0.000
90	0.000									

Logarithms of cosines

$\theta \rightarrow \log \cos \theta$

θ	.0	.1	.2	.3	.4	.5	.6	.7	.8	.9
0	.0000	000	000	000	000	000	000	000	000	000
1	.000	000	000	000	000	000	000	000	000	000
2	0.000	000	000	000	000	000	000	000	$\bar{1}$.999	$\bar{1}$.999
3	$\bar{1}$.999	999	999	999	999	999	999	999	999	999
4	.999	999	999	999	999	999	999	999	998	998
5	$\bar{1}$.998	998	998	998	998	998	998	998	998	998
6	.998	998	997	997	997	997	997	997	997	997
7	.997	997	997	996	996	996	996	996	996	996
8	.996	996	996	995	995	995	995	995	995	995
9	.995	994	994	994	994	994	994	994	994	993
10	$\bar{1}$.993	993	993	993	993	993	993	992	992	992
11	.992	992	992	991	991	991	991	991	991	991
12	.990	990	990	990	990	990	989	989	989	989
13	.989	989	988	988	988	988	988	987	987	987
14	.987	987	987	986	986	986	986	986	985	985
15	$\bar{1}$.985	985	985	984	984	984	984	983	983	983
16	.983	983	982	982	982	982	982	981	981	981
17	.981	980	980	980	980	979	979	979	979	978
18	.978	978	978	977	977	977	977	976	976	976
19	.976	975	975	975	975	974	974	974	974	973
20	$\bar{1}$.973	973	972	972	972	972	971	971	971	970
21	.970	970	970	969	969	969	968	968	968	967
22	.967	967	967	966	966	966	965	965	965	964
23	.964	964	963	963	963	962	962	962	961	961
24	.961	960	960	960	959	959	959	958	958	958
25	$\bar{1}$.957	957	957	956	956	955	955	955	954	954
26	.954	953	953	953	952	952	951	951	951	950
27	.950	949	949	949	948	948	948	947	947	946
28	.946	946	945	945	944	944	943	943	943	942
29	.942	941	941	941	940	940	939	939	938	938
30	$\bar{1}$.938	937	937	936	936	935	935	934	934	934
31	.933	933	932	932	931	931	930	930	929	929
32	.928	928	927	927	927	926	926	925	925	924
33	.924	923	923	922	922	921	921	920	920	919
34	.919	918	918	917	917	916	915	915	914	914
35	$\bar{1}$.913	913	912	912	911	911	910	910	909	909
36	.908	907	907	906	906	905	905	904	903	903
37	.902	902	901	901	900	899	899	898	898	897
38	.897	896	895	895	894	894	893	892	892	891
39	.891	890	889	889	888	887	887	886	886	885
40	$\bar{1}$.884	884	883	882	882	881	880	880	879	878
41	.878	877	876	876	875	874	874	873	872	872
42	.871	870	870	869	868	868	867	866	866	865
43	.864	863	863	862	861	861	860	859	858	858
44	.857	856	855	855	854	853	852	852	851	850

θ	.0	.1	.2	.3	.4	.5	.6	.7	.8	.9
45	$\bar{1}$.849	849	848	847	846	846	845	844	843	843
46	.842	841	840	839	839	838	837	836	835	835
47	.834	833	832	831	831	830	829	828	827	826
48	.826	825	824	823	822	821	820	820	819	818
49	.817	816	815	814	813	813	812	811	810	809
50	$\bar{1}$.808	807	806	805	804	804	803	802	801	800
51	.799	798	797	796	795	794	793	792	791	790
52	.789	788	787	786	785	784	783	782	781	780
53	.799	778	777	776	775	774	773	772	771	770
54	.769	768	767	766	765	764	763	762	761	760
55	$\bar{1}$.759	758	756	755	754	753	752	751	750	749
56	.748	746	745	744	743	742	741	740	738	737
57	.736	735	734	733	731	730	729	728	727	725
58	.724	723	722	721	719	718	717	716	714	713
59	.712	711	709	708	707	705	704	703	702	700
60	$\bar{1}$.699	698	696	695	694	692	691	690	688	687
61	.686	684	683	681	680	679	677	676	674	673
62	.672	670	669	667	666	664	663	661	660	659
63	.657	656	654	653	651	650	648	646	645	643
64	.642	640	639	637	636	634	632	631	629	628
65	$\bar{1}$.626	624	623	621	619	618	616	614	613	611
66	.609	608	606	604	602	601	599	597	595	594
67	.592	590	588	586	585	583	581	579	577	575
68	.574	572	570	568	566	564	562	560	558	556
69	.554	552	550	548	546	544	542	540	538	536
70	$\bar{1}$.534	532	530	528	526	523	521	519	517	515
71	.513	510	508	506	504	501	499	497	495	492
72	.490	488	485	483	481	478	476	473	471	468
73	.466	463	461	458	456	453	451	448	446	443
74	.440	428	435	432	430	427	424	421	419	416
75	$\bar{1}$.413	410	407	404	402	399	396	393	390	387
76	.384	381	378	374	371	368	365	362	359	355
77	.352	349	345	342	339	335	332	328	325	321
78	.318	314	311	307	303	300	296	292	288	284
79	.281	277	273	269	265	261	257	252	248	244
80	$\bar{1}$.240	235	231	227	222	218	213	208	204	199
81	.194	190	185	180	175	170	165	159	154	149
82	.144	138	133	127	121	116	110	104	098	092
83	.086	080	073	067	060	054	047	040	033	026
84	$\bar{1}$.019	012	005	$\bar{2}$.997	$\bar{2}$.989	$\bar{2}$.982	$\bar{2}$.974	$\bar{2}$.966	$\bar{2}$.957	$\bar{2}$.949
85	$\bar{2}$.940	932	923	913	904	895	885	875	865	854
86	.844	833	821	810	798	786	773	760	747	733
87	.719	704	689	673	657	640	622	603	584	564
88	.543	521	497	472	446	418	388	356	321	283
89	$\bar{2}$.242	196	145	987	020	$\bar{3}$.941	$\bar{3}$.844	$\bar{3}$.719	$\bar{3}$.543	$\bar{3}$.242

Logarithms of tangents

$\theta \rightarrow \log \tan \theta$

θ	.0	.1	.2	.3	.4	.5	.6	.7	.8	.9
0		$\bar{3}.242$	$\bar{3}.543$	$\bar{3}.719$	$\bar{3}.844$	$\bar{3}.941$	$\bar{2}.020$	$\bar{2}.087$	$\bar{2}.145$	$\bar{2}.196$
1	$\bar{2}.242$	283	321	356	388	418	446	472	497	521
2	.543	564	585	604	622	640	657	674	689	705
3	.719	734	747	761	774	786	799	811	822	834
4	.845	855	866	876	886	896	906	915	924	933
5	$\bar{2}.942$	951	959	967	976	984	991	999	$\bar{1}.007$	$\bar{1}.014$
6	$\bar{1}.022$	029	036	043	050	057	063	070	076	083
7	.089	095	102	108	114	119	125	131	137	142
8	.148	153	159	164	169	174	180	185	190	195
9	.200	205	209	214	219	224	228	233	237	242
10	$\bar{1}.246$	251	255	259	264	268	272	276	280	285
11	.289	293	297	301	305	308	312	316	320	324
12	.327	331	335	339	342	346	349	353	356	360
13	.363	367	370	374	377	380	384	387	390	394
14	.397	400	403	406	410	413	416	419	422	425
15	$\bar{1}.428$	431	434	437	440	443	446	449	452	455
16	.457	460	463	466	469	472	474	477	480	483
17	.485	488	491	493	496	499	501	504	507	509
18	.512	514	517	519	522	525	527	530	532	535
19	.537	539	542	544	547	549	552	554	556	559
20	$\bar{1}.561$	563	566	568	570	573	575	577	580	582
21	.584	586	589	591	593	595	598	600	602	604
22	.606	609	611	613	615	617	619	621	624	626
23	.628	630	632	634	636	638	640	642	644	647
24	.649	651	653	655	657	659	661	663	665	667
25	$\bar{1}.669$	671	673	675	677	678	680	682	684	686
26	.688	690	692	694	696	698	700	702	703	705
27	.707	709	711	713	715	716	718	720	722	724
28	.726	728	729	731	733	735	737	738	740	742
29	.744	746	747	749	751	753	754	756	758	760
30	$\bar{1}.761$	763	765	767	768	770	772	774	775	777
31	.779	780	782	784	786	787	789	791	792	794
32	.796	797	799	801	803	804	806	808	809	811
33	.813	814	816	817	819	821	822	824	826	827
34	.829	831	832	834	836	837	839	840	842	844
35	$\bar{1}.845$	847	848	850	852	853	855	856	858	860
36	.861	863	864	866	868	869	871	872	874	876
37	.877	879	880	882	883	885	887	888	890	891
38	.893	894	896	897	899	901	902	904	905	907
39	.908	910	911	913	915	916	918	919	921	922
40	$\bar{1}.924$	925	927	928	930	931	933	935	936	938
41	.939	941	942	944	945	947	948	950	951	953
42	.954	956	957	959	961	962	964	965	967	968
43	.970	971	973	974	976	977	979	980	982	983
44	.985	986	988	989	991	992	994	995	997	998

θ	.0	.1	.2	.3	.4	.5	.6	.7	.8	.9
45	0.000	002	003	005	006	008	009	011	012	014
46	.015	017	018	020	021	023	024	026	027	029
47	.030	032	033	035	036	038	039	041	043	044
48	.046	047	049	050	052	053	055	056	058	059
49	.061	062	064	065	067	069	070	072	073	075
50	0.076	078	079	081	082	084	085	087	089	090
51	.092	093	095	096	098	099	101	103	104	106
52	.107	109	110	112	113	115	117	118	120	121
53	.123	124	126	128	129	131	132	134	136	137
54	.139	140	142	144	145	147	148	150	152	153
55	0.155	156	158	160	161	163	164	166	168	169
56	.171	173	174	176	178	179	181	183	184	186
57	.187	189	191	192	194	196	197	199	201	203
58	.204	206	208	209	211	213	214	216	218	220
59	.221	223	225	226	228	230	232	233	235	237
60	0.239	240	242	244	246	247	249	251	253	254
61	.256	258	260	262	263	265	267	269	271	272
62	.274	276	278	280	282	284	285	287	289	291
63	.293	295	297	298	300	302	304	306	308	310
64	.312	314	316	318	320	322	323	325	327	329
65	0.331	333	335	337	339	341	343	345	347	349
66	.351	353	356	358	360	362	364	366	368	370
67	.372	374	376	379	381	383	385	387	389	391
68	.394	396	398	400	402	405	407	409	411	414
69	.416	418	420	423	425	427	430	432	434	437
70	0.439	441	444	446	448	451	453	456	458	461
71	.463	465	468	470	473	475	478	481	483	486
72	.488	491	493	496	499	501	504	507	509	512
73	.515	517	520	523	526	528	531	534	537	540
74	.543	545	548	551	554	557	560	563	566	569
75	0.572	575	578	581	584	587	590	594	597	600
76	.603	606	610	613	616	620	623	626	630	633
77	.637	640	644	647	651	652	658	661	665	669
78	.673	676	680	684	688	692	695	699	703	707
79	.711	715	720	724	728	732	736	.741	745	749
80	0.754	758	763	767	772	776	781	786	791	795
81	.800	805	810	815	820	826	831	836	841	847
82	.852	858	863	869	875	881	886	892	898	905
83	.911	917	924	930	937	943	950	957	964	971
84	0.978	986	993	1.001	1.009	1.016	1.024	1.033	1.041	1.049
85	1.058	067	076	085	094	104	114	124	134	145
86	.155	166	178	189	201	214	226	239	253	266
87	.281	295	311	326	343	360	378	396	415	436
88	.457	479	503	528	554	582	612	644	679	717
89	1.758	804	855	913	980	2.059	2.156	2.281	2.457	2.758

Square roots from 1 to 10

$x \to \sqrt{x}$

x	0	1	2	3	4	5	6	7	8	9
1.0	1.00	1.00	1.01	1.01	1.02	1.02	1.03	1.03	1.04	1.04
1.1	1.05	1.05	1.06	1.06	1.07	1.07	1.08	1.08	1.09	1.09
1.2	1.10	1.10	1.10	1.11	1.11	1.12	1.12	1.13	1.13	1.14
1.3	1.14	1.14	1.15	1.15	1.16	1.16	1.17	1.17	1.17	1.18
1.4	1.18	1.19	1.19	1.20	1.20	1.20	1.21	1.21	1.22	1.22
1.5	1.22	1.23	1.23	1.24	1.24	1.24	1.25	1.25	1.26	1.26
1.6	1.26	1.27	1.27	1.28	1.28	1.28	1.29	1.29	1.30	1.30
1.7	1.30	1.31	1.31	1.32	1.32	1.32	1.33	1.33	1.33	1.34
1.8	1.34	1.35	1.35	1.35	1.36	1.36	1.36	1.37	1.37	1.37
1.9	1.38	1.38	1.39	1.39	1.39	1.40	1.40	1.40	1.41	1.41
2.0	1.41	1.42	1.42	1.42	1.43	1.43	1.44	1.44	1.44	1.45
2.1	1.45	1.45	1.46	1.46	1.46	1.47	1.47	1.47	1.48	1.48
2.2	1.48	1.49	1.49	1.49	1.50	1.50	1.50	1.51	1.51	1.51
2.3	1.52	1.52	1.52	1.53	1.53	1.53	1.54	1.54	1.54	1.55
2.4	1.55	1.55	1.56	1.56	1.56	1.57	1.57	1.57	1.57	1.58
2.5	1.58	1.58	1.59	1.59	1.59	1.60	1.60	1.60	1.61	1.61
2.6	1.61	1.62	1.62	1.62	1.62	1.63	1.63	1.63	1.64	1.64
2.7	1.64	1.65	1.65	1.65	1.66	1.66	1.66	1.66	1.67	1.67
2.8	1.67	1.68	1.68	1.68	1.69	1.69	1.69	1.69	1.70	1.70
2.9	1.70	1.71	1.71	1.71	1.71	1.72	1.72	1.72	1.73	1.73
3.0	1.73	1.73	1.74	1.74	1.74	1.75	1.75	1.75	1.75	1.76
3.1	1.76	1.76	1.77	1.77	1.77	1.77	1.78	1.78	1.78	1.79
3.2	1.79	1.79	1.79	1.80	1.80	1.80	1.81	1.81	1.81	1.81
3.3	1.82	1.82	1.82	1.82	1.83	1.83	1.83	1.84	1.84	1.84
3.4	1.84	1.85	1.85	1.85	1.85	1.86	1.86	1.86	1.87	1.87
3.5	1.87	1.87	1.88	1.88	1.88	1.88	1.89	1.89	1.89	1.89
3.6	1.90	1.90	1.90	1.91	1.91	1.91	1.91	1.92	1.92	1.92
3.7	1.92	1.93	1.93	1.93	1.93	1.94	1.94	1.94	1.94	1.95
3.8	1.95	1.95	1.95	1.96	1.96	1.96	1.96	1.97	1.97	1.97
3.9	1.97	1.98	1.98	1.98	1.98	1.99	1.99	1.99	1.99	2.00
4.0	2.00	2.00	2.00	2.01	2.01	2.01	2.01	2.02	2.02	2.02
4.1	2.02	2.03	2.03	2.03	2.03	2.04	2.04	2.04	2.04	2.05
4.2	2.05	2.05	2.05	2.06	2.06	2.06	2.06	2.07	2.07	2.07
4.3	2.07	2.08	2.08	2.08	2.08	2.09	2.09	2.09	2.09	2.10
4.4	2.10	2.10	2.10	2.10	2.11	2.11	2.11	2.11	2.12	2.12
4.5	2.12	2.12	2.13	2.13	2.13	2.13	2.14	2.14	2.14	2.14
4.6	2.14	2.15	2.15	2.15	2.15	2.16	2.16	2.16	2.16	2.17
4.7	2.17	2.17	2.17	2.17	2.18	2.18	2.18	2.18	2.19	2.19
4.8	2.19	2.19	2.20	2.20	2.20	2.20	2.20	2.21	2.21	2.21
4.9	2.21	2.22	2.22	2.22	2.22	2.22	2.23	2.23	2.23	2.23
5.0	2.24	2.24	2.24	2.24	2.24	2.25	2.25	2.25	2.25	2.26
5.1	2.26	2.26	2.26	2.26	2.27	2.27	2.27	2.27	2.28	2.28
5.2	2.28	2.28	2.28	2.29	2.29	2.29	2.29	2.30	2.30	2.30
5.3	2.30	2.30	2.31	2.31	2.31	2.31	2.32	2.32	2.32	2.32
5.4	2.32	2.33	2.33	2.33	2.33	2.33	2.34	2.34	2.34	2.34

x	0	1	2	3	4	5	6	7	8	9
5.5	2.35	2.35	2.35	2.35	2.35	2.36	2.36	2.36	2.36	2.36
5.6	2.37	2.37	2.37	2.37	2.37	2.38	2.38	2.38	2.38	2.39
5.7	2.39	2.39	2.39	2.39	2.40	2.40	2.40	2.40	2.40	2.41
5.8	2.41	2. 41	2.41	2.41	2.42	2.42	2.42	2.42	2.42	2.43
5.9	2.43	2.43	2.43	2.44	2.44	2.44	2.44	2.44	2.45	2.45
6.0	2.45	2.45	2.45	2.46	2.46	2.46	2.46	2.46	2.47	2.47
6.1	2.47	2.47	2.47	2.48	2.48	2.48	2.48	2.48	2.49	2.49
6.2	2.49	2.49	2.49	2.50	2.50	2.50	2.50	2.50	2.51	2.51
6.3	2.51	2.51	2.51	2.52	2.52	2.52	2.52	2.52	2.53	2.53
6.4	2.53	2.53	2.53	2.54	2.54	2.54	2.54	2.54	2.55	2.55
6.5	2.55	2.55	2.55	2.56	2.56	2.56	2.56	2.56	2.57	2.57
6.6	2.57	2.57	2.57	2.57	2.58	2.58	2.58	2.58	2.58	2.59
6.7	2.59	2.59	2.59	2.59	2.60	2.60	2.60	2.60	2.60	2.61
6.8	2.61	2.61	2.61	2.61	2.62	2.62	2.62	2.62	2.62	2.62
6.9	2.63	2.63	2.63	2.63	2.63	2.64	2.64	2.64	2.64	2.64
7.0	2.65	2.65	2.65	2.65	2.65	2.66	2.66	2.66	2.66	2.66
7.1	2.66	2.67	2.67	2.67	2.67	2.67	2.68	2.68	2.68	2.68
7.2	2.68	2.69	2.69	2.69	2.69	2.69	2.69	2.70	2.70	2.70
7.3	2.70	2.70	2.71	2.71	2.71	2.71	2.71	2.71	2.72	2.72
7.4	2.72	2.72	2.72	2.73	2.73	2.73	2.73	2.73	2.73	2.74
7.5	2.74	2.74	2.74	2.74	2.75	2.75	2.75	2.75	2.75	2.75
7.6	2.76	2.76	2.76	2.76	2.76	2.77	2.77	2.77	2.77	2.77
7.7	2.77	2.78	2.78	2.78	2.78	2.78	2.79	2.79	2.79	2.79
7.8	2.79	2.79	2.80	2.80	2.80	2.80	2.80	2.81	2.81	2.81
7.9	2.81	2.81	2.81	2.82	2.82	2.82	2.82	2.82	2.82	2.83
8.0	2.83	2.83	2.83	2.83	2.84	2.84	2.84	2.84	2.84	2.84
8.1	2.85	2.85	2.85	2.85	2.85	2.85	2.86	2.86	2.86	2.86
8.2	2.86	2.87	2.87	2.87	2.87	2.87	2.87	2.88	2.88	2.88
8.3	2.88	2.88	2.88	2.89	2.89	2.89	2.89	2.89	2.89	2.90
8.4	2.90	2.90	2.90	2.90	2.91	2.91	2.91	2.91	2.91	2.91
8.5	2.92	2.92	2.92	2.92	2.92	2.92	2.93	2.93	2.93	2.93
8.6	2.93	2.93	2.94	2.94	2.94	2.94	2.94	2.94	2.95	2.95
8.7	2.95	2.95	2.95	2.95	2.96	2.96	2.96	2.96	2.96	2.96
8.8	2.97	2.97	2.97	2.97	2.97	2.97	2.98	2.98	2.98	2.98
8.9	2.98	2.98	2.99	2.99	2.99	2.99	2.99	2.99	3.00	3.00
9.0	3.00	3.00	3.00	3.00	3.01	3.01	3.01	3.01	3.01	3.01
9.1	3.02	3.02	3.02	3.02	3.02	3.02	3.03	3.03	3.03	3.03
9.2	3.03	3.03	3.04	3.04	3.04	3.04	3.04	3.04	3.05	3.05
9.3	3.05	3.05	3.05	3.05	3.06	3.06	3.06	3.06	3.06	3.06
9.4	3.07	3.07	3.07	3.07	3.07	3.07	3.08	3.08	3.08	3.08
9.5	3.08	3.08	3.09	3.09	3.09	3.09	3.09	3.09	3.10	3.10
9.6	3.10	3.10	3.10	3.10	3.10	3.11	3.11	3.11	3.11	3.11
9.7	3.11	3.12	3.12	3.12	3.12	3.12	3.12	3.13	3.13	3.13
9.8	3.13	3.13	3.13	3.14	3.14	3.14	3.14	3.14	3.14	3.14
9.9	3.15	3.15	3.15	3.15	3.15	3.15	3.16	3.16	3.16	3.16

Square roots from 10 to 100

$x \to \sqrt{x}$

x	.0	.1	.2	.3	.4	.5	.6	.7	.8	.9
10	3.16	3.18	3.19	3.21	3.22	3.24	3.26	3.27	3.29	3.30
11	3.32	3.33	3.35	3.36	3.38	3.39	3.41	3.42	3.44	3.45
12	3.46	3.48	3.49	3.51	3.52	3.54	3.55	3.56	3.58	3.59
13	3.61	3.62	3.63	3.65	3.66	3.67	3.69	3.70	3.71	3.73
14	3.74	3.75	3.77	3.78	3.79	3.81	3.82	3.83	3.85	3.86
15	3.87	3.89	3.90	3.91	3.92	3.94	3.95	3.96	3.97	3.99
16	4.00	4.01	4.02	4.04	4.05	4.06	4.07	4.09	4.10	4.11
17	4.12	4.14	4.15	4.16	4.17	4.18	4.20	4.21	4.22	4.23
18	4.24	4.25	4.27	4.28	4.29	4.30	4.31	4.32	4.34	4.35
19	4.36	4.37	4.38	4.39	4.40	4.42	4.43	4.44	4.45	4.46
20	4.47	4.48	4.49	4.51	4.52	4.53	4.54	4.55	4.56	4.57
21	4.58	4.59	4.60	4.62	4.63	4.64	4.65	4.66	4.67	4.68
22	4.69	4.70	4.71	4.72	4.73	4.74	4.75	4.76	4.77	4.79
23	4.80	4.81	4.82	4.83	4.84	4.85	4.86	4.87	4.88	4.89
24	4.90	4.91	4.92	4.93	4.94	4.95	4.96	4.97	4.98	4.99
25	5.00	5.01	5.02	5.03	5.04	5.05	5.06	5.07	5.08	5.09
26	5.10	5.11	5.12	5.13	5.14	5.15	5.16	5.17	5.18	5.19
27	5.20	5.21	5.22	5.22	5.23	5.24	5.25	5.26	5.27	5.28
28	5.29	5.30	5.31	5.32	5.33	5.34	5.35	5.36	5.37	5.38
29	5.39	5.39	5.40	5.41	5.42	5.43	5.44	5.45	5.46	5.47
30	5.48	5.49	5.50	5.50	5.51	5.52	5.53	5.54	5.55	5.56
31	5.57	5.58	5.59	5.59	5.60	5.61	5.62	5.63	5.64	5.65
32	5.66	5.67	5.67	5.68	5.69	5.70	5.71	5.72	5.73	5.74
33	5.74	5.75	5.76	5.77	5.78	5.79	5.80	5.81	5.81	5.82
34	5.83	5.84	5.85	5.86	5.87	5.87	5.88	5.89	5.90	5.91
35	5.92	5.92	5.93	5.94	5.95	5.96	5.97	5.97	5.98	5.99
36	6.00	6.01	6.02	6.02	6.03	6.04	6.05	6.06	6.07	6.07
37	6.08	6.09	6.10	6.11	6.12	6.12	6.13	6.14	6.15	6.16
38	6.16	6.17	6.18	6.19	6.20	6.20	6.21	6.22	6.23	6.24
39	6.24	6.25	6.26	6.27	6.28	6.28	6.29	6.30	6.31	6.32
40	6.32	6.33	6.34	6.35	6.36	6.36	6.37	6.38	6.39	6.40
41	6.40	6.41	6.42	6.43	6.43	6.44	6.45	6.46	6.47	6.47
42	6.48	6.49	6.50	6.50	6.51	6.52	6.53	6.53	6.54	6.55
43	6.56	6.57	6.57	6.58	6.59	6.60	6.60	6.61	6.62	6.63
44	6.63	6.64	6.65	6.66	6.66	6.67	6.68	6.69	6.69	6.70
45	6.71	6.72	6.72	6.73	6.74	6.75	6.75	6.76	6.77	6.77
46	6.78	6.79	6.80	6.80	6.81	6.82	6.83	6.83	6.84	6.85
47	6.86	6.86	6.87	6.88	6.88	6.89	6.90	6.91	6.91	6.92
48	6.93	6.94	6.94	6.95	6.96	6.96	6.97	6.98	6.99	6.99
49	7.00	7.01	7.01	7.02	7.03	7.04	7.04	7.05	7.06	7.06
50	7.07	7.08	7.09	7.09	7.10	7.11	7.11	7.12	7.13	7.13
51	7.14	7.15	7.16	7.16	7.17	7.18	7.18	7.19	7.20	7.20
52	7.21	7.22	7.22	7.23	7.24	7.25	7.25	7.26	7.27	7.27
53	7.28	7.29	7.29	7.30	7.31	7.31	7.32	7.33	7.33	7.34
54	7.35	7.36	7.36	7.37	7.38	7.38	7.39	7.40	7.40	7.41

x	.0	.1	.2	.3	.4	.5	.6	.7	.8	.9
55	7.42	7.42	7.42	7.44	7.44	7.45	7.46	7.46	7.47	7.48
56	7.48	7.49	7.50	7.50	7.51	7.52	7.52	7.53	7.54	7.54
57	7.55	7.56	7.56	7.57	7.58	7.58	7.59	7.60	7.60	7.61
58	7.62	7.62	7.63	7.64	7.64	7.65	7.66	7.66	7.67	7.67
59	7.68	7.69	7.69	7.70	7.71	7.71	7.72	7.73	7.73	7.74
60	7.75	7.75	7.76	7.77	7.77	7.78	7.78	7.79	7.80	7.80
61	7.81	7.82	7.82	7.83	7.84	7.84	7.85	7.85	7.86	7.87
62	7.87	7.88	7.89	7.89	7.90	7.91	7.91	7.92	7.92	7.93
63	7.94	7.94	7.95	7.96	7.96	7.97	7.97	7.98	7.99	7.99
64	8.00	8.01	8.01	8.02	8.02	8.03	8.04	8.04	8.05	8.06
65	8.06	8.07	8.07	8.08	8.09	8.09	8.10	8.11	8.11	8.12
66	8.12	8.13	8.14	8.14	8.15	8.15	8.16	8.17	8.17	8.18
67	8.19	8.19	8.20	8.20	8.21	8.22	8.22	8.23	8.23	8.24
68	8.25	8.25	8.26	8.26	8.27	8.28	8.28	8.29	8.29	8.30
69	8.31	8.31	8.32	8.32	8.33	8.34	8.34	8.35	8.35	8.36
70	8.37	8.37	8.38	8.38	8.39	8.40	8.40	8.41	8.41	8.42
71	8.43	8.43	8.44	8.44	8.45	8.46	8.46	8.47	8.47	8.48
72	8.49	8.49	8.50	8.50	8.51	8.51	8.52	8.53	8.53	8.54
73	8.54	8.55	8.56	8.56	8.57	8.57	8.58	8.58	8.59	8.60
74	8.60	8.61	8.61	8.62	8.63	8.63	8.64	8.64	8.65	8.65
75	8.66	8.67	8.67	8.68	8.68	8.69	8.69	8.70	8.71	8.71
76	8.72	8.72	8.73	8.73	8.74	8.75	8.75	8.76	8.76	8.77
77	8.77	8.78	8.79	8.79	8.80	8.80	8.81	8.81	8.82	8.83
78	8.83	8.84	8.84	8.85	8.85	8.86	8.87	8.87	8.88	8.88
79	8.89	8.89	8.90	8.91	8.91	8.92	8.92	8.93	8.93	8.94
80	8.94	8.95	8.96	8.96	8.97	8.97	8.98	8.98	8.99	8.99
81	9.00	9.01	9.01	9.02	9.02	9.03	9.03	9.04	9.04	9.05
82	9.06	9.06	9.07	9.07	9.08	9.08	9.09	9.09	9.10	9.10
83	9.11	9.12	9.12	9.13	9.13	9.14	9.14	9.15	9.15	9.16
84	9.17	9.17	9.18	9.18	9.19	9.19	9.20	9.20	9.21	9.21
85	9.22	9.22	9.23	9.24	9.24	9.25	9.25	9.26	9.26	9.27
86	9.27	9.28	9.28	9.29	9.30	9.30	9.31	9.31	9.32	9.32
87	9.33	9.33	9.34	9.34	9.35	9.35	9.36	9.36	9.37	9.38
88	9.38	9.39	9.39	9.40	9.40	9.41	9.41	9.42	9.42	9.43
89	9.43	9.44	9.44	9.45	9.46	9.46	9.47	9.47	9.48	9.48
90	9.49	9.49	9.50	9.50	9.51	9.51	9.52	9.52	9.53	9.53
91	9.54	9.54	9.55	9.56	9.56	9.57	9.57	9.58	9.58	9.59
92	9.59	9.60	9.60	9.61	9.61	9.62	9.62	9.63	9.63	9.64
93	9.64	9.65	9.65	9.66	9.66	9.67	9.67	9.68	9.69	9.69
94	9.70	9.70	9.71	9.71	9.72	9.72	9.73	9.73	9.74	9.74
95	9.75	9.75	9.76	9.76	9.77	9.77	9.78	9.78	9.79	9.79
96	9.80	9.80	9.81	9.81	9.82	9.82	9.83	9.83	9.84	9.84
97	9.85	9.85	9.86	9.86	9.87	9.87	9.88	9.88	9.89	9.89
98	9.90	9.90	9.91	9.91	9.92	9.92	9.93	9.93	9.94	9.94
99	9.95	9.95	9.96	9.96	9.97	9.97	9.98.	9.98	9.99	9.99

Squares

$x \rightarrow x^2$

x	0	1	2	3	4	5	6	7	8	9
1.0	1.00	1.02	1.04	1.06	1.08	1.10	1.12	1.14	1.17	1.19
1.1	1.21	1.23	1.25	1.28	1.30	1.32	1.35	1.37	1.39	1.42
1.2	1.44	1.46	1.49	1.51	1.54	1.56	1.59	1.61	1.64	1.66
1.3	1.69	1.72	1.74	1.77	1.80	1.82	1.85	1.88	1.90	1.93
1.4	1.96	1.99	2.02	2.04	2.07	2.10	2.13	2.16	2.19	2.22
1.5	2.25	2.28	2.31	2.34	2.37	2.40	2.43	2.46	2.50	2.53
1.6	2.56	2.59	2.62	2.66	2.69	2.72	2.76	2.79	2.82	2.86
1.7	2.89	2.92	2.96	2.99	3.03	3.06	3.10	3.13	3.17	3.20
1.8	3.24	3.28	3.31	3.35	3.39	3.42	3.46	3.50	3.53	3.57
1.9	3.61	3.65	3.69	3.72	3.76	3.80	3.84	3.88	3.92	3.96
2.0	4.00	4.04	4.08	4.12	4.16	4.20	4.24	4.28	4.33	4.37
2.1	4.41	4.45	4.49	4.54	4.58	4.62	4.67	4.71	4.75	4.80
2.2	4.84	4.88	4.93	4.97	5.02	5.06	5.11	5.15	5.20	5.24
2.3	5.29	5.34	5.38	5.43	5.48	5.52	5.57	5.62	5.66	5.71
2.4	5.76	5.81	5.86	5.90	5.95	6.00	6.05	6.10	6.15	6.20
2.5	6.25	6.30	6.35	6.40	6.45	6.50	6.55	6.60	6.66	6.71
2.6	6.76	6.81	6.86	6.92	6.97	7.02	7.08	7.13	7.18	7.24
2.7	7.29	7.34	7.40	7.45	7.51	7.56	7.62	7.67	7.73	7.78
2.8	7.84	7.90	7.95	8.01	8.07	8.12	8.18	8.24	8.29	8.35
2.9	8.41	8.47	8.53	8.58	8.64	8.70	8.76	8.82	8.88	8.94
3.0	9.00	9.06	9.12	9.18	9.24	9.30	9.36	9.42	9.49	9.55
3.1	9.61	9.67	9.73	9.80	9.86	9.92	9.99	10.05	10.11	10.18
3.2	10.24	10.30	10.37	10.43	10.50	10.56	10.63	10.69	10.76	10.82
3.3	10.89	10.96	11.02	11.09	11.16	11.22	11.29	11.36	11.42	11.49
3.4	11.56	11.63	11.70	11.76	11.83	11.90	11.97	12.04	12.11	12.18
3.5	12.25	12.32	12.39	12.46	12.53	12.60	12.67	12.74	12.82	12.89
3.6	12.96	13.03	13.10	13.18	13.25	13.32	13.40	13.47	13.54	13.62
3.7	13.69	13.76	13.84	13.91	13.99	14.06	14.14	14.21	14.29	14.36
3.8	14.44	14.52	14.59	14.67	14.75	14.82	14.90	14.98	15.05	15.13
3.9	15.21	15.29	15.37	15.44	15.52	15.60	15.68	15.76	15.84	15.92
4.0	16.00	16.08	16.16	16.24	16.32	16.40	16.48	16.56	16.65	16.73
4.1	16.81	16.89	16.97	17.06	17.14	17.22	17.31	17.39	17.47	17.56
4.2	17.64	17.72	17.81	17.89	17.98	18.06	18.15	18.23	18.32	18.40
4.3	18.49	18.58	18.66	18.75	18.84	18.92	19.01	19.10	19.18	19.27
4.4	19.36	19.45	19.54	19.62	19.71	19.80	19.89	19.98	20.07	20.16
4.5	20.25	20.34	20.43	20.52	20.61	20.70	20.79	20.88	20.98	21.07
4.6	21.16	21.25	21.34	21.44	21.53	21.62	21.72	21.81	21.90	22.00
4.7	22.09	22.18	22.28	22.37	22.47	22.56	22.66	22.75	22.85	22.94
4.8	23.04	23.14	23.23	23.33	23.43	23.52	23.62	23.72	23.81	23.91
4.9	24.01	24.11	24.21	24.30	24.40	24.50	24.60	24.70	24.80	24.90
5.0	25.00	25.10	25.20	25.30	25.40	25.50	25.60	25.70	25.81	25.91
5.1	26.01	26.11	26.21	26.32	26.42	26.52	26.63	26.73	26.83	26.94
5.2	27.04	27.14	27.25	27.35	27.46	27.56	27.67	27.77	27.88	27.98
5.3	28.09	28.20	28.30	28.41	28.52	28.62	28.73	28.84	28.94	29.05
5.4	29.16	29.27	29.38	29.48	29.59	29.70	29.81	29.92	30.03	30.14

x	0	1	2	3	4	5	6	7	8	9
5.5	30.25	30.36	30.47	30.58	30.69	30.80	30.91	31.02	31.14	31.25
5.6	31.36	31.47	31.58	31.70	31.81	31.92	32.04	32.15	32.26	32.38
5.7	32.49	32.60	32.72	32.83	32.95	33.06	33.18	33.29	33.41	33.52
5.8	33.64	33.76	33.87	33.99	34.11	34.22	34.34	34.46	34.57	34.69
5.9	34.81	34.93	35.05	35.16	35.28	35.40	35.52	35.64	35.76	35.88
6.0	36.00	36.12	36.24	36.36	36.48	36.60	36.72	36.84	36.97	37.09
6.1	37.21	37.33	37.45	37.58	37.70	37.82	37.95	38.07	38.19	38.32
6.2	38.44	38.56	38.69	38.81	38.94	39.06	39.19	39.31	39.44	39.56
6.3	39.69	39.82	39.94	40.07	40.20	40.32	40.45	40.58	40.70	40.83
6.4	40.96	41.09	41.22	41.34	41.47	41.60	41.73	41.86	41.99	42.12
6.5	42.25	42.38	42.51	42.64	42.77	42.90	43.03	43.16	43.30	43.43
6.6	43.56	43.69	43.82	43.96	44.09	44.22	44.36	44.49	44.62	44.76
6.7	44.89	45.02	45.16	45.29	45.43	45.56	45.70	45.83	45.97	46.10
6.8	46.24	46.38	46.51	46.65	46.79	46.92	47.06	47.20	47.33	47.47
6.9	47.61	47.75	47.89	48.02	48.16	48.30	48.44	48.58	48.72	48.86
7.0	49.00	49.14	49.28	49.42	49.56	49.70	49.84	49.98	50.13	50.27
7.1	50.41	50.55	50.69	50.84	50.98	51.12	51.27	51.41	51.55	51.70
7.2	51.84	51.98	52.13	52.27	52.42	52.56	52.71	52.85	53.00	53.14
7.3	53.29	53.44	53.58	53.73	53.88	54.02	54.17	54.32	54.46	54.61
7.4	54.76	54.91	55.06	55.20	55.35	55.50	55.65	55.80	55.95	56.10
7.5	56.25	56.40	56.55	56.70	56.85	57.00	57.15	57.30	57.46	57.61
7.6	57.76	57.91	58.06	58.22	58.37	58.52	58.68	58.83	58.98	59.14
7.7	59.29	59.44	59.60	59.75	59.91	60.06	60.22	60.37	60.53	60.68
7.8	60.84	61.00	61.15	61.31	61.47	61.62	61.78	61.94	62.09	62.25
7.9	62.41	62.57	62.73	62.88	63.04	63.20	63.36	63.52	63.68	63.84
8.0	64.00	64.16	64.32	64.48	64.64	64.80	64.96	65.12	65.29	65.45
8.1	65.61	65.77	65.93	66.10	66.26	66.42	66.59	66.75	66.91	67.08
8.2	67.24	67.40	67.57	67.73	67.90	68.06	68.23	68.39	68.56	68.72
8.3	68.89	69.06	69.22	69.39	69.56	69.72	69.89	70.06	70.22	70.39
8.4	70.56	70.73	70.90	71.06	71.23	71.40	71.57	71.74	71.91	72.08
8.5	72.25	72.42	72.59	72.76	72.93	73.10	73.27	73.44	73.62	73.79
8.6	73.96	74.13	74.30	74.48	74.65	74.82	75.00	75.17	75.34	75.52
8.7	75.69	75.86	76.04	76.21	76.39	76.56	76.74	76.91	77.09	77.26
8.8	77.44	77.62	77.79	77.97	78.15	78.32	78.50	78.68	78.85	79.03
8.9	79.21	79.39	79.57	79.74	79.92	80.10	80.28	80.46	80.64	80.82
9.0	81.00	81.18	81.36	81.54	81.72	81.90	82.08	82.26	82.45	82.83
9.1	82.81	82.99	83.17	83.36	83.54	83.72	83.91	84.09	84.27	84.46
9.2	84.64	84.82	85.01	85.19	85.38	85.56	85.75	85.93	86.12	86.30
9.3	86.49	86.68	86.86	87.05	87.24	87.42	87.61	87.80	87.98	88.17
9.4	88.36	88.55	88.74	88.92	89.11	89.30	89.49	89.68	89.87	90.06
9.5	90.25	90.44	90.63	90.82	91.01	92.10	91.39	91.58	91.78	91.97
9.6	92.16	92.35	92.54	92.74	92.93	93.12	93.32	93.51	93.70	93.90
9.7	94.09	94.28	94.48	94.67	94.87	95.06	95.26	95.45	95.65	95.84
9.8	96.04	96.24	96.43	96.63	96.83	97.02	97.22	97.42	97.61	97.81
9.9	98.01	98.21	98.41	98.60	98.80	99.00	99.20	99.40	99.60	99.80

Reciprocals

$x \rightarrow \frac{1}{x}$

x	0	1	2	3	4	5	6	7	8	9
1.0	1.000	0.990	980	971	962	952	943	935	926	917
1.1	0.909	901	893	885	877	870	862	855	847	840
1.2	.833	826	820	813	806	800	794	787	781	775
1.3	.769	763	758	752	746	741	735	730	725	719
1.4	.714	709	704	699	694	690	685	680	676	671
1.5	0.667	662	658	654	649	645	641	637	633	629
1.6	.625	621	617	613	610	606	602	599	595	592
1.7	.588	585	581	578	575	571	568	565	562	559
1.8	.556	552	549	546	543	541	538	535	532	529
1.9	.526	524	521	518	515	513	510	508	505	503
2.0	0.500	498	495	493	490	488	485	483	481	478
2.1	.476	474	472	469	467	465	463	461	459	457
2.2	.455	452	450	448	446	444	442	441	439	437
2.3	.435	433	431	429	427	426	424	422	420	418
2.4	.417	415	413	412	410	408	407	405	403	402
2.5	0.400	398	397	395	394	392	391	389	388	386
2.6	.385	383	382	380	379	377	376	375	373	372
2.7	.370	369	368	366	365	364	362	361	360	358
2.8	.357	356	355	353	352	351	350	348	347	346
2.9	.345	344	342	341	340	339	338	337	336	334
3.0	0.333	332	331	330	329	328	327	326	325	324
3.1	.323	322	321	319	318	317	316	315	314	313
3.2	.313	312	311	310	309	308	307	306	305	304
3.3	.303	302	301	300	299	299	298	297	296	295
3.4	.294	293	292	292	291	290	289	288	287	287
3.5	0.286	285	284	283	282	282	281	280	279	279
3.6	.278	277	276	275	275	274	273	272	272	271
3.7	.270	270	269	268	267	267	266	265	265	264
3.8	.263	262	262	261	260	260	259	258	258	257
3.9	.256	256	255	254	254	253	253	252	251	251
4.0	0.250	249	249	248	248	247	246	246	245	244
4.1	.244	243	243	242	242	241	240	240	239	239
4.2	.238	238	237	236	236	235	235	234	234	233
4.3	.233	232	231	231	230	230	229	229	228	228
4.4	.227	227	226	226	225	225	224	224	223	223
4.5	0.222	222	221	221	220	220	219	219	218	218
4.6	.217	217	216	216	216	215	215	214	214	213
4.7	.213	212	212	211	211	211	210	210	209	209
4.8	.208	208	207	207	207	206	206	205	205	204
4.9	.204	204	203	203	202	202	202	201	201	200
5.0	0.200	200	199	199	198	198	198	197	197	196
5.1	.196	196	195	195	195	194	194	193	193	193
5.2	.192	192	192	191	191	190	190	190	189	189
5.3	.189	188	188	188	187	187	187	186	186	186
5.4	.185	185	185	184	184	183	183	183	182	182

x	0	1	2	3	4	5	6	7	8	9
5.5	0.182	181	181	181	181	180	180	180	179	179
5.6	.179	178	178	178	177	177	177	176	176	176
5.7	.175	175	175	175	174	174	174	173	173	173
5.8	.172	172	172	172	171	171	171	170	170	170
5.9	.169	169	169	169	168	168	168	168	167	167
6.0	0.167	166	166	166	166	165	165	165	164	164
6.1	.164	164	163	163	163	163	162	162	162	162
6.2	.161	161	161	161	160	160	160	159	159	159
6.3	.159	158	158	158	158	157	157	157	157	156
6.4	.156	156	156	156	155	155	155	155	154	154
6.5	0.154	154	153	153	153	153	152	152	152	152
6.6	.152	151	151	151	151	150	150	150	150	149
6.7	.149	149	149	149	148	148	148	148	147	147
6.8	.147	147	147	146	146	146	146	146	145	145
6.9	.145	145	145	144	144	144	144	143	143	143
7.0	0.143	143	142	142	142	142	142	141	141	141
7.1	.141	141	140	140	140	140	140	139	139	139
7.2	.139	139	139	138	138	138	138	138	137	137
7.3	.137	137	137	136	136	136	136	136	136	135
7.4	.135	135	135	135	134	134	134	134	134	134
7.5	0.133	133	133	133	133	132	132	132	132	132
7.6	.132	131	131	131	131	131	131	130	130	130
7.7	.130	130	130	129	129	129	129	129	129	128
7.8	.128	128	128	128	128	127	127	127	127	127
7.9	.127	126	126	126	126	126	126	125	125	125
8.0	0.125	125	125	125	124	124	124	124	124	124
8.1	.123	123	123	123	123	123	123	122	122	122
8.2	.122	122	122	122	121	121	121	121	121	121
8.3	.120	120	120	120	120	120	120	119	119	119
8.4	.119	119	119	119	118	118	118	118	118	118
8.5	0.118	118	117	117	117	117	117	117	117	116
8.6	.116	116	116	116	116	116	115	115	115	115
8.7	.115	115	115	115	114	114	114	114	114	114
8.8	.114	114	113	113	113	113	113	113	113	112
8.9	.112	112	112	112	112	112	112	111	111	111
9.0	0.111	111	111	111	111	110	110	110	110	110
9.1	.110	110	110	110	109	109	109	109	109	109
9.2	.109	109	108	108	108	108	108	108	108	108
9.3	.108	107	107	107	107	107	107	107	107	106
9.4	.106	106	106	106	106	106	106	106	105	105
9.5	0.105	105	105	105	105	105	105	104	104	104
9.6	.104	104	104	104	104	104	104	103	103	103
9.7	.103	103	103	103	103	103	102	102	102	102
9.8	.102	102	102	102	102	102	101	101	101	101
9.9	.101	101	101	101	101	101	100	100	100	100

Random numbers

	01	02	03	04	05	06	07	08	09	10	11	12	13	14	15	16	17	18	19	20	21	22	23	24	25
01	49	06	36	74	56	67	90	50	57	31	76	47	11	78	13	59	48	77	61	56	93	97	24	71	41
02	94	48	24	90	58	30	90	59	91	81	75	99	65	41	04	95	73	16	99	58	89	10	72	81	98
03	22	36	16	68	46	78	59	98	27	06	60	72	89	40	98	41	01	96	38	36	94	88	15	44	17
04	85	78	75	97	50	95	40	60	46	08	87	37	42	71	54	83	49	33	37	95	58	40	41	27	19
05	09	42	03	69	83	70	03	72	29	89	77	71	51	59	41	33	86	32	78	43	04	41	07	66	55
06	91	49	43	39	47	62	58	93	55	59	78	96	25	16	56	46	71	92	36	01	99	85	76	54	59
07	61	18	26	08	35	43	17	10	69	49	77	33	95	63	63	79	48	60	08	15	52	38	87	25	56
08	43	63	67	96	86	98	79	16	21	60	58	34	41	31	74	55	50	34	29	81	21	64	43	61	04
09	06	16	64	05	02	99	19	06	02	17	54	06	85	13	75	16	64	62	93	46	57	40	68	40	14
10	41	31	75	63	35	11	00	90	22	61	95	26	43	99	86	01	35	57	02	44	56	63	21	89	51
11	96	38	77	32	85	80	25	90	11	06	86	01	88	77	37	60	33	87	68	52	16	18	11	75	78
12	02	48	82	21	98	75	48	24	60	89	57	74	40	71	49	51	79	08	89	11	93	38	74	48	01
13	47	42	98	31	16	27	54	34	61	77	68	30	39	34	60	63	74	59	23	93	10	66	49	11	06
14	60	17	51	97	12	87	94	88	28	59	97	15	75	61	15	12	28	77	31	62	18	50	22	32	25
15	38	29	21	03	29	78	75	17	16	60	61	18	99	64	48	06	04	36	81	31	23	05	91	95	93
16	86	15	69	42	67	80	04	07	44	30	96	69	18	29	50	24	69	16	00	25	68	81	42	77	33
17	71	27	21	23	34	77	49	88	15	38	25	48	13	96	65	89	72	72	71	70	03	36	32	10	57
18	48	25	02	57	00	43	53	63	89	53	82	61	14	12	38	13	07	88	27	95	00	85	13	64	48
19	52	02	19	34	95	61	72	84	01	71	66	36	71	09	40	76	95	74	18	39	74	82	90	86	72
20	75	86	92	92	64	73	56	30	06	50	49	62	94	45	95	41	07	52	61	73	42	24	18	43	96
21	78	99	82	39	31	94	86	25	20	59	05	47	25	83	09	60	83	47	03	89	85	06	99	82	24
22	96	19	62	03	10	00	07	78	64	02	84	16	02	91	00	89	34	07	67	68	32	20	28	00	76
23	69	19	30	69	75	52	67	38	05	80	56	74	90	97	11	95	59	22	75	10	28	64	73	63	38
24	49	37	22	85	60	15	42	25	52	60	75	99	15	59	69	64	41	64	46	44	79	07	88	73	47
25	77	48	59	84	81	19	46	04	60	39	48	27	80	77	10	27	11	85	03	74	15	62	65	72	35
26	12	58	53	40	48	73	90	96	89	94	06	68	39	29	40	80	19	34	97	45	81	03	98	69	85
27	81	12	85	84	55	63	26	35	96	13	08	61	19	80	45	38	81	37	56	00	45	63	98	68	88
28	75	50	34	41	93	21	44	41	72	27	90	56	55	17	08	04	15	11	56	92	41	46	09	86	65
29	20	88	39	83	63	60	35	34	66	42	63	81	58	80	40	81	63	31	30	55	40	08	68	98	90
30	47	71	24	29	40	70	06	31	47	09	69	81	57	21	63	48	37	87	80	97	70	35	23	12	58
31	99	36	96	23	70	79	21	78	85	68	44	52	16	09	15	58	56	35	07	84	61	38	28	91	30
32	70	74	56	74	59	40	73	82	91	31	44	80	78	58	19	09	03	28	17	33	29	12	60	35	58
33	32	20	70	38	26	31	49	89	77	69	69	12	21	69	77	31	07	51	34	39	26	28	04	84	81
34	59	53	08	62	23	22	33	92	42	32	11	50	21	88	92	06	13	49	51	37	90	53	83	87	58
35	63	35	19	27	99	57	50	55	52	70	46	93	39	04	53	52	19	31	95	65	04	65	63	64	19
36	45	35	41	02	83	30	05	94	31	34	75	03	41	34	87	58	90	48	15	60	13	16	28	04	86
37	02	89	69	66	05	63	79	72	20	67	33	78	98	64	43	77	70	04	32	72	41	10	01	42	42
38	02	23	43	79	86	11	27	81	38	99	25	42	15	75	78	40	07	87	49	88	31	11	47	19	78
39	94	32	34	76	68	95	44	34	47	87	51	47	04	68	75	46	09	90	08	74	13	71	73	31	90
40	47	84	14	48	29	70	21	72	37	27	79	97	87	35	48	16	87	51	01	55	98	29	10	14	36
41	34	12	79	01	61	62	30	28	92	77	14	59	78	97	09	76	44	93	35	56	70	30	07	03	05
42	80	19	42	68	02	40	90	13	20	15	73	78	85	64	21	13	78	02	98	44	33	20	47	84	01
43	69	00	43	02	59	78	71	25	87	92	08	39	09	08	44	89	70	70	70	56	68	51	17	79	61
44	96	28	74	94	31	02	16	33	40	19	94	28	89	74	77	16	15	03	40	39	10	21	75	12	51
45	52	73	55	47	51	12	52	70	50	81	84	13	48	69	59	89	55	82	06	24	46	66	92	62	54
46	48	81	62	34	28	48	65	20	43	15	58	25	86	52	66	91	93	66	70	91	86	05	80	62	65
47	16	24	75	75	05	78	22	86	00	59	92	33	84	79	59	41	64	57	69	28	02	38	48	07	03
48	62	15	92	43	75	35	46	34	12	15	25	73	28	01	57	26	40	46	76	51	36	82	63	87	31
49	28	73	77	77	01	00	71	34	74	64	23	36	12	96	08	84	65	57	36	50	98	47	43	54	57
50	09	99	28	75	30	85	16	94	34	42	09	56	14	64	40	68	27	30	31	41	63	41	94	38	64

Answers

Exercise 1a (p. 1)

1. $3(b + c)$
2. $3r(r + h)$
3. $2(m + 2m^2 - 3)$
4. $x(7x^2 - y - 3)$
5. $y(3x^2 + 5xy + 2x + 1)$
6. $7(a + c)$
7. $(b + 3)(x + y)$
8. $t(u + v - t)$
9. $(r - h)(m + r - h)$
10. $(x - y)(u + v + p + r)$

Exercise 1b (p. 2)

1. $(x + y)(a + 3b)$
2. $(a + 2b)(7 + x)$
3. $(x + 5)(x + 2)$
4. $(p + r)(q + s)$
5. $(a - 9)(a + 3)$
6. $(4m - 1)(2 + n)$
7. $(x - 2)(5x + 3)$
8. $(a - c)(b + d)$
9. $(2b - 5)(a + 1)$
10. $(3m - 1)(1 + 2m)$

Exercise 1c (p. 2)

1. $(a + c)(b - m)$
2. $(3x + 2)(3 - x)$
3. $(x - y)(2a - 3b)$
4. $(x - 7)(x - 2)$
5. $(a - b)(5 - c)$
6. $(q + 4r)(3p - y)$
7. $(a - 3)(a - 3)$
8. $(2s + 5t)(p - r)$
9. $(x - 6)(x - 1)$
10. $(3k + 1)(1 - h)$

Exercise 1d (p. 3)

1. $(a + b)(6 + m)$
2. $(p + q)(r + s)$
3. $(3 + y)(5 - x)$
4. $(a - b)(c + d)$
5. $(a + x)(x - y)$
6. $(d - m)(a + c)$
7. $(x - 3)(x - 5)$
8. $(2a + 3y)(4 + 5b)$
9. $(a - b)(3 + c)$
10. $(t + 3s)(1 + 2z)$

Exercise 1e (p. 3)

1. $(m + n)(x + y)$
2. $(x - y)(a + b)$
3. $(u + v)(h - k)$
4. $(a - b)(u - v)$
5. $(a + 2b)(m + n)$
6. $(c - d)(x + 2y)$
7. no factors
8. $(a - 2x)(b - 2y)$
9. $(m - n)(a + 1)$
10. no factors
11. $(a + 1)(a^2 + 1)$
12. $(h + k)(2m - 3n)$
13. $(x - y)(3s + 5t)$
14. $(ax + y)(bx + y)$
15. $(h - 2m)(k + 3n)$
16. no factors
17. $(g + h)(2k - 3l)$
18. $(f + 2g)(2h - k)$
19. no factors
20. $(h + 2k)(l - 3m)$
21. $(e - 2f)(3c - 2d)$
22. $(x - 2n)(y + 3n)$
23. $(a + 2b)(b - 2c)$
24. no factors
25. $(4u - v)(2v + 3w)$
26. $(m - 2n)(n + 3p)$
27. $(3x + 2y)(y - a)$
28. $3(a - u)(b + v)$
29. no factors
30. $(2c + 3d)(4e - f)$
31. $(mu + v)(nu - v)$
32. $5(m - n)(x - y)$
33. $(3a - c)(b - 3d)$
34. $(2a - 5c)(3b + 2d)$
35. $2a(m + n)(u - v)$
36. $(am + 2)(bm - 3)$
37. $2(2a + b)(x + 2y)$
38. $(7m - x)(3n + y)$
39. no factors
40. $(2a - 3m)(m + 2n)$
41. $(5u - 1)(2v + 1)$
42. $(a + m)(am - n)$
43. $(xy + a)(2x - y)$
44. $(1 - 5a)(1 + 3x)$
45. $(d + 2xy)(2dx - 3y)$

Exercise 1f (p. 4)

1. $(x + 5)(x + 1)$
2. $(x + 11)(x + 1)$
3. $(a + 13)(a + 1)$
4. $(b + 7)(b + 1)$
5. $(y + 8)(y + 1)$
6. $(z + 4)(z + 2)$
7. $(c + 5)(c + 3)$
8. $(d + 11)(d + 2)$
9. $(n + 6)(n + 2)$
10. $(r + 5)(r + 4)$
11. $(s + 8)(s + 2)$
12. $(t + 4)(t + 4)$

Exercise 1g (p. 5)

1. $(x - 3)(x - 1)$
2. $(y - 2)(y - 1)$
3. $(z - 17)(z - 1)$
4. $(a - 7)(a - 1)$
5. $(b - 3)(b - 2)$
6. $(c - 6)(c - 1)$
7. $(d - 7)(d - 2)$
8. $(n - 5)(n - 2)$
9. $(p - 8)(p - 3)$
10. $(q - 7)(q - 3)$
11. $(f - 14)(f - 2)$
12. $(x - 5)(x - 5)$

Exercise 1h (p. 5)

1. $(x + 5)(x - 1)$
2. $(a - 5)(a + 1)$
3. $(x + 7)(x - 1)$
4. $(b - 7)(b + 1)$
5. $(n + 2)(n - 1)$
6. $(r - 3)(r + 1)$
7. $(x - 11)(x + 1)$
8. $(y + 13)(y - 1)$
9. $(x - 5)(x + 3)$
10. $(x - 15)(x + 1)$
11. $(s + 6)(s - 1)$
12. $(t - 6)(t + 1)$
13. $(u - 3)(u + 2)$
14. $(v + 3)(v - 2)$
15. $(z + 5)(z - 4)$
16. $(c - 10)(c + 2)$
17. $(x + 7)(x - 7)$
18. $(x + 2)(x - 2)$

Exercise 1i (p. 6)

1. $a^2 + 8a + 16$
2. $b^2 - 6b + 9$
3. $25 + 10c + c^2$
4. $4 - 4d + d^2$
5. $1 + 2m + m^2$
6. $4n^2 + 4n + 1$
7. $9x^2 + 6xy + y^2$
8. $u^2 - 4uv + 4v^2$

9 $25h^2 - 10hk + k^2$ 10 $p^2 + 8pq + 16q^2$
11 $4a^2 + 12ad + 9d^2$ 12 $9b^2 - 30bc + 25c^2$
13 $49e^2 - 28ef + 4f^2$ 15 $100x^2 - 20x + 1$
15 $1 + 24y + 144y^2$ 16 $9a^2 + 42ab + 49b^2$
17 $c^2 - 16cd + 64d^2$ 18 $81u^2 + 18uv + v^2$

Exercise 1j (p. 6)

1 10 201 2 9801 3 10 609
4 9 604 5 1 002 001 6 998 001
7 1 010 025 8 992 016 9 990 025
10 5 184 11 6 889 12 6 241

Exercise 1k (p. 6)

1 $(a + 5)^2$ 2 $(b + 4)^2$
3 $(c + 3)^2$ 4 $(d + 10)^2$
5 $(m - 3)^2$ 6 $(n - 6)^2$
7 $(x - 2)^2$ 8 $(y - 1)^2$
9 $(z + 8)^2$ 10 $(k - 7)^2$
11 $(2 - b)^2$ 12 $(9 + d)^2$
13 $(x + 3y)^2$ 14 $(2u - 3)^2$
15 $(1 - a)^2$ 16 $(5n - 3v)^2$
17 $(3a - 4b)^2$ 18 $(11 - y)^2$

Exercise 1l (p. 7)

1 $(x + 1)(x - 1)$ 2 $(1 + y)(1 - y)$
3 $(2m - n)(2m + n)$ 4 $(u + 4v)(u - 4v)$
5 $(1 - ab)(1 + ab)$ 6 $(3 - 2c)(3 + 2c)$
7 $(2d + 3e)(2d - 3e)$ 8 $3(1 - f)(1 + f)$
9 $4(g + 1)(g - 1)$ 10 $(2h + 5)(2h - 5)$
11 $(5k - 4)(5k + 4)$ 12 $(7m - n)(7m + n)$
13 $(pq - 3)(pq + 3)$ 14 $(5 + uv)(5 - uv)$
15 $(9 - w)(9 + w)$ 16 $(10x + 1)(10x - 1)$
17 $4(2y + z)(2y - z)$ 18 $(4h - k)(4h + k)$
19 $(2c + 7d)(2c - 7d)$
20 $(e + 2f)(e - 2f)$
21 $(6a + 7b)(6a - 7b)$
22 $5(c - 3d)(c + 3d)$
23 $(xy + z)(xy - z)$ 24 $(10 - w)(10 + w)$

Exercise 1m (p. 8)

1 9 200 2 13 600 3 288
4 9 600 5 10 600 6 400
7 2 600 8 224 9 1 008 000
10 994 000 11 125.6 mm^2 12 2 640 cm^3

Exercise 1n (p. 10)

1 $(a + 3)(a + 5)$ 2 $(b - 2)(b - 5)$
3 $(c - 3)(c + 7)$ 4 $(d + 2)(d - 7)$
5 $(e - 2)(e + 4)$ 6 $(w + 2)(w + 3)$
7 $(x + 6)(x - 1)$ 8 $(y - 2)(y - 3)$
9 $(z - 6)(z + 1)$ 10 $(d + 1)(2d + 1)$
11 $(e - 1)(2e - 1)$ 12 $(f - 1)(2f + 1)$
13 $(a + 2)(a + 5)$ 14 $(a + 2b)(a + 5b)$
15 $(ab + 2)(ab + 5)$ 16 $(x + 3y)(x - 5y)$
17 $(m - 2)(m + 12)$ 18 $(n + 2)(n - 12)$
19 $(u - 4)(u - 6)$ 20 $(v - 3)(v - 8)$
21 $(m - 3)(m + 7)$ 22 $(m - 3n)(m + 7n)$
23 $(mn - 3)(mn + 7)$ 24 $(3a - 1)(a - 1)$
25 $(3b - 2)(b + 1)$ 26 $(x + 3)(2x - 1)$
27 $(y - 3)(2y + 1)$ 28 $(z - 1)(2z - 3)$
29 $(1 + 2m)(1 + m)$ 30 $(5 + n)(3 - n)$
31 $(1 - 4u)(1 + 2u)$ 32 $(u - 2v)(u + 4v)$
33 $(a - 4b)(a + 9b)$ 34 $(a - 3b)(a + 12b)$
35 $(a - 2b)(a + 18b)$ 36 $2(b - 2)(b - 3)$
37 $(c + 7)(c - 11)$ 38 $(7 - d)(11 + d)$
39 $3(e - 2)(e + 3)$ 40 $(f + 1)(3f - 1)$
41 $(a + b)(a + 3b)$ 42 $(1 + x)(1 + 3x)$
43 $(g - 2)(2g - 1)$ 44 $(h + 1)(2h + 3)$
45 $(h + 2k)(3h + k)$ 46 $(3x + 2)(4x - 7)$
47 $(a + 30)(a - 5)$ 48 $(b + 10)(b + 15)$
49 $(c - 3)(3c - 2)$ 50 $(d + 3)(3d - 2)$

Exercise 1o (p. 10)

1 $(e - 2)(5e + 1)$ 2 $(f + 1)(7f + 3)$
3 $(5 - a)(7 - a)$ 4 $(5 - b)(7 + b)$
5 $(35 + c)(1 + c)$ 6 $5(7 - d)(1 + d)$
7 $(a + b)(3a + 2b)$ 8 $(m + 2n)(3m - n)$
9 $(u + 2v)(3u + v)$ 10 $(2n - 3)(3n + 1)$
11 $(v + 3)(7v + 1)$ 12 $(2y - 1)(2y - 5)$
13 $(h - 9)(2h + 3)$ 14 $(k - 3)(2k - 9)$
15 $(xy + 5)(xy - 6)$ 16 $(uv + 2)(2uv - 3)$
17 $(1 - 2a)(5 + 3a)$ 18 $(p - 5)(10p + 9)$
19 $(2q - 5)(5q - 9)$ 20 $(a - 1)(8a - 9)$
21 $(2b - 3)(4b - 3)$ 22 $(c - 3)(8c + 3)$
23 $(2d - 1)(4d - 9)$ 24 $(e - 3)(8e - 25)$
25 $(2f - 5)(4f - 15)$ 26 $(3ab - 1)(4ab + 5)$
27 $(2m + n)(6m - 5n)$
28 $(3t - 2)(4t - 1)$
29 $(xy - 1)(12xy + 1)$ 30 $(p + q)(24p - 23q)$

Practice exercise P1.1 (p. 11)

1 $-3g(1 + 4r)$ 2 $2(k + k^3 + 1)$
3 $(c + 2)(2c + 7)$ 4 $x(6 + 3y - x)$
5 $rd(d - 2 + 4r)$
6 $\dfrac{3y - 2}{3y + 1}$

7. $2b(2c + 3cd + 1)$
8. $-5(x^2 - 2x + 2)$ or $5(-x^2 + 2x - 2)$
9. $3t(t^2 - t - s)$
10. $4(l - b)(h - 2k)$
11. $(y + 1)(xy + x + 2b)$
12. $(3 - f)(8g - 6 + 2f)$
13. $(g + h)(3f - 2)$
14. $(R - h)(R - h + 2)$
15. $n(2n - 3)(n + 3m)$
16. $(x + y)^2$
17. $\frac{a(5a - 1)}{2}$
18. $\frac{y^2}{2(y + 2)}$
19. $\frac{3(3x - 2)}{x + 2}$
20. $\frac{3c - 2b^2}{2 - c}$

Practice exercise P1.2 (p. 11)

1. $2x^2 + 13x + 21$
2. $3x^2 - 3x$
3. $ab + a + 3b + 3$
4. $cd + c - 2d - 2$
5. $ab + 3a + 4b + 12$
6. $2xy + 4x + 3y + 6$
7. $cd + 5c - 2d - 10$
8. $xy - dx - cy + cd$
9. $2fg - 2f - 3g + 3$
10. $4k^2 + 4k - 3$
11. $3km + 12kn + nm + 4n^2$
12. $2p^2 - 5sp - 3s^2$
13. $3m^2 + 12m + 4$
14. $2x^2 + 7xy + 3y^2$
15. $d^2 - dr - 4r^2$
16. $d^2 + 4d + 4$
17. $d^2 + 7d + 10$
18. $6 - 5t + t^2$
19. $3n^2 - 2n + 1$
20. $3a^2 + 7ab + 2b^2$
21. $9 + 6g + g^2$
22. $m^2 - 4mn + 4n^2$
23. $t^2 - 6t + 9$
24. $4d^2 + 4d + 1$
25. $4x^2 - 12xy + 9y^2$
26. $9 - 12y + 4y^2$
27. $-2x^2 - 5x + 3$
28. $12x^2 + 10x - 12$
29. $9x^2 + 6x + 1$
30. $4x^2 - 25$
31. $n^2 - 1$
32. $9v^2 - 1$
33. $x^2 - 4y^2$
34. $5x^2 - 2x + 10$
35. $3x^2 - 6xy$
36. 0

Practice exercise P1.3 (p. 11)

	(a)	(b)	(c)
1	1	6	8
2	4	7	3
3	1	0	−4
4	1	−4	3
5	1	0	−9
6	1	0	−25
7	1	8	16
8	4	0	−9
9	16	0	−1
10	4	0	−1
11	3	−8	−3
12	2	7	−4
13	1	4	4
14	3	2	−1
15	1	−6	9
16	1	14	49
17	1	2	1
18	1	−2	1
19	1	−8	16
20	1	6	9

Practice exercise P1.4 (p. 12)

1. 1
2. 16
3. $8x$
4. x
5. 9
6. 9

Practice exercise P1.5 (p. 12)

1. 361
2. 961
3. 168
4. 9604
5. 725
6. 6600

Practice exercise P1.6 (p. 12)

1. $(2a - b)(m - 3n)$
2. $(x + 5)(x - 3)$
3. $(h - 1)(h - r)$
4. $(a + b)(c + 2d)$
5. $(3u - w)(x - 2y)$
6. $(r + t)(2t - 3s)$
7. $(5 - v)(3 + w)$
8. $a(1 - 2b)(ab - 3bc)$
9. $2a(c - 2d)(3x + 2y)$
10. $(5y - 3)(y - 2)$

Practice exercise P1.7 (p. 12)

1. $(a + 2)(a + 1)$
2. $(b - 3)(b - 2)$
3. $(c + 3)(c + 2)$
4. $(f + 6)(f - 2)$
5. $(p - 1)(p + 8)$
6. $(r - 2)(r - 10)$
7. $(x + 8)^2$
8. $(m - 6)^2$
9. $(k + 11)^2$
10. $(1 - y)^2$
11. $(x + 7y)^2$
12. $(9 + n)^2$
13. $(2t - 3)(t - 2)$
14. $(d + 5)(d + 1)$
15. $2(b - 2)(b - 3)$
16. $4(h + 1)(h - 4)$
17. $(g - 3)(g - 4)$
18. $(j + 6)(j + 2)$
19. $3(v - 1)(v + 3)$
20. $2(w + 1)(w - 6)$
21. $4(y - 1)^2$
22. $(2y + 1)(y + 3)$
23. $(x - 9y)(x - 6y)$
24. $(2y - 1)(y + 3)$
25. $(2y - 3)(y - 1)$
26. $(2x - y)(2x - 3y)$
27. $(x - y)(4x + 3y)$
28. $(4x - y)(x - 3y)$
29. $(p + 4)^2$
30. $(r - 5)$
31. $2(t + 3)^2$
32. $(d + 3f)^2$
33. $2(b - 2)^2$
34. $(3y + 1)(y + 3)$
35. $(y + 8)(y + 1)$
36. $3(y + 1)^2$
37. $3(y - 1)^2$
38. $(3x + 1)(x - 3)$
39. $(3x + 1)(2x - 3)$
40. $(3x - 1)(2x + 3)$
41. $6(m + 1)(m - 1)$
42. $6(m + 1)^2$
43. $(3m - 2)(2m - 3)$
44. $(3y + 1)(y + 3)$
45. $(x - 3 - 2y)(x - 3 + 2y)$
46. $2(2c - 3d)(2c + 3d)$
47. $(2y - 5)(y - 3)$
48. $3(3x - 2y)(3x + 2y)$
49. $(3y - 1)^2$
50. $(x - 5)(x + 2)$

Practice exercise P1.8 (p. 12)

1. 4682 2. 3686.84 3. $4752\ \text{cm}^3$
4. 4 5. 10

Practice exercise P1.9 (p. 12)

1. $x + 4$ 2. $\frac{x-2}{x-3}$
3. $\frac{2x-1}{x-2}$ 4. $\frac{b}{1-6b}$

Exercise 2a (p. 13)

1. (a) $\begin{pmatrix} 15 & 10 \\ 3\frac{1}{2} & 4\frac{1}{2} \\ 18 & 20 \end{pmatrix}$ (b) $\begin{pmatrix} 15 \\ 3\frac{1}{2} \\ 18 \end{pmatrix}$
   (c) $(10, 4\frac{1}{2}, 20)$ (d) $(18, 20)$
2. (a) 2×5 (b) 2×1 (c) 3×3
   (d) 3×2 (e) 1×4 (f) 2×2
3. (a) $\begin{pmatrix} a & b & c & d & e \\ f & g & h & i & j \\ k & l & m & n & o \end{pmatrix}$
   (b) $\begin{pmatrix} a & b & c & d \\ e & f & g & h \end{pmatrix}$
   (c) $\begin{pmatrix} a & b \\ c & d \\ e & f \\ g & h \end{pmatrix}$
   (d) $\begin{pmatrix} a & b & c & d \\ e & f & g & h \\ i & j & k & l \\ m & n & o & p \end{pmatrix}$
4. (a) (w) (b) $\begin{pmatrix} w & x \\ y & z \end{pmatrix}$
   (c) $\begin{pmatrix} k & l & m \\ p & q & r \\ x & y & z \end{pmatrix}$ (d) see 3(d)
5. (a) 6 (b) 6 (c) 12
   (d) mn (e) x^2 (f) 4
6. (a) a (b) g (c) f (d) d
7. (a) 2nd row, 3rd col (b) 4th row, 1st col
   (c) 1st row, 2nd col (d) 4th row, 3rd col
   (e) 3rd row, 1st col (f) 2nd row, 2nd col
   (g) 1st row, 3rd col (h) 3rd row, 2nd col
8. $\begin{pmatrix} 11 & 3 & 5 \\ 8 & 0 & 2 \end{pmatrix}$

Exercise 2b (p. 15)

1. $\begin{pmatrix} 3 & 1 \\ 7 & 5 \end{pmatrix}$ 2. $\begin{pmatrix} 4 & 3 \\ 5 & 1 \\ 6 & 4 \end{pmatrix}$ 3. $\begin{pmatrix} 1 & 1 & -5 \\ 1 & 0 & 0 \end{pmatrix}$
4. impossible 5. $\begin{pmatrix} -2 \\ 2 \\ -2 \end{pmatrix}$ 6. $\begin{pmatrix} 8.4 & 1.0 \\ 0.2 & 10.5 \end{pmatrix}$
7. impossible 8. impossible
9. $\begin{pmatrix} 12 & -2 & 2 \\ 7 & 5 & 3 \\ 7 & 8 & 8 \end{pmatrix}$ 10. $\begin{pmatrix} 1 & -1 \\ 2 & 2 \\ 0 & -2 \end{pmatrix}$
11. $\begin{pmatrix} -2.2 & -1.7 \\ 0.9 & -1.9 \end{pmatrix}$ 12. $\begin{pmatrix} -3 \\ 2 \end{pmatrix}$
13. impossible 14. $\begin{pmatrix} 0 & 6 \\ 5 & 7 \end{pmatrix}$ 15. $\begin{pmatrix} 1 & -10 \\ 0 & 11 \end{pmatrix}$

Exercise 2c (p. 15)

1. (a) $\begin{pmatrix} 9 \\ 6 \end{pmatrix}$ (b) $\begin{pmatrix} 8 & 4 \\ 0 & 4 \end{pmatrix}$
   (c) $(-2, -4, -16)$ (d) $\begin{pmatrix} 3 & 4 \\ 1 & 7 \end{pmatrix}$
   (e) $\begin{pmatrix} 12 & 11 \\ 2 & 17 \end{pmatrix}$ (f) $\begin{pmatrix} 0 & 5 \\ 2 & 11 \end{pmatrix}$
2. $n = 3$
3. (a) $\text{M} = \begin{pmatrix} -2 & 3 \\ 0 & 1 \end{pmatrix}$ (b) $\text{M} = \begin{pmatrix} -3 & 3 \\ 2 & 1 \end{pmatrix}$
4. $x = 2, y = -3$
5. $p = -4, q = 11$

Exercise 2d (p. 16)

1. (a) 28 (b) -39 (c) 1
   (d) 35 (e) 2 (f) 0
2. (a) 156 (b) impossible
   (c) impossible (d) 0
3. $\$(70, 80)\begin{pmatrix} 5 \\ 3 \end{pmatrix} = \$590$
4. \$27.80, \$27.50
5. $\$(2, 1.2)\begin{pmatrix} 400 & 305 \\ 600 & 420 \end{pmatrix} = \$(1\,520, 1\,114)$

Exercise 2e (p. 17)

1. $\begin{pmatrix} 12 \\ 14 \end{pmatrix}$ 2. $\begin{pmatrix} 14 & 5 \\ 13 & 5 \end{pmatrix}$
3. $\begin{pmatrix} 15 & 5 \\ 11 & 4 \end{pmatrix}$ 4. $\begin{pmatrix} 1 & 0 \\ 0 & 1 \end{pmatrix}$
5. $\begin{pmatrix} 9 & 0 \\ 0 & 9 \end{pmatrix}$ 6. $\begin{pmatrix} 10 \\ 13 \\ 10 \end{pmatrix}$
7. $\begin{pmatrix} 4 & -4 \\ 0 & 16 \end{pmatrix}$ 8. $\begin{pmatrix} 0 & 4 \\ 7 & 7 \end{pmatrix}$
9. $\begin{pmatrix} 7 & 4 & -5 \\ 1 & -3 & 0 \end{pmatrix}$ 10. $\begin{pmatrix} 1 & 4 & -5 \\ 4 & 19 & -17 \end{pmatrix}$

Exercise 2f (p. 19)

1. $\frac{1}{9}\begin{pmatrix} 2 & -3 \\ -1 & 6 \end{pmatrix}$ 2. $\frac{1}{9}\begin{pmatrix} 3 & -3 \\ -2 & 5 \end{pmatrix}$

3 $-\frac{1}{2}\begin{pmatrix} 1 & -3 \\ -2 & 4 \end{pmatrix}$

4 $-\frac{2}{3}\begin{pmatrix} 3 & -3 \\ -5 & 4\frac{1}{2} \end{pmatrix}$

5 $\begin{pmatrix} 2 & -3 \\ -1 & 2 \end{pmatrix}$

6 $-\begin{pmatrix} 1 & -2 \\ -2 & 3 \end{pmatrix}$

7 singular

8 $-\frac{1}{17}\begin{pmatrix} 4 & -3 \\ 1 & -5 \end{pmatrix}$

9 $\frac{1}{60}\begin{pmatrix} -6 & 9 \\ -8 & 2 \end{pmatrix}$

10 $-\frac{1}{8}\begin{pmatrix} 0 & 4 \\ 2 & -5 \end{pmatrix}$

11 $-\frac{1}{2}\begin{pmatrix} 3 & -1 \\ -16 & 4 \end{pmatrix}$

12 $\begin{pmatrix} 0 & 3 \\ -2 & 0 \end{pmatrix}$

Exercise 2g (p. 20)

1 $\begin{pmatrix} x \\ y \end{pmatrix} = \begin{pmatrix} 3 \\ 1 \end{pmatrix}$

2 $\begin{pmatrix} x \\ y \end{pmatrix} = \begin{pmatrix} 3 \\ 2 \end{pmatrix}$

3 $\begin{pmatrix} x \\ y \end{pmatrix} = \begin{pmatrix} -1 \\ 5 \end{pmatrix}$

4 $\begin{pmatrix} x \\ y \end{pmatrix} = \begin{pmatrix} 1 \\ 3 \end{pmatrix}$

5 $\begin{pmatrix} x \\ y \end{pmatrix} = \begin{pmatrix} 2\frac{1}{2} \\ \frac{1}{2} \end{pmatrix}$

6 $\begin{pmatrix} x \\ y \end{pmatrix} = \begin{pmatrix} \frac{1}{2} \\ -\frac{1}{2} \end{pmatrix}$

7 $\begin{pmatrix} a \\ b \end{pmatrix} = \begin{pmatrix} 3 \\ 1 \end{pmatrix}$

8 $\begin{pmatrix} u \\ v \end{pmatrix} = \begin{pmatrix} 12 \\ 6 \end{pmatrix}$

9 $\begin{pmatrix} s \\ t \end{pmatrix} = \begin{pmatrix} \frac{1}{3} \\ \frac{1}{6} \end{pmatrix}$

10 $\begin{pmatrix} c \\ d \end{pmatrix} = \begin{pmatrix} \frac{1}{2} \\ 5 \end{pmatrix}$

Exercise 2h (p. 20)

1 $\begin{pmatrix} -7 & -6 \\ 0 & 12 \end{pmatrix}$

2 (a) 2 (b) $\frac{1}{2}\begin{pmatrix} 4 & 6 \\ 1 & 2 \end{pmatrix}$

3 14, $\frac{1}{14}\begin{pmatrix} 3 & 4 \\ -5 & -2 \end{pmatrix}$

4 (a) $x = 3$ (b) $\frac{1}{7}\begin{pmatrix} 3 & 2 \\ 4 & 5 \end{pmatrix}$

5 $k = 5$

6 (a) $\begin{pmatrix} -6 & 15 & 0 \\ 2 & -5 & 0 \\ -4 & 10 & 0 \end{pmatrix}$ (b) −11

7 (a) $\begin{pmatrix} 14 \\ -4 \\ 7 \end{pmatrix}$ (b) $\begin{pmatrix} 0 & 12 & 15 & 3 \\ 0 & 3 & 3 & 0 \end{pmatrix}$

8 $m = 5, n = 3$

9 $P = \begin{pmatrix} -1 & 0 \\ 1 & 2 \end{pmatrix}$

10 (a) $\begin{pmatrix} 5 & -2 \\ 0 & -4 \end{pmatrix}$ (b) $p = 9, q = -6$

11 $a = 0, b = -3, c = -\frac{1}{2}$

12 $a = 5, b = 2$

13 (a) $\begin{pmatrix} 4 & 0 \\ 5 & 9 \end{pmatrix}$ (b) $r = -4$
(c) $r = -2$

14 (a) $\frac{1}{41}\begin{pmatrix} 7 & 4 \\ -5 & 3 \end{pmatrix}$ (b) $x = 2, y = -1$

15 $\begin{pmatrix} 5 & 3 \\ 3 & 2 \end{pmatrix}\begin{pmatrix} x \\ y \end{pmatrix} = \begin{pmatrix} 3 \\ -1 \end{pmatrix}$, $x = 9$ and $y = -14$

Practice exercise P2.1 (p. 21)

1 (a) 2×1 (b) 3×3
(c) 1×3 (d) 3×2

2 (a) $\begin{pmatrix} 0.7 & -0.7 \\ 1.6 & 3.3 \end{pmatrix}$ (b) $\begin{pmatrix} 0.3 & 4 \\ 0.2 & -2.7 \end{pmatrix}$
(c) $\begin{pmatrix} 1.5 & 9 \\ 2.7 & 0.9 \end{pmatrix}$ (d) $\begin{pmatrix} -0.4 & 2 \\ -1.4 & -6 \end{pmatrix}$

3 The number of columns in the first matrix is different from the numbers of rows in the second matrix.

4 $\begin{pmatrix} 0 & 3 \\ -3 & 2 \end{pmatrix}$ 7 $n = 2$ 9 $x = 2, y = 10$

Practice exercise P2.2 (p. 21)

1 (a) 36 (b) 27
(c) 11 (d) $8x + 5y + 9z$

2 15.55 kg

3 (a) $\begin{pmatrix} 10 & 10 \\ 23 & 4 \end{pmatrix}$ (b) $\begin{pmatrix} -8 & 7 \\ 2 & 22 \end{pmatrix}$

4 (a) $\begin{pmatrix} 52 \\ 28 \end{pmatrix}$ (b) $\begin{pmatrix} -62 & -74 \\ 23 & 29 \end{pmatrix}$
(c) $\begin{pmatrix} -58 & 128 & 7 \\ -16 & 38 & 1 \end{pmatrix}$ (d) $\begin{pmatrix} 108 & 64 \\ 29 & 2 \end{pmatrix}$

Practice exercise P2.3 (p. 22)

1 (a) $\begin{pmatrix} -3 & -4 \\ 2 & -3 \end{pmatrix}$ (b) $\begin{pmatrix} 1 & -1 \\ -2\frac{1}{2} & 3 \end{pmatrix}$
(c) $\begin{pmatrix} -2\frac{1}{2} & -4 \\ -\frac{1}{2} & -1 \end{pmatrix}$ (d) $\begin{pmatrix} 0.2 & -0.4 \\ -0.7 & 0.9 \end{pmatrix}$
(e) $\begin{pmatrix} 1.2 & -1.4 \\ -0.2 & 0.4 \end{pmatrix}$ (f) $\begin{pmatrix} -0.2 & 0.3 \\ -0.2 & 0.8 \end{pmatrix}$

2 (a) $k = 9$ (b) $\begin{pmatrix} 1 & \frac{4}{9} \\ \frac{1}{3} & \frac{1}{3} \end{pmatrix}$

3 $p = 3$

Practice exercise P2.4 (p. 22)

1 $x = 3, y = 9$
2 $x = 2, y = 8$
3 $x = 1, y = 8$
4 $x = 3, y = 3$
5 $x = -6, y = 7$
6 $x = -8, y = 2$
7 $x = 5, y = 8$
8 $x = -1, y = 5$

Practice exercise P2.5 (p. 22)

1 (a) $\begin{pmatrix} -3 & 5 \\ -2 & 3 \end{pmatrix}$ (b) $\begin{pmatrix} 11 \\ 3 \end{pmatrix}$

2 $x = -3, y = 4$

3 (a) $5x + 7y = 375$
$5x + 2y = 155$
(b) Cost of pencil = $13.40
Cost of pen = $44

4 (a) $9a + 5b = 799$
$3a + 14b = 661$
(b) Weight of a block of metal A = 71 grams
Weight of a block of metal B = 32 grams

Exercise 3a (p. 24)

1 0.940 2 −0.342 3 −2.75
4 0.454 5 0.990 6 −0.788
7 −19.1 8 −0.616 9 0.399
10 −0.114 11 −0.948 12 −1.12
13 −0.227 14 0.877 15 0.028
16 −21.5

Exercise 3b (p. 25)

1 36° 2 144° 3 75°
4 105° 5 66.9°, 113.1° 6 117°
7 160° 8 25°, 155° 9 98°
10 148° 11 115.4° 12 162.5°
13 115.2° 14 56.4°, 123.6° 15 144.1°
16 73.6°, 106.4°

Exercise 3c (p. 26)

1 −0.940 2 −0.342 3 2.75
4 −0.602 5 −0.707 6 −0.577
7 −0.936 8 0.184 9 −0.863
10 0.253 11 1.75 12 −0.297

Exercise 3d (p. 26)

1 194.8°, 345.2° 2 290°
3 248.2° 4 221.7°
5 248.6°, 291.4° 6 329.7°
7 289° 8 253.4°
9 308.4° 10 242.7°

Exercise 3e (p. 27)

1 −0.766 2 −0.342 3 0.577
4 0.839 5 −0.707 6 0.577
7 0.936 8 −0.936 9 0.648
10 −0.253

Exercise 3f (p. 28)

1 194.8°, 345.2°, −14.8°, −165.2°
2 70°, 290°, −70°, −290°
3 68.2°, 248.2°, −111.8°, −291.8°
4 138.3°, 221.7°, −138.3°, −221.7°
5 248.6°, 291.4°, −68.6°, −111.4°
6 149.7°, 329.7°, −30.3°, −210.3°
7 109°, 289°, −71°, −251°
8 106.6°, 253.4°, −106.6°, −253.4°
9 51.6°, 308.4°, −51.6°, −308.4°
10 62.7°, 242.7°, 117.3°, 297.3°

Exercise 3g (p. 32)

1 $x = 3, y = 3\sqrt{2}$ 2 $x = y = \frac{5\sqrt{2}}{2}$
3 $x = 5, y = 5\sqrt{3}$
4 $x = \frac{7\sqrt{3}}{3}, y = \frac{14\sqrt{3}}{3}$
5 $x = 2\frac{1}{2}, y = 2\frac{1}{2}\sqrt{3} = \frac{5\sqrt{3}}{2}$
6 $x = 14, y = 7\sqrt{3}$ 7 $x = 2, y = 2\sqrt{2}$
8 $x = 12, y = 6\sqrt{3}$ 9 $x = \sqrt{2}, y = 2$
10 $x = \frac{1}{2}\sqrt{3}, y = 1\frac{1}{2}$ 11 $x = \frac{\sqrt{3}}{3}, y = \frac{2\sqrt{3}}{3}$
12 $x = 5, y = \frac{5\sqrt{3}}{3}$ 13 $x = 4, y = 4\sqrt{3}$
14 $x = y = \frac{5\sqrt{2}}{2}$ 15 $x = 3\sqrt{2}, y = 6$
16 $x = 3, y = 3\sqrt{2}$ 17 $x = 2\sqrt{2}, y = \sqrt{2}$
18 $x = 2\sqrt{3}, y = 6$ 19 $x = 3, y = 2\sqrt{3}$
20 $x = 2\sqrt{3}, y = 2\sqrt{6}$

Exercise 3h (p. 33)

1 $20\sqrt{3}$ m 2 20 m
3 $\sqrt{3}$ cm 4 $2(2 - \sqrt{3})$ cm
5 $6\sqrt{3}$ cm 6 $14\sqrt{3}$ cm, 462 cm^2
7 $10\sqrt{3}$ m 8 $50(3 + \sqrt{3})$ m
9 $20(\sqrt{3} - 1)$ m 10 $8(3 - \sqrt{3})$ m

Practice exercise P3.1 (p. 34)

1 −0.985 2 −0.766 3 −5.67
4 0.956 5 0.530 6 0.287

Practice exercise P3.2 (p. 34)

1 75°, 105° 2 77°, 283° 3 67°, 247°
4 263°, 277° 5 123°, 237° 6 147°, 327°

Practice exercise P3.3 (p. 34)

1 B
2 $$\cos 30° = \frac{a}{12}$$
$$\text{but } \cos 30° = \frac{\sqrt{3}}{2}$$
$$\text{so } \frac{a}{12} = \frac{\sqrt{3}}{2}$$
$$a = \frac{12\sqrt{3}}{2} = 6\sqrt{3} \text{ cm}$$
3 $\frac{28\sqrt{3}}{3}$ m

4 (a) 7.5 cm
(b)
$$\cos 30 = \frac{AB}{15}$$
$$\text{but } \cos 30° = \frac{\sqrt{3}}{2}$$
$$\text{so } \frac{AB}{15} = \frac{\sqrt{3}}{2}$$
$$AB = \frac{15\sqrt{3}}{2}\text{ cm}$$

5 $\frac{7\sqrt{3}}{2}$ m

6 (a) 54 m (b) $18\sqrt{3}$ m

7 $\frac{13\sqrt{3}}{2}$ cm

8 $16\sqrt{3}$ m

9 $4.6\sqrt{2}$ m

10 (a) $15\sqrt{3}$ cm (b) 30 cm (c) 60°

Exercise 4a (p. 35)

1 −2, 4 2 0, −6 3 0, 7
4 0, −3 5 6, −4 6 −5, 3
7 2, −2 8 −5, 5 9 9 twice
10 −1 twice

Exercise 4b (p. 36)

1 $-3\frac{1}{3}$, 4 2 $\frac{3}{4}$ twice 3 $-\frac{1}{2}$ twice
4 $\frac{1}{2}, -\frac{2}{3}$ 5 $-1\frac{2}{3}$ twice 6 0, $1\frac{1}{3}$
7 $-3, \frac{5}{7}$ 8 $2\frac{3}{4}$ twice 9 0, $-4\frac{1}{2}$
10 0, $-3\frac{3}{4}$

Exercise 4c (p. 36)

1 1, 2 2 −2, −3 3 2, −1
4 1, −3 5 2, 5 6 0, 4
7 0, −5 8 −3, −4 9 2, −4
10 1 twice 11 1, 4 12 0, 9
13 ±3 14 ±5 15 −1, 9
16 5, −7 17 3 twice 18 −4 twice
19 0, 4 20 ±2

Exercise 4d (p. 37)

1 6, 9 2 −3, 18 3 0, $2\frac{1}{2}$
4 1, $1\frac{1}{2}$ 5 3, $-\frac{1}{2}$ 6 0, $-\frac{1}{3}$
7 9, −10 8 −8, 9 9 −1, $-\frac{1}{3}$
10 $\frac{1}{3}$ twice 11 2, $-\frac{1}{3}$ 12 $2\frac{1}{2}$ twice
13 $-\frac{2}{3}$ twice 14 3, $-\frac{1}{4}$ 15 4, −11
16 0, $\frac{3}{7}$ 17 5, $\frac{1}{2}$ 18 $\frac{1}{2}, -\frac{1}{3}$
19 $\frac{1}{3}, -2\frac{1}{2}$ 20 $\pm 1\frac{3}{4}$ 21 $\pm 3\frac{1}{2}$
22 $\frac{3}{4}, -2\frac{1}{2}$ 23 $2\frac{1}{2}, -1\frac{1}{3}$ 24 $2\frac{1}{3}, 1\frac{1}{6}$

Exercise 4e (p. 37)

1 5, −1 2 9, 5 3 −1, −5
4 3, −7 5 $1 \pm \sqrt{2}$ 6 $-4 \pm \sqrt{3}$
7 $1\frac{1}{2}, 2\frac{1}{2}$ 8 0, 12 9 $4 \pm \sqrt{10}$
10 $-4\frac{2}{3}, -5\frac{1}{3}$ 11 $8 \pm \sqrt{3}$ 12 $\frac{2}{5}, 1\frac{3}{5}$
13 $\frac{1}{2}, -2\frac{1}{2}$ 14 $-7 \pm \sqrt{6}$ 15 $\frac{1}{3}, -1$
16 $-9 \pm \sqrt{3}$ 17 $6 \pm \sqrt{5}$ 18 0, 5
19 $-10 \pm \sqrt{8}$ 20 $4\frac{1}{3}, 7\frac{2}{3}$

Exercise 4f (p. 38)

1 $a^2 + 8a + 16 = (a + 4)^2$
2 $b^2 + 10b + 25 = (b + 5)^2$
3 $c^2 - 4c + 4 = (c - 2)^2$
4 $d^2 - 6d + 9 = (d - 3)^2$
5 $x^2 + 5x + 6\frac{1}{4} = (x + 2\frac{1}{2})^2$
6 $y^2 - 3y + 2\frac{1}{4} = (y - 1\frac{1}{2})^2$
7 $z^2 - 7z + 12\frac{1}{4} = (z - 3\frac{1}{2})^2$
8 $m^2 + 2m + 1 = (m + 1)^2$
9 $n^2 - n + \frac{1}{4} = (n - \frac{1}{2})^2$
10 $u^2 - \frac{1}{2}u + \frac{1}{16} = (u - \frac{1}{4})^2$
11 $h^2 + \frac{2}{3}h + \frac{1}{9} = (h + \frac{1}{3})^2$
12 $k^2 - 1\frac{1}{3}k + \frac{4}{9} = (k - \frac{2}{3})^2$
13 $g^2 - 4\frac{2}{3}g + 5\frac{4}{9} = (g - 2\frac{1}{3})^2$
14 $a^2 + \frac{3}{5}a + \frac{9}{100} = (a + \frac{3}{10})^2$
15 $b^2 - \frac{4}{5}b + \frac{4}{25} = (b - \frac{2}{5})^2$
16 $m^2 - 8m + 16 = (m - 4)^2$
17 $m^2 - 8mn + 16n^2 = (m - 4n)^2$
18 $a^2 - 6ad + 9d^2 = (a - 3d)^2$
19 $x^2 + 10xy + 25y^2 = (x + 5y)^2$
20 $m^2 + 3mn + 2\frac{1}{4}n^2 = (m + 1\frac{1}{2}n)^2$

Exercise 4g (p. 39)

1 3, −7 2 −3, 4 3 $2 \pm \sqrt{6}$
4 $-1 \pm \sqrt{3}$ 5 −2 twice 6 $5 \pm \sqrt{10}$
7 $-5 \pm \sqrt{3}$ 8 3 twice 9 $-3 \pm \sqrt{2}$
10 $\frac{3 \pm \sqrt{5}}{2}$ 11 2, 3 12 −1, −4
13 $\frac{5 \pm \sqrt{17}}{2}$ 14 $\frac{-5 \pm \sqrt{17}}{2}$
15 $4 \pm \sqrt{17}$ 16 $\frac{1 \pm \sqrt{5}}{2}$
17 $\frac{-1 \pm \sqrt{13}}{2}$ 18 3, −10
19 5 twice 20 $6 \pm \sqrt{35}$

Exercise 4h (p. 40)

1 −2, −3 2 1, 4 3 5, −1

4 $-1, -1\frac{1}{2}$ 5 $1, \frac{1}{3}$ 6 $2, -\frac{1}{3}$
7 $1, -\frac{2}{5}$ 8 $-2, \frac{1}{4}$
9 $-\frac{2}{3}, -1\frac{1}{2}$ 10 $5, -\frac{2}{3}$
11 −2.62, −0.38 12 3.24, −1.24
13 0.39, −3.89 14 2.14, −0.47
15 1.58, −0.38 16 0.22, −1.82
17 −1.77, −0.57 18 2.82, 1.18
19 2.39, 0.28 20 −1.13, 0.53

Exercise 4i (p. 42)

1 $x = 1$ or -3 2 $x = 0.5$ or -2
3 $x = 0$ or 3
4

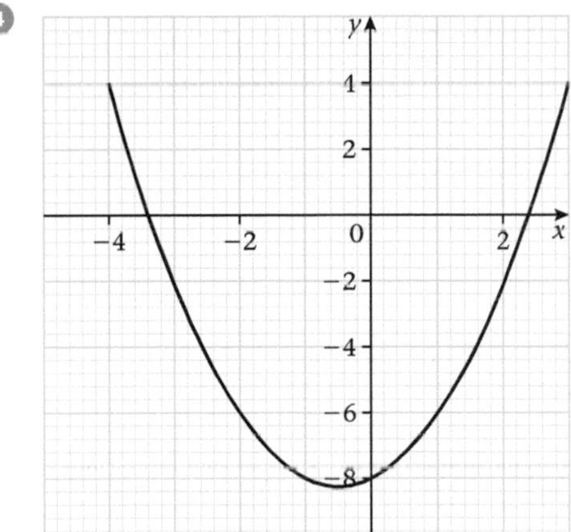

$x = 2.4$ or -3.4

5

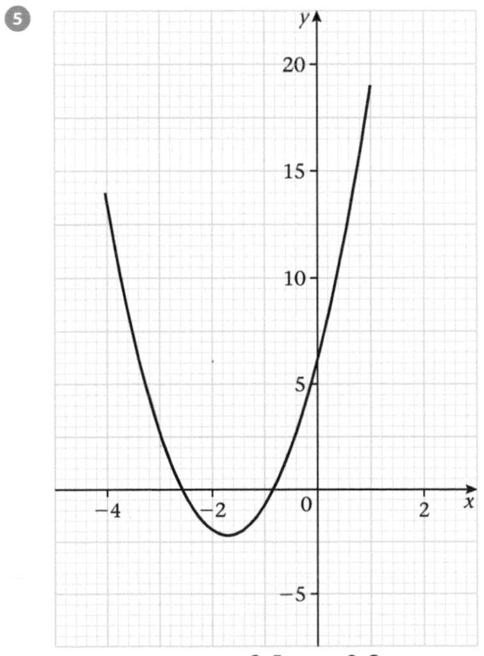

x −2.5 or −0.8

6 (a) $x = 3$ or -7 (b) $x = -4$ (twice)
(c) $x = 5$ or -3 (d) imaginary roots

7 (a)

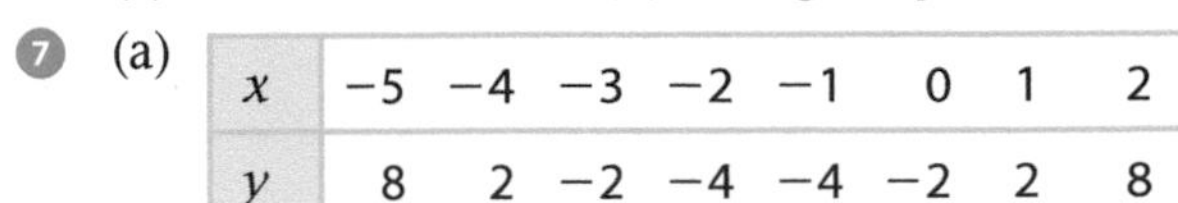

x	−5	−4	−3	−2	−1	0	1	2
y	8	2	−2	−4	−4	−2	2	8

(b)

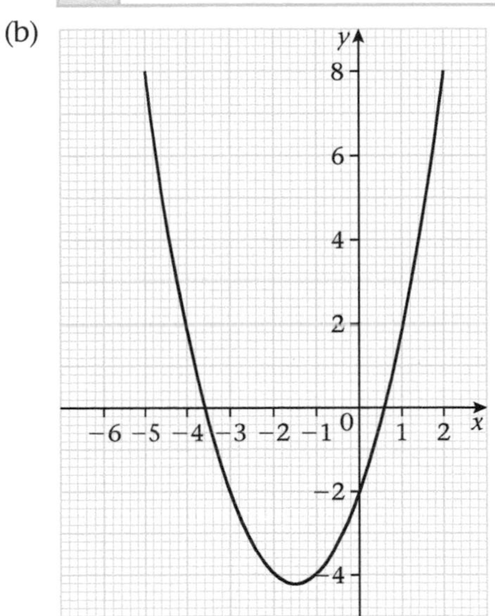

$x = 0.6$ or -3.6

8 (a)

x	−1	0	1	2	3	4
y	25	9	1	1	9	25

(b)

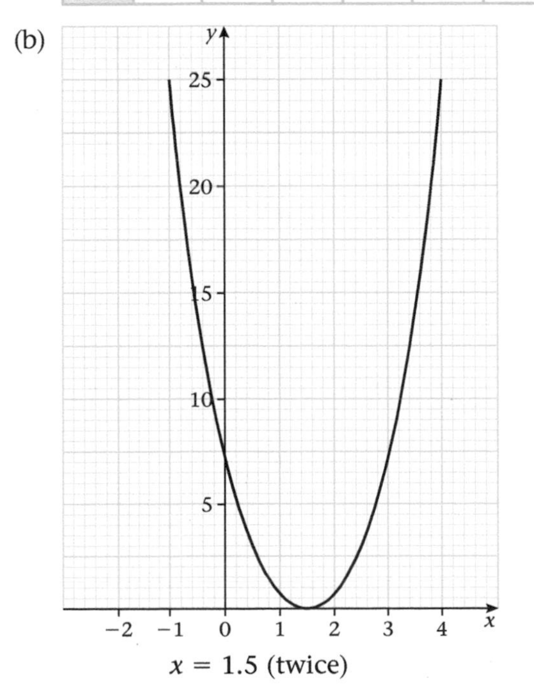

$x = 1.5$ (twice)

9 (a)

x	−4	−3	−2	−1	0	1	2
$f(x)$	6	1	2	−3	−2	1	6

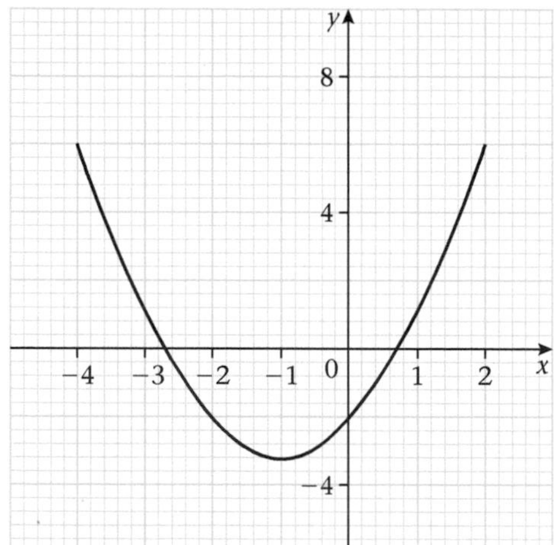

(b) $x = 0.7$ or -2.7

10 (a)

x	−2	−1	0	1	2	3	4	5	6
$f(x)$	−13	1	11	17	19	17	11	1	−13

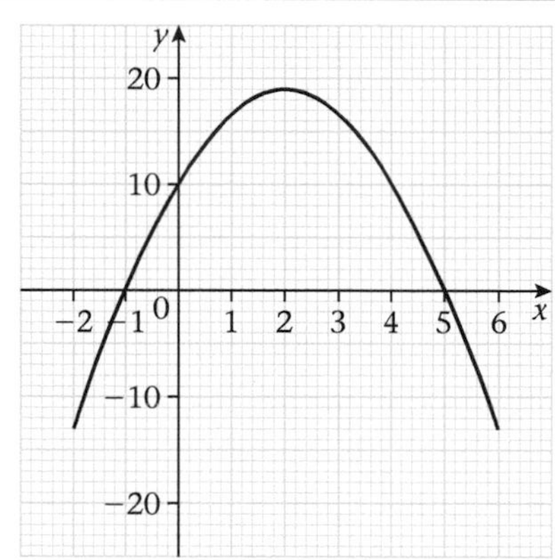

(b) $x = 5.1$ or -1.1

Exercise 4j (p. 44)

1 9, 5 or −5, −9
2 9 m by 5 m
3 8, 5 or −5, −8
4 8 m square, 5 m square
5 15 years, 9 years
6 5 years
7 9 or −10
8 6
9 4 cm
10 3 m
11 5 cm
12 5 years
13 7 or 24
14 12, 13 or −13, −12
15 14, 16 or −16, −14
16 2 or 11

Exercise 4k (p. 46)

1 (a) $(1, 1), (1\frac{2}{3}, 2\frac{1}{3})$ (b) $(5, 3), (-3, -5)$
(c) $(3, -1), (-3, 1)$ (d) $(1, \frac{2}{3})$ twice
(e) $(3, -2), (\frac{1}{3}, 6)$ (f) $(0, \frac{1}{2}), (1, 3)$
2 (a) $(8, 1), (\frac{1}{3}, 24)$
(b) $(2, 1), (-1\frac{1}{3}, -1\frac{1}{2})$
(c) $(6, 1), (-2\frac{1}{2}, -2\frac{2}{5})$
(d) $(3, -\frac{1}{2}), (-\frac{2}{3}, 2\frac{1}{4})$
3 (a) $(6, -3)$ (b) $(5, -2)$
(c) $(2, -1)$ (d) $(5, -3)$
4 $q + p = 2(q - p)$, $pq = 675$; son: 15 years, father: 45 years
5 8 and $1\frac{1}{2}$

Exercise 4l (p. 47)

1 (a) $x = 2.7$ or -1.3
(b) when $x = 3.4$, $y = 5.2$;
when $x = -0.9$, $y = -3$
(c) $x = 3.4$ or -0.9
(d) 3.3, −1.9
(e) $x = 2.3$ or -0.9
2 (a) $x = 0.4$ or 4.6 (b) $x = 0.7$ or 4.3
3 when $x = 0$, $y = 0$; when $x = 2$, $y = 2$
4 (a) $x = 0.4$ or 2.6 (b) $x = -0.6$ or 3.6
5 (a) $x = 0$ or 4 (b) $x = 3.3$ or -1

Practice exercise P4.1 (p. 48)

1 −1, 3
2 0, 1
3 −2 twice
4 5 twice
5 1, 8
6 10, 2
7 −4, 5
8 −2, 5

Practice exercise P4.2 (p. 48)

1 $d^2 + 6d + 9 = (d + 3)^2$
2 $b^2 - 5b + 6\frac{1}{4} = (d - 2\frac{1}{4})^2$
3 $a^2 + 3a + 2\frac{1}{4} = (a + 1\frac{1}{2})^2$
4 $-(x^2 - 16x + 64) = -(x - 8)^2$
5 $3(v^2 + 2v + 1) = 3(v + 1)^2$
6 $-2(5w + w^2 + 6\frac{1}{4}) = -2(w + 2\frac{1}{2})^2$
7 $2(t - 3\frac{1}{2}t + 3\frac{1}{16}) = -2(t - 1\frac{3}{4})^2$
8 $-4(y^2 - 2y + 1) = -4(y + 1)^2$

Practice exercise P4.3 (p. 48)

1. $(p + 3\frac{1}{2})^2 - 20\frac{1}{4}$
2. $(m - 6)^2 - 11$
3. $(c + 2\frac{1}{2})^2 - \frac{1}{4}$
4. $(x + 5\frac{1}{2})^2 - 74\frac{1}{2}$
5. $2(b - 1\frac{1}{4})^2 + 13\frac{9}{16}$
6. $-(r - 6)^2 - 56$
7. $-(f + 2)^2 + 16$
8. $4(c - 1\frac{1}{2})^2 - 27$
9. $-2(y + 1)^2 + 6$
10. $-4(h - 1\frac{1}{2})^2 + 7$

Practice exercise P4.4 (p. 48)

1. $1 \pm \sqrt{6}$
2. $-3 \pm 2\sqrt{3}$
3. $-1 \pm \sqrt{5}$
4. $\frac{1 \pm \sqrt{5}}{2}$
5. $3 \pm \sqrt{3}$
6. $-2 \pm \sqrt{10}$
7. $\frac{-5 \pm \sqrt{185}}{4}$
8. $-8 \pm \sqrt{50}$
9. $6 \pm \sqrt{11}$
10. $\frac{3 \pm 3\sqrt{5}}{4}$

Practice exercise P4.5 (p. 48)

1. 3.45, −1.45
2. 0.46, −6.46
3. 0.24, −4.24
4. 1.62, −0.62
5. 4.73, 1.27
6. 1.16, −5.16
7. 2.15, −4.65
8. −0.93, −15.07
9. 9.32, 2.68
10. 2.43, −0.93

Practice exercise 4.6 (p. 49)

1. (a) All the terms of the equation are positive so the solutions must both be negative.
   (b) No, since the constant term is negative, the roots must have different signs.
2. (a) No real roots
   (b) One repeated root
   (c) Two distinct roots
3. 8.165 seconds

Practice exercise 4.7 (p. 49)

1.

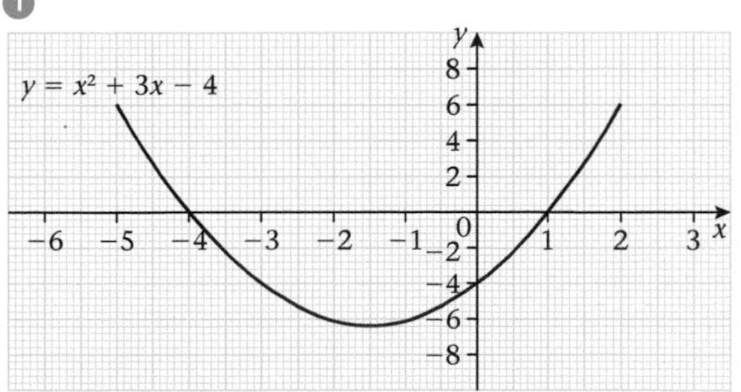

$x = -4$ or 1

2.

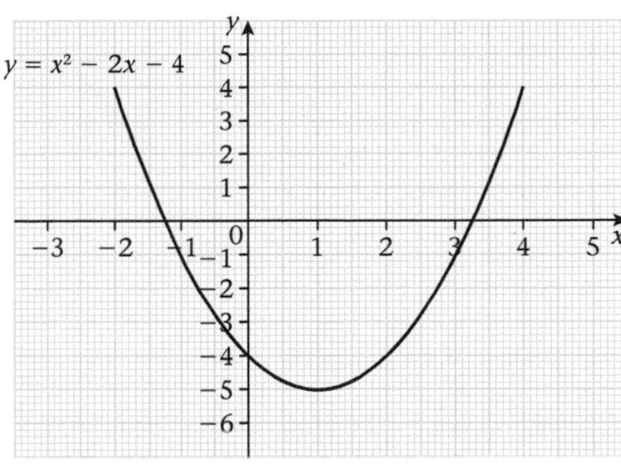

$x = -1.2$ or 3.2

3.

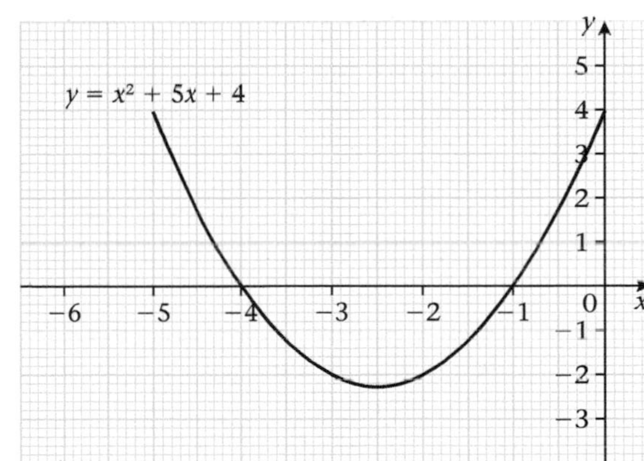

$x = -4$ or -1

4.

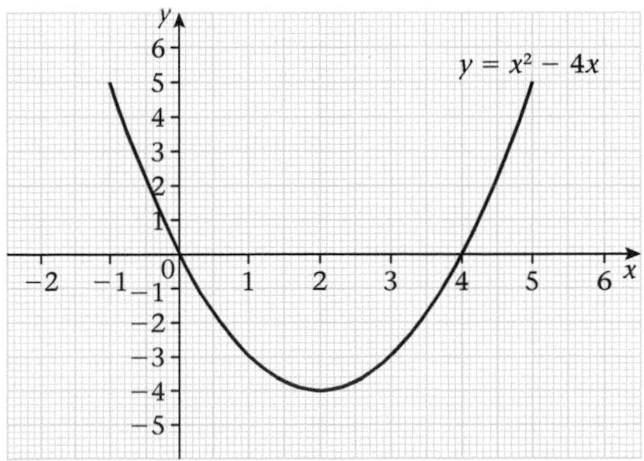

$x = 0$ or 4

5 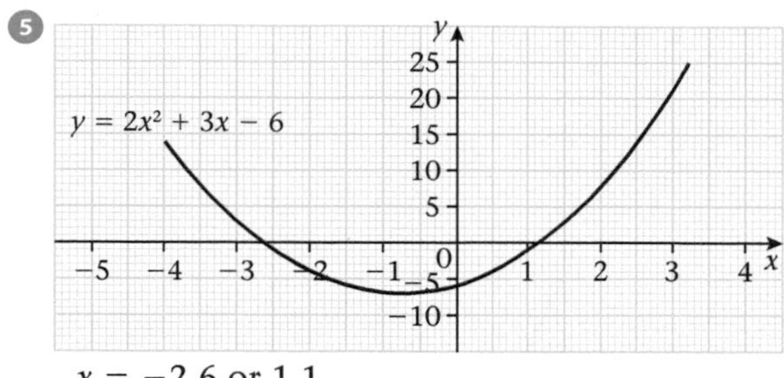

$x = -2.6$ or 1.1

Practice exercise P4.8 (p. 49)

1 (a)

x	−2	−1	0	1	2	3	4	1.5
x^2	4	1	0	1	4	9	16	2.25
$-3x$	6	3	0	−3	−6	−9	−12	−4.5
-2	−2	−2	−2	−2	−2	−2	−2	−2
y	8	2	−2	−4	−4	−2	2	−4.25

(b) 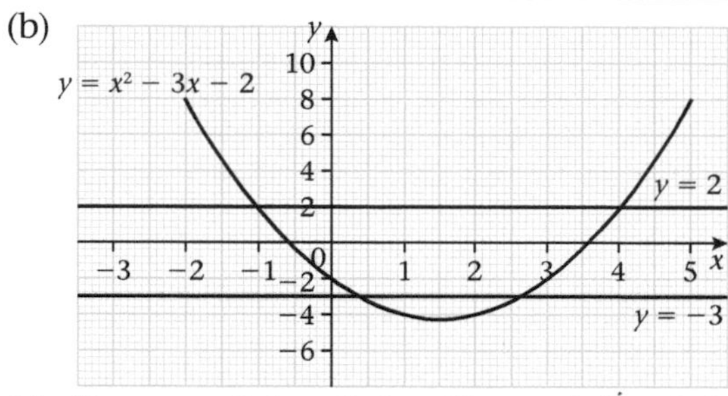

(c) (i) $x = -0.6$ or 3.6 (ii) $x = 0.4$ or 2.6
(iii) $x = -1$ or 4

2 (a) $p = -1$

(b)

x	−2	−1	0	1	2	3	4
y	−4	1	4	5	4	1	−4

(c) 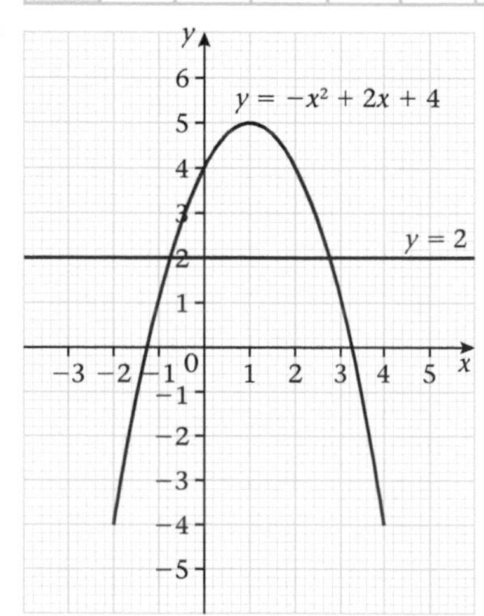

(d) (i) $x = -1.2$ or 3.2
(ii) $x = -0.7$ or 2.7

3 (a)

(b) (i) $t = 0$ seconds and $t = 6$ seconds
(ii) $t = 2.2$ seconds and $t = 3.8$ seconds
(iii) $t = 1.5$ seconds and $t = 4.5$ seconds

Practice exercise P4.9 (p. 50)

1 9
2 12, 16
3 10 cm²
4 6 or 30
5 $2n^2 - 15n = 4230$; 50
6 7
7 9
8 45 cm, 9 cm

Practice exercise P4.10 (p. 50)

1 $(-1.25, -12), (3, 5)$
2 $(1, 2)$ twice
3 $(1\frac{8}{11}, -\frac{1}{11}), (-1, -1)$
4 $(0, 5), (4, -3)$
5 $(5, 6), (2, 15)$
6 $(4, -1)$, twice
7 $(-2.5, 6), (3, -5)$
8 $(-\frac{1}{7}, 1\frac{3}{7}), (1, -2)$
9 $(-\frac{3}{4}, 2\frac{3}{4})$ twice
10 $(-1, 4), (2, -2)$
11 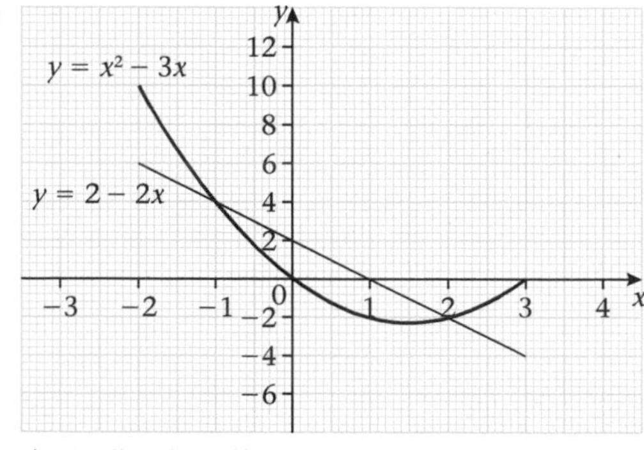

$(-1, 4), (2, -2)$

12

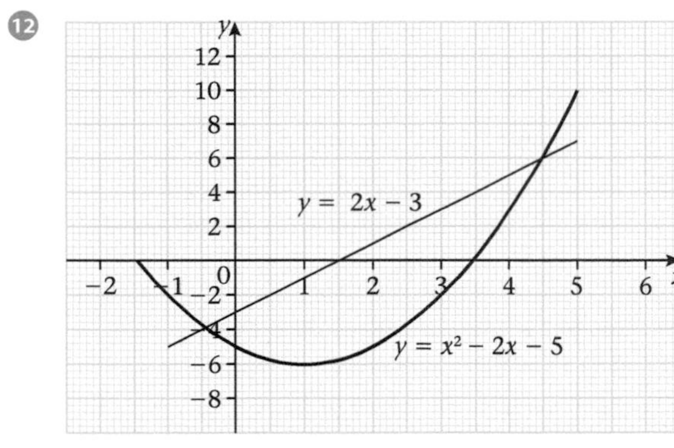

(−0.4, 4, −3.9), (4, 4, 5.9)

Practice exercise P4.11 (p. 50)

1 $x^2 - 6x - 4 = 0$ 2 $3x^2 + 2x + 1 = 0$
3 $4x^2 + 2 = 0$

Practice exercise P4.12 (p. 50)

1 $m = g + 24$, $gm = 180$; 6 years and 30 years
2 (a) 10 (b) $48
6 (a) (1, 2); the curve and the line meet at a single point so the line is a tangent to the curve.
(b)

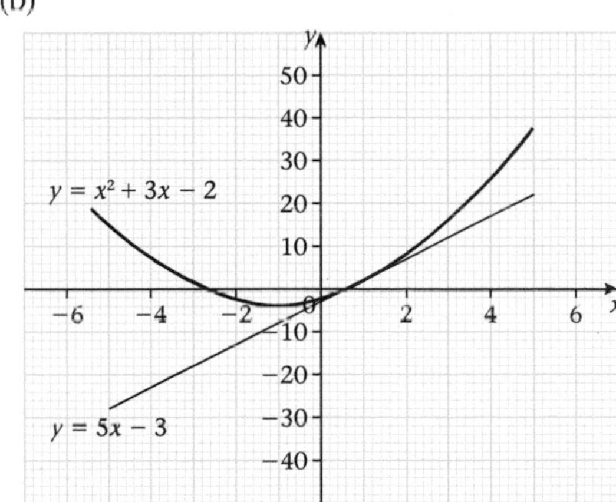

Exercise 5a (p. 52)

1 (a) (i) arc AE (ii) $A\hat{C}E$
(b) (i) arc BC (ii) $B\hat{A}C$, $B\hat{D}C$
(c) (i) arc EDC (ii) $E\hat{B}C$
2 (a) 50° (b) 32° (c) 23°
(d) 33° (e) 79° (f) 117°
(g) 60° (h) 22° (i) 124°
3 106° 4 31°
5 107°, 107°, 73°, 73° 6 120°
7 70° 8 60° each, equilateral

9 $B\hat{C}E = 2x°$ (*ext. angle of cyclic quad.*)
$2x° + x° + 60° = 180°$ (*angle sum of* △BCE)
$x = 40$
10 (b) $S\hat{Q}P = 45°$, $S\hat{R}Z = 90°$

Exercise 5b (p. 55)

1 (a) 18° (b) 45° (c) 63° (d) 70°
2 (a) 36° (b) 42° (c) 124° (d) 70°
3 (a) 15 cm (b) 5 cm
4 (a) 43° (b) 63° (c) 48° (d) 32°
5 4.5 cm 6 3.3 cm
7 141° 8 8 cm
9 (a) (i) $(90 - x)°$, (ii) $(90 - x)°$
(b) $O\hat{B}A = A\hat{B}C$
10 (a) 90° (b) $(90 - x)°$
(c) $A\hat{D}B = x°$ (*sum of angles of* △$A\hat{D}B$)
∴ $B\hat{A}T = A\hat{D}B = x°$

Exercise 5c (p. 57)

1 (a) 36° (b) 57° (c) 46° (d) 12 cm
2 $T\hat{A}O = T\hat{B}O = 90°$ (*radius* ⊥ *tangent*)
∴ TAOB is cyclic (*opposite angles are supplementary*)
3 43° 4 106°
5 59° or 121° 10 134°

Exercise 5d (p. 59)

1 (a) $A\hat{B}Y$, $A\hat{C}Y$ (b) $C\hat{B}Y$, $C\hat{A}Y$
(c) $B\hat{C}Y$ (d) $B\hat{A}Y$
(e) 58° (f) 112°
(g) 55° (h) 80°
2 (a) 56° (b) 52° (c) 43°
(d) 102° (e) 78°
3 $P\hat{Y}X$ 4 84°
5 72° 6 82°, 49°, 49°
7 54° 8 101°, 74°
9 40° 10 34°, 77°, 69°
11 50°, 60°, 70° 12 44°, 108°

Practice exercise P5.1 (p. 60)

1 $360 - 2a$
2 $B\hat{O}D = 2a$ (*angle at centre = 2 × angle at circumference*)
reflex $B\hat{O}D = 2c$ (*same reason*)
∴ $B\hat{O}D$ + reflex = 360° (*angles at a point*)
∴ $a + c = 180°$
∴ $B\hat{A}D + B\hat{C}D = 180°$

3 $B\hat{O}D = 2B\hat{A}D$
because the angle at the centre =
2 × the angle at the circumference.
$B\hat{C}D = 180° - B\hat{A}D$
because opposite angles of cyclic quad. are supplementary.
Adding the two equations gives
$B\hat{C}D + B\hat{O}D = 2B\hat{A}D + 180° - B\hat{A}D$
$= 180° + B\hat{A}D$, as required.

Practice exercise P5.2 (p. 61)

Note: other reasons may be possible.

1 $A\hat{B}C = 90°$ *(angle in a semicircle)*
$B\hat{C}A = 57°$ *(sum of angles of △)*
2 $A\hat{O}C = 60°$ *(angles at a point)*
Angle from AC at circumference = 30°
(angle at centre = 2 × angle at circumference)
$A\hat{B}C = 150°$ *(opp. angles of cyclic quad.)*
3 $C\hat{A}B = 43°$ *(in same segment as $C\hat{D}B$)*
4 $C\hat{B}D = 54°$ *(sum of angles of △)*
$C\hat{A}D = 54°$ *(in same segment as $C\hat{B}D$)*
5 $A\hat{C}B = 90°$ *(angle in a semicircle)*
∴ $A\hat{C}O = 25°$
$O\hat{B}C = 65°$ *(isos. △OBC)*
$A\hat{O}C = 130°$ *(angle at centre = 2 × angle at circumference)*
6 $A\hat{C}D = 90°$ *(angle in a semicircle)*
$C\hat{E}D = 75°$ *(sum of angles of △)*
$C\hat{D}A = 36°$ *(sum of angles of △)*
∴ $B\hat{D}A = 21°$
$A\hat{B}D = 90°$ *(angle in a semicircle)*
$B\hat{A}D = 69°$ *(sum of angles of △)*
∴ $B\hat{A}C = 15°$

Practice exercise P5.3 (p. 61)

Note: other reasons may be possible.

1 $A\hat{B}C = 73°$ *(opp. angles of cyclic quad.)*
$B\hat{C}D = 99°$ *(same reason)*
2 $B\hat{C}D = 64°$ *(angles on str. line)*
$D\hat{A}D = 116°$ *(opp. angles of cyclic quad.)*
$C\hat{D}X = 69°$ *(opp. interior angle of cyclic quad.)*
3 $D\hat{A}B = 108°$ *(opp. angles of cyclic quad.)*
$A\hat{B}C = 96°$ *(same reason)*
4 $A\hat{B}C = 67°$ *(angle at centre = 2 × angle at circumference)*
$C\hat{D}X = 67°$ *(= exterior angle of cyclic quad.)*
5 $B\hat{A}C = 36°$ *(in same segment as $C\hat{D}B$)*
$A\hat{B}C = 98°$ *(sum of angles of △)*
$A\hat{D}B = 46°$ *(in same segment as $B\hat{C}A$)*
$A\hat{B}D = 49°$ *(sum of angles of △)*
$B\hat{E}A = 95°$ *(sum of angles of △)*
6 $A\hat{D}C = 141°$ *(corresponding angles)*
$A\hat{B}C = 39°$ *(opp. angles of cyclic quad.)*

Practice exercise P5.4 (p. 61)

Note: other reasons may be possible.

1 $B\hat{A}O = 90°$ *(tangent ⊥ radius)*
$A\hat{O}B = 61°$ *(sum of angles of △)*
2 Angle in alternate segment to $T\hat{A}B = 70°$
$A\hat{O}B = 140°$ *(angle at centre = 2 × angle at circumference)*
3 $A\hat{B}C = 79°$ *(alternate segment)*
$B\hat{C}T = 63°$ *(alternate segment)*
4 $T\hat{A}O = T\hat{B}O = 90°$ *(tangent ⊥ radius)*
$T\hat{A}O + T\hat{B}O = 180°$
$A\hat{T}B + A\hat{O}B = 180°$ *(angles of quad.)*
∴ $A\hat{T}B = 24°$
5 $C\hat{A}D = 68°$ *(alternate segment)*
$A\hat{D}C = 90°$ *(angle in a semicircle)*
$A\hat{B}C = 90°$ *(same reason)*
6 $A\hat{C}B = 55°$ *(angle at centre = 2 × angle at circumference)*
$B\hat{C}T = 52°$ *(angles on str. line)*
7 $T\hat{C}O = 90°$ *(tangent ⊥ radius)*
$T\hat{O}C = 65°$ *(sum of angles of △)*
$O\hat{B}C = 57.5°$ *(sum of angles of isos. △)*
$O\hat{B}A = 57.5°$ *(symmetry)*
∴ $A\hat{B}C = 115°$
$O\hat{C}B = 57.5$ *(base angle of isos. △)*
8 $B\hat{C}A = 60°$ *(alternate segment)*
$B\hat{A}C = 60°$ *(isos. △)*
$A\hat{B}C = 60°$ *(sum of angles of △)*

Practice exercise P5.5 (p. 62)

1 (a) $180° - x - y$
(b) In △ABD, $B\hat{A}D = x$, $B\hat{D}A = y$ and $A\hat{B}D = 180° - x - y$
In △ACD, $C\hat{A}D = x$, $D\hat{C}A = y$ and $A\hat{D}C = 180° - x - y$
The angles of the two △s are the same ∴ the △s are similar.
2 $A\hat{C}E = 180° - x - y$ *(sum of angles of △ACE)*
$C\hat{B}D = \frac{1}{2}(x + y)$

3 $\hat{AOE} = \hat{DAB}$ (*corresponding angles*)
$\hat{DAB} = \hat{ACB}$ (*alternate segment*)
$\therefore \hat{AOE} = \hat{ACB}$

Exercise 6a (p. 64)

1

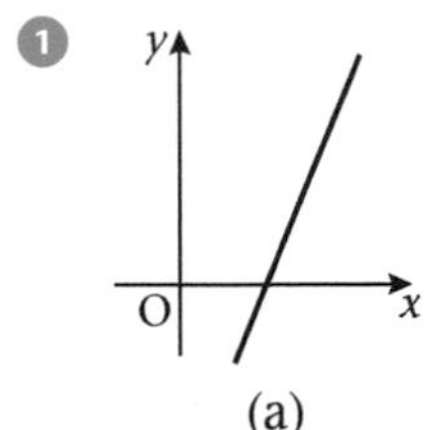

(a)

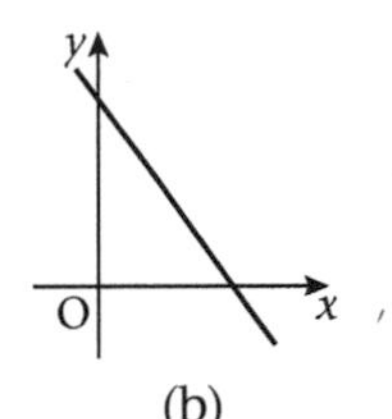

(b)

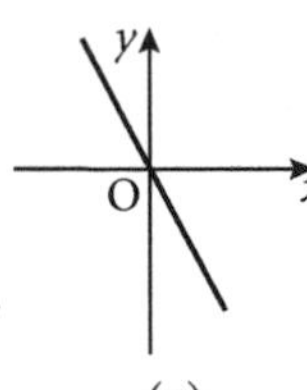

(c)

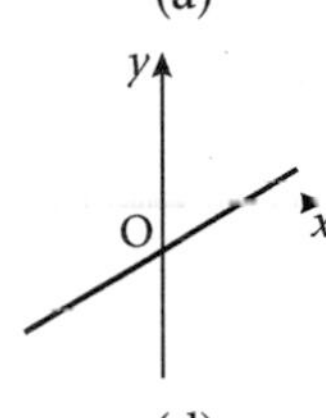

(d)

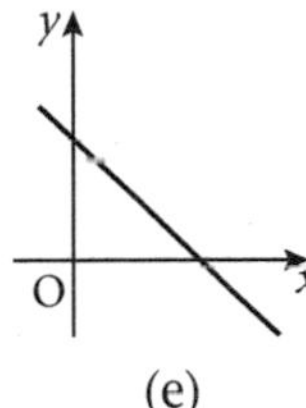

(e)

2 (a) 2 (b) $\frac{1}{3}$ (c) $\frac{5}{4}$ (d) $-\frac{3}{7}$ (e) $\frac{4}{7}$

3 (a) (0, −2), (1, 0) (b) (0, 1), (−3, 0)
(c) (0, −6), (10, 0) (d) $(0, \frac{2}{3}), (\frac{1}{2}, 0)$
(e) $(0, \frac{4}{5}), (\frac{1}{2}, 0)$

4 (a) 1 (b) 2 (c) $\frac{3}{4}$ (d) $-\frac{5}{4}$

Exercise 6b (p. 66)

1 perpendicular
2 parallel
3 perpendicular
4 parallel
5 parallel
6 perpendicular

Exercise 6c (p. 67)

1 (2, 3)
2 $(\frac{5}{2}, \frac{1}{2})$
3 (−2, 3)
4 $(-\frac{1}{2}, \frac{11}{2})$
5 (1, 1)
6 $(\frac{1}{2}, -\frac{11}{2})$

Exercise 6d (p. 68)

1 $2\sqrt{2}$
2 $\sqrt{122}$
3 $2\sqrt{2}$
4 $\sqrt{26}$
5 $2\sqrt{13}$
6 $\sqrt{26}$

Exercise 6e (p. 69)

1 (a) (i) $\frac{1}{2}$ (ii) $2y - x = 1$
(iii) (5, 3) (iv) $2\sqrt{5}$
(b) (i) −3 (ii) $y + 3x = 1$
(iii) (1, −2) (iv) $4\sqrt{10}$
(c) (i) $\frac{1}{4}$ (ii) $4y = x$
(iii) (0, 0) (iv) $2\sqrt{17}$
(d) (i) $-\frac{15}{8}$ (ii) $8y + 15x = -26$
(iii) $(-2, \frac{1}{2})$ (iv) 17
(e) (i) 4 (ii) $y - 4x = 10$
(iii) (−2, 2) (iv) $2\sqrt{17}$
(f) (i) $\frac{1}{2}$ (ii) $2y - x + 1 = 0$
(iii) (−5, −3) (iv) $2\sqrt{5}$

2 $5\sqrt{2}$
3 (b) 30
4 (a) 13 (b) $8\sqrt{2}$
5 (a) $2y + 3x = 12$ (b) 6 (c) 12

Practice exercise P6.1 (p. 69)

1 2
2 $-\frac{1}{3}$
3 $-\frac{1}{2}$
4 $\frac{4}{3}$
5 0
6 undefined
7 undefined
8 −6
9 4
10 −4

Practice exercise P6.2 (p. 69)

1 $\frac{3}{7}$
2 1
3 $1\frac{2}{3}$
4 1

Practice exercise P6.3 (p. 70)

1 Gradient = $\frac{\text{change in } y}{\text{change in } x} = \frac{7 - 10}{1 - 2} = 3$

2 OP = 4

3 Equation of str. line: $y = mx + c$
At (2, 10) and with gradient = 3,
$10 = 3 \times 2 + c$
$\therefore c = 4$ = the intercept on y-axis

Practice exercise P6.4 (p. 70)

1

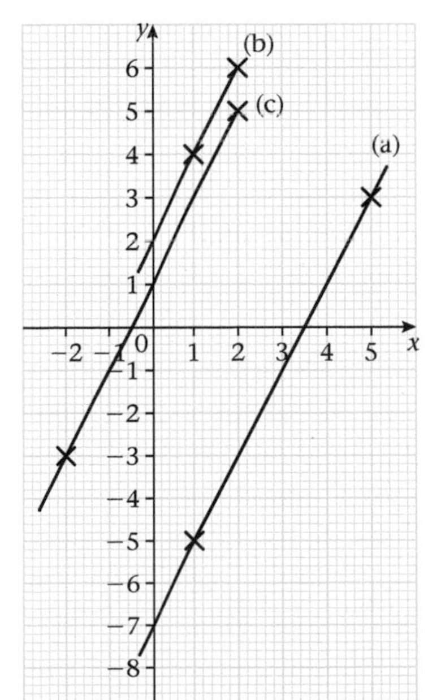

2 (a) -7 (b) 2 (c) 1
3 (a) 2 (b) 2 (c) 2
4 The lines are parallel.

Practice exercise P6.5 (p. 70)

1 (a)

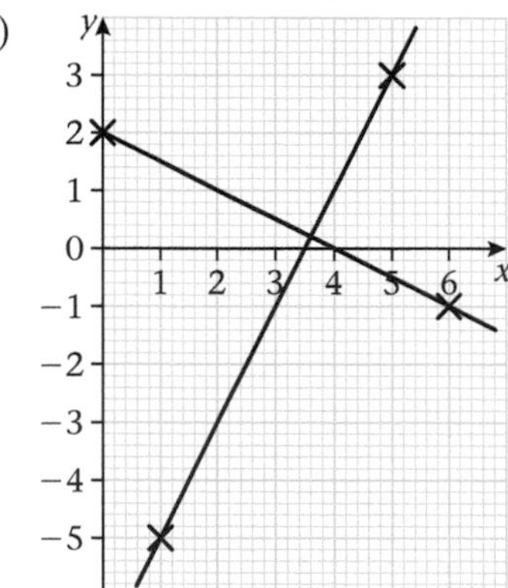

(b)

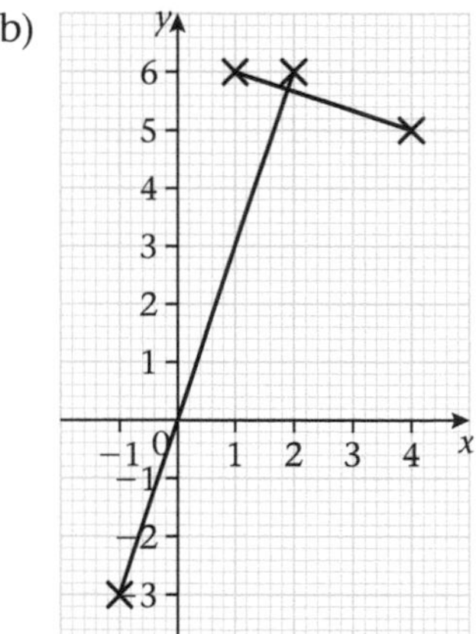

2 (a) $2, -\frac{1}{2}$
(b) $3, -\frac{1}{3}$
3 $m_1m_2 = -1$
4 The lines are perpendicular.

Practice exercise P6.6 (p. 70)

1 (a) (i) 1 (ii) 1
(b) (i) -1 (ii) 1
(c) (i) 0 (ii) 2
(d) (i) $\frac{1}{2}$ (ii) 0
(e) (i) $\frac{1}{2}$ (ii) -1
(f) (i) undefined
(ii) does not intersect with y-axis
2 Lines (d) and (e) are parallel; their gradients are the same.
3 Lines (a) and (b) are perpendicular; the product of their gradients is -1.

Practice exercise P6.7 (p. 70)

1

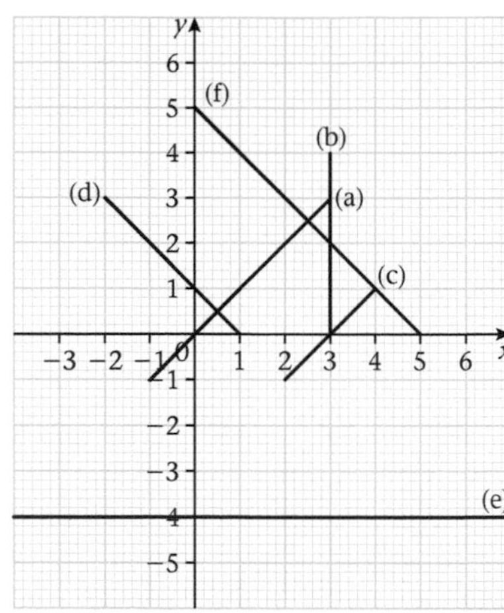

2 (a) (i) 1 (ii) 0
(b) (i) undefined
(ii) does not intersect with y-axis
(c) (i) 1 (ii) -3 (d) (i) -1 (ii) 1
(e) (i) 0 (ii) -4 (f) (i) -1 (ii) 5
3 Lines (a) and (c) are parallel.
Lines (d) and (f) are parallel.
Line (a) is perpendicular to lines (d) and (f).
Line (c) is perpendicular to lines (d) and (f).

Practice exercise P6.8 (p. 70)

1 (a)

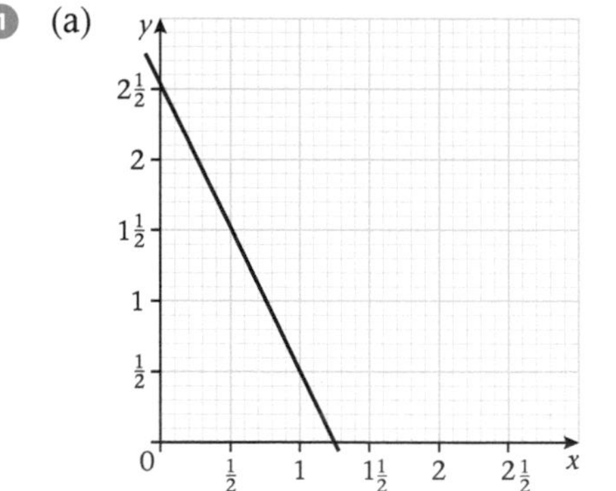

(b) 4

2 (a) (i) 7 (ii) 5 (iii) $-\frac{1}{2}$ (iv) m
(b) greatest: (i), least (iii)
(c) (i) (ii)

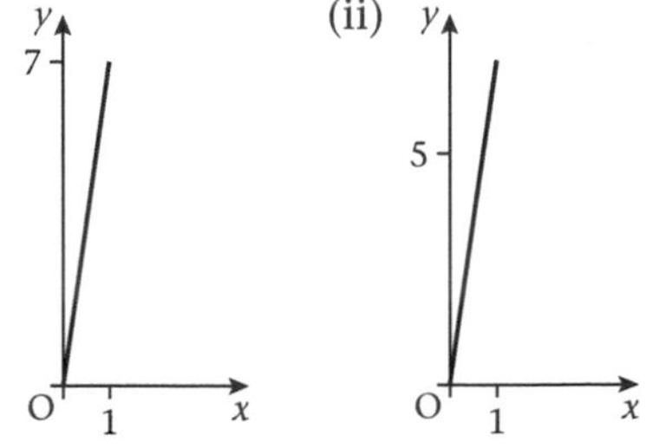

(iii)
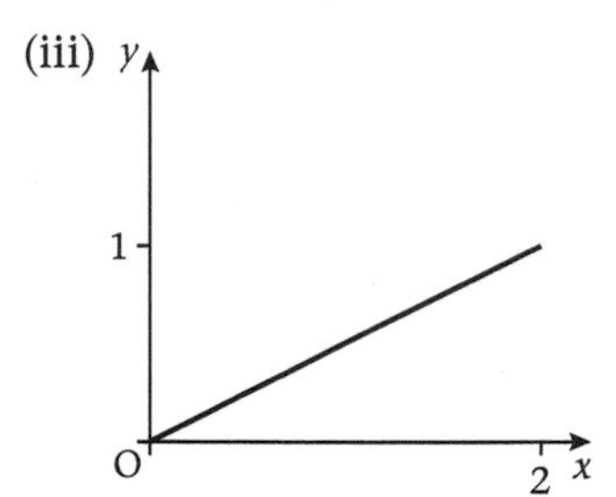

(iv)
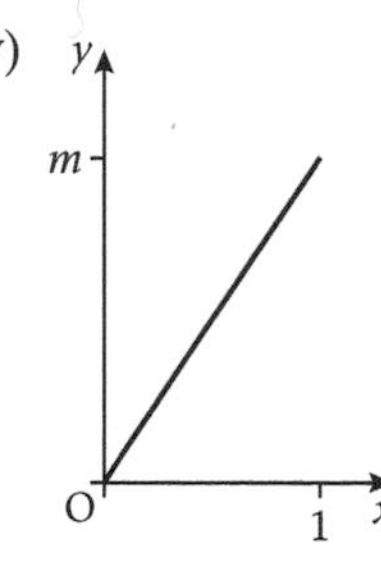

Practice exercise P6.9 (p. 71)

1. $y = -\frac{1}{4}x + 5\frac{1}{2}$
2. $y = -\frac{3}{4}x + \frac{1}{4}$
3. $y = -\frac{3}{4}x + 2$
4. $y = 6x + 9$
5. $y = 2\frac{1}{3}x$
6. $y = -\frac{3}{7}x + -4\frac{5}{7}$
7. $y = -\frac{5}{16}x + 4\frac{3}{16}$
8. $y = x + 2$
9. $y = -2\frac{1}{3}x$
10. $y + x = 3$
11. $y = 3x - 4$
12. $y = 2\frac{1}{2}x$
13. $y + x = 4$
14. $y + 2x = -5$
15. $y = 3x - 10$

Practice exercise P6.10 (p. 71)

1. (a) (i) $-\frac{1}{2}$ (ii) 6
   (b) $y - 2x + 7$
2. (a) $y = -\frac{2}{7}x$
   (b) (i) $y = -\frac{2}{7}x + 2\frac{2}{7}$
   (ii) $(0, 2\frac{2}{7})$

Practice exercise P6.11 (p. 71)

1. (a) 2 (b) 0
2. (a) 2 (b) 3
3. (a) 2 (b) $-1\frac{1}{2}$
4. (a) $-\frac{1}{2}$ (b) 0

Practice exercise P6.12 (p. 71)

1. (1, 4)
2. (2, 3)
3. $(-\frac{1}{2}, 5\frac{1}{2})$
4. (−1, 2)
5. (−2, 3)
6. $(2\frac{1}{2}, 2\frac{1}{2})$
7. $(0, \frac{1}{2})$
8. $(1, -\frac{1}{2})$
9. $(2, \frac{1}{2})$
10. $(-1\frac{1}{2}, 0)$

Practice exercise P6.13 (p. 71)

1. (a) (i) $(2\frac{1}{2}, 3)$
   (ii) $\left(\frac{X + 5}{2}\right)$, 3
   (b) They are both equal to half the y-coordinate of T.
   (c) $\left(\frac{x_1}{2}, \frac{y_1}{2}\right), \left(\frac{X + x_1}{2}, \frac{y_1}{2}\right)$
   (d) It is parallel to the x-axis and half the length of OF.
2. (a) (2, 3)
   (b) (i) $(3, 1\frac{1}{2})$ (ii) $(3, 1\frac{1}{2})$
   (c) The diagonals bisect each other.
   (d) (6, 3)
   (e) B(4, 4), C(−2, 4)

Practice exercise P6.14 (p. 72)

1. 5.66 units
2. 5 units
3. 2.83 units
4. 4.24 units
5. 6 units

Practice exercise P6.15 (p. 72)

1. (a) −3 (b) $y = -3x + 1$
   (c) (1, −2) (d) 12.65 units
2. (a) $-2\frac{2}{7}$ (b) $y = -2\frac{2}{7}x - 2\frac{6}{7}$
   (c) $(\frac{1}{2}, -4)$ (d) 17.46 units
3. (a) $\frac{1}{3}$ (b) $y = \frac{1}{3}x + 4\frac{1}{3}$
   (c) (5, 6) (d) 6.32 units
4. (a) −3 (b) $y = -3x + 11$
   (c) (3, 2) (d) 6.32 units
5. (a) 2 (b) $y = 2x - 7$
   (c) (3, −1) (d) 8.94 units
6. (a) Let d be the distance of the point (x_1, y_1) from the origin.
   Using Pythagoras,
   $$d^2 = x_1^2 + y_1^2$$
   $$\therefore d = \sqrt{x_1^2 + y_1^2}$$
   (b) (i) The locus of a point a fixed distance from a point is a circle.
   (ii) (0, 0)
   (iii) 10 units

Revision exercise 1 (p. 73)

1. $p = 3, q = 2$
2. (a) $(y - b)(3a + 2x)$ (b) $(s - t)(d - r)$
   (c) $(q - r)(p + 8)$ (d) $(x + y)(5 + k)$
3. (a) $(x - 4)(x + 1)$ (b) $(3y + 1)(2y + 3)$
   (c) $(3a - 8)(a - 1)$ (d) $(2d - 3)(d + 5)$
4. (a) $d = 5, \frac{2}{3}$ (b) $m = \frac{1}{2}, -4$
5. (a) $a(a + 3)$ (b) $(3p - 2q)^2$
   (c) $(4x + 1)(4x - 1)$ (d) $2(y + 2)^2$
6. (a) $x = 0, 5$ (b) $a = -\frac{1}{2}, -5$
   (c) $t = \frac{4}{3}, -\frac{1}{2}$ (d) $x = -2, \frac{1}{2}$
7. −0.18, −2.82

8 (a)

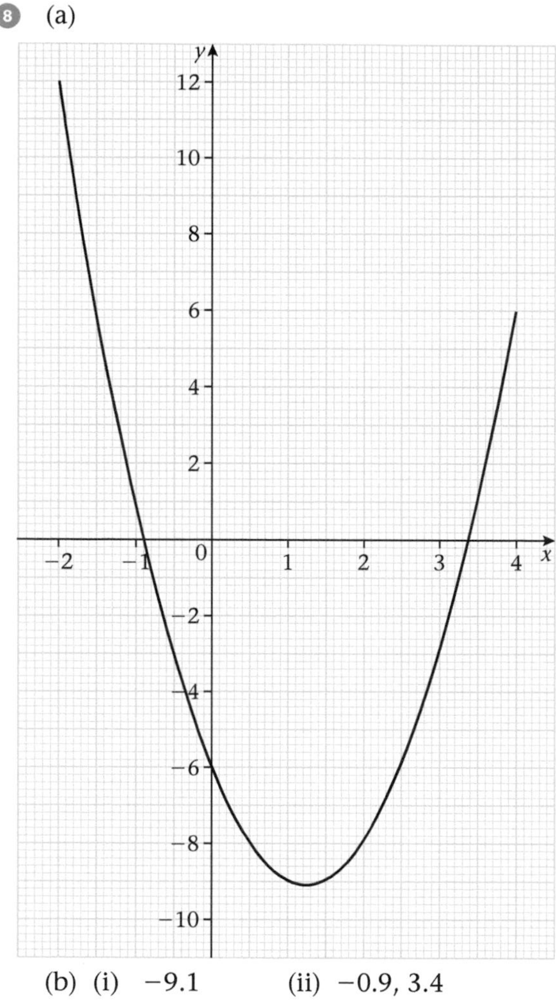

(b) (i) −9.1 (ii) −0.9, 3.4

9 7 **10** 11 cm

Revision test 1 (p. 73)

1 C **2** B **3** C **4** B **5** C

6 (a) $a(5a + b)$ (b) $(a - b)(3x - y)$
(c) $(a + 1)(a + 6)$ (d) $2b(b - 4)(b - 7)$
(e) $(3c + 5)(5c + 2)$ (f) $(8d - 3)(d + 5)$

7 (a) $x = 0, 4$ (b) $y = 2, 8$
(c) $x = -\frac{2}{3}, -4$ (d) $x = -\frac{3}{2}, 7$

8 (a) $x = \frac{13}{3}, \frac{5}{3}$ (b) $x = \pm 3$
(c) $x = -2.2, -7.8$ (d) $x = 0.4, -2.4$

9 12 m, 18 m **10** 1.62, −0.62

Revision exercise 2 (p. 74)

1 $\begin{pmatrix} 1 \\ 2 \end{pmatrix}$

2 $P + Q = \begin{pmatrix} 1 & -1 \\ 3 & 2 \end{pmatrix}$ $3(P + Q) = \begin{pmatrix} 3 & -3 \\ 9 & 6 \end{pmatrix}$
$3P + 3Q = \begin{pmatrix} 6 & -3 \\ 0 & 3 \end{pmatrix} + \begin{pmatrix} -3 & 0 \\ 9 & 3 \end{pmatrix} = \begin{pmatrix} 3 & -3 \\ 9 & 6 \end{pmatrix}$

3 (a) $\begin{pmatrix} 6 \\ 8 \end{pmatrix}$ (b) $\begin{pmatrix} 4 \\ -5 \end{pmatrix}$ $\begin{pmatrix} 3 \\ -2 \end{pmatrix}$

4 (a) $x = \frac{7}{2}, y = \frac{7\sqrt{3}}{2}$ (b) $x = 8, y = 8\sqrt{2}$
(c) $x = 5, y = 5\sqrt{2}$ (d) $x = 3\sqrt{3}, y = 6\sqrt{3}$

5 $\begin{pmatrix} 3 & 1 & 2 \\ 2 & 2 & 2 \\ 1 & 3 & 2 \end{pmatrix}$

6 $a = -2, b = -1$

7 (a) 0.67 (b) −0.88 (c) −0.11
(d) −0.98 (e) −0.69 (f) 0.86

8 (a) −148°, −32°, 212°, 328°
(b) −283°, −77°, 77°, 283°
(c) −250°, −70°, 110°, 290°

9 $14\frac{1}{3}$ cm

Revision test 2 (p. 74)

1 D **2** B **3** D **4** C **5** C

6 (a) $-\frac{12}{13}$

7 (a) $\begin{pmatrix} 1 & 1 \\ 4 & 4 \end{pmatrix}$ (b) $\begin{pmatrix} 3 & 5 \\ 0 & -2 \end{pmatrix}$

8 (a) BC = 10 cm, AC = $10\sqrt{2}$ cm
CD = $10(\sqrt{3} - 1)$ cm
CE = $5(\sqrt{3} - 1)$ cm
CAD = 15°
(b) $\sin 15° = \frac{\sqrt{2}(\sqrt{3} - 1)}{4}$

9 (a) $\begin{pmatrix} 2 & -1 \\ 1 & 3 \end{pmatrix}\begin{pmatrix} x \\ y \end{pmatrix} = \begin{pmatrix} 7 \\ 0 \end{pmatrix}$ (b) 7
(c) $\frac{1}{7}\begin{pmatrix} 3 & 1 \\ -1 & 2 \end{pmatrix}$ (d) $x = 3, y = -1$

10 2.7 cm

Revision exercise 3 (p. 75)

1 (a) parallel (b) perpendicular
(c) parallel (d) perpendicular

2 (a) (i) $6\sqrt{2}$ (ii) (5, 3) (iii) −1
(b) (i) $\sqrt{2}$ (ii) (−3.5, 3.5) (iii) 1
(c) (i) $7\sqrt{5}$ (ii) (1, −1.5) (iii) $-\frac{1}{2}$

3 90°, 34°, 124°, 34°, 22°

4 (a) $T\hat{D}A$, $A\hat{B}D$, $A\hat{C}D$
(b) $A\hat{B}C$
(c) $\frac{1}{2}(180 - z)°$

5 61° **7** $2\sqrt{26}$

8 $2y + x - 3 = 0$

9 (a) 90°, 146°, 90°, 34° (b) kite

Revision test 3 (p. 76)

1 B 2 D 3 B 4 A 5 C

6 (a) $2x + 7y = 0$
(b) $2x + 7y = 56$
(c) $(28, 0)$

7 $-2, 1, 3$

8 $\text{O}\widehat{\text{B}}\text{F} = 38°$, $\text{O}\widehat{\text{B}}\text{D} = 25°$, $\text{B}\widehat{\text{F}}\text{D} = 128°$

9 $\text{P}\widehat{\text{T}}\text{Q} = 36°$

10 (a) $\text{C}\widehat{\text{T}}\text{P} = (90 - a)°$, $\text{A}\widehat{\text{T}}\text{O} = a°$
(b) $\text{C}\widehat{\text{T}}\text{A} = 90° \Rightarrow$ CA is a diameter $\Rightarrow$ AC and BT intersect at centre of circle.

General revision test A (p. 77)

1 D 2 C 3 B 4 A 5 D

6 C 7 B 8 B 9 C

10 (a) $(x + 1)(x^2 + 1)$
(b) $(2 + b)(1 + b)(1 - b)$
(c) $(x + p)(x - q)$

11 (a) $\begin{pmatrix} 1 & -1 \\ 0 & 1 \end{pmatrix}$ (b) $\text{M}^{-1} = \frac{1}{2}\begin{pmatrix} 1 & 1 \\ 0 & 1 \end{pmatrix}$

12 (a) $3c(c^2 - 5)$
(b) $(2x - 5)^2$
(c) $2(y^2 + 4)(y + 2)(y - 2)$

13 (a) $x = 6.8$; $y = 9.5$
(b) $x = 4.3$; $y = 2.1$

14 (a) $\frac{7}{2}, -3$ (b) $-0.6, -13.4$
(c) $8.9, -26.9$

15 $a = \frac{2}{3}$, $3y = 2x + 6$, $2y + 3x + 4 = 0$

16 (a) 63° (b) 27° (c) 105°

Exercise 7a (p. 80)

1 $\frac{m}{v}$ 2 $\frac{4x}{5y}$

3 No simpler form

4 $\frac{b + c}{d + e}$ 5 $\frac{a + b}{a + c}$

6 No simpler form

7 $\frac{h - k}{k}$ 8 $\frac{1}{3dnv}$ 9 $-\frac{c}{d}$

10 $-\frac{a + b}{b}$ 11 $\frac{x}{x - y}$ 12 $\frac{4cd^2}{5e^2}$

13 No simpler form

14 $\frac{c - d}{c}$ 15 $\frac{m + n}{m - n}$ 16 $\frac{c + 3}{c + 2}$

17 $\frac{d + 3}{d - 4}$ 18 $\frac{y}{x - y}$

19 No simpler form 20 No simpler form

21 $\frac{h + k}{h - k}$ 22 $\frac{x + 3y}{x - y}$ 23 $-\frac{x + 3}{x + 5}$

24 $-\frac{3a + m}{a + m}$ 25 $-\frac{a + 4}{2a + 1}$ 26 $\frac{a - n}{a + n}$

27 $\frac{w + u - v}{w - u + v}$ 28 $\frac{a + b + c}{b - a - c}$ 29 $\frac{b + c}{b - c}$

30 $\frac{a - b}{c - a}$

Exercise 7b (p. 80)

1 $\frac{a}{e}$ 2 $\frac{8n^2}{9c^2}$ 3 $\frac{n}{3}$

4 $\frac{4}{3u}$ 5 $\frac{a}{2}$ 6 $\frac{3u^3v}{16m^2}$

7 $\frac{a + 2}{a}$ 8 $\frac{m + 3}{m}$ 9 $-2a$

10 $\frac{d(d - 2)}{2}$ 11 $\frac{b}{2}$ 12 $\frac{n - 3}{n}$

13 $-\frac{n}{m}$ 14 $\frac{a + 2b}{a + 3b}$ 15 $2c$

Exercise 7c (p. 82)

1 $\frac{8a + 9c}{6abc}$ 2 $\frac{5c - a + b}{c}$ 3 $\frac{3c - 14b}{30bc}$

4 $\frac{7}{6(x + y)}$ 5 $\frac{2}{a - 2b}$ 6 $\frac{3a - b}{a - b}$

7 $\frac{x + 4y}{x + 2y}$ 8 $\frac{u - 9v}{2u + 3v}$ 9 $\frac{6a - b}{2(2a + b)}$

10 $\frac{19mn}{6(m^2 + n^2)}$ 11 $\frac{3}{2(2x - y)}$ 12 $\frac{u(u - 2v)}{v^2}$

13 $\frac{5a + 7}{(a + 1)(a + 2)}$ 14 $\frac{3x^2 + 2x + 4}{(x - 1)(x + 2)}$

15 $\frac{4m}{(m - n)^2}$ 16 $\frac{2a - b}{(a - 2b)^2}$ 17 $\frac{3}{c - d}$

18 $\frac{2}{4m + 3n}$

Exercise 7d (p. 83)

1 $\frac{1}{5}$ 2 1 3 $8\frac{1}{4}$

4 d 5 $\frac{4a + 5}{a + 10}$ 6 $-\frac{5}{3m}$

7 $\frac{3w - 8}{8w - 5}$ 8 $\frac{5 - a}{7a + 4}$

Exercise 7e (p. 84)

1 3 2 4 3 -7

4 0 5 $2\frac{1}{2}$ 6 $6\frac{2}{3}$

7 $0, -2$ 8 $-4, -3$ 9 ± 1

10 6

Exercise 7f (p. 84)

1 5, −2 2 −6 3 0, ±3 4 0, ±2

Exercise 7g (p. 85)

1 1
2 $2\frac{1}{5}$
3 3, −1
4 3, $-\frac{2}{3}$
5 −2, $1\frac{1}{3}$
6 $-\frac{1}{3}$, 2
7 −1, −2
8 5, $-\frac{1}{3}$
9 0, 2
10 6, $-1\frac{1}{3}$
11 1
12 $-\frac{1}{4}$
13 5, $8\frac{1}{2}$
14 3, $1\frac{3}{4}$
15 −2, $1\frac{3}{7}$
16 3, $-7\frac{3}{8}$
17 3, $1\frac{1}{19}$
18 $2\frac{1}{2}$
19 2, −3
20 $-1\frac{1}{6}$

Exercise 7h (p. 85)

1 $\frac{6 - x^2 + 10x}{2x}$
2 $\frac{3x + 2}{4}$
3 $\frac{5}{6(x - 2)}$
4 $\frac{2y - 5x}{4}$
5 $\frac{6m + 1}{2m + 3}$
6 10
7 (a) −3 (b) $x < 4$ (but not −3)
8 (a) $\frac{3(a - 6)}{(a + 4)(a - 2)}$ (b) $a = 6$
9 (a) 0, 3 (b) $k = -2$
10 $\frac{y + z}{y - z}$
11 $\frac{3(2a - 1)}{2(a - 1)}$, 1
12 $-\frac{a + 2b}{b(a + b)}$
13 $\frac{2}{x}$
14 $x = \frac{1}{2}$

Exercise 7i (p. 87)

1 $\frac{1}{a^2}$
2 $\frac{1}{b}$
3 $\frac{1}{c^{\frac{2}{3}}}$
4 $\frac{x}{y}$
5 $\frac{1}{xy}$
6 $\frac{b^3}{a^2}$
7 $\frac{a}{b^3}$
8 $\frac{1}{a^3b^3}$
9 $\frac{2}{x^{\frac{1}{2}}}$
10 $\frac{3}{y^{\frac{2}{3}}}$
11 $\frac{9b^2}{a^2}$
12 $n^{\frac{1}{3}}$

Exercise 7j (p. 87)

1 $18a^3$
2 ±2
3 5
4 4
5 4
6 $\frac{1}{4}$
7 $\frac{1}{27}$
8 $\pm\frac{1}{3}$
9 $\pm 5a$
10 $\frac{2}{a}$
11 ±8
12 2
13 $\frac{1}{100}$
14 $\pm 1\frac{1}{4}$
15 $\frac{3}{a^2}$
16 $\frac{1}{9a^2}$
17 ±9
18 $\pm\frac{1}{a}$
19 9
20 9
21 9
22 $\frac{1}{3}$
23 ±0.2
24 $18a$
25 $\pm\frac{1}{64}$
26 4
27 $\frac{1}{25}$
28 1
29 0.09
30 $\frac{6}{a}$
31 $\frac{2}{9a}$
32 $2\frac{1}{4}$
33 ±2
34 $\frac{2}{a^2}$
35 2
36 $\frac{12a^4}{b}$
37 $\frac{1}{5}$
38 $4x^5$

Exercise 7k (p. 88)

1 $x = 4$
2 $x = 27$
3 $a = \frac{1}{2}$
4 $a = \pm\frac{1}{3}$
5 $x = 3$
6 $x = \frac{1}{25}$
7 $n = \pm\frac{1}{27}$
8 $r = -\frac{1}{2}$
9 $x = 4$
10 $x = 2$

Exercise 7l (p. 89)

1 (a) (i) I (b) (i) I (c) (i) I
(d) (i) E (ii) 4 (e) (i) E (ii) −3
2 (a) $p = 1, q = \frac{1}{2}$ (b) $p = 1, q = -3$
(c) $p = 3, q = -1; p = -1, q = 3$

Exercise 7m (p. 90)

1 $25xg$
2 $\$4y$
3 $15t$ km
4 (b) $B = 56\frac{2}{3}$
5 (b) \$11.20 (c) £5
6 $C = 7n$
7 $d = 4s$
8 $a = 0.8b$
9 (a) $x = 2\frac{1}{2}y$ (b) 25 (c) 5.6
10 (a) $P = \frac{3}{8}Q$ (b) 6 (c) 6.4
11 (a) $y = \frac{1}{5}x$ (b) $y = 8$ (c) $x = 50$
12 $p = 25q$
13 (a) $x = \frac{5}{3}y$ (b) $x = 20$
14 225 cm^2
15 (b) $H = \frac{13}{15}V$ (or, by graph, $H = 0.9V$)
(c) $V = 7.5$ (d) $H = 3.9$

Exercise 7n (p. 92)

1 $a = 2$
2 (a) $A = \frac{1}{2}B^3$ (b) $A = 108$ (c) $B = 3$
3 (a) $P = \frac{5}{2}\sqrt{Q}$ (b) $P = 7\frac{1}{2}$ (c) $Q = \frac{9}{16}$
4 $y = 4x^2$, $y = 1$ when $x = \frac{1}{2}$
5 8.8 amps
6 35 km
7 (a) x increases by 21%
(b) x decreases by 19%
8 x increases by 20%
9 72.8%
10 $27\frac{3}{4}$%

Exercise 7o (p. 94)

1 $d \propto \frac{1}{t}$
2 (a) inversely (b) $n \propto \frac{1}{l}$
3 (a) $l = \frac{A}{b}$ (b) $b = \frac{A}{l}$ (c) inversely
4 $R = \frac{32}{T}$
5 $y = 1$
6 $Q = \frac{4}{5}$
7 $R = \frac{k}{r^2}$
8 $v = 2.25$
9 $Y = 4$
10 11.25 km

Exercise 7p (p. 95)

1. (a) $x = 5yz$ (b) 120
2. (a) $x = \frac{6y}{z}$ (b) 7
3. (a) $p = \frac{6q}{r^2}$ (b) $p = 37\frac{1}{2}$
4. $h = \frac{kV}{r^2}$
5. (a) $A = 5$ (b) $C = 8$
   (c) A decreases by 1%
6. (a) $x = 37\frac{1}{2}$ (b) $y = \pm 4$
   (c) x is doubled
7. $x \propto z^3$
8. $x \propto z^4$
9. $A \propto \frac{1}{C}$
10. 0.261 kg

Exercise 7q (p. 96)

1. $x = y - 8$
2. $x = y + 3$
3. $a = b - c, c = b - a$
4. $x = \frac{y}{3}$
5. $x = 4y$
6. $a = \frac{b}{c}, c = \frac{b}{a}$
7. $a = \frac{n}{5x}, x = \frac{n}{5a}$
8. $x = 9y$
9. $x = \frac{y}{2}$
10. $x = \frac{y - 11}{6}$
11. $a = \frac{b + c}{5}$
12. $x = 13 - y, y = 13 - x$
13. $q = 2p, p = \frac{1}{2}q$
14. $x = \frac{d + y}{1}, y = 2x - d$
15. $d = \frac{p}{4}$ or $\frac{1}{4}p$
16. $\theta = T - 273$
17. $V = \frac{W}{I}, I = \frac{W}{V}$
18. $r = \frac{A\pi}{l}, l = \frac{A}{\pi r}$
19. $l = \frac{V}{bh}, b = \frac{V}{lh}, h = \frac{V}{lb}$
20. $r = \frac{A}{2\pi h}, h = \frac{A}{2\pi r}$

Exercise 7r (p. 97)

1. (a) $x = \frac{y + 9}{2}$ (b) $x = 7$
2. (a) $x = \frac{d - y}{3}$ (b) $x = -4$
3. (a) $h = \frac{A}{2\pi r}$ (b) $h = 6$
4. (a) $b = \frac{2A}{h}$ (b) $b = 15$
5. (a) $r = \frac{w - 59}{3}$ (b) 8 h
6. (a) $P = \frac{100I}{RT}$ (b) $850
7. (a) $R = \frac{100I}{PT}$ (b) $3\frac{1}{3}$%
8. (a) $d = \frac{c}{p}$ (b) 1.9
9. $x \propto \frac{1}{y^2}$
10. $y = k + \frac{h}{x^2}$

Exercise 7s (p. 98)

1. $b - a$
2. $a - b$
3. $\frac{b}{a}$
4. $\frac{c}{a + d}$
5. $\frac{b}{1 - a}$
6. $\frac{a}{b}$
7. $\frac{a}{c - b}$
8. $a(c - b)$
9. $\frac{ab}{a + b}$
10. $a + b$
11. $\frac{c}{a} - b$
12. $\frac{bc}{a - b}$
13. $\frac{ab}{a + c}$
14. $\frac{6ab}{5}$
15. $\frac{bc}{a}$
16. $\frac{b - a}{c}$
17. $\frac{b(3a - 2)}{2a + 3}$
18. a^2
19. $\frac{a^2}{2}$
20. $\frac{a^2}{4}$
21. $2a^2$
22. $4a^2$
23. $\frac{b^2}{a^2}$
24. $\frac{b^2}{a}$
25. ab^3
26. $\pm a^2$
27. $\pm\sqrt{a}$
28. $b^2 - a$
29. $(b - a)^2$
30. $\pm\sqrt{b^2 - a^2}$
31. $\pm 2a\sqrt{2}$
32. $a - \frac{b}{2}$
33. $\frac{1 + b^2}{a^2}$
34. $\left(\frac{1 + b}{a}\right)^2$
35. $\frac{a}{a - b}$
36. 0
37. $b - a$
38. $a^2 + ab + b^2$
39. $\frac{-a + b}{2}$
40. $\pm\frac{a}{b}\sqrt{b^2 - y^2}$

Exercise 7t (p. 99)

1. $r = \frac{C}{2\pi}$
2. $W = \frac{P - b}{a}$
3. $N = PD - 2, \quad D = \frac{N + 2}{P}$
4. $T = \frac{100(A - P)}{PR}, \quad P = \frac{100A}{100 + RT}$
5. $s = \frac{v^2 - u^2}{2a}, \quad u = \pm\sqrt{v^2 - 2as}$
6. $n = \frac{2s}{a + l}, \quad l = \frac{2s}{n} - a$
7. $h = \frac{S}{2\pi r} - r$
8. $b = \frac{kv}{k + rt}$
9. $r = \sqrt{\frac{S}{4\pi}}$
10. $r = \frac{V}{\pi h^2} + \frac{h}{3}$
11. $W = \frac{aLP}{h - aL}$
12. $R = \frac{2aE + rL}{L}$

13 $K = \dfrac{mu^2}{T + 5mg}$
14 $h = \dfrac{2D^2}{3}$
15 $r = \dfrac{3p}{t - s}$
16 $q = \dfrac{bp}{a - cp}$
17 $v = \pm\sqrt{\dfrac{2gxH}{m} + u^2}$
18 $x = \dfrac{QR^2 + P}{a - R^2b}$
19 $M = \dfrac{4\pi^2 I}{HT^2}$
20 $u = \pm\sqrt{v^2 - \dfrac{2A}{m}}$

Exercise 7u (p. 99)

1 (a) $a = \sqrt{\dfrac{3V}{h}}$ (b) $a = 4\frac{1}{2}$
2 (a) $r = \sqrt{\dfrac{3V}{\pi h}}$ (b) $r = 2\frac{1}{2}$ cm
3 (a) $a = \sqrt{h^2 - b^2}$ (b) $a = 30$
4 (a) $d = \dfrac{2(S_n - an)}{n(n - 1)}$ (b) $d = -1\frac{1}{2}$
5 (a) $v = \pm\sqrt{\dfrac{2(E - mgh)}{m}}$
(b) $v = 14$ m/s
6 (a) $h = \dfrac{3V}{\pi r^2} - 2r$ (b) $h = 3.4$ cm
7 (a) $L = \dfrac{8S^2}{3d} + d$ (b) $L = 16.06$ m
8 (a) $d = \dfrac{21T}{4b^2} + \dfrac{3b}{5}$ (b) $d = 70$ m

Practice exercise P7.1 (p. 100)

1 $\dfrac{5x + 20}{6}$
2 $\dfrac{x + 4}{6}$
3 $\dfrac{x^2 + 8x + 16}{6}$
4 $1\frac{1}{2}$
5 $\dfrac{7x + 24}{x^2 + 7x + 12}$
6 $\dfrac{-x}{x^2 + 7x + 12}$
7 $\dfrac{12}{x^2 + 7x + 12}$
8 $\dfrac{3x + 12}{4x + 12}$

Practice exercise P7.2 (p. 100)

1 $\frac{1}{3}$
2 $\dfrac{2a^2 + 5a - 3}{6}$
3 $\dfrac{3w^2 + 7w - 7}{w^2 + 3w + 2}$
4 $\dfrac{3m}{2m + 12}$
5 $\dfrac{3d^2 - 8d - 6}{d^2 + d - 6}$
6 $\dfrac{9v + 5}{3}$
7 $\dfrac{y - 3}{3}$
8 $-\dfrac{p - 2}{2p + 1}$
9 $\dfrac{3abc - 2c^2 + a^2}{a^2b^2c^2}$
10 $\dfrac{2 - x - x^2}{6}$
11 $\dfrac{2x - 26}{5x + 10}$
12 $\dfrac{3m - 3n - 10}{6}$
13 $\dfrac{y + 18}{y^2 + y - 6}$
14 $-\dfrac{5h^2 + 44h + 21}{6h^2 + 24h + 3}$
15 $\dfrac{7y - 9 - 7y^2}{6y}$

Practice exercise P7.3 (p. 100)

1 (a) 2 (b) −3 (c) 12 (d) 36
(e) −7 (f) 4 (g) 5 (h) 1
(i) $\frac{1}{2}$ (j) −1 (k) 2 (l) $\frac{1}{5}$
2 (a) 42 (b) 10
3 (a) 20 (b) 8 (c) −2
(d) 25 (e) 11
4 −7
5 (a) −10 (b) 20 (c) $\frac{2}{11}$
6 $T = \dfrac{2F - 4}{F}$
7 4.4 m
8 4.5% p.a.
9 (a) $y = x + nd$ (b) 12 weeks

Practice exercise P7.4 (p. 101)

1 $C = K - 273$
2 $b = \dfrac{P - 2l}{2}$
3 $s = t - \dfrac{r}{2}$, $r = 2t - 2s$
4 $T = \dfrac{100I}{PR}$, $R = \dfrac{100I}{PT}$, $P = \dfrac{100I}{RT}$
5 $u = v - at$, $t = \dfrac{v - u}{a}$
6 $u = \dfrac{2s - a^2t}{2t}$, $t = \dfrac{2s}{2u + a^2}$
7 $n = -\dfrac{3 + 5p}{p - 2}$

Practice exercise P7.5 (p. 101)

1 $8x^3$ 2 $16x^6$ 3 1 4 $64x^9$

Practice exercise P7.6 (p. 101)

1 $4a^7$
2 $p^5r^5q^3$
3 $15bd^5$
4 $6xy^5z^3$
5 $\dfrac{3jk^2}{h^3}$
6 $4ac^2$
7 $8a^9f^3$
8 $27h^5k^5n^6$

Practice exercise P7.7 (p. 101)

1 x^{-5} 2 $x^{\frac{1}{5}}$ 3 $x^{-\frac{1}{2}}$ 4 $x^{\frac{5}{2}}$

Practice exercise P7.8 (p. 101)

1 $\pm\frac{1}{3}$
2 $\frac{1}{2}$
3 $\pm 4b^3$
4 $1\frac{1}{3}$
5 $6x^2$
6 $54x^2$

7. $\frac{\sqrt{6}}{36}$ 8. $\frac{1}{3}$ 9. $8k^3$
10. $2\frac{2}{3}$ 11. a^{-3} 12. $-27f^6$
13. x^4y 14. $\frac{r^3}{8}$ 15. $\frac{a^9x^5y}{2}$
16. $6h^{-5}$ 17. $4y^2$ 18. v^2
19. $3^{\frac{1}{2}}$ 20. $m^{-\frac{1}{3}}$

Practice exercise P7.9 (p. 101)

1. (a) E (b) $x = \frac{1}{5}$
2. (a) I
   (b) $\text{RHS} = 3 + 3z + 3y - 2 - 4z$
   $= 3y + 1 - z$
   $= \text{LHS}$
3. (a) E (b) $d = 0$
4. (a) I
   (b) $\text{LHS} = c^2 - c - 6$
   $= \text{RHS}$
5. (a) E (b) $h = 3$
6. (a) E (b) $y = 0$
7. (a) E (b) $m = 2$
8. (a) I
   (b) $\text{LHS} = \frac{(2t - 1)(1 + t) - 2t(t + \frac{1}{2})}{2t(1 - t)}$
   $= -\frac{1}{2t(1 - t)}$
   $= \text{RHS}$
9. (a) E (b) $y = -4$
10. (a) I
    (b) $\text{LHS} = 2(8 - 2f - f^2 - 3f + 2$
    $= 18 - 7f - 2f^2$
    $\text{RHS} = 2 - 7f + 6f^2 - 8f^2 + 16$
    $= 18 - 7f - 2f^2$
    $= \text{LHS}$
11. (a) E (b) $a = -3$
12. (a) E (b) $b = 2$

Practice exercise P7.10 (p. 102)

1. $\text{RHS} = 4 + 3x$
   $= 3x + 4$
   $= \text{LHS}$
2. $\text{LHS} = 3a - 2b + 1$
   $\text{RHS} = b + 5 + 3a - 3b - 4$
   $= 3a - 2b + 1$
   $= \text{LHS}$
3. $\text{LHS} = 2d^2 + d$
   $\text{RHS} = d^2 + d^2 + d$
   $= 2d^2 + d$
   $= \text{LHS}$
4. $\text{LHS} = 2y^2 + 2xy + y + x$
   $\text{RHS} = 3y^2 + x + 2xy - y^2 + y$
   $= 2y^2 + 2xy + y + x$
   $= \text{LHS}$
5. $\text{LHS} = 3x + 3 - 2x + 2$
   $= x + 5$
   $= \text{RHS}$
6. $\text{LHS} = 2a - 4b + b$
   $= 2a - 3b$
   $\text{RHS} = 3a - 3b - a$
   $= 2a - 3b$
   $= \text{LHS}$
7. $\text{LHS} = m^2 + mn + 2n$
   $\text{RHS} = m^2 + mn + 2n$
   $= \text{LHS}$
8. $\text{LHS} = jk + kj + k^2 + j^2$
   $= k^2 + 2kj + j^2$
   $\text{RHS} = 2k^2 + 2j^2 + 2kj - j^2 - k^2$
   $= k^2 + 2kj + j^2$
   $= \text{LHS}$
9. $\text{LHS} = \frac{(x + 1)(x + 3)}{x(x + 3)}$
   $= \frac{x + 1}{x}$
   $= \text{RHS}$
10. $\text{LHS} = mn - mp + np - nm + pm - np$
    $= 0$
    $= \text{RHS}$

Practice exercise P7.11 (p. 102)

1. $c = 2, d = 4$
2. $p = -1, q = 2$

Practice exercise P7.12 (p. 102)

1. (a) $C \propto r$ (b) $A \propto r^2$
   (c) $A \propto C$ (d) $l \propto r$
2. $V \propto r^3$
3. 2.8 seconds
4. $y = \frac{ac}{b}$

Exercise 8a (p. 104)

1. $\frac{x}{\sin 80°} = \frac{4}{\sin 30°}$ or $x = \frac{4 \sin 80°}{\sin 30°}$

The answers to questions 2–8 have been rounded to 3 s.f.

2. 13.0 cm 3. 14.5 cm 4. 21.5 cm
5. 48.8° 6. 41.6°
7. B = 26.5°, C = 38.5°, c = 44.7 m
8. A = 30°, a = 23.7 m, b = 20.5 m

Exercise 8b (p. 106)

1. $a = 3.78$ cm 2. $b = 1.49$ cm
3. $c = 1.58$ m 4. $a = 2.31$ m
5. $a = 16.6$ cm 6. $b = 4.18$ m
7. $c = 5.73$ cm 8. $b = 12.1$ cm

Exercise 8c (p. 107)

1. $a = 4.97$ cm B = 47.4° C = 66.6°
2. $c = 9.87$ cm B = 29.2° A = 76.8°
3. $c = 12.2$ m A = 25.4° B = 15.6°
4. $b = 375$ m C = 32.1° A = 52.9°
5. $a = 9.08$ m B = 69.3° C = 52.6°
6. $b = 9.45$ cm C = 28.7° A = 25.3°
7. $c = 3.23$ cm B = 28.1° A = 126.2°
8. $a = 65.3$ m B = 26.2° C = 13.6°

Exercise 8d (p. 108)

1. (a) $\frac{2}{3}$ (b) 7 cm
2. (a) 117° (b) 9.40 cm
3. $x = 6.36$, $\theta = 61.4°$
4. (a) 120.4° (b) 29.8°
5. $-\frac{23}{40}$, 4.24 cm
6. $\frac{1}{10}$, 7.42 m
7. 7.63 cm, 3.12 cm
8. $y = 5$
9. 4 : 1 : 1
10. (a) 82.8° (b) 8.89 cm

Exercise 8e (p. 110)

The answers are given to 3 s.f. as a check on calculation. Round each answer to the nearest whole number to give a suitable degree of accuracy.

1. (a) 17.6 cm² (b) 4.22 cm² (c) 4.53 cm² (d) 16.2 cm²
2. (a) 28.7 cm² (b) 26.5 cm² (c) 15.9 cm² (d) 31.1 cm²
3. (a) 26.7 m² (b) 5.80 m² (c) 17.6 m² (d) 38.1 m²
4. 48 cm² 5. (a) 15 cm² (b) 11 cm²
6. 72 cm² 7. 43°

Practice exercise P8.1 (p. 112)

1. 13.1 cm 2. 24°
3. R = 89° $p = 2.4$ cm $q = 9.6$ cm
4. 51.9 m 5. 6.1 mm 6. 26°
7. A = 84° B = 56° C = 40°
8. $a = 12.7$ cm B = 34° C = 17°

Practice exercise P8.2 (p. 112)

1. 7.4 cm, 10.4 cm 2. 14.8 km
3. BT = 61 km, CT = 131 km
4. (a) 42° (b) 34 km (c) 144°

Practice exercise P8.3 (p. 112)

1. 13.7 cm² 2. 205.2 cm²
3. (a) 27° (b) 8.9 cm 4. 162.9 cm²
5. (a) 7.2 cm² (b) 6.8 cm (c) 41° (d) 26.2 cm²
6. (a) P = 42.5°, Q = 17°, R = 120.5° (b) 4.8 cm²

Exercise 9a (p. 114)

1. (a) 120 cm³ (b) 616 cm³ (c) 480 cm³ (d) 48 cm³ (e) 77 cm³ (f) 420 cm³
2. (a) 158 cm² (b) 484 cm² (c) 528 cm²
3. 108 times 4. 16 cm
5. (a) 5 000 cm³ (b) 5 500 cm³ (c) 7 500 cm³
6. 1.44 tonnes
7. (a) 27 cm³ (b) $1\frac{2}{3}$ g/cm³ (c) 19 g
8. 42π cm² 9. 396 cm² 10. 9 cm
11. (a) 275 cm³ (b) 1.09 g/cm³
12. 1.55 g/cm³ 13. 9 cm 14. 64
15. (a) 16π cm³ (b) 5 cm (c) 20π cm² (d) 288°

Exercise 9b (p. 116)

1. (a) 4 190 cm³, 1 260 cm² (b) 2 140 cm³, 804 cm² (c) 16.8 cm³, 37.7 cm² (d) 191 cm³, 191 cm²
2. 7.06 kg 3. 27 cm 4. 1 728
5. (a) 6.20 cm (b) 483 cm²

Exercise 9c (p. 119)

1. 38 125 cm³
2. (a) 3 000 litres (b) 1.224 m³
3. 218π cm³
4. (a) 500 cm³ (b) 468 cm³
5. $121\frac{1}{2}$ tonnes 6. 57.6 kg
7. (a) 7.08 litres (b) 3.24 kg
8. 67.0 cm³ 9. 352 cm³
10. 52.3% 11. 16.5 kg
12. (a) 195π cm² (b) 700π cm³
13. cone: 185 *ml*, frustum: 4 815 *ml*
14. 32.25 m³ 15. 12.7 cm
16. 9 cm 17. $2\frac{1}{4}$ cm
18. $3\frac{25}{77}$ 19. 1 cm 20. 0.343 kg

Exercise 9d (p. 123)

1 6.29 m, 4.42 m   2 54.3°
3 (a) 34.9 m (b) 19°
4 (a) 32.2 m (b) 60.8° (c) 74.8°
5 (a) 4.5 cm (b) 65°
6 87°

Practice exercise P9.1 (p. 123)

1 volume = 504 cm^3, surface area = 402 cm^2
2 volume = 168 cm^3, surface area = 357 cm^2
3 volume = 40.3π cm^3, surface area = 38.4π cm^2
4 volume = 9200 m^3, surface area = 2120 m^2
5 volume = 47.8 cm^3, surface area = 86.7 cm^2
6 (a) 2.51 litres (b) 16.0 cm
7 12 cm   8 823 cm^3 9 160 cm
10 135   11 7.82 cm
12 volume = 59.8 cm^3, surface area = 77.1 cm^2

Practice exercise P9.2 (p. 124)

1 1450 cm^3
2 767 000 cm^3 or 0.767 m^3
3 468 cm^3

Practice exercise P9.3 (p. 124)

1 12.6 m
2 (a) 61.3° (b) 2.1 cm
3 (a) 130 m (b) 151 m

Exercise 10a (p. 127)

1 (a) $x = \frac{1}{2}$ (b) $x = 0$
2 (a) (i) 6 (ii) -2 (b) -5 (c) $x = 3$
3 (a) (i) 2 (ii) -4 (b) 4 (c) $x = -1$
4 (c) (i) 3 (ii) -1 (d) $x = 1\frac{1}{2}$ (e) $2\frac{1}{4}$
5 (a) 1.2 (b) 0.6 (c) -0.8
6 (a) $x = 1$ (b) 8 (c) 1
7 (a) 4 (b) 0 (c) -4
8 (a) (i) 5 (ii) 1 (iii) -7 (b) $x = 1\frac{1}{4}$
9 (a) 2 (b) 0 (c) -3 (d) -4
10 (a) -4 (b) -2 (c) 0 (d) 3

Exercise 10b (p. 130)

1 (a) 2 km (b) 36 km/h
(c) 1 h 24 min (d) 10 km
2 (a) 1 h, 48 km
(b) (i) 82 km, (ii) 36 km
(c) (i) 1 h 15 min, (ii) 42 min
3 (a) 10 min (b) 11.5 km (c) 15 km/h
(d) 18 km/h (e) 50 min
4 (a) No (b) 2.39 p.m. (c) 145 km
(d) 50 km/h (e) 80 km/h (f) 30 km/h
(g) 70 km/h (h) 20 km/h
5 (a) 50 m (b) 100 m
(c) 6 km/h (d) 30
6 (a) A (b) 11 m (c) 6 m (d) $8\frac{1}{3}$ m/s

Exercise 10c (p. 132)

1 (b) the 200 m record
2 (b) (i) 12.5 km (ii) 28.3 km
(c) (i) 8.8 min (ii) 25.6 min
3 14:58
4 12:05
5 20 min, 25 min, 30 min
6 (a) 09:17 (b) 09:05, 09:29
7 (a) 11:41 (b) 31.1 km
8 the first man by 12 min
9 4.9 m   10 12 km/h

Exercise 10d (p. 136)

1 (a) (1) acceleration from rest to 20 m/s in 10 s
(2) deceleration from 20 m/s to 15 m/s in 5 s
(3) constant speed of 15 m/s for 5 s
(4) deceleration from 15 m/s to rest in 5 s
(b) (1) constant speed of 26 km/h for 1 min
(2) deceleration from 26 km/h to 10 km/h in $\frac{1}{2}$ min
(3) acceleration from 10 km/h to 30 km/h in $\frac{1}{2}$ min
(4) deceleration from 30 km/h to rest in 0.6 min
2 (a) in m/s^2: 2, -1, 0, -3
(b) in km/h per min: 0, -32, 40, -50
3 (a) 300 m (b) 0.9 km
4 (a) 0.7 m/s^2 (b) 920 m
5 (a) 2 s (b) 9 m/s^2 (c) 225 m
6 (a) $v = 30$ (b) $\frac{5}{6}$ m/s^2
7 (a) $1\frac{3}{4}$ m/s^2, 14 m (b) $v = 11$, $\frac{2}{3}$ m/s^2
8 (a) 29.2 m/s (b) 26.8 m/s (c) 26 m/s
(d) 27.3 m/s (e) 28.6 m/s (f) 33.8 m/s
9 (a) 32 m/s (b) 312 m (c) 39 m/s
10 (a) 14 m/s (b) $25\frac{1}{2}$ m/s (c) 18.1 m/s

Practice exercise P10.1 (p. 138)

1 (a)

x	−4	−3	−2	−1	0	1	2	3	4
x^2	16	9	4	1	0	1	4	9	16
y	8	4.5	2	0.5	0	0.5	2	4.5	8

(b), (c)

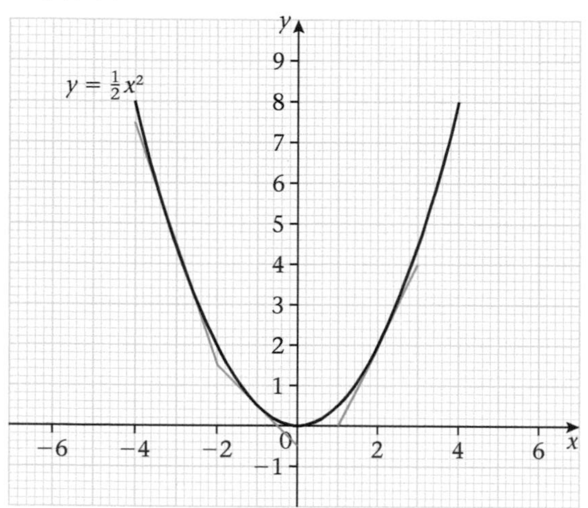

(d) (i) −3 (ii) −1 (iii) 0 (iv) 2

2 (a) (i) At $x = -1$, gradient $= -4$
At $x = 3$, gradient $= 4$

(ii)

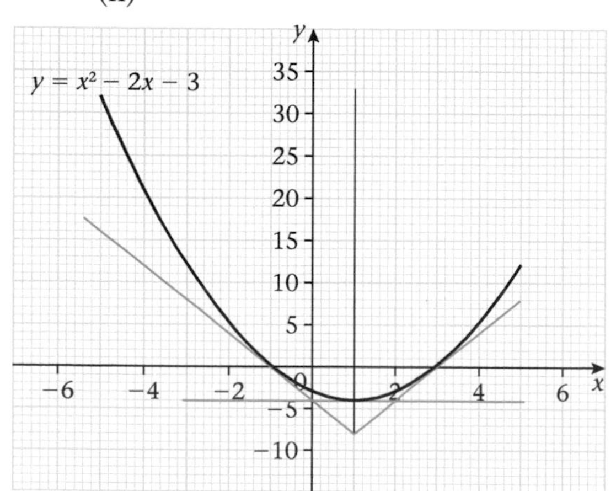

Minimum at (1, −4)

(iii) see graph, $x = 1$

(b) (i) At $x = -2$, gradient $= 5$
At $x = 2$, gradient $= -3$

(ii)

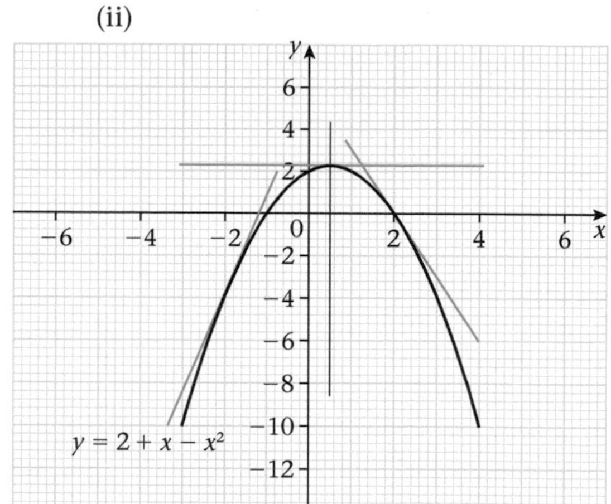

Maximum at (0.5, 2.25)

(iii) see graph, $x = 0.5$

3 (a)

x	−4	−3	−2	−1	0	1	2	3	4	5
x^2	16	9	4	1	0	1	4	9	16	25
$-4x$	16	12	8	4	0	−4	−8	−12	−16	−20
$+1$	1	1	1	1	1	1	1	1	1	1
y	33	22	13	6	1	−2	−3	−2	1	6

(b)

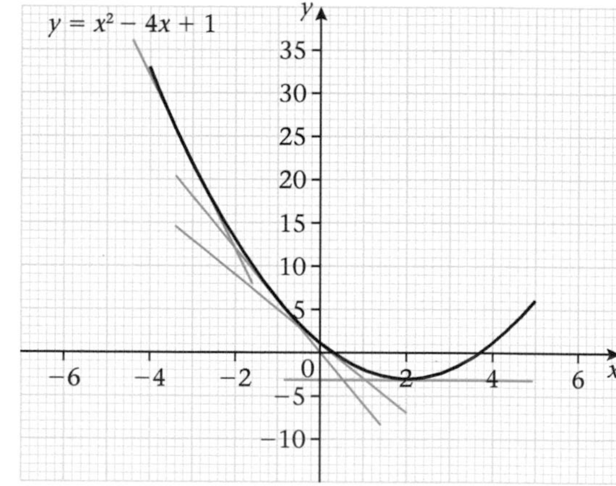

(c) $x = 0.3$ or 3.7

(d) see graph

(e) (i) −10 (ii) −6 (iii) −4
(iv) 0

(f) (2, −3), minimum

Practice exercise P10.2 (p. 139)

1 A leaves from a point 40 km from home at $t = 0$ minutes and travels at a constant speed of 96 km/h, arriving at home 25 minutes later.
B leaves from home at $t = 0$ minutes and travels 10 km in 5 minutes at a speed of 120 km/h and then travels 30 km in 6 minutes at a speed of 300 km/h. B is then 40 km from home. B is stationary for 4 minutes and then travels the 40 km home in 10 minutes at a constant speed of 240 km/h.

2 (a) OA: The object moves 6 m away from O in 2 minutes at a constant speed of 0.05 m/s.
AB: The object remains stationary for 2 minutes.
BC: The object travels the 6 m back to O in 1 minute at a constant speed of 0.1 m/s.
(b) (i) 12 m (ii) 0.04 m/s

Exercise 11a (p. 140)

1 (a) 1 (b) $14\frac{3}{5}$ (c) undefined
(d) -1 (e) $3b - \frac{2}{b}$ (f) $\frac{2}{b} - 3b$
(g) $\frac{2}{x} - 3x$ (h) $6x - \frac{1}{x}$ (i) $\frac{1}{x} - 6x$

2 126

3 (a) 1 (b) $\frac{9}{5}$ (c) $\frac{3}{2}$ (d) 1, 3

4 (a) (i) 1 (ii) 1 (iii) $(a - 1)^2$
(b) (i) 1 (ii) $-1, 3$ (iii) 2

5 3

6 (a) $y = \frac{1}{5}(1 - 2x)$ (b) $y = \frac{-2x}{3} - 4$
(c) $y = x^2 + x - 3$ (d) $y = -\frac{1}{2}(x - 3)$
(e) $y = \frac{7 - x^2}{x + 1}$

Exercise 11b (p. 143)

1 (a) (i) yes, f is a one-to-one mapping
(ii) f^{-1} is the function 'divided by 2'
(b) (i) yes, if there is only one town in the domain for each country in the range; no, if there is more than one town in the domain for each country in the range
(ii) if g^{-1} exists then g^{-1} is the function 'is the country in which ... is the town'
(c) (i) yes, different solids have different volumes
(ii) h^{-1} is the function 'is the solid of volume ...'

2 (a) 3, 4, 5, 6, 7
(b) yes, f^{-1}: $x \rightarrow x - 5$; $f^{-1}(3) = -2$, $f^{-1}(4) = -1$, etc.

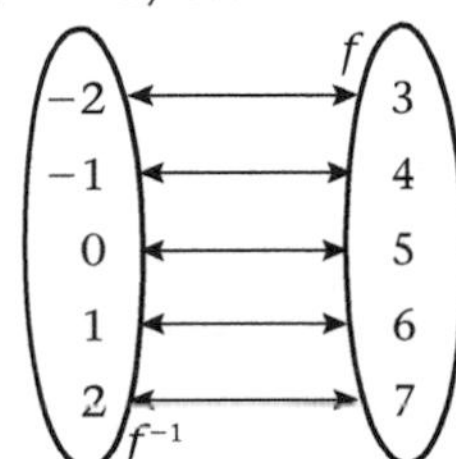

3 (a) (i) the set of real numbers
(ii) $f^{-1}(x) = \frac{x}{5}$
(b) (i) the set of real numbers
(ii) $f^{-1}(x) = 2(x - 1)$
(c) (i) the set of real numbers
(ii) $f^{-1}(x) = 4x + 3$
(d) (i) the set of real numbers
(ii) inverse does not exist
(e) (i) $x < 3, x > 3$
(ii) $f^{-1}(x) = \frac{3x + 5}{x - 2}$

4 (a) $\frac{4x - 3}{x - 5}$ (b) $\frac{15}{8}$ (c) $x = 4$

5 $x = 4$

Exercise 11c (p. 144)

1 (a) 3, 5, does not exist
(b) domain: {1, 2}, range: {3, 5}

2 (a)

x	$f(x)$	$gf(x)$
−2	0	−1
−1	1	2
0	2	5
1	3	8
2	4	11

(c)

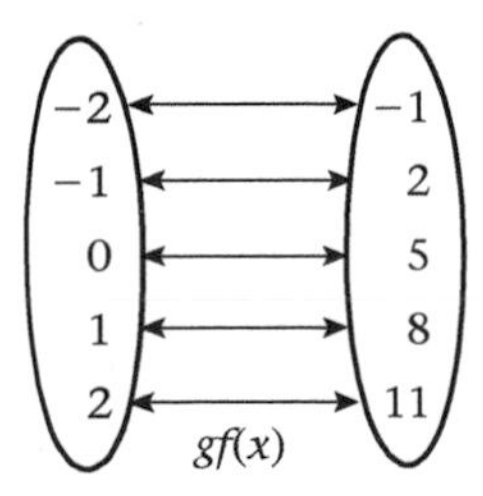

3 (a) $f^{-1}: x \to \frac{1-x}{2}$; $g^{-1}: x \to \frac{x}{3}$

(b) (i) $gf: x \to 3(1-2x)$

(ii) $f^{-1}g^{-1}: x \to \frac{3-x}{6}$

(iii) $(gf)^{-1}: x \to \frac{3-x}{6}$

4 (a) (i) $\frac{x}{3}$ (ii) $4(x-1)$

(iii) $\sqrt{x}$ (iv) $\frac{3x}{4} + 3\sqrt{3}$

(v) $\frac{3x}{4} + 1$ (vi) $3x^2$

(vii) $\frac{\sqrt{x}}{3}$ (viii) $\sqrt{4(x-1)}$

(ix) $4(\sqrt{x}+1)$

5 (a) 0, 35 (b) 3, 5 (c) 1,7

Practice exercise P11.1 (p. 145)

1 (a) is not a function because not all the members of the first set are connected.
(b) and (d) are not functions because one member of the first set is connected to more than one member of the second set.
(c) is a function because each member of the first set is connected to one and only one member of the second set.

2 (a) (i)

x	−3	−2	−1	0	1	2	3
y	19	12	7	4	3	4	7

(ii) (−3, 19), (−2, 12), (−1, 7), (0, 4), (1, 3), (2, 4), (3, 7)

(b) (i)

x	−2	−1	0	1	2	3
$h(x)$	0.6	0.25	−0.333…	−1.5	−5	*undefined*

(ii) (−2, 0.6), (−1, 0.25), (0, 0.333…), (1, −1.5), (2, −5), undefined at $x = 3$

3 (a) (i) (−3, 4), (−1, 2), (2, −1)

(ii) not a function, not all members of first set connected

(b) (i) (15, 5), (15, 3), (12, 4), (12, 3), (12, 2), (9, 3), (6, 3), (6, 2), (3, 3)

(ii) not a function, some of the members of the first set are connected to more than one member of the second set

(c) (i) (−3, 7), (−2, 2), (−1, −1), (0, −2), (1, −1), (2, 2), (3, 7)

(ii) function, each member of the first set is connected to one and only one member of the second set

(iii) many-to-one

(d) (i) (2, 10), (2, 12), (2, 14), (2, 16), (2, 18), (2, 20), (3, 12), (3, 18), (5, 10), (5, 20)

(ii) not a function, not all members of first set connected

Practice exercise P11.2 (p. 145)

1 (a) (i) 1 (ii) 9 (iii) $-(w-2)^2$

(b) (i) 2 (ii) 1, 4 (iii) 1.6, −5.6

2 (a)

x	−3	−2	−1	0	1	2	3
$g: x$	4	*undefined*	−3	$-1\frac{1}{4}$	$-\frac{2}{3}$	$-\frac{3}{8}$	$-\frac{1}{5}$

(b) $(-1, -3)$, $(0, -1\frac{1}{4})$, $(1, -\frac{2}{3})$, $(2, -\frac{3}{8})$, $(3, -\frac{1}{5})$

(c)

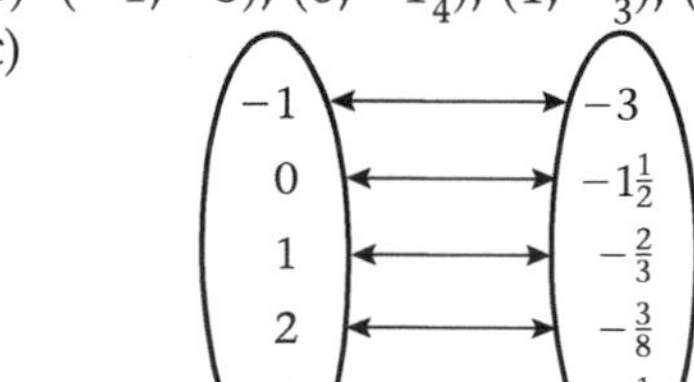

3 (a) (1, m), (2, n), (3, m), (4, n)

(b)

N: 1, 2, 3, 4 → L: m, n

(c) yes, each member of the first set is connected to one and only one member of the second set

Practice exercise P11.3 (p. 146)

1 (a) (i) yes, there is only one town in the domain for each country in the range

(ii) f^{-1} is the function 'is the country of which … is the capital'

(b) (i) yes, g is a one-to-one function

(ii) g^{-1} is the function 'is 5 less than'

(c) (i) yes if there is only one street in the domain for each town in the range; no if there is more than one street in the domain for each town in the range

(ii) if h^{-1} exists then h^{-1} is the function 'is the town in which … is a street

2 (a)

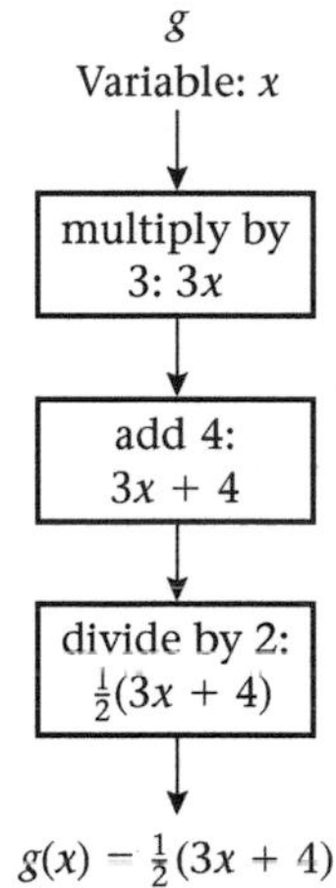

(b)

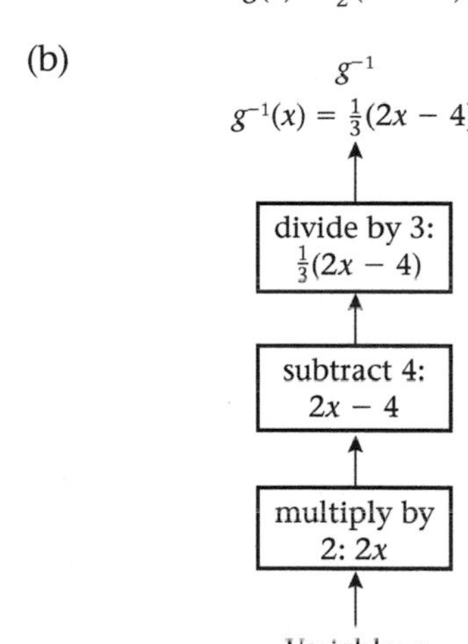

3 (a) $f^{-1}(x) = \dfrac{3x + 5}{2 - x}$

(b) $x = -3$

(c) $x = 8$

4 (a) (i)

x	−2	−1	0	1	2	3	4
$f(x)$	9	7	5	3	1	−1	−3

(ii) $f^{-1}(x) = \dfrac{5 - x}{2}$

(iii) none

x	−2	−1	0	1	2	3	4
$f(x)$	8	1	−2	−1	4	13	26

(ii) $f^{-1}(x)$ does not exist

(iii) none

x	−2	−1	0	1	2	3	4
$f(x)$	1	1	−1	−5	−11	−19	−29

(ii) $f^{-1}(x)$ does not exist

(iii) none

x	−2	−1	0	1	2	3	4
$f(x)$	−4	3	$\frac{2}{3}$	$\frac{1}{5}$	0	$-\frac{1}{9}$	$-\frac{2}{11}$

(ii) $f^{-1}(x) = \dfrac{2 - 3x}{2x + 1}$

(iii) $x = -1.5$

Practice exercise P11.4 (p. 146)

1 (a) (i) $gf(x) = 4x + 1$
(ii) $fg(x) = 4x + 4$
(b) (i) $gf(x) = x^2 - 3x + 10$
(ii) $fg(x) = x^2 - 2$

2 (a) $f(3) = 8$, $ff(3) = 13$, $ff(x) = x + 10$
(b) $f^{-1}(x) = x - 5$; $f^{-1}(3) = -2$;
$f^{-1}f^{-1}(3) = -5$
(c) $f^{-1}f^{-1}(x) = x - 10$

3 (a) $f(x) = x^2 = w$
$g(w) = w + 5 = x^2 + 5$
so $gf(x) = x^2 + 5$
(b) $fg(x) = x^2 + 10x + 25$;
$gf(2) = 9$; $fg(2) = 49$

4 (a) $f^{-1} : x \to \dfrac{2 - x}{3}$;
$g^{-1} : x \to \dfrac{x}{2}$
(b) (i) $4 - 6x$ (ii) $\dfrac{4 - x}{6}$
(iii) $\dfrac{4 - x}{6}$

5 (a) $\dfrac{x - 5}{2}$
(b) $f(7) = 9$; $f^{-1}(7) = 1$; $f^{-1}f(7) = 2$; $ff^{-1}(7) = -3$; $f^{-1}f^{-1}(7) = -2$
(c) $f^{-1}f(x) = x - 5$; $ff^{-1}(x) = x - 10$; the composite of the function and its inverse is not commutative

6 (a) $x = \pm 3$
(b) $x = -2, 1$

Exercise 12a (p. 149)

1. (a) 0.891 (b) 0.656 (c) −0.695
   (d) −0.5 (e) −0.530 (f) −0.643
2. 44°
3. 136°, −224°, −316°
4. 27°
5. 42°, 138°
6. 242.6°, 297.4°, −62.6°, −117.4°
7. (a) tan θ (b) −tan θ
8.

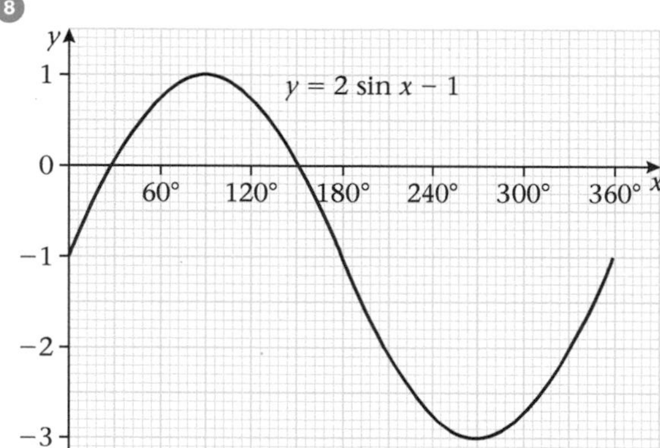

9. (a) 48.6°, 131.4° (b) 11.5°, 168.5°
10. $y = -2$; 210°, 330°
11.

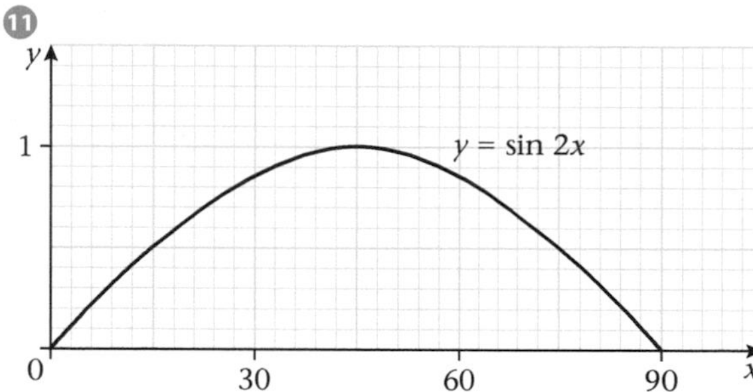

Exercise 12b (p. 152)

1. (a) 0.454 (b) −0.755 (c) −0.719
   (d) 0.866 (e) 0.848 (f) −0.707
2. 66.4°
3. −315°, −45°, 45°, 315°
4. 25°
5. 48°
6. 151.6°, 208.4°
7. (a) −tan θ (b) tan θ
8.

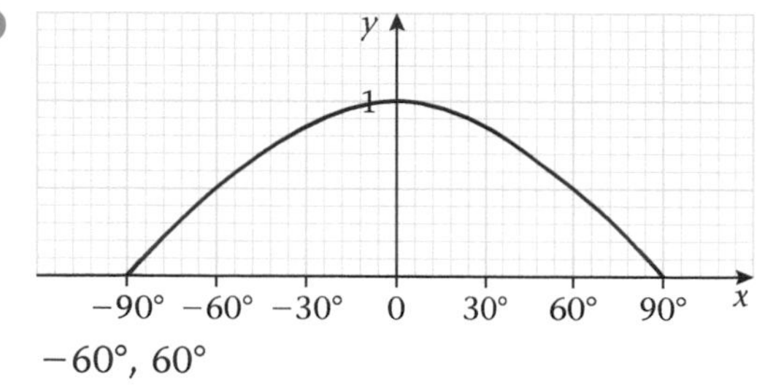

   −60°, 60°

9. $y = 2 \cos x + 1$

10. (a) 104° (b) 143°
11. $y = -1$; 180°

Exercise 12c (p. 153)

1. (a) 1.327 (b) −0.869 (c) 3.487
   (d) −0.900 (e) −2.050 (f) 0.900
2. (a) 60°, 240° (b) 145°, 325°
3. 35°
4. 48°
5. −150°, 30°
6. −41.3°, −221.3°, 138.7°, 318.7°
7. (a) −1 (b) 1
8.

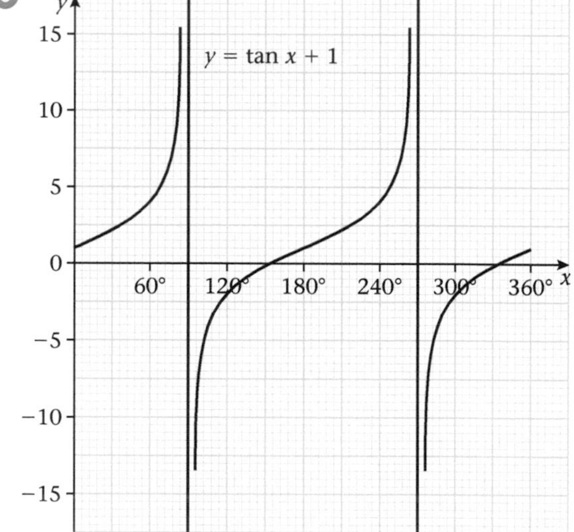

9. (a) 154°, 334° (b) 122°, 302°
10. $y = -2$; 108°, 288°

Practice exercise P12.1 (p. 154)

1. (a) 30°, 150° (b) 60°, 120° (c) 67°, 113°
   (d) 52°, 128° (e) 48°, 132° (f) 21°, 159°
2. (a) 63°, 297° (b) 45°, 315°
   (c) 120°, 240° (d) 84°, 276°
3. 24°, 204°
4. 23°
5. 51°
6. 73°
7. 45°, 225°
8. 53°, 127°
9. 127°, 223°
10. 248°

Practice exercise P12.2 (p. 154)

1. −1
2. −1
3. −1

Practice exercise P12.3 (p. 154)

1.

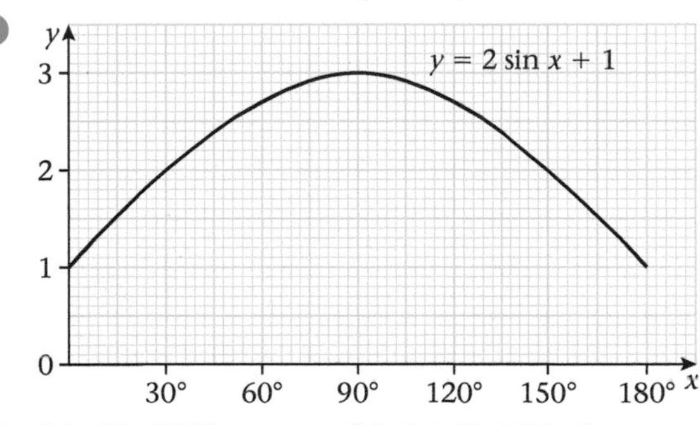

2. (a) 0°, 180° (b) 14.5°, 165.5°
3. $y = 3$; $x = 90°$
4.

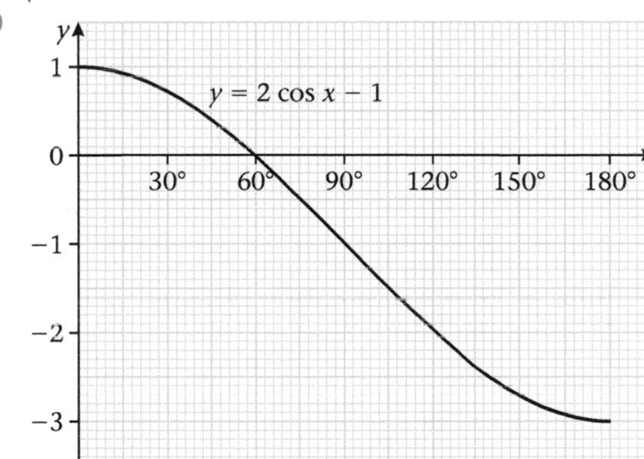

5. (a) 90° (b) 36.9°
6. $y = -2$; $x = 120°$
7.

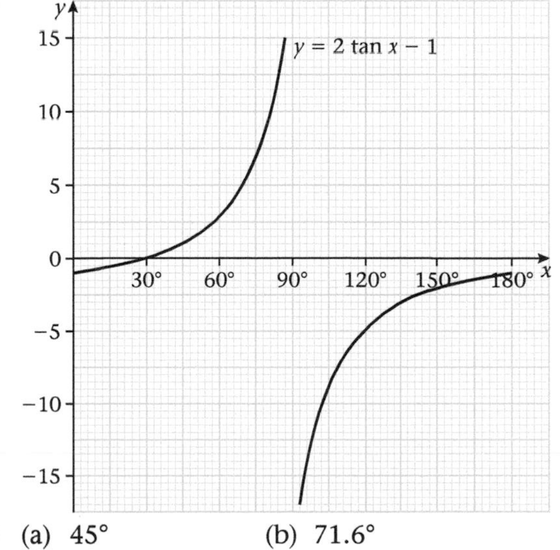

8. (a) 45° (b) 71.6°
9. $y = -1$; $x = 0°$ or 180°

Revision exercise 4 (p. 156)

1. (a) $\dfrac{x + y}{x - y}$ (b) $\dfrac{1}{x + 3}$
2. 1 552 cm³
3. $k = 15$, $V = 3\,240$
4. length = 12 cm, volume = 576 cm³
5. (a) y^3 (b) 1
6. height = 3.9 m, distance = 3.2 m
7. 2.3 cm
8. (b) $13\frac{1}{3}$ cm

Revision test 4 (p. 156)

1. D
2. A
3. B
4. B
5. D
6. 864 cm³
7. 10 453 cm³
8. $\frac{9}{2}$, 4
9. (a) 4, −5 (b) $-\frac{5}{2}$
10. (a) $\dfrac{3(b - 5)}{2(b + 1)(b - 3)}$

    (b) $b = 5$

Revision exercise 5 (p. 157)

1. $x = 39°$, $y = 47°$, $z = 94°$
2. $a = 7.6$ cm, $\widehat{A} = 102.4°$, $\widehat{B} = 42.1°$
3. (b) 14 cm² (d) 4.0 cm
4. (a) 36.9° (b) 1.05 cm
5. (a) 6 cm 5 mm (b) 41 cm²
7.

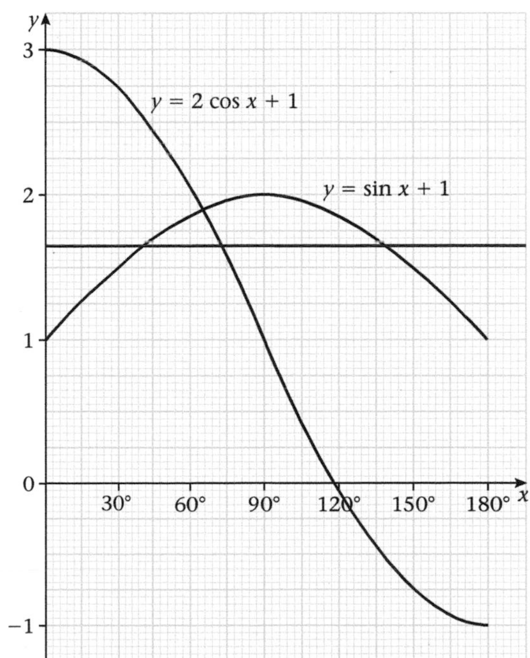

8 (a) 63.5° (b) 40.5°, 139.5°

9 (a) 2.4 cm (b) 40.6°

10 (a) 73.7° (b) 9 mm
(c) 48 cm^2 (d) 16.3 cm^2

Revision test 5 (p. 158)

1 D

2 C

3 A

4 B

5 B

6 (a) 13.0 cm^2
(b) 60.2 cm^2

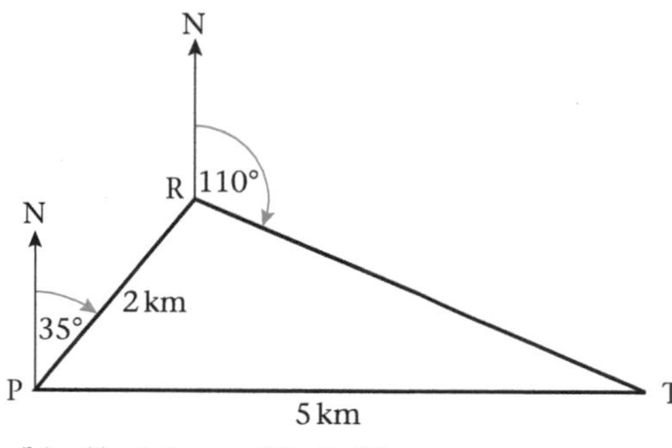

(b) (i) 4.1 m (ii) 8.7°

8 (b) 57.8°

9

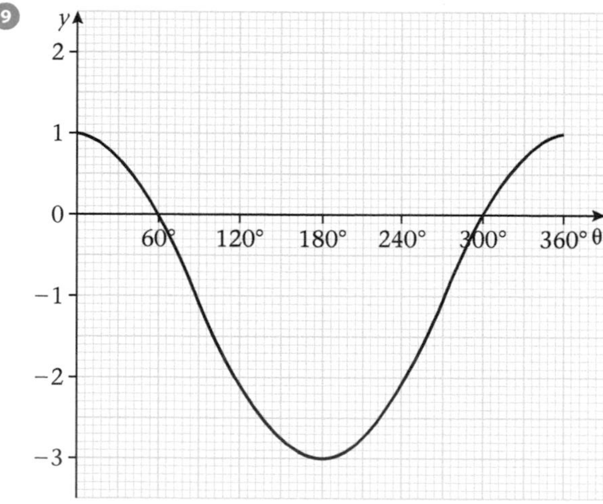

10 41°, 319°

Revision exercise 6 (p. 159)

1 (a) 25 cm (b) 5 km
(c) 2 hour (d) $1\frac{1}{2}$ hour
(e) $1\frac{1}{2}$ hour (f) 6 hours
(g) 10 km/h

2 21 km

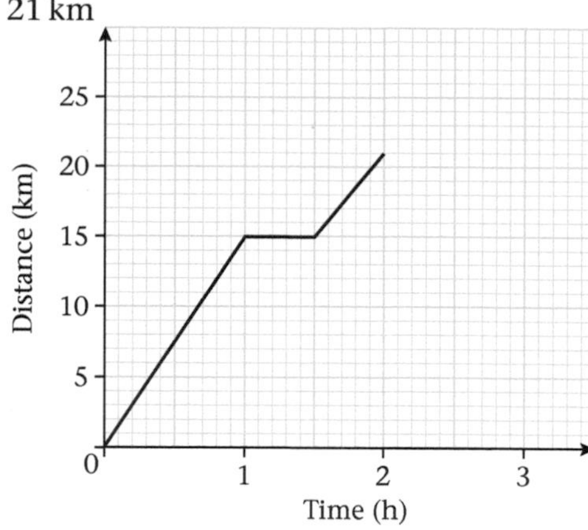

3

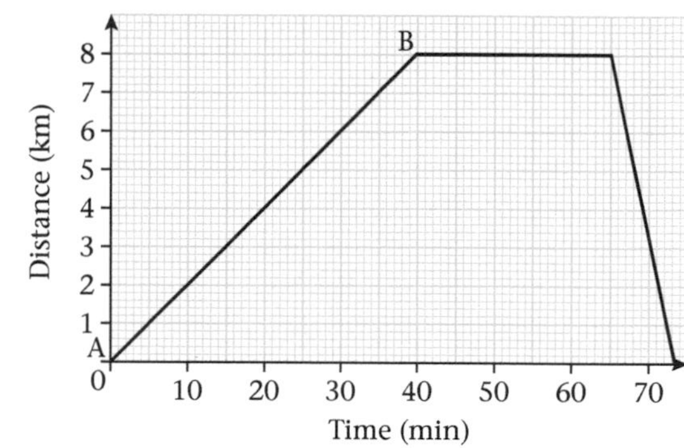

4 (a) 12 km/h (b) 48 km/h (c) 12.8 km/h

5 (a) 3.5 (b) −2.25 (c) $y = -1.5$

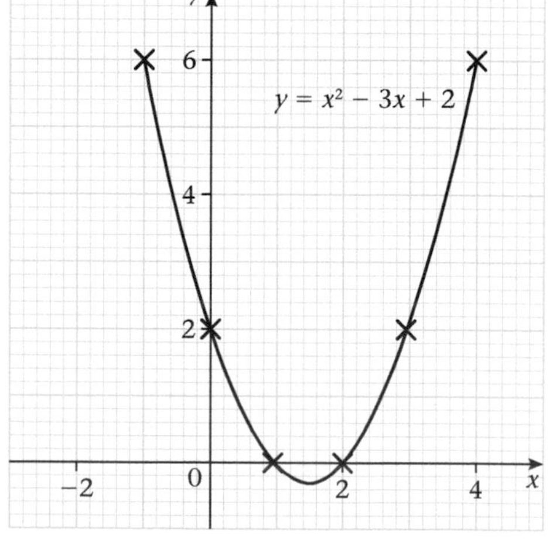

(a) −0.25 (b) $y = 1.5$ (c) 3

7 uniform acceleration from rest to 40 km/h in 15 min;
constant speed of 40 km/h for 45 min;
uniform acceleration from 40 km/h to V km/h in 10 min;

uniform deceleration from V km/h to rest in 12 min

8 (a) 160 km/h^2 (b) 54 km/h (c) $48\frac{7}{30}$ km

9 (i) $\frac{8x+1}{6}$ (ii) $\frac{4x}{3}$
(iii) $\frac{3x}{4}$ (iv) $\frac{6x-1}{8}$

10 (i) $x = 3$ (ii) $\frac{3x+2}{x+3}$
$x = \frac{8}{5}$

Revision test 6 (p. 160)

1 C 2 C 3 A 4 D 5 D

6 (a) $2x + 7y = 0$
(b) $2x + 7y = 56$
(c) (28, 0)

7 (a)

t	0	1	2	3	4	5
v	0	5	16	33	56	85

(b)

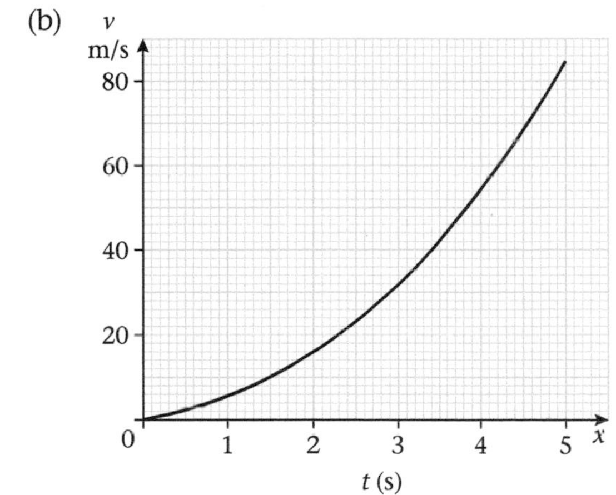

(c) (i) 14 m/s^2 (ii) 80 m (iii) 44 m

8 (a) 23
(b) (i) $x = \frac{5}{2}, x = 3$
(ii) $\frac{5-3x}{3-x}$, 4

General revision test B (p. 161)

1 C 2 D 3 B 4 D 5 A
6 B 7 A 8 C 9 C 10 A

11 (a) $\frac{1}{a+2}$ (b) -2

12 (a) 24.5 m (b) 24.0°
(c) 5.7 m (d) 69.7 m^2

13 1 : 8

14

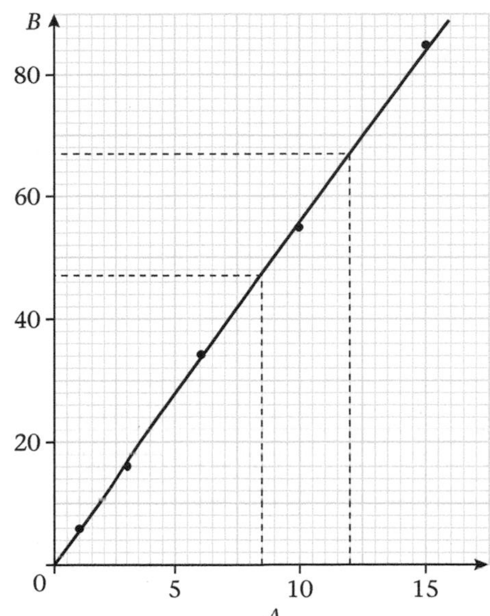

$B = 5.6\,A$, that is, the graph is a straight line.
(a) $A = 8.5$ (b) $B = 67$

15 (a)

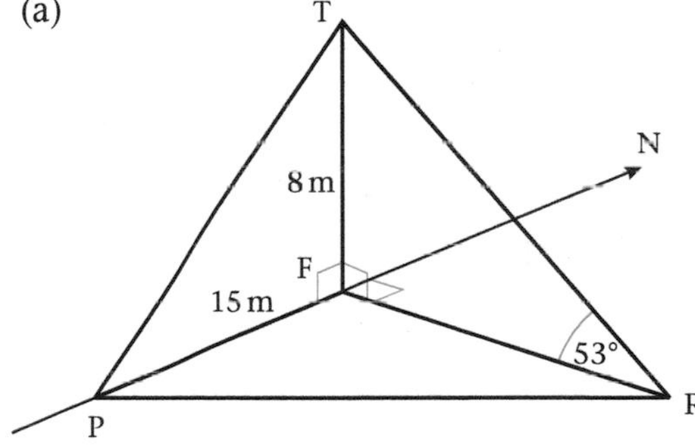

(b) (i) 28° (ii) 6.0 m (iii) 16.2 m

16 (a)

x	22.5	45	60	75	90
$f(x)$	1.12	2	1.60	0.5	−1

(b)

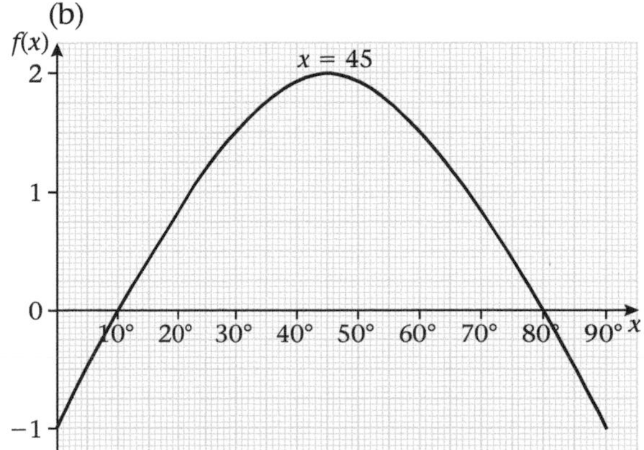

(c) $x = 45°$

(d) (i) $f_{max} = 2$

(ii) 10°, 80°

(iii) $x < 10$, $x > 80$

17 (a)

(b) (i) $v = 12\frac{1}{4}$ m/s, $t = \frac{7}{2}$ s

(ii) 3 m/s², −4 m/s²

(iii) 57 m

18

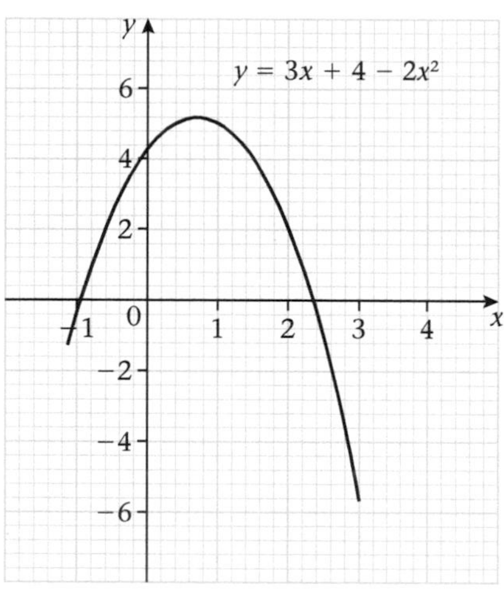

(i) 25

(ii) 5

 **19** (a)

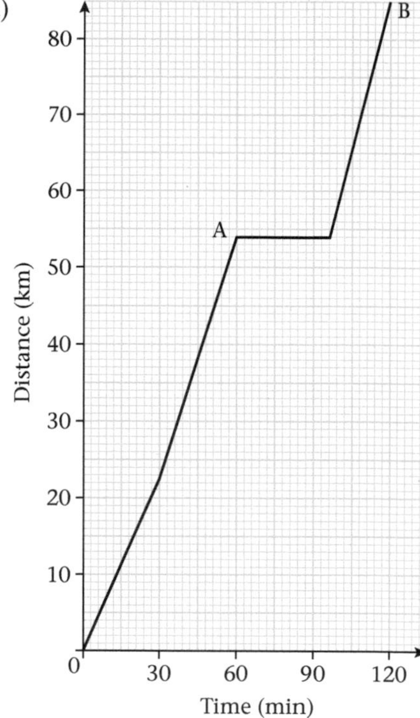

(b) 32 km

(c) 58 km/h

Exercise 13a (p. 166)

1 (a) 28%

(b) 20 students

(c) 6%

2 (a)

Class interval	Frequency	Cumulative frequency
1–10	2	2
11–20	7	9
21–30	9	18
31–40	11	29
41–50	13	42
51–60	16	58
61–70	16	74
71–80	15	89
81–90	8	97
91–100	3	100

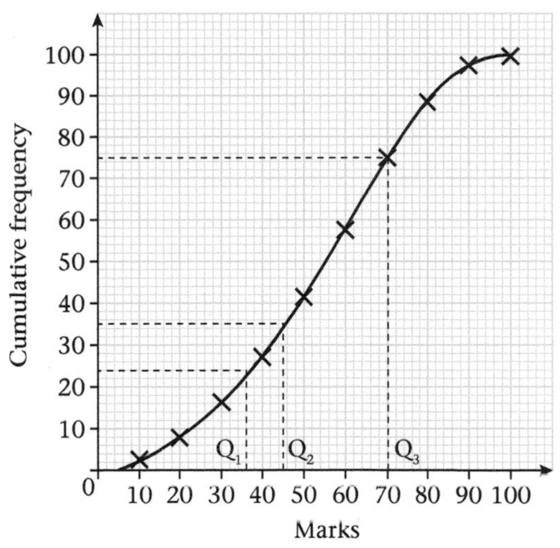

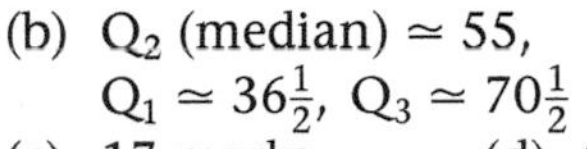
(b) Q_2 (median) $\simeq$ 55,
$Q_1 \simeq 36\frac{1}{2}$, $Q_3 \simeq 70\frac{1}{2}$

(c) 17 marks (d) 65%

3 (a)

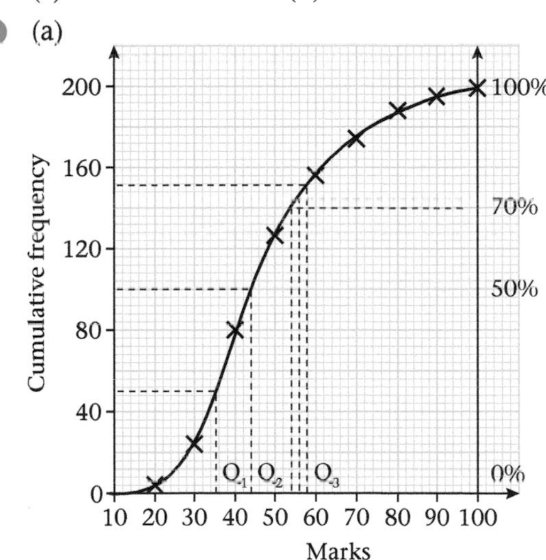

(b) 44,11 (c) 28% (d) 54

4 (a)

Class interval	Frequency	Cumulative frequency
551–560	2	2
561–570	2	4
571–580	4	8
581–590	7	15
591–600	15	30
601–610	8	38
611–620	5	43
621–630	5	48
631–640	2	50

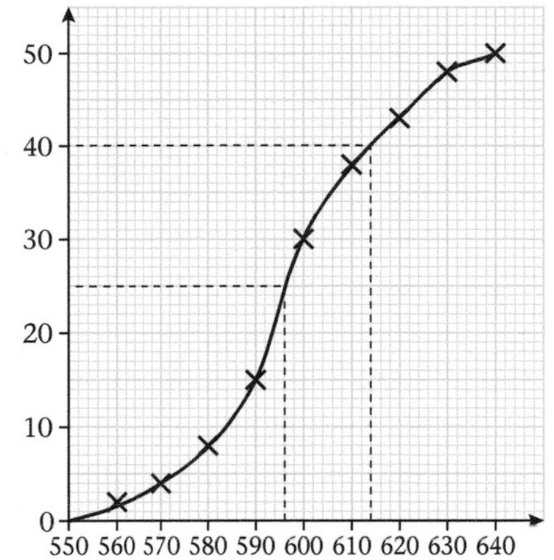

(b) median $\simeq$ 596, 80th percentile $\simeq$ 614

5 (a)

Class interval	Frequency	Cumulative frequency
2–4	3	3
5–7	9	12
8–10	32	44
11–13	46	90
14–16	6	96
17–19	4	100

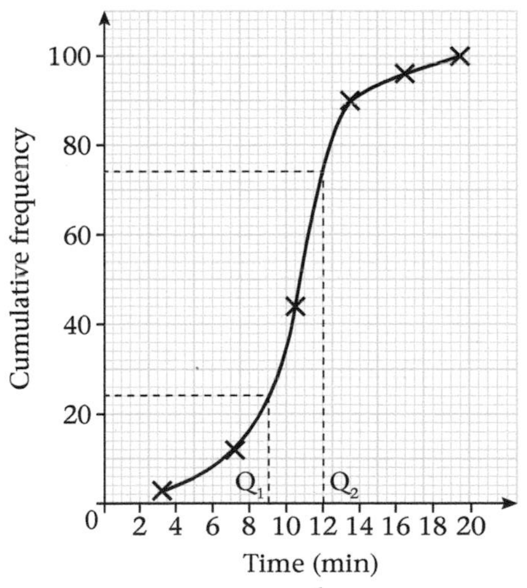

(b) 26 (c) $2\frac{1}{2}$ minutes

Exercise 13b (p. 168)

1 $\frac{7}{25}$ **2** $\frac{19}{40}$

3 $\frac{1}{5000}$ **4** $\frac{2}{25}$

5 (a) $\frac{1}{2}$ (b) $\frac{3}{8}$ (c) $\frac{9}{16}$
(d) $\frac{5}{16}$ (e) $\frac{1}{4}$ (f) $\frac{1}{8}$

6 (a) $\frac{2}{9}$ (b) $\frac{1}{3}$ (c) $\frac{4}{9}$
(d) 0 (e) $\frac{7}{9}$ (f) $\frac{2}{3}$

7 (a) $\frac{1}{6}$ (b) $\frac{1}{6}$ (c) 0
(d) $\frac{1}{2}$ (e) $\frac{1}{3}$ (f) $\frac{2}{3}$

8 (a) $\frac{1}{26}$ (b) $\frac{12}{13}$ (c) $\frac{2}{13}$ (d) $\frac{3}{13}$

9 (a) $\frac{5}{27}$ (b) $\frac{11}{25}$

10 (a) $\frac{1}{52}$ (b) $\frac{1}{52}$ (c) $\frac{1}{13}$ (d) $\frac{1}{26}$
(e) $\frac{1}{4}$ (f) $\frac{3}{13}$ (g) $\frac{1}{2}$ (h) 0

Exercise 13c (p. 170)

1 (a) $\frac{11}{36}$ (b) $\frac{1}{6}$

2 (a) $\frac{1}{25}$ (b) $\frac{1}{5}$ (c) $\frac{4}{25}$ (d) $\frac{3}{5}$

3 $\frac{7}{18}$

4 (a) $\frac{16}{25}$ (b) $\frac{8}{25}$

5 (a) $\frac{1}{2}$ (b) (i) 0 (ii) $\frac{19}{112}$

6 (a) $\frac{8}{27}$ (b) $\frac{19}{27}$ (c) $\frac{12}{27}$

7 (a) $\frac{3}{8}$ (b) $\frac{7}{8}$

8 (a) $\frac{1}{2}$ (b) $\frac{7}{39}$

9 (a) $\frac{3}{11}$ (b) $\frac{4}{11}$

10 (a) $\frac{2}{13}$ (b) $\frac{1}{102}$

Practice exercise P13.1 (p. 171)

1 (a) 13 (b) 17 (c) 9 (d) 8 (e) 4 (f) 14

2 (a)

Marks	Frequency	Cumulative frequency
11–20	2	2
21–30	4	6
31–40	9	15
41–50	19	34
51–60	28	62
61–70	13	75
71–80	2	80

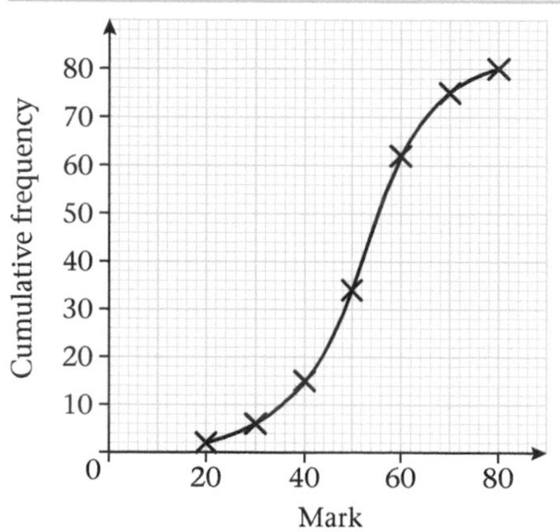

(b) median = 52,
semi-interquartile range = 8

(c) 57

3 (a)

Height (cm)	Frequency	Cumulative frequency
0–10	11	11
11–20	67	78
21–30	138	216
31–40	167	383
41–50	129	512
51–60	70	582
61–70	14	596
71–80	4	600

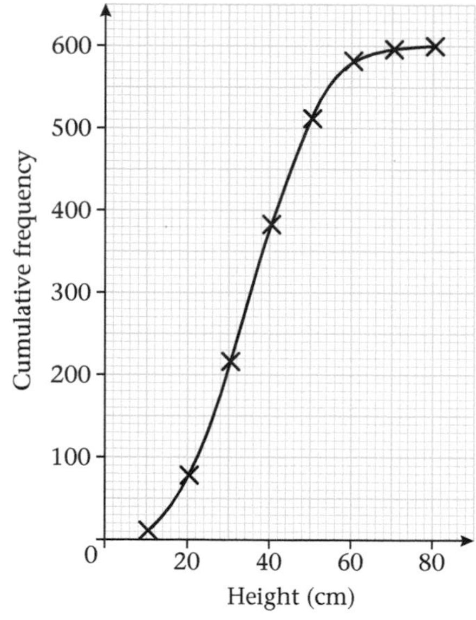

(b) median = 36 cm, semi-interquartile range = 9.5 cm

(c) 140

Practice exercise P13.2 (p. 172)

1 (a) $\frac{1}{9}$ (b) $\frac{4}{9}$ (c) $\frac{16}{75}$ (d) $\frac{12}{25}$

2 (a) $\frac{64}{16\,575}$ (b) $\frac{11}{850}$ (c) $\frac{1}{5\,525}$ (d) $\frac{22}{425}$

3 (a) 0.028 9 (b) 0.971 1 (c) 0.282 2

4 (a) $\frac{8}{125}$ (b) $\frac{54}{125}$ (c) $\frac{71}{125}$

5 (a) $\frac{9}{20}$ or 0.45 (b) $\frac{17}{100}$ or 0.17 (c) $\frac{19}{50}$ or 0.38
(d) $\frac{289}{1\,000}$ or 0.0289 (e) $\frac{68}{2\,475}$

Exercise 14a (p. 174)

1

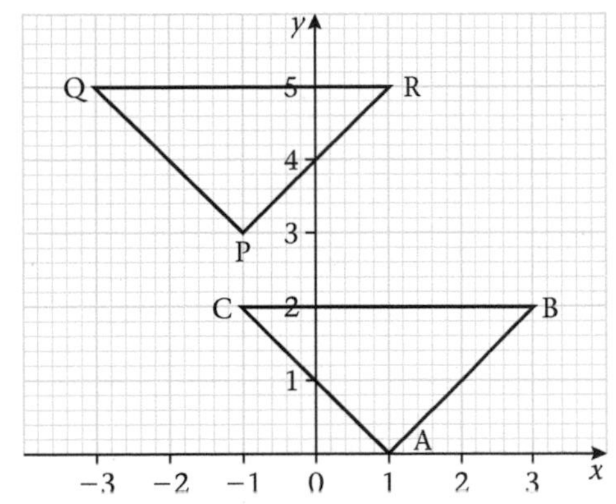

Glide axis is $x = 0$ (y-axis)

2

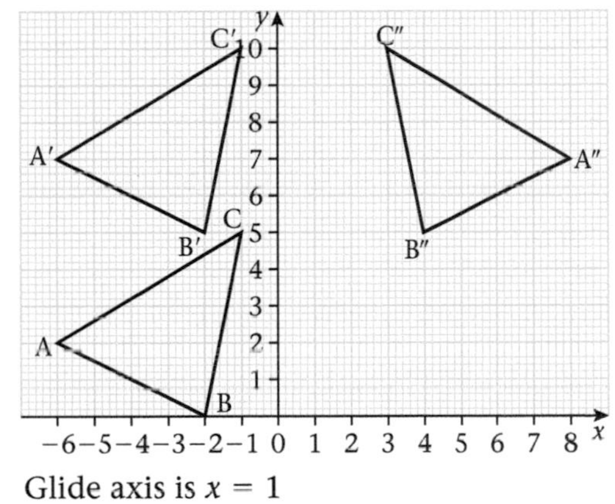

Glide axis is $x = 1$

Exercise 14b (p. 175)

1

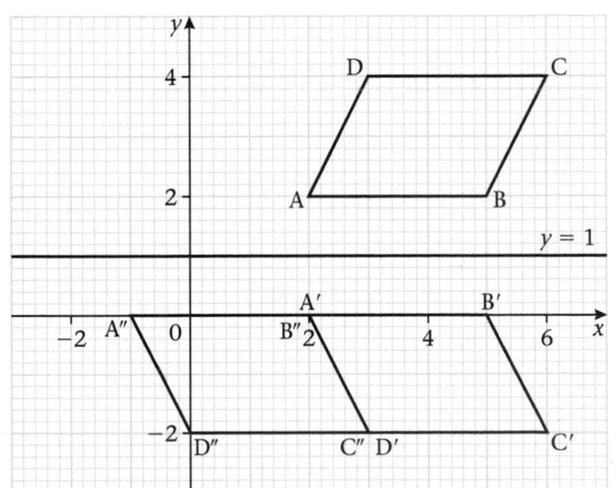

A″(−1, 0), B″(2, 0), C″(3, −2), D″(0, −2)

2

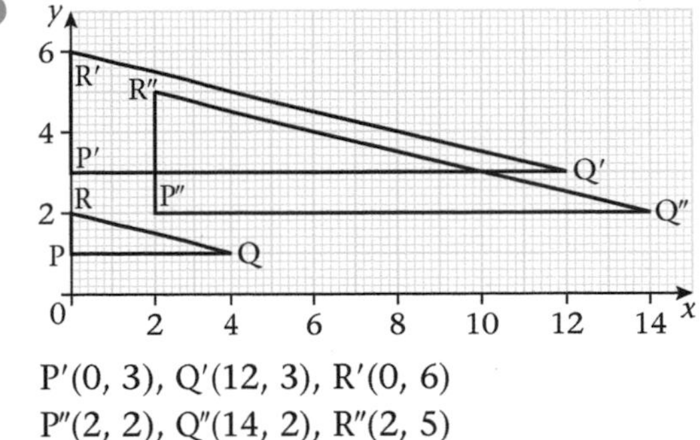

P′(0, 3), Q′(12, 3), R′(0, 6)
P″(2, 2), Q″(14, 2), R″(2, 5)

3

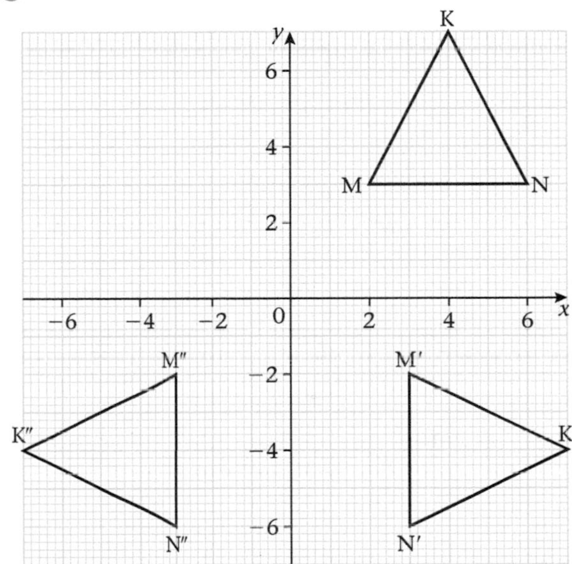

K′(7, −4), m′(3, −2), N′(3, −6)
K″(−7, −4), m″(−3, −2), N″(−3, −6)

4

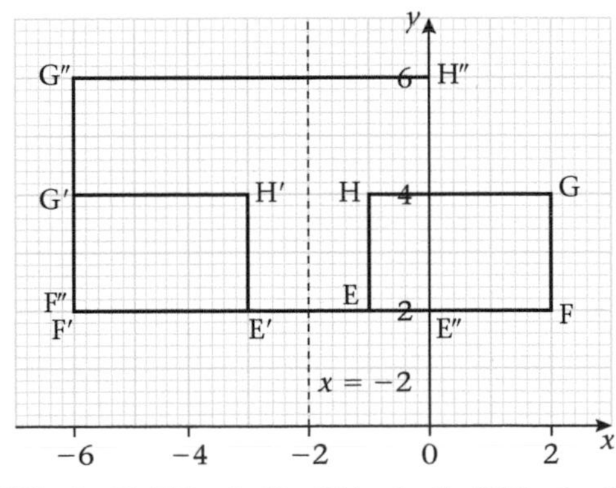

E′(−3, 2), F′(−6, 2), G′(−6, 4), H′(−3, 4)
E″(0, 2), F″(−6, 2), G″(−6, 6), H″(0, 6)

5 (a)

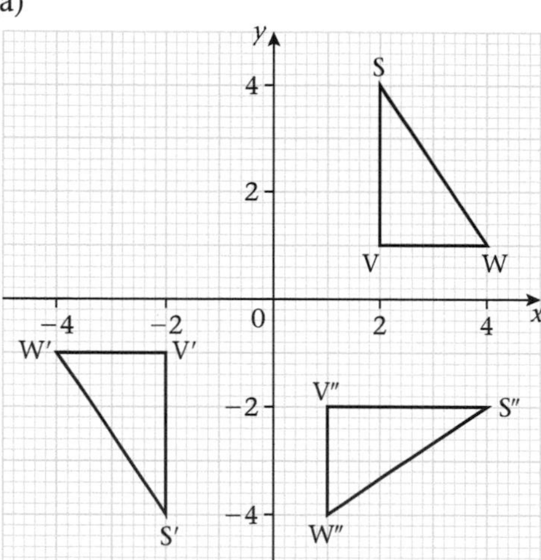

(b)

(c) $\triangle S''V''W''$ and $\triangle S_2V_2W_2$ coincide, that is, S″, V″ and W″ have the same coordinates as S_2, V_2 and W_2.

6 (a) and (b)(i)

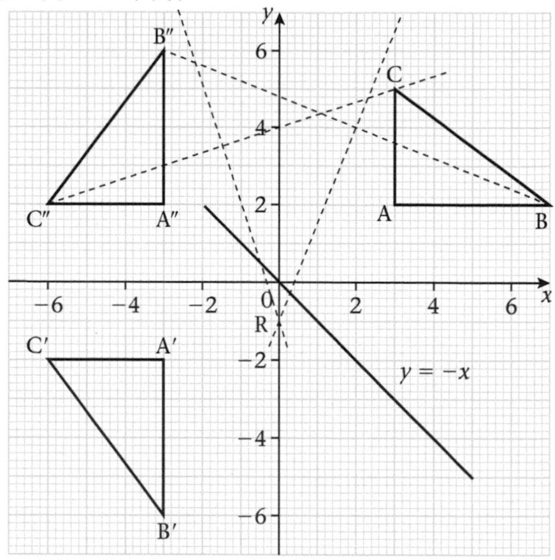

(ii) Centre of rotation R (0, −1)

(iii) Clockwise rotation of 90° about (0, −1).

7 (a) A′(2, 3)

(b) translation by the vector $\begin{pmatrix} -0.5 \\ 1.5 \end{pmatrix}$

(c) (i) $OB' = \begin{pmatrix} 4 \\ 1 \end{pmatrix}$

(ii)

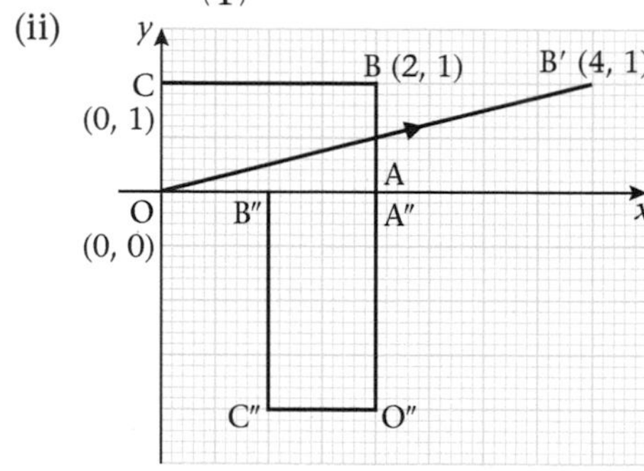

Exercise 14c (p. 179)

1

Transformation	Matrix
identity	$\begin{pmatrix} 1 & 0 \\ 0 & 1 \end{pmatrix}$
reflection in the x-axis	$\begin{pmatrix} 1 & 0 \\ 0 & -1 \end{pmatrix}$
reflection in the y-axis	$\begin{pmatrix} -1 & 0 \\ 0 & 1 \end{pmatrix}$
rotation of 180° about (0, 0)	$\begin{pmatrix} -1 & 0 \\ 0 & -1 \end{pmatrix}$

2 (a) $\begin{pmatrix} 0 & -1 \\ 1 & 0 \end{pmatrix}$ (b) $\begin{pmatrix} 0 & 1 \\ -1 & 0 \end{pmatrix}$
3 (1, −1), (4, −2), (7, −3)
4 P(−1, 5)
5 (0, 0), (−1, 1), (−1, −1)
6 (0, 0), (−1, −1), (−1, 1)
7 (a) (2, −2), (1, 1), (−2, 2), (−1, −1)
(b) (−2, 2), (−1, −1), (2, −2), (1, 1)
8 (a) $\begin{pmatrix} 7 \\ -5 \end{pmatrix}$
(b) (i) 270°, (iii) $\begin{pmatrix} 0 & 1 \\ -1 & 0 \end{pmatrix}$ (c) $y = -1$

Exercise 14d (p. 181)

1 A′(−3, −3), B′(−6, −9), C′(−12, −6)
2 $\begin{pmatrix} 1\frac{1}{2} & 0 \\ 0 & 1\frac{1}{2} \end{pmatrix}$
3 (0, 0), $(4\frac{1}{2}, 0)$, (0, 3), $(4\frac{1}{2}, 3)$
4 An enlargement of scale factor 4 with the origin as centre; $\begin{pmatrix} 4 & 0 \\ 0 & 4 \end{pmatrix}$

Exercise 14e (p. 183)

1 X′(−5, −2), Y′(−6, −5), Z′(−1, −4)
2 X′(−4, 10), Y′(−10, 12), Z′(−8, 2)
3 (a) A′(15, 30) (b) A′(11, 14) (c) A′$(1, \frac{2}{3})$
4 (a) 90° (b) $\begin{pmatrix} 0 & -1 \\ 1 & 0 \end{pmatrix}$
(c) $\begin{pmatrix} -1 \\ -7 \end{pmatrix}$ (d) $x + y = -1$
5 (a) 2 (b) $\begin{pmatrix} 2 & 0 \\ 0 & 2 \end{pmatrix}$
(c) $\begin{pmatrix} -3 \\ -7 \end{pmatrix}$ (d) $\begin{pmatrix} -1\frac{1}{2} \\ -3\frac{1}{2} \end{pmatrix}$
(e) $h = 3, y = 7$
6 Enlargement, scale factor 4 with origin as centre followed by a translation of vector $\begin{pmatrix} -3 \\ 1 \end{pmatrix}$; (−7, 17)
7 (a) (b) (c) (d)

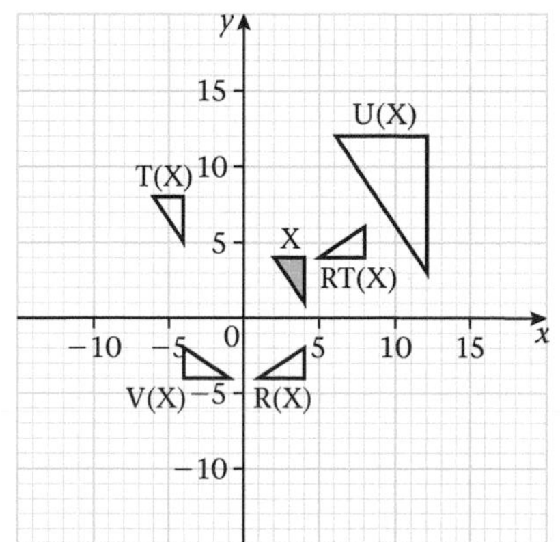

(b) U is an enlargement of scale factor 3 with the origin as centre.
(e) V is a reflection in the line $y = x$.

8 (a) (i) $\begin{pmatrix} -1 & 0 \\ 0 & -1 \end{pmatrix}$
(ii) A′(−1, −3), B(−2, −4)
(b) (i) $M = \begin{pmatrix} 1 & 0 \\ 0 & -1 \end{pmatrix}$, $T = \begin{pmatrix} 2 & 0 \\ 0 & 2 \end{pmatrix}$
(ii) $K = \begin{pmatrix} 2 & 0 \\ 0 & -2 \end{pmatrix}$
(iii) $K' = -\frac{1}{4}\begin{pmatrix} -2 & 0 \\ 0 & -2 \end{pmatrix}$
(iv) $(x, y) = (\frac{3}{2}, 1)$

Practice exercise P14.1 (p. 184)

1

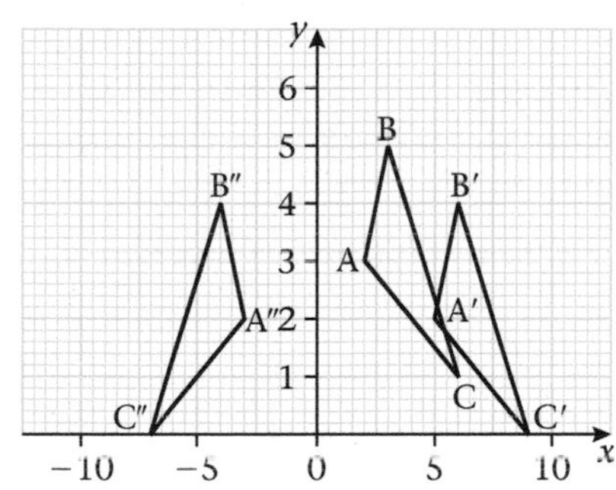

2

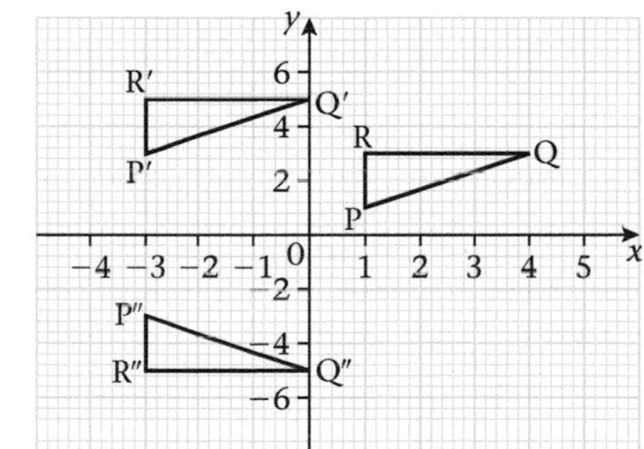

3 (a)

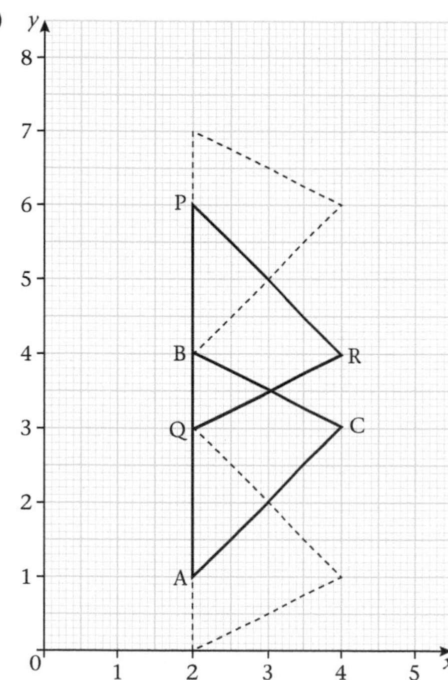

(b) $y = 2$

4

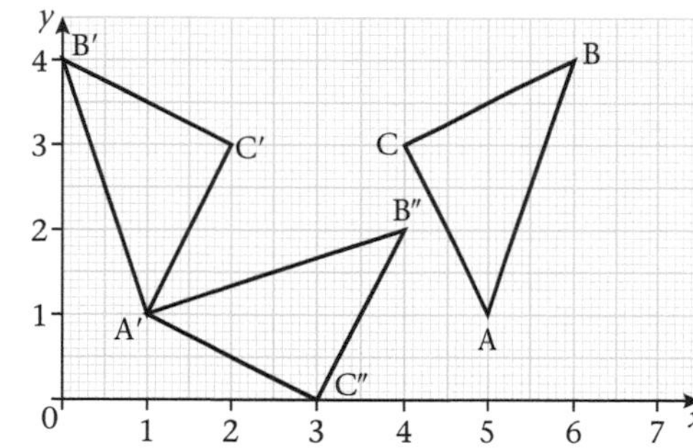

5

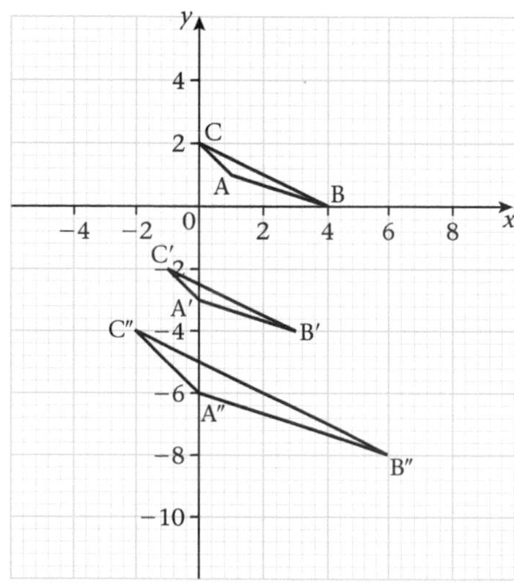

A′(0, −3), B′(3, −4), C′(−1, −2)
A″(0, −6), B″(6, −8), C″(−2, −4)

Practice exercise P14.2 (p. 185)

1 A′(1, 0), B′(−5, 5), C′(−7, −2)

2 (a) Rotation of 90° clockwise about the origin
(b) A′(1, −2), B′(1, −3), C′(3, −2)

3 A′(6, 0), B′(0, 6); $\begin{pmatrix} 3 & 0 \\ 0 & 3 \end{pmatrix}$

4 (a) Reflection in the x-axis
(b) A′(2, −5), B′(−1, −3)
(c) $\begin{pmatrix} 1 & 0 \\ 0 & -1 \end{pmatrix}$, (−5, 10)

5 (a) $\begin{pmatrix} \frac{17}{13} & -\frac{18}{13} \\ 0 & 1 \end{pmatrix}$
(b)

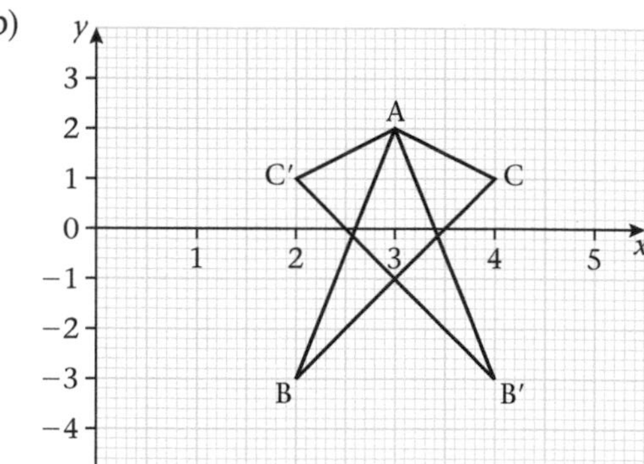

(c) Reflection in $x = 3$

Practice exercise P14.3 (p. 185)

1 (a) Enlargement, scale factor 3 with the origin as the centre, followed by a translation by vector $\begin{pmatrix} -2 \\ 1 \end{pmatrix}$.
(b) (4, −14)

2 (a) $\begin{pmatrix} x' \\ y' \end{pmatrix} = \left(\begin{pmatrix} x \\ y \end{pmatrix} + \begin{pmatrix} 2 \\ 4 \end{pmatrix}\right)\begin{pmatrix} -1 & 0 \\ 0 & 1 \end{pmatrix}$
(b) (−5, 2)

3 (a) $\begin{pmatrix} 0 & -1 \\ 1 & 0 \end{pmatrix}$
(b) $\begin{pmatrix} 1 & 0 \\ 0 & -1 \end{pmatrix}$
(c) (i) (3, 4)
(ii) (−3, −4)
(iii) (3, −4)

4 (a) $\begin{pmatrix} -1 & 0 \\ 0 & -1 \end{pmatrix}$

(b) A′(−6, −2), B′(2, 4)

(c) $\begin{pmatrix} \frac{1}{2} & 0 \\ 0 & \frac{1}{2} \end{pmatrix}$

(d) $\begin{pmatrix} -\frac{1}{2} & 0 \\ 0 & -\frac{1}{2} \end{pmatrix}$

(e) $\begin{pmatrix} -2 & 0 \\ 0 & -2 \end{pmatrix}$

(f) (10, −4)

5 P′(12, 3), Q′(−5, 2), R′(11, 4)

Exercise 15a (p. 187)

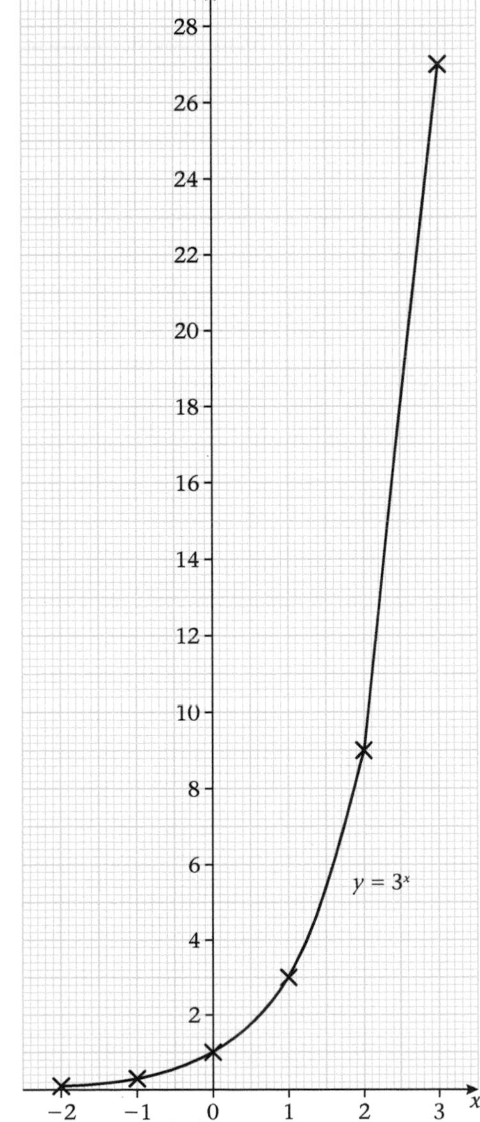

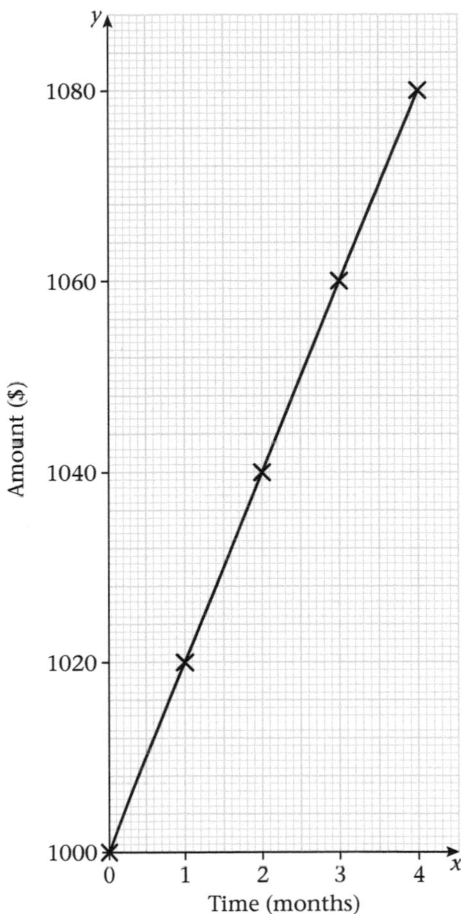

3 (a)

year	1985	1990	1995	2000	2005
pop'n	8 000	8 280	8 570	8 870	9 180

(b)

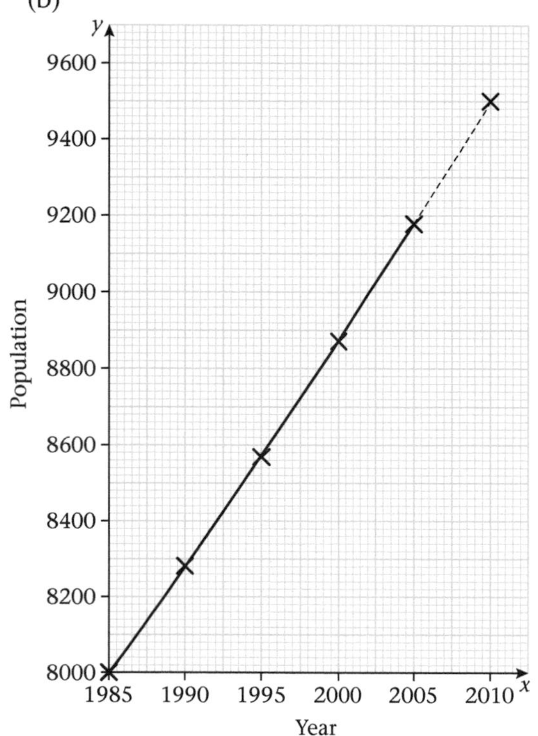

Fig. A95

(c) 9 500

4 (c)

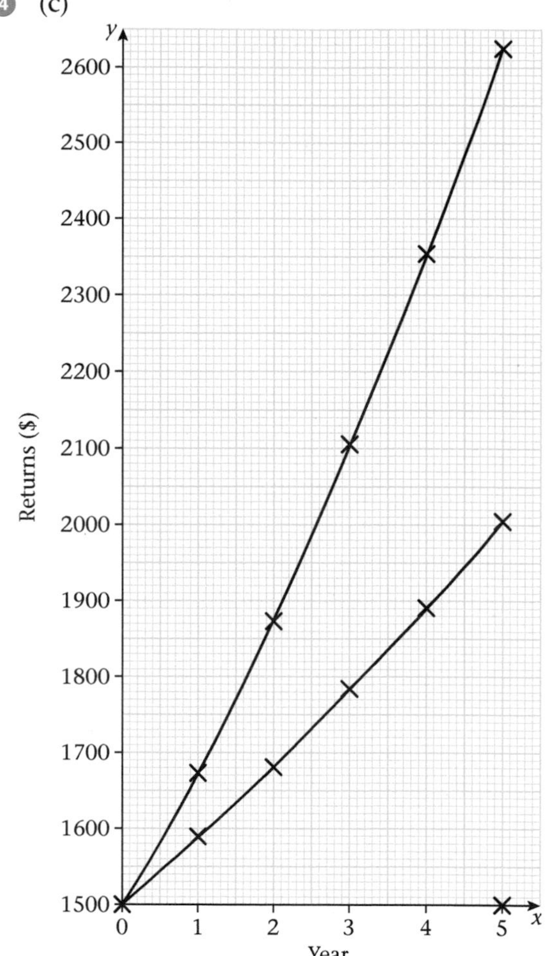

Exercise 15b (p. 190)

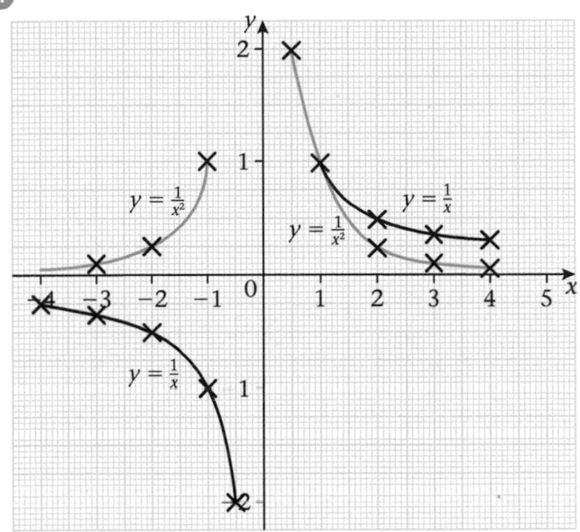

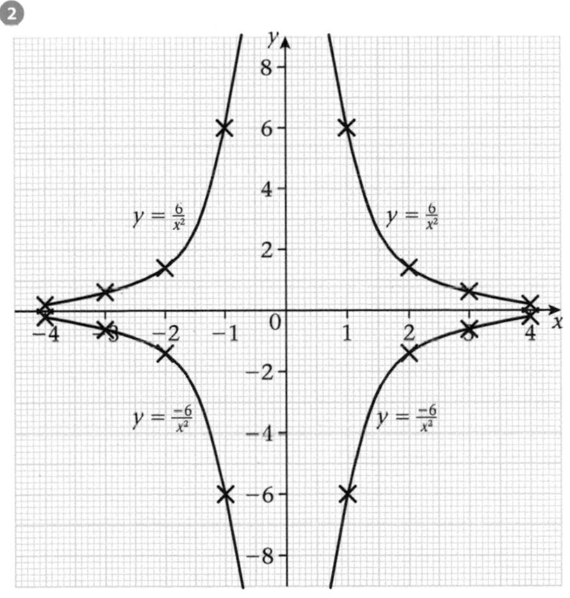

3 -1, 1.3; $3x^2 - x - 4 = 0$

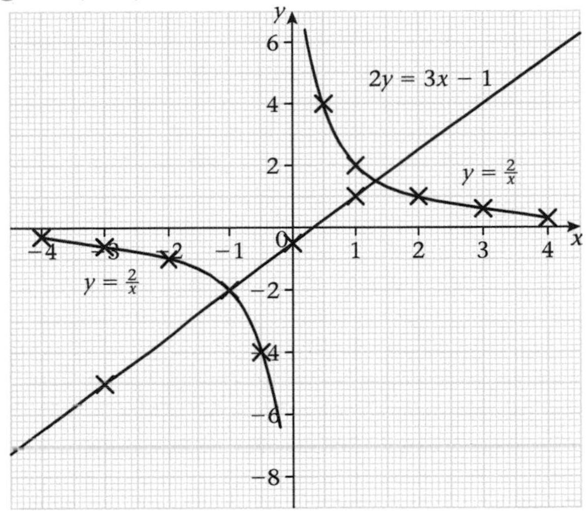

4 (a) $-\frac{1}{2}$, 1; $2x^2 - x - 1 = 0$

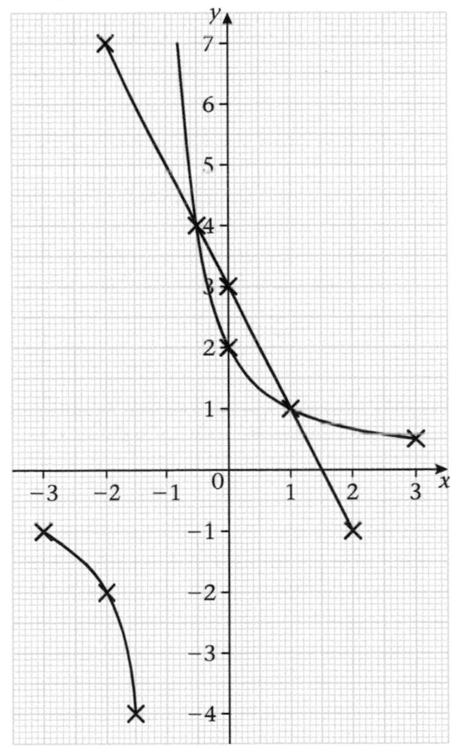

(b) -2.7, 1; $3x^2 + 5x - 8 = 0$

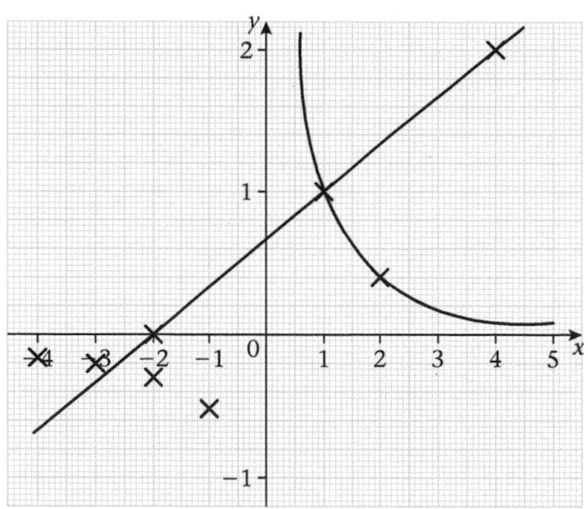

(c) 2.1; $x^3 - x - 5 = 0$

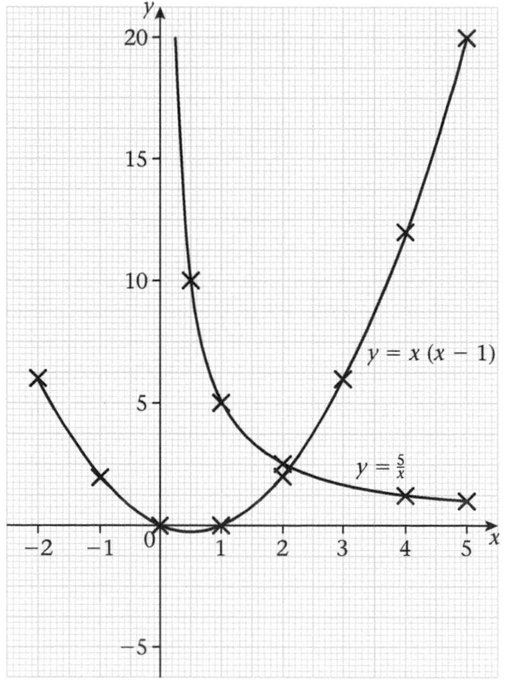

Exercise 15c (p. 191)

1

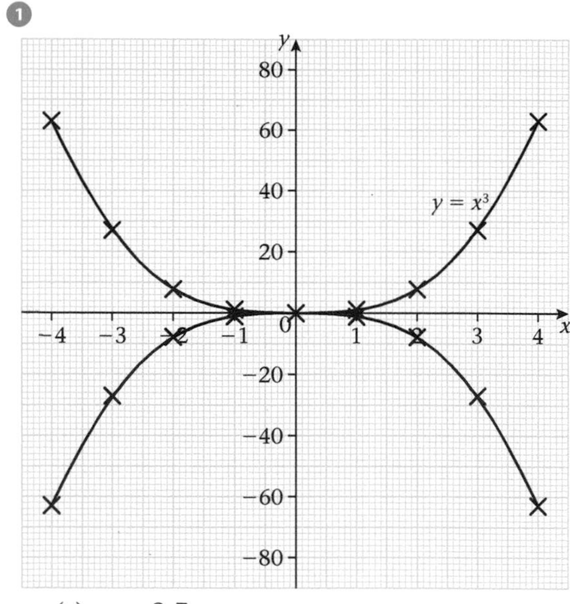

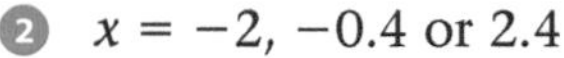
(c) $x \simeq 3.7$

2 $x = -2, -0.4$ or 2.4

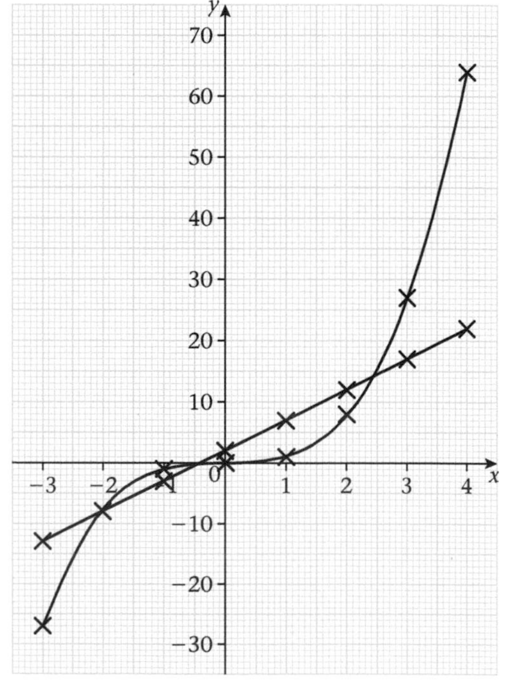

3 $x \simeq -2.5, 0.7, 1.8$

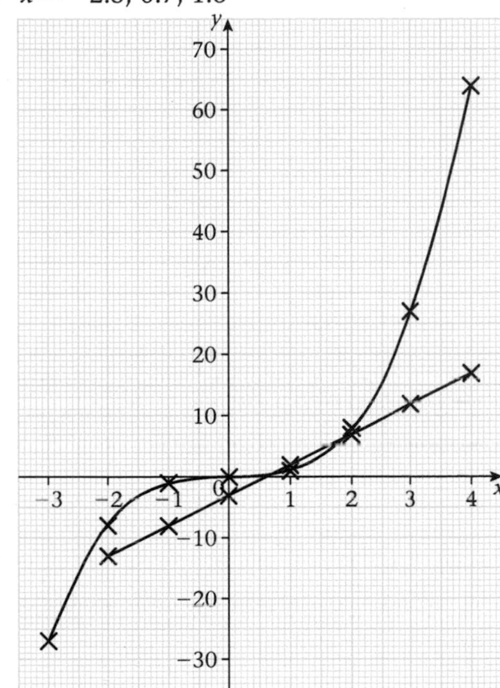

4 $x \simeq 1.4$

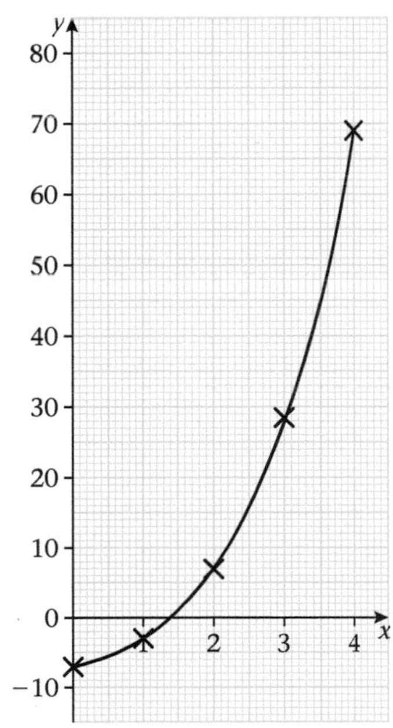

5 (a) $x = 0, 1$ or 3

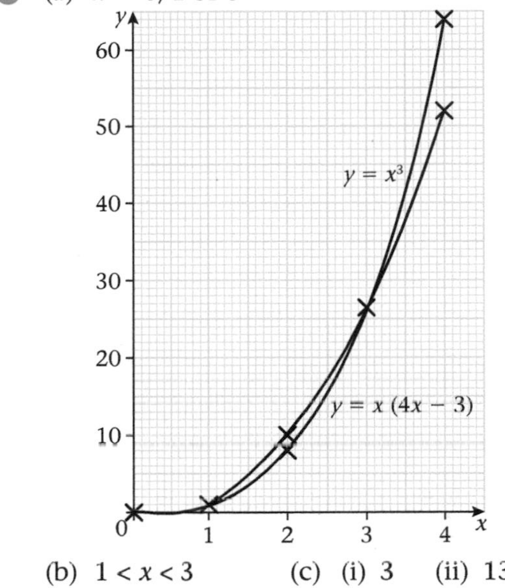

(b) $1 < x < 3$ (c) (i) 3 (ii) 13

Exercise 15d (p. 193)

1

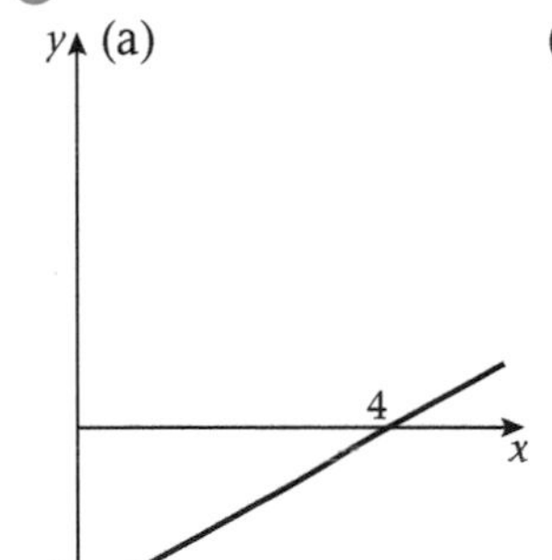

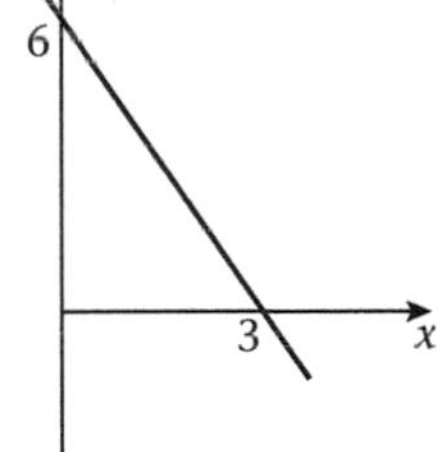

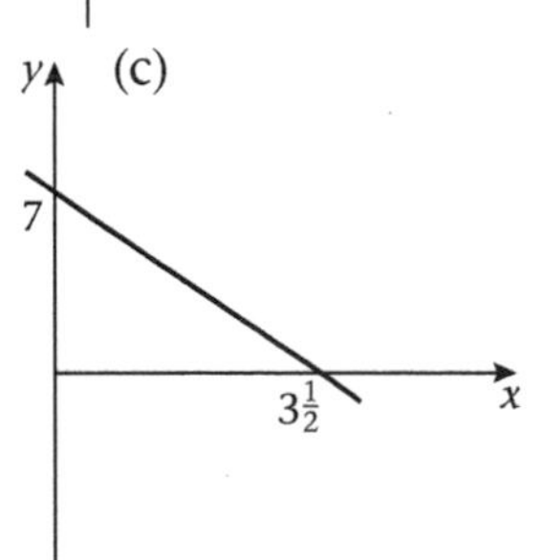

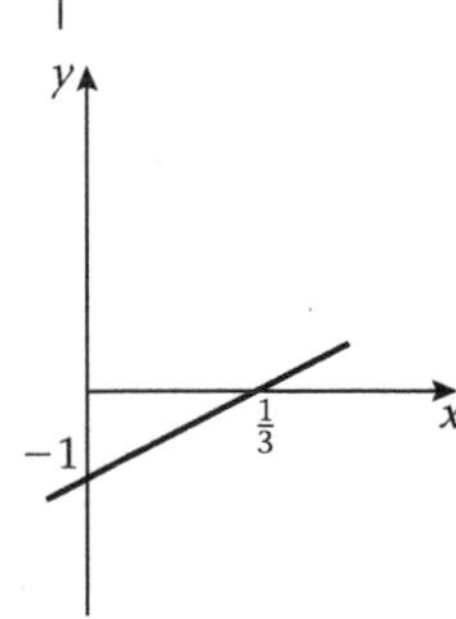

2 (a) $5x - 6y = 30$ (b) $2x + 3y = 18$
(c) $x + 2y = -8$ (d) $4x - y = -12$

3 -11

(a)

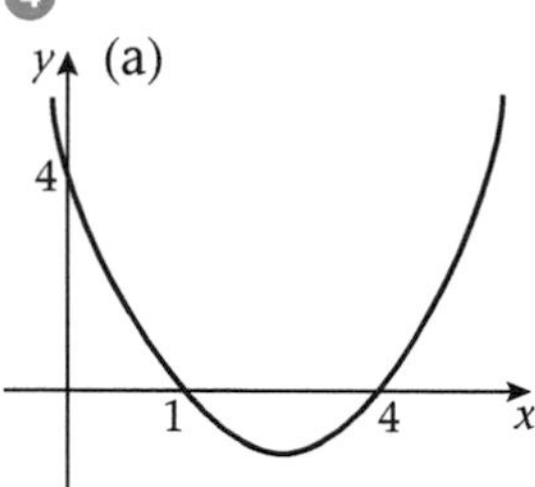

(b)

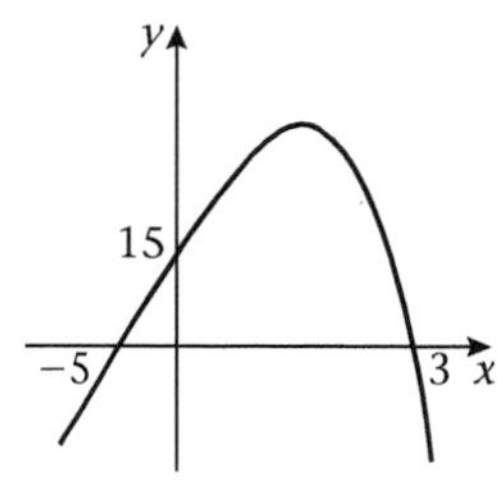

(c)

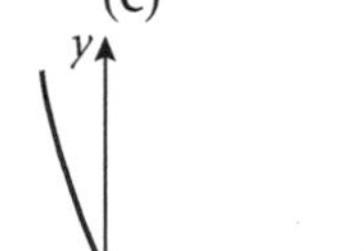

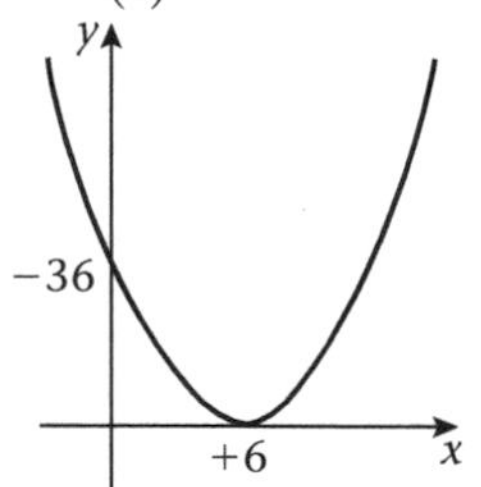

(d)

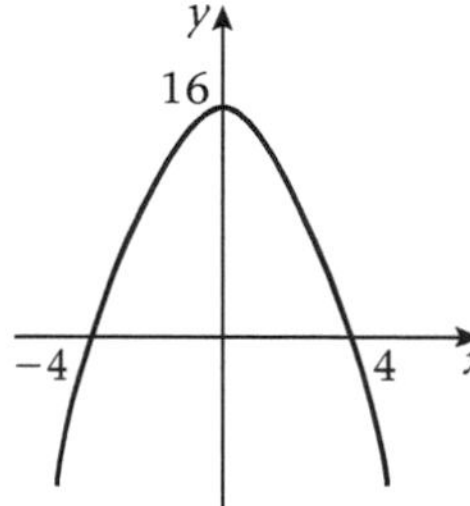

(e)

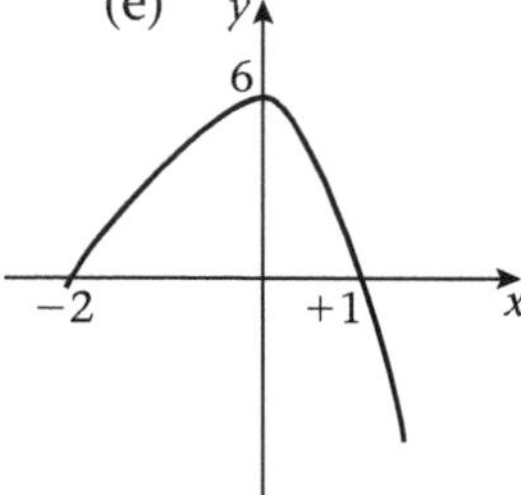

(f)

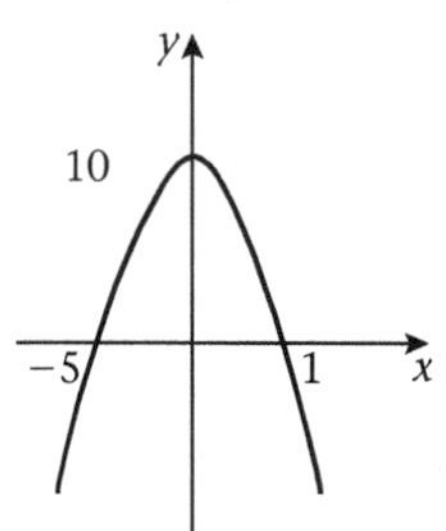

5 (a) $x = 3$
(b) $\tan O\widehat{M}N = 2$

6 $y = x^2 - x - 2$

Practice exercise P15.1 (p. 195)

1 (a) $(-1, -4), (0, -2), (1, 0)$
(b) y is 2 less than twice x
(c) $y = 2x - 2$

2 (a) $(-2, 3), (-1, 0), (0, -1), (1, 0), (2, 3)$
(b) y is 1 less than x squared
(c) $y = x^2 - 1$

Practice exercise P15.2 (p. 195)

1 (a)

x	−2	−1	0	1	2	3
y	−7	−5	−3	−1	1	3

(b)

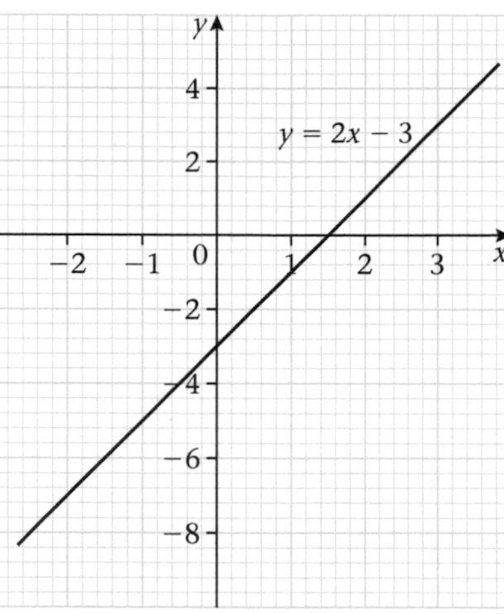

2 (a)

x	−3	−2	−1	0	1	2	3
y	10	5	2	1	2	5	10

(b)

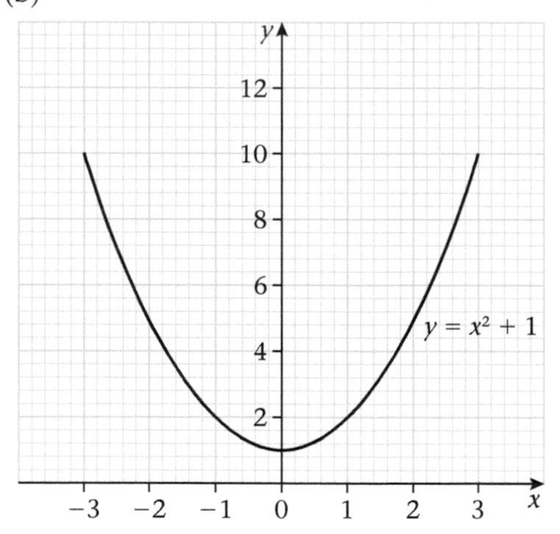

3 (a)

x	−3	−2	−1	0	1	2	3
y	19	12	7	4	3	4	7

(b)

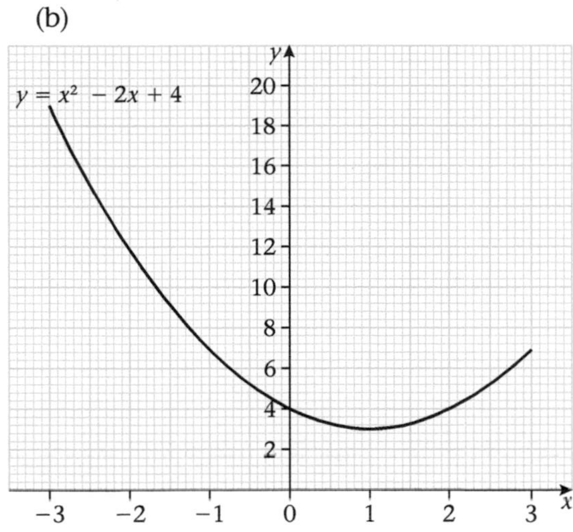

Practice exercise P15.3 (p. 195)

1 (a)

x	−3	−2	−1	0	1	2	3
y	$\frac{1}{64}$	$\frac{1}{16}$	$\frac{1}{4}$	1	4	16	16

(b)

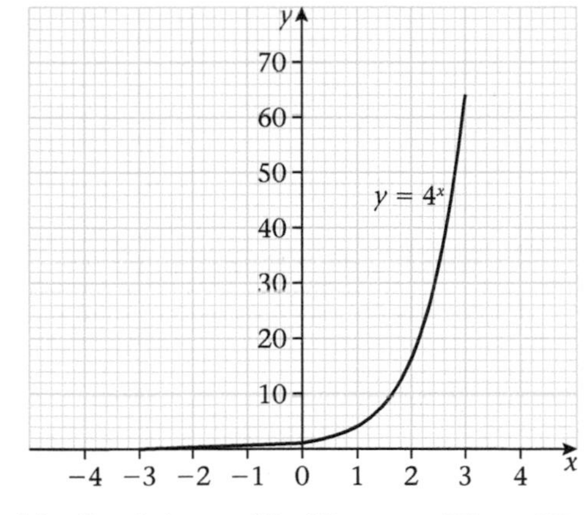

(c) (i) 1.4 (ii) 22 (iii) 89

2 (a)

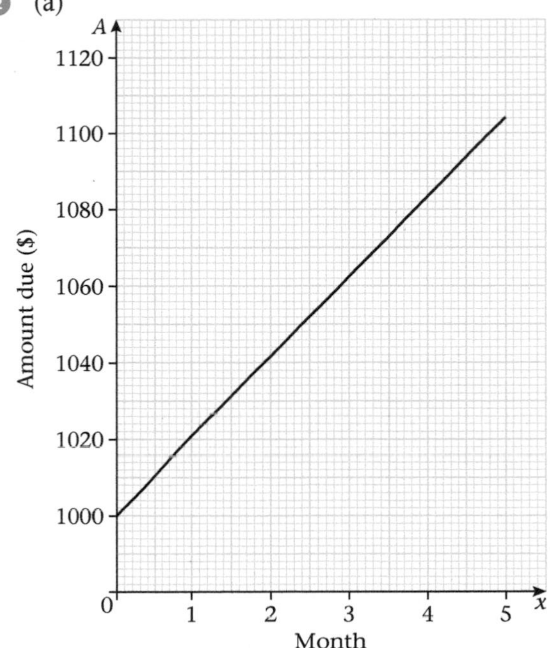

(b) $1104

Practice exercise P15.4 (p. 195)

1

x	−4	−3	−2	−1	$-\frac{1}{2}$	0	$\frac{1}{2}$	1	2	3	4
y	$-\frac{1}{8}$	$-\frac{2}{9}$	$-\frac{1}{2}$	−2	−8		−8	−2	$-\frac{1}{2}$	$-\frac{2}{9}$	$-\frac{1}{8}$

2

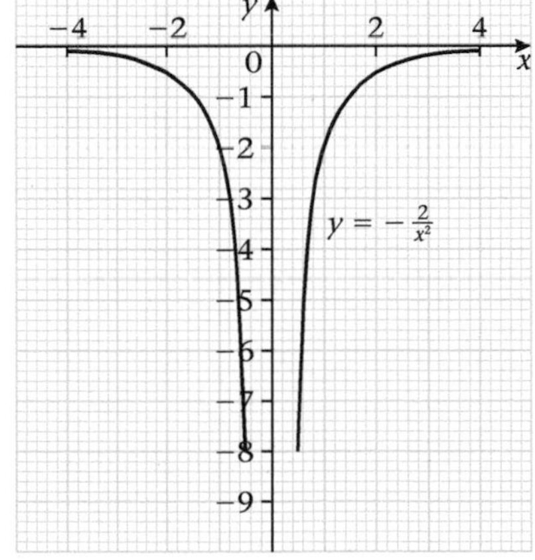

Practice exercise P15.5 (p. 195)

1 (a)

x	−3	−2	−1	0	1	2	3
y	−54	−16	−2	0	2	16	54

(b), (c)

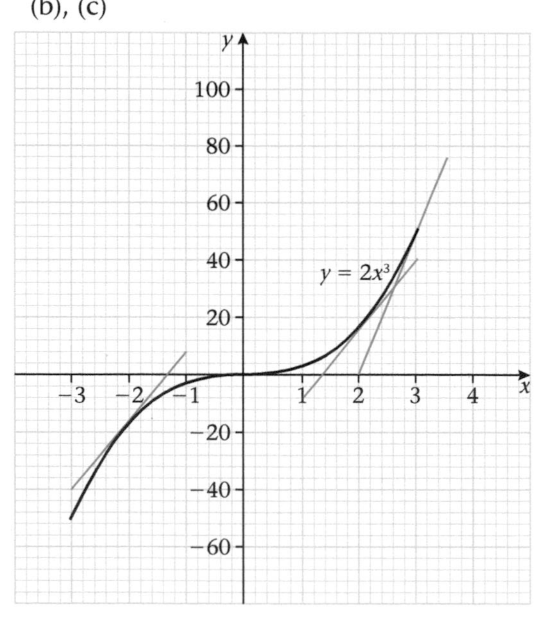

(d) (i) 24 (ii) 0 (iii) 24 (iv) 54

2 (a)

x	−6	−5	−4	−3	−2	−1	0	1	2	3	4	5	6
y	−108	−62.5	−32	−13.5	−4	−0.5	0	0.5	4	13.5	32	62.5	108

(b), (c)

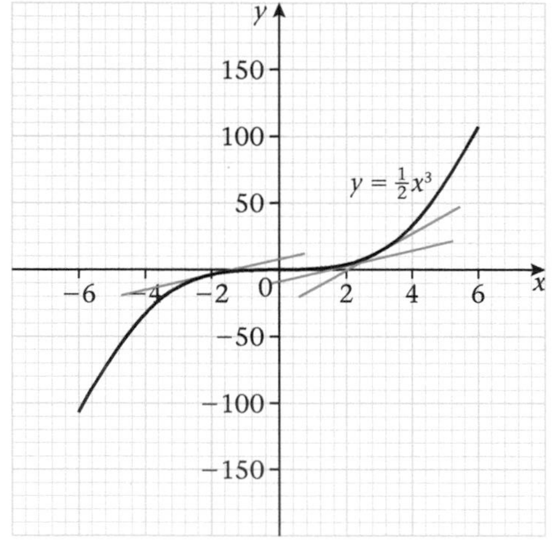

(d) (i) 6 (ii) 0 (iii) 6 (iv) 13.5

Practice exercise P15.6 (p. 195)

1

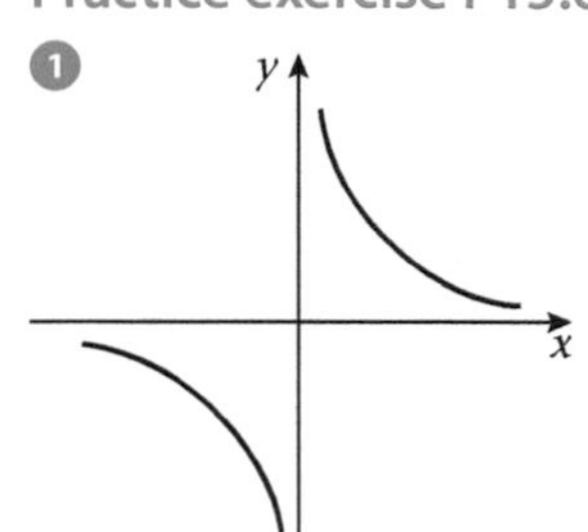

2

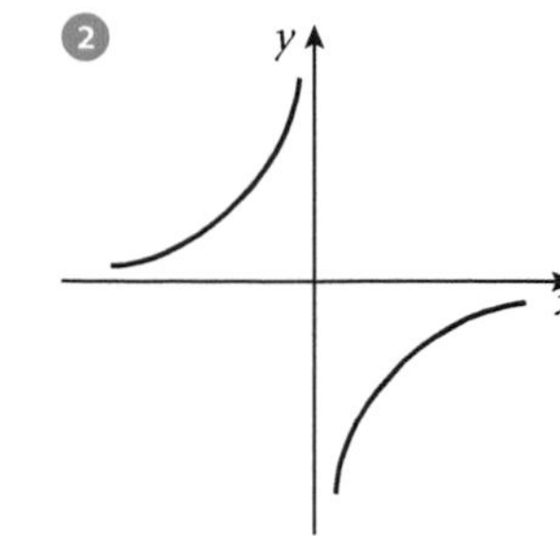

3

4

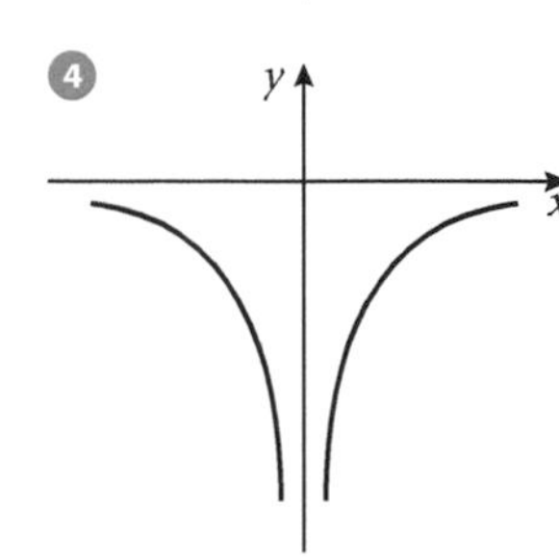

5

6

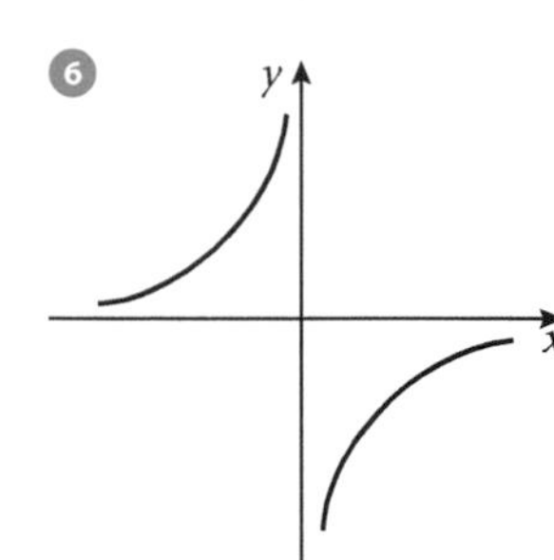

7

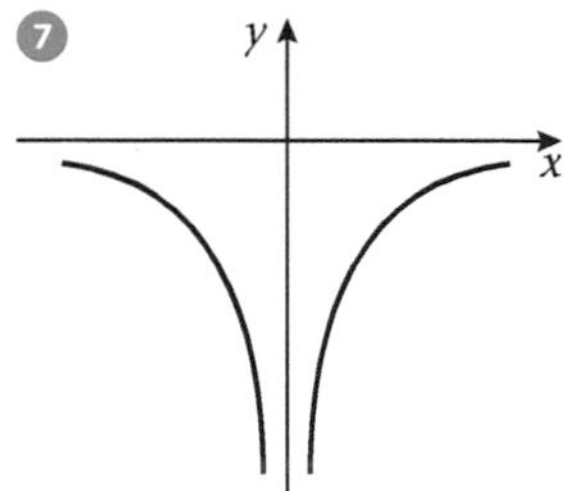

8

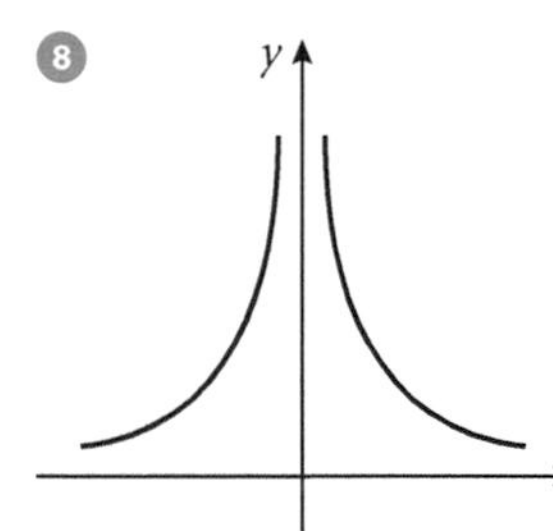

Exercise 16a (p. 197)

1

(a)

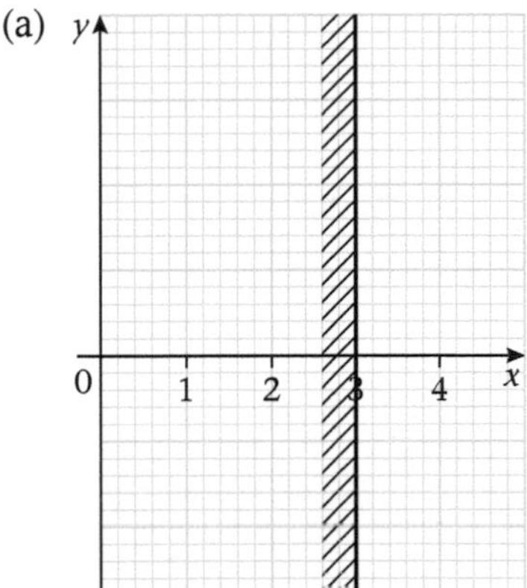

(b)

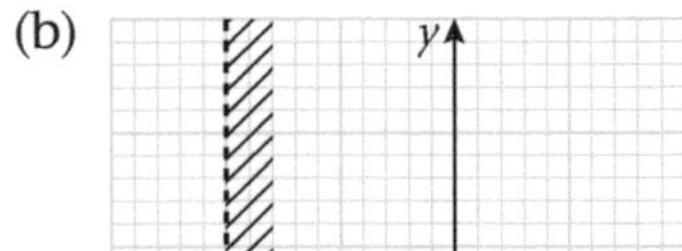

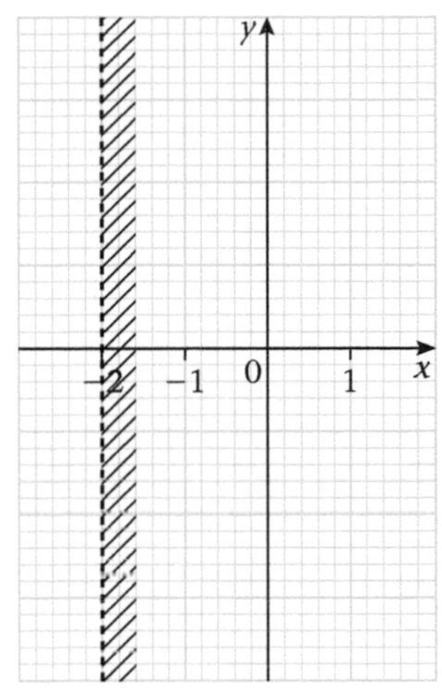

(c)

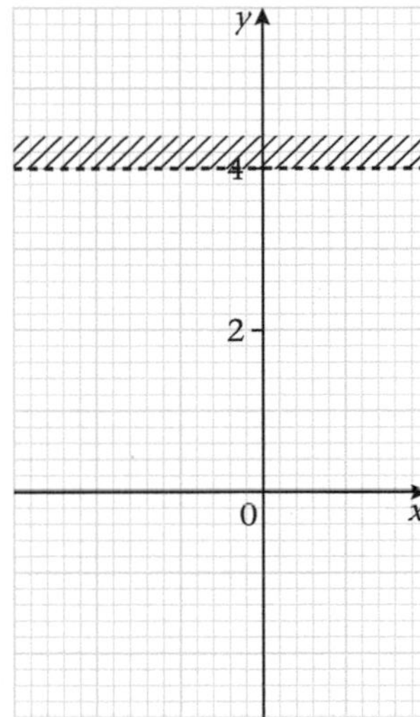

(d)

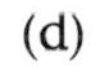

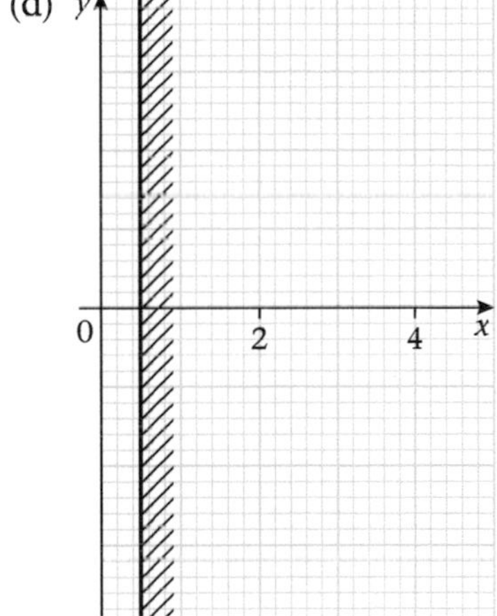

(e)

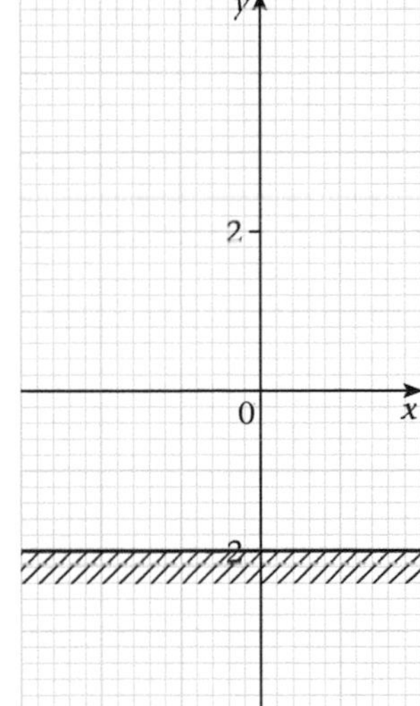

(f)

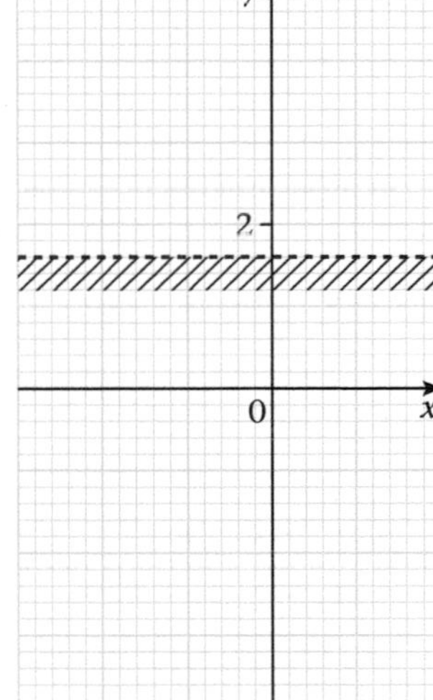

2 (a) (b) (c) (d)

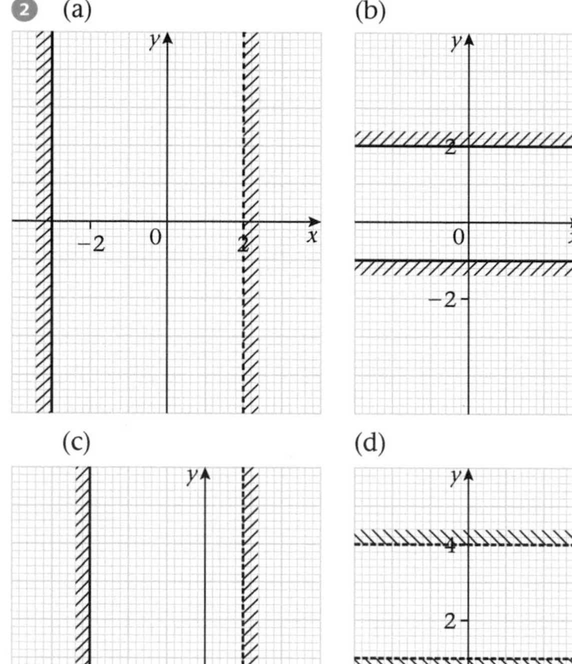

3 (a)

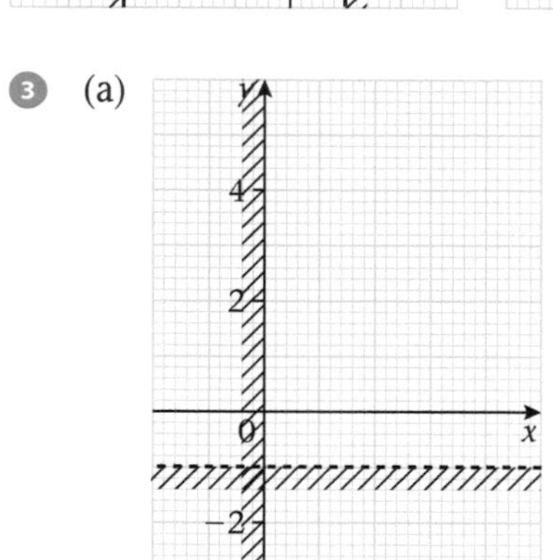

(b)

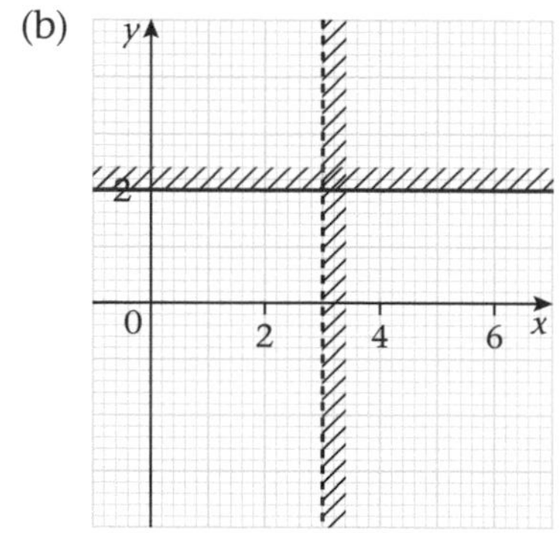

(c)

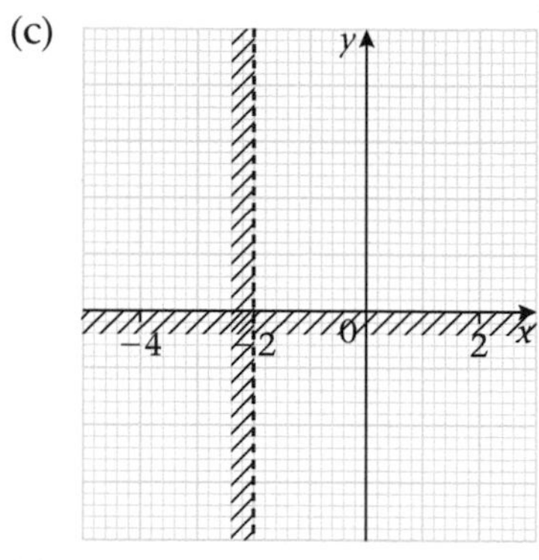

(d)

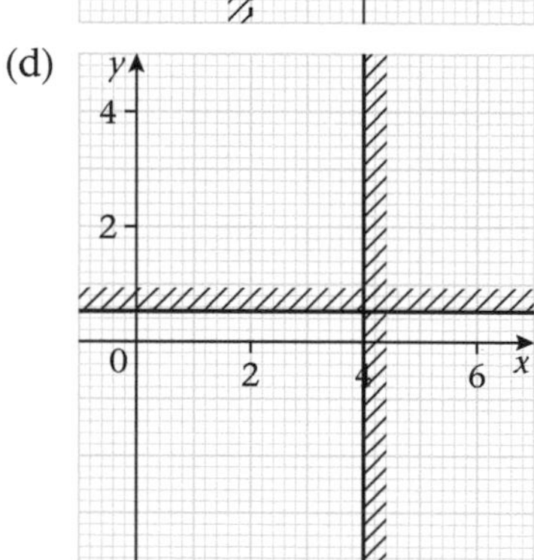

4 (a) (i) $y = 4$
(ii) $\{(x, y): y < 4\}$
(b) (i) $x = -1$
(ii) $\{(x, y): x \leqslant -1\}$
(c) (i) $x = -1$ and $x = 3$
(ii) $\{(x, y): -1 \leqslant x < 3\}$
(d) (i) $x = 2$ and $y = -1$
(ii) $\{(x, y): x \leqslant 2 \text{ and } y \geqslant -1\}$
(e) (i) $x = 1$ and $y = \frac{3}{2}$
(ii) $\{(x, y): x < 1 \text{ and } 2y < 3\}$

5 (a)

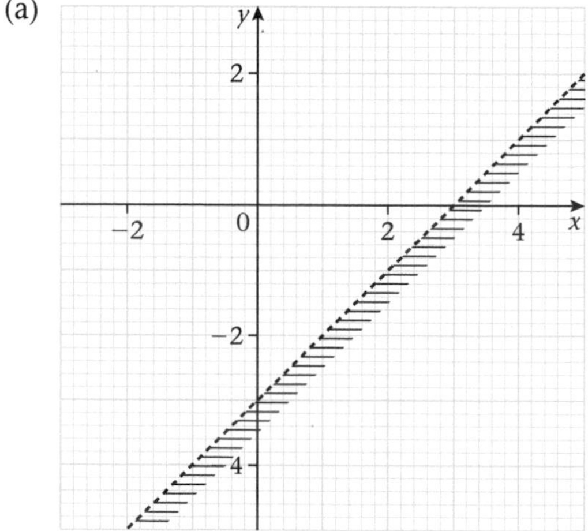

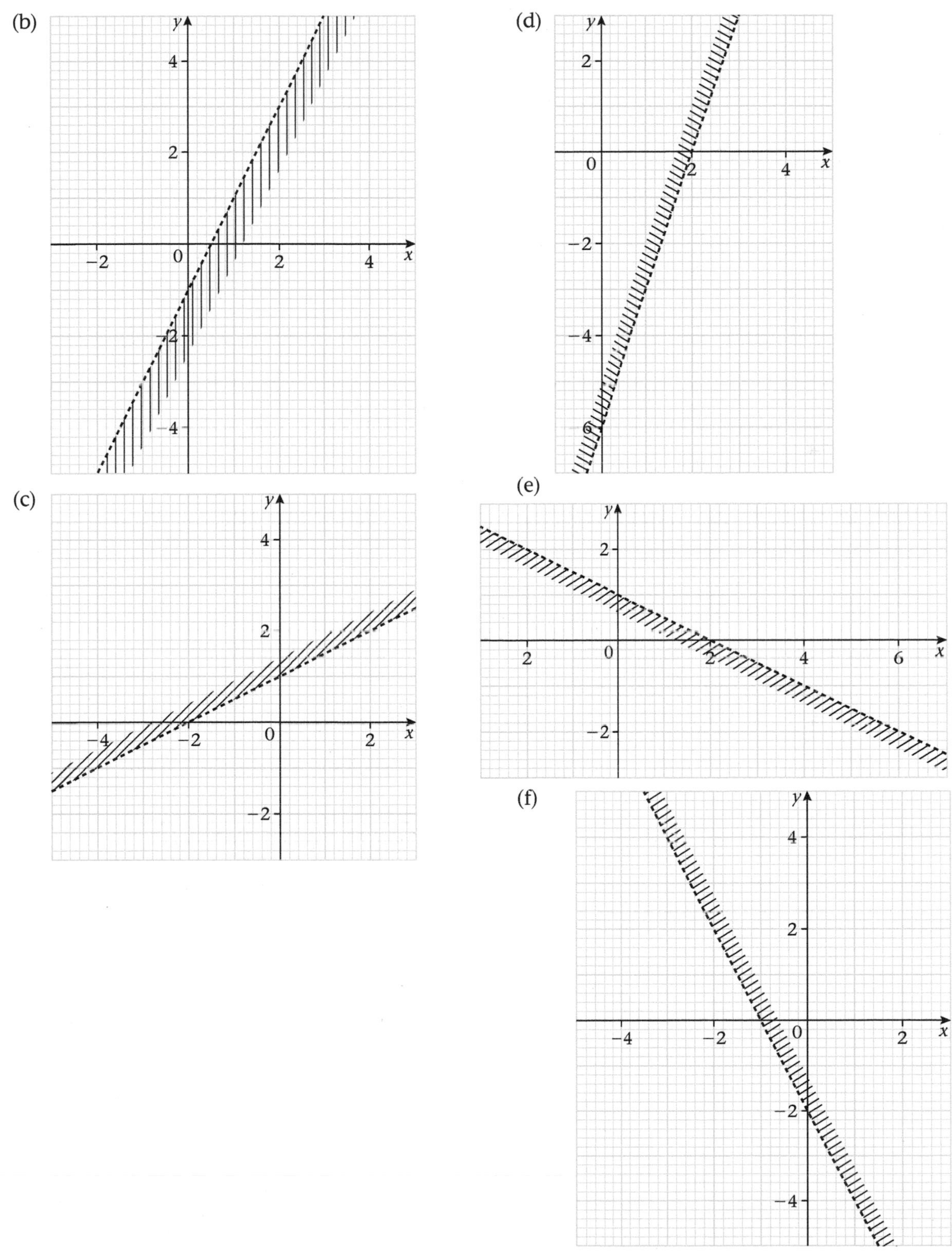
(b)
(c)
(d)
(e)
(f)

6 (a) (i) $2y = x$
(ii) $\{(x, y): 2y \leq x\}$
(b) (i) $y + 2x = 0$
(ii) $\{(x, y): y + 2x < 0\}$
(c) (i) $y = x + 1$
(ii) $\{(x, y): y \geq x + 1\}$
(d) (i) $y - 3x = 3$
(ii) $\{(x, y): y - 3x < 3\}$
(e) (i) $y + 2x = 2$
(ii) $\{(x, y): y + 2x \geq 2\}$

Exercise 16b (p. 200)

1 (a) $x = 3$
(b) $x + 2y \leq 5$, $x + y > 2$, $x < 3$
(c) $\{(1, 2), (2, 1)\}$

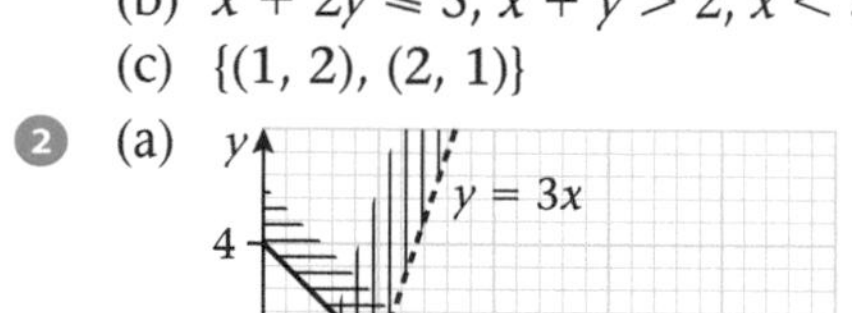

2 (a)

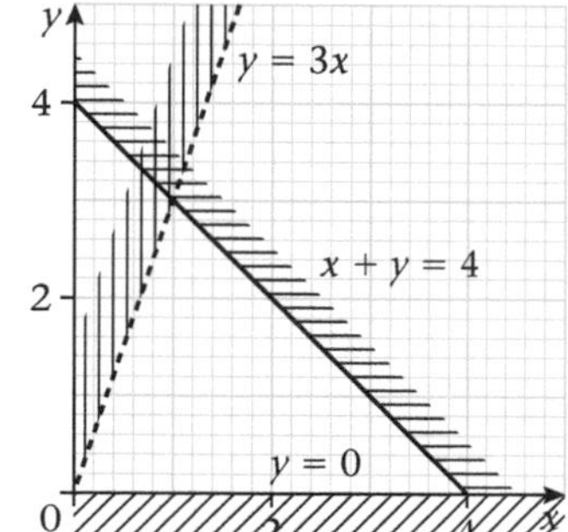

(b)

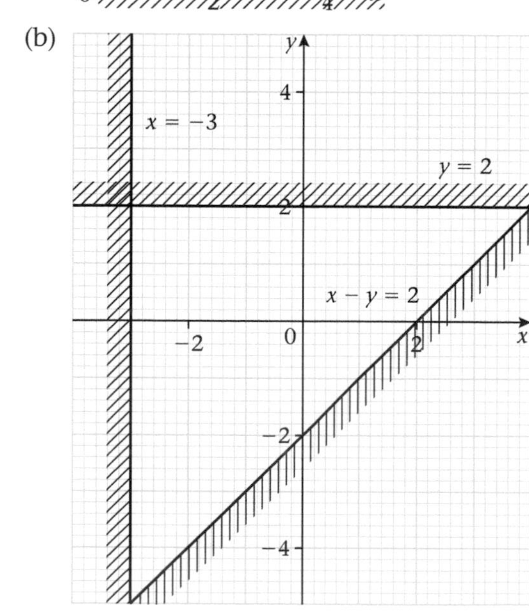

(c)

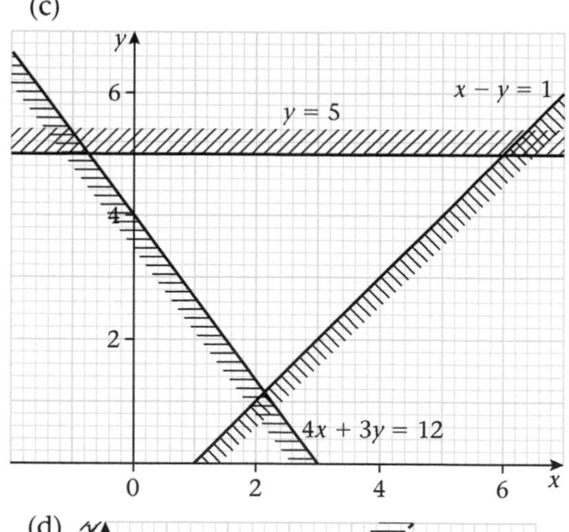

(d)

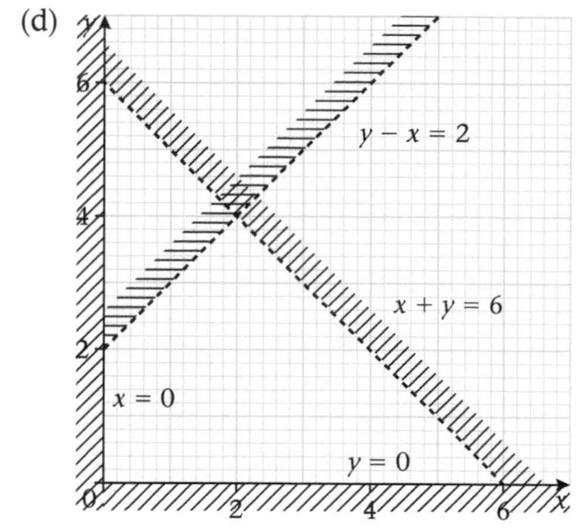

(e)

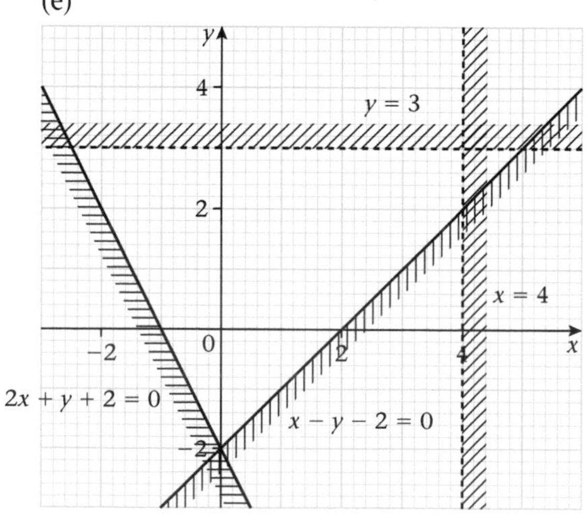

3 (a)

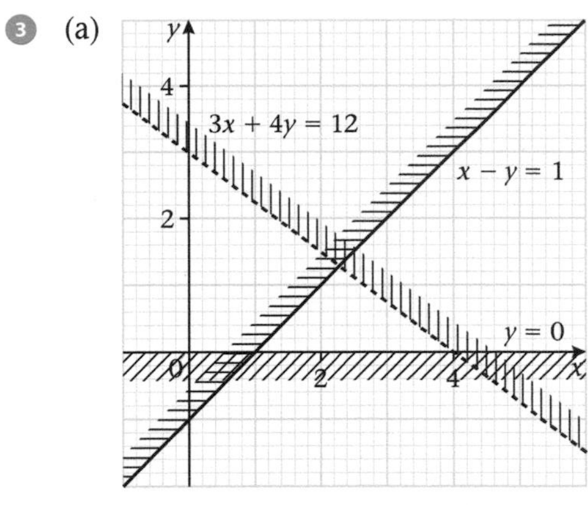

{(1, 0), (2, 0), (3, 0), (2, 1)}

(b)

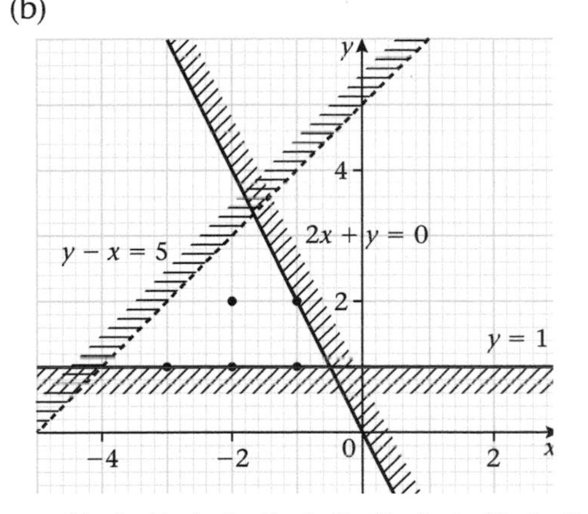

{(−1, 1), (−2, 1), (−3, 1), (−1, 2), (−2, 2)}

(c)

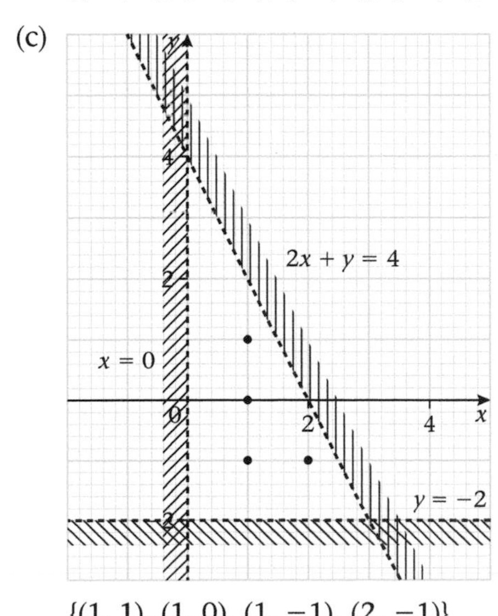

{(1, 1), (1, 0), (1, −1), (2, −1)}

(d)

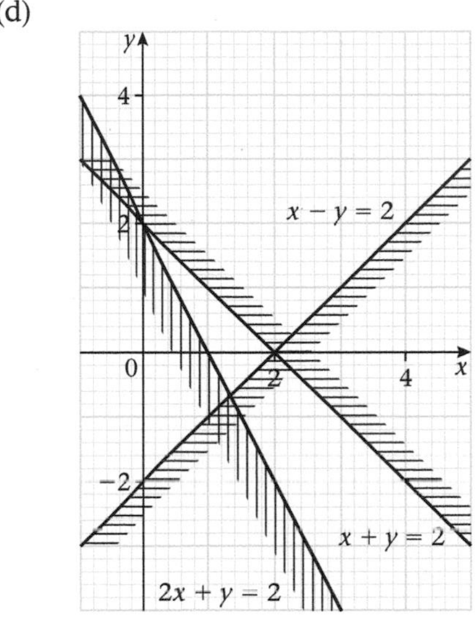

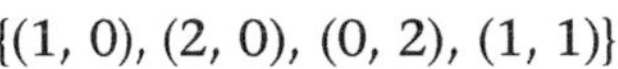

{(1, 0), (2, 0), (0, 2), (1, 1)}

(e)

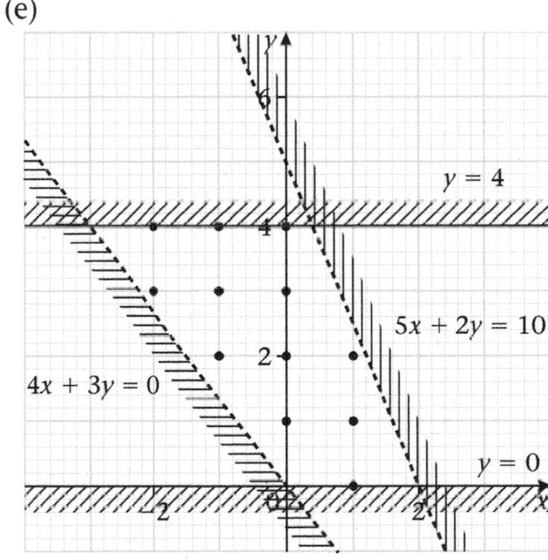

{(1, 1), (1, 2), (0, 1), (0, 2), (0, 3), (−1, 2), (−1, 3), (−2, 3), (−2, 4), (−1, 4), (0, 4), (1, 0)}

4 $x \geqslant 1$, $x + y < 5$, $3y \geqslant x$

5 $y > 2$, $y > 3x$, $x + y \geqslant -2$

6 (a) N(0, 5)

(b) $3y = -7x$ (or $7x + 3y = 0$)

(c) $x \leqslant 0$, $3y + 7x \geqslant 0$, $2y < x + 10$

Exercise 16c (p. 203)

1 14 buses, 12 minibuses

2 (a) two ways: 3 notebooks, 5 pencils or 4 notebooks, 3 pencils

(b) yes, the 2nd way: 60 cents change

3 (a) 93: either (72, 21) or (73, 20)
(b) 60 cheap, 30 expensive

4 (a) (i) (96, 34) (ii) (140, 20)
(b) $13 000

5 (a) $4x + y \geqslant 20$, $4x + 3y \geqslant 30$
(b)

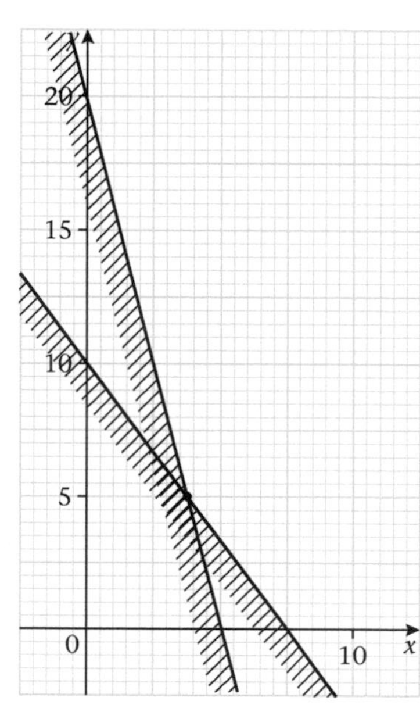

(c) 4 Feelgood pills and 5 Getbetta pills, $13.00

6 15 lorries (5 Landmasters, 10 Sandrovers)

7 (a) 37 (10, 27)
(b) either (10, 27) or (11, 25)
(c) $18.80

8 (a) 9
(b) 3 of A and 6 of B

Practice exercise P16.1 (p. 204)

1

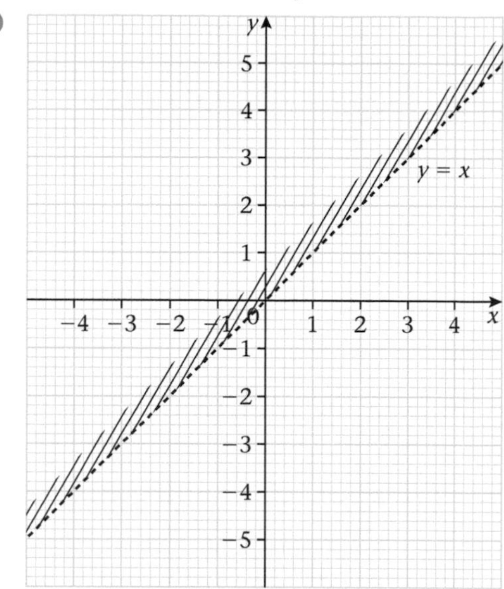

The points on the line satisfy the equality $y = x$ but are *not* included in the solution set of the inequality $y < x$.

2

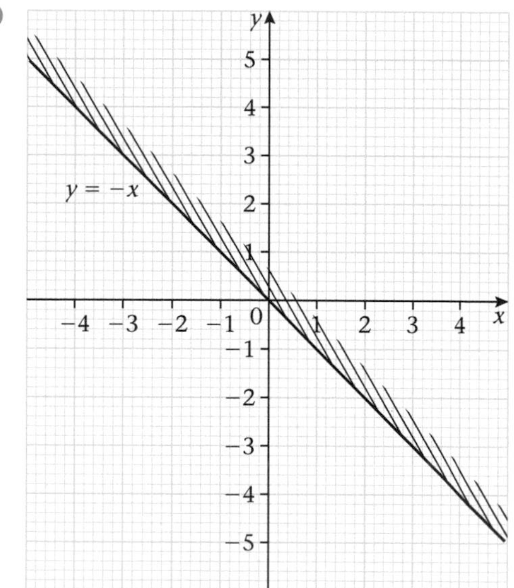

The points on the line satisfy the equality $y = -x$ and are included in the solution set of the inequality $y \leqslant x$.

3

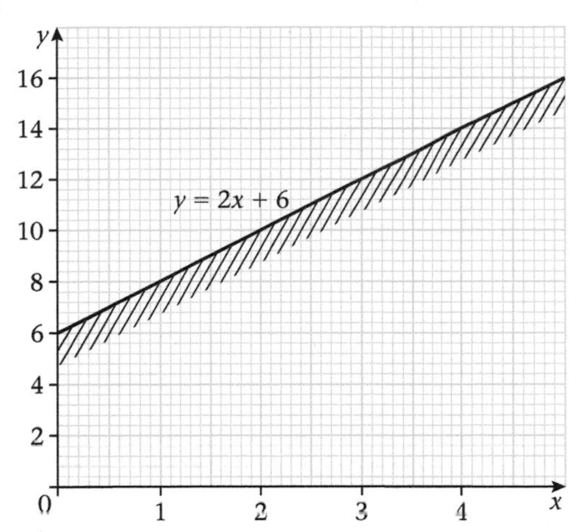

The points on the line satisfy the equality $y = 2x + 6$ and are included in the solution set of the inequality $y \geqslant 2x + 6$.

Practice exercise P16.2 (p. 204)

1

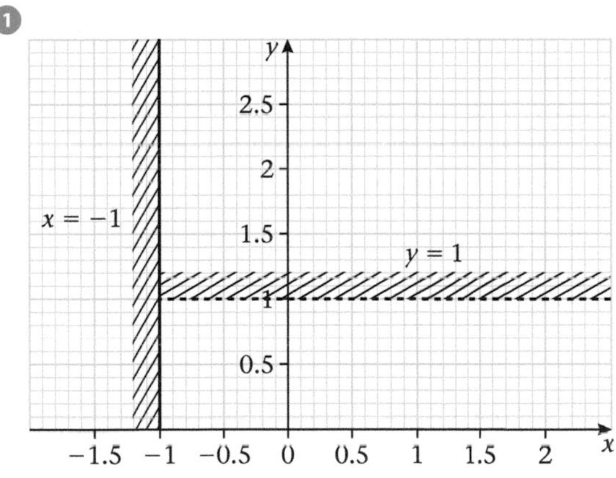

2

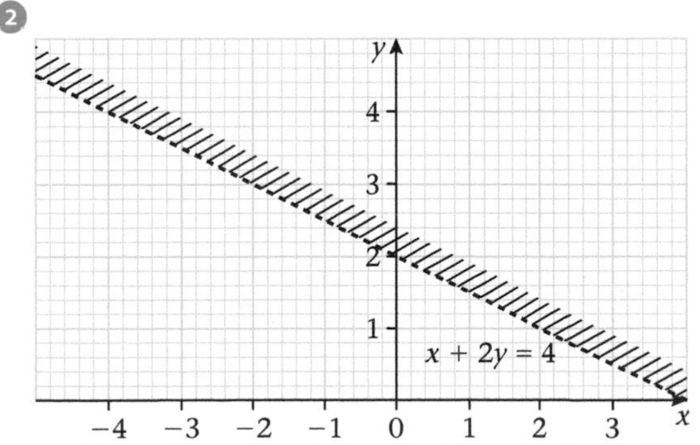

3

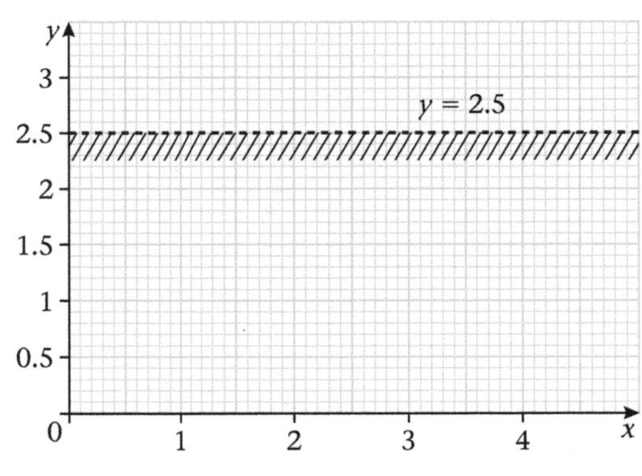

4

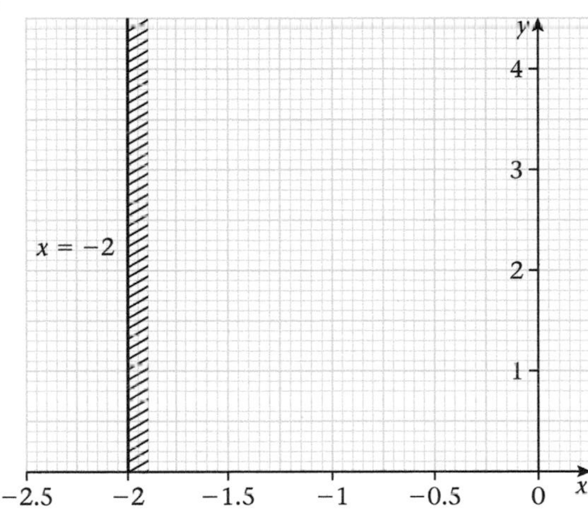

5

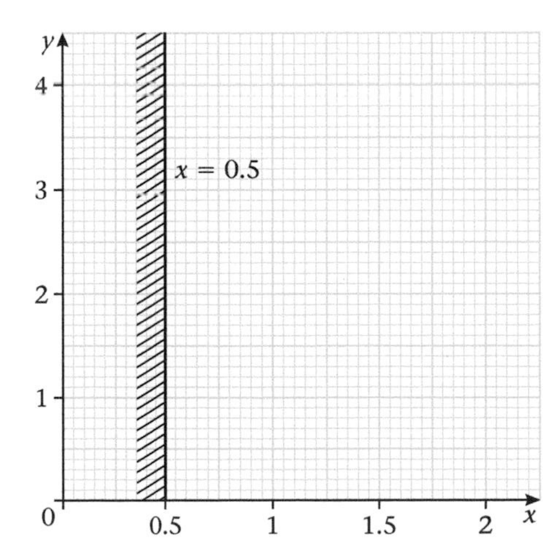

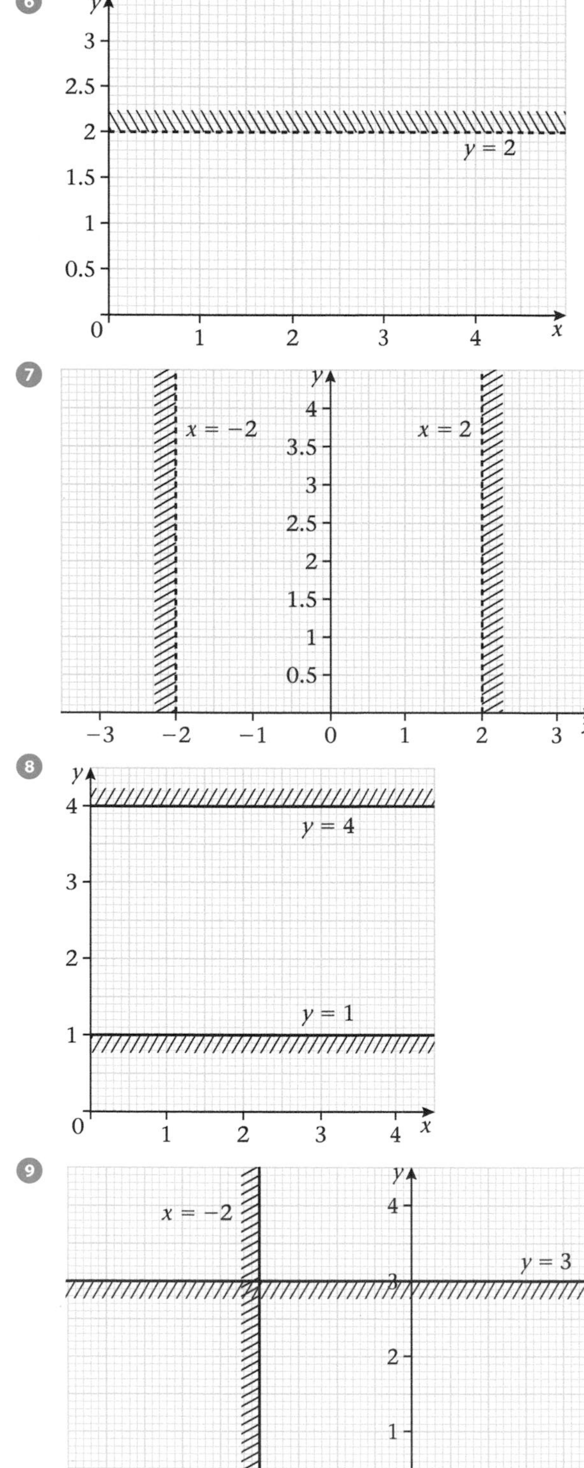

10

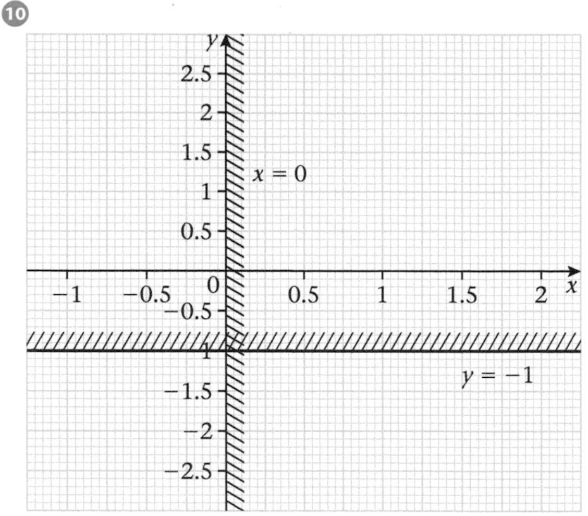

Practice exercise P16.3 (p. 205)

1. (a) $4y = 5x$ (b) $4y \geqslant 5x$
2. (a) $y + 2x = 6$ (b) $y + 2x \leqslant 6$

Practice exercise P16.4 (p. 205)

1. (a)

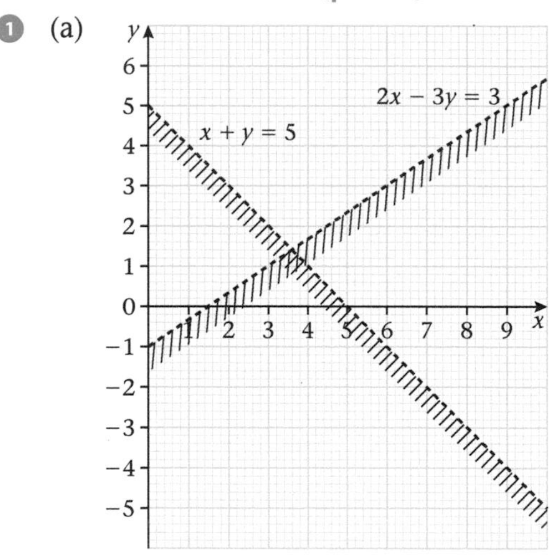

(b)

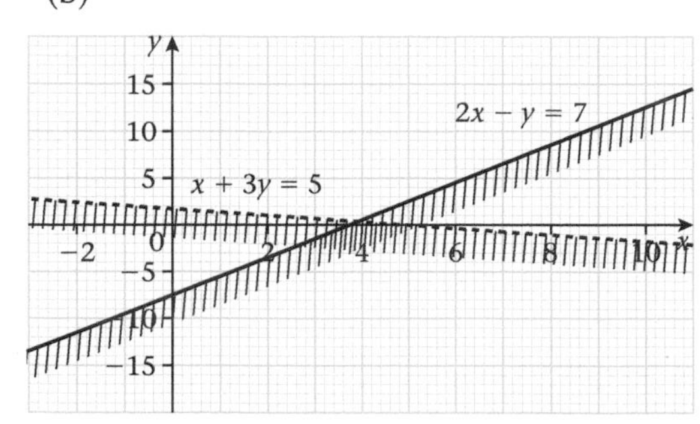

2 (a), (b)

(c) {(3, 1), (3, 2), (3, 3), (4, 1), (4, 2), (5, 1)}
(d) 7, 8, 9, 9, 10, 11, respectively
(e) (5, 1)

3 (a) $x + 2y \leqslant 12$; $y \geqslant 2$; $2x + y \geqslant 6$
(b) A(0, 6), B(8, 2), C(2, 2)
(c) 18, 22, 10, respectively
(d) B

4

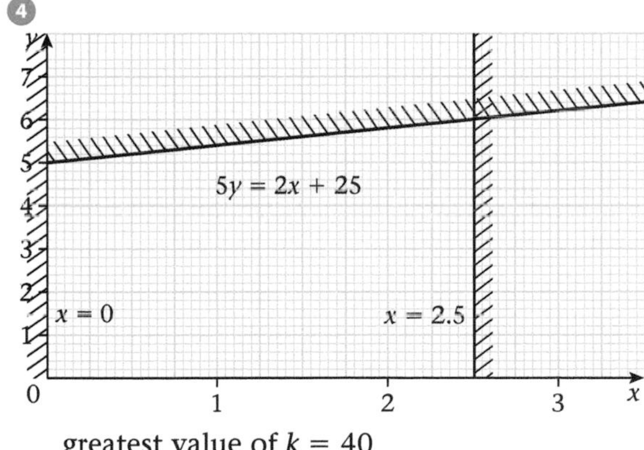

greatest value of $k = 40$

Practice exercise P16.5 (p. 205)

1 (a) $h + s \leqslant 36$; $h \geqslant 20$, $h \geqslant 2s$
(b) $h + s = 36$; $h = 20$, $h = 2s$
(c)

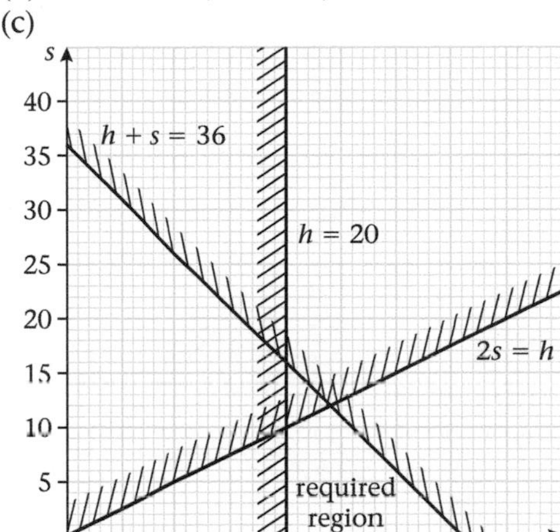

2 (a), (b)

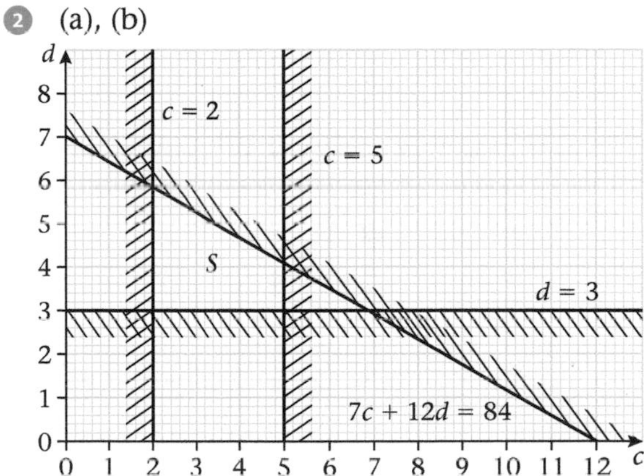

3 (a) (i) $c \geqslant 15$
(ii) $p > 25$
(iii) $45 \leqslant c + p < 60$
(b)

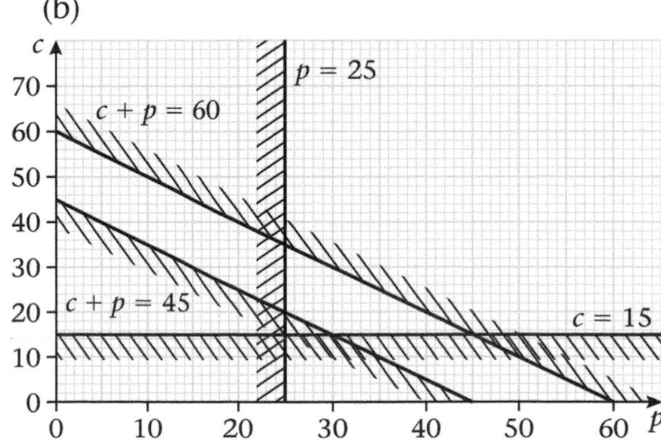

4 In this graph, x represents the number of cartons of orange juice and y represents the number of cartons of grapefruit juice.

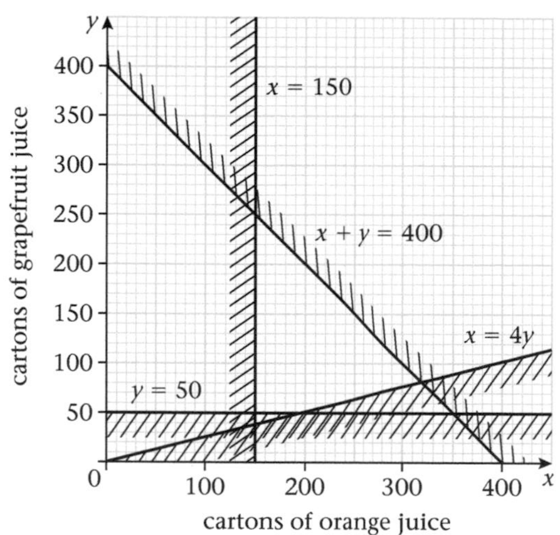

For maximum profit (of \$1056) he should take 320 cartons of orange juice and 80 cartons of grapefruit juice.

5 In this graph, m represents the number of minibuses and t represents the number of taxis.

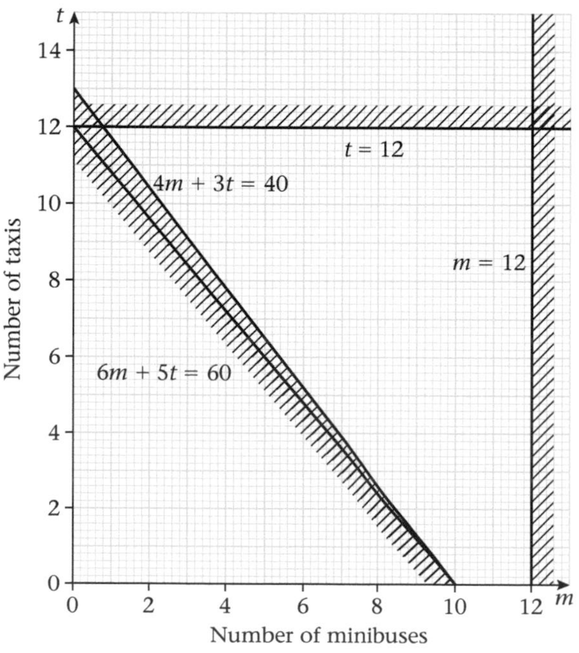

The smallest number of vehicles they must hire is 10 minibuses and no taxis.

6 (a) In this graph, t represents the number of kilograms of tomatoes and c represents the number of kilograms of carrots.

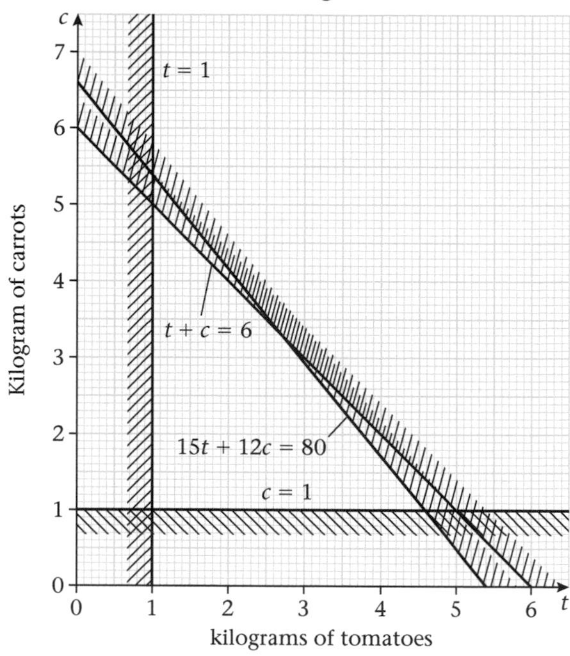

(b) The customer buys 1 kg of carrots and 4.53 kg of tomatoes.

(c) \$51.33

7 (a)

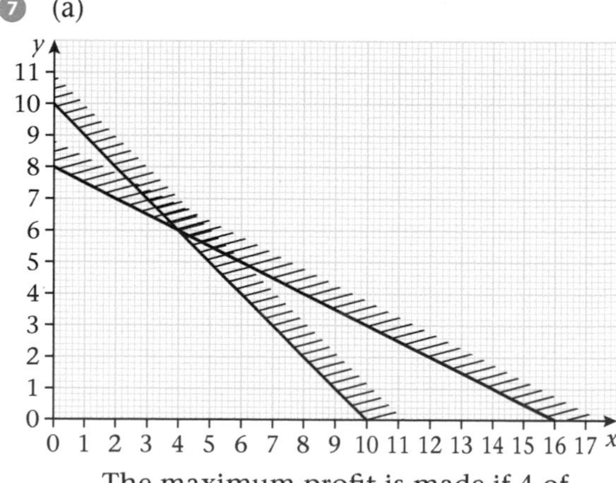

The maximum profit is made if 4 of item X and 6 of item Y are made.

(b) \$144

8

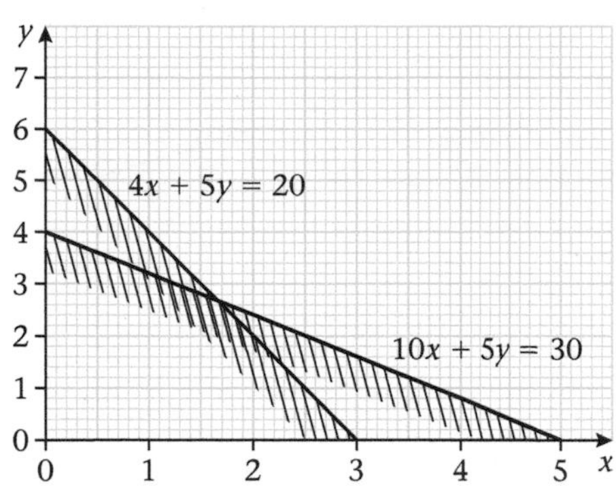

Assuming the cook buys and uses whole packets of mix X and mix Y, the cheapest way (at $17) to get the correct mix is to use 1 packet of mix X and 4 packets of mix Y.

Exercise 17a (p. 208)

1 great circles
2 $\alpha = 30°$ 3 arc AXB
4 (a) 8 cm (b) $64\pi\ \text{cm}^2$
5 (a) 5 cm (b) 3 cm (c) 4 cm (d) 1 cm

Exercise 17b (p. 210)

1 (a) 90°W (b) 90°E (c) 135°E (d) 120°W
2 105° 3 two
4 (a) 90° (b) 130°
5 Turkmenistan
7 (a) 161°W (b) 52°S
8 (a) 70°N, 40°W (b) 70°N, 0°E
(c) 70°N, 80°E (d) 0°N, 80°E
(e) 30°S, 80°E (f) 30°S, 40°W
10 6370 km

Exercise 17c (p. 212)

1 (a) 2330 km (b) 1260 n.mi.
2 5670 km
3 2100 n.mi.
4 4110 km
5 4700 km
6 5820 n.mi.
7 (a) 378 km
(b) geographical and social factors often prevent roads from taking the shortest distance between two points.
8 1500 n.mi. 9 6000 n.mi.
10 2.38° (2°23′)

Exercise 17d (p. 213)

1 10.07 cm 2 147.3 cm
3 2400 km
4 (a) 29 900 km (b) 16 050 n.mi.
5 (a) 2820 n.mi. (b) 19 810 n.mi.
6 2225 n.mi. 7 1600 km
8 27.2 n.mi. 9 1820 km
10 925 n.mi.

Exercise 17e (p. 215)

1 (a) 11 472 km (b) 7778 km
2 6.9°S 3 52.1°N
4 5°5′S 5 320 knots
6 (a) 61.5°S (b) 6 p.m.
7 (a) 1500 km (b) 52.6°W
8 (a) 46 (b) 33°E
9 (a) 5560 km (b) 8450 km
10 (a) 2400 n.mi. (b) 3690 n.mi.

Practice exercise P17.1 (p. 215)

1 5800 km (2 s.f.)
2 15 000 km (2 s.f.)
3 (8.6°S, 95°E)
4 (22°S, 124.4°W)
5 (30.2°N, 32°E)
6 1000 km (2 s.f.)
7 6200 km (2 s.f.)
8 4000 km (3 s.f.)
9 74.5°E
10 (a) $\text{Arc AB} = \frac{25 + 35}{360} \times 2\pi R$
$= \frac{60 \times 2}{360} \times \pi R$
$= \frac{1}{3}\pi R$
(b) 3600 n.mi.

Practice exercise P17.2 (p. 216)

1 (a) (63°N, 50°W) or (57°S, 50°W)
(b) 6700 km (2 s.f.)
2 (a) 5300 km (b) 000 km
3 (a) (49.5°N, 20°W) (b) (49.5°N, 27.9°W)
(c) 493 km/h
4 (0°N, 10.75°E)
5 (a) 21 600 n.i. (b) 277.8 km/h

Exercise 18a (p. 218)

1 $\begin{pmatrix} -9 \\ 2 \end{pmatrix}$

2

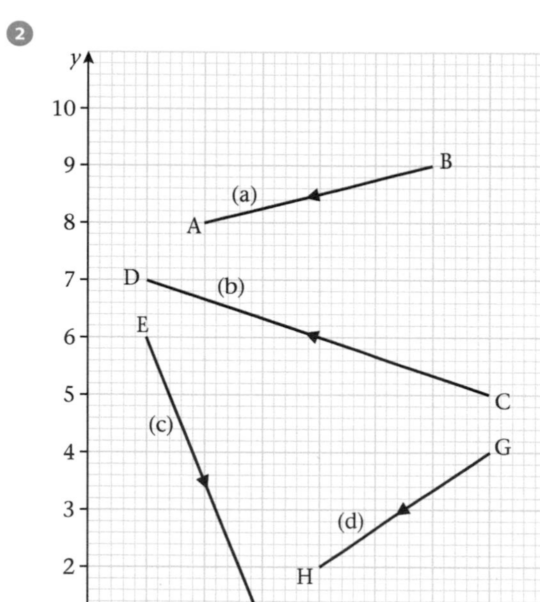
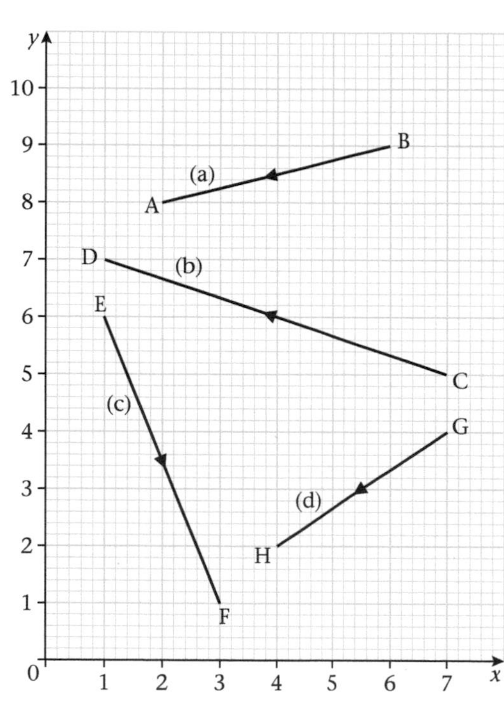

3

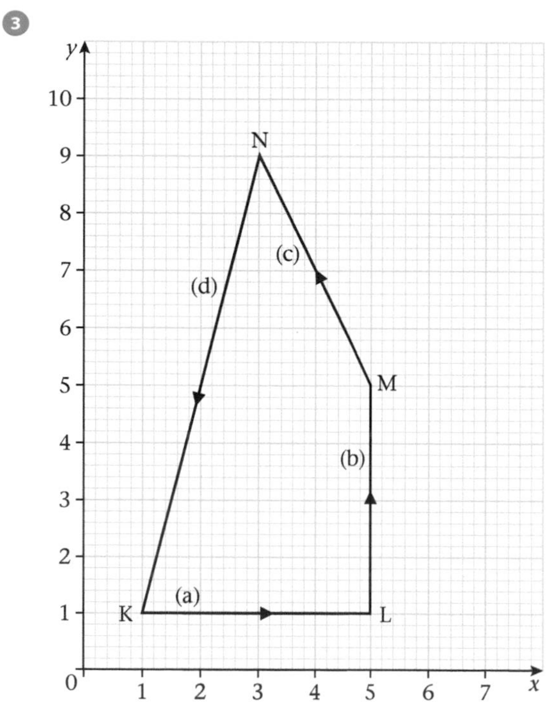

4 (a)

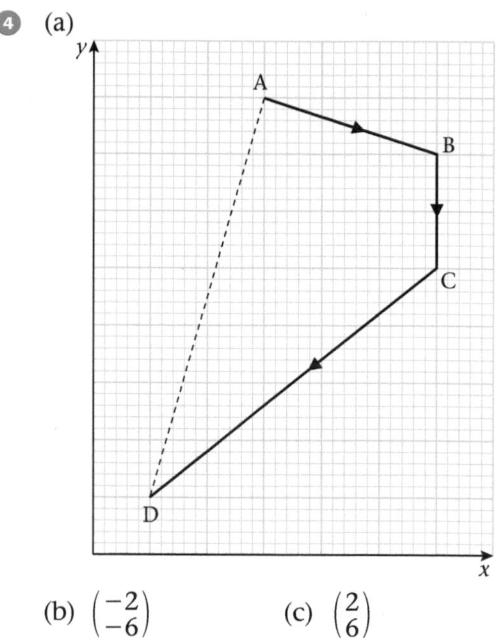

(b) $\begin{pmatrix} -2 \\ -6 \end{pmatrix}$ (c) $\begin{pmatrix} 2 \\ 6 \end{pmatrix}$

5 (a)

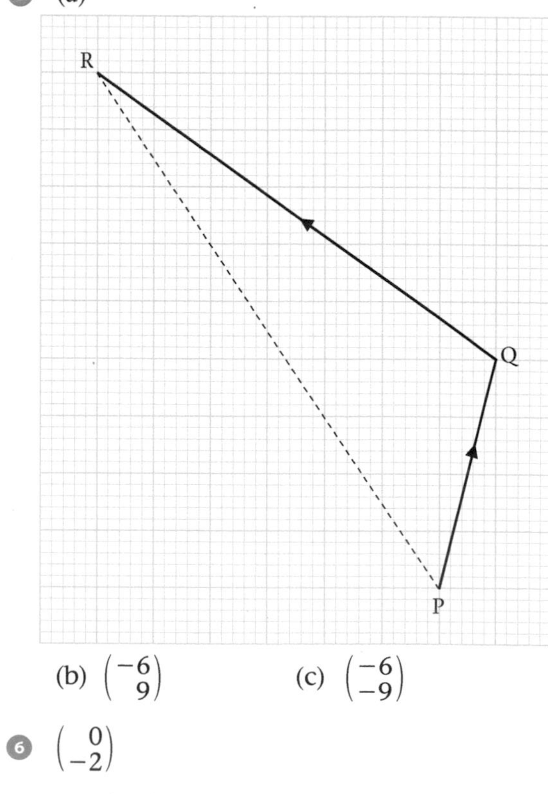

(b) $\begin{pmatrix} -6 \\ 9 \end{pmatrix}$ (c) $\begin{pmatrix} -6 \\ -9 \end{pmatrix}$

6 $\begin{pmatrix} 0 \\ -2 \end{pmatrix}$

7 (a) $\begin{pmatrix} 3 \\ -1 \end{pmatrix}$ (b) $\begin{pmatrix} -4 \\ 5 \end{pmatrix}$ (c) $\begin{pmatrix} 8 \\ 6 \end{pmatrix}$ (d) $\begin{pmatrix} -2 \\ -7 \end{pmatrix}$

8 $\begin{pmatrix} 8 \\ -5 \end{pmatrix}$

9 (a) $\begin{pmatrix} 7 \\ 5 \end{pmatrix}$ (b) $\sqrt{74}$

10 $\overrightarrow{PR} = \begin{pmatrix} -3 \\ 7 \end{pmatrix}$, $\overrightarrow{RP} = \begin{pmatrix} 3 \\ -7 \end{pmatrix}$

Exercise 18b (p. 219)

1 (a) $\begin{pmatrix} 2 \\ 4 \end{pmatrix}$ (b) $\begin{pmatrix} 4 \\ -4 \end{pmatrix}$ (c) $\begin{pmatrix} -4 \\ 3 \end{pmatrix}$ (d) $\begin{pmatrix} -4 \\ 0 \end{pmatrix}$

2 (a) $\begin{pmatrix} 2 \\ 5 \end{pmatrix}$ (b) $\begin{pmatrix} 5 \\ 3 \end{pmatrix}$ (c) $\begin{pmatrix} 6 \\ 0 \end{pmatrix}$
(d) $\begin{pmatrix} 5 \\ -3 \end{pmatrix}$ (e) $\begin{pmatrix} 2 \\ -5 \end{pmatrix}$ (f) $\begin{pmatrix} 4 \\ 0 \end{pmatrix}$

3 (a)

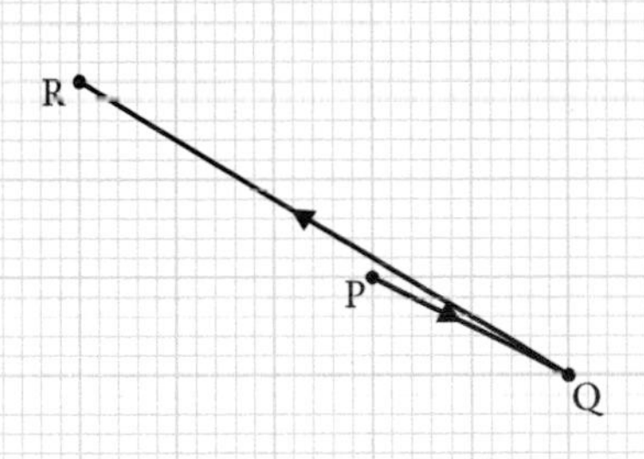

(b) $\overrightarrow{PR} = \begin{pmatrix} -3 \\ 2 \end{pmatrix}$

4 (a) $\begin{pmatrix} -2 \\ 1 \end{pmatrix}$ (b) $\begin{pmatrix} 4 \\ 2 \end{pmatrix}$ (c) $\begin{pmatrix} 2 \\ 4 \end{pmatrix}$ (d) $\begin{pmatrix} 2 \\ -5 \end{pmatrix}$
(e) $\begin{pmatrix} 0 \\ -3 \end{pmatrix}$ (f) $\begin{pmatrix} 6 \\ -2 \end{pmatrix}$ (g) $\begin{pmatrix} 2 \\ -1 \end{pmatrix}$ (h) $\begin{pmatrix} 0 \\ 1 \end{pmatrix}$
(i) $\begin{pmatrix} 4 \\ -5 \end{pmatrix}$ (j) $\begin{pmatrix} 4 \\ -1 \end{pmatrix}$

5 $\begin{pmatrix} -3 \\ -7 \end{pmatrix}$

6 (a) $\begin{pmatrix} 4 \\ -3 \end{pmatrix}$ (b) (2, 4) (c) (−6, 3)

Exercise 18c (p. 221)

1 (a) 5 (b) 13 (c) 10
(d) 7 (e) 2 (f) 17

2 (a) $\begin{pmatrix} -8 \\ 1 \end{pmatrix}$ (b) $\begin{pmatrix} 2 \\ -3 \end{pmatrix}$ (c) $\begin{pmatrix} 0 \\ -2 \end{pmatrix}$
(d) $\begin{pmatrix} 6 \\ 5 \end{pmatrix}$ (e) $\begin{pmatrix} 4 \\ 0 \end{pmatrix}$ (f) $\begin{pmatrix} -7 \\ -3 \end{pmatrix}$

3 15 units

4 $\begin{pmatrix} -4 \\ 4 \end{pmatrix}$

5 $\begin{pmatrix} 5 \\ -3 \end{pmatrix}$, $\sqrt{34}$ units

6 (a) $\begin{pmatrix} 1 \\ -1 \end{pmatrix}$ (b) $\begin{pmatrix} 9 \\ -3 \end{pmatrix}$ (c) $\begin{pmatrix} 8 \\ 5 \end{pmatrix}$ (d) $\begin{pmatrix} 1 \\ 4 \end{pmatrix}$

7 (a) $\overrightarrow{AC}' = \begin{pmatrix} 10 \\ -7 \end{pmatrix}$, where $\overrightarrow{BC}' = -\overrightarrow{BC}$

(b)

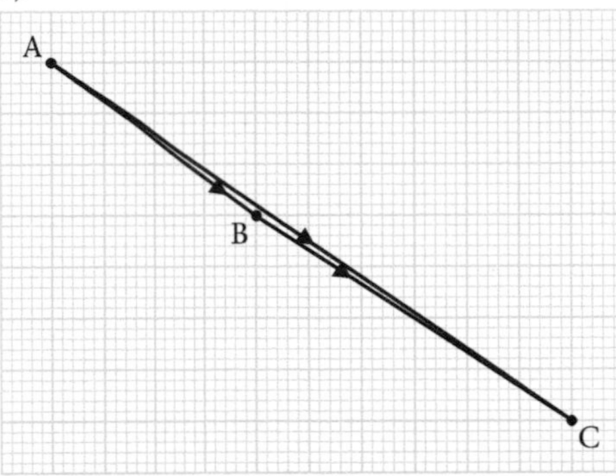

8 (a) $\begin{pmatrix} 13 \\ -5 \end{pmatrix}$ (b) $\begin{pmatrix} 1 \\ -1 \end{pmatrix}$ (c) $\begin{pmatrix} -13 \\ 5 \end{pmatrix}$
(d) $\sqrt{2}$ units (e) $\sqrt{194}$ units

9 (a) $\begin{pmatrix} 7 \\ -2 \end{pmatrix}$ (b) $\begin{pmatrix} 9 \\ 6 \end{pmatrix}$ (c) $\begin{pmatrix} 10 \\ 1 \end{pmatrix}$ (d) $\begin{pmatrix} 2 \\ 8 \end{pmatrix}$
(e) $\begin{pmatrix} 1 \\ -8 \end{pmatrix}$ (f) $\begin{pmatrix} -5 \\ 10 \end{pmatrix}$ (g) $\begin{pmatrix} 11 \\ -4 \end{pmatrix}$ (h) $\begin{pmatrix} 3 \\ 0 \end{pmatrix}$

10 (a) $\begin{pmatrix} 3 \\ 12 \end{pmatrix}$ (b) $\begin{pmatrix} 3 \\ 12 \end{pmatrix}$
(c) $\overrightarrow{OQ} - \overrightarrow{OP} = \overrightarrow{PQ}$
(d) always true since if
$\overrightarrow{OQ} - \overrightarrow{OP} = \overrightarrow{PQ}$ then $\overrightarrow{OP} + \overrightarrow{PQ} = \overrightarrow{OQ}$

Exercise 18d (p. 222)

1 (a) $\begin{pmatrix} 9 \\ 21 \end{pmatrix}$ (b) $\begin{pmatrix} 20 \\ 5 \end{pmatrix}$
(c) $\begin{pmatrix} -48 \\ 0 \end{pmatrix}$ (d) $\begin{pmatrix} -6 \\ -10 \end{pmatrix}$
(e) $\begin{pmatrix} 4 \\ -16 \end{pmatrix}$ (f) $\begin{pmatrix} -4 \\ 16 \end{pmatrix}$
(g) $\begin{pmatrix} 12 \\ 3 \end{pmatrix}$ (h) $\begin{pmatrix} 12 \\ 3 \end{pmatrix}$

2 (a) $\begin{pmatrix} -4 \\ -1 \end{pmatrix}$ (b) $\begin{pmatrix} 2 \\ 0 \end{pmatrix}$
(c) $\begin{pmatrix} -2 \\ -1 \end{pmatrix}$ (d) $\begin{pmatrix} 4 \\ 1 \end{pmatrix}$

3 (a) $\begin{pmatrix} -8 \\ 2 \end{pmatrix}$ (b) $\begin{pmatrix} -10 \\ 15 \end{pmatrix}$
(c) $\begin{pmatrix} -6 \\ 0 \end{pmatrix}$ (d) $\begin{pmatrix} \frac{1}{2} \\ -2 \end{pmatrix}$

4 (a) $\begin{pmatrix} 0 \\ 11 \end{pmatrix}$ (b) $\begin{pmatrix} 8 \\ 2 \end{pmatrix}$ (c) $\begin{pmatrix} 0 \\ 3 \end{pmatrix}$ (d) $\begin{pmatrix} -8 \\ 15 \end{pmatrix}$

5 (a) $\begin{pmatrix} 4 \\ 0 \end{pmatrix}$ (b) $\begin{pmatrix} 0 \\ 1 \end{pmatrix}$ (c) $\begin{pmatrix} -5 \\ -3 \end{pmatrix}$ (d) $\begin{pmatrix} 10 \\ 1 \end{pmatrix}$

6 (a) $\begin{pmatrix} 1 \\ 2 \end{pmatrix}$ (b) $\begin{pmatrix} 4 \\ 1 \end{pmatrix}$ (c) $\begin{pmatrix} 8 \\ 2 \end{pmatrix}$ (d) $\begin{pmatrix} 0 \\ -1 \end{pmatrix}$

7 (a) $\begin{pmatrix}21\\30\end{pmatrix}$ (b) $\begin{pmatrix}21\\30\end{pmatrix}$
(c) $\begin{pmatrix}-5\\-65\end{pmatrix}$ (d) $\begin{pmatrix}-5\\-65\end{pmatrix}$
(e) $\begin{pmatrix}-10\\-6\end{pmatrix}$ (f) $\begin{pmatrix}-10\\-6\end{pmatrix}$
(g) $\begin{pmatrix}-28\\0\end{pmatrix}$ (h) $\begin{pmatrix}-28\\0\end{pmatrix}$

8

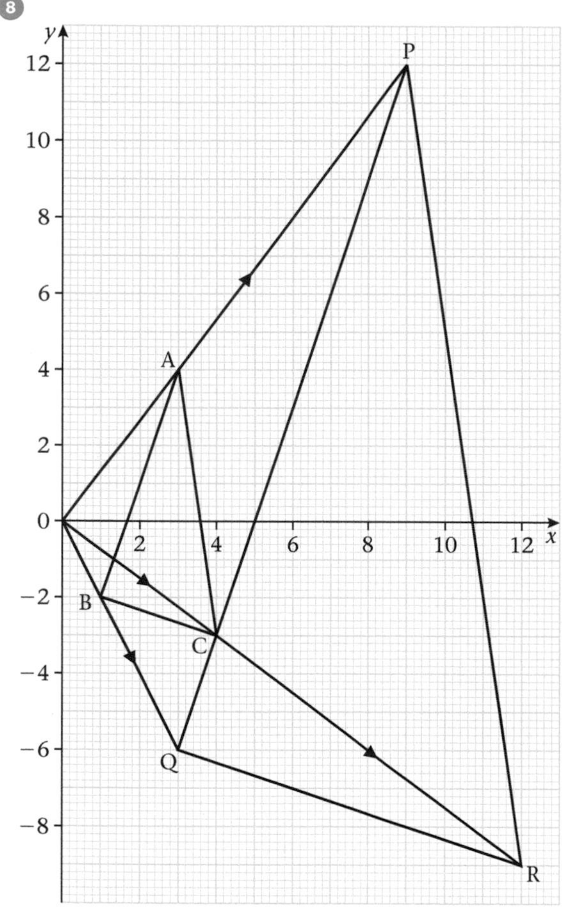

△PQR is an enlargement of △ABC, scale factor 3
△PQR = 9△ABC

9

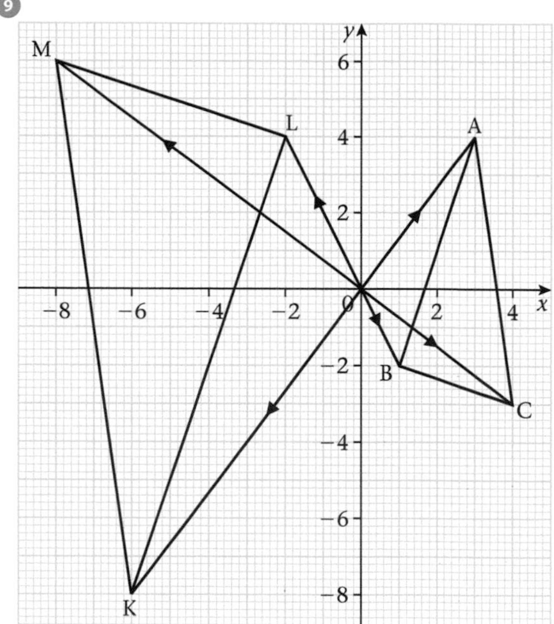

K(−6, −8), L(−2, 4), M(−8, 6)
△KLM is an enlargement of △ABC, scale factor 2
△KLM = 4△ABC

Exercise 18e (p. 226)

1 (a) $\begin{pmatrix}-5\\-9\end{pmatrix}$ (b) $\begin{pmatrix}5\\9\end{pmatrix}$

2 (a) $\begin{pmatrix}3\\8\end{pmatrix}$ (b) $\begin{pmatrix}10\\3\end{pmatrix}$ (c) $\begin{pmatrix}2\\3\end{pmatrix}$
(d) $\begin{pmatrix}4\\2\end{pmatrix}$ (e) $\begin{pmatrix}7\\-5\end{pmatrix}$ (f) $\begin{pmatrix}-6\\-5\end{pmatrix}$

3 (a) $\begin{pmatrix}-1\\3\end{pmatrix}$ (b) $\begin{pmatrix}8\\-1\end{pmatrix}$ (c) $\begin{pmatrix}6\\1\end{pmatrix}$
(d) $\begin{pmatrix}3\\-5\end{pmatrix}$ (e) $\begin{pmatrix}5\\4\end{pmatrix}$ (f) $\begin{pmatrix}9\\-4\end{pmatrix}$
(g) $\begin{pmatrix}5\\4\end{pmatrix}$ (h) $\begin{pmatrix}8\\-1\end{pmatrix}$ (i) $\begin{pmatrix}-3\\5\end{pmatrix}$
(j) $\begin{pmatrix}-8\\1\end{pmatrix}$

4 (a)

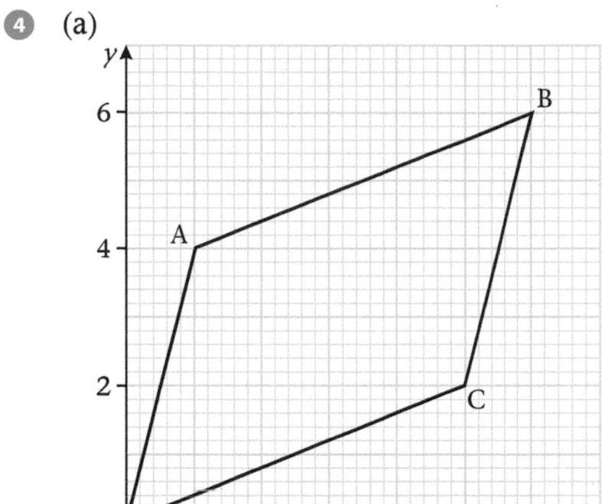

A(1, 4), B(6, 6) C(5, 2)

(b) (i) $\begin{pmatrix} 6 \\ 6 \end{pmatrix}$ (ii) $\begin{pmatrix} -4 \\ 2 \end{pmatrix}$

5 $\overrightarrow{QR} = \begin{pmatrix} 4 \\ -1 \end{pmatrix}$ and $\overrightarrow{PS} = \begin{pmatrix} 8 \\ -2 \end{pmatrix} = 2\begin{pmatrix} 4 \\ -1 \end{pmatrix}$
$\Rightarrow$ QR $\parallel$ PS

6 AB = BC = CD = DA [= $\sqrt{125}$] $\Rightarrow$ ABCD is a rhombus

7 $\overrightarrow{OA} = \overrightarrow{CB}\left[= \begin{pmatrix} 4 \\ 0 \end{pmatrix}\right]$
$\Rightarrow$ OA $\parallel$ CB and OA = CB
$\Leftrightarrow$ OABC is a parallelogram

8 $\overrightarrow{PQ} = \overrightarrow{SR}\left[= \begin{pmatrix} 6 \\ 3 \end{pmatrix}\right]$
$\Rightarrow$ PQ $\parallel$ SR and PQ = SR
$\Leftrightarrow$ PQRS is a parallelogram, diagonals intersect at (7, 5)

9 Q(12, 8), diagonals intersect at (6, 4)

10 (a) $\begin{pmatrix} 8 \\ 6 \end{pmatrix}$ (b) 10 (c) P(−7, −10)

Exercise 18f (p. 228)

1 (a) $\overrightarrow{PR}$ (b) $\overrightarrow{PS}$ (c) $\overrightarrow{PT}$ (d) $\overrightarrow{PT}$
(e) $\overrightarrow{PS}$ (f) $\overrightarrow{PS}$ (g) $\overrightarrow{PT}$ (h) $\overrightarrow{PS}$

2 $\frac{1}{2}(\mathbf{a} + \mathbf{b})$

3 $\overrightarrow{XY} = \mathbf{b} - \mathbf{a}$, $\overrightarrow{YZ} = \mathbf{c} - \mathbf{b}$, $\overrightarrow{ZX} = \mathbf{a} - \mathbf{c}$

4 (a) $2\mathbf{b} - \mathbf{a}$ (b) $\mathbf{a} + \mathbf{b}$

5 (a) $\mathbf{t}$ (b) $-\mathbf{r}$ (c) $-\frac{1}{4}\mathbf{r}$
(d) $\mathbf{t} - \frac{1}{4}\mathbf{r}$ (e) $\frac{3}{4}\mathbf{r} + \mathbf{t}$

6 (a) $\mathbf{b} - \mathbf{a}$ (b) $\frac{1}{2}\mathbf{b}$ (c) $\frac{1}{2}\mathbf{b} - \mathbf{a}$

7 (a) $2\mathbf{a}$ (b) $\mathbf{b} - \mathbf{a}$
(c) $1\frac{1}{2}\mathbf{a} - \frac{1}{2}\mathbf{b}$ (d) $1\frac{1}{2}\mathbf{a} + \frac{1}{2}\mathbf{b}$

8 (a) $2\mathbf{x}$ (b) $\mathbf{x} + \mathbf{y}$ (c) $\mathbf{x} + \mathbf{y}$
(d) $\mathbf{x} + 2\mathbf{y}$ (e) $2\mathbf{x} + 2\mathbf{y}$ (f) $2\mathbf{x} + \mathbf{y}$

9 (a) $\frac{4}{5}\mathbf{x}$ (b) $\frac{1}{2}(\mathbf{x} + \mathbf{y})$ (c) $1\frac{1}{2}\mathbf{y}$
(d) $\frac{1}{2}\mathbf{y} - \frac{3}{10}\mathbf{x}$ (e) $\mathbf{y} - \frac{1}{2}\mathbf{x}$

12 (a) $\mathbf{b} - \mathbf{a}$, $\frac{1}{2}\mathbf{a}$, $\frac{1}{2}\mathbf{b}$, $\frac{1}{2}(\mathbf{b} - \mathbf{a})$
(b) MN $\parallel$ AB and MN = $\frac{1}{2}$AB

15 (a) $\overrightarrow{RP}$ $\mathbf{a} - \mathbf{b}$, $\overrightarrow{QS} = 3\mathbf{a} - \mathbf{b}$

16 (a) (i) $\mathbf{b} - \mathbf{a}$, (ii) $\frac{1}{4}\mathbf{a} + \frac{3}{4}\mathbf{b}$, (iii) $\frac{1}{2}\mathbf{a} - \mathbf{b}$
(b) $\frac{1}{2}h\mathbf{a} + (1 - h)\mathbf{b}$
(c) $h = \frac{2}{5}$, $k = \frac{4}{5}$ (d) $\frac{1}{5}\mathbf{a} + \frac{3}{5}\mathbf{b}$

17 (a) (i) $\frac{1}{2}\mathbf{a}$, (ii) $\mathbf{c}$,
(iii) $\mathbf{c} - \mathbf{a}$, (iv) $\frac{2}{7}(\mathbf{c} - \mathbf{a})$
(b) $\frac{3}{14}\mathbf{a} + \frac{2}{7}\mathbf{c}$ (c) $\frac{1}{2}\mathbf{a} + p\mathbf{c}$
(d) $\frac{3}{14}q\mathbf{a} + \frac{2}{7}q\mathbf{c}$
(c) $p = \frac{2}{3}$, $q = 2\frac{1}{3}$, AY : YB = 2 : 1

18 (a) $\frac{1}{2}(\mathbf{x} + \mathbf{y})$ (b) $\frac{1}{2}h\mathbf{x} + \frac{1}{2}h\mathbf{y}$
(c) $\frac{1}{2}\mathbf{y} - \mathbf{x}$ (d) $(1 - k)\mathbf{x} + \frac{1}{2}k\mathbf{y}$
(e) $h = k = \frac{2}{3}$
(f) The lines meet each other at a single point and divide each other in the ratio 2 : 1.

Exercise 18g (p. 231)

1 95 km/h, 030° 2 60 km/h E
3 94 km/h, 033° 4 480 km/h, 037°

Practice exercise P18.1 (p. 232)

1 (a) $\begin{pmatrix} 1 \\ 9 \end{pmatrix}$ (b) $\begin{pmatrix} 1 \\ 11 \end{pmatrix}$ (c) $\begin{pmatrix} 13 \\ -4 \end{pmatrix}$

2 $\overrightarrow{AB} = \begin{pmatrix} -2 \\ -6 \end{pmatrix}$, $\overrightarrow{CB} = \begin{pmatrix} -2 \\ -3 \end{pmatrix}$

3

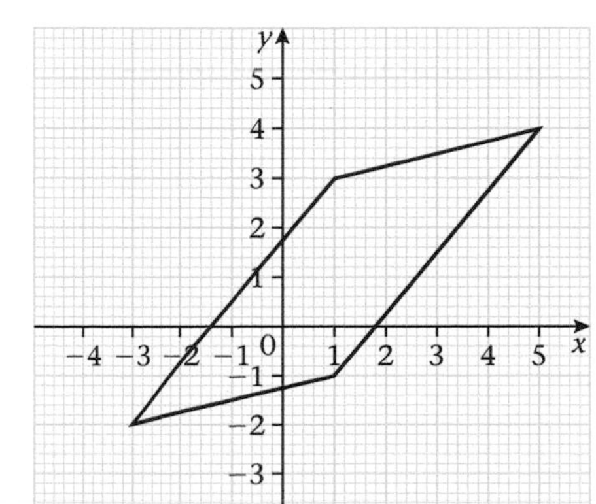

$\overrightarrow{PQ} = \begin{pmatrix} 4 \\ 1 \end{pmatrix}$, $\overrightarrow{RS} = \begin{pmatrix} -4 \\ -1 \end{pmatrix}$

4 (a)

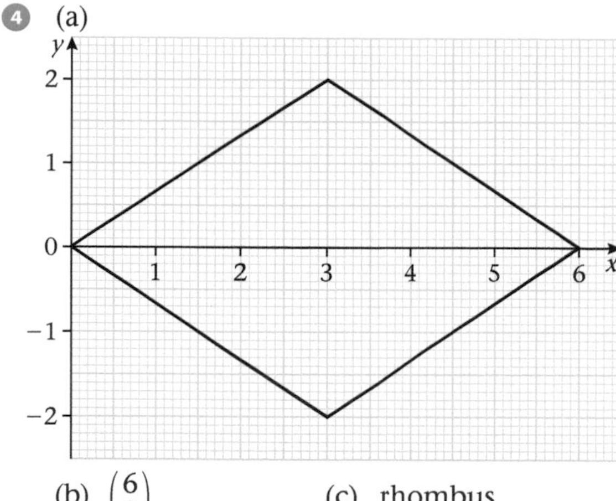

(b) $\begin{pmatrix}6\\0\end{pmatrix}$ (c) rhombus

5 $\overrightarrow{OB} = \begin{pmatrix}x_1\\y_1\end{pmatrix}$ and $\overrightarrow{OC} = \begin{pmatrix}x_2\\y_2\end{pmatrix}$.
Since △ADE is an enlargement by scale factor 2 and centre the origin,
$\overrightarrow{OD} = \begin{pmatrix}2x_1\\2y_1\end{pmatrix}$ and $\overrightarrow{OE} = \begin{pmatrix}2x_2\\2y_2\end{pmatrix}$
$\overrightarrow{BC} = \begin{pmatrix}x_2 - x_1\\y_2 - y_1\end{pmatrix}$ and $\overrightarrow{DE} = \begin{pmatrix}2x_2 - x_1\\2y_2 - y_1\end{pmatrix}$
But $\begin{pmatrix}2x_2 - x_1\\2y_2 - y_1\end{pmatrix} = 2\begin{pmatrix}x_2 - x_1\\y_2 - y_1\end{pmatrix}$,
so the length of DE = 2 × the length of BC or BC = $\frac{1}{2}$DE.

6 $\overrightarrow{OR} = \begin{pmatrix}-4\\3\end{pmatrix}$, $\overrightarrow{OS} = \begin{pmatrix}0\\13\end{pmatrix}$.
$\overrightarrow{RS} = -\begin{pmatrix}-4\\3\end{pmatrix} + \begin{pmatrix}0\\13\end{pmatrix} = \begin{pmatrix}4\\10\end{pmatrix} = 2\begin{pmatrix}2\\5\end{pmatrix}$
Therefore PQ is parallel to RS.

7 (a) $\overrightarrow{OA} = \begin{pmatrix}3\\2\end{pmatrix}$, $\overrightarrow{OB} = \begin{pmatrix}5\\3\end{pmatrix}$, $\overrightarrow{OC} = \begin{pmatrix}4\\-1\end{pmatrix}$
(b) 3.61 units, 5.83 units, 4.12 units

8 Let $\overrightarrow{AB}$ = **a** and $\overrightarrow{AC}$ = **c** therefore
$\overrightarrow{BC} = -\mathbf{a} + \mathbf{c}$.
Since D, E and F are the midpoints of the sides of △ABC,
$\overrightarrow{AD} = \frac{1}{2}\mathbf{a}$, $\overrightarrow{AF} = \frac{1}{2}\mathbf{c}$ and $\overrightarrow{AE} = \mathbf{a} + \frac{1}{2}(-\mathbf{a} + \mathbf{c})$.
Thus FE $= -\frac{1}{2}\mathbf{c} + \mathbf{a} + \frac{1}{2}(-\mathbf{a} + \mathbf{c}) = \frac{1}{2}\mathbf{a}$,
DE $= -\frac{1}{2}\mathbf{a} + \mathbf{a} + \frac{1}{2}(-\mathbf{a} + \mathbf{c}) = \frac{1}{2}\mathbf{c}$ and
DF $= -\frac{1}{2}\mathbf{a} + \frac{1}{2}\mathbf{c} = \frac{1}{2}(-\mathbf{a} + \mathbf{c})$.
So FE is parallel to and half the length of AB, DE is parallel to and half the length of AC and DF is parallel to and half the length of BC. As △ABC is an equilateral triangle, △DEF must also be an equilateral triangle.

Revision exercise 19a (p. 233)

1 (a) (i) relatives/family members
(ii) female relatives/family members
(iii) male relatives/family members
(b) (i) the letters of the alphabet from a to g
(ii) vowels
(iii) consonants
(c) (i) shapes with straight sides
(ii) triangles
(iii) shapes with four or more sides
(d) (i) integers from 1 to 9
(ii) even numbers
(iii) odd numbers

2 (a) finite (b) infinite
(c) finite (d) finite
(e) empty (f) finite
(g) finite (h) empty
(i) empty (j) infinite

3 (a) (i) 7
(ii) {0}, {1}, {2}, {0, 1}, {0, 2}, {1, 2}, {0, 1, 2}
(b) (i) 15
(ii) {*m*}, {*e*}, {*t*}, {*r*}, {*m*, *e*}, {*m*, *t*}, {*m*, *r*}, {*e*, *t*}, {*e*, *r*}, {*t*, *r*}, {*m*, *e*, *t*}, {*m*, *e*, *r*}, {*m*, *t*, *r*}, {*e*, *t*, *r*}, {*m*, *e*, *t*, *r*}

4 (a) (i) Y′ (ii) S′ (iii) Y ∩ H (iv) S′ ∩ Y′
(b) *All unsuccessful men are unhappy.*
All unhappy men are old.

5

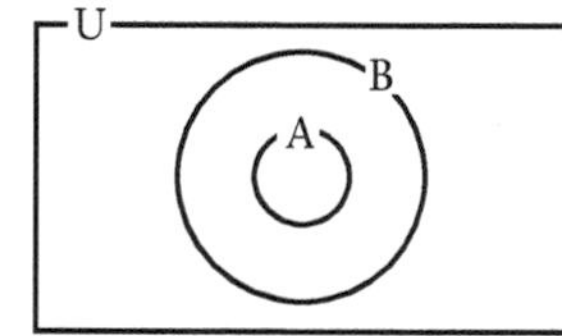

Revision exercise 19b (p. 234)

1 (a) {1, 3, 7, 21}
(b) {2, 3, 11}
(c) {−3, −2, −1, 0, 1, ...}
(d) {−5, −6, −7, ...}
(e) {−3, −2, −1, 0, 1, 2, 3}
(f) {2}

2 (a) equal
(b) equivalent
(c) equivalent
(d) not equivalent
(e) equal
(f) not equivalent

Revision exercise 19c (p. 234)

1 (a)

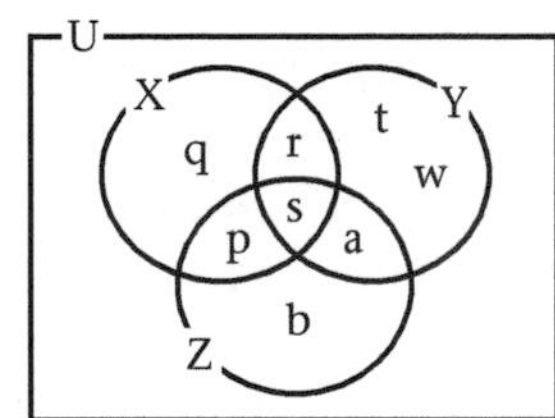

(b) (i) {p, r, s, a, b}

(ii) {r, s, p}

2 (a) $P \cap R$

(i)

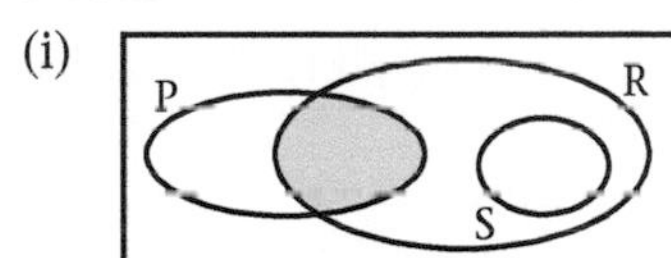

(ii)

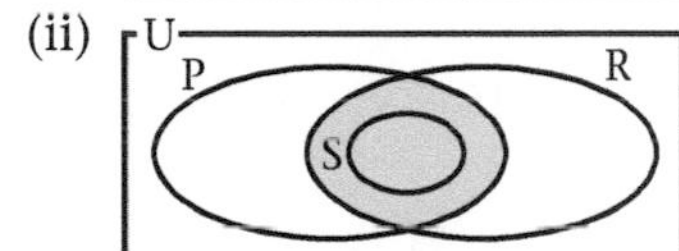

(iii)

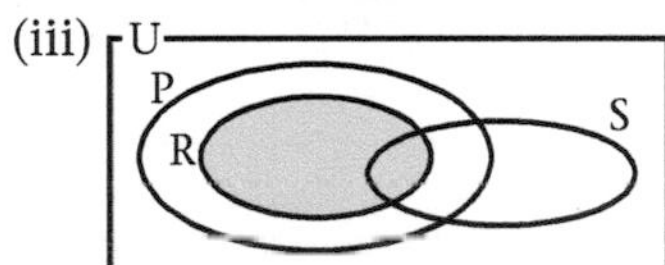

(b) $R \cup S$

(i)

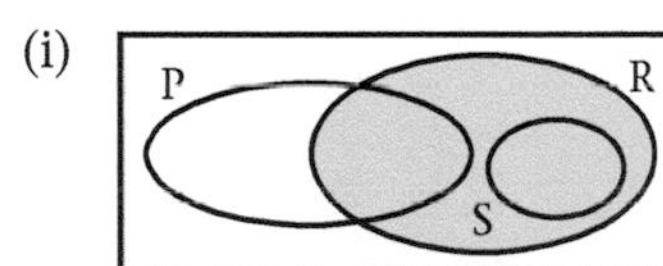

(ii)

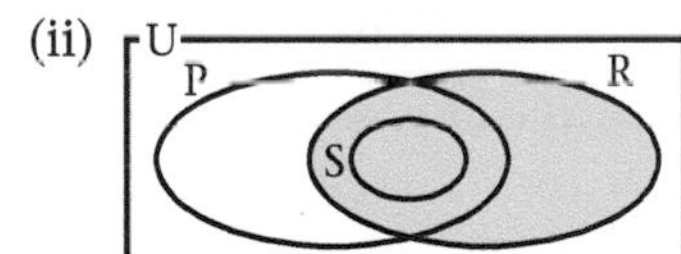

(iii)

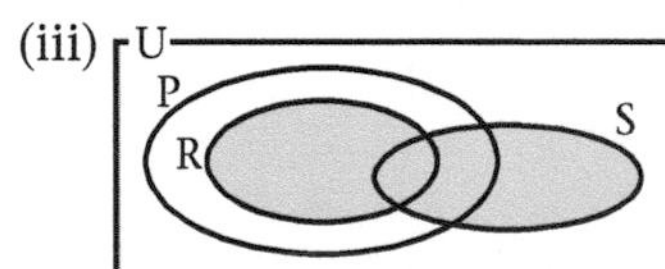

(c) $(S \cup P)'$

(i)

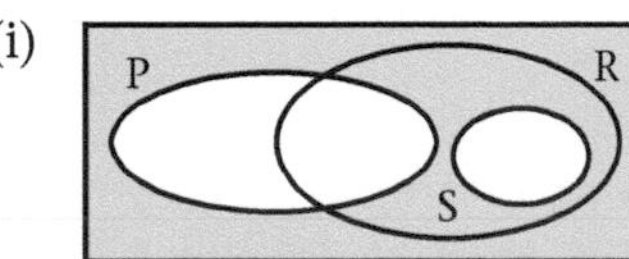

(ii)

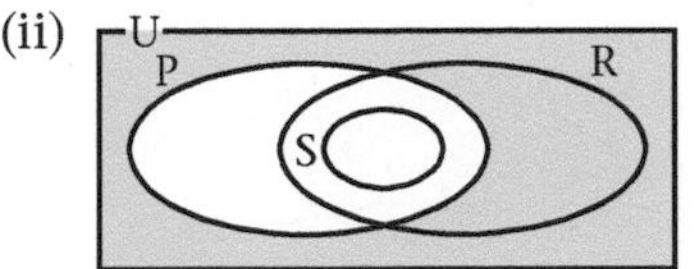

(iii)

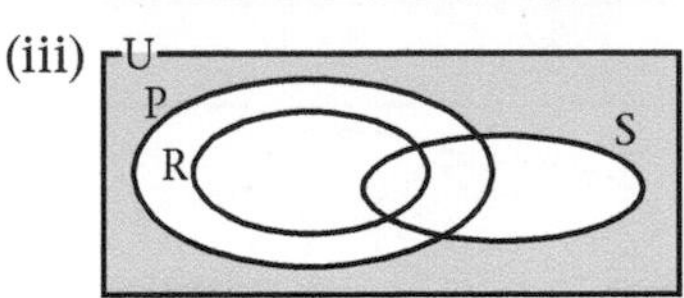

(d) $(P \cup R) \cap S$

(i)

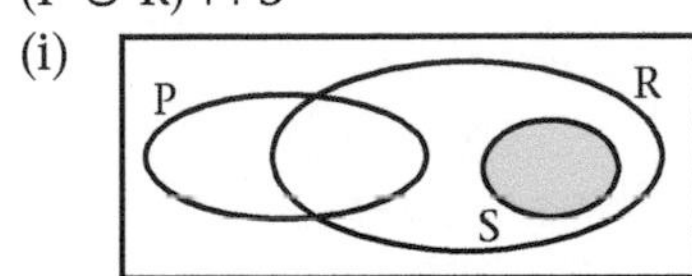

(ii)

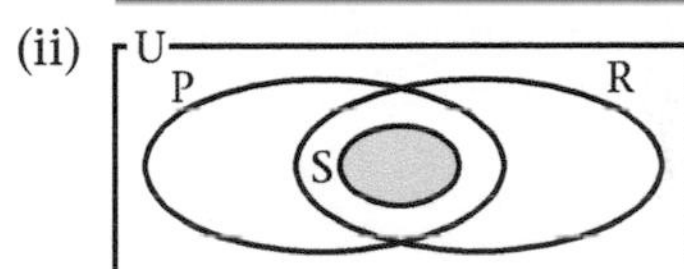

(iii)

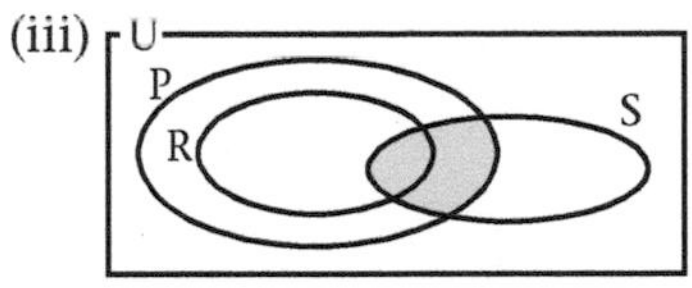

(e) $(P \cap R) \cup S$

(i)

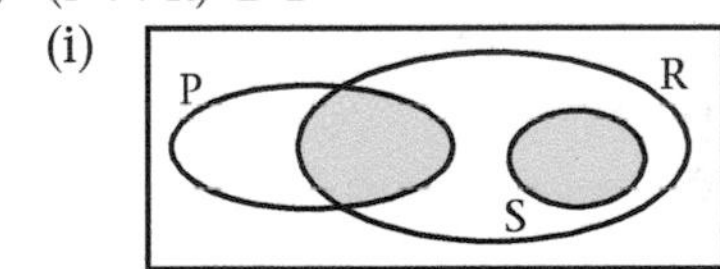

(ii)

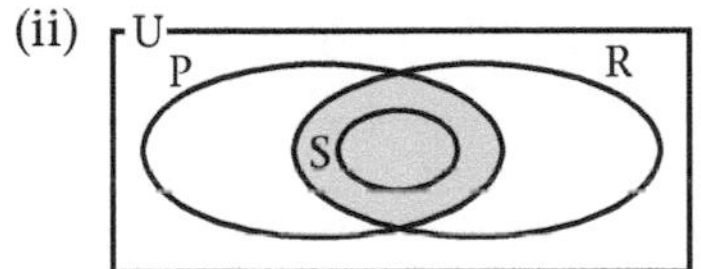

(iii)

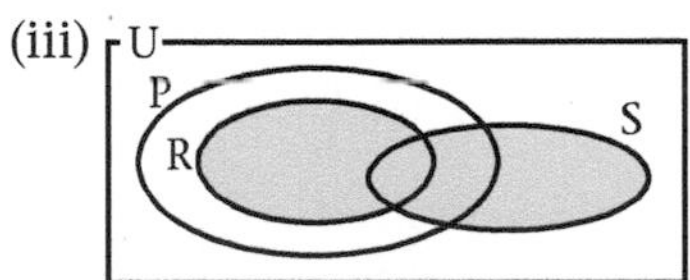

(f) $P \cup (R \cap S')$

(i)

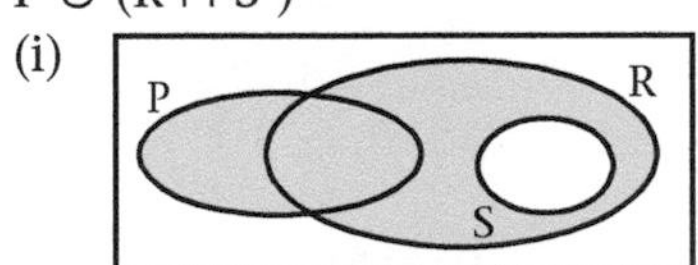

(ii)

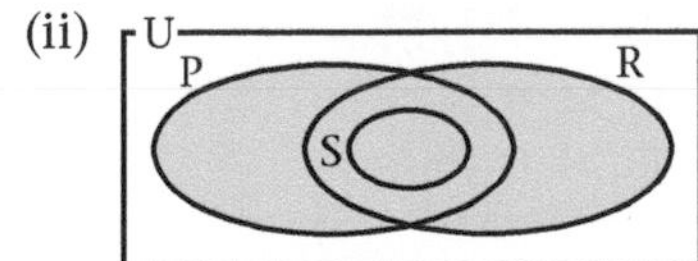

(iii)

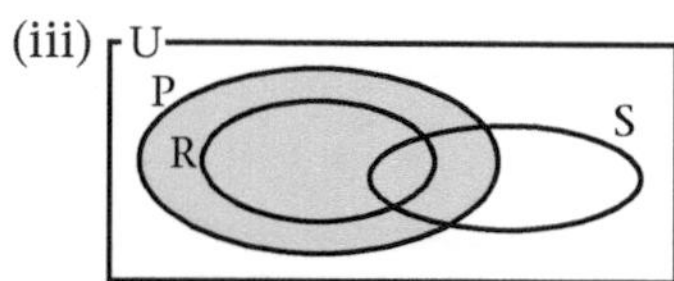

(g) P ∩ (R ∪ S′)

(i)

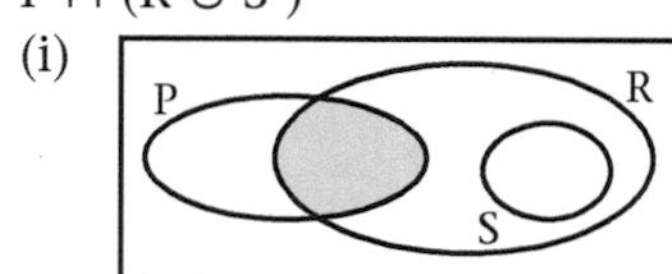

(ii)

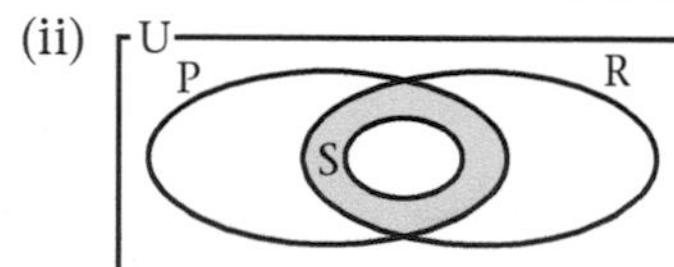

(iii)

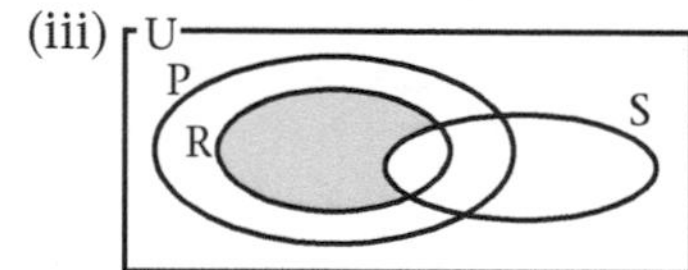

3 (a) 6 (b) 28 (c) 0 (d) 15

4 (a) (i) U = {3, 4, …, 28, 29}

(ii) A = {17, 19, 21, 23, 25, 27, 29}
B = {3, 5, 7, 9, 11, 13, 17, 19, 23, 29}

(iii) A ∩ B = {17, 19, 23, 29]

(iv) (A ∩ B)′ = {3, 4, …, 15, 16, 18, 20, 21, 22, 24, 25, 26, 27, 28}

(v) A ∪ B = {3, 5, 7, 9, 11, 13, 17, 19, 21, 23, 25, 27, 29}

(vi) (A ∪ B)′ = {4, 6, 8, 10, 12, 14, 15, 16, 18, 20, 22, 24, 26, 28}

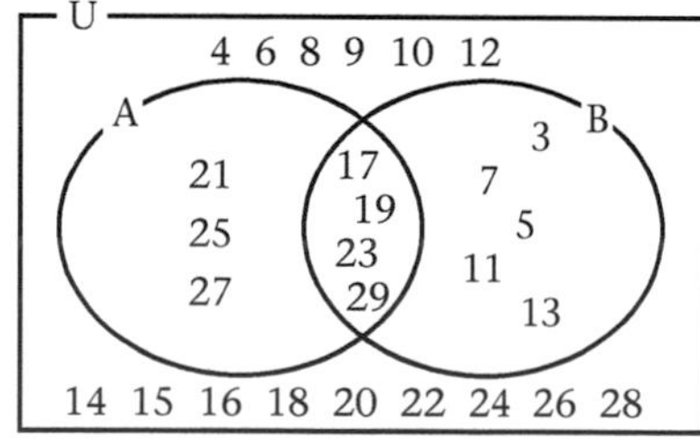

(b) (i) U = {3, 4, …, 28, 29}

(ii) F = {3, 4, 5, 6, 8, 10, 12, 15, 20, 24}
P = {3, 5, 7}
M = {5, 10, 15}

(iii) F ∩ P = {3, 5}
F ∩ M = {5, 10, 15}
P ∩ M = {5}
F ∩ P ∩ M = {5}

(iv) (F ∩ P)′ = {4, 6, 7, …, 29}
(F ∩ M)′ = {3, 4, 6, 7, 8, 9, 11, 12, 13, 14, 16, 17, …, 29}
(P ∩ M)′ = {3, 4, 6, 7, …, 29}
(F ∩ P ∩ M)′ = {3, 4, 6, 7, …, 29}

(v) F ∪ P = {3, 4, 5, 6, 7, 8, 10, 12, 15, 20, 24}
F ∪ M = {3, 4, 5, 6, 8, 10, 12, 15, 20, 24}
P ∪ M = {3, 5, 7, 10, 15}
F ∪ P ∪ M = {3, 4, 5, 6, 7, 8, 10, 12, 15, 20, 24}

(vi) (F ∪ P)′ = {9, 11, 13, 14, 16, 17, 18, 19, 21, 22, 23, 25, 26, 27, 28, 29}
(F ∪ M)′ = {7, 9, 11, 13, 14, 16, 17, 18, 19, 21, 22, 23, 25, 26, 27, 28, 29}
(P ∪ M)′ = {4, 6, 8, 9, 11, 12, 13, 14, 16, 17, 18, 19, 20, 21, 22, 23, 24, 25, 26, 27, 28, 29}
(F ∪ P ∪ M)′ = {9, 11, 13, 14, 16, 17, 18, 19, 21, 22, 23, 25, 26, 27, 28, 29}

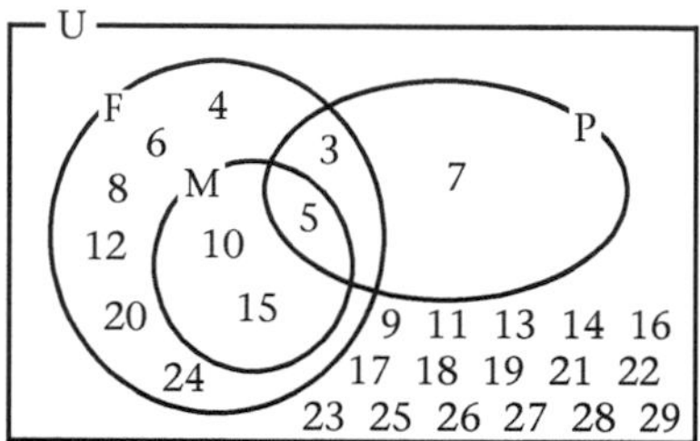

5 (a) H ∩ K′

(b) (L ∩ N) ∪ (L ∩ M′)

6 (a) (i) Possible answers include:
A = {prime factors of 210}
B = {even numbers between 0 and 10}

(ii) A ∩ C = {2}

(iii) (A ∩ C′) = {3, 4, 5, 6, 7, 8}

(iv) A ∪ B = {2, 3, 4, 5, 6, 7, 8}

(v) (A ∪ B)′ = { }

(b) (i) Possible answers include:
X = {multiples of 3 less than 30}
Y = {multiples of 6 less than 30}
W = {multiples of 9 less than 30}

(ii) X ∩ Y = {6, 12, 18, 24}
X ∩ W = {9, 18, 27}
Y ∩ W = {18}
X ∩ Y ∩ W = {18}

(iii) $(X \cap Y)' = \{3, 9, 15, 21, 27\}$
$(X \cap W)' = \{3, 6, 12, 15, 21, 24\}$
$(Y \cap W)' = \{3, 6, 9, 12, 15, 21, 24, 27\}$
$(X \cap Y \cap W)' = \{3, 6, 9, 12, 15, 21, 24, 27\}$

(iv) $X \cup Y = \{3, 6, 9, 12, 15, 18, 21, 24, 27\}$
$X \cup W = \{3, 6, 9, 12, 15, 18, 21, 24, 27\}$
$Y \cup W = \{6, 9, 12, 18, 24, 27\}$
$X \cup Y \cup W = \{3, 6, 9, 12, 15, 18, 21, 24, 27\}$

(v) $(X \cup Y)' = \{\ \}$
$(X \cup W)' = \{\ \}$
$(Y \cup W)' = \{3, 15, 21\}$
$(X \cup Y \cup W)' = \{\ \}$

7 (a) $m = 40 - x$ (b) 45
(c) 15
(d) $x = 12, r = 3, y = 57$

8 C = {students who use computers}
R = {students who read}
A = {clever students}
Some students both use computers and read therefore $C \cap R$.
All clever students read therefore $A \subset R$.
Some clever students use computers therefore $A \cap C$.

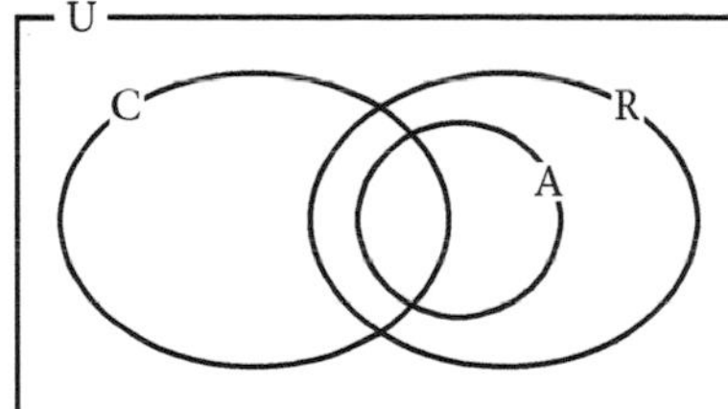

9 (a) (i) All geometry books are mathematics books.
Some but not all mathematics books have pictures.
Some but not all books with pictures are mathematics books.
Geometry books do not have pictures.

(ii) $G \subset M, M \cap P, M' \cap P', G \cap P = \{\ \}$

(b) (i) All geometry books are mathematics books.
Some but not all mathematics books are geometry books.
All mathematics books have pictures.
Some but not all books with pictures are mathematics books.

(ii) $G \subset M, M' \cap G = \{\ \}, M \subset P, P' \cap M = \{\ \}$

(c) (i) Mathematics books do not have pictures.
All geometry books are mathematics.
Some but not all mathematics books are geometry books.

(ii) $M \cap P = \{\ \}, G \subset M, M' \cap G = \{\ \}$

(d) (i) Some but not all mathematics books have pictures.
Some but not all geometry books have pictures.
Some books with pictures are not mathematics books.
All geometry books are mathematics books.

(ii) $M \cap P, G \cap P, P \cap M', G \subset M$

10 (a) (i) $A = \{x: x \text{ is a multiple of } 6\}$
$B = \{y: y \text{ is a multiple of } 16\}$
$A \cap B = \{\text{common multiples of 6 and 16}\}$

(ii) $A = \{6, 12, 18, 24, 30, 36, 42, 48, \ldots\}$
$B = \{16, 32, 48, \ldots\}$
$A \cap B = \{48, \ldots\}$

(iii) 48

(b) (i) $A = \{x: x \text{ is a perfect square}\}$
$B = \{x: 15 < x < 40, x \text{ is a natural number}\}$
$A \cap B = \{\text{perfect squares greater than 15 and less than 40}\}$

(ii) $A = \{1, 4, 9, 16, 25, 36, 49, \ldots\}$
$B = \{16, 17, \ldots, 38, 39\}$
$A \cap B = \{16, 25, 36\}$

(iii) 16

11 (a)

U
2
X
Y
Z
6
3
5
8

(b) $n(U) = n(X \cup Y \cup Z) + n(X \cup Y \cup Z)'$ or $n(X \cup Y) + n(X \cup Y)'$

(c) 24 (d) $Z \subset Y$

12 (a) (i) U = {boxes with straight edges}
T = {boxes made of glass}
W = {boxes made of wood}
G = {gift boxes}
$G \subset T$

(ii)

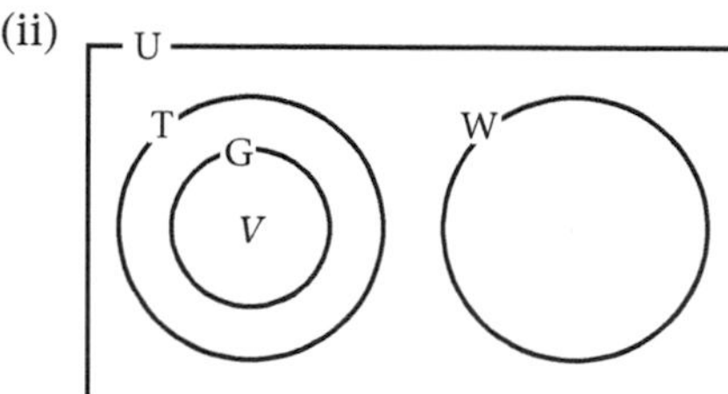

(iii) $V \in G$

(b) U = {fifth formers}
G = {students studying Geography}
M = {students studying Mathematics}
B = {students studying Biology}
$G \cap M, B \cap M, G \cap B \cap M = \{\ \}$

(ii)

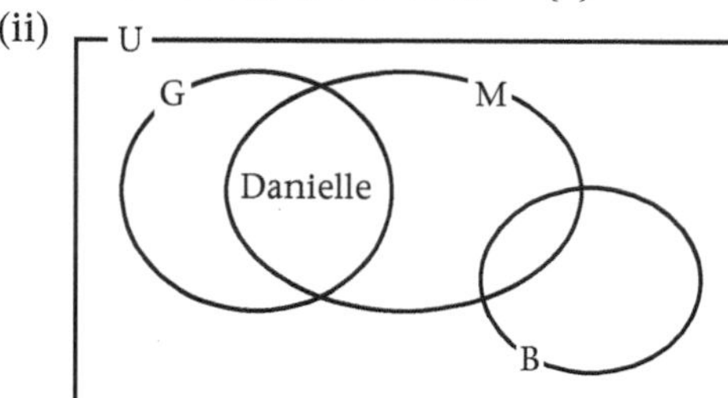

(iii) Danielle $\in (G \cap M)$
Tim $\in (G \cup B)'$

Exercise 20a (p. 238)

1. $(3c - d)(3m - 4n)$
2. $(a + b)(c + 2d)$
3. $(x + 4a)(x - 4a)$
4. $(2c + 5d)(2c - 5d)$
5. $(a + 2)(a - 5)$
6. $(a + 2b)(a - 5b)$
7. $(ab + 2)(ab - 5)$
8. $5(m - 3n)(m + 3n)$
9. $(h - 2k)(m - 2n)$
10. $(2a + b)(a + 4b)$
11. $(3m - 1)(m - 3)$
12. $(cd - 9)(cd + 9)$
13. $(4x + 3am)(4x - 3am)$
14. $(2n + 3)(3n + 2)$
15. $(2a + 5)^2$
16. $9(h - 2k)(h + 2k)$
17. $(a - 2b)(a - 4b)$
18. $(5abc + 3d)(5abc - 3d)$
19. $\left(\frac{m}{3} - \frac{n}{2}\right)\left(\frac{m}{3} + \frac{n}{2}\right)$
20. $(3s + t)(u - 2v)$
21. $(3x - 2y)^2$
22. $(4d - 1)(3d + 2)$
23. $(x^2 + y)(x^2 - y)$
24. $(2mn - 3)(5mn + 4)$
25. $(4 + n^2)(2 + n)(2 - n)$
26. $(a + b + c)(a + b - c)$
27. $(x + m - n)(x - m + n)$
28. $2(m + 2n)(a + b)$
29. $(2 + 5h)(1 - 3h)$
30. $(2a - b)(m - 3n)$
31. $(a - 6b)(a - 9b)$
32. $(m - 18n)(m + 3n)$
33. $(c - 2d - 3e)(c - 2d + 3e)$
34. $(3x + 2y)(4x + 9y)$
35. $(h - k)(3h - 4k)$
36. $2a(3x + 2y)(c - 2d)$
37. $(2a - 9x)(3a + 4x)$
38. $(5a + 2m + 4n)(5a - 2m - 4n)$
39. $(7a - 4b)(3a + 4b)$
40. $(x - y)(x + y - 4)$

Exercise 20b (p. 239)

1. 2 400
2. 75.6
3. 308
4. 10 560

Exercise 20c (p. 239)

1. $\frac{a + b}{a}$
2. $\frac{m - n}{m + n}$
3. $2x + 1$
4. $\frac{x - 2}{x + 3}$
5. $-\frac{x + 5}{x + 3}$
6. $\frac{3 - a}{3 + a}$
7. $-\frac{a + b}{ab}$
9. $\frac{1}{x - 2}$

Exercise 20d (p. 240)

1. $\frac{1}{x^3}$
2. $\frac{x}{y}$
3. $\frac{1}{xy}$
4. $\frac{b^3}{a^2}$
5. $\frac{a^2}{b^6}$
6. $\frac{3}{x^{\frac{1}{2}}}$

Exercise 20e (p. 240)

1. 9
2. $\frac{1}{4}$
3. a^2
4. d
5. $\frac{8}{25}$
6. 4
7. 4
8. $\pm\frac{1}{8}$
9. $\pm 1\frac{1}{4}$
10. ± 3
11. 4
12. ± 0.3
13. 1
14. ± 2
15. $\pm\frac{64}{27}$
16. $\frac{25}{9}$
17. $\frac{2}{a^2}$
18. 1
19. $\pm\frac{1}{32}$
20. $\frac{6}{x^2}$
21. $\pm\frac{27}{64}$
22. $\pm\frac{2}{a^3}$
23. $\pm\frac{64}{27}$
24. $\frac{4a}{9b}$
25. $\frac{15b}{a}$

Exercise 20f (p. 240)

1. 9
2. 8
3. $\pm\frac{1}{4}$
4. 2
5. $\pm\frac{1}{27}$
6. ± 8

Exercise 20g (p. 241)

1. (a) $l \propto \frac{1}{b}$ or $l = \frac{k}{b}$
   (b) $r \propto \frac{1}{\sqrt{c}}$ or $r = \frac{k}{\sqrt{c}}$
   (c) $G \propto \frac{m_1 m_2}{d^2}$ or $G = \frac{k m_1 m_2}{d^2}$
2. $y = 20x$, $x = 1\frac{3}{4}$
3. $A = 6$, $M = 17\frac{1}{2}$
4. $P = 10\frac{1}{2}$, $Q = 16$
5. $D = 135$, $V = 5\frac{1}{4}$
6. (a) $P = \frac{3}{4}Q^2$ (b) $P = 75$ (c) $Q = 5$
7. (a) $xy = 30$ (b) $x = 2\frac{1}{2}$ (c) $y = 1\frac{1}{2}$
8. (a) $M = \frac{5}{8}R^3$ (b) $M = 625$ (c) $R = 1.6$
9. (a) $\sqrt{Y} = \frac{2}{3}Z$ (b) $Y = 100$ (c) $Z = 6$
10. $A = 30$, $B = 7\frac{1}{2}$
11. $P = 13\frac{1}{2}$, $R = 5$
12. $x \propto z^2$
13. $x \propto \frac{1}{z^2}$
14. (a) 12 h (b) 2 h (c) 1 h
15.

x	10	15	20	25	30	35
y	$\frac{1}{2}$	$\frac{1}{3}$	$\frac{1}{4}$	$\frac{1}{5}$	$\frac{1}{6}$	$\frac{1}{7}$

Exercise 20h (p. 242)

1. 7 cm
2. 11
3. 528 cm^2
4. 1 169.7
5. (a) 314.2 (b) 45.5
6. (a) 99c (b) 49 words
7. (a) \$138.70 (b) 151 kWh
8. (a) 30 km (b) 98 m

Exercise 20i (p. 243)

1. $T = \frac{100I}{PR}$
2. $h = \frac{4V}{\pi d^2}$, $d = 2\sqrt{\frac{V}{\pi h}}$
3. $t = \frac{mv - mu}{F}$
4. $M = N - RD$
5. $N = \sqrt{\frac{T - a}{b}}$
6. $r = \sqrt[3]{\frac{3V}{4\pi}}$
7. $h = \frac{S - 2\pi r^2}{2\pi r}$
8. $a = \frac{x}{\sqrt{1 - w^2}}$, $w = \sqrt{1 - \frac{x^2}{a^2}} = \frac{\sqrt{(a^2 - x^2)}}{a}$
9. $f = \frac{uv}{u + v}$, $u = \frac{vf}{v - f}$
10. $a = \frac{2s}{n} - (n - 1)d$
11. $Q = \frac{P(y + 3x)}{x - 3y}$
12. $s = \frac{u^2 - v^2}{2a}$, $u = \pm\sqrt{v^2 + 2as}$
13. $u = \pm\sqrt{v^2 - \frac{2Eg}{m}}$
14. $p = \frac{P(1 - eT)}{1 - et}$
15. $W = \frac{Tgx}{gx + v^2}$
16. $g = \frac{4\pi^2 l}{t^2}$
17. $h = \frac{V}{\pi r^2} - \frac{2}{3}r$
18. $y = \frac{(3x - 2)}{(x - 1)}$
19. (a) $T = \frac{100}{R}\left(\frac{A}{P} - 1\right)$ (b) 8 years
20. (a) $r = \frac{C}{2\pi}$ (b) $V = \frac{C^2 h}{4\pi}$

Exercise 20k (p. 245)

1. (a) $y = 2x$ (b) $x + y = 5$
   (c) $y < 2x$, $x + y \leqslant 5$, $y \geqslant 0$
2. $x > -4$, $x + y < -5$, $y \geqslant x - 2$
3. $x \geqslant 0$, $y > 1$, $x + y \leqslant 4$, $x + 2y \leqslant 6$
4. (a)

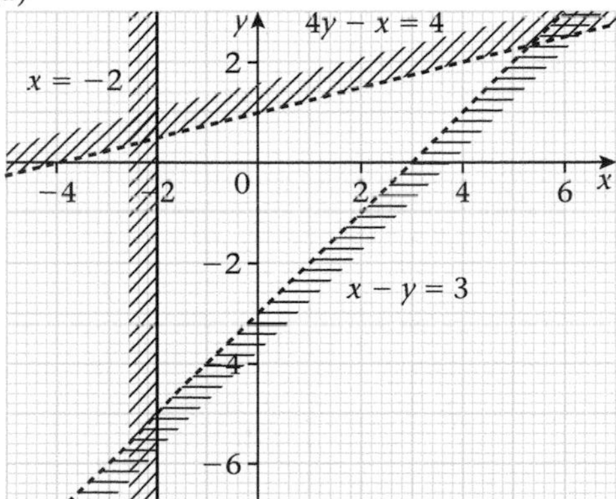

(b)

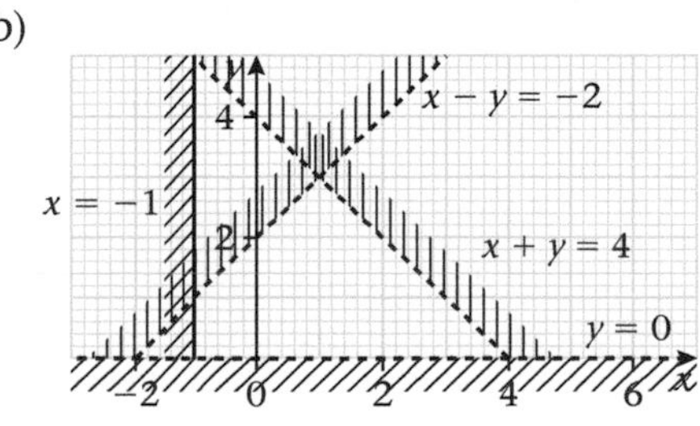

(c)

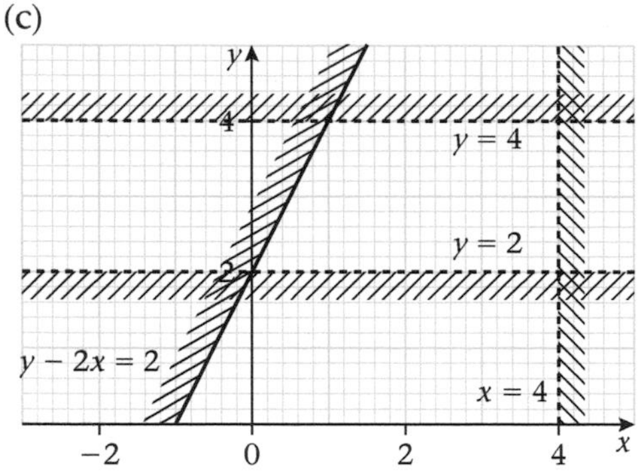

(d)

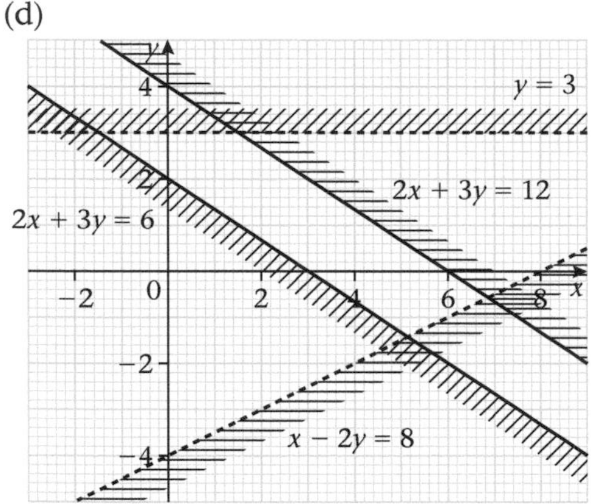

5 (a)

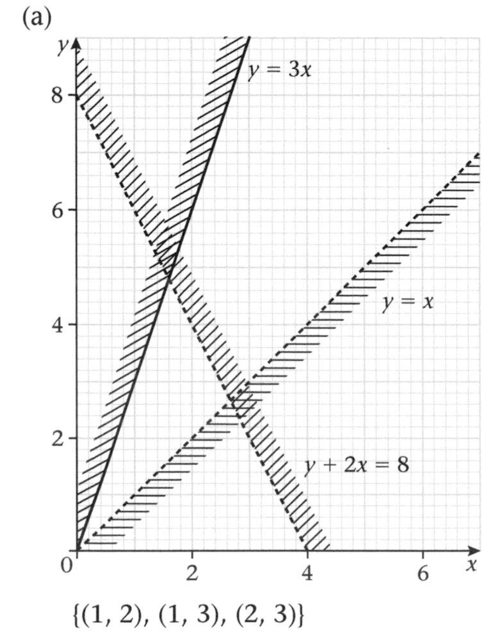

{(1, 2), (1, 3), (2, 3)}

(b)

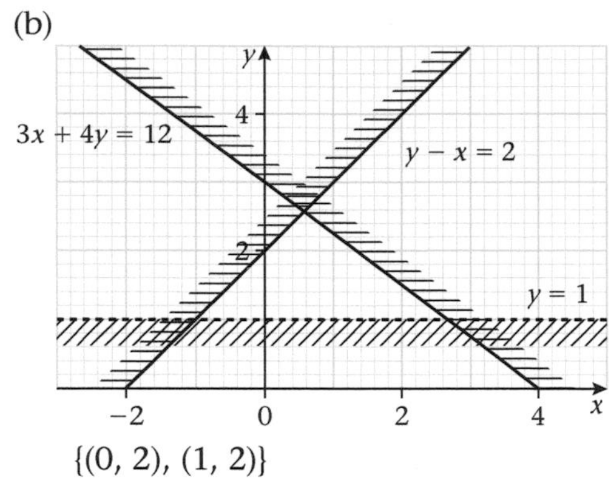

{(0, 2), (1, 2)}

6

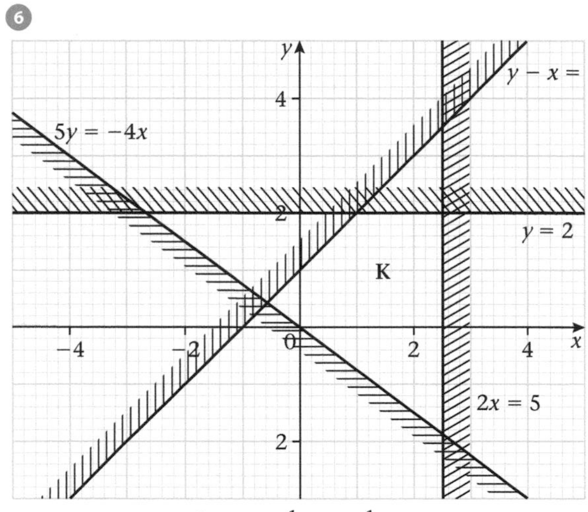

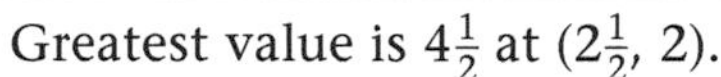

Greatest value is $4\frac{1}{2}$ at $(2\frac{1}{2}, 2)$.

7

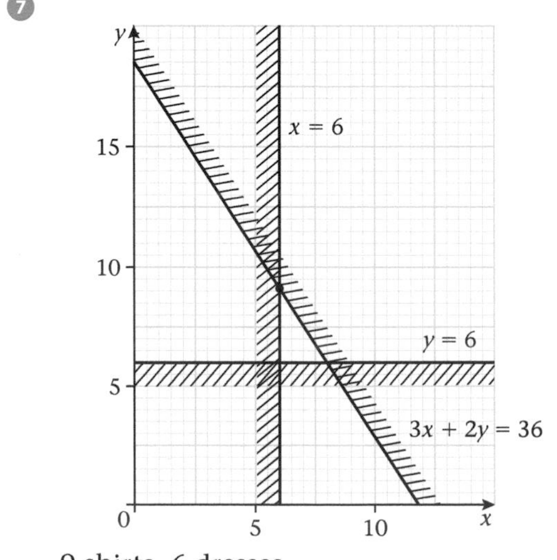

9 shirts, 6 dresses

8 (a)

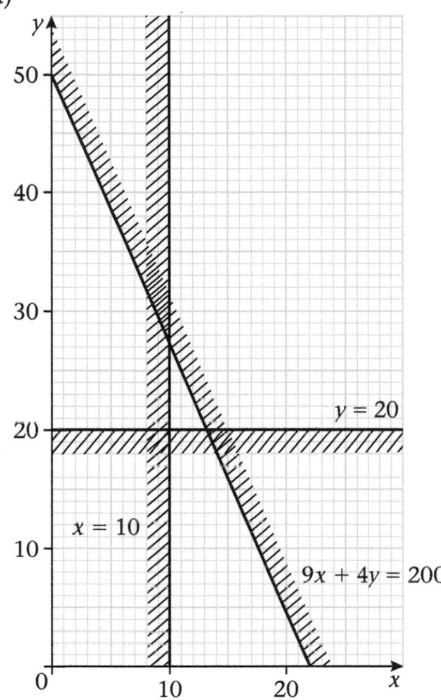

(b) maximum 37 (10 large, 27 small)

(c) (i) greatest profit from either (10 large, 27 small) or (11 large, 25 small)

(ii) $2.35

9 (a)

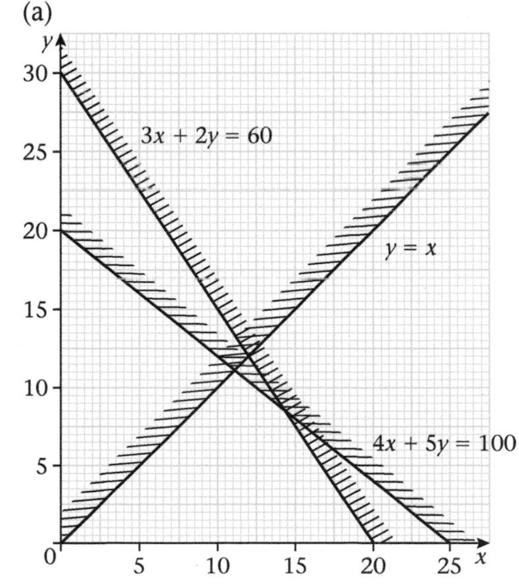

(b) (i) 15A and 7B

(ii) 12A and 10B

10 (a) $g + c \leqslant 50$; $g + 2c \leqslant 80$

(b)

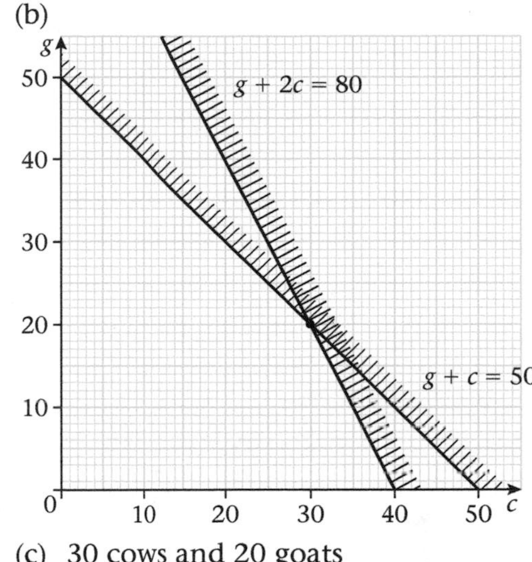

(c) 30 cows and 20 goats

Exercise 20l (p. 247)

1 3, −5	**2** 2, 1	**3** −2, −6
4 5, 0	**5** −3, 4	**6** 5, twice
7 0, −1	**8** $-3, -\frac{2}{3}$	**9** $-\frac{3}{4}$, twice
10 $3\frac{1}{2}, -2$	**11** 3, 5	**12** $\frac{1}{4}, -1\frac{1}{2}$
13 $5, 2\frac{1}{2}$	**14** $-1\frac{2}{3}, \frac{2}{5}$	**15** $-2\frac{1}{3}, \frac{1}{4}$

Exercise 20m (p. 247)

1 3, 7	**2** −1, −2	**3** 2, −3
4 5, −2	**5** 1, −2	**6** 0, −3
7 $2, 1\frac{1}{2}$	**8** 3, twice	**9** $-2, -\frac{3}{4}$
10 $5, \frac{2}{3}$	**11** 2, −2	**12** $\frac{3}{2}$, twice
13 0, 4	**14** $\frac{9}{4}, -\frac{2}{3}$	**15** $-1, -2\frac{1}{2}$
16 −5, twice	**17** $4, -\frac{2}{3}$	**18** $1\frac{1}{2}, \frac{3}{4}$
19 $-\frac{3}{2}$, twice	**20** $1\frac{1}{2}, -1\frac{3}{4}$	

Exercise 20n (p. 248)

1 (a) $v^2 + \frac{1}{4}v + (\frac{1}{8})^2 = (v + \frac{1}{8})^2$

(b) $c^2 - 1\frac{1}{2}c + (-\frac{3}{4})^2 = (c - \frac{3}{4})^2$

(c) $w^2 - \frac{3}{4}w + (-\frac{3}{8})^2 = (w - \frac{3}{8})^2$

2 (a) 4, −10 (b) 3.65, −1.65

3 (a) 14.2, −0.2 (b) 0.56, − 3.56

(c) 2.18, 0.15 (d) −1.18, −3.82

Exercise 20o (p. 248)

1. 2.73, −0.73
2. 2.41, −0.41
3. 3.24, −1.24
4. 0.41, −2.41
5. 1, −3
6. −1, 5
7. 3.73, 0.27
8. 5.73, 2.27
9. −1.59, −4.41
10. 5.24, 0.76
11. 1.87, −5.87
12. −0.59, −3.41
13. 1, 2
14. 5.14, −2.14
15. 0.19, −5.19
16. 2.79, −1.79
17. 1.71, 0.29
18. $-\frac{1}{2}$, 1
19. 2.29, −0.29
20. 9.47, 0.53
21. 2.37, −3.37
22. 3, −2
23. 6.46, −0.46
24. 5.08, −1.08
25. 3, $-1\frac{1}{2}$
26. 6.41, 3.59
27. 8, −5
28. 1.87, 0.13
29. 5, 7
30. 5.14, −2.14
31. 2.11, −7.11
32. 1, $\frac{2}{3}$
33. 2.32, −0.32
34. 2.76, 0.24
35. 5.10, −1.10

Exercise 20p (p. 250)

1

x	−2	−1	0	1	2	3	4	5
y	14	7	2	−1	−2	−1	2	7

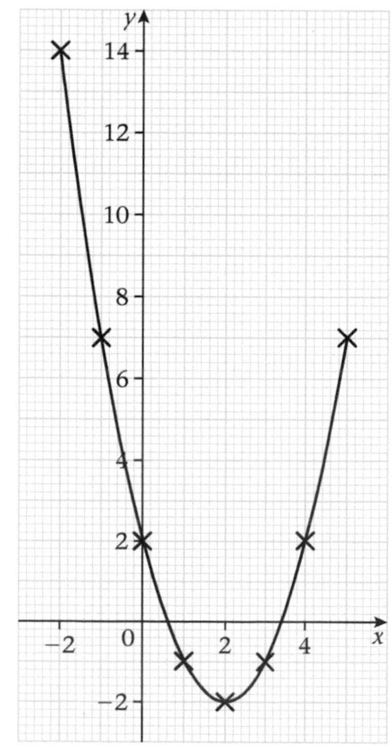

(a) $x = 3.4$ or 0.6 (b) −2

2

x	−2	−1	0	1	2	3	4
y	25	10	1	−2	1	10	25

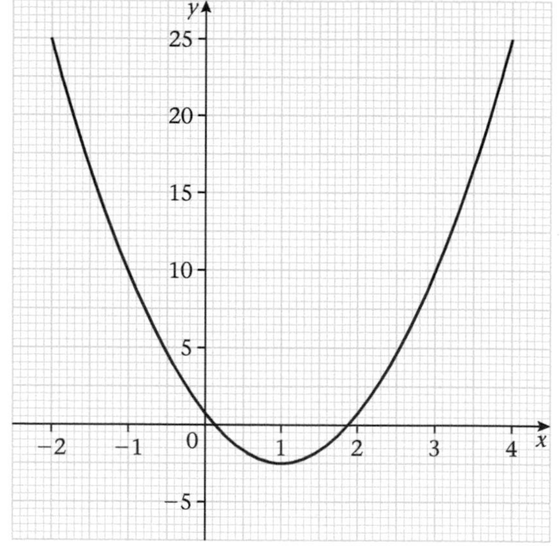

$x = 0.18, 1.82$

3

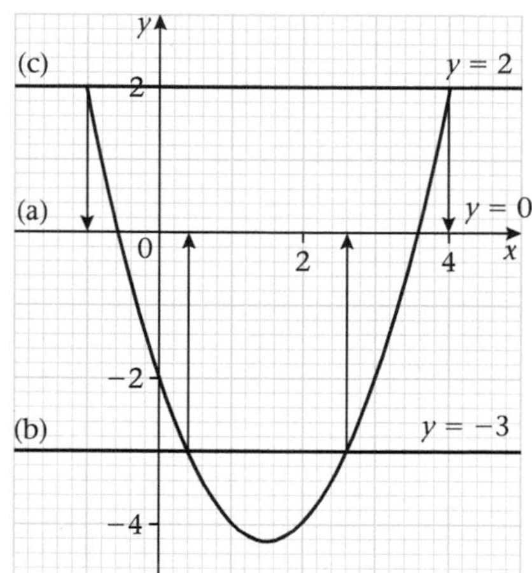

(a) 3.56, −0.56
(b) 2.62, 0.38
(c) 4, −1

4

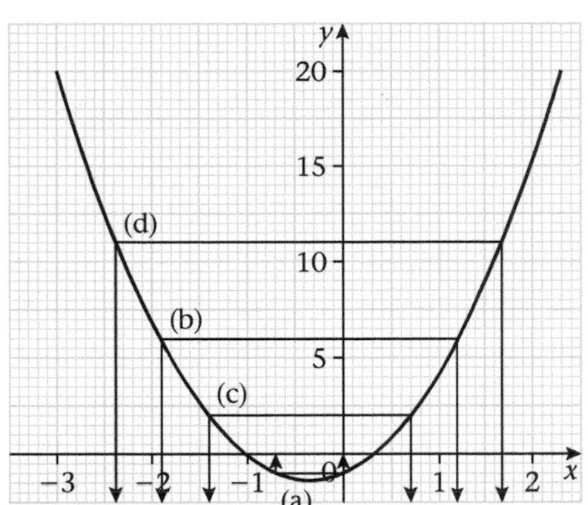

(a) 0, −0.67
(b) 1.23, −1.9
(c) 0.72, −1.39
(d) 1.69, −2.36

5

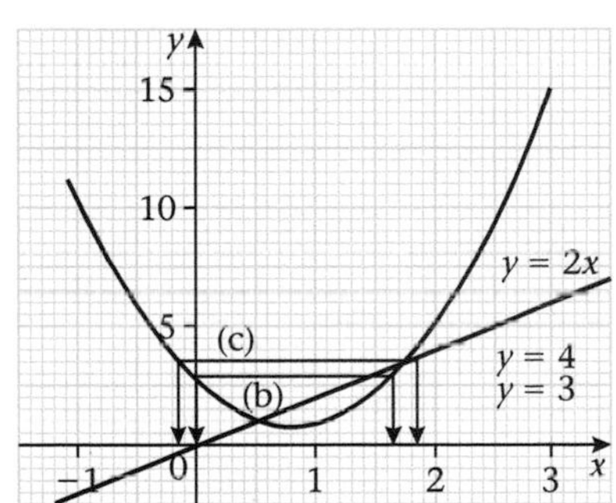

(a) imaginary roots
(b) 0, 1.67
(c) 1.85, −0.18

6 $x = 1.77$ or 0.57 (see above)

7

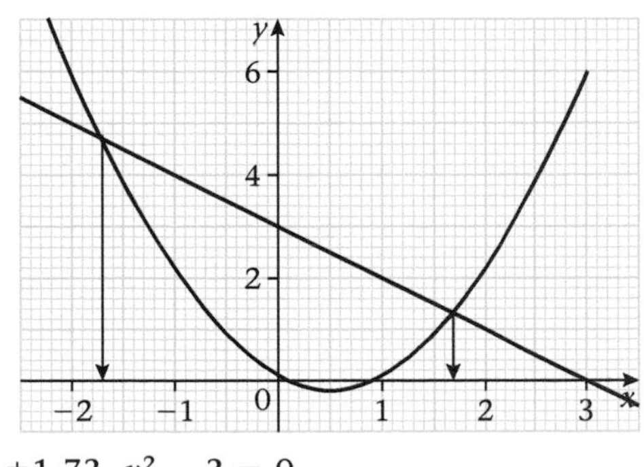

± 1.73, $x^2 - 3 = 0$

8

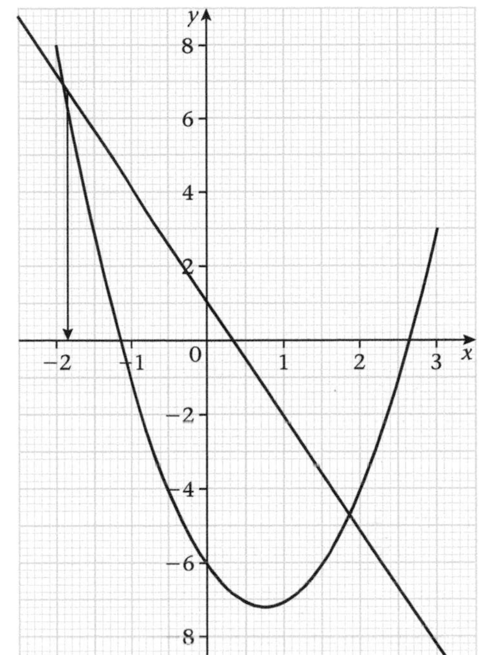

± 1.87

9

x	−2	−1.5	1	−0.5	0	0.5
y	−4	−1.75	0	1.25	2	2.25

x	1	1.5	2	2.5	3
y	2	1.25	0	−1.25	−4

(a)

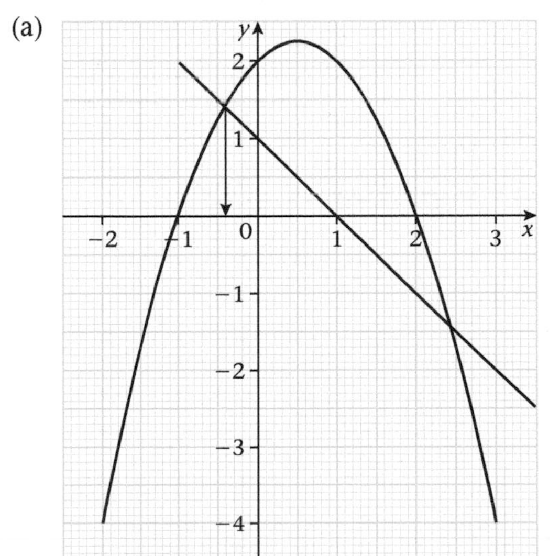

(b) $y_{max} = 2.25$ when $x = 0.5$
(d) 2.4, −0.4

10

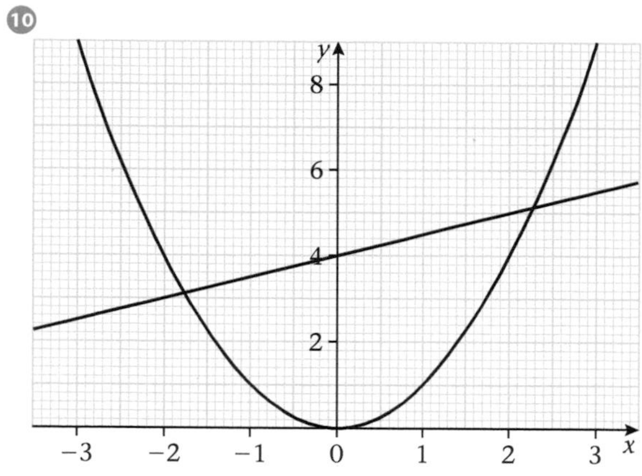

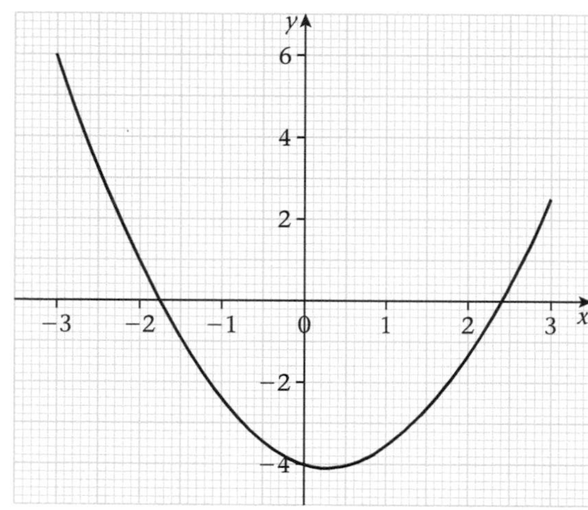

2.27, −1.77

Exercise 20q (p. 251)

1 (0, 5), (4, −3)
2 (3, 5), $(-1\frac{1}{4}, -12)$
3 (1, 2)
4 (3, −5), $(-2\frac{1}{2}, 6)$
5 (2, −1)
6 (1, −2), $(-\frac{1}{7}, 1\frac{3}{7})$
7 (−4, 3)
8 (4, −1)
9 (−1, −1), $(1\frac{8}{11}, -\frac{1}{11})$
10 (5, 6), (2, 15)
11 (2, 4)
12 (4, 1), (−8, −3)
13 (1, 2), $(-2\frac{3}{7}, -\frac{2}{7})$
14 $(1\frac{1}{5}, -1), (-1\frac{2}{15}, \frac{2}{3})$
15 (2, 1), $(-1\frac{1}{3}, -1\frac{1}{2})$

Exercise 20r (p. 251)

1 15
2 12, 15
3 18 cm
4 17 cm, 4 cm
5 10 yr
6 2 m
7 5
8 3 cm
9 6 years ago
10 12 cm, 2 cm

Exercise 20s (p. 252)

1 (a) (i) $x = -15$
(b) (i) $R = \frac{60 - A}{2\pi h}$ (ii) $R = \frac{7}{6}$

2 (a) (i) $\frac{8y - 3y^2 - 1}{2(y^2 - 1)}$
(b) (i) 21 (ii) −5

3 (a) (i) $y = 9x^2$ (ii) 324
(b) (i) $p = 5$ (ii) $y = \frac{5}{x + 3}$

4 (a) $x = 1, 6$ (b) (i) $x > -2$
(ii)

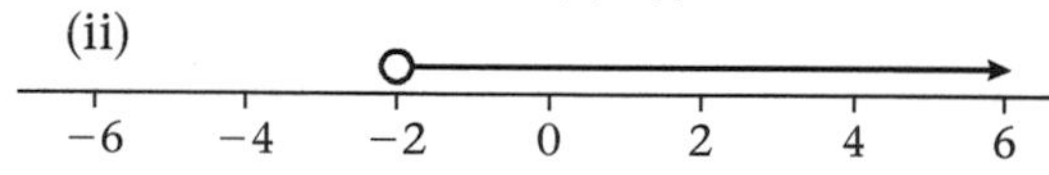

5 (a) $50 \leqslant x \leqslant 80$, $25 \leqslant y \leqslant 60$, $x + y \leqslant 120$
(b)

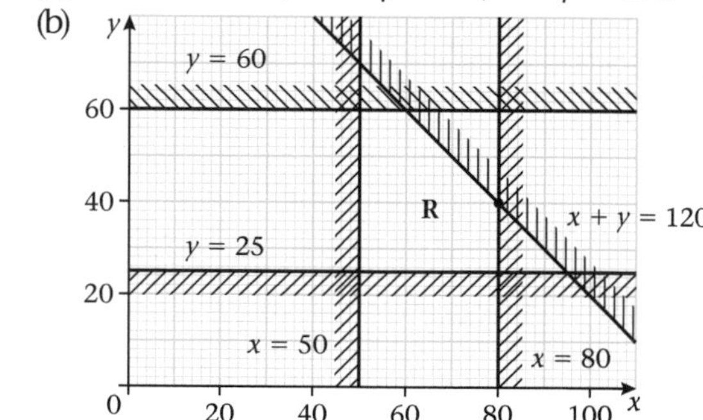

(c) (i) $P = \frac{3x}{2} + \frac{11y}{10}$
(ii) maximum when $x = 80$ and $y = 40$; \$164.

6 (a) $3 \leqslant x \leqslant 10$; $5 < y < 12$
$x + y \geqslant 18$
(b) to (d)

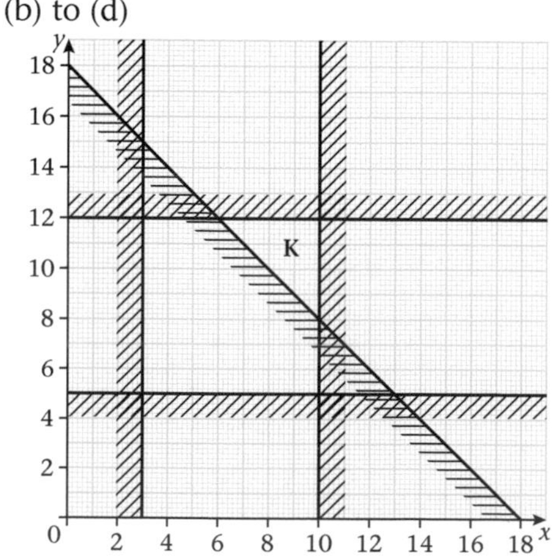

(e) (7,11), (8,10), (8,11), (9,9), (9,10), (9,11), (10,8), (10,9), (10,10), (10,11)

7 (a) $x_{min} = 1, y_{min} = 2$
(b) $x \geqslant 1, y \geqslant 2, y \geqslant 2x, x + y \leqslant 12$
(c)

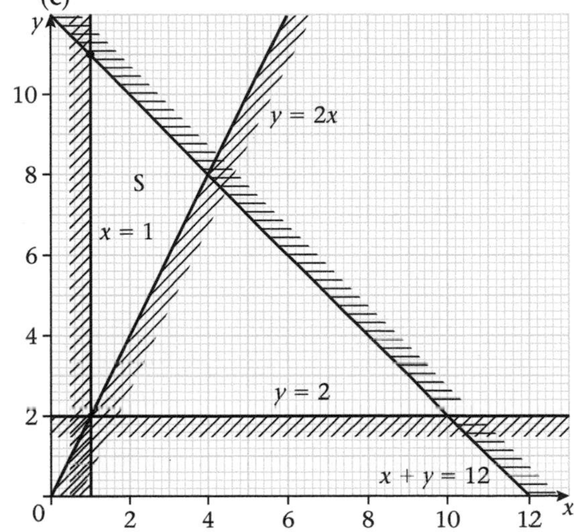

(d) when $x = 1, y = 11$

8 (a) $x < -2$

(b) (i) $(x + 1)(x + 2)$ (c) $p = 9$
(ii) $(x + 3)(x - 2 + a)$

9 (i) $-4 \leqslant x \leqslant 3$ (ii) $-3, 2$ (iii) $-3.4, 2.4$

10 (a) $-2.5, 1$ (b) (ii) $-2, 0.5$
(c) (ii) $-2.6, 0.6$

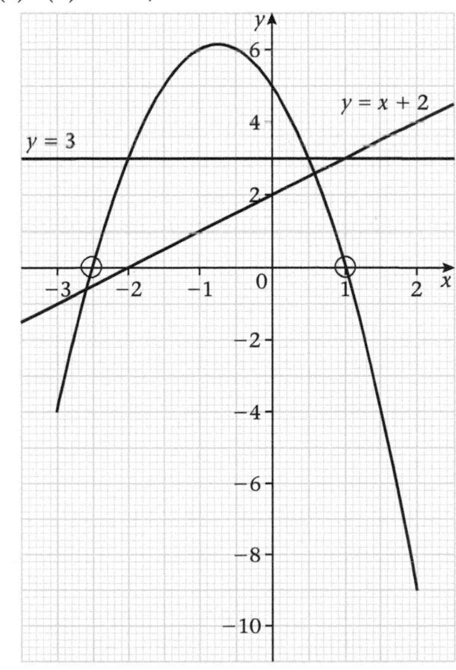

11 (i) $(3x - 5)^2 - 4$

13 (a) $\dfrac{3x - y}{3}$ (b) $a = \frac{2}{3}, b = -4$

14 (a) $3r(r + 2R)$ (b) $(k - n)(m - n)$
(c) $2(a - 3)(a + 3)$ (d) $(5y - 4)(y + 1)$

15 (a) $\frac{11}{12} + 3(x - \frac{1}{6})^2$
(b) minimum, since $3(x - \frac{1}{6})^2 > 0$
(c) $\frac{11}{12}$ at $x = \frac{1}{6}$

16 (a) $v = -2, w = 2$
(b) $y = -1, z = 2; y = \frac{23}{17}, z = \frac{-26}{17}$

17 (a) (i) $t = \sqrt{\dfrac{2S}{g}}$ (ii) 3.2
(b) (i) $2(x + \frac{5}{4})^2 - \frac{33}{8}$
(ii) 0.19, −2.69

18 (a) (i) $\dfrac{x^2y}{5z^4}$ (ii) $\dfrac{3(2b + 3)}{5}$
(iii) $\dfrac{2(h - 1)}{(h + 3)(h + 1)(3h + 1)}$
(c) $m = \frac{5}{2}, n = \frac{1}{2}$

Revision test Ch. 20 (p. 254)

1 C	2 D	3 B	4 A	5 C
6 D	7 D	8 C	9 B	10 A
11 A	12 D	13 A	14 D	15 B
16 B	17 B	18 A	19 C	20 D

Revision exercise 21a (p. 257)

1 (a) \$80 (b) 25%
2 (a) \$14 990 (b) 5 996 (c) \$2 822.50
3 15%
4 (a) \$62.25 (b) 20.75%

Revision exercise 21b (p. 257)

1 (a) \$27 (b) \$4.73
2 (a) 9.4% (b) 2.7%
3 6.7%

Revision exercise 21c (p. 257)

1 (a) \$17 795
(b) \$1845
2 (a) Mortgage A costs \$235 500.
Mortgage B costs \$267 000.
Mortgage A costs less.
(b) They make 57% on mortgage A and 78% on mortgage B.

Revision exercise 21d (p. 258)

1. (a) \$89.61 (b) \$20.46
2. (a) \$147.07 (b) \$689.43
3. Tariff A costs \$106, tariff B costs \$114.50 and tariff C costs \$105. Tariff C is cheapest.

Revision exercise 21e (p. 258)

1. (a) \$662.20 (b) \$900
2. (a) \$1362 (b) \$2465.50
3. \$12 644.35
4. (a) \$799.62 (b) \$471.30
5. (a) 35% (b) \$32 803.18
6. (a) \$600 (b) \$150 (c) 5

Revision exercise 21f (p. 259)

1. (a) \$3 663 (b) \$4 972.80 (c) \$5 705.40
2. \$1 222.50

Revision exercise 21g (p. 259)

1. (a) US\$194.40 (b) EC\$534.60 (c) Ja\$12 101.40 (d) Bds\$388.80
2. (a) EC\$4 394.25 (b) Bds\$2 991.61
3. 7747.44 Mexican Pesos

Exercise 22a (p. 260)

1. (a) $\{(-1, -1), (0, 2), (1, 5), (2, 8)\}$
   (c) (i) $p = -4, q = 3, r = -\frac{2}{3}$
   (ii) yes, one-to-one mapping from elements in the domain to elements in the range
2. (a) $f^{-1}: x \rightarrow \dfrac{2}{7 - x}$
   (b) (i) $6\frac{1}{2}$ (ii) $\frac{1}{2}$
   (c) (i) $\frac{3}{10}$ (ii) 8
3. (a) $fg(x) = x - 2$ (b) $gf(x) = x - \frac{2}{3}$
   (c) (i) -3 (ii) 2
   (d) (i) 1 (ii) $-\frac{5}{3}$
4. (a) $-4 \leqslant f(x) \leqslant 8$
   (b) (i) $\frac{5}{3}$ (ii) $-2, 2$

Exercise 22b (p. 260)

1. $2y = x + 2$
2. $5y + 4x = 3$
3. (a) $\frac{2}{3}$ (b) $(-\frac{3}{2}, 1)$ (c) $\sqrt{13}$
4. (a) $2y = 3x - 6$ (b) $(0, -3), (2, 0)$
5. (a) $y + x = 2$
   (b) (i) $x + y = 0$ or $y = -x$;
   (ii) $y + x = 5$

Exercise 22c (p. 262)

1.

x	-2	-1	0	1	2	3	4
y	25	11	3	1	5	15	31

   (b) 0.92 (c) $x = 0.83$
   (d) $x = 2, x = -0.3$
2. (a) (i) 2.75 (ii) -5.25
   (b) (i) 4.7 or -1.7 (ii) -1 or 4
   (c) $x = 1.5$
3. (a) -2.25 (b) 4.65 or -0.65
   (c) $x = 2, 4$
4. (a) $x > -2$ (b) $(-2, -3)$
5. (a) $x < \frac{1}{2}$ (b) $-1 < x < 2$
6. (a) (i) $2\sqrt{5}$ (ii) $\frac{1}{2}$ (iii) (4, 2)
   (b) $y + 2x - 10 = 0$
8. (a) (4, 0) and (−2, 0) (b) (0, 8)
   (c) maximum
   (d) $x = 1$
   (e) $y = 9$
9. (a) (0, 0) and (5, 0)
   (b) $y = 0$ (the x-axis), $x = 2\frac{1}{2}$
10. (2.64, 0.76), (−1.14, −1.75)
11. (−1, −1), (−3, 1)

Exercise 22d (p. 265)

1. (a) 5 (b) 145 km (c) 58 km/h
   (d) 36 km/h (e) 100 km/h (f) 70 km/h
   (g) 10 km
2. (a) 280 m (b) 4 min (c) 12:17 hrs
   (d) 400 m (e) 80 m/min (f) 120 m
3. (a) 12:35 hrs (b) 12:30 hrs (c) 12:37 hrs
4. (a) 4 600 km (b) 400 km
5. (a) 17.9 km/h (b) 2 min
6. (a) 9 km/h per min (b) 3.4 km
7. (a) 5 m/s^2 (b) $\frac{1}{9}$ m/s^2 (c) 172 m
8. (a) (i) constant speed of 120 km/h for 3 min
   (ii) uniform acceleration from 120 km/h to 150 km/h in 2 min
   (iii) uniform deceleration from 150 km/h to rest in 1 min
   (b) (ii) 10 km/h per min
   (iii) −150 km/h per min
   (c) $11\frac{3}{4}$ km
9. (a) 1.5 m/s^2 (b) (i) 225 (ii) 6 675 m
10. (a) 19 m/s (b) 24.95 m/s
11. (b) 227 min (c) 2 639 km

12 (a) $0\,m/s^2$ (b) $-6\,m/s^2$ (c) 116 m
13 (b) $-6\,m/s^2$
14 (a) $m = 2, n = 18$ (c) 0.3 s and 2.7 s
(d) $12\,m/s^2$
15 (a) $a = 16, b = 0$ (c) 14.3 m/s
(d) (i) $4\,m/s^2$ (ii) $-6\,m/s^2$

Revision test (Ch. 22) (p. 267)

1 C 2 D 3 A 4 C 5 A
6 C 7 B 8 C 9 A 10 D
11 B 12 C 13 B 14 C 15 A
16 C 17 D 18 D 19 B 20 A

Exercise 23a (p. 270)

1 (a) $g - 95°, h - 33°$
(b) $i = j = 63°, k = 108°, l = 9°, m = 45°$
(c) $n = 226°, p = 113°, q = 67°$
(d) $r = 60°, s = 80°, t = 28°, u = 72°$
(e) $w = 90°, x = 42°, y = 12°$
(f) $z = 62°$
2 7 cm, 6 cm, 5 cm
3 3 cm, 4 cm, 6 cm
4 AY = 4 cm, BX = 5 cm
5 (a) $h = 90°, i = 90°, j = 40°, k = 50°$
(b) $m = n = 61°$
(c) $p = 23°, q = 67°, r = 67°, s = 23°$
(d) $t = u = 90°, v = 113°$
(e) $w = 128°, x = 64°$
(f) $y = 57°, z = 66°$
6 62°, 56°, 62°
7 (a) 62° (b) 120° (c) 58° (d) 106°
8 (a) $f = 56°, g = 68°, h = 56°$
(b) $x = 74°, y = 72°, z = 34°$
(c) $p = 38°$ (d) $r = 86°$ (e) $q = 25°$
9 $y = 2x - 180°$
10 (a) 58° (b) 72° (c) 26° (d) 19°

Revision test (Chapter 23) (p. 272)

1 A 2 A 3 C 4 C 5 A
6 D 7 C 8 B 9 D 10 A
11 C 12 B

Exercise 24a (p. 274)

1 (a) $\tan \alpha = \frac{y}{z} = \frac{z}{p} = \frac{x}{q}$
(b) $\sin \alpha = \frac{y}{x} = \frac{z}{q} = \frac{x}{y + p}$
(c) $\cos \alpha = \frac{z}{x} = \frac{p}{q} = \frac{q}{y + p}$
2 (a) 51.3° (b) 26.7°
(c) 66.4° (d) 53.1°
3 (a) 3.75 cm (b) 7.5 cm, 19.3°
4 (a) 80.4° (b) 2.96 cm
5 $\frac{7 \sin 70°}{21} = \frac{1}{3} \sin 70° = 0.313$
6 32° 7 43° 8 9.2 cm

Exercise 24b (p. 275)

1 (a) $5, 5\sqrt{2}$ (b) $2\sqrt{2}, 2\sqrt{2}$
(c) $\frac{7\sqrt{2}}{2}, \frac{7\sqrt{2}}{2}$ (d) $6, 3\sqrt{3}$
(e) $2\sqrt{3}, 4\sqrt{3}$ (f) $2\frac{1}{2}, 2\frac{1}{2}\sqrt{3}$
(g) $10, 5\sqrt{3}$ (h) $4, 4\sqrt{3}$
2 (a) 12 (b) $6\sqrt{2}$
(c) $6\sqrt{3}$ (d) 8
(e) $2\sqrt{6}$ (f) $\sqrt{6}$
(g) $3\sqrt{2} - \sqrt{6}$ (h) $4\sqrt{6}$

Exercise 24c (p. 277)

1 −0.819 2 −0.574 3 1.43
4 −0.342 5 0.819 6 −0.839
7 0.998 8 0.163 9 0.352
10 13.6 11 −0.833 12 −0.678
13 213.4°, 326.6°, −33.4°, −146.6°
14 ±76.4°, ±283.6°
15 79.1°, 259.1°, −100.9°, −280.9°
16 208.3°, 331.7°, −28.3°, −151.7°
17 39.1°, 140.9°, −320.9°
18 149.1°, 329.1°, −30.9°, −210.9°
19 136°, 316°, −44°, −224°
20 ±149.5°, ±210.5° 21 ±136.7°, ±223.3°
22 71.2°, 251.2°, −108.8°, −288.8°
23 30.2°, 149.8°, −210.2°, −329.8°
24 ±127.5°, ±232.5°

Exercise 24d (p. 278)

1 4.36 m 2 (a) 5.6 cm (b) 6.5 cm
3 (a) (i) $20\sqrt{13}$ cm (ii) 50 cm
(b) (i) 11.2° (ii) 16.2°
4 (a) 7.81 km (b) 085.2°
5 $47\frac{1}{2}$ m (45.9 m + 1.6 m)
6 (a) 800 m (b) $400\sqrt{3}$ m
7 (a) DC = w m, BD = $\frac{\sqrt{3}\,w}{2}$ m (b) 11 m

8 (a) (i) 70° (ii) 4.1 cm (iii) 11 cm
(b) (i) 10.4 cm (ii) 147.4 cm^2
9 17m
10 10 m
11 (a) (i) 5.7 km (ii) 1.8 km (b) 030°
12 (a) 24 cm (b) 32 cm

Exercise 24e (p. 281)

1 (a) 6.93 cm ($\sqrt{48}$ cm) (b) 35°21′
2 (a) 11 cm (b) 39°31′
3 (a) 14.5° (b) 9°
4 4.48 m
5 6°54′, 17°38′
6 2.87 m, 60.5°
7 46°41′
8 53°8′
9 (a) 33°4′ (b) $6\frac{6}{11}$ cm
10 (a)

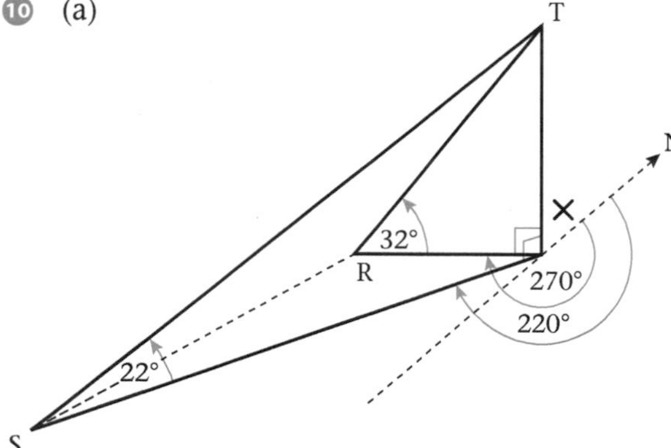

(b) (i) RS = 98.6 m (ii) 180°

11 (a)

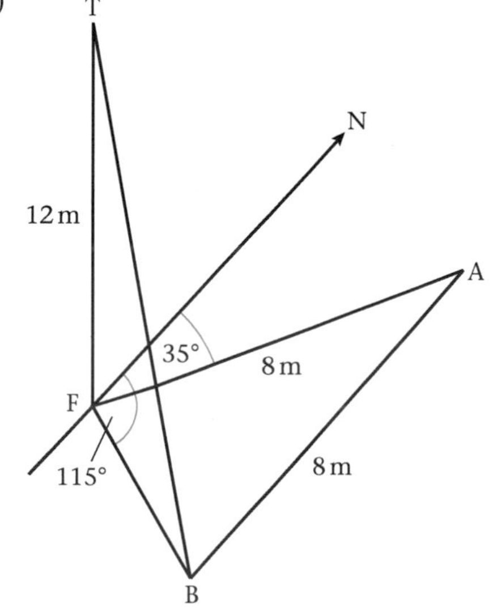

(b) (i) 2.8 m (ii) 77°

Exercise 24f (p. 284)

1 5.6 cm
2 20.4 cm
3 16°25′
4 4.36 cm
5 4.42 m
6 61.8 cm
7 $\widehat{B} = 57.1°$, $\widehat{C} = 78.5°$
8 60°
9 112.4°
10 240.5 m
11 135 m
12 $d = 12.3$ cm, $\alpha = 50°3'$
13 $x = 8.90$ m, $\theta = 107.7°$
14 (a) 20 cm (b) 68°
15 110.8°
16 1041 km, 017°52′
17 (a) 7.0 m
(b) 2.8 m
(c) 23.6°
18 (c) (i) 4.7 cm (ii) 17 cm

Exercise 24g (p. 286)

1 155°, −205°, −335°
2 ±144°, ±216°
3 30°, 210°, −150°, −330°
5 41.8°, 138.2°
6

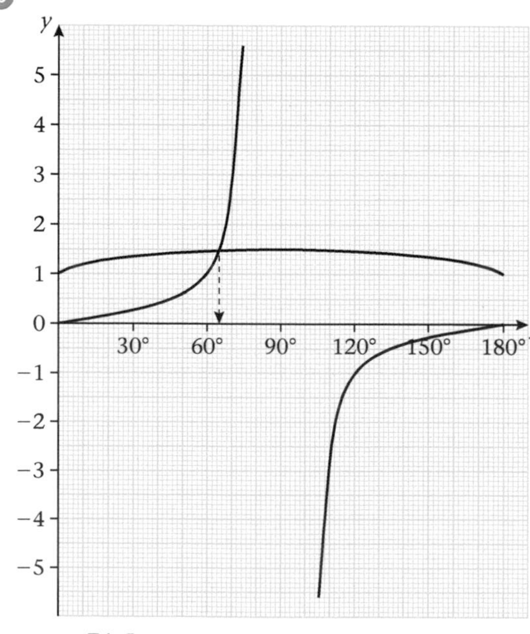

$x = 71.5$

7 (a)

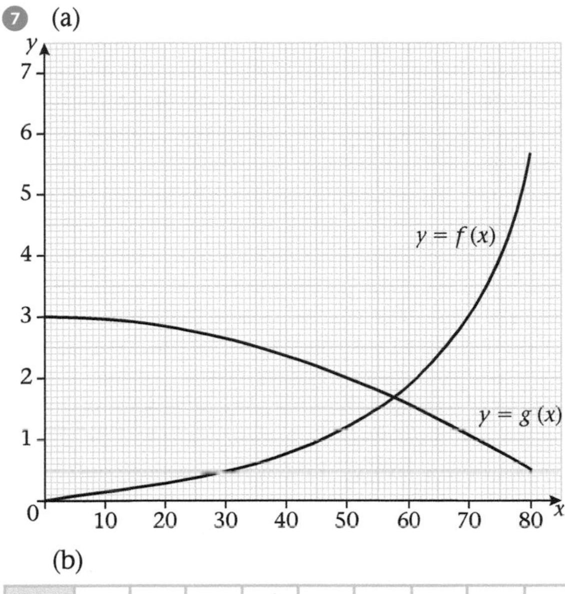

(b)

$x°$	0	10	20	30	40	50	60	70	80
$g(x)$	3	2.95	2.82	2.60	2.30	1.93	1.5	1.03	0.52

(d) (i) 58° (ii) 1.0 radians
(e) 122°, −238°, −302°

8 (a)

$x°$	−60	−30	0	30	60	90	120	150
$F(x)$	−3.60	−3.23	−2	−0.23	1.60	3	3.60	3.23

(b)

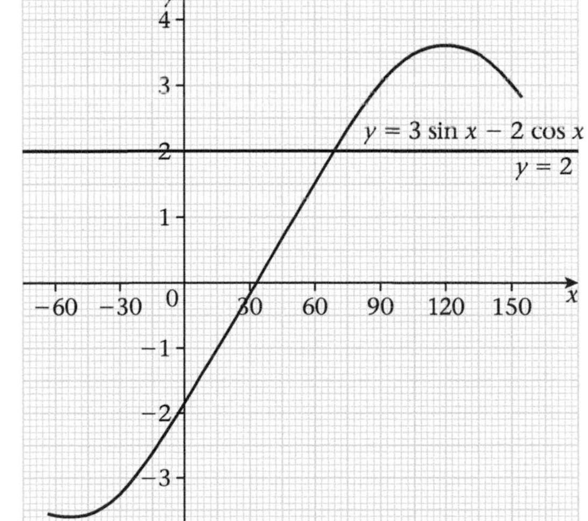

(d) (i) 68° or 69° (ii) 123° or 124°
(e) (i) 1.2 radians (ii) 2.1 or 2.2 radians

Exercise 24h (p. 289)

1 2 000 km 2 7 110 km 3 4 140 n.mi.
4 6 120 n.mi. 5 17°28′S
6 15 900 km, 2 610 km
7 6 667 km 8 5 590 km
9 (a) 3 715 km
(b) (i) 28°N (ii) 13 111 km
10 3 835 n.mi.
11 (a) 5 659 n.mi. (b) 3 792 n.mi.
12 (a) 9 666 km (b) (8°S, 70°W)
(c) 1.67×10^4 km
13 (a) (i) 9 600 km (ii) 20 100 km
(b) (60°N, 22°W)
14 (b) (24°S, 35°W)
(c) 3 940 km
15 (a) (10°N, 60°W) (b) (40°N, 20°W)

Revision test (Chapter 24) (p. 291)

1 A 2 C 3 C 4 B 5 B
6 C 7 C 8 B 9 C 10 D
11 D 12 C 13 C 14 A 15 B
16 B 17 D 18 A 19 A 20 A

Exercise 25a (p. 295)

1 (a) $\begin{pmatrix} 7 & 9 \\ -4 & 3 \end{pmatrix}$ (b) $\begin{pmatrix} 12 \\ 9 \end{pmatrix}$
(c) $\begin{pmatrix} 3 & -7 \\ 0 & 6 \\ -5 & 7 \end{pmatrix}$ (d) $\begin{pmatrix} -10 & -7 \\ 3 & 1 \end{pmatrix}$
(e) $\begin{pmatrix} 5 \\ 10 \end{pmatrix}$ (f) $\begin{pmatrix} 0 & -3 \\ -8 & 6 \end{pmatrix}$

2 (a) $\begin{pmatrix} -3 & 15 \\ 6 & 9 \end{pmatrix}$ (b) $\begin{pmatrix} 12 & 0 \\ 8 & -16 \end{pmatrix}$
(c) $\begin{pmatrix} 2 & -10 \\ -4 & -6 \end{pmatrix}$ (d) $\begin{pmatrix} 3 & 0 \\ 2 & -4 \end{pmatrix}$
(e) $\begin{pmatrix} 4 & 10 \\ 8 & -2 \end{pmatrix}$ (f) $\begin{pmatrix} -19 & 5 \\ -10 & 27 \end{pmatrix}$

3 (a) (29) (b) (−8)
(c) $(3x - y)$ (d) $(4x + 5y + 6z)$

4 $x = 8, y = -5$

5 (a) $\begin{pmatrix} -7 \\ -10 \end{pmatrix}$ (b) (17, 4, 1)
(c) $\begin{pmatrix} 4 & 3 \\ -2 & -11 \end{pmatrix}$ (d) $\begin{pmatrix} -2 & 24 \\ 2 & -5 \end{pmatrix}$
(e) $\begin{pmatrix} 33 & 24 \\ -8 & 1 \end{pmatrix}$ (f) $\begin{pmatrix} 2 & 3 & 0 \\ 5 & -1 & 6 \end{pmatrix}$

6 (a) $\begin{pmatrix} 13 & 2 \\ 6 & 4 \end{pmatrix}$ (b) $x = 7, y = 4$

7 $a = 0, b = -1$

8 (a) $\begin{pmatrix}4 & -5\\3 & 5\end{pmatrix}$ (b) $\begin{pmatrix}13 & -15\\-3 & 6\end{pmatrix}$

9 $x = 3, y = -2$

10 (a) $\begin{matrix} & J & F\\ E & 45 & 30\\ W & 40 & 38\end{matrix}$ or $\begin{matrix} & E & W\\ J & 45 & 40\\ F & 30 & 38\end{matrix}$

(b) $\begin{matrix}E & W\end{matrix}$ $Rate(6 \quad 5)$ $\begin{matrix}J & F\end{matrix}$ $\begin{pmatrix}45 & 30\\40 & 38\end{pmatrix}$

or $\begin{matrix}E & W\end{matrix}$ $\begin{pmatrix}45 & 40\\30 & 38\end{pmatrix}$ $Rate$ $\begin{pmatrix}6\\5\end{pmatrix}$

(c) Electricity : \$450; Water : \$390

Exercise 25b (p. 297)

1 $7, \frac{1}{7}\begin{pmatrix}-3 & -2\\-4 & -5\end{pmatrix}$

2 $x = -8\frac{1}{2}$

3 (a) $\begin{pmatrix}4 & 4\\1 & 7\end{pmatrix}$ (b) $x = 4, y = 7$

4 (a) $\begin{pmatrix}7 & -3\\6 & -3\end{pmatrix}$ (b) $m = 1\frac{1}{2}$ (c) $n = -2$

5 $a = 2, \frac{1}{2}\begin{pmatrix}3 & -4\\1 & 2\end{pmatrix}$

6 $a = 4, b = -3, c = 1$

7 (a) $\begin{pmatrix}11 & -8\\3 & 0\end{pmatrix}$ (b) $\frac{1}{2}\begin{pmatrix}4 & -5\\-2 & 3\end{pmatrix}$

(c) $\frac{1}{3}\begin{pmatrix}1 & 2\\3 & 4\end{pmatrix}$

8 $k = -\frac{1}{2}$ or 6

9 (a) $-\frac{1}{7}\begin{pmatrix}3 & -4\\-1 & -1\end{pmatrix}$ (b) $\begin{pmatrix}x\\y\end{pmatrix} = \begin{pmatrix}-5\\2\end{pmatrix}$

10 $\begin{pmatrix}2 & 5\\1 & -1\end{pmatrix}\begin{pmatrix}x\\y\end{pmatrix} = \begin{pmatrix}7\\7\end{pmatrix}, \begin{pmatrix}x\\y\end{pmatrix} = \begin{pmatrix}6\\-1\end{pmatrix}$

11 (a) 5 (b) $\begin{pmatrix}\frac{1}{10} & -\frac{3}{5}\\\frac{1}{5} & \frac{4}{5}\end{pmatrix}$

(c) $x = -1, y = 6$

12 (a) $x = 4, y = 5$

(b) $3x - 2y = 2$

$5x - 3y = 5$

13 (a) $x = 0, y = 0$ (b) (i) -1

14 (a) $3b + 2t = 23$

$2b + 3t = 22$

(b) $\begin{pmatrix}3 & 2\\2 & 3\end{pmatrix}\begin{pmatrix}b\\t\end{pmatrix} = \begin{pmatrix}22\\23\end{pmatrix}$

(c) $\frac{1}{5}\begin{pmatrix}3 & -2\\-2 & 3\end{pmatrix}$; $b = 5, t = 4$

Exercise 25c (p. 298)

1 (a) $\begin{pmatrix}15\\20\end{pmatrix}$ (b) $\begin{pmatrix}-9\\3\end{pmatrix}$ (c) $\begin{pmatrix}-2\\0\end{pmatrix}$ (d) $\begin{pmatrix}6\\3\end{pmatrix}$

(e) $\begin{pmatrix}-7\\-4\end{pmatrix}$ (f) $\begin{pmatrix}7\\4\end{pmatrix}$ (g) $\begin{pmatrix}5\\12\end{pmatrix}$ (h) $\begin{pmatrix}-3\\6\end{pmatrix}$

(i) $\begin{pmatrix}-1\\24\end{pmatrix}$ (j) $\begin{pmatrix}36\\11\end{pmatrix}$

2 (a) 5 (b) $\sqrt{10}$ (c) 4

(d) $\sqrt{17}$ (e) $\sqrt{2}$ (f) 5

3 (a) unit vector (b) unit vector

(c) $\sqrt{2}$ (d) $\sqrt{5}$

(e) unit vector (f) unit vector

4 (a) $\begin{pmatrix}-8\\6\end{pmatrix}$ (b) 10

5

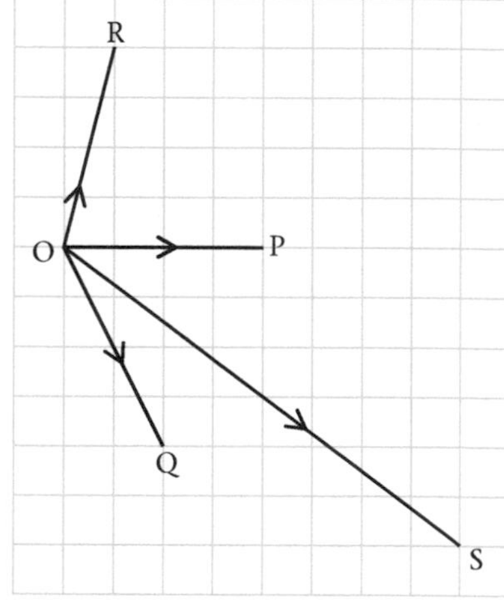

6 (b) (i) $\begin{pmatrix}-1\\-4\end{pmatrix}$, (ii) $\begin{pmatrix}3\\-3\end{pmatrix}$, (iii) $\begin{pmatrix}5\\5\end{pmatrix}$

7 (b) (i) $\begin{pmatrix}5\\6\end{pmatrix}$, (ii) $\begin{pmatrix}1\\-4\end{pmatrix}$, (iii) $\begin{pmatrix}-4\\3\end{pmatrix}$

8 (a) $2\mathbf{a} - 5\mathbf{b}$ (b) $\overrightarrow{PR} = -\overrightarrow{PQ}$, Q, P, R lie on a straight line

9 (a) $\begin{pmatrix}10\\-12\end{pmatrix}$ (b) $\begin{pmatrix}5\\-6\end{pmatrix}$ (c) $\begin{pmatrix}1\\3\end{pmatrix}$

10 (a) $\mathbf{b}$ (b) $-\frac{1}{2}\mathbf{a}$ (c) $\frac{1}{2}\mathbf{a} + \frac{1}{2}\mathbf{b}$

(d) $\frac{1}{2}\mathbf{b}$ (e) $\frac{1}{2}\mathbf{a} + \mathbf{b}$ (f) $\mathbf{b} - \frac{1}{2}\mathbf{a}$

11 (a) 5 (b) $q = 7$

12 $h = 2, k = 9$

13 $m = -1, n = -10$

14 (a) A′(4, 4), B′(3, 6), C′(4, 7) (b) 5

15 (a) $\overrightarrow{XY} = \frac{1}{2}\mathbf{a} + \frac{1}{2}\mathbf{c}$, $\overrightarrow{CZ} = \mathbf{a} - \frac{1}{2}\mathbf{c}$

(b) $\overrightarrow{XP} = \frac{1}{2}h\mathbf{a} + \frac{1}{2}h\mathbf{c}$

(c) $\overrightarrow{CP} = k\mathbf{a} - \frac{1}{2}k\mathbf{c}$ (d) $h = \frac{2}{3}, k = \frac{1}{3}$

16 (a) (i) $\begin{pmatrix}5\\0\end{pmatrix}$ (ii) $\begin{pmatrix}3\\4\end{pmatrix}$ (b) $\begin{pmatrix}1\\7\end{pmatrix}$

17 (a) $\sqrt{17}$ (b) $\frac{1}{4}$

18 (a) (i) $\begin{pmatrix}1\\5\end{pmatrix}$ (ii) $\begin{pmatrix}5\\3\end{pmatrix}$ (b) $\begin{pmatrix}3\\4\end{pmatrix}$

19 (a) $\overrightarrow{OF} = 2u + v$; $\overrightarrow{OH} = -u + 5v$

(b) $\overrightarrow{HF} = 3u - 4v$ (c) $|\overrightarrow{HF}| = 5$

20 (a)

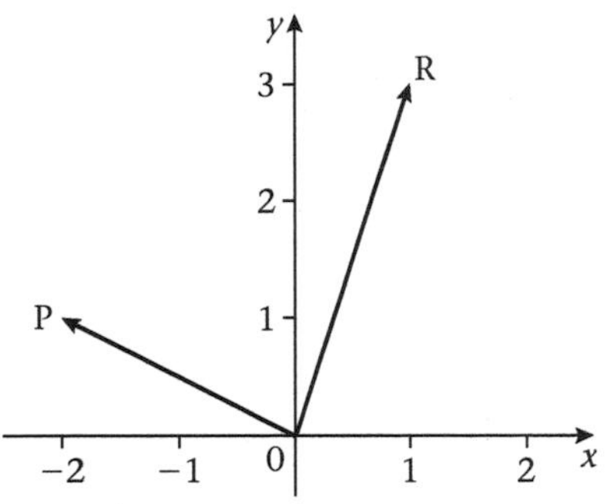

(b) $PR = \begin{pmatrix} 3 \\ 2 \end{pmatrix}$;

(c) a = 7, b = 7

21 (a) (i) $\begin{pmatrix} -6 \\ -2\frac{1}{2} \end{pmatrix}$, (ii) 13

(b) (i) $3\mathbf{a} + 3\mathbf{b}$, (ii) $-\mathbf{a} + 3\mathbf{b}$

(c) $4\mathbf{b} + m(3\mathbf{a} - \mathbf{b}) = n(4\mathbf{a})$

(d) $m = 4, n = 3$

22 (a) $\overrightarrow{OS} = \begin{pmatrix} 1 \\ 0 \end{pmatrix}, \overrightarrow{OT} = \begin{pmatrix} 5 \\ 0 \end{pmatrix}, \overrightarrow{OV} = \begin{pmatrix} 4 \\ 4 \end{pmatrix}$

(b) $\overrightarrow{OW} = \begin{pmatrix} 14 \\ 12 \end{pmatrix}$

23 (a) (i) $\overrightarrow{AC} = \begin{pmatrix} 0 \\ -4 \end{pmatrix}, \overrightarrow{CB} = \begin{pmatrix} -6 \\ 0 \end{pmatrix}, \overrightarrow{AB} = \begin{pmatrix} -6 \\ -4 \end{pmatrix}$

(ii) $|\overrightarrow{AC}| = 4, |\overrightarrow{CB}| = 6, |\overrightarrow{AB}| = 2\sqrt{13}$

(b) (i) $\mathbf{q} - \mathbf{p}$

Exercise 25d (p. 303)

1 Reflection in the line $x = y$

2 (a) (5, −3)

(b) (−4, −1)

(c) (1, 2)

3 (a)

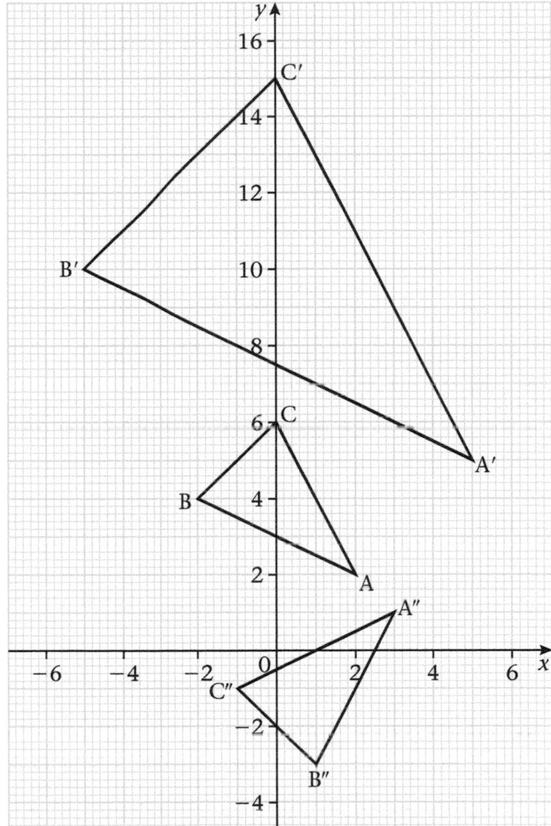

(b) Enlargement, centre at origin, factor k, $k = 2\frac{1}{2}$

(c) A″(3, 1), B″(1, −3), C″(−1, −1)

4 (a) $\begin{pmatrix} 1 & 0 \\ 4 & 1 \end{pmatrix}$ (b) $\begin{pmatrix} 1 & 0 \\ -4 & 1 \end{pmatrix}$

5 (a) (i) (−7, 3), (ii) (5, 1)

(b) Rotation of 90° anticlockwise about (2, 2)

6 (a) Reflection in the line $y = -x$

(b) $\begin{pmatrix} 0 & -1 \\ -1 & 0 \end{pmatrix}$

(c) Reflection in the line $x = 1$

(d) Rotation of 90° clockwise about (1, −1)

7 (a) (ii) $\begin{pmatrix} 0 & 1 \\ 1 & 1 \end{pmatrix}$

(b) (i) $\overrightarrow{OC} = \begin{pmatrix} 0 \\ 4 \end{pmatrix}, \overrightarrow{OD} = \begin{pmatrix} 4 \\ 8 \end{pmatrix}$

(ii) $\begin{pmatrix} 2 & 0 \\ 0 & 2 \end{pmatrix}$

8. (a) (i) $(5, -8)$ (ii) $(2,1)$
   (b) $W = V^{-1}$ since B is the image of A under V, and A is the image of B under W.

9. (a)

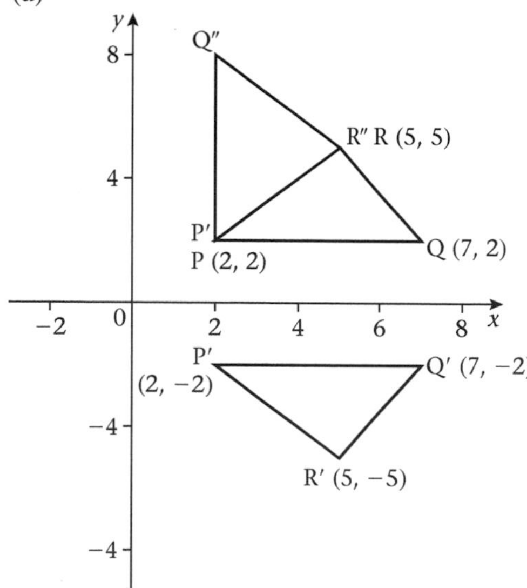

   (b) $P' = \begin{pmatrix} 2 \\ -2 \end{pmatrix}$, $Q' = \begin{pmatrix} 7 \\ -2 \end{pmatrix}$, $R' = \begin{pmatrix} 5 \\ -5 \end{pmatrix}$
   (c) $P'' = \begin{pmatrix} 2 \\ 2 \end{pmatrix}$, $Q'' = \begin{pmatrix} 2 \\ 7 \end{pmatrix}$, $R'' = \begin{pmatrix} 5 \\ 5 \end{pmatrix}$
   (d) $\begin{pmatrix} 0 & 1 \\ 1 & 0 \end{pmatrix}$
   (e) reflection in $y = x$

10. (a) (i) $\begin{pmatrix} 3 & 3 \\ 0 & 0 \end{pmatrix}$
   (ii) enlargement, centre at (0, 0), scale factor 3
   (iii) $\frac{1}{9}$ or 1 : 9
   (b) (i) $X'(-4, 5)$, $Y'(3, 2)$, $Z'(1, 4)$
   (ii) $\begin{pmatrix} 0 & -1 \\ -1 & 0 \end{pmatrix}$

11. (a) $\overrightarrow{OK} = \begin{pmatrix} 2 \\ 0 \end{pmatrix}$, $\overrightarrow{OL} = \begin{pmatrix} 2 \\ 2 \end{pmatrix}$, $\overrightarrow{ON} = \begin{pmatrix} 0 \\ 2 \end{pmatrix}$
   (b) $OK' = \begin{pmatrix} 6 \\ 0 \end{pmatrix}$, $OL' = \begin{pmatrix} 6 \\ 6 \end{pmatrix}$, $ON' = \begin{pmatrix} 0 \\ 6 \end{pmatrix}$
   (c) enlargement, centre of enlargement (0,0), scale factor 3
   (d)

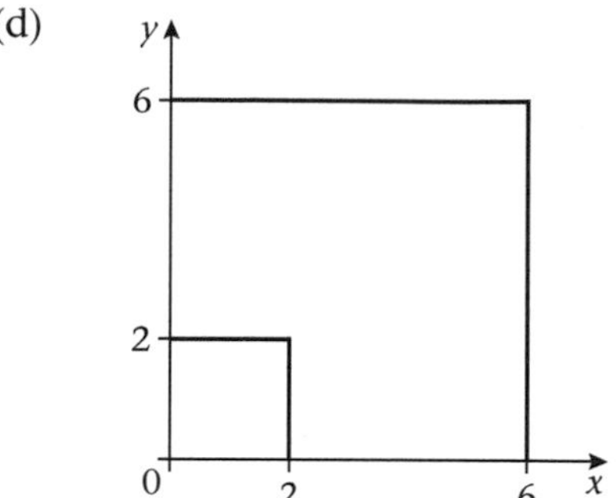

   (e) $y = x$

12. (a) (i) $A'(-2,0)$, $B'(-6,3)$, $C'(-6,0)$
   (ii) reflection in the y-axis
   (b) (i) $A''(2,0)$, $B''(6,-3)$, $C''(6,0)$
   (ii) rotation about O through 180°
   (c) (i) $\begin{pmatrix} 1 & 0 \\ 0 & -1 \end{pmatrix}$
   (ii) reflection in the x-axis

Revision test (Chapter 25) (p. 305)

1 B	2 C	3 A	4 C	5 C
6 D	7 A	8 D	9 D	10 C
11 B	12 B	13 D	14 A	15 C
16 B	17 A	18 D		

Exercise 26 (p. 308)

1. (a) Dwight: 15.75 Vashan: 14.42 (b) Dwight: 16.5 Vashan: 14.5
   (c) Dwight: 12 Vashan: 10 (d) Dwight: 0.58 Vashan: 0.42

2. (b)

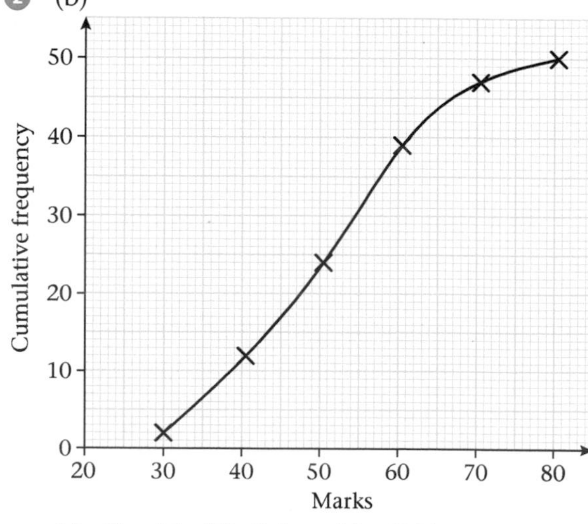

   (c) (i) 51 (ii) 9.5 (d) 68%

3 (a)

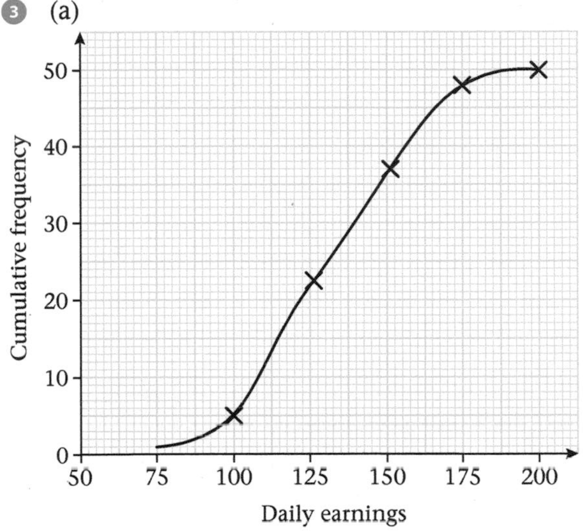

(b) (i) $\frac{3}{35}$ (ii) $\frac{3}{7}$

4 (a)

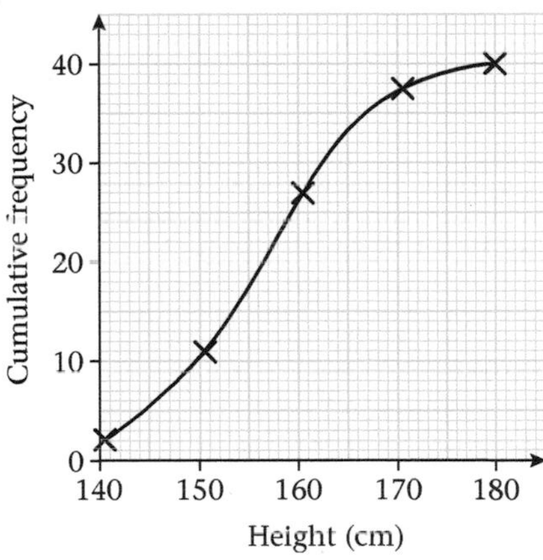

(b) 115 cm (c) $\frac{17}{40}$

5 (a)

Cumulative Frequency
8
21
38
58
77
91
97
100

(b)

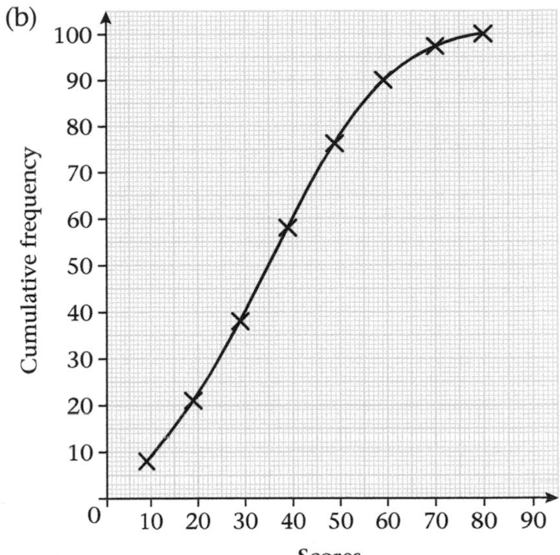

(c) (i) 30 ± 1 (ii) 53 ± 1 (iii) $\frac{3}{10}$

6 (a)

Time (minutes)	Frequency	Cumulative frequency
81–90	3	3
91–100	18	21
101–110	36	57
111–120	49	106
121–130	11	117
131–140	3	120

(b)

(c) median = 114 minutes,
interquartile range = 15 minutes

(d) 84

7 (a) 40 (b) 30–40 cm
(c) (i) 31–40 cm (i) 40.5 cm
(d)

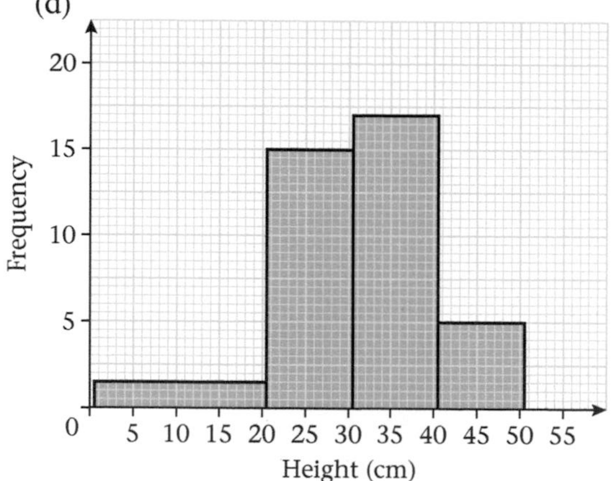

(e) $\frac{17}{40}$ or 0.425

8 (a)

Marks	Frequency
10–14	2
15–19	2
20–24	5
25–29	8
30–34	4
35–39	3

(b)

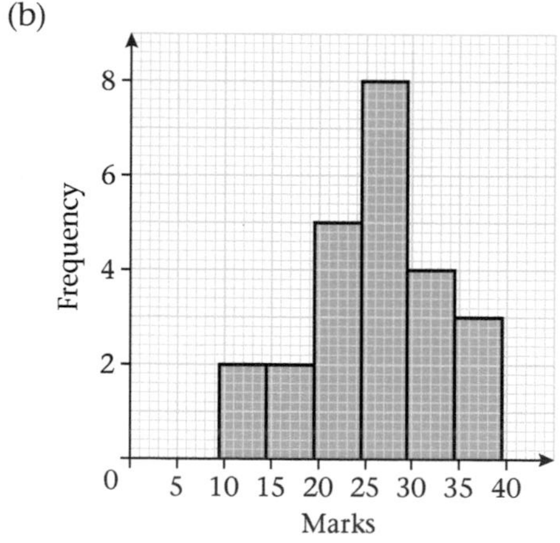

(c) $\frac{19}{24}$ or 0.79

9 (a) (i) 60
(ii) 14.5 cm
(b)

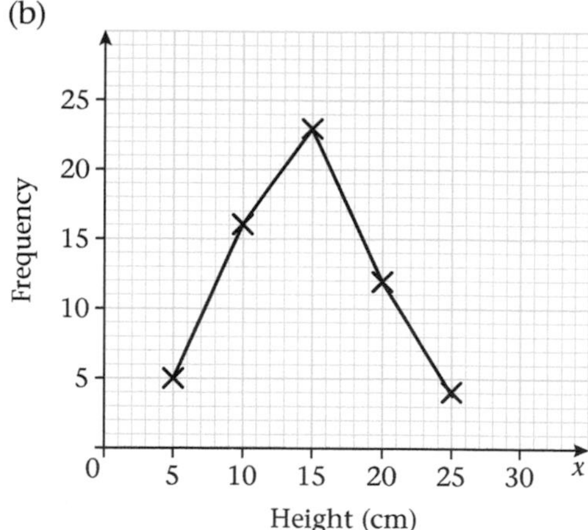

(c) $\frac{7}{20}$ or 0.35

Revision test Chapter 26 (p. 310)

1 C 2 A 3 C 4 D 5 C
6 B 7 D 8 D 9 A 10 D
11 A 12 A 13 D 14 B 15 D
16 B 17 A 18 A

Practice examination I: Paper 1 (p. 312)

1 C 2 D 3 B 4 A 5 C
6 C 7 D 8 A 9 B 10 C
11 A 12 A 13 D 14 D 15 A
16 C 17 D 18 C 19 A 20 D
21 D 22 B 23 C 24 C 25 B
26 A 27 C 28 C 29 D 30 D
31 A 32 B 33 C 34 B 35 B
36 C 37 B 38 C 39 B 40 D
41 D 42 A 43 B 44 D 45 C
46 D 47 D 48 B 49 D 50 D
51 C 52 B 53 C 54 C 55 A
56 C 57 A 58 A 59 D 60 A

Practice examination I: Paper 2 (p. 317)

1 (a) 3 (b) 5×10^3
(c) 0.7 kg

2 (a)

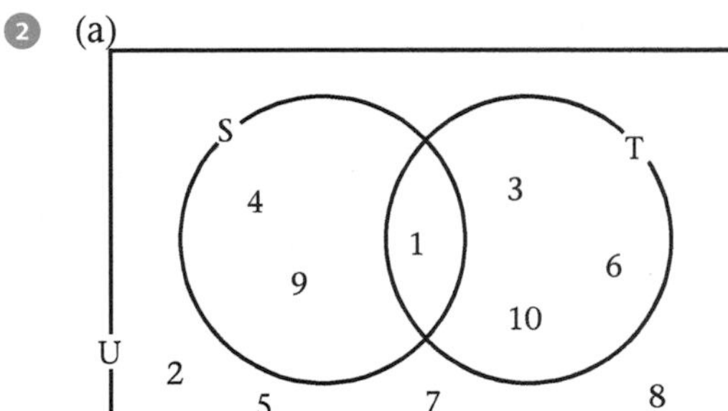

(b) (i) {2, 5, 7, 8} (ii) 7

3 (a) $-\frac{1}{3}$ (b) $3y + x = 5$ (c) $(7, -\frac{2}{3})$

4 (a) 5 (b) $y = \dfrac{8}{\sqrt{x}}$

(c) $(4a - 3)(4a + 3)$, 37×43

5 (a) $(-6, -21)$ (b) $(-2, 3)$

(c) enlargement scale factor -3, centre O (0, 0).

6 (a) (i) \$890 (ii) 30%

(b) (i) 16 000 cm² (ii) 57 000 cm³

7 (a) (i) 15 cm (ii) 19.2 cm (iii) 38.7°

(b) (diagram is not the actual size)

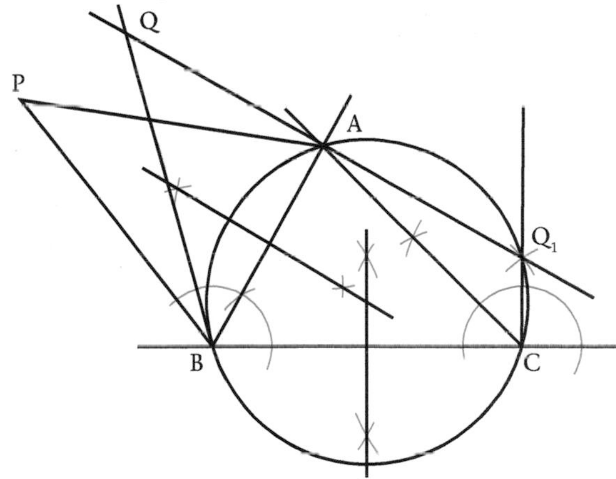

(b) (i) 9.0 cm (vi) 16.7 cm *or* 3.8 cm

8 (a) (i)

Days	Cumulative frequency
1–5	113
6–10	238
11–15	386
16–20	542
21–25	684
26–30	781

(ii)

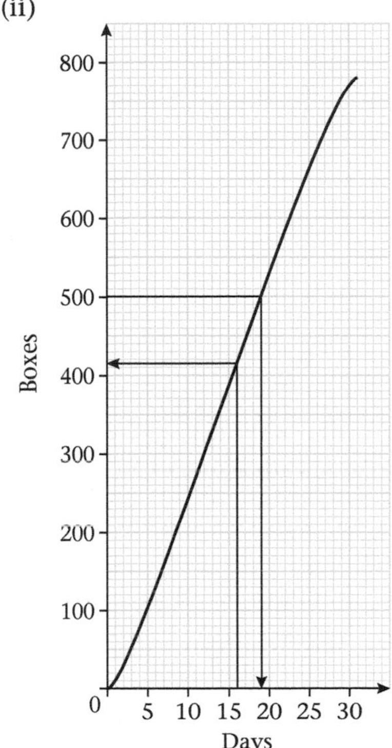

(iii) 414 boxes

(iv) 19 days

(b) $\frac{11}{12}$

9 (a)

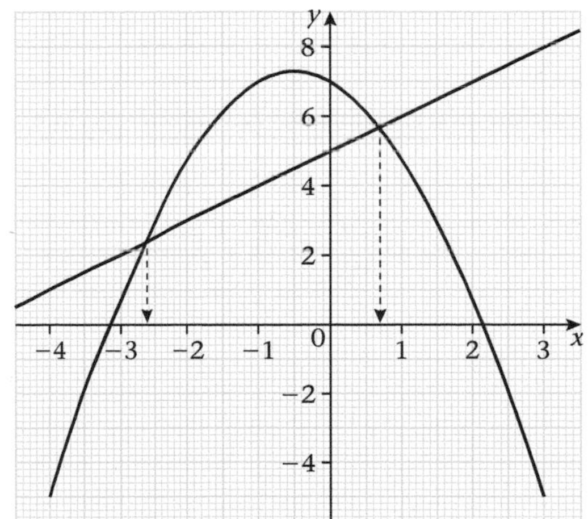

(b) 7.25

(c) (i) -3.2, 2.2

(ii) -2.7, 0.7

10 (a) $3r + 2p \leqslant 1200$; $r + 2p \leqslant 800$

(b)

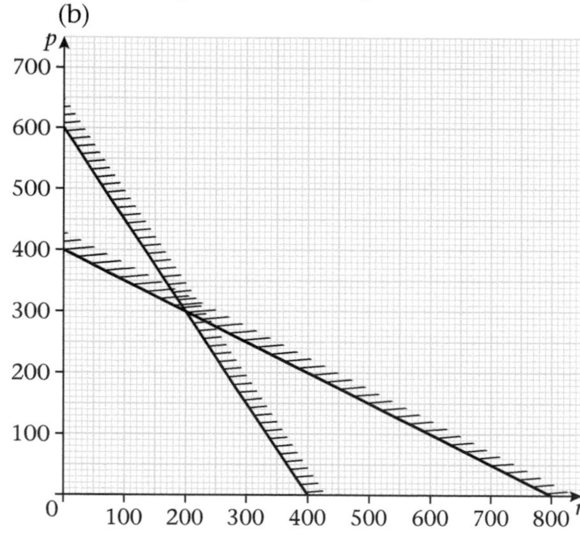

(c) (i) $r = 200$, $p = 300$ (ii) \$6500

11 (a) (i) 139 m (ii) 139 m (iii) 10100 m^2
(b) (i) 320° (ii) 140° (iii) 056°

12 (a) 1710 n.mi. (b) 3 975 km
(c) 53°40'S

13 (a) (i) $\begin{pmatrix} -1 \\ -4 \end{pmatrix}$ (ii) 270°; (0, 5)
(iii) $y = x - 1$ (b) (i) $(-2, -7)$

14 (a) (i)

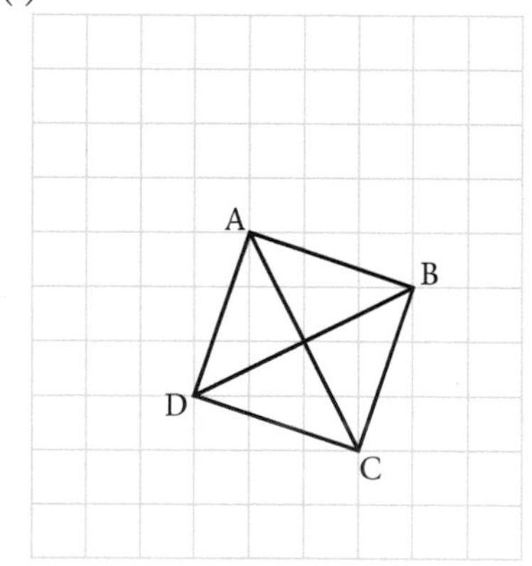

(ii) $p = -1$, $q = -3$

(iii) $\begin{pmatrix} 3 \\ -1 \end{pmatrix}$ (iv) $\begin{pmatrix} 4 \\ 2 \end{pmatrix}$ (b) 2, $\frac{2}{7}$

Practice examination II: Paper 1 (p. 321)

1 A	**2** C	**3** D	**4** A	**5** D
6 A	**7** B	**8** C	**9** B	**10** A
11 A	**12** D	**13** C	**14** C	**15** D
16 B	**17** C	**18** D	**19** C	**20** D
21 B	**22** B	**23** D	**24** A	**25** D
26 C	**27** D	**28** B	**29** D	**30** D
31 A	**32** C	**33** B	**34** C	**35** B
36 C	**37** A	**48** A	**39** D	**40** B
41 D	**42** C	**43** A	**44** B	**45** C
46 A	**47** D	**48** B	**49** C	**50** B
51 A	**52** C	**53** D	**54** B	**55** D
56 A	**57** B	**48** A	**59** C	**60** B

Practice examination II: Paper 2 (p. 325)

1 (a) 20
(b) (i) \$90 (ii) \$6 000
(iii) \$507.50

2 (a) $w = 7$
(b) $y = 14$
(c) $x = 1$ or 6

3 (a) (*note*: figure is half actual size)

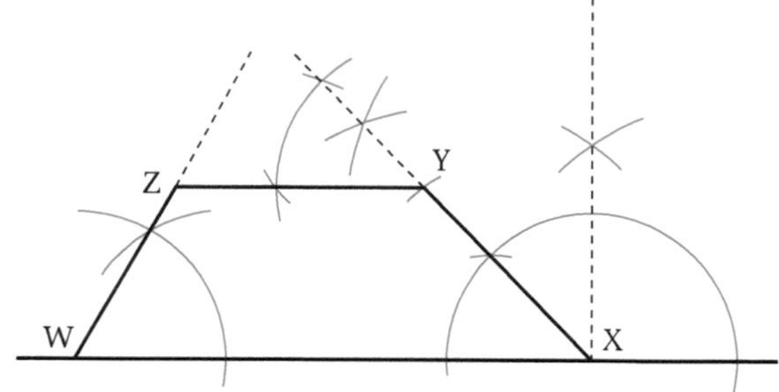

(b) 4.1 cm

4 (a) (i) 7 cm (ii) 68.25 cm^2
(iii) 546 cm^3 (iv) 8.53 m^3

5 (a)

(b) 58 kg
(c) 57.9 kg

6 (a) $(1, \frac{7}{2})$
(b) $\frac{3}{4}$
(c) $6y + 8x = 29$

7 (a) (i) $V \subset B$

(ii)

(b) (i) $\overrightarrow{PS} = \begin{pmatrix} 4 \\ -4 \end{pmatrix}$, $\overrightarrow{PT} = \begin{pmatrix} -2 \\ -6 \end{pmatrix}$

(ii) $\overrightarrow{TS} = \begin{pmatrix} 6 \\ 2 \end{pmatrix}$

8 (a) (i) $7, \frac{4}{3}$

(b) (i) $3m = n$

(ii) $\$\left(2m + \frac{3n}{2}\right)$

(iii) $2m + \frac{3n}{2} = 65$

(iv) $m = 10, n = 30$

9 (a) $-2(x - \frac{5}{4})^2 + \frac{73}{8}$

(b) (i) $\frac{73}{8}$ (ii) $\frac{5}{4}$

(c) 3.4 or −0.9

(d)

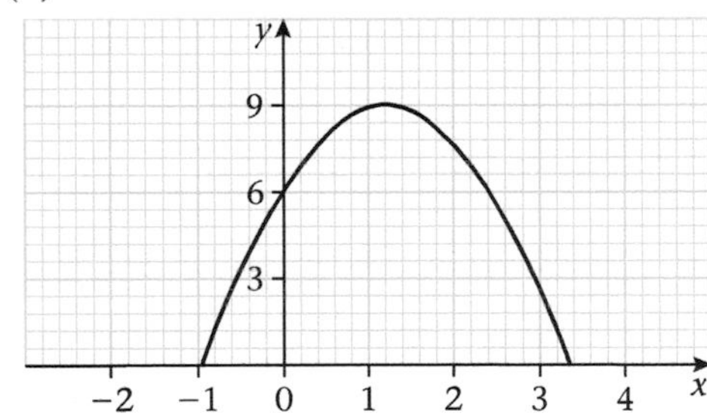

10 (a) (i) constant speed of 20 km/h from 09:00 hrs to 09:05 hrs

(ii) constant acceleration from 09:05 hrs to 09:30 hrs

(iii) constant speed of 60 km/h from 09:30 hrs to 09:40 hrs

(b) (iv) acceleration = 3 km/h per min

(v) acceleration = 0

(c) $19\frac{3}{8}$ km

(d) 52 km/h

11 (a) (i) 20.7 cm (ii) 20.4 cm

(b) (ii) 59.5 cm

12 (a) (i)

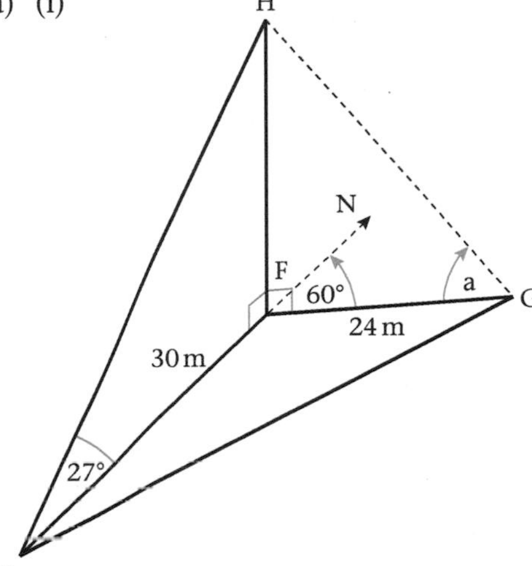

(ii) 15.3 m (iii) 32.5° (iv) 47 m

(b) (i) enlargement, scale factor $-\frac{1}{2}$, centre of enlargement at $(-2, 0)$

(ii) $\frac{1}{2}$

(iii) centre of rotation at $(-\frac{3}{2}, 0)$ through 180°

13 (a) (i) G(3, 1), H(3, 2), J(4, 3), K(6, 1)
G′(1, 3), H′(2, 3), J′(3, 4), K′(1, 6)

(ii) $\begin{pmatrix} 0 & 1 \\ 1 & 0 \end{pmatrix}$ (iii) $\begin{pmatrix} 0 & 1 \\ 1 & 0 \end{pmatrix}$

(b) (i) G″(3, −1), H″(3, −2), J″(4, −3), K″(6, −1)

(ii)

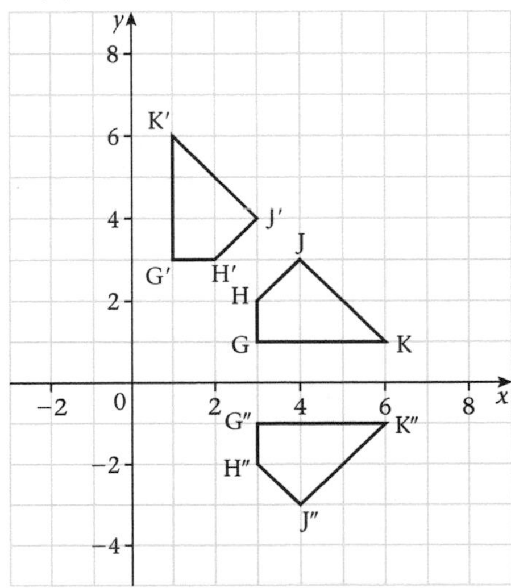

(iii) Reflection in x-axis

 (a) (i)

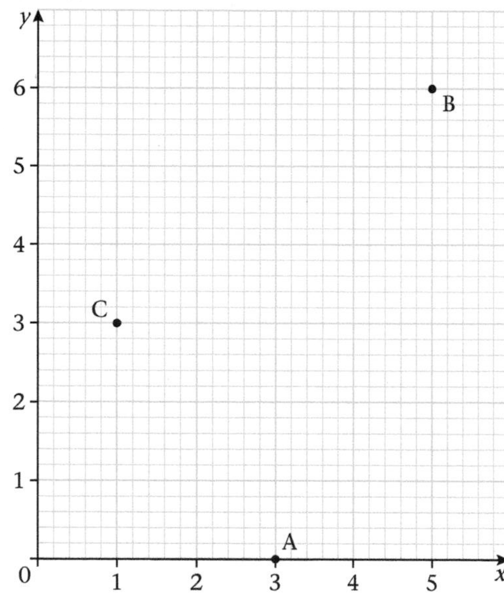

(ii) $\overrightarrow{OA} = \begin{pmatrix} 3 \\ 0 \end{pmatrix}$ $\overrightarrow{OB} = \begin{pmatrix} 5 \\ 6 \end{pmatrix}$ $\overrightarrow{OC} = \begin{pmatrix} 1 \\ 3 \end{pmatrix}$

(iii) $\overrightarrow{AB} = 2\,\overrightarrow{OC}$ (iv) trapezium

(b) (i) $\frac{1}{7}\begin{pmatrix} 3 & 1 \\ -1 & 2 \end{pmatrix}$

(ii) $x = 2, y = -1$

Index

acceleration 133–136, 263–264
addition: fractions 81–83; matrices 14–15, 293–294; vectors 218–220, 226, 298
airspeed 231
algebra 79–102, 237–256; functions 140–146, 260; matrices 18–19; revision test 254–256; undefined fractions 83–84, 189–190
alternate segments 57–60, 270
angles: negative 27–28; obtuse 23–25; positive 26–27; reflex 25–26; special 28–33, 275–277; supplementary 51
antilogarithm tables 334
area 113–123; cones 113–115, 117–119; cuboids 113, 118; cylinders 113, 117–119; frustrums 117–118; hemispheres 117; parallelograms 110; prisms 113; problems 109–111; pyramids 113–115, 117; solids 113–123; spheres 116–117, 119; surface area 113–123; trapeziums 110; travel graphs 134, 263; triangles 109; trigonometry 109–111; units 330

bearings 231

change of subject: formulae 95–100, 243–244
circles 51–62, 270–272; alternate segments 57–60, 270; cyclic quadrilaterals 51–54, 270; geometric proofs 51–62; tangents 54–60, 270–272
circumference of the earth 210
column matrices 13, 16
column vectors 217
combined transformations 173–177
common factors 1–3
completing the square 38–39, 248
composite functions 143–145
compound interest 186
cones 113–115, 117–119
congruent transformations 174, 301
constant ratios: variation 89
consumer arithmetic 257–259
coordinate geometry 63–72; distance between two points 67–69; gradients 63–66; graph sketching 63–65; linear functions 63; mid-points of lines 66–67; parallel lines 65–66; perpendicular lines 65–66; straight lines 63–72; zero gradient 64
cosine 149–152, 286–287; curve 150–152, 286; function 149–152, 286–287; rule 105–109, 283–284; tables 336, 339; trigonometric ratios 23–33, 274–277
course of an aircraft 231
cube roots 86
cubic functions 190–191
cuboids 113, 118
cumulative frequency 164–167; curves 307–310
cyclic quadrilaterals 51–54, 270
cylinders 113, 117–119

deceleration 134–136, 263
determinants of matrices 18–19
difference of two squares 7–8
direct variation (proportion) 89–92, 240–241; constant ratios 89; non-linear quantities 91–92; sketch graphs 90
displacements: vectors 224, 227
distance: area under graph 134, 263; between two points 67–69, 214–215; distance-time graphs 128–133, 263; great circles 210–212, 215; parallels of latitude 212–215
divisibility tests 331
division: fractions 80–81

earth 208–210
enlargements 174–175, 180–183, 302
equal matrices 15
equations 244–245; fractions 84–86; graphical solutions 40–44, 46–47, 248–250; inverse functions 142–143, 192–193; linear equations 244; quadratic equations 35–50, 244, 247–250; simultaneous equations 19–20, 45–47, 250–251
equator 209–211, 287
exercises *see* revision exercises and tests
experimental probability 167
exponential functions 186–188
expressions: fractions 85; quadratic 1–11
extrapolation 189

factorisation 1–12, 237–239; common factors 1–3; definition of factors 4; difference of two squares 7–8; grouping 1–3, 8–11; highest common factor 1; perfect squares 6–7, 238–239; products/sums method 4–6; quadratic expressions 1–11; trial and error method 4–6
flow charts 142
formulae 95–100, 242–244; change of subject 95–100, 243–244; mensuration 331–332; quadratic equations 39–40, 248
fractions 79–88; addition 81–83; division 80–81; equations 84–86; expressions 85; indices 86–88, 240; lowest terms 79–80; multiplication 80–81; simplifying 79–80; subtraction 81–83; undefined 83–84, 189–190
frequency distributions 164, 307
frustrums 117–118
functions 140–146, 147–154, 260; composite 143–145; cosine 149–152, 286–287; inverses 140–143, 192–193; notation 140; one-to-one 141; sine 147–149, 286–287; tangent 152–154, 286–287

geometry: alternate segments 57–60, 270; coordinate 63–72; cyclic quadrilaterals 51–54, 270; proofs 51–62, 270–272; revision test 272–273; tangents 54–56, 125–128, 270–272; transformations 173–185, 301–305; vectors 226–230

glide reflections 173–174, 301
gradients: curves 125–128; distance-time graphs 129; speed-time graphs 133; straight lines 63–66; travel graphs 129, 133, 263
graphs 63–72, 125–139, 186–195, 260–267; coordinate geometry 63–72; cubic functions 190–191; distance-time 128–133, 263; exponential functions 186–188; extrapolation 189; gradients of curves 125–128; hyperbolas 188; interpolation 189; inverse functions 192–193; linear functions 191–192; maximum/minimum values 127–128; parabolas 192; quadratic equations 40–44, 46–47, 248–250; quadratic functions 192; sketching 63–65, 90, 131–133, 191–193, 261–262; solving equations 40–44, 46–47, 248–250; speed-time 133–138, 263–264; straight-lines 63–72, 191–192, 260–261; travel 128–138, 263–267; trigonometric functions 286; variation 90–91, 93
great circles 207–212, 215, 287
Greenwich Meridian 209–210
groundspeed 231
grouping: factorisation 1–3, 8–11
growth: exponential 186–187

HCF *see* highest common factor
hemispheres 117
highest common factor (HCF) 1
hollow objects 118
hyperbolas 188

identities: equations 88–89, 244–245
identity matrix 18, 178, 181, 296–298
imaginary roots 42
indices 86–88, 240
inequalities 196–206; linear programming 201–204, 245–247; two variables 196–201
interest: compound 186
interpolation 189
interquartile range 165, 307
invariant lines/points 178, 180
inverse functions 140–143, 192–193
inverse matrices 18–19, 181, 296–298
inverse variation (proportion) 93–95, 241
isometric transformations 174, 301

joint variation (proportion) 241–242

kilometres 210–211
knots 214

latitude and longitude 207–216, 287–291; earth 208–210; equator 209–211, 287; great circles 207–212, 215, 287; Greenwich Meridian 209–210; kilometres 210–211; knots 214; lines 208, 287; meridians 209, 287; nautical miles 211–214; parallels of latitude 209, 212–214, 287; polar axis 209; shortest distance between points 214–215; sketches as aids 211; small circles 207–209, 215; speed 214
laws of indices 240
LCM *see* lowest common multiple
linear equations 244
linear functions 63, 191–192
linear programming 201–204, 245–247
lines of symmetry 127
logarithm tables 333, 338–340
longitude and latitude 207–216, 287–291
lowest common multiple (LCM) 81, 85
lowest terms: fractions 79–80

magnitude of vectors 220, 298
matrices 13–22, 177–184, 293–298; addition 14–15, 293–294; algebra 18–19; column 13, 16; determinants 18–19; equal 15; identity 18, 178, 181, 296–298; inverses 18–19, 181, 296–298; multiplication 15–17, 178–179, 294–295; null 18; operators 19–20; orders 13–14, 293–294; row 13, 16; scalars 15, 293–294; simultaneous equations 19–20; square 13; subtraction 14–15, 293–294; transformations 177–184, 302; two by two 18–19; zero 18
maximum values: cosine curve 151; graphs 127–128; linear programming 201–203; sine curve 148
median 165, 307
mensuration 330–345; formulae 331–332; tables 333–345; units 330–331
meridians 209, 287
minimum values: cosine curve 151; graphs 127–128; linear programming 201–203; sine curve 148
modulus of a vector 220
multiplication: fractions 80–81; matrices 15–17, 178–179, 294–295; vectors 221–223, 298

naming vectors 217–218
nautical miles 211–214
null matrix 18

obtuse angles 23–25
ogives (cumulative frequency curves) 164–165, 307–310
one-to-one functions 141
orders of matrices 13–14, 293–294
outcome tables: probability 168–171

parabolas 192
parallel lines 65–66
parallelogram law 223
parallelograms: area 110
parallels of latitude 209, 212–215, 287
percentiles 165–167
perfect squares 6–7, 238–239
periodic functions 148, 151, 153
perpendicular lines 65–66
polar axis 209
position vectors 177, 224–226
Practice examinations: I 312–320; answers 428–430; II 321–329; answers 430–432
prisms 113
probability 167–171; experimental 167; outcome tables 168–171; set language 167–169; theoretical 167; tree diagrams 168–171
product/sums factorisation method 4–6
proofs: geometry 51–62
proportion (variation) 89–95, 240–242
pyramids 113–115, 117
Pythagoras' theorem 67–68, 114, 224, 277–280

quadrants of Cartesian plane 26–27
quadratic equations 35–50, 244, 247–250; completing the square 38–39, 248; graphical solutions 40–44, 46–47, 248–250; imaginary roots 42; non-rational roots 37–38, 248; roots 35, 37–38, 42, 248; second degree 35; simultaneous equations 45–47, 250–251; solving by formula 39–40, 248; variables 35; word problems 44–45, 251
quadratic expressions 1–11
quadratic functions 192
quadrilaterals: cyclic 51–54, 270; parallelograms 110, 223; trapeziums 110
quartiles 165

radius: earth 210; parallels of latitude 212–213
random numbers table 345
ratios: trigonometric 23–33, 274–277
reciprocals table 344
reflections 173–175, 178–180, 182–183, 301–302
reflex angles 25–26
relations 260
relative velocity 230–231
resultant vectors 223–224
revision course 233–311
revision exercises and tests: chapters 1-6 73–78; answers 361–363; chapters 7-12 156–163; answers 375–378; chapter 19 233–236; answers 408–412; chapter 20 254–256;

answers 419; chapter 21 257–259; answers 419–420; chapter 22 267–269; answers 421; chapter 23 272–273; answers 421; chapter 24 291–292; answers 423; chapter 25 305–306; answers 426; chapter 26 310–311; answers 428
right-angled triangles 277–280
roots of equations 35, 37–38, 42, 248
rotations 177–178, 182, 301–302
rounding off 110
row matrices 13, 16

scalars: matrices 15, 293–294; vectors 221–223, 298
scale factors: enlargements 174–175, 180–183
second degree equations 35
semi-interquartile range 165, 307
sets: probability 167–169; revision exercises 233–236
SI units 330–331
sign changes: simplifying fractions 79
similar triangles 108
simplifying: expressions 85; fractions 79–80
simultaneous equations: graphical solutions 46–47; matrices 19–20; quadratic equations 45–47, 250–251; substitution method 45–46, 250–251
sine: curve 148–149, 286; function 147–149, 286–287; rule 103–107, 282–283; tables 335, 338; trigonometric ratios 23–33, 274–277
sketching graphs: curves 192–193, 261–262; straight lines 63–65, 191–192; travel 131–133; variation 90
small circles 207–209, 215
solids 113–124
solutions of equations 244; graphical solutions 40–44, 46–47, 248–250; solving by formula 39–40, 248
speed: air/ground speed 231; distance-time graphs 129; knots 214; speed-time graphs 133–138, 263–264
spheres 116–117, 119
square matrices 13
square roots 86; tables 341–342
square-based pyramids 113
squares: completing the square 38–39, 248; difference of two squares 7–8; perfect 6–7, 238–239; tables 343
statistics 164–172, 307–311; cumulative frequency 164–167, 307–310; frequency distributions 164, 307; interquartile range 165, 307; median 165, 307; percentiles 165–167; probability 167–171; quartiles 165; revision test 310–311; semi-interquartile range 165, 307
straight lines 63–72, 191–192, 260–261
subjects of formulae 95–100, 243–244
substitution method: simultaneous equations 45–46, 250–251
subtraction: fractions 81–83; matrices 14–15, 293–294; vectors 220–221, 298
supplementary angles 51
surface area 113–123
symbols: meanings 332
symmetry: lines 127

tables 333–345; antilogarithms 334; cosines of angles 336; logarithms 333, 338–340; logarithms of cosines 339; logarithms of sines 338; logarithms of tangents 340; random numbers 345; reciprocals 344; sines of angles 335; square roots 341–342; squares 343; tangents of angles 337
tangent of angle: tables 337, 340; tangent curve 153–154; tangent function 152–154, 286–287; trigonometric ratios 23–33, 274–277
tangents to curves 54–60, 125–128, 270–272
tests *see* revision exercises and tests
theoretical probability 167
three-dimensional solids 113–124; composite solids 116–121; cones 113–115, 117–119; cuboids 113, 118; cylinders 113, 117–119; frustrums 117–118; hemispheres 117; hollow objects 118; prisms 113; problems 121–123, 280–282; pyramids 113–115, 117; spheres 116–117, 119
time: distance-time graphs 128–133, 263; speed-time graphs 133–138, 263–264
track of an aircraft 231
transformations 173–185, 301–305; combined 173–176, 181–184; enlargements 174–175, 180–183, 302; glide reflections 173–174, 301; invariant lines/points 178, 180; isometric 174, 301; matrices 177–184, 302; ordering 182; reflections 173–175, 178–180, 182–183, 301–302; rotations 177–178, 182, 301–302; translations 173–174, 177, 181–182, 301–302; vectors 173, 177
translations 173–174, 177, 181–182, 301–302; vectors 224, 227
trapeziums 110
travel graphs 128–138, 263–267; acceleration 133–136, 263–264; area under graph 134, 263; deceleration 134–136, 263; distance-time graphs 128–133, 263; drawing graphs 131–133; speed-time graphs 133–138, 263–264
tree diagrams 168–171
trial and error factorisation method 4–6
triangle law: vectors 223
triangles: area calculation 109; right-angled triangles 277–280; similar triangles 108; solving completely 109
trigonometric functions 147–154, 286–287; cosine curve 150–152, 286; cosine function 149–152, 286–287; periodic functions 148, 151, 153; sine curve 148–149, 286; sine function 147–149, 286–287; tangent curve 153–154; tangent function 152–154, 286–287
trigonometric ratios 23–33, 274–277; Cartesian quadrants 26–27; negative angles 27–28; obtuse angles 23–25; positive angles 26–27; reflex angles 25–26; special angles 28–33, 275–277
trigonometry 23–34, 103–112, 147–154, 274–292; area problems 109–111; cosine rule 105–109, 283–284; formulae 331–332; functions 147–154, 286–287; Pythagoras' theorem 277–280; ratios 23–33, 274–277; revision test 291–292; right-angled triangles 277–280; sine rule 103–107, 282–283; three-dimensional problems 280–282
turning points 127
two by two matrices 18–19

undefined fractions 83–84, 189–190
unit vectors 220
units: mensuration 330–331

variables: definition 35
variation (proportion) 89–95, 240–242; direct 89–92, 240–241; inverse 93–95, 241; joint 241
vectors 173, 217–232, 298–301; addition 218–220, 226, 298; air travel example 231; column 217; diagrams 231; displacements 224, 227; geometry 226–230; magnitude 220, 298; modulus 220; multiplication 221–223, 298; naming 217–218; parallelogram law 223; position 177, 224–226; resultant 223–224; scalars 221–223, 298; subtraction 220–221, 298; translations 173, 177; triangle law 223; unit 220; velocity 230
velocity: relative 230–231; vectors 230
volumes 113–123; cones 113–115, 117–119; cuboids 113, 118; cylinders 113, 117–119; frustrums 117–118; hemispheres 117; hollow objects 118; prisms 113; pyramids 113–115, 117; solids 113–123; spheres 116–117, 119; units 330

word problems 44–45, 251–254

zero gradient 64
zero matrix 18